Universitext

Universitext

Universitext is a series of textbooks that presents material from a wide variety of mathematical disciplines at master's level and beyond. The books, often well class-tested by their author, may have an informal, personal, even experimental approach to their subject matter. Some of the most successful and established books in the series have evolved through several editions, always following the evolution of teaching curricula, into very polished texts.

Thus as research topics trickle down into graduate-level teaching, first textbooks written for new, cutting-edge courses may make their way into *Universitext.*

For further volumes:
www.springer.com/series/223

Alexandr A. Borovkov

Probability Theory

Edited by K.A. Borovkov

Translated by O.B. Borovkova and P.S. Ruzankin

Springer

Alexandr A. Borovkov
Sobolev Institute of Mathematics and
Novosibirsk State University
Novosibirsk, Russia

Translation from the 5th edn. of the Russian language edition:
'Teoriya Veroyatnostei' by Alexandr A. Borovkov

ISSN 0172-5939 ISSN 2191-6675 (electronic)
Universitext
ISBN 978-1-4471-5200-2 ISBN 978-1-4471-5201-9 (eBook)
DOI 10.1007/978-1-4471-5201-9
Springer London Heidelberg New York Dordrecht

Library of Congress Control Number: 2013941877

Mathematics Subject Classification: 60-XX, 60-01

Printed on acid-free paper

Springer is part of Springer Science+Business Media (www.springer.com)

Foreword

The present edition of the book differs substantially from the previous one. Over the period of time since the publication of the previous edition the author has accumulated quite a lot of ideas concerning possible improvements to some chapters of the book. In addition, some new opportunities were found for an accessible exposition of new topics that had not appeared in textbooks before but which are of certain interest for applications and reflect current trends in the development of modern probability theory. All this led to the need for one more revision of the book. As a result, many methodological changes were made and a lot of new material was added, which makes the book more logically coherent and complete. We will list here only the main changes in the order of their appearance in the text.

• Section 4.4 "Expectations of Sums of a Random Number of Random Variables" was significantly revised. New sufficient conditions for Wald's identity were added. An example is given showing that, when summands are non-identically distributed, Wald's identity can fail to hold even in the case when its right-hand side is well-defined. Later on, Theorem 11.3.2 shows that, for identically distributed summands, Wald's identity is always valid whenever its right-hand side is well-defined.

• In Sect. 6.1 a criterion of uniform integrability of random variables is constructed, which simplifies the use of this notion. For example, the criterion directly implies uniform integrability of weighted sums of uniformly integrable random variables.

• Section 7.2, which is devoted to inversion formulas, was substantially expanded and now includes assertions useful for proving integro-local theorems in Sect. 8.7.

• In Chap. 8, integro-local limit theorems for sums of identically distributed random variables were added (Sects. 8.7 and 8.8). These theorems, being substantially more precise assertions than the integral limit theorems, do not require additional conditions and play an important role in investigating large deviation probabilities in Chap. 9.

• A new chapter was written on probabilities of large deviations of sums of random variables (Chap. 9). The chapter provides a systematic and rather complete exposition of the large deviation theory both in the case where the Cramér condition (rapid decay of distributions at infinity) is satisfied and where it is not. Both integral and integro-local theorems are obtained. The large deviation principle is established.

• Assertions concerning the case of non-identically distributed random variables were added in Chap. 10 on "Renewal Processes". Among them are renewal theorems as well as the law of large numbers and the central limit theorem for renewal processes. A new section was written to present the theory of generalised renewal processes.

• An extension of the Kolmogorov strong law of large numbers to the case of non-identically distributed random variables having the first moment only was added to Chap. 11. A new subsection on the "Strong law of large numbers for generalised renewal processes" was written.

• Chapter 12 on "Random walks and factorisation identities" was substantially revised. A number of new sections were added: on finding factorisation components in explicit form, on the asymptotic properties of the distribution of the suprema of cumulated sums and generalised renewal processes, and on the distribution of the first passage time.

• In Chap. 13, devoted to Markov chains, a section on "The law of large numbers and central limit theorem for sums of random variables defined on a Markov chain" was added.

• Three new appendices (6, 7 and 8) were written. They present important auxiliary material on the following topics: "The basic properties of regularly varying functions and subexponential distributions", "Proofs of theorems on convergence to stable laws", and "Upper and lower bounds for the distributions of sums and maxima of sums of independent random variables".

As has already been noted, these are just the most significant changes; there are also many others. A lot of typos and other inaccuracies were fixed. The process of creating new typos and misprints in the course of one's work on a book is random and can be well described mathematically by the Poisson process (for the definition of Poisson processes, see Chaps 10 and 19). An important characteristic of the quality of a book is the intensity of this process. Unfortunately, I am afraid that in the two previous editions (1999 and 2003) this intensity perhaps exceeded a certain acceptable level. Not renouncing his own responsibility, the author still admits that this may be due, to some extent, to the fact that the publication of these editions took place at the time of a certain decline of the publishing industry in Russia related to the general state of the economy at that time (in the 1972, 1976 and 1986 editions there were much fewer such defects).

Before starting to work on the new edition, I asked my colleagues from our laboratory at the Sobolev Institute of Mathematics and from the Chair of Probability Theory and Mathematical Statistics at Novosibirsk State University to prepare lists of any typos and other inaccuracies they had spotted in the book, as well as suggested improvements of exposition. I am very grateful to everyone who provided me with such information. I would like to express special thanks to I.S. Borisov, V.I. Lotov, A.A. Mogul'sky and S.G. Foss, who also offered a number of methodological improvements.

I am also deeply grateful to T.V. Belyaeva for her invaluable assistance in typesetting the book with its numerous changes. Without that help, the work on the new edition would have been much more difficult.

A.A. Borovkov

Foreword to the Third and Fourth Editions

This book has been written on the basis of the Russian version (1986) published by "Nauka" Publishers in Moscow. A number of sections have been substantially revised and several new chapters have been introduced. The author has striven to provide a complete and logical exposition and simpler and more illustrative proofs. The 1986 text was preceded by two earlier editions (1972 and 1976). The first one appeared as an extended version of lecture notes of the course the author taught at the Department of Mechanics and Mathematics of Novosibirsk State University. Each new edition responded to comments by the readers and was completed with new sections which made the exposition more unified and complete.

The readers are assumed to be familiar with a traditional calculus course. They would also benefit from knowing elements of measure theory and, in particular, the notion of integral with respect to a measure on an arbitrary space and its basic properties. However, provided they are prepared to use a less general version of some of the assertions, this lack of additional knowledge will not hinder the reader from successfully mastering the material. It is also possible for the reader to avoid such complications completely by reading the respective Appendices (located at the end of the book) which contain all the necessary results.

The first ten chapters of the book are devoted to the basics of probability theory (including the main limit theorems for cumulative sums of random variables), and it is best to read them in succession. The remaining chapters deal with more specific parts of the theory of probability and could be divided into two blocks: random processes in discrete time (or random sequences, Chaps. 12 and 14–16) and random processes in continuous time (Chaps. 17–21).

There are also chapters which remain outside the mainstream of the text as indicated above. These include Chap. 11 "Factorisation Identities". The chapter not only contains a series of very useful probabilistic results, but also displays interesting relationships between problems on random walks in the presence of boundaries and boundary problems of complex analysis. Chapter 13 "Information and Entropy" and Chap. 19 "Functional Limit Theorems" also deviate from the mainstream. The former deals with problems closely related to probability theory but very rarely treated in texts on the discipline. The latter presents limit theorems for the convergence of processes generated by cumulative sums of random variables to the Wiener and Poisson processes; as a consequence, the law of the iterated logarithm is established in that chapter.

The book has incorporated a number of methodological improvements. Some parts of it are devoted to subjects to be covered in a textbook for the first time (for example, Chap. 16 on stochastic recursive sequences playing an important role in applications).

The book can serve as a basis for third year courses for students with a reasonable mathematical background, and also for postgraduates. A one-semester (or two-trimester) course on probability theory might consist (there could be many variants) of the following parts: Chaps. 1–2, Sects. 3.1–3.4, 4.1–4.6 (partially), 5.2 and 5.4 (partially), 6.1–6.3 (partially), 7.1, 7.2, 7.4–7.6, 8.1–8.2 and 8.4 (partially), 10.1, 10.3, and the main results of Chap. 12.

For a more detailed exposition of some aspects of Probability Theory and the Theory of Random Processes, see for example [2, 10, 12–14, 26, 31].

While working on the different versions of the book, I received advice and help from many of my colleagues and friends. I am grateful to Yu.V. Prokhorov, V.V. Petrov and B.A. Rogozin for their numerous useful comments which helped to improve the first variant of the book. I am deeply indebted to A.N. Kolmogorov whose remarks and valuable recommendations, especially of methodological character, contributed to improvements in the second version of the book. In regard to the second and third versions, I am again thankful to V.V Petrov who gave me his comments, and to P. Franken, with whom I had a lot of useful discussions while the book was translated into German.

In conclusion I want to express my sincere gratitude to V.V. Yurinskii, A.I. Sakhanenko, K.A. Borovkov, and other colleagues of mine who also gave me their comments on the manuscript. I would also like to express my gratitude to all those who contributed, in one way or another, to the preparation and improvement of the book.

A.A. Borovkov

For the Reader's Attention

The numeration of formulas, lemmas, theorems and corollaries consists of three numbers, of which the first two are the numbers of the current chapter and section. For instance, Theorem 4.3.1 means Theorem 1 from Sect. 3 of Chap. 4. Section 6.2 means Sect. 2 of Chap. 6.

The sections marked with an asterisk may be omitted in the first reading.

The symbol □ at the end of a paragraph denotes the end of a proof or an important argument, when it should be pointed out that the argument has ended.

The symbol $:=$, systematically used in the book, means that the left-hand side is *defined* to be given by the right-hand side. The relation $=:$ has the opposite meaning: the right-hand side is defined by the left-hand side.

The reader may find it useful to refer to the **Index of Basic Notation** and **Subject index**, which can be found at the end of this book.

Introduction

1. It is customary to set the origins of Probability Theory at the 17th century and relate them to combinatorial problems of games of chance. The latter can hardly be considered a serious occupation. However, it is games of chance that led to problems which could not be stated and solved within the framework of the then existing mathematical models, and thereby stimulated the introduction of new concepts, approaches and ideas. These new elements can already be encountered in writings by P. Fermat, D. Pascal, C. Huygens and, in a more developed form and somewhat later, in the works of J. Bernoulli, P.-S. Laplace, C.F. Gauss and others. The above-mentioned names undoubtedly decorate the genealogy of Probability Theory which, as we saw, is also related to some extent to the vices of society. Incidentally, as it soon became clear, it is precisely this last circumstance that can make Probability Theory more attractive to the reader.

The first text on Probability Theory was Huygens' treatise *De Ratiociniis in Ludo Alea* ("On Ratiocination in Dice Games", 1657). A bit later in 1663 the book *Liber de Ludo Aleae* ("Book on Games of Chance") by G. Cardano was published (in fact it was written earlier, in the mid 16th century). The subject of these treatises was the same as in the writings of Fermat and Pascal: dice and card games (problems within the framework of Sect. 1.2 of the present book). As if Huygens foresaw future events, he wrote that if the reader studied the subject closely, he would notice that one was not dealing just with a game here, but rather that the foundations of a very interesting and deep theory were being laid. Huygens' treatise, which is also known as the first text introducing the concept of mathematical expectation, was later included by J. Bernoulli in his famous book *Ars Conjectandi* ("The Art of Conjecturing"; published posthumously in 1713). To this book is related the notion of the so-called Bernoulli scheme (see Sect. 1.3), for which Bernoulli gave a cumbersome (cf. our Sect. 5.1) but mathematically faultless proof of the first limit theorem of Probability Theory, the Law of Large Numbers.

By the end of the 19th and the beginning of the 20th centuries, the natural sciences led to the formulation of more serious problems which resulted in the development of a large branch of mathematics that is nowadays called Probability Theory. This subject is still going through a stage of intensive development. To a large extent,

Probability Theory owes its elegance, modern form and a multitude of achievements to the remarkable Russian mathematicians P.L. Chebyshev, A.A. Markov, A.N. Kolmogorov and others.

The fact that increasing our knowledge about nature leads to further demand for Probability Theory appears, at first glance, paradoxical. Indeed, as the reader might already know, the main object of the theory is randomness, or uncertainty, which is due, as a rule, to a lack of knowledge. This is certainly so in the classical example of coin tossing, where one cannot take into account all the factors influencing the eventual position of the tossed coin when it lands.

However, this is only an apparent paradox. In fact, there are almost no exact deterministic quantitative laws in nature. Thus, for example, the classical law relating the pressure and temperature in a volume of gas is actually a result of a probabilistic nature that relates the number of collisions of particles with the vessel walls to their velocities. The fact is, at typical temperatures and pressures, the number of particles is so large and their individual contributions are so small that, using conventional instruments, one simply cannot register the random deviations from the relationship which actually take place. This is not the case when one studies more sparse flows of particles—say, cosmic rays—although there is no qualitative difference between these two examples.

We could move in a somewhat different direction and name here the uncertainty principle stating that one cannot simultaneously obtain exact measurements of any two conjugate observables (for example, the position and velocity of an object). Here randomness is not entailed by a lack of knowledge, but rather appears as a fundamental phenomenon reflecting the nature of things. For instance, the lifetime of a radioactive nucleus is essentially random, and this randomness cannot be eliminated by increasing our knowledge.

Thus, uncertainty was there at the very beginning of the cognition process, and it will always accompany us in our quest for knowledge. These are rather general comments, of course, but it appears that the answer to the question of when one should use the methods of Probability Theory and when one should not will always be determined by the relationship between the degree of precision we want to attain when studying a given phenomenon and what we know about the nature of the latter.

2. In almost all areas of human activity there are situations where some experiments or observations can be repeated a large number of times under the same conditions. Probability Theory deals with those experiments of which the result (expressed in one way or another) may vary from trial to trial. The events that refer to the experiment's result and which may or may not occur are usually called *random events*.

For example, suppose we are tossing a coin. The experiment has only two outcomes: either heads or tails show up, and before the experiment has been carried out, it is impossible to say which one will occur. As we have already noted, the reason for this is that we cannot take into account all the factors influencing the final position of the coin. A similar situation will prevail if you buy a ticket for each lottery draw and try to predict whether it will win or not, or, observing the operation of a complex machine, you try to determine in advance if it will have failed before or

Fig. 1 The plot of the relative frequencies n_h/n corresponding to the outcome sequence *htthtthhhthht* in the coin tossing experiment

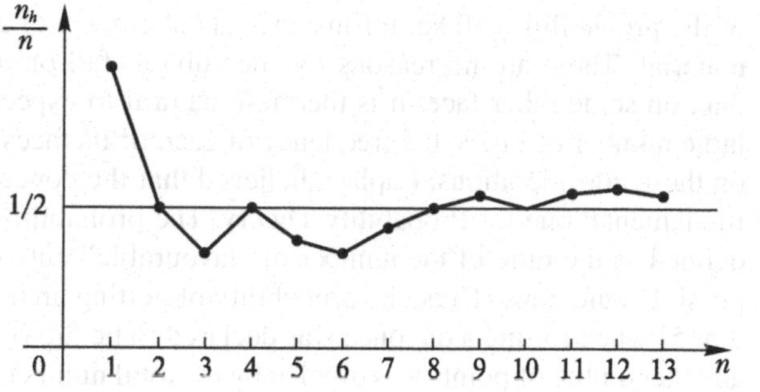

after a given time. In such situations, it is very hard to find any laws when considering the results of individual experiments. Therefore there is little justification for constructing any theory here.

However, if one turns to a *long* sequence of repetitions of such an experiment, an interesting phenomenon becomes apparent. While individual results of the experiments display a highly "irregular" behaviour, the average results demonstrate stability. Consider, say, a long series of repetitions of our coin tossing experiment and denote by n_h the number of heads in the first n trials. Plot the ratio n_h/n versus the number n of conducted experiments (see Fig. 1; the plot corresponds to the outcome sequence *htthtthhhthh*, where h stands for heads and t for tails, respectively).

We will then see that, as n increases, the polygon connecting the consecutive points $(n, n_h/n)$ very quickly approaches the straight line $n_h/n = 1/2$. To verify this observation, G.L. Leclerc, comte de Buffon,[1] tossed a coin 4040 times. The number of heads was 2048, so that the relative frequency n_h/n of heads was 0.5069. K. Pearson tossed a coin 24,000 times and got 12,012 heads, so that $n_h/n = 0.5005$.

It turns out that this phenomenon is universal: *the relative frequency of a certain outcome in a series of repetitions of an experiment under the same conditions tends towards a certain number* $p \in [0, 1]$ *as the number of repetitions grows*. It is an objective law of nature which forms the foundation of Probability Theory.

It would be natural to define the probability of an experiment outcome to be just the number p towards which the relative frequency of the outcome tends. However, such a definition of probability (usually related to the name of R. von Mises) has proven to be inconvenient. First of all, in reality, each time we will be dealing not with an infinite sequence of frequencies, but rather with finitely many elements thereof. Obtaining the entire sequence is unfeasible. Hence the frequency (let it again be n_h/n) of the occurrence of a certain outcome will, as a rule, be different for each new series of repetitions of the same experiment.

This fact led to intense discussions and a lot of disagreement regarding how one should define the concept of probability. Fortunately, there was a class of phenomena that possessed certain "symmetry" (in gambling, coin tossing etc.) for which one could compute *in advance*, prior to the experiment, the expected numerical values

[1]The data is borrowed from [15].

of the probabilities. Take, for instance, a cube made of a sufficiently homogeneous material. There are no reasons for the cube to fall on any of its faces more often than on some other face. It is therefore natural to expect that, when rolling a die a large number of times, the frequency of each of its faces will be close to 1/6. Based on these considerations, Laplace believed that the concept of *equiprobability* is the fundamental one for Probability Theory. The probability of an event would then be defined as the ratio of the number of "favourable" outcomes to the total number of possible outcomes. Thus, the probability of getting an odd number of points (e.g. 1, 3 or 5) when rolling a die once was declared to be 3/6 (i.e. the number of faces with an odd number of points was divided by the total number of all faces). If the die were rolled ten times, then one would have 6^{10} in the denominator, as this number gives the total number of equally likely outcomes and calculating probabilities reduces to counting the number of "favourable outcomes" (the ones resulting in the occurrence of a given event).

The development of the mathematical theory of probabilities began from the instance when one started defining probability as the ratio of the number of favourable outcomes to the total number of equally likely outcomes, and this approach is nowadays called "classical" (for more details, see Chap. 1).

Later on, at the beginning of the 20th century, this approach was severely criticised for being too restrictive. The initiator of the critique was R. von Mises. As we have already noted, his conception was based on *postulating* stability of the frequencies of events in a long series of experiments. That was a confusion of physical and mathematical concepts. No passage to the limit can serve as justification for introducing the notion of "probability". If, for instance, the values n_h/n were to converge to the limiting value $1/2$ in Fig. 1 too slowly, that would mean that nobody would be able to find the value of that limit in the general (non-classical) case. So the approach is clearly vulnerable: it would mean that Probability Theory would be applicable only to those situations where *frequencies have a limit*. But why frequencies would have a limit remained unexplained and was not even discussed.

In this relation, R. von Mises' conception has been in turn criticised by many mathematicians, including A.Ya. Khinchin, S.N. Bernstein, A.N. Kolmogorov and others. Somewhat later, another approach was suggested that proved to be fruitful for the development of the mathematical theory of probabilities. Its general features were outlined by S.N. Bernstein in 1908. In 1933 a rather short book "Foundations of Probability Theory" by A.N. Kolmogorov appeared that contained a complete and clear exposition of the axioms of Probability Theory. The general construction of the concept of probability based on Kolmogorov's axiomatics removed all the obstacles for the development of the theory and is nowadays universally accepted.

The creation of an axiomatic Probability Theory provided a solution to the sixth Hilbert problem (which concerned, in particular, Probability Theory) that had been formulated by D. Hilbert at the Second International Congress of Mathematicians in Paris in 1900. The problem was on the axiomatic construction of a number of physical sciences, Probability Theory being classified as such by Hilbert at that time.

An axiomatic foundation separates the mathematical aspect from the physical: one no longer needs to explain *how and where* the concept of probability comes from. The concept simply becomes a primitive one, its properties being described by *axioms* (which are essentially the axioms of Measure Theory). However, the problem of how the probability thus introduced is related (and can be applied) to the real world remains open. But this problem is mostly removed by the remarkable fact that, under the axiomatic construction, the desired fundamental property that the frequencies of the occurrence of an event converge to the probability of the event does take place and is a precise mathematical result. (For more details, see Chaps. 2 and 5.)[2]

We will begin by defining probability in a somewhat simplified situation, in the so-called *discrete case*.

[2]Much later, in the 1960s A.N. Kolmogorov attempted to develop a fundamentally different approach to the notions of probability and randomness. In that approach, the measure of randomness, say, of a sequence $0, 1, 0, 0, 1, \dots$ consisting of 0s and 1s (or some other symbols) is the complexity of the algorithm describing this sequence. The new approach stimulated the development of a number of directions in contemporary mathematics, but, mostly due to its complexity, has not yet become widely accepted.

Contents

Chapter 1
Discrete Spaces of Elementary Events

Abstract Section 1.1 introduces the fundamental concept of probability space, along with some basic terminology and properties of probability when it is easy to do, i.e. in the simple case of random experiments with finitely or at most countably many outcomes. The classical scheme of finitely many equally likely outcomes is discussed in more detail in Sect. 1.2. Then the Bernoulli scheme is introduced and the properties of the binomial distribution are studied in Sect. 1.3. Sampling without replacement from a large population is considered, and convergence of the emerging hypergeometric distributions to the binomial one is formally proved. The inclusion-exclusion formula for the probabilities of unions of events is derived and illustrated by some applications in Sect. 1.4.

1.1 Probability Space

To mathematically describe experiments with random outcomes, we will first of all need the notion of the *space of elementary events* (or *outcomes*) corresponding to the experiment under consideration. We will denote by Ω any set such that each result of the experiment we are interested in can be uniquely specified by the elements of Ω.

In the simplest experiments we usually deal with *finite* spaces of elementary outcomes. In the coin tossing example we considered above, Ω consists of two elements, "heads" and "tails". In the die rolling experiment, the space Ω is also finite and consists of 6 elements. However, even for tossing a coin (or rolling a die) one can arrange such experiments for which finite spaces of elementary events will not suffice. For instance, consider the following experiment: a coin is tossed until heads shows for the first time, and then the experiment is stopped. If t designates tails in a toss and h heads, then an "elementary outcome" of the experiment can be represented by a sequence $(tt\dots th)$. There are infinitely many such sequences, and all of them are different, so there is no way to describe unambiguously all the outcomes of the experiment by elements of a finite space.

Consider finite or countably infinite spaces of elementary events Ω. These are the so-called *discrete* spaces. We will denote the elements of a space Ω by the letter ω and call them *elementary events* (or *elementary outcomes*).

A.A. Borovkov, *Probability Theory*, Universitext,
DOI 10.1007/978-1-4471-5201-9_1, © Springer-Verlag London 2013

The notion of the space of elementary events itself is mathematically undefinable: it is a primitive one, like the notion of a point in geometry. The specific nature of Ω will, as a rule, be of no interest to us.

Any subset $A \subseteq \Omega$ will be called an *event* (the event A occurs if any of the elementary outcomes $\omega \in A$ occurs).

The *union* or *sum* of two events A and B is the event $A \cup B$ (which may also be denoted by $A + B$) consisting of the elementary outcomes which belong to at least one of the events A and B. The *product* or *intersection* AB (which is often denoted by $A \cap B$ as well) is the event consisting of all elementary events belonging to both A and B. The *difference* of the events A and B is the set $A - B$ (also often denoted by $A \setminus B$) consisting of all elements of A not belonging to B. The set Ω is called the *certain* event. The empty set $\varnothing$ is called the *impossible* event. The set $\overline{A} = \Omega - A$ is called the *complementary* event of A. Two events A and B are *mutually exclusive* if $AB = \varnothing$.

Let, for instance, our experiment consist in rolling a die twice. Here one can take the space of elementary events to be the set consisting of 36 elements (i, j), where i and j run from 1 to 6 and denote the numbers of points that show up in the first and second roll respectively. The events $A = \{i + j \le 3\}$ and $B = \{j = 6\}$ are mutually exclusive. The product of the events A and $C = \{j \text{ is even}\}$ is the event $(1, 2)$. Note that if we were interested in the events related to the first roll only, we could consider a smaller space of elementary events consisting of just 6 elements $i = 1, 2, \ldots, 6$.

One says that the *probabilities of elementary events* are given if a nonnegative real-valued function $\mathbf{P}$ is given on Ω such that $\sum_{\omega \in \Omega} \mathbf{P}(\omega) = 1$ (one also says that the *function* $\mathbf{P}$ *specifies a probability distribution on* Ω).

The *probability of an event A is the number*

$$\mathbf{P}(A) := \sum_{\omega \in A} \mathbf{P}(\omega).$$

This definition is consistent, for the series on the right hand side is absolutely convergent.

We note here that specific numerical values of the function $\mathbf{P}$ will also be of no interest to us: this is just an issue of the practical value of the model. For instance, it is clear that, in the case of a symmetric die, for the outcomes $1, 2, \ldots, 6$ one should put $\mathbf{P}(1) = \mathbf{P}(2) = \cdots = \mathbf{P}(6) = 1/6$; for a symmetric coin, one has to choose the values $\mathbf{P}(h) = \mathbf{P}(t) = 1/2$ and not any others. In the experiment of tossing a coin until heads shows for the first time, one should put $\mathbf{P}(h) = 1/2$, $\mathbf{P}(th) = 1/2^2$, $\mathbf{P}(tth) = 1/2^3, \ldots$. Since $\sum_{n=1}^{\infty} 2^{-n} = 1$, the function $\mathbf{P}$ given in this way on the outcomes of the form $(t \ldots th)$ will define a probability distribution on Ω. For example, to calculate the probability that the experiment stops on an even step (that is, the probability of the event composed of the outcomes $(th), (ttth), \ldots$), one should consider the sum of the corresponding probabilities which is equal to

$$\sum_{n=1}^{\infty} 2^{-2n} = \frac{1}{4} \times \frac{4}{3} = \frac{1}{3}.$$

In the experiments mentioned in the Introduction, where one had to guess when a device will break down—before a given time (the event A) or after it, quantitative estimates of the probability $\mathbf{P}(A)$ can usually only be based on the results of the experiments themselves. The methods of estimating unknown probabilities from observation results are studied in Mathematical Statistics, the subject-matter of which will be exemplified somewhat later by a problem from this chapter.

Note further that by no means can one construct models with discrete spaces of elementary events for all experiments. For example, suppose that one is measuring the energy of particles whose possible values fill the interval $[0, V]$, $V > 0$, but the set of points of this interval (that is, the set of elementary events) is continuous. Or suppose that the result of an experiment is a patient's electrocardiogram. In this case, the result of the experiment is an element of some functional space. In such cases, more general schemes are needed.

From the above definitions, making use of the absolute convergence of the series $\sum_{\omega \in A} \mathbf{P}(\omega)$, one can easily derive the following properties of probability:

(1) $\mathbf{P}(\varnothing) = 0$, $\mathbf{P}(\Omega) = 1$.
(2) $\mathbf{P}(A + B) = \sum_{\omega \in A \cup B} \mathbf{P}(\omega) = \sum_{\omega \in A} \mathbf{P}(\omega) + \sum_{\omega \in B} \mathbf{P}(\omega) - \sum_{\omega \in A \cap B} \mathbf{P}(\omega) = \mathbf{P}(A) + \mathbf{P}(B) - \mathbf{P}(AB)$.
(3) $\mathbf{P}(\overline{A}) = 1 - \mathbf{P}(A)$.

This entails, in particular, that, for *disjoint* (mutually exclusive) events A and B,

$$\mathbf{P}(A + B) = \mathbf{P}(A) + \mathbf{P}(B).$$

This property of the *additivity of probability* continues to hold for an arbitrary number of disjoint events $A_1, A_2, \ldots$: if $A_i A_j = \varnothing$ for $i \neq j$, then

$$\mathbf{P}\left(\bigcup_{k=1}^{\infty} A_k\right) = \sum_{k=1}^{\infty} \mathbf{P}(A_k). \tag{1.1.1}$$

This follows from the equality

$$\mathbf{P}\left(\bigcup_{k=1}^{n} A_k\right) = \sum_{k=1}^{n} \mathbf{P}(A_k)$$

and the fact that $\mathbf{P}(\bigcup_{k=n+1}^{\infty} A_k) \to 0$ as $n \to \infty$. To prove the last relation, first enumerate the elementary events. Then we will be dealing with the sequence $\omega_1, \omega_2, \ldots$; $\bigcup \omega_k = \Omega$, $\mathbf{P}(\bigcup_{k>n} \omega_k) = \sum_{k>n} \mathbf{P}(\omega_k) \to 0$ as $n \to \infty$. Denote by n_k the number of events A_j such that $\omega_k \in A_j = A_{n_k}$; $n_k = 0$ if $\omega_k A_j = \varnothing$ for all j. If $n_k \le N < \infty$ for all k, then the events A_j with $j > N$ are empty and the desired relation is obvious. If $N_s := \max_{k \le s} n_k \to \infty$ as $s \to \infty$, then one has $\bigcup_{j>n} A_j \subset \bigcup_{k>s} \omega_k$ for $n > N_s$, and therefore

$$\mathbf{P}\left(\bigcup_{j>n} A_j\right) \le \mathbf{P}\left(\bigcup_{k>s} \omega_k\right) = \sum_{k>s} \mathbf{P}(\omega_k) \to 0 \quad \text{as } s \to \infty.$$

The required relation is proved.

For arbitrary A and B, one has $\mathbf{P}(A+B) \le \mathbf{P}(A) + \mathbf{P}(B)$. A similar inequality also holds for the sum of an arbitrary number of events:

$$\mathbf{P}\left(\bigcup_{k=1}^{\infty} A_k\right) \le \sum_{k=1}^{\infty} \mathbf{P}(A_k).$$

This follows from (1.1.1) and the representation of $\bigcup A_k$ as the union $\bigcup A_k \overline{B}_k$ of disjoint events $A_k \overline{B}_k$, where $B_k = \bigcup_{j<k} A_j$. It remains to note that $\mathbf{P}(A_k \overline{B}_k) \le \mathbf{P}(A_k)$.

Now we will consider several important special cases.

1.2 The Classical Scheme

Let Ω consist of n elements and all the outcomes be equally likely, that is $\mathbf{P}(\omega) = 1/n$ for any $\omega \in \Omega$. In this case, the probability of any event A is defined by the formula

$$\mathbf{P}(A) := \frac{1}{n}\{\text{number of elements of } A\}.$$

This is the so-called *classical definition of probability* (the term *uniform discrete distribution* is also used).

Let a set $\{a_1, a_2, \dots, a_n\}$ be given, which we will call the *general population*. A *sample of size* k from the general population is an ordered sequence $(a_{j_1}, a_{j_2}, \dots, a_{j_k})$. One can form this sequence as follows: the first element a_{j_1} is chosen from the whole population. The next element a_{j_2} we choose from the general population without the element a_{j_1}; the element a_{j_3} is chosen from the general population without the elements a_{j_1} and a_{j_2}, and so on. Samples obtained in such a way are called *samples without replacement*. Clearly, one must have $k \le n$ in this case. The number of such samples of size k coincides with the number of arrangements of k elements from n:

$$(n)_k := n(n-1)(n-2)\cdots(n-k+1).$$

Indeed, according to the sampling process, in the first position we can have any element of the general population, in the second position any of the remaining $(n-1)$ elements, and so on. We could prove this more formally by induction on k.

Assign to each of the samples without replacement the probability $1/(n)_k$. Such a sample will be called *random*. This is clearly the classical scheme.

Calculate the probability that $a_{j_1} = a_1$ and $a_{j_2} = a_2$. Since the remaining $k-2$ positions can be occupied by any of the remaining $n-2$ elements of the general population, the number of samples without replacement having elements a_1 and a_2

in the first two positions equals $(n-2)_{k-2}$. Therefore the probability of that event is equal to

$$\frac{(n-2)_{k-2}}{(n)_k} = \frac{1}{n(n-1)}.$$

One can think of a sample without replacement as the result of sequential sampling from a collection of enumerated balls placed in an urn. Sampled balls are not returned back to the urn.

However, one can form a sample in another way as well. One takes a ball out of the urn and memorises it. Then the ball is returned to the urn, and one again picks a ball from the urn; this ball is also memorised and put back to the urn, and so on. The sample obtained in this way is called a *sample with replacement*. At each step, one can pick any of the n balls. There are k such steps, so that the total number of such samples will be n^k. If we assign the probability of $1/n^k$ to each sample, this will also be a classical scheme situation.

Calculate, for instance, the probability that, in a sample with replacement of size $k \le n$, all the elements will be different. The number of samples of elements without repetitions is the same as the number of samples without replacement, i.e. $(n)_k$. Therefore the desired probability is $(n)_k/n^k$.

We now return to sampling without replacement for the general population $\{a_1, a_2, \ldots, a_n\}$. We will be interested in the number of samples of size $k \le n$ which differ from each other in their composition only. The number of samples without replacement of size k which have the same composition and are only distinguished by the order of their elements is $k!$ Hence the number of samples of different composition equals

$$\frac{(n)_k}{k!} = \binom{n}{k}.$$

This is the *number of combinations of* k *items chosen from a total of* n for $0 \le k \le n$.[1] If the initial sample is random, we again get the classical probability scheme, for the probability of each new sample is

$$\frac{k!}{(n)_k} = \frac{1}{\binom{n}{k}}.$$

Let our urn contain n balls, of which n_1 are black and $n - n_1$ white. We sample k balls without replacement. What is the probability that there will be exactly k_1 black balls in the sample? The total number of samples which differ in the composition is, as was shown above, $\binom{n}{k}$. There are $\binom{n_1}{k_1}$ ways to choose k_1 black balls from the totality of n_1 black balls. The remaining $k - k_1$ white balls can be chosen from the totality of $n - n_1$ white balls in $\binom{n-n_1}{k-k_1}$ ways. Note that clearly any collection of black balls can be combined with any collection of white balls. Therefore the total

[1] In what follows, we put $\binom{n}{k} = 0$ for $k < 0$ and $k > n$.

number of samples of size k which differ in composition and contain exactly k_1 black balls is $\binom{n_1}{k_1}\binom{n-n_1}{k-k_1}$. Thus the desired probability is equal to

$$P_{n_1,n}(k_1,k) = \binom{n_1}{k_1}\binom{n-n_1}{k-k_1} \Big/ \binom{n}{k}.$$

The collection of numbers $P_{n_1,n}(0,k)$, $P_{n_1,n}(1,k), \dots, P_{n_1,n}(k,k)$ forms the so-called *hypergeometric distribution*. From the derived formula it follows, in particular, that, for any $0 < n_1 < n$,

$$\sum_{k_1=0}^{k} \binom{n_1}{k_1}\binom{n-n_1}{k-k_1} = \binom{n}{k}.$$

Example 1.2.1 In the 1980s, a version of a lottery called "Sportloto 6 out of 49" had became rather popular in Russia. A gambler chooses six from the totality of 49 sports (designated just by numbers). The prize amount is determined by how many sports he guesses correctly from another group of six sports, to be drawn at random by a mechanical device in front of the public. What is the probability that the gambler correctly guesses all six sports? A similar question could be asked about five sports, and so on.

It is not difficult to see that this is nothing else but a problem on the hypergeometric distribution where the gambler has labelled as "white" six items in a general population consisting of 49 items. Therefore the probability that, of the six items chosen at random, k_1 will turn out to be "white" (i.e. will coincide with those labelled by the gambler) is equal to $P_{6,49}(k_1,k)$, where the sample size k equals 6. For example, the probability of guessing all six sports correctly is

$$P_{6,49}(6,6) = \binom{49}{6}^{-1} \approx 7.2 \times 10^{-8}.$$

In connection with the hypergeometric distribution, one could comment on the nature of problems in Probability Theory and Mathematical Statistics. Knowing the composition of the general population, we can use the hypergeometric distribution to find out what chances different compositions of the sample would have. This is a typical *direct* problem of probability theory. However, in the natural sciences one usually has to solve *inverse* problems: how to determine the nature of general populations from the composition of random samples. Generally speaking, such inverse problems form the subject matter of Mathematical Statistics.

1.3 The Bernoulli Scheme

Suppose one draws a sample with replacement of size r from a general population consisting of two elements $\{0, 1\}$. There are 2^r such samples. Let p be a number in the interval [0, 1]. Define a nonnegative function $\mathbf{P}$ on the set Ω of all samples in the following way: if a sample ω contains exactly k ones, then $\mathbf{P}(\omega) = p^k(1-p)^{r-k}$. To verify that $\mathbf{P}$ is a probability, one has to prove the equality

$$\mathbf{P}(\Omega) = 1.$$

It is easy to see that k ones can be arranged in r places in $\binom{r}{k}$ different ways. Therefore there is the same number of samples containing exactly k ones. Now we can compute the probability of Ω:

$$\mathbf{P}(\Omega) = \sum_{k=0}^{r} \binom{r}{k} p^k (1-p)^{r-k} = \big(p + (1-p)\big)^r = 1.$$

The second equality here is just the binomial formula. At the same time we have found that the probability $P(k,r)$ that the sample contains exactly k ones is:

$$P(k,r) = \binom{r}{k} p^k (1-p)^{r-k}.$$

This is the so-called *binomial distribution*. It can be considered as the distribution of the number of "successes" in a series of r trials with two possible outcomes in each trial: 1 ("success") and 0 ("failure"). Such a series of trials with probability $\mathbf{P}(\omega)$ defined as $p^k(1-p)^{r-k}$, where k is the number of successes in ω, is called the *Bernoulli scheme*. It turns out that the trials in the Bernoulli scheme have the independence property which will be discussed in the next chapter.

It is not difficult to verify that the probability of having 1 at a fixed place in the sample (say, at position s) equals p. Indeed, having removed the item number s from the sample, we obtain a sample from the same population, but of size $r-1$. We will find the desired probability if we multiply the probabilities of these truncated samples by p and sum over all "short" samples. Clearly, we will get p. This is why the number p in the Bernoulli scheme is often called the success probability.

Arguing in the same way, we find that the probability of having 1 at k fixed positions in the sample equals p^k.

Now consider how the probabilities $P(k,r)$ of various outcomes behave as k varies. Let us look at the ratio

$$R(k,r) := \frac{P(k,r)}{P(k-1,r)} = \frac{p}{1-p}\,\frac{r-k+1}{k} = \frac{p}{1-p}\left(\frac{r+1}{k} - 1\right).$$

It clearly monotonically decreases as k increases, the value of the ratio being less than 1 for $k/(r+1) < p$ and greater than 1 for $k/(r+1) > p$. This means that the probabilities $P(k,r)$ first increase and then, for $k > p(r+1)$, decrease as k increases.

The above enables one to estimate, using the quantities $P(k,r)$, the probabilities

$$Q(k,r) = \sum_{j=0}^{k} P(j,r)$$

that the number of successes in the Bernoulli scheme does not exceed k. Namely, for $k < p(r+1)$,

$$Q(k,r) = P(k,r)\left(1 + \frac{1}{R(k,r)} + \frac{1}{R(k,r)R(k-1,r)} + \cdots\right)$$
$$\le P(k,r)\frac{R(k,r)}{R(k,r)-1} = P(k,r)\frac{(r+1-k)p}{(r+1)p-k}.$$

It is not difficult to see that this bound will be rather sharp if the numbers k and r are large and the ratio $k/(pr)$ is not too close to 1. In that case the sum

$$1 + \frac{1}{R(k,r)} + \frac{1}{R(k,r)R(k-1,r)} + \cdots$$

will be close to the sum of the geometric series

$$\sum_{j=0}^{\infty} R^{-j}(k,r) = \frac{R(k,r)}{R(k,r)-1},$$

and we will have the approximate equality

$$Q(k,r) \approx P(k,r)\frac{(r+1-k)p}{(r+1)p-k}. \tag{1.3.1}$$

For example, for $r = 30$, $p = 0.7$ and $k = 16$ one has $rp = 21$ and $P(k,r) \approx 0.023$. Here the ratio $\frac{(r+1-k)p}{(r-1)p-1}$ equals $15 \times 0.7/5.7 \approx 1.84$. Hence the right hand side of (1.3.1) estimating $Q(k,r)$ is approximately equal to $0.023 \times 1.84 \approx 0.042$. The true value of $Q(k,r)$ for the given values of r, p and k is 0.040 (correct to three decimals).

Formula (1.3.1) will be used in the example in Sect. 5.2.

Now consider a general population composed of n items, of which n_1 are of the first type and $n_2 = n - n_1$ of the second type. Draw from it a sample without replacement of size r.

Theorem 1.3.1 *Let n and n_1 tend to infinity in such a way that $n_1/n \to p$, where p is a number from the interval $[0,1]$. Then the following relation holds true for the hypergeometric distribution*:

$$P_{n_1,n}(r_1,r) \to P(r_1,r).$$

Proof Divide both the numerator and denominator in the formula for $P_{n_1,n}(r_1,r)$ (see Sect. 1.2) by n^r. Putting $r_2 = r - r_1$ and $n_2 := n - n_1$, we get

$$P_{n_1,n}(r_1,r) = \frac{r!(n-r)!}{n!} \frac{n_1!}{r_1!(n_1-r_1)} \frac{n_2!}{r_2!(n_2-r_2)}$$

$$= \frac{r!}{r_1!r_2!} \frac{\frac{n_1}{n}(\frac{n_1}{n} - \frac{1}{n})(\frac{n_1}{n} - \frac{2}{n}) \cdots (\frac{n_1}{n} - \frac{r_1-1}{n})}{\frac{n}{n}(1 - \frac{1}{n}) \cdots (1 - \frac{r-1}{n})}$$
$$\times \frac{n_2}{n}\left(\frac{n_2}{n} - \frac{1}{n}\right) \cdots \left(\frac{n_2}{n} - \frac{r_2 - 1}{n}\right)$$
$$\to \binom{r}{r_1} p^{r_1}(1-p)^{r_2} = P(r_1, r)$$

as $n \to \infty$. The theorem is proved. □

For sufficiently large n, $P_{n_1,n}(r_1, r)$ is close to $P(r_1, r)$ by the above theorem. Therefore the Bernoulli scheme can be thought of as sampling without replacement from a very large general population consisting of items of two types, the proportion of items of the first type being p.

In conclusion we will consider two problems.

Imagine n bins in which we place at random r enumerated particles. Each particle can be placed in any of the n bins, so that the total number of different allocations of r particles to n bins will be n^r. Allocation of particles to bins can be thought of as drawing a sample with replacement of size r from a general population of n items. We will assume that we are dealing with the classical scheme, where the probability of each outcome is $1/n^r$.

(1) What is the probability that there are exactly r_1 particles in the k-th bin? The remaining $r - r_1$ particles which did not fall into bin k are allocated to the remaining $n - 1$ bins. There are $(n-1)^{r-r_1}$ different ways in which these $r - r_1$ particles can be placed into $n - 1$ bins. Of the totality of r particles, one can choose $r - r_1$ particles which did not fall into bin k in $\binom{r}{r-r_1}$ different ways. Therefore the desired probability is

$$\binom{r}{r-r_1} \frac{(n-1)^{r-r_1}}{n^r} = \binom{r}{r-r_1} \frac{1}{n}^{r_1} \left(1 - \frac{1}{n}\right)^{r-r_1}.$$

This probability coincides with $P(r_1, r)$ in the Bernoulli scheme with $p = 1/n$.

(2) Now let us compute the probability that at least one bin will be empty. Denote this event by A. Let A_k mean that the k-th bin is empty, then

$$A = \bigcup_{k=1}^{n} A_k.$$

To find the probability of the event A, we will need a formula for the probability of a sum (union) of events. We cannot make use of the additivity of probability, for the events A_k are not disjoint in our case.

1.4 The Probability of the Union of Events. Examples

Let us return to an arbitrary discrete probability space.

Theorem 1.4.1 *Let $A_1, A_2, \ldots, A_n$ be events. Then*

$$\mathbf{P}\left(\bigcup_{i=1}^{n} A_i\right) = \sum_{i=1}^{n} \mathbf{P}(A_i) - \sum_{i<j} \mathbf{P}(A_i A_j)$$
$$+ \sum_{i<j<k} \mathbf{P}(A_i A_j A_k) - \cdots + (-1)^{n-1}\mathbf{P}(A_1 \cdots A_n).$$

Proof One has to make use of induction and the property of probability that

$$\mathbf{P}(A+B) = \mathbf{P}(A) + \mathbf{P}(B) - \mathbf{P}(AB)$$

which we proved in Sect. 1.1. For $n = 2$ the assertion of the theorem is true. Suppose it is true for any $n-1$ events $A_1, \ldots, A_{n-1}$. Then, setting $B = \bigcup_{i=1}^{n-1} A_i$, we get

$$\mathbf{P}\left(\bigcup_{i=1}^{n} A_i\right) = \mathbf{P}(B + A_n) = \mathbf{P}(B) + \mathbf{P}(A_n) - \mathbf{P}(A_n B).$$

Substituting here the known values

$$\mathbf{P}(B) = \mathbf{P}\left(\bigcup_{i=1}^{n-1} A_i\right) \quad \text{and} \quad \mathbf{P}(A_n B) = \mathbf{P}\left(\bigcup_{i=1}^{n-1} (A_i A_n)\right),$$

we obtain the assertion of the theorem. □

Now we will turn to the second problem about bins (see the end of Sect. 1.3) and find the probability of the event A that at least one bin is empty. We represented A in the form $\bigcup_{k=1}^{n} A_k$, where A_k denotes the event that all the r particles miss the k-th bin. One has

$$\mathbf{P}(A_k) = \frac{(n-1)^r}{n^r} = \left(1 - \frac{1}{n}\right)^r, \quad k \le n.$$

The event $A_k A_l$ means that all r particles are allocated to $n-2$ bins with labels differing from k and l, and therefore

$$\mathbf{P}(A_k A_l) = \frac{(n-2)^r}{n^r} = \left(1 - \frac{2}{n}\right)^r, \quad k, l \le n.$$

Similarly,

$$\mathbf{P}(A_k A_l A_m) = \frac{n-3^r}{n^r} = \left(1 - \frac{3}{n}\right)^r, \quad k, l, m \le n,$$

and so on. The probability of the event A is equal by Theorem 1.4.1 to

$$\mathbf{P}(A) = n\left(1 - \frac{1}{n}\right)^r - \binom{n}{2}\left(1 - \frac{2}{n}\right)^r + \cdots$$

$$= \sum_{j=1}^{n} (-1)^{j-1} \binom{n}{j} \left(1 - \frac{j}{n}\right)^r.$$

Discussion of this problem will be continued in Example 4.1.5.

As an example of the use of Theorem 1.4.1 we consider one more problem having many varied applications. This is the so-called *matching problem.*

Suppose n items are arranged in a certain order. They are rearranged at random (all $n!$ permutations are equally likely). What is the probability that at least one element retains its position?

There are $n!$ different permutations. Let A_k denote the event that the k-th item retains its position. This event is composed of $(n-1)!$ outcomes, so its probability equals

$$\mathbf{P}(A_k) = \frac{(n-1)!}{n!}.$$

The event $A_k A_l$ means that the k-th and l-th items retain their positions; hence

$$\mathbf{P}(A_k A_l) = \frac{(n-2)!}{n!}, \quad \ldots, \quad \mathbf{P}(A_1 \cdots A_k) = \frac{(n-(n-1))!}{n!} = \frac{1!}{n!}.$$

Now $\bigcup_{k=1}^{n} A_k$ is precisely the event that at least one item retains its position. Therefore we can make use of Theorem 1.4.1 to obtain

$$\mathbf{P}\left(\bigcup_{k=1}^{n} A_k\right) = \binom{n}{1}\frac{(n-1)!}{n!} - \binom{n}{2}\frac{(n-2)!}{n!} + \binom{n}{3}\frac{(n-3)!}{n!} - \cdots + \frac{(-1)^{n-1}}{n!}$$

$$= 1 - \frac{1}{2!} + \frac{1}{3!} - \cdots + \frac{(-1)^{n-1}}{n!}$$

$$= 1 - \left(1 - 1 + \frac{1}{2!} - \frac{1}{3!} + \cdots + \frac{(-1)^n}{n!}\right).$$

The last expression in the parentheses is the first $n+1$ terms of the expansion of e^{-1} into a series. Therefore, as $n \to \infty$,

$$\mathbf{P}\left(\bigcup_{k=1}^{n} A_k\right) \to 1 - e^{-1}.$$

Chapter 2
An Arbitrary Space of Elementary Events

Abstract The chapter begins with the axiomatic construction of the probability space in the general case where the number of outcomes of an experiment is not necessarily countable. The concepts of algebra and sigma-algebra of sets are introduced and discussed in detail. Then the axioms of probability and, more generally, measure are presented and illustrated by several fundamental examples of measure spaces. The idea of extension of a measure is discussed, basing on the Carathéodory theorem (of which the proof is given in Appendix 1). Then the general elementary properties of probability are discussed in detail in Sect. 2.2. Conditional probability given an event is introduced along with the concept of independence in Sect. 2.3. The chapter concludes with Sect. 2.4 presenting the total probability formula and the Bayes formula, the former illustrated by an example leading to the introduction of the Poisson process.

2.1 The Axioms of Probability Theory. A Probability Space

So far we have been considering problems in which the set of outcomes had at most countably many elements. In such a case we defined the probability $\mathbf{P}(A)$ using the probabilities $\mathbf{P}(\omega)$ of elementary outcomes ω. It proved to be a function defined on all the subsets A of the space Ω of elementary events having the following properties:

(1) $\mathbf{P}(A) \ge 0$.
(2) $\mathbf{P}(\Omega) = 1$.
(3) For disjoint events $A_1, A_2, \dots$

$$\mathbf{P}\Big(\bigcup A_j\Big) = \sum \mathbf{P}(A_j).$$

However, as we have already noted, one can easily imagine a problem in which the set of all outcomes is uncountable. For example, choosing a point at random from the segment $[t_1, t_2]$ (say, in an experiment involving measurement of temperature) has a continuum of outcomes, for any point of the segment could be the result of the experiment. While in experiments with finite or countable sets of outcomes any collection of outcomes was an event, this is not the case in this example. We will

A.A. Borovkov, *Probability Theory*, Universitext,
DOI 10.1007/978-1-4471-5201-9_2, © Springer-Verlag London 2013

encounter serious difficulties if we treat any subset of the segment as an event. Here one needs to select a *special class of subsets* which will be treated as events.

Let the space of elementary events Ω be an arbitrary set, and $\mathcal{A}$ be a system of subsets of Ω.

Definition 2.1.1 $\mathcal{A}$ is called an *algebra* if the following conditions are met:

A1. $\Omega \in \mathcal{A}$.

A2. *If* $A \in \mathcal{A}$ *and* $B \in \mathcal{A}$, *then*

$$A \cup B \in \mathcal{A}, \quad A \cap B \in \mathcal{A}.$$

A3. *If* $A \in \mathcal{A}$ *then* $\overline{A} \in \mathcal{A}$.

It is not hard to see that in condition A2 it suffices to require that only one of the given relations holds. The second relation will be satisfied automatically since

$$\overline{A \cap B} = \overline{A} \cup \overline{B}.$$

An algebra $\mathcal{A}$ is sometimes called a *ring* since there are two operations defined on $\mathcal{A}$ (addition and multiplication) which do not lead outside of $\mathcal{A}$. An algebra $\mathcal{A}$ is a *ring with identity*, for $\Omega \in \mathcal{A}$ and $A\Omega = \Omega A = A$ for any $A \in \mathcal{A}$.

Definition 2.1.2 A class of sets $\mathfrak{F}$ is called a *sigma-algebra* (σ*-algebra*, or σ*-ring*, or *Borel field of events*) if property A2 is satisfied for any sequences of sets:

A2$'$. *If* $\{A_n\}$ *is a sequence of sets from* $\mathfrak{F}$, *then*

$$\bigcup_{n=1}^{\infty} A_n \in \mathfrak{F}, \qquad \bigcap_{n=1}^{\infty} A_n \in \mathfrak{F}.$$

Here, as was the case for A2, it suffices to require that only one of the two relations be satisfied. The second relation will follow from the equality

$$\overline{\bigcap_n A_n} = \bigcup_n \overline{A}_n.$$

Thus an algebra is a class of sets which is closed under a *finite* number of operations of taking complements, unions and intersections; a σ-algebra is a class of sets which is closed under a *countable* number of such operations.

Given a set Ω and an algebra or σ-algebra $\mathfrak{F}$ of its subsets, one says that we are given a *measurable space* $\langle \Omega, \mathfrak{F} \rangle$.

For the segment [0, 1], all the sets consisting of a finite number of segments or intervals form an algebra, but not a σ-algebra.

Consider all the σ-algebras on $[0,1]$ containing all intervals from that segment (there is at least one such σ-algebra, for the collection of all the subsets of a given set clearly forms a σ-algebra). It is easy to see that the intersection of all such σ-algebras (i.e. the collection of all the sets which belong simultaneously to all the σ-algebras) is again a σ-algebra. It is the *smallest σ-algebra containing all intervals* and is called the *Borel σ-algebra*. Roughly speaking, the Borel σ-algebra could be thought of as the collection of sets obtained from intervals by taking countably many unions, intersections and complements. This is a rather rich class of sets which is certainly sufficient for any practical purposes. The elements of the Borel σ-algebra are called *Borel sets*. Everything we have said in this paragraph equally applies to systems of subsets of the whole real line.

Along with the intervals (a,b), the one-point sets $\{a\}$ and sets of the form $(a,b]$, $[a,b]$ and $[a,b)$ (in which a and b can take infinite values) are also Borel sets. This assertion follows, for example, from the representations of the form

$$\{a\} = \bigcap_{n=1}^{\infty}(a-1/n, a+1/n), \qquad (a,b] = \bigcap_{n=1}^{\infty}(a, b+1/n).$$

Thus all countable sets and countable unions of intervals and segments are also Borel sets.

For a given class $\mathcal{B}$ of subsets of Ω, one can again consider the intersection of all σ-algebras containing $\mathcal{B}$ and obtain in this way the *smallest σ-algebra containing $\mathcal{B}$*.

Definition 2.1.3 The smallest σ-algebra containing $\mathcal{B}$ is called the *σ-algebra generated by* $\mathcal{B}$ and is denoted by $\sigma(\mathcal{B})$.

In this terminology, the Borel σ-algebra in the n-dimensional Euclidean space $\mathbb{R}^n$ is the σ-algebra generated by rectangles or balls. If Ω is countable, then the σ-algebra generated by the elements $\omega \in \Omega$ clearly coincides with the σ-algebra of all subsets of Ω.

As an exercise, we suggest the reader to describe the algebra and the σ-algebra of sets in $\Omega = [0,1]$ generated by: (a) the intervals $(0,1/3)$ and $(1/3,1)$; (b) the semi-open intervals $(a,1]$, $0<a<1$; and (c) individual points.

To formalise a probabilistic problem, one has to find an appropriate measurable space $\langle \Omega, \mathfrak{F} \rangle$ for the corresponding experiment. The symbol Ω denotes the set of elementary outcomes of the experiment, while the algebra or σ-algebra $\mathfrak{F}$ specifies a class of events. All the remaining subsets of Ω which are not elements of $\mathfrak{F}$ *are not events*. Rather often it is convenient to define the class of events $\mathfrak{F}$ as the σ-algebra generated by a certain algebra $\mathcal{A}$.

Selecting a specific algebra or σ-algebra $\mathfrak{F}$ depends, on the one hand, on the nature of the problem in question and, on the other hand, on that of the set Ω. As we will see, one cannot always define probability in such a way that it would make sense for *any* subset of Ω.

We have already noted in Chap. 1 that, in probability theory, one uses, along with the usual set theory terminology, a somewhat different terminology related to the fact that the subsets of Ω (belonging to $\mathfrak{F}$) are interpreted as events. The set Ω itself is often called the *certain event*. By axioms A1 and A2, the empty set $\emptyset$ also belongs to $\mathfrak{F}$; it is called the *impossible event*. The event $\overline{A}$ is called the *complement event* or simply the *complement* of A. If $A \cap B = \varnothing$, then the events A and B are called *mutually exclusive* or *disjoint*.

Now it remains to introduce the notion of probability. Consider a space Ω and a system $\mathcal{A}$ of its subsets which forms an *algebra* of events.

Definition 2.1.4 A probability on $\langle \Omega, \mathcal{A} \rangle$ is a real-valued function defined on the sets from $\mathcal{A}$ and having the following properties:

P1. $\mathbf{P}(A) \ge 0$ for any $A \in \mathcal{A}$.
P2. $\mathbf{P}(\Omega) = 1$.
P3. If a sequence of events $\{A_n\}$ is such that $A_i A_j = \varnothing$ for $i \neq j$ and $\bigcup_1^\infty A_n \in \mathcal{A}$, then

$$\mathbf{P}\Biggl(\bigcup_{n=1}^{\infty} A_n\Biggr) = \sum_{n=1}^{\infty} \mathbf{P}(A_n). \tag{2.1.1}$$

These properties can be considered as an *axiomatic* definition of probability.

An equivalent to axiom P3 is the requirement of additivity (2.1.1) for *finite collections* of events A_j plus the following *continuity axiom*.

P3′. *Let $\{B_n\}$ be a sequence of events such that $B_{n+1} \subset B_n$ and $\bigcap_{n=1}^\infty B_n = B \in \mathcal{A}$. Then $\mathbf{P}(B_n) \to \mathbf{P}(B)$ as $n \to \infty$.*

Proof of the equivalence Assume P3 is satisfied and let $B_{n+1} \subset B_n$, $\bigcap_n B_n = B \in \mathcal{A}$. Then the sequence of the events B, $C_k = B_k \overline{B}_{k+1}$, $k = 1, 2, \ldots$, consists of disjoint events and $B_n = B + \bigcup_{k=n}^\infty C_k$ for any n. Now making use of property P3 we see that the series $\mathbf{P}(B_1) = \mathbf{P}(B) + \sum_{k=n}^\infty \mathbf{P}(C_k)$ is convergent, which means that

$$\mathbf{P}(B_n) = \mathbf{P}(B) + \sum_{k=n}^{\infty} \mathbf{P}(C_k) \to \mathbf{P}(B)$$

as $n \to \infty$. This is just the property P3′.

Conversely, if A_n is a sequence of disjoint events, then

$$\mathbf{P}\Biggl(\bigcup_{k=1}^{\infty} A_k\Biggr) = \mathbf{P}\Biggl(\bigcup_{k=1}^{n} A_k\Biggr) + \mathbf{P}\Biggl(\bigcup_{k=n+1}^{\infty} A_k\Biggr)$$

and one has

$$\sum_{k=1}^{\infty} \mathbf{P}(A_k) = \lim_{n\to\infty} \sum_{k=1}^{n} \mathbf{P}(A_k) = \lim_{n\to\infty} \mathbf{P}\Biggl(\bigcup_{k=1}^{n} A_k\Biggr)$$

$$= \lim_{n\to\infty} \left\{ \mathbf{P}\left(\bigcup_{k=1}^{\infty} A_k \right) - \mathbf{P}\left(\bigcup_{k=n+1}^{\infty} A_k \right) \right\} = \mathbf{P}\left(\bigcup_{k=1}^{\infty} A_k \right).$$

The last equality follows from P3′. □

Definition 2.1.5 A triple $\langle \Omega, \mathcal{A}, \mathbf{P} \rangle$ is called a *wide-sense probability space.* If an algebra $\mathfrak{F}$ is a σ-algebra ($\mathfrak{F} = \sigma(\mathfrak{F})$), then condition $\bigcup_{n=1}^{\infty} A_n \in \mathfrak{F}$ in axiom P3 (for a probability on $\langle \Omega, \mathfrak{F} \rangle$) will be automatically satisfied.

Definition 2.1.6 A triple $\langle \Omega, \mathfrak{F}, \mathbf{P} \rangle$, where $\mathfrak{F}$ is a σ-algebra, is called a *probability space.*

A probability $\mathbf{P}$ on $\langle \Omega, \mathfrak{F} \rangle$ is also sometimes called a *probability distribution* on Ω or just a *distribution* on Ω (on $\langle \Omega, \mathfrak{F} \rangle$).

Thus defining a probability space means defining a countably additive nonnegative measure on a measurable space such that the measure of Ω is equal to one. In this form the axiomatics of Probability Theory was formulated by A.N. Kolmogorov. The system of axioms we introduced is incomplete and consistent.

Constructing a probability space $\langle \Omega, \mathfrak{F}, \mathbf{P} \rangle$ is the *basic stage* in creating a mathematical model (formalisation) of an experiment.

Discussions on *what* should one understand by probability have a long history and are related to the desire to connect the definition of probability with its "physical" nature. However, because of the complexity of the latter, such attempts have always encountered difficulties not only of mathematical, but also of philosophical character (see the Introduction). The most important stages in this discussion are related to the names of Borel, von Mises, Bernstein and Kolmogorov. The emergence of Kolmogorov's axiomatics separated, in a sense, the mathematical aspect of the problem from all the rest. With this approach, the "physical interpretation" of the notion of probability appears in the form of a theorem (the strong law of large numbers, see Chaps. 5 and 7), by virtue of which the relative frequency of the occurrence of a certain event in an increasingly long series of independent trials approaches (in a strictly defined sense) the probability of this event.

We now consider examples of the most commonly used measurable and probability spaces.

1. *Discrete measurable spaces.* These are spaces $\langle \Omega, \mathfrak{F} \rangle$ where Ω is a finite or countably infinite collection of elements, and the σ-algebra $\mathfrak{F}$ usually consists of all the subsets of Ω. *Discrete probability spaces* constructed on discrete measurable spaces were studied, with concrete examples, in Chap. 1.

2. *The measurable space* $\langle \mathbb{R}, \mathfrak{B} \rangle$, where $\mathbb{R}$ is the real line(or a part of it) and $\mathfrak{B}$ is the σ-algebra of Borel sets. The necessity of considering such spaces arises in situations where the results of observations of interest may assume any values in $\mathbb{R}$.

Example 2.1.1 Consider an experiment consisting of choosing a point "at random" from the interval [0, 1]. By this we will understand the following. The set of elementary outcomes Ω is the interval [0, 1]. The σ-algebra $\mathfrak{F}$ will be taken to be the class of subsets B for which the notion of length (Lebesgue measure) $\mu(B)$ is defined—for example, the σ-algebra $\mathfrak{B}$ of Borel measurable sets. To "conduct a trial" means to choose a point $\omega \in \Omega = [0, 1]$, the probability of the event $\omega \in B$ being $\mu(B)$. All the axioms are clearly satisfied for the probability space $\langle [0, 1], \mathfrak{B}, \mu \rangle$. We obtain the so-called *uniform distribution* on $[0, 1]$.

Why did we take the σ-algebra of Borel sets $\mathfrak{B}$ to be our $\mathfrak{F}$ in this example? If we considered on $\Omega = [0, 1]$ the σ-algebra generated by "individual" points of the interval, we would get the sets of which the Lebesgue measure is either 0 or 1. In other words, the obtained sets would be either very "dense" or very "thin" (countable), so that the intervals (a, b) for $0 < b - a < 1$ do not belong to this σ-algebra.

On the other hand, if we considered on $\Omega = [0, 1]$ the σ-algebra of all subsets of Ω, it would be impossible to define a probability measure on it in such a way that $\mathbf{P}([a, b]) = b - a$ (i.e. to get the uniform distribution).[1]

Turning back to the uniform distribution $\mathbf{P}$ on $\Omega = [0, 1]$, it is easy to see that it is impossible to define this distribution using the same approach as we used to define a probability on a discrete space of elementary events (i.e. by defining the probabilities of elementary outcomes ω). Since in this example the ωs are individual points from $[0, 1]$, we clearly have $\mathbf{P}(\omega) = 0$ for any ω.

3. *The measurable space* $\langle \mathbb{R}^n, \mathfrak{B}^n \rangle$ is used in the cases when observations are vectors. Here $\mathbb{R}^n$ is the n-dimensional Euclidean space($\mathbb{R}^n = \mathbb{R}_1 \times \cdots \times \mathbb{R}^n$, where $\mathbb{R}_1, \ldots, \mathbb{R}_n$ are n copies of the real line), $\mathfrak{B}^n$ is the σ-algebra of Borel sets in $\mathbb{R}^n$, i.e. the σ-algebra generated by the sets $B = B_1 \times \cdots \times B^n$, where $B_i \subset \mathbb{R}_i$ are Borel sets on the line. Instead of $\mathbb{R}^n$ we could also consider some measurable part $\Omega \in \mathfrak{B}^n$ (for example a cube or ball), and instead of $\mathfrak{B}^n$ the restriction of $\mathfrak{B}^n$ onto Ω. Thus, similarly to the last example one can construct a probability space for choosing a point at random from the cube $\Omega = [0, 1]^n$. We put here $\mathbf{P}(\omega \in B) = \mu(B)$, where $\mu(B)$ is the Lebesgue measure (volume) of the set B. Instead of the cube $[0, 1]^n$ we could consider any other cube, for example $[a, b]^n$, but in this case we would have to put

$$\mathbf{P}(\omega \in B) = \mu(B)/\mu(\Omega) = \mu(B)/(b - a)^n.$$

This is the *uniform distribution on a cube.*

In Probability Theory one also needs to deal with more complex probability spaces. What to do if the result of the experiment is an infinite random sequence? In this case the space $\langle \mathbb{R}^\infty, \mathfrak{B}^\infty \rangle$ is often the most appropriate one.

4. *The measurable space* $\langle \mathbb{R}^\infty, \mathfrak{B}^\infty \rangle$, where

$$\mathbb{R}^\infty = \prod_{j=1}^{\infty} \mathbb{R}_j$$

[1] See e.g. [28], p. 80.

is the space of all sequences $(x_1, x_2, \ldots)$ (the direct product of the spaces $\mathbb{R}_j$), and $\mathfrak{B}^\infty$ the σ-algebra generated by the sets of the form

$$\left(\prod_{k=1}^{N} B_{j_k}\right) \times \left(\prod_{\substack{j \neq j_k \\ k \leq N}} \mathbb{R}_j\right); \quad B_{j_k} \in \mathfrak{B}_{j_k},$$

for any $N, j_1, \ldots, j_N$, where $\mathfrak{B}_j$ is the σ-algebra of Borel sets from $\mathbb{R}_j$.

5. If an experiment results, say, in a continuous function on the interval $[a, b]$ (a trajectory of a moving particle, a cardiogram of a patient, etc.), then the probability spaces considered above turn out to be inappropriate. In such a case one should take Ω to be the space $C(a, b)$ of all continuous functions on $[a, b]$ or the space $\mathbb{R}^{[a,b]}$ of all functions on $[a, b]$. The problem of choosing a suitable σ-algebra here becomes somewhat more complicated and we will discuss it later in Chap. 18.

Now let us return to the definition of a probability space.

Let a triple $\langle \Omega, \mathcal{A}, \mathbf{P} \rangle$ be a wide-sense probability space ($\mathcal{A}$ is an algebra). As we have already seen, to each algebra $\mathcal{A}$ there corresponds a σ-algebra $\mathfrak{F} = \sigma(\mathcal{A})$ generated by $\mathcal{A}$. The following question is of substantial interest: does the probability measure $\mathbf{P}$ on $\mathcal{A}$ define a measure on $\mathfrak{F} = \sigma(\mathcal{A})$? And if so, does it define it in a unique way? In other words, to construct a probability space $\langle \Omega, \mathcal{A}, \mathbf{P} \rangle$, is it sufficient to define the probability just on some algebra $\mathcal{A}$ generating $\mathfrak{F}$ (i.e. to construct a wide-sense probability space $\langle \Omega, \mathcal{A}, \mathbf{P} \rangle$, where $\sigma(\mathcal{A}) = \mathfrak{F}$)? An answer to this important question is given by the Carathéodory theorem.

The measure extension theorem *Let $\langle \Omega, \mathcal{A}, \mathbf{P} \rangle$ be a wide-sense probability space. Then there exists a unique probability measure $\mathbf{Q}$ defined on $\mathfrak{F} = \sigma(\mathcal{A})$ such that*

$$\mathbf{Q}(A) = \mathbf{P}(A) \quad \textit{for all } A \in \mathcal{A}.$$

Corollary 2.1.1 *Any wide-sense probability space $\langle \Omega, \mathcal{A}, \mathbf{P} \rangle$ automatically defines a probability space $\langle \Omega, \mathfrak{F}, \mathbf{P} \rangle$ with $\mathfrak{F} = \sigma(\mathcal{A})$.*

We will make extensive use of this fact in what follows. In particular, it implies that to define a probability measure on the measurable space $\langle \mathbb{R}, \mathfrak{B} \rangle$, it suffices to define the probability on intervals.

The proof of the Carathéodory theorem is given in Appendix 1.

In conclusion of this section we will make a general comment. Mathematics differs qualitatively from such sciences as physics, chemistry, etc. in that it does not always base its conclusions on empirical data with the help of which a naturalist tries to answer his questions. Mathematics develops in the framework of an initial construction or system of axioms with which one describes an object under study. Thus mathematics and, in particular, Probability Theory, studies the nature of the phenomena around us in a methodologically different way: one studies not the phenomena themselves, but rather the *models* of these phenomena that have been created based on human experience. The value of a particular model is determined by the agreement of the conclusions of the theory with our observations and therefore depends on the choice of the axioms characterising the object.

In this sense axioms P1, P2, and the additivity of probability look indisputable and natural (see the remarks in the Introduction on desirable properties of probability). Countable additivity of probability and the property A2′ of σ-algebras are more delicate and less easy to intuit (as incidentally are a lot of other things related to the notion of infinity). Introducing the last two properties was essentially brought about by the possibility of constructing a meaningful mathematical theory. Numerous applications of Probability Theory developed from the system of axioms formulated in the present section demonstrate its high efficiency and purposefulness.

2.2 Properties of Probability

1. $\mathbf{P}(\varnothing)=0$. This follows from the equality $\varnothing+\Omega=\Omega$ and properties P2 and P3 of probability.

2. $\mathbf{P}(\overline{A})=1-\mathbf{P}(A)$, since $A+\overline{A}=\Omega$ and $A\cap\overline{A}=\varnothing$.

3. If $A\subset B$, then $\mathbf{P}(A)\le\mathbf{P}(B)$. This follows from the relation $\mathbf{P}(A)+\mathbf{P}(\overline{A}B)=\mathbf{P}(B)$.

4. $\mathbf{P}(A)\le 1$ (by properties 3 and P2).

5. $\mathbf{P}(A\cup B)=\mathbf{P}(A)+\mathbf{P}(B)-\mathbf{P}(AB)$, since $A\cup B=A+(B-AB)$ and $\mathbf{P}(B-AB)=\mathbf{P}(B)-\mathbf{P}(AB)$.

6. $\mathbf{P}(A\cup B)\le\mathbf{P}(A)+\mathbf{P}(B)$ follows from the previous property.

7. The formula

$$\mathbf{P}\left(\bigcup_{j=1}^{n}A_j\right)=\sum_{k=1}^{n}\mathbf{P}(A_k)-\sum_{k<l}\mathbf{P}(A_kA_l)+\sum_{k<l<m}\mathbf{P}(A_kA_lA_m)-\cdots+(-1)^{n-1}\mathbf{P}(A_1\ldots A_n)$$

has already been proved and used for discrete spaces Ω. Here the reader can prove it in exactly the same way, using induction and property 5.

Denote the sums on the right hand side of the last formula by $Z_1, Z_2, \ldots, Z_n$, respectively. Then statement 7 for the event $B_n=\bigcup_{j=1}^{n}A_j$ can be rewritten as $\mathbf{P}(B_n)=\sum_{j=1}^{n}(-1)^{j-1}Z_j$.

8. An important addition to property 7 is that *the sequence* $\sum_{j=1}^{k}(-1)^{j-1}Z_j$ *approximates* $\mathbf{P}(B_n)$ *by turns from above and from below as k grows*, i.e.

$$\begin{aligned}&\mathbf{P}(B_n)-\sum_{j=1}^{2k-1}(-1)^{j-1}Z_j\le 0,\\&\mathbf{P}(B_n)-\sum_{j=1}^{2k}(-1)^{j-1}Z_j\ge 0,\quad k=1,2,\ldots\end{aligned}\tag{2.2.1}$$

This property can also be proved by induction on n. For $n = 2$ this property is ascertained in 5. Let (2.2.1) be valid for any events $A_1, \dots, A_{n-1}$ (i.e. for any B_{n-1}). Then by 5 we have

$$\mathbf{P}(B_n) = \mathbf{P}(B_{n-1} \cup A_n) = \mathbf{P}(B_{n-1}) + \mathbf{P}(A_n) - \mathbf{P}\left(\bigcup_{j=1}^{k-1} A_j A_n\right),$$

where, in view of (2.2.1) for $k = 1$,

$$\sum_{j=1}^{n=1} \mathbf{P}(A_j) - \sum_{i<j}^{n-1} \mathbf{P}(A_i A_j) \le \mathbf{P}(B_{n-1}) \le \sum_{j=1}^{n-1} \mathbf{P}(A_j),$$

$$\mathbf{P}\left(\bigcup_{j=1}^{n-1} A_j A_n\right) \le \sum_{j=1}^{n-1} \mathbf{P}(A_j A_n).$$

Hence, for $B_n = B_{n-1} \cup A_n$, we get

$$\mathbf{P}(B_n) \le \sum_{j=1}^{n} \mathbf{P}(A_j),$$

$$\mathbf{P}(B_n) = \mathbf{P}(B_{n-1}) + \mathbf{P}(A_n) - \mathbf{P}(B_{n-1} A_n)$$
$$\ge \sum_{j=1}^{n} \mathbf{P}(A_j) - \sum_{i<j}^{n-1} \mathbf{P}(A_i A_j) - \sum_{i=1}^{n-1} \mathbf{P}(A_i A_n) = \sum_{j=1}^{n} \mathbf{P}(A_n) - \sum_{i<j}^{n} \mathbf{P}(A_i A_j).$$

This proves (2.2.1) for $k = 1$. For $k = 2, 3, \dots$ the proof is similar.

9. If A_n is a monotonically increasing sequence of sets (i.e. $A_n \subset A_{n+1}$) and $A = \bigcup_{n=1}^{\infty} A_n$, then

$$\mathbf{P}(A) = \lim_{n \to \infty} \mathbf{P}(A_n). \tag{2.2.2}$$

This is a different form of the continuity axiom equivalent to P3′.

Indeed, introducing the sets $B_n = A - A_n$, we get $B_{n+1} \subset B_n$ and $\bigcap_{n=1}^{\infty} B_n = \varnothing$. Therefore, by the continuity axiom,

$$\mathbf{P}(A - A_n) = \mathbf{P}(A) - \mathbf{P}(A_n) \to 0$$

as $n \to \infty$. The converse assertion that (2.2.2) implies the continuity axiom can be obtained in a similar way. □

2.3 Conditional Probability. Independence of Events and Trials

We will start with examples. Let an experiment consist of three tosses of a fair coin. The probability that heads shows up only once, i.e. that one of the elementary events htt, tht, or tth occurs, is equal in the classical scheme to $3/8$. Denote this event by A. Now assume that we know in addition that the event $B = \{\textit{the number of heads is odd}\}$ has occurred.

What is the probability of the event A given this additional information? The event B consists of four elementary outcomes. The event A is constituted by three outcomes from the event B. In the framework of the classical scheme, it is natural to define the new probability of the event A to be $3/4$.

Consider a more general example. Let a classical scheme with n outcomes be given. An event A consists of r outcomes, an event B of m outcomes, and let the event AB have k outcomes. Similarly to the previous example, it is natural to define the probability of the event A given the event B has occurred as

$$\mathbf{P}(A|B) = \frac{k}{m} = \frac{k/n}{m/n}.$$

The ratio is equal to $\mathbf{P}(AB)/\mathbf{P}(B)$, for

$$\mathbf{P}(A|\,B) = \frac{k}{n}, \qquad \mathbf{P}(B) = \frac{m}{n}.$$

Now we can give a general definition.

Definition 2.3.1 Let $\langle \Omega, \mathfrak{F}, \mathbf{P}\rangle$ be a probability space and A and B be arbitrary events. If $\mathbf{P}(B) > 0$, the *conditional probability* of the event A given B has occurred is denoted by $\mathbf{P}(A|B)$ and is defined by

$$\mathbf{P}(A|B) := \frac{\mathbf{P}(AB)}{\mathbf{P}(B)}.$$

Definition 2.3.2 Events A and B are called *independent* if

$$\mathbf{P}(AB) = \mathbf{P}(A)\,\mathbf{P}(B).$$

Below we list several properties of independent events.

1. If $\mathbf{P}(B) > 0$, then the independence of A and B is equivalent to the equality

$$\mathbf{P}(A|B) = \mathbf{P}(A).$$

The proof is obvious.

2. If A and B are independent, then $\overline{A}$ and B are also independent.

Indeed,

$$\begin{aligned}\mathbf{P}(\overline{A}B) &= \mathbf{P}(B - AB)\\ &= \mathbf{P}(B) - \mathbf{P}(AB) = \mathbf{P}(B)\bigl(1 - \mathbf{P}(A)\bigr) = \mathbf{P}(\overline{A})\mathbf{P}(B).\end{aligned}$$

3. Let the events A and B_1 and the events A and B_2 each be independent, and assume $B_1B_2 = \varnothing$. Then the events A and $B_1 + B_2$ are independent.

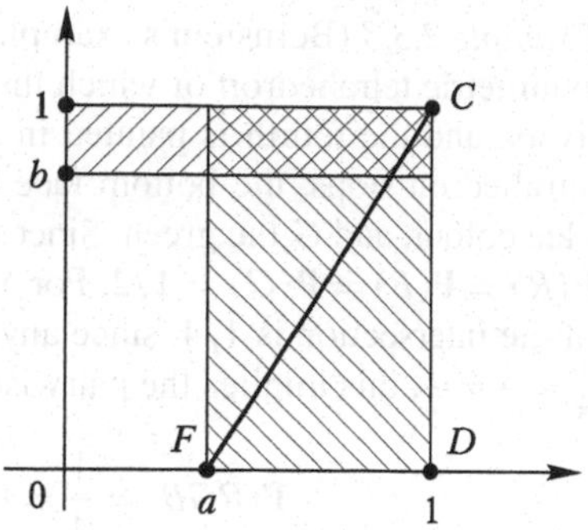

Fig. 2.1 Illustration to Example 2.3.2: the *dashed rectangles* represent the events A and B

The property is proved by the following chain of equalities:

$$\begin{aligned}\mathbf{P}\big(A(B_1+B_2)\big) &= \mathbf{P}(AB_1+AB_2) = \mathbf{P}(AB_1)+\mathbf{P}(AB_2)\\ &= \mathbf{P}(A)\big(\mathbf{P}(B_1)+\mathbf{P}(B_2)\big) = \mathbf{P}(A)\mathbf{P}(B_1+B_2).\end{aligned}$$

As we will see below, the requirement $B_1B_2 = \varnothing$ is essential here.

Example 2.3.1 Let event A mean that heads shows up in the first of two tosses of a fair coin, and event B that tails shows up in the second toss. The probability of each of these events is $1/2$. The probability of the intersection AB is

$$\mathbf{P}(AB) = \frac{1}{4} = \frac{1}{2}\cdot\frac{1}{2} = \mathbf{P}(A)\mathbf{P}(B).$$

Therefore the events A and B are independent.

Example 2.3.2 Consider the uniform distribution on the square $[0,1]^2$ (see Sect. 2.1). Let A be the event that a point chosen at random is in the region on the right of an abscissa a and B the event that the point is in the region above an ordinate b.

Both regions are hatched in Fig. 2.1. The event AB is squared in the figure. Clearly, $\mathbf{P}(AB) = \mathbf{P}(A)\mathbf{P}(B)$, and hence the events A and B are independent.

It is also easy to verify that if B is the event that the chosen point is inside the triangle FCD (see Fig. 2.1), then the events A and B will already be dependent.

Definition 2.3.3 Events $B_1, B_2, \ldots, B_n$ are *jointly independent* if, for any $1 \le i_1 < i_2 < \cdots < i_r \le n$, $r = 2, 3, \ldots, n$,

$$\mathbf{P}\Bigg(\bigcap_{k=1}^{r} B_{j_k}\Bigg) = \prod_{k=1}^{r} \mathbf{P}(B_{i_k}).$$

Pairwise independence is not sufficient for joint independence of n events, as one can see from the following example.

Example 2.3.3 (Bernstein's example) Consider the following experiment. We roll a symmetric tetrahedron of which three faces are painted red, blue and green respectively, and the fourth is painted in all three colours. Event R means that when the tetrahedron stops, the bottom face has the red colour on it, event B that it has the blue colour, and G the green. Since each of the three colours is present on two faces, $\mathbf{P}(R)=\mathbf{P}(B)=\mathbf{P}(G)=1/2$. For any two of the introduced events, the probability of the intersection is $1/4$, since any two colours are present on one face only. Since $\frac{1}{4}=\frac{1}{2}\times\frac{1}{2}$, this implies the pairwise independence of all three events. However,

$$\mathbf{P}(RGB)=\frac{1}{4}\neq\mathbf{P}(R)\mathbf{P}(B)\mathbf{P}(G)=1/8. \quad \square$$

Now it is easy to construct an example in which property 3 of independent events does not hold when $B_1B_2\neq\varnothing$.

An example of a sequence of jointly independent events is given by the series of outcomes of trials in the Bernoulli scheme.

If we assume that each outcome was obtained as a result of a *separate trial*, then we will find that any event related to a fixed trial will be independent of any event related to other trials. In such cases one speaks of a sequence of *independent trials*.

To give a general definition, consider two arbitrary experiments G_1 and G_2 and denote by $\langle\Omega_1,\mathfrak{F}_1,\mathbf{P}_1\rangle$ and $\langle\Omega_2,\mathfrak{F}_2,\mathbf{P}_2\rangle$ the respective probability spaces. Consider also the "compound" experiment G with the probability space $\langle\Omega,\mathfrak{F},\mathbf{P}\rangle$, where $\Omega=\Omega_1\times\Omega_2$ is the direct product of the spaces Ω_1 and Ω_2, and the σ-algebra $\mathfrak{F}$ is generated by the direct product $\mathfrak{F}_1\times\mathfrak{F}_2$ (i.e. by the events $B=B_1\times B_2$, $B_1\in\mathfrak{F}_1$, $B_2\in\mathfrak{F}_2$).

Definition 2.3.4 We will say that the *trials G_1 and G_2 are independent* if, for any $B=B_1\times B_2$, $B_1\in\mathfrak{F}_1$, $B_2\in\mathfrak{F}_2$ one has

$$\mathbf{P}(B)=\mathbf{P}_1(B_1)\mathbf{P}_2(B_2)=\mathbf{P}(B_1\times\Omega_2)\mathbf{P}(\Omega_1\times B_2).$$

Independence of n trials $G_1,\dots,G_n$ is defined in a similar way, using the equality

$$\mathbf{P}(B)=\mathbf{P}_1(B_1)\cdots\mathbf{P}_n(B_n),$$

where $B=B_1\times\dots\times B_n$, $B_k\in\mathfrak{F}_k$, and $\langle\Omega_k,\mathfrak{F}_k,\mathbf{P}_k\rangle$ is the probability space corresponding to the experiment G_k, $k=1,\dots,n$.

In the Bernoulli scheme, the probability of any sequence of outcomes consisting of r zeros and ones and containing k ones is equal to $p^k(1-p)^{r-k}$. Therefore the Bernoulli scheme may be considered as a result of r independent trials in each of which one has 1 (success) with probability p and 0 (failure) with probability $1-p$. Thus, the probability of k successes in r independent trials equals $\binom{r}{k}p^k(1-p)^{r-k}$.

The following assertion, which is in a sense converse to the last one, is also true: any sequence of identical independent trials with two outcomes makes up a Bernoulli scheme.

In Chap. 3 several remarks will be given on the relationship between the notions of independence we introduced here and the common notion of causality.

2.4 The Total Probability Formula. The Bayes Formula

Let A be an event and $B_1, B_2, \dots, B_n$ be mutually exclusive events having positive probabilities such that

$$A \subset \bigcup_{j=1}^{n} B_j.$$

The sequence of events $B_1, B_2, \dots$ can be infinite, in which case we put $n = \infty$. The following *total probability formula* holds true:

$$\mathbf{P}(B) = \sum_{j=1}^{n} \mathbf{P}(B_j)\mathbf{P}(A|B_j).$$

Proof It follows from the assumptions that

$$A = \bigcup_{j=1}^{n} B_j A.$$

Moreover, the events $AB_1, AB_2, \dots, AB_n$ are disjoint, and hence

$$\mathbf{P}(A) = \sum_{j=1}^{n} \mathbf{P}(AB_j) = \sum_{j=1}^{n} \mathbf{P}(B_j)\mathbf{P}(A|B_j).$$

□

Example 2.4.1 In experiments with colliding electron-positron beams, the probability that during a time unit there will occur j collisions leading to the birth of new elementary particles is equal to

$$p_j = \frac{e^{-\lambda}\lambda^j}{j!}, \quad j = 0, 1, \dots,$$

where λ is a positive parameter (this is the so-called Poisson distribution, to be considered in more detail in Chaps. 3, 5 and 19). In each collision, different groups of elementary particles can appear as a result of the interaction, and the probability of each group is fixed and does not depend on the outcomes of other collisions. Consider one such group, consisting of two μ-mesons, and denote by p the probability of its appearance in a collision. What is the probability of the event A_k that, during a time unit, k pairs of μ-mesons will be born?

Assume that the event B_j that there were j collisions during the time unit has occurred. Given this condition, we will have a sequence of j independent trials, and the probability of having k pairs of μ-mesons will be $\binom{j}{k} p^k (1-p)^{j-k}$. Therefore by the total probability formula,

$$\mathbf{P}(A_k) = \sum_{j=k}^{\infty} \mathbf{P}(B_j)\mathbf{P}(A_k|B_j) = \sum_{j=k}^{\infty} \frac{e^{-\lambda}\lambda^j}{j!} \frac{j!}{k!(j-k)!} p^k (1-p)^{j-k}$$

$$= \frac{e^{-\lambda} p^k \lambda^k}{k!} \sum_{j=0}^{\infty} \frac{(\lambda(1-p))^j}{j!} = \frac{e^{-\lambda p}(\lambda p)^k}{k!}.$$

Thus we again obtain a Poisson distribution, but this time with parameter λp.

The solution above was not formalised. A formal solution would first of all require the construction of a probability space. The space turns out to be rather complex in this example. Denote by Ω_j the space of elementary outcomes in the Bernoulli scheme corresponding to j trials, and let ω_j denote an element of Ω_j. Then one could take Ω to be the collection of all pairs $\{(j, \omega_j)\}_{j=0}^{\infty}$, where the number j indicates the number of collisions, and ω_j is a sequence of "successes" and "failures" of length j ("success" stands for the birth of two μ-mesons). If ω_j contains k "successes", one has to put

$$\mathbf{P}\big((j, \omega_j)\big) = p_j p^k (1-p)^{j-k}.$$

To get $\mathbf{P}(A_k)$, it remains to sum up these probabilities over all ω_j containing k successes and all $j \ge k$ (the idea of the total probability formula is used here tacitly when splitting A_k into the events (j, ω_j)).

The fact that the number of collisions is described here by a Poisson distribution could be understood from the following circumstances related to the nature of the physical process. Let $B_j(t,u)$ be the event that there were j collisions during the time interval $[t, t+u)$. Then it turns out that:

(a) the pairs of events $B_j(v,t)$ and $B_k(v+t,u)$ related to non-overlapping time intervals are independent for all v, t, u, j, and k;
(b) for small Δ the probability of a collision during the time Δ is proportional to Δ:

$$\mathbf{P}\big(B_1(t, \Delta)\big) = \lambda \Delta + o(\Delta),$$

and, moreover, $\mathbf{P}(B_k(t, \Delta)) = o(\Delta)$ for $k \ge 2$.

Again using the total probability formula with the hypotheses $B_j(v,t)$, we obtain for the probabilities $p_k(t) = \mathbf{P}(B_k(v,t))$ the following relations:

$$\begin{aligned} p_k(t+\Delta) &= \sum_{j=0}^{k} p_j(t)\,\mathbf{P}\big(B_k(v, t+\Delta) \bigm| B_j(v,t)\big) \\ &= \sum_{j=0}^{k} p_j(t)\mathbf{P}\big(B_{k-j}(v+t, \Delta)\big) = o(\Delta) + p_{k-1}(t)\big(\lambda\Delta + o(\Delta)\big) \\ &= p_k(t)\big(1 - \lambda\Delta - o(\Delta)\big), \quad k \ge 1; \\ p_0(t+\Delta) &= p_0(t)\big(1 - \lambda\Delta - o(\Delta)\big). \end{aligned}$$

Transforming the last equation, we find that

$$\frac{p_0(t+\Delta) - p_0(t)}{\Delta} = -\lambda p_0(t) + o(1).$$

Therefore the derivative of p_0 exists and is given by

$$p_0'(t) = -\lambda p_0(t).$$

In a similar way we establish the existence of

$$p_k'(t) = \lambda p_{k-1}(t) - \lambda p_k(t), \quad k \ge 1. \tag{2.4.1}$$

Now note that since the functions $p_k(t)$ are continuous, one should put $p_0(0) = 1$, $p_k(0) = 0$ for $k \ge 1$. Hence

$$p_0(t) = e^{-\lambda t}.$$

Using induction and substituting into (2.4.1) the function $p_{k-1}(t) = \frac{(\lambda t)^{k-1} e^{-\lambda t}}{(k-1)!}$, we establish (it is convenient to make the substitution $p_k = e^{-\lambda t} u_k$, which turns (2.4.1) into $u_k' = \frac{\lambda(\lambda t)^{k-1}}{(k-1)!}$) that

$$p_k(t) = \frac{(\lambda t)^k e^{-\lambda t}}{k!}, \quad k = 0, 1, \ldots$$

This is the Poisson distribution with parameter λt.

To understand the construction of the probability space in this problem, one should consider the set Ω of all non-decreasing step-functions $x(t) \ge 0$, $t \ge 0$, taking values $0, 1, 2, \ldots$. Any such function can play the role of an elementary outcome: its jump points indicate the collision times, the value $x(t)$ itself will be the number of collisions during the time interval $(0, t)$. To avoid a tedious argument related to introducing an appropriate σ-algebra, for the purposes of our computations we could treat the probability as given on the *algebra* $\mathcal{A}$ (see Sect. 2.1) generated by the sets $\{x(t) = k\}$, $t \ge 0$; $k = 0, 1, \ldots$ (note that all the events considered in this problem are just of such form). The above argument shows that one has to put

$$\mathbf{P}\big(x(v+t) - x(v) = k\big) = \frac{(\lambda t)^k e^{-\lambda t}}{k!}.$$

(See also the treatment of Poisson processes in Chap. 19.) □

By these examples we would like not only to illustrate the application of the total probability formula, but also to show that the construction of probability spaces in real problems is not always a simple task.

Of course, for each particular problem, such constructions are by no means necessary, but we would recommend to carry them out until one acquires sufficient experience.

Assume that events A and $B_1, \ldots, B_n$ satisfy the conditions stated at the beginning of this section. If $\mathbf{P}(A) > 0$, then under these conditions the following *Bayes' formula* holds true:

$$\mathbf{P}(B_j | A) = \frac{\mathbf{P}(B_j)\mathbf{P}(A | B_j)}{\sum_{k=1}^{n} \mathbf{P}(B_k)\mathbf{P}(A | B_k)}.$$

This formula is simply an alternative way of writing the equality

$$\mathbf{P}(B_j|A) = \frac{\mathbf{P}(B_j A)}{\mathbf{P}(A)},$$

where in the numerator one should make use of the definition of conditional probability, and in the denominator, the total probability formula. In Bayes' formula we can take $n = \infty$, just as for the total probability formula.

Example 2.4.2 An item is manufactured by two factories. The production volume of the first factory is k times the production of the second one. The proportion of defective items for the first factory is P_1, and for the second one P_2. Now assume that the items manufactured by the factories during a certain time interval were mixed up and then sent to retailers. What is the probability that you have purchased an item produced by the second factory given the item proved to be defective?

Let B_1 be the event that the item you have got came from the first factory, and B_2 from the second. It easy to see that

$$\mathbf{P}(B_1) = \frac{1}{1+k}, \qquad \mathbf{P}(B_2) = \frac{k}{1+k}.$$

These are the so-called *prior* probabilities of the events B_1 and B_2. Let A be the event that the purchased item is defective. We are given conditional probabilities $\mathbf{P}(A|B_1) = P_1$ and $\mathbf{P}(A|B_2) = P_2$. Now, using Bayes' formula, we can answer the posed question:

$$\mathbf{P}(B_2|A) = \frac{\frac{k}{1+k}P_2}{\frac{1}{1+k}P_1 + \frac{k}{1+k}P_2} = \frac{kP_2}{P_1 + kP_2}.$$

Similarly, $\mathbf{P}(B_1|A) = \frac{P_1}{P_1+kP_2}$. □

The probabilities $\mathbf{P}(B_1|\,A)$ and $\mathbf{P}(B_2|\,A)$ are sometimes called *posterior* probabilities of the events B_1 and B_2 respectively, after the event A has occurred.

Example 2.4.3 A student is suggested to solve a numerical problem. The answer to the problem is known to be one of the numbers $1, \ldots, k$. Solving the problem, the student can either find the correct way of reasoning or err. The training of the student is such that he finds a correct way of solving the problem with probability p. In that case the answer he finds coincides with the right one. With the complementary probability $1 - p$ the student makes an error. In that case we will assume that the student can give as an answer any of the numbers $1, \ldots, k$ with equal probabilities $1/k$.

We know that the student gave a correct answer. What is the probability that his solution of the problem was correct?

Let B_1 (B_2) be the event that the student's solution was correct (wrong). Then, by our assumptions, the prior probabilities of these events are $\mathbf{P}(B_1) = p$,

$\mathbf{P}(B_2) = 1 - p$. If the event A means that the student got a correct answer, then

$$\mathbf{P}(A|B_1) = 1, \qquad \mathbf{P}(A|B_2) = 1/k.$$

By Bayes' formula the desired posterior probability $\mathbf{P}(B_1|\,A)$ is equal to

$$\mathbf{P}(B_1|A) = \frac{\mathbf{P}(B_1)\mathbf{P}(A|\,B_1)}{\mathbf{P}(B_1)\mathbf{P}(A|B_1) + \mathbf{P}(B_2)\mathbf{P}(A|B_2)} = \frac{p}{p + \frac{1-p}{k}} = \frac{1}{1 + \frac{1-p}{kp}}.$$

Clearly, $\mathbf{P}(B_1|A) > \mathbf{P}(B_1) = p$ and $\mathbf{P}(B_1|A)$ is close to 1 for large k.

Chapter 3
Random Variables and Distribution Functions

Abstract Section 3.1 introduces the formal definitions of random variable and its distribution, illustrated by several examples. The main properties of distribution functions, including a characterisation theorem for them, are presented in Sect. 3.2. This is followed by listing and briefly discussing the key univariate distributions. The second half of the section is devoted to considering the three types of distributions on the real line and the distributions of functions of random variables. In Sect. 3.3 multivariate random variables (random vectors) and their distributions are introduced and discussed in detail, including the two key special cases: the multinomial and the normal (Gaussian) distributions. After that, the concepts of independence of random variables and that of classes of events are considered in Sect. 3.4, establishing criteria for independence of random variables of different types. The theorem on independence of sigma-algebras generated by independent algebras of events is proved with the help of the probability approximation theorem. Then the relationships between the introduced notions are extensively discussed. In Sect. 3.5, the problem of existence of infinite sequences of random variables is solved with the help of Kolmogorov's theorem on families of consistent distributions, which is proved in Appendix 2. Section 3.6 is devoted to discussing the concept of integral in the context of Probability Theory (a formal introduction to Integration Theory is presented in Appendix 3). The integrals of functions of random vectors are discussed, including the derivation of the convolution formulae for sums of independent random variables.

3.1 Definitions and Examples

Let $\langle \Omega, \mathfrak{F}, \mathbf{P} \rangle$ be an arbitrary probability space.

Definition 3.1.1 A *random variable* ξ is a measurable function $\xi = \xi(\omega)$ mapping $\langle \Omega, \mathfrak{F} \rangle$ into $\langle \mathbb{R}, \mathfrak{B} \rangle$, where $\mathbb{R}$ is the set of real numbers and $\mathfrak{B}$ is the σ-algebra of all Borel sets, i.e. a function for which the inverse image $\xi^{(-1)}(B) = \{\omega : \xi(\omega) \in B\}$ of any Borel set $B \in \mathfrak{B}$ is a set from the σ-algebra $\mathfrak{F}$.

A.A. Borovkov, *Probability Theory*, Universitext,
DOI 10.1007/978-1-4471-5201-9_3, © Springer-Verlag London 2013

For example, when tossing a coin once, Ω consists of two points: heads and tails. If we put 1 in correspondence to heads and 0 to tails, we will clearly obtain a random variable.

The number of points showed up on a die will also be a random variable.

The distance between the origin to a point chosen at random in the square $[0 \le x \le 1, 0 \le y \le 1]$ will also be a random variable, since the set $\{(x, y) : x^2 + y^2 < t\}$ is measurable. The reader might have already noticed that in these examples it is very difficult to come up with a non-measurable function of ω which would be related to any real problem. This is often the case, but not always. In Chap. 18, devoted to random processes, we will be interested in sets which, generally speaking, are not events and which require special modifications to be regarded as events.

As we have already mentioned above, it follows from the definition of a random variable that, for any set B from the σ-algebra $\mathfrak{B}$ of Borel sets on the real line,

$$\xi^{(-1)}(B) = \{\omega : \xi(\omega) \in B\} \in \mathfrak{F}.$$

Hence one can define a probability $\mathbf{F}_\xi(B) = \mathbf{P}(\xi \in B)$ on the measurable space $\langle \mathbb{R}, \mathfrak{B} \rangle$ which generates the probability space $\langle \mathbb{R}, \mathfrak{B}, \mathbf{F}_\xi \rangle$.

Definition 3.1.2 The probability $\mathbf{F}_\xi(B)$ is called the *distribution of the random variable* ξ.

Putting $B = (-\infty, x)$ one obtains the function

$$F_\xi(x) = \mathbf{F}_\xi(-\infty, x) = \mathbf{P}(\xi < x)$$

defined on the whole real line which is called the *distribution function*[1] *of the random variable* ξ.

We will see below that the distribution function of a random variable completely specifies its distribution and is often used to describe the latter.

Where it leads to no confusion, we will write just $\mathbf{F}$, $F(x)$ instead of $\mathbf{F}_\xi$, $F_\xi(x)$, respectively. More generally, in what follows, as a rule, we will be using boldface letters $\mathbf{F}$, $\mathbf{G}$, $\mathbf{I}$, $\boldsymbol{\Phi}$, $\mathbf{K}$, $\boldsymbol{\Pi}$, etc. to denote distributions, and the standard font letters F, G, I, Φ, ... to denote the respective distribution functions.

Since a random variable ξ is a mapping of Ω into $\mathbb{R}$, one has $\mathbf{P}(|\xi| < \infty) = 1$. Sometimes, it is also convenient to consider along with such random variables random variables which can assume the values $\pm\infty$ (they will be measurable mappings of Ω into $\mathbb{R} \cup \{\pm\infty\}$). If $\mathbf{P}(|\xi| = \infty) > 0$, we will call such random variables $\xi(\omega)$ *improper*. Each situation where such random variables appear will be explicitly noted.

Example 3.1.1 Consider the Bernoulli scheme with success probability p and sample size k (see Sect. 3.3). As we know, the set of elementary outcomes Ω in this case

[1] In the English language literature, the distribution function is conventionally defined as $F_\xi(x) = \mathbf{P}(\xi \le x)$. The only difference is that, with the latter definition, F will be right-continuous, cf. property F3 below.

is the set of all k-tuples of zeros and ones. Take the σ-algebra $\mathfrak{F}$ to be the system of all subsets of Ω. Define a random variable on Ω as follows: to each k-tuple of zeros and ones we relate the number of ones in this tuple.

The probability of r successes is, as we already know,

$$P(r,k) = \binom{k}{r} p^r (1-p)^{k-r}.$$

Therefore the distribution function $F(x)$ of our random variable will be defined as

$$F(x) = \sum_{r<x} P(r,k).$$

Here the summation is over all integers r which are less than x. If $x \le 0$ then $F(x) = 0$, and if $x > k$ then $F(x) = 1$.

Example 3.1.2 Suppose we choose a point at random from the segment $[a,b]$, i.e. the probability that the chosen point is in a subset of $[a,b]$ is taken to be proportional to the Lebesgue measure of this subset. Here, Ω is the segment $[a,b]$, the σ-algebra $\mathfrak{F}$ is the class of Borel subsets of $[a,b]$. Define a random variable ξ by

$$\xi(\omega) = \omega, \qquad \omega \in [a,b],$$

i.e. the value of the random variable is equal to the number from $[a,b]$ we have chosen. It is a measurable function. If $x \le a$, then $F(x) = \mathbf{P}(\xi < x) = 0$. Let $x \in (a,b]$. Then $\{\xi < x\}$ means that the point is in the interval $[a,x)$. The probability of this event is proportional to the length of the interval, hence

$$F(x) = \mathbf{P}(\xi < x) = \frac{x-a}{b-a}.$$

If $x > b$, then clearly $F(x) = 1$. Finally, we find that

$$F(x) = \begin{cases} 0, & x < a, \\ \frac{x-a}{b-a}, & a \le x \le b, \\ 1, & x > b. \end{cases} \qquad (3.1.1)$$

This distribution function defines the so-called *uniform distribution* on the interval $[a,b]$.

If $\mu(B)$ is the Lebesgue measure on $\langle \mathbb{R}, \mathfrak{B} \rangle$, then, as we will see in the next section, it is not hard to show that in this case $\mathbf{F}_\xi(B) = \mu(B \cap [a,b])/(b-a)$.

3.2 Properties of Distribution Functions. Examples

3.2.1 The Basic Properties of Distribution Functions

Let $F(x)$ be the distribution function of a random variable ξ. Then $F(x)$ has the following properties:

F1. *Monotonicity*: *if* $x_1 < x_2$, *then* $F(x_1) \le F(x_2)$.
F2. $\lim_{x\to-\infty} F(x) = 0$ *and* $\lim_{x\to\infty} F(x) = 1$.
F3. *Left-continuity*: $\lim_{x\uparrow x_0} F(x) = F(x_0)$.

Proof Since for $x_1 \le x_2$ one has $\{\xi < x_1\} \subseteq \{\xi < x_2\}$, F1 immediately follows from property 3 of probability (see Sect. 3.2.2).

To prove F2, consider two number sequences $\{x_n\}$ and $\{y_n\}$ such that $\{x_n\}$ is decreasing and $x_n \to -\infty$, while $\{y_n\}$ is increasing and $y_n \to \infty$. Put $A_n = \{\xi < x_n\}$ and $B_n = \{\xi < y_n\}$. Since x_n tends monotonically to $-\infty$, the sequence of sets A_n decreases monotonically to $\bigcap A_n = \varnothing$. By the continuity axiom (see Sect. 3.2.1), $\mathbf{P}(A_n) \to 0$ as $n \to \infty$ or, which is the same, $\lim_{n\to\infty} F(x_n) = 0$. This and the monotonicity of $F(x)$ imply that

$$\lim_{x\to-\infty} F(x) = 0.$$

Since the sequence $\{y_n\}$ tends monotonically to ∞, the sequence of sets B_n increases to $\bigcup B_n = \Omega$, and hence (see property 9 in Sect. 3.2.2) $\mathbf{P}(B_n) \to 1$. This implies, as above, that

$$\lim_{n\to\infty} F(y_n) = 1, \qquad \lim_{x\to\infty} F(x) = 1.$$

Property F3 is proved in a similar way. Let $\{x_n\}$ be an increasing sequence with $x_n \uparrow x_0$,

$$A = \{\xi < x_0\}, \qquad A_n = \{\xi < x_n\}.$$

The sequence of sets A_n also increases, and $\bigcup A_n = A$. Therefore, $\mathbf{P}(A_n) \to \mathbf{P}(A)$. This means that

$$\lim_{x\uparrow x_0} F(x) = F(x_0). \qquad \square$$

It is not hard to see that the function F would be *right-continuous* if we put $F(x) = \mathbf{P}(\xi \le x)$.

With our definition, the function F is generally speaking not right-continuous, since by the continuity axiom

$$\begin{aligned} F(x+0) - F(x) &= \lim_{n\to\infty} \left(F\left(x + \frac{1}{n}\right) - F(x) \right) \\ &= \lim_{n\to\infty} \mathbf{P}\left(x \le \xi < x + \frac{1}{n} \right) = \mathbf{P}\left(\bigcap_{n=1}^{\infty} \left\{ \xi \in \left[x, x + \frac{1}{n} \right) \right\} \right) \\ &= \mathbf{P}(\xi = x). \end{aligned}$$

This means that $F(x)$ is continuous if and only if $\mathbf{P}(\xi = x) = 0$ for any x. Examples 3.1.1 and 3.1.2 show that both continuous and discontinuous $F(x)$ are quite common.

From the above relations it also follows that

$$\mathbf{P}(x \le \xi \le y) = \mathbf{F}_\xi\big([x, y]\big) = F(y+0) - F(x).$$

Theorem 3.2.1 *If a function* $F(x)$ *has properties* F1, F2 *and* F3, *then there exist a probability space* $\langle \Omega, \mathfrak{F}, \mathbf{P} \rangle$ *and a random variable* ξ *such that* $F_\xi(x) = F(x)$.

Proof First we construct a probability space $\langle \Omega, \mathfrak{F}, \mathbf{P} \rangle$. Take Ω to be the real line $\mathbb{R}$, $\mathfrak{F}$ the σ-algebra $\mathfrak{B}$ of Borel sets. As we already know (see Sect. 3.2.1), to construct a probability space $\langle \mathbb{R}, \mathfrak{B}, \mathbf{P} \rangle$ it suffices to define a probability on the algebra $\mathcal{A}$ generated, say, by the semi-intervals of the form $[\cdot,\cdot)$ (then $\sigma(\mathcal{A}) = \mathfrak{B}$). An arbitrary element of the algebra $\mathcal{A}$ has the form of a finite union of disjoint semi-intervals:

$$A = \bigcup_{i=1}^{n} [a_i, b_i), \quad a_i < b_i$$

(the values of a_i and b_i can be infinite). We define

$$\mathbf{P}(A) = \sum_{i=1}^{n} \big(F(b_i) - F(a_i)\big).$$

It is absolutely clear that axioms P1 and P2 are satisfied by virtue of F1 and F2. It remains to verify the countable additivity, or continuity, of $\mathbf{P}$ on the algebra $\mathcal{A}$. Let $B_n \in \mathcal{A}$, $B_{n+1} \subset B_n$, $\bigcap_{n=1}^{\infty} B_n = B \in \mathcal{A}$. One has to show that $\mathbf{P}(B_n) \to \mathbf{P}(B)$ as $n \to \infty$ or, which is the same, that $\mathbf{P}(B_n\overline{B}) \to 0$ $(B_n\overline{B} \in \mathcal{A})$. To this end, it suffices to prove that, for any fixed N, $\mathbf{P}(B_n\overline{B}C_N) \to 0$, where $C_N = [-N, N)$. Indeed, for any given $\varepsilon > 0$, by virtue of F2 we can choose an N such that $\mathbf{P}(\overline{C}_N) < \varepsilon$. Then $\mathbf{P}(B_n\overline{B}\,\overline{C}_N) \le \mathbf{P}(\overline{C}_N) < \varepsilon$ and

$$\limsup_{n\to\infty} \mathbf{P}(B_n\overline{B}) \le \limsup_{n\to\infty} \mathbf{P}(B_n\overline{B}C_N) + \varepsilon.$$

Since ε is arbitrary, the convergence $\mathbf{P}(B_n\overline{B}C_N) \to 0$ as $n \to \infty$ implies the required convergence $\mathbf{P}(B_n\overline{B}) \to 0$. It follows that we can assume that the sets B_n are bounded ($B_n \subset [-N, N)$ for some $N < \infty$). Moreover, we can assume without loss of generality that B is the empty set.

By the above remarks, B_n admits the representation

$$B_n = \bigcup_{i=1}^{k_n} [a_i^n, b_i^n), \quad k_n < \infty,$$

where a_i^n, b_i^n are finite. Further note that, for a given $\varepsilon > 0$ and any semi-interval $[a, b)$, one can always find an embedded interval $[a, b - \delta)$, $\delta > 0$, such that $\mathbf{P}([a, b - \delta)) \ge \mathbf{P}([a, b)) - \varepsilon$. This follows directly from property F3: $F(b - \delta) \to F(b)$ as $\delta \downarrow 0$. Hence, for a given $\varepsilon > 0$ and set B_n, there exist $\delta_i^n > 0$, $i = 1, \ldots, k_n$, such that

$$\widetilde{B}_n = \bigcup_{i=1}^{k_n} \big[a_i^n, b_i^n - \delta_i^n\big) \subset B_n, \qquad \mathbf{P}(\widetilde{B}_n) > \mathbf{P}(B_n) - \varepsilon 2^{-n}.$$

Now add the right end points of the semi-intervals to the set $\widetilde{B}_n$ and consider the closed bounded set

$$K_n = \bigcup_{i=1}^{k_n} [a_i^n, b_i^n - \delta_i^n].$$

Clearly,

$$\widetilde{B}_n \subset K_n \subset B_n, \qquad K = \bigcap_{n=1}^{\infty} K_n = \varnothing,$$

$$\mathbf{P}(B_n - K_n) = \mathbf{P}(B_n \overline{K}_n) \le \varepsilon 2^{-n}.$$

It follows from the relation $K = \varnothing$ that $K_n = \varnothing$ for all sufficiently large n. Indeed, all the sets K_n belong to the closure $[C_N] = [N, -N]$ which is compact. The sets $\{\Delta_n = [C_N] - K_n\}_{n=1}^{\infty}$ form an *open covering* of $[C_N]$, since

$$\bigcup_n \Delta_n = [C_N]\Big(\bigcup_n \overline{K}_n\Big) = [C_N]\Big(\overline{\bigcap_n K_n}\Big) = [C_N].$$

Thus, by the Heine–Borel lemma there exists a *finite subcovering* $\{\Delta_n\}_{n=1}^{n_0}$, $n_0 < \infty$, such that $\bigcup_{n=1}^{n_0} \Delta_n = [C_N]$ or, which is the same, $\bigcap_{n=1}^{n_0} K_n = \varnothing$. Therefore

$$\begin{aligned}
\mathbf{P}(B_{n_0}) &= \mathbf{P}\Bigg(B_{n_0}\overline{\Bigg(\bigcap_{n=1}^{n_0} K_n\Bigg)}\Bigg) = \mathbf{P}\Bigg(B_{n_0}\Bigg(\bigcup_{n=1}^{n_0} \overline{K}_n\Bigg)\Bigg) \\
&= \mathbf{P}\Bigg(\bigcup_{n=1}^{n_0} B_{n_0}\overline{K}_n\Bigg) \le \mathbf{P}\Bigg(\bigcup_{n=1}^{n_0} B_n\overline{K}_n\Bigg) \le \sum_{n=1}^{n_0} \varepsilon 2^{-n} < \varepsilon.
\end{aligned}$$

Thus, for a given $\varepsilon > 0$ we found an n_0 (depending on ε) such that $\mathbf{P}(B_{n_0}) < \varepsilon$. This means that $\mathbf{P}(B_n) \to 0$ as $n \to \infty$. We proved that axiom P3 holds.

So we have constructed a probability space. It remains to take ξ to be the identity mapping of $\mathbb{R}$ onto itself. Then

$$F_\xi(x) = \mathbf{P}(\xi < x) = \mathbf{P}(-\infty, x) = F(x). \qquad \square$$

The model of the *sample probability space* based on the assertion just proved is often used in studies of distribution functions.

Definition 3.2.1 A probability space $\langle \Omega, \mathfrak{F}, \mathbf{F} \rangle$ is called a *sample space* for a random variable $\xi(\omega)$ if Ω is a subset of the real line $\mathbb{R}$ and $\xi(\omega) \equiv \omega$.

The probability $\mathbf{F} = \mathbf{F}_\xi$ is called, in accordance with Definition 3.1.1 from Sect. 3.1, the *distribution* of ξ. We will write this as

$$\xi \in \mathbf{F}. \tag{3.2.1}$$

It is obvious that constructing a sample probability space is always possible. It suffices to put $\Omega = \mathbb{R}$, $\mathfrak{F} = \mathfrak{B}$, $\mathbf{F}(B) = \mathbf{P}(\xi \in B)$. For integer-valued variables

ξ the space $\langle \Omega, \mathfrak{F} \rangle$ can be chosen in a more "economical" way by taking $\Omega = \{\ldots, -1, 0, \ldots\}$.

Since by Theorem 3.2.1 the distribution function $F(x)$ of a random variable ξ uniquely specifies the distribution $\mathbf{F}$ of this random variable, along with (3.2.1) we will also write $\xi \in F$.

Now we will give examples of some of the most common distributions.

3.2.2 The Most Common Distributions

1. *The degenerate distribution* $\mathbf{I}_a$. The distribution $\mathbf{I}_a$ is defined by

$$\mathbf{I}_a(B) = \begin{cases} 0 & \text{if } a \in B, \\ 1 & \text{if } a \notin B. \end{cases}$$

This distribution is concentrated at the point a: if $\xi \in \mathbf{I}_a$, then $\mathbf{P}(\xi = a) = 1$. The distribution function of $\mathbf{I}_a$ has the form

$$F(x) = \begin{cases} 0 & \text{for } x \le a, \\ 1 & \text{for } x > a. \end{cases}$$

The next two distributions were described in Examples 3.1.1 and 3.1.2 of Sect. 3.1.

2. *The binomial distribution* $\mathbf{B}_p^n$. By the definition, $\xi \in \mathbf{B}_p^n$ ($n > 0$ is an integer, $p \in (0, 1)$) if $\mathbf{P}(\xi = k) = \binom{n}{k} p^k (1-p)^{n-k}$, $0 \le k \le n$. The distribution $\mathbf{B}_p^1$ will be denoted by $\mathbf{B}_p$.

3. *The uniform distribution* $\mathbf{U}_{a,b}$. If $\xi \in \mathbf{U}_{a,b}$, then

$$\mathbf{P}(\xi \in B) = \frac{\mu(B \cap [a, b])}{\mu([a, b])},$$

where μ is the Lebesgue measure. We saw that this distribution has distribution function (3.1.1).

The next distribution plays a special role in probability theory, and we will encounter it many times.

4. *The normal distribution* $\mathbf{\Phi}_{\alpha,\sigma^2}$ (the normal or Gaussian law). We will write $\xi \in \mathbf{\Phi}_{\alpha,\sigma^2}$ if

$$\mathbf{P}(\xi \in B) = \mathbf{\Phi}_{\alpha,\sigma^2}(B) = \frac{1}{\sigma\sqrt{2\pi}} \int_B e^{-(u-\alpha)^2/(2\sigma^2)}\, du. \tag{3.2.2}$$

The distribution $\mathbf{\Phi}_{\alpha,\sigma^2}$ depends on two parameters: α and $\sigma > 0$. If $\alpha = 0, \sigma = 1$, the normal distribution is called *standard*. The distribution function of $\mathbf{\Phi}_{0,1}$ is equal to

$$\Phi(x) = \mathbf{\Phi}_{0,1}\big((-\infty, x)\big) = \frac{1}{\sqrt{2\pi}} \int_{-\infty}^{x} e^{-u^2/2}\, du.$$

The distribution function of $\mathbf{\Phi}_{\alpha,\sigma^2}$ is obviously equal to $\Phi((x-\alpha)/\sigma)$, so that the parameters α and σ have the meaning of the "location" and "scale" of the distribution.

The fact that formula (3.2.2) defines a distribution follows from Theorem 3.2.1 and the observation that the function $\Phi(x)$ (or $\Phi((x-a)/\sigma)$) satisfies properties F1–F3, since $\Phi(-\infty)=0$, $\Phi(\infty)=1$, and $\Phi(x)$ is continuous and monotone. One could also directly use the fact that the integral in (3.2.2) is a countably additive set function (see Sect. 3.6 and Appendix 3).

5. *The exponential distribution* $\mathbf{\Gamma}_\alpha$. The relation $\xi \in \mathbf{\Gamma}_\alpha$ means that ξ is nonnegative and

$$\mathbf{P}(\xi \in B) = \mathbf{\Gamma}_\alpha(B) = \alpha \int_{B\cap(0,\infty)} e^{-\alpha u}\,du.$$

The distribution function of $\xi \in \mathbf{\Gamma}_\alpha$ clearly has the form

$$\mathbf{P}(\xi < x) = \begin{cases} 1-e^{-\alpha x} & \text{for } x \ge 0, \\ 0 & \text{for } x < 0. \end{cases}$$

The exponential distribution is a special case of the gamma distribution $\mathbf{\Gamma}_{\alpha,\lambda}$, to be considered in more detail in Sect. 7.7.

6. A discrete analogue of the exponential distribution is called the *geometric distribution*. It has the form

$$\mathbf{P}(\xi = k) = (1-p)p^k, \quad p \in (0,1),\ k = 0, 1, \ldots$$

7. *The Cauchy distribution* $\mathbf{K}_{\alpha,\sigma}$. As was the case with the normal distribution, this distribution depends on two parameters α and σ which are also location and scale parameters. If $\xi \in \mathbf{K}_{\alpha,\sigma}$ then

$$\mathbf{P}(\xi \in B) = \frac{1}{\pi\sigma}\int_B \frac{du}{1+((u-a)/\sigma)^2}.$$

The distribution function $K(x)$ of $\mathbf{K}_{0,1}$ is

$$K(x) = \frac{1}{\pi}\int_{-\infty}^{x} \frac{du}{1+u^2}.$$

The distribution function of $\mathbf{K}_{\alpha,\sigma}$ is equal to $K((x-\alpha)\sigma)$. All the remarks made for the normal distribution continue to hold here.

Example 3.2.1 Suppose that there is a source of radiation at a point (α,σ), $\sigma > 0$, on the plane. The radiation is registered by a detector whose position coincides with the x-axis. An emitted particle moves in a random direction distributed uniformly over the circle. In other words, the angle η between this direction and the vector $(0,-1)$ has the uniform distribution $\mathbf{U}_{-\pi,\pi}$ on the interval $[-\pi,\pi]$. Observation results are the coordinates $\xi_1, \xi_2, \ldots$ of the points on the x-axis where the particles interacted with the detector. What is the distribution of the random variable $\xi = \xi_1$?

To find this distribution, consider a particle emitted at the point (α,σ) given that the particle hit the detector (i.e. given that $\eta \in [-\pi/2, \pi/2]$). It is clear that the conditional distribution of η given the last event (of which the probability is $\mathbf{P}(\eta \in [-\pi/2,\pi/2]) = 1/2$) coincides with $\mathbf{U}_{-\pi/2,\pi/2}$. Since $(\xi-\alpha)/\sigma = \tan\eta$, one obtains that

$$\mathbf{P}(\xi < x) = \mathbf{P}(\alpha + \sigma \tan\eta < x)$$
$$= \mathbf{P}\left(\frac{\eta}{\pi} < \frac{1}{\pi}\arctan\frac{x-\alpha}{\sigma}\right) = \frac{1}{2} + \frac{1}{\pi}\arctan\frac{x-\alpha}{\sigma}.$$

Recalling that $(\arctan u)' = 1/(1+u^2)$, we have

$$\arctan x = \int_0^x \frac{du}{1+u^2} = \int_{-\infty}^x \frac{du}{1+u^2} - \frac{\pi}{2},$$
$$\mathbf{P}(\xi < x) = \frac{1}{\pi}\int_{-\infty}^{(x-\alpha)/\sigma} \frac{du}{1+u^2} = K\left(\frac{x-\alpha}{\sigma}\right).$$

Thus the coordinates of the traces on the x-axis of the particles emitted from the point (α, σ) have the Cauchy distribution $\mathbf{K}_{\alpha,\sigma}$.

8. *The Poisson distribution* $\mathbf{\Pi}_\lambda$. We will write $\xi \in \mathbf{\Pi}_\lambda$ if ξ assumes nonnegative integer values with probabilities

$$\mathbf{P}(\xi = m) = \frac{\lambda^m}{m!}e^{-\lambda}, \quad \lambda > 0,\ m = 0, 1, 2, \ldots$$

The distribution function, as in Example 3.1.1, has the form of a sum:

$$F(x) = \begin{cases} \sum_{m<x} \frac{\lambda^m}{m!} e^{-\lambda} & \text{for } x > 0, \\ 0 & \text{for } x \le 0. \end{cases}$$

3.2.3 The Three Distribution Types

All the distributions considered in the above examples can be divided into two types.

I. Discrete Distributions

Definition 3.2.2 The distribution of a random variable ξ is called *discrete* if ξ can assume only finitely or countably many values $x_1, x_2, \ldots$ so that

$$p_k = \mathbf{P}(\xi = x_k) > 0, \qquad \sum p_k = 1.$$

A discrete distribution $\{p_k\}$ can obviously always be defined on a discrete probability space. It is often convenient to characterise such a distribution by a table:

Values	x_1	x_2	x_3	$\ldots$
Probabilities	p_1	p_2	p_3	$\ldots$

The distributions $\mathbf{I}_a$, $\mathbf{B}_p^n$, $\mathbf{\Pi}_\lambda$, and the geometric distribution are discrete. The derivative of the distribution function of such a distribution is equal to zero everywhere except at the points $x_1, x_2, \ldots$ where $F(x)$ is discontinuous, the jumps being

$$F(x_k + 0) - F(x_k) = p_k.$$

An important class of discrete distributions is formed by lattice distributions.

Definition 3.2.3 We say that random variable ξ has a *lattice distribution* with span h if there exist a and h such that

$$\sum_{k=-\infty}^{\infty} \mathbf{P}(\xi = a + kh) = 1. \tag{3.2.3}$$

If h is the greatest number satisfying (3.2.3) and the number a lies in the interval $[0, h)$ then these numbers are called the *span* and the *shift*, respectively, of the lattice.

If $a = 0$ and $h = 1$ then the distribution is called *arithmetic*. The same terms will be used for random variables.

Obviously the greatest common divisor (g.c.d.) of all possible values of an arithmetic random variable equals 1.

II. Absolutely Continuous Distributions

Definition 3.2.4 The distribution $\mathbf{F}$ of a random variable ξ is said to be *absolutely continuous*[2] if, for any Borel set B,

$$\mathbf{F}(B) = \mathbf{P}(\xi \in B) = \int_B f(x)\,dx, \tag{3.2.4}$$

where $f(x) \geq 0$, $\int_{-\infty}^{\infty} f(x)\,dx = 1$.

The function $f(x)$ in (3.2.4) is called the *density* of the distribution.

It is not hard to derive from the proof of Theorem 3.2.1 (to be more precise, from the theorem on uniqueness of the extension of a measure) that the above definition of absolute continuity is equivalent to the representation

$$F_\xi(x) = \int_{-\infty}^{x} f(u)\,du$$

for all $x \in R$. Distribution functions with this property are also called absolutely continuous.

---

[2]The definition refers to absolute continuity with respect to the *Lebesgue measure*. Given a measure μ on $\langle \mathbb{R}, \mathfrak{B} \rangle$ (see Appendix 3), a distribution $\mathbf{F}$ is called *absolutely continuous with respect to* μ if, for any $B \in \mathfrak{B}$, one has

$$\mathbf{F}(B) = \int_B f(x)\mu(dx).$$

In this sense discrete distributions are also absolutely continuous, but with respect to the counting measure m. Indeed, if one puts $f(x_k) = p_k$, $m(B) = \{$*the number of points from the set* $(x_1, x_2, \ldots)$ *which are in* $B\}$, then

$$\mathbf{F}(B) = \sum_{x_k \in B} p_k = \sum_{x_k \in B} f(x_k) = \int_B f(x)m(dx)$$

(see Appendix 3).

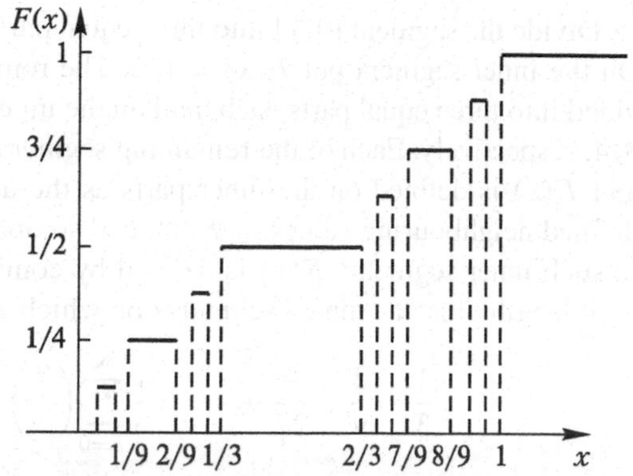

Fig. 3.1 The plot shows the result of the first three steps in the construction of the Cantor function

The function $f(x)$ is determined by the above equalities up to its values on a set of Lebesgue measure 0. For this function, the relation $f(x) = \frac{dF(x)}{dx}$ holds[3] almost everywhere (with respect to the Lebesgue measure).

The distributions $\mathbf{U}_{a,b}$, $\mathbf{\Phi}_{\alpha,\sigma^2}$, $\mathbf{K}_{\alpha,\sigma}$ and $\mathbf{\Gamma}_\alpha$ are absolutely continuous. The density of the normal distribution with parameters $\alpha$and σ is equal to

$$\phi_{\alpha,\sigma^2}(x) = \frac{1}{\sqrt{2\pi}\sigma} e^{-(x-\alpha)^2/(2\sigma^2)}.$$

From their definitions, one could easily derive the densities of the distributions $\mathbf{U}_{a,b}$, $\mathbf{K}_{\alpha,\sigma}$ and $\mathbf{\Gamma}_\alpha$ as well. The density of $\mathbf{K}_{\alpha,\sigma}$ has a shape resembling that of the normal density, but with "thicker tails" (it vanishes more slowly as $|x| \to \infty$).

We will say that a distribution $\mathbf{F}$ *has an atom* at point x_1 if $\mathbf{F}(\{x_1\}) > 0$. We saw that any discrete distribution consists of atoms but, for an absolutely continuous distribution, the probability of hitting a set of zero Lebesgue measure is zero. It turns out that there exists yet a *third* class of distributions which is characterised by the negation of both mentioned properties of discrete and absolutely continuous distributions.

III. Singular Distributions

Definition 3.2.5 A distribution $\mathbf{F}$ is said to be *singular* (with respect to Lebesgue measure) if it has no atoms and is concentrated on a set of zero Lebesgue measure.

Because a singular distribution has no atoms, its distribution function is continuous. An example of such a distribution function is given by the famous Cantor function of which the whole variation is concentrated on the interval $[0, 1]$: $F(x) = 0$ for $x \le 0$, $F(x) = 1$ for $x \ge 1$. It can be constructed as follows (the construction process is shown in Fig. 3.1).

[3]The assertion about the "almost everywhere" uniqueness of the function f follows from the Radon–Nikodym theorem (see Appendix 3).

Divide the segment [0, 1] into three equal parts [0, 1/3], [1/3, 2/3], and [2/3, 1]. On the inner segment put $F(x) = 1/2$. The remaining two segments are again divided into three equal parts each, and on the inner parts one sets $F(x)$ to be 1/4 and 3/4, respectively. Each of the remaining segments is divided in turn into three parts, and $F(x)$ is defined on the inner parts as the arithmetic mean of the two already defined neighbouring values of $F(x)$, and so on. At the points which do not belong to such inner segments $F(x)$ is defined by continuity. It is not hard to see that the total length of such "inner" segments on which $F(x)$ is constant is equal to

$$\frac{1}{3} + \frac{2}{9} + \frac{4}{27} + \cdots = \frac{1}{3} \sum_{k=0}^{\infty} \left(\frac{2}{3}\right)^k = \frac{1}{3} \frac{1}{1 - 2/3} = 1,$$

so that the function $F(x)$ grows on a set of measure zero but has no jumps.

From the construction of the Cantor distribution we see that $dF(x)/dx = 0$ almost everywhere.

It turns out that these three types of distribution exhaust all possibilities.

More precisely, there is a theorem belonging to Lebesgue[4] stating that any distribution function $F(x)$ can be represented in a unique way as a sum of three components: discrete, absolutely continuous, and singular. Hence an arbitrary distribution function cannot have more than a countable number of jumps (which can also be observed directly: we will count all the jumps if we first enumerate all the jumps which are greater than 1/2, then the jumps greater than 1/3, then greater than 1/4, etc.). This means, in particular, that $F(x)$ is everywhere continuous except perhaps at a countable or finite set of points.

In conclusion of this section we will list several properties of distribution functions and densities that arise when forming new random variables.

3.2.4 *Distributions of Functions of Random Variables*

For a given function $g(x)$, to find the distribution of $g(\xi)$ we have to impose some measurability requirements on the function. The function $g(x)$ is called *Borel* if the inverse image

$$g^{-1}(B) = \{x : g(x) \in B\}$$

of any Borel set B is again a Borel set. For such a function g the distribution function of the random variable $\eta = g(\xi)$ equals

$$F_{g(\xi)}(x) = \mathbf{P}\big(g(\xi) < x\big) = \mathbf{P}\big(\xi \in g^{-1}(-\infty, x)\big).$$

If $g(x)$ is continuous and strictly increasing on an interval (a, b) then, on the interval $(g(a), g(b))$, the inverse function $y = g^{(-1)}(x)$ is defined as the solution to

[4] See Sect. 3.5 in Appendix 3.

the equation $g(y) = x$.[5] Since g is a monotone mapping we have

$$\{g(\xi) < x\} = \{\xi < g^{(-1)}(x)\} \quad \text{for } x \in (g(a), g(b)).$$

Thus we get the following representation for $F_{g(\xi)}$ in terms of F_ξ: for $x \in (g(a), g(b))$,

$$F_{g(\xi)}(x) = \mathbf{P}(\xi < g^{-1}(x)) = F_\xi(g^{-1}(x)). \tag{3.2.5}$$

Putting $g = F_\xi$ we obtain, in particular, that *if F_ξ is continuous and strictly increasing on (a, b) and $F(a) = 0$, $F(b) = 1$ ($-a$ and b may be ∞) then*

$$F_\xi(g^{(-1)}(x)) \equiv x$$

for $x \in [0, 1]$ and therefore the random variable $\eta = F_\xi(\xi)$ is uniformly distributed over $[0, 1]$.

Definition 3.2.6 *The quantile transform $F^{(-1)}(f)$ of an arbitrary distribution* **F** with the distribution function $F(x)$ is the "generalised" inverse of the function F

$$F^{(-1)}(y) := \sup\{x : F(x) < y\} \quad \text{for } y \in (0, 1];$$
$$F^{(-1)}(0) := \inf\{x : F(x) > 0\}.$$

In mathematical statistics, the number $F^{(-1)}(y)$ is called the *quantile* of order y of the distribution **F**. The function $F^{(-1)}$ has a discontinuity of size $b - a$ at a point y if (a, b) is the interval on which F is constant and such that $F(x) = y \in [0, 1)$.

Roughly speaking, the plot of the function $F^{(-1)}$ can be obtained from that of the function $F(x)$ on the (x, y) plane in the following way: rotate the (x, y) plane in the counter clockwise direction by 90°, so that the x-axis becomes the ordinate axis, but the y-axis becomes the abscissa axis directed to the left. To switch to normal coordinates, we have to reverse the direction of the new x-axis.

Further, if x is a point of continuity and a point of growth of the function F (i.e., $F(x)$ is a point of continuity of $F^{(-1)}$) then $F^{(-1)}(y)$ is the unique solution of the equation $F(x) = y$ and the equality $F(F^{(-1)}(y)) = y$ holds.

In some cases the following statement proves to be useful.

Theorem 3.2.2 *Let $\eta \in \mathbf{U}_{0,1}$. Then, for any distribution* **F**,

$$f^{(-1)}(\eta) \in \mathbf{F}.$$

Proof If $F(x) > y$ then $F^{(-1)}(y) = \sup\{v : F(v) < y\} < x$, and vice versa: if $F(x) < y$ then $F^{(-1)}(y) \geq x$ (recall that $F(x)$ is left-continuous). Therefore the following inclusions are valid for the sets in the (x, y) plane:

$$\{y < F(x)\} \subset \{F^{(-1)}(y) < x\} \subset \{y \leq F(x)\}.$$

---

[5]For an arbitrary non-decreasing function g, the inverse function $g^{(-1)}(x)$ is defined by the equation

$$g^{(-1)}(y) := \inf\{x : g(x) \geq y\} = \sup\{x : g(x) < y\}.$$

Substituting $\eta \in \mathbf{U}_{0,1}$ in place of y in these relations yields that, for any x, such inclusions hold for the respective events, and hence

$$\mathbf{P}\big(F^{(-1)}(\eta) < x\big) = \mathbf{P}\big(\eta < F(x)\big) = F(x).$$

The theorem is proved. □

Thus we have obtained an important method for constructing random variables with prescribed distributions from uniformly distributed random variables. For instance, if $\eta \in \mathbf{U}_{0,1}$ then $\xi = -(1/\alpha)\ln\eta \in \boldsymbol{\Gamma}_\alpha$.

In another special case, when $g(x) = a + bx$, $b > 0$, from (3.2.5) we get $F_{g(\xi)} = F_\xi((x-a)/b)$. We have already used this relation to some extent when considering the distributions $\boldsymbol{\Phi}_{\alpha,\sigma^2}$ and $\mathbf{K}_{\alpha,\sigma}$.

If a function g is strictly increasing and differentiable (the inverse function $g^{(-1)}$ is defined in this case), and ξ has a density $f(x)$, then there exists a density for $g(\xi)$ which is equal to

$$f_{g(\xi)}(y) = f\big(g^{(-1)}(y)\big)\big(g^{(-1)}(y)\big)' = f(x)\frac{dx}{dy},$$

where $x = g^{(-1)}(y)$, $y = g(x)$. A similar argument for decreasing g leads to the general formula

$$f_{g(\xi)}(y) = f(x)\left|\frac{dx}{dy}\right|.$$

For $g(x) = a + bx$, $b \neq 0$, one obtains

$$f_{a+b\xi}(y) = \frac{1}{|b|} f\left(\frac{y-a}{b}\right).$$

3.3 Multivariate Random Variables

Let $\xi_1, \xi_2, \ldots, \xi_n$ be random variables given on a common probability space $\langle \Omega, \mathfrak{F}, \mathbf{P}\rangle$. To each ω, these random variables put into correspondence an n-dimensional vector $\xi(\omega) = (\xi_1(\omega), \xi_2(\omega), \ldots, \xi_n(\omega))$.

Definition 3.3.1 A mapping $\Omega \to \mathbb{R}^n$ given by random variables $\xi_1, \xi_2, \ldots, \xi_n$ is called a *random vector* or *multivariate random variable*.

Such a mapping $\Omega \to \mathbb{R}^n$ is a measurable mapping of the space $\langle \Omega, \mathbf{F}\rangle$ into the space $\langle \mathbb{R}^n, \mathfrak{B}^n\rangle$, where $\mathfrak{B}^n$ is the σ-algebra of Borel sets in $\mathbb{R}^n$. Therefore, for Borel sets B, the function $\mathbf{P}_\xi(B) = \mathbf{P}(\xi \in B)$ is defined.

Definition 3.3.2 The function $\mathbf{F}_\xi(B)$ is called the *distribution of the vector* ξ.

The function

$$F_{\xi_1\ldots\xi_n}(x_1, \ldots, x_n) = \mathbf{P}(\xi_1 < x_1, \ldots, \xi_n < x_n)$$

is called the *distribution function* of the random vector $(\xi_1, \dots, \xi_n)$ or *joint distribution function* of the random variables $\xi_1, \dots, \xi_n$.

The following properties of the distribution functions of random vectors, analogous to properties F1–F3 in Sect. 3.2, hold true.

FF1. *Monotonicity*: "Multiple" differences of the values of the function $F_{\xi_1 \dots \xi_n}$, which correspond to probabilities of hitting arbitrary "open at the right" parallelepipeds, are nonnegative. For instance, in the two-dimensional case this means that, for any $x_1 < x_2$, $y_1 < y_2$ (the points (x_1, y_1) and (x_2, y_2) being the "extreme" vertices of the parallelepiped),

$$F_{\xi_1,\xi_2}(x_2, y_2) - F_{\xi_1,\xi_2}(x_2, y_1) - \bigl(F_{\xi_1,\xi_2}(x_1, y_2) - F_{\xi_1,\xi_2}(x_1, y_1)\bigr) \ge 0.$$

This double difference is nothing else but the probability of hitting the "semi-open" parallelepiped $[x_1, x_2) \times [y_1, y_2)$ by ξ.

In other words, the *differences*

$$F_{\xi_1,\xi_2}(t, y_2) - F_{\xi_1,\xi_2}(t, y_1) \quad \text{for } y_1 < y_2$$

must be monotone in t. (For this to hold, the monotonicity of the function $F_{\xi_1,\xi_2}(t, y_1)$ is not sufficient.)

FF2. The second property can be called *consistency*.

$$\lim_{x_n \to \infty} F_{\xi_1 \dots \xi_n}(x_1, \dots, x_n) = F_{\xi_1 \dots \xi_{n-1}}(x_1, \dots, x_{n-1}),$$
$$\lim_{x_n \to -\infty} F_{\xi_1 \dots \xi_n}(x_1, \dots, x_n) = 0.$$

FF3. *Left-continuity*:

$$\lim_{x_n' \uparrow \infty} F_{\xi_1 \dots \xi_n}\bigl(x_1, \dots, x_n'\bigr) = F_{\xi_1 \dots \xi_n}(x_1, \dots, x_n).$$

That the limits in properties FF2 and FF3 are taken in the last variable is inessential, for one can always renumber the components of the vectors.

One can prove these properties in the same way as in the one-dimensional case. As above, *any function $F(x_1, \dots, x_n)$ possessing this collection of properties will be the distribution function of a (multivariate) random variable.*

As in the one-dimensional case, when considering random vectors $\xi = (\xi_1, \dots, \xi_n)$, we can make use of the simplest sample model of the probability space $\langle \Omega, \mathbf{F}, \mathbf{P} \rangle$. Namely, let Ω coincide with $\mathbb{R}^n$ and $\mathbf{F} = \mathfrak{B}^n$ be the σ-algebra of Borel sets. We will complete the construction of the required probability space if we put $\mathbf{F}(B) = \mathbf{F}_\xi(B) = \mathbf{P}(\xi \in B)$ for any $B \in \mathfrak{B}^n$. It remains to define the random variable as the value of the elementary event itself, i.e. to put $\xi(\omega) = \omega$, where ω is a point in $\mathbb{R}^n$.

It is not hard to see that the distribution function $F_{\xi_1 \dots \xi_n}$ uniquely determines the distribution $\mathbf{F}_\xi(B)$. Indeed, $F_{\xi_1 \dots \xi_n}$ defines a probability on the σ-algebra $\mathcal{A}$ generated by rectangles $\{a_i \le x_i < b_i;\ i = 1, \dots, n\}$. For example, in the two-dimensional case

$$\begin{aligned}
&\mathbf{P}(a_1 \le \xi_1 < b_1, a_2 \le \xi_2 < b_2) \\
&\quad = \mathbf{P}(\xi_1 < b_1, a_2 \le \xi_2 < b_2) - \mathbf{P}(\xi_1 < a_1, a_2 \le \xi_2 < b_2) \\
&\quad = \left[F_{\xi_1,\xi_2}(b_1, b_2) - F_{\xi_1,\xi_2}(b_1, a_2)\right] - \left[F_{\xi_1,\xi_2}(a_1, b_2) - F_{\xi_1,\xi_2}(a_1, a_2)\right].
\end{aligned}$$

But $\mathfrak{B}^n = \sigma(\mathcal{A})$, and it remains to make use of the measure extension theorem (see Sect. 3.2.1).

Thus *from a distribution function $F_{\xi_1 \ldots \xi_n} = F$ one can always construct a sample probability space $\langle \mathbb{R}^n, \mathfrak{B}^n, \mathbf{F}_\xi \rangle$ and a random variable $\xi(\omega) \equiv \omega$ on it so that the latter will have the prescribed distribution $\mathbf{F}_\xi$.*

As in the one-dimensional case, we say that the distribution of a random vector is discrete if the random vector assumes at most a countable set of values.

The distribution of a random vector will be *absolutely continuous* if, for any Borel set $B \subset \mathbb{R}^n$,

$$\mathbf{F}_\xi(B) = \mathbf{P}(\xi \in B) = \int_B f(x)\,dx,$$

where clearly $f(x) \ge 0$ and $\int_\Omega f(x)\,dx = 1$.

This definition can be replaced with an equivalent one requiring that

$$F_{\xi_1 \ldots \xi_n}(x_1, \ldots, x_n) = \int_{-\infty}^{x_1} \cdots \int_{-\infty}^{x_n} f(t_1, \ldots, t_n)\,dt_1 \cdots dt_n. \tag{3.3.1}$$

Indeed, if (3.3.1) holds, we define a countably additive set function

$$\mathbf{Q}(B) = \int_B f(x)\,dx$$

(see properties of integrals in Appendix 3), which will coincide on rectangles with $\mathbf{F}_\xi$. Consequently, $\mathbf{F}_\xi(B) = \mathbf{Q}(B)$.

The function $f(x)$ is called the *density of the distribution* of ξ or *density of the joint distribution* of $\xi_1, \ldots, \xi_n$. The equality

$$\frac{\partial^n}{\partial x_1 \cdots \partial x_n} F_{\xi_1 \ldots \xi_n}(x_1, \ldots, x_n) = f(x_1, \ldots, x_n)$$

holds for this function almost everywhere.

If a random vector ξ has density $f(x_1, \ldots, x_n)$, then clearly any "subvector" $(\xi_{k_1} \ldots \xi_{k_n})$, $k_i \le n$, also has a density equal (let for the sake of simplicity $k_i = i$, $i = 1, \ldots, s$) to

$$f(x_1, \ldots, x_s) = \int f(x_1, \ldots, x_n)\,dx_{s+1} \cdots dx_n.$$

Let continuously differentiable functions $y_i = g_i(x_1, \ldots, x_n)$ be given in a region $A \subset \mathbb{R}^n$. Suppose they are univalently resolvable for $x_1, \ldots, x_n$: there exist functions $x_i = g_i^{(-1)}(y_1, \ldots, y_n)$, and the Jacobian $J = |\partial x_i / \partial y_i| \ne 0$ in A. Denote by B the image of A in the range of $(y_1, \ldots, y_n)$. Suppose further that a random vector $\xi = (\xi_1, \ldots, \xi_n)$ has a density $f_\xi(x)$. Then $\eta_i = g_i(\xi_1, \ldots, \xi_n)$ will be random variables with a joint density which, at a point $(y_1, \ldots, y_n) \in B$, is equal to

$$f_n(y_1, \ldots, y_n) = f_\xi(x_1, \ldots, x_n)|J|; \tag{3.3.2}$$

moreover

$$\mathbf{P}(\xi \in A) = \int_A f_\xi(x_1, \dots, x_n)\,dx_1 \cdots dx_n = \int_B f_\xi(x_1, \dots, x_n)|J|\,dy_1 \cdots dy_n$$
$$= \int_B f_\eta(y_1, \dots, y_n)\,dy_1 \cdots dy_n = \mathbf{P}(\eta \in B). \tag{3.3.3}$$

This is clearly an extension to the multi-dimensional case of the property of densities discussed at the end of Sect. 3.2. Formula (3.3.3) for integrals is well-known in calculus as the change of variables formula and could serve as a proof of (3.3.2).

The distribution $\mathbf{F}_\xi$ of a random vector ξ is called *singular* if the distribution has no atoms ($\mathbf{F}_\xi(\{x\}) = 0$ for any $x \in \mathbb{R}^n$) and is concentrated on a set of zero Lebesgue measure.

Consider the following two important examples of multivariate distributions (we continue the list of the most common distribution from Sect. 3.2).

9. *The multinomial distribution* $\mathbf{B}_p^n$. We use here the same symbol $\mathbf{B}_p^n$ as we used for the binomial distribution. The only difference is that now by p we understand a vector $p = (p_1, \dots, p_r)$, $p_j \ge 0$, $\sum_{j=1}^r p_j = 1$, which could be interpreted as the collection of probabilities of disjoint events A_j, $\bigcup A_j = \Omega$. For an integer-valued random vector $\nu = (\nu_1, \dots, \nu_r)$, we will write $\nu \Subset B$ if for $k = (k_1, \dots, k_r)$, $k_j \ge 0$, $\sum_{j=1}^r k_j = n$ one has

$$\mathbf{P}(\nu = k) = \frac{n!}{k_1! \cdots k_r!} p_1^{k_1} \cdots p_r^{k_r}. \tag{3.3.4}$$

On the right-hand side we have a term from the expansion of the polynomial $(p_1 + \cdots + p_r)^n$ into powers of $p_1, \dots, p_r$. This explains the name of the distribution. If p is a number, then evidently $\mathbf{B}_p^n = \mathbf{B}_{(p,1-p)}^n$, so that the binomial distribution is a multinomial distribution with $r = 2$.

The numbers ν_j could be interpreted as the frequencies of the occurrence of events A_j in n independent trials, the probability of occurrence of A_j in a trial being p_j. Indeed, the probability of any fixed sequence of outcomes containing $k_1, \dots, k_r$ outcomes $A_1, \dots, A_r$, respectively, is equal to $p_1^{k_1} \cdots p_r^{k_r}$, and the number of different sequences of this kind is equal to $n!/k_1! \cdots k_r!$ (of $n!$ permutations we leave only those which differ by more than merely permutations of elements inside the groups of $k_1, \dots, k_r$ elements). The result will be the probability (3.3.4).

Example 3.3.1 The simplest model of a chess tournament with two players could be as follows. In each game, independently of the outcomes of the past games, the 1st player wins with probability p, loses with probability q, and makes a draw with probability $1 - p - q$. In that case the probability that, in n games, the 1st player wins i and loses j games ($i + j \le n$), is

$$p(n; i, j) = \frac{n!}{i!\,j!\,(n-i-j)!} p^i q^j (1 - p - q)^{n-i-j}.$$

Suppose that the tournament goes on until one of the players wins N games (and thereby wins the tournament). If we denote by η the duration of the tournament (the number of games played before its end) then

$$\mathbf{P}(\eta = n) = \sum_{i=0}^{N-1} p(n-1; N-1, i)p + \sum_{i=0}^{N-1} p(n-1; i, N-1)q.$$

10. *The multivariate normal* (or *Gaussian*) *distribution* $\mathbf{\Phi}_{\alpha,\sigma^2}$. Let $\alpha = (\alpha_1, \dots, \alpha_r)$ be a vector and $\sigma^2 = \|\sigma_{ij}\|$, $i, j = 1, \dots, r$, a symmetric positive definite matrix, and $A = \|a_{ij}\|$ the matrix inverse to $\sigma^2 = A^{-1}$. We will say that a vector $\xi = (\xi_1, \dots, \xi_r)$ has the *normal distribution*: $\xi \in \mathbf{\Phi}_{\alpha,\sigma^2}$, if it has the density

$$\varphi_{\alpha,\sigma^2}(x) = \frac{\sqrt{|A|}}{(2\pi)^{r/2}} \exp\left\{-\frac{1}{2}(x-\alpha)A(x-\alpha)^T\right\}.$$

Here T denotes transposition:

$$xAx^T = \sum a_{ij}x_i x_j.$$

It is not hard to verify that

$$\int \varphi_{\alpha,\sigma^2}(x)\,dx_1 \cdots dx_r = 1$$

(see also Sect. 7.6).

3.4 Independence of Random Variables and Classes of Events

3.4.1 Independence of Random Vectors

Definition 3.4.1 Random variables $\xi_1, \dots, \xi_n$ are said to be *independent* if

$$\mathbf{P}(\xi_1 \in B_1, \dots, \xi_n \in B_n) = \mathbf{P}(\xi_1 \in B_1) \cdots \mathbf{P}(\xi_n \in B_n) \tag{3.4.1}$$

for any Borel sets $B_1, \dots, B_n$ on the real line.

One can introduce the notion of a *sequence* of independent random variables. The random variables from the sequence $\{\xi_n\}_{n=1}^{\infty}$ given on a probability space $\langle\Omega, \mathfrak{F}, \mathbf{P}\rangle$, are independent if (3.4.1) holds for any integer n so that the independence of a sequence of random variables reduces to that of any finite collection of random variable from this sequence. As we will see below, for a sequence of independent random variables, any two events related to disjoint groups of random variables from the sequence are independent.

Another possible definition of independence of random variables follows from the assertion below.

Theorem 3.4.1 *Random variables $\xi_1, \dots, \xi_n$ are independent if and only if*

$$F_{\xi_1 \dots \xi_n}(x_1, \dots, x_n) = F_{\xi_1}(x_1) \cdots F_{\xi_n}(x_n).$$

The proof of the theorem is given in the third part of the present section.

An important criterion of independence in the case when the distribution of $\xi = (\xi_1, \dots, \xi_n)$ is absolutely continuous is given in the following theorem.

Theorem 3.4.2 *Let random variables $\xi_1, \dots, \xi_n$ have densities $f_1(x), \dots, f_n(x)$, respectively. Then for the independence of $\xi_1, \dots, \xi_n$ it is necessary and sufficient that the vector $\xi = (\xi_1, \dots, \xi_n)$ has a density $f(x_1, \dots, x_n)$ which is equal to*

$$f(x_1, \dots, x_n) = f_1(x_1) \cdots f_n(x_n).$$

Thus, if it turns out that the density of ξ equals the product of densities of ξ_j, that will mean that the random variables ξ_j are independent.

We leave it to the reader to verify, using this theorem, that the components of a normal vector $(\xi_1, \dots, \xi_n)$ are independent if and only if $a_{ij} = 0, \sigma_{ij} = 0$ for $i \neq j$.

Proof of Theorem 3.4.2 If the distribution function of the random variable ξ_i is given by

$$F_{\xi_i}(x_i) = \int_{-\infty}^{x_i} f_i(t_i)\, dt_i$$

and ξ_i are independent, then the joint distribution function will be defined by the formula

$$\begin{aligned} F_{\xi_1 \dots \xi_n}(x_1, \dots, x_n) &= F_{\xi_1}(x_1) \cdots F_{\xi_n}(x_n) \\ &= \int_{-\infty}^{x_1} f_1(t_1)\, dt_1 \cdots \int_{-\infty}^{x_n} f_n(t_n)\, dt_n \\ &= \int_{-\infty}^{x_1} \cdots \int_{-\infty}^{x_n} f_1(t_1) \cdots f_n(t_n)\, dt_1 \cdots dt_n. \end{aligned}$$

Conversely, assuming that

$$F_{\xi_1 \dots \xi_n}(x_1, \dots, x_n) = \int_{-\infty}^{x_1} \cdots \int_{-\infty}^{x_n} f_1(t_1) \cdots f_n(t_n)\, dt_1 \cdots dt_n,$$

we come to the equality

$$F_{\xi_1 \dots \xi_n}(x_1, \dots, x_n) = F_{\xi_1}(x_1) \cdots F_{\xi_n}(x_n).$$

The theorem is proved. □

Now consider the discrete case. Assume for the sake of simplicity that the components of ξ may assume only integral values. Then for the independence of ξ_j it is necessary and sufficient that, for all $k_1, \dots, k_n$,

$$\mathbf{P}(\xi_1 = k_1, \dots, \xi_n = k_n) = \mathbf{P}(\xi_1 = k_1) \cdots \mathbf{P}(\xi_n = k_n).$$

Verifying this assertion causes no difficulties, and we leave it to the reader.

The notion of independence is very important for Probability Theory and will be used throughout the entire book. Assume that we are formalising a practical problem (constructing an appropriate probability model in which various random variables are to be present). How can one find out whether the random variables (or events) to appear in the model are independent? In such situations it is a *justified rule to consider events and random variables with no causal connection as independent.*

The detection of "probabilistic" independence in a mathematical model of a random phenomenon is often connected with a deep understanding of its physical essence.

Consider some simple examples. For instance, it is known that the probability of a new-born child to be a boy (event A) has a rather stable value $\mathbf{P}(A) = 22/43$. If B denotes the condition that the child is born on the day of the conjunction of Jupiter and Mars, then, under the assumption that the position of the planets does not determine individual fates of humans, the conditional probability $\mathbf{P}(A|B)$ will have the same value: $\mathbf{P}(A|B) = 22/43$. That is, the actual counting of the frequency of births of boys under these specific astrological conditions would give just the value 22/43. Although such a counting might never have been carried out at a sufficiently large scale, we have no grounds to doubt its results.

Nevertheless, one should not treat the connection between "mathematical" and causal independence as an absolute one. For instance, by Newton's law of gravitation the flight of a missile undoubtedly influences the simultaneous flight of another missile. But it is evident that in practice one can ignore this influence. This example also shows that independence of events and variables in the concrete and relative meaning of this term does not contradict the principle of the universal interdependence of all events.

It is also interesting to note that the formal definition of independence of events or random variables is much wider than the notion of real independence in the sense of affiliation to causally unrelated phenomena. This follows from the fact that "mathematical" independence can take place in such cases when one has no reason for assuming no causal relation. We illustrate this statement by the following example. Let η be a random variable uniformly distributed over $[0, 1]$. Then in the expansion of η into a binary fraction

$$\eta = \frac{\xi_1}{2} + \frac{\xi_2}{4} + \frac{\xi_3}{8} + \cdots$$

the random variables ξ_k will be independent (see Example 11.3.1), although they all have a related origin.

One can see that this circumstance only enlarges the area of applicability of all the assertions we obtain below under the formal condition of independence.[6]

The notion of independence of random variables is closely connected with that of *independence of σ-algebras*.

3.4.2 Independence of Classes of Events

Let $\langle \Omega, \mathfrak{F}, \mathbf{P} \rangle$ be a probability space and $\mathcal{A}_1$ and $\mathcal{A}_2$ classes of events from the σ-algebra $\mathfrak{F}$.

---

[6]For a more detailed discussion of connections between causal and probabilistic independence, see [24], from where we borrowed the above examples.

Definition 3.4.2 The classes of events $\mathcal{A}_1$ and $\mathcal{A}_2$ are said to be *independent* if, for any events A_1 and A_2 such that $A_1 \in \mathcal{A}_1$ and $A_2 \in \mathcal{A}_2$, one has

$$\mathbf{P}(A_1 A_2) = \mathbf{P}(A_1)\mathbf{P}(A_2).$$

The following definition introduces the notion of independence of a sequence of classes of events.

Definition 3.4.3 Classes of events $\{\mathcal{A}_n\}_{n=1}^{\infty}$ are *independent* if, for any collection of integers $n_1, \ldots, n_k$,

$$\mathbf{P}\left(\bigcap_{j=1}^{k} A_{n_j}\right) = \prod_{j=1}^{k} \mathbf{P}(A_{n_j})$$

for any $A_{n_j} \in \mathcal{A}_{n_j}$.

For instance, in a sequence of independent trials, the sub-σ-algebras of events related to different trials will be independent. The independence of a sequence of algebras of events also reduces to the independence of any finite collection of algebras from the sequence. It is clear that subalgebras of events of independent algebras are also independent.

Theorem 3.4.3 *σ-algebras $\mathfrak{A}_1$ and $\mathfrak{A}_2$ generated, respectively, by independent algebras of events $\mathcal{A}_1$ and $\mathcal{A}_2$ are independent.*

Before proving this assertion we will obtain an approximation theorem which will be useful for the sequel. By virtue of the theorem, any event A from the σ-algebra $\mathfrak{A}$ generated by an algebra $\mathcal{A}$ can, in a sense, be approximated by events from $\mathcal{A}$. To be more precise, we introduce the "distance" between events defined by

$$d(A, B) = \mathbf{P}(A\overline{B} \cup \overline{A}B) = \mathbf{P}(A\overline{B}) + \mathbf{P}(\overline{A}B) = \mathbf{P}(A - B) + \mathbf{P}(B - A).$$

This distance possesses the following properties:

$$\begin{aligned} d(\overline{A}, \overline{B}) &= d(A, B), \\ d(A, C) &\le d(A, B) + d(B, C), \\ d(AB, CD) &\le d(A, C) + d(B, D), \\ \big|\mathbf{P}(A) - \mathbf{P}(B)\big| &\le d(A, B). \end{aligned} \tag{3.4.2}$$

The first relation is obvious. The triangle inequality follows from the fact that

$$\begin{aligned} d(A, C) = \mathbf{P}(A\overline{C}) + \mathbf{P}(\overline{A}C) &= \mathbf{P}(A\overline{C}B) + \mathbf{P}(A\overline{C}\overline{B}) + \mathbf{P}(\overline{A}CB) + \mathbf{P}(\overline{A}C\overline{B}) \\ &\le \mathbf{P}(\overline{C}B) + \mathbf{P}(A\overline{B}) + \mathbf{P}(\overline{A}B) + \mathbf{P}(C\overline{B}) = d(A, B) + d(B, C). \end{aligned}$$

The third relation in (3.4.2) can be obtained in a similar way by enlarging events under the probability sign. Finally, the last inequality in (3.4.2) is a consequence of the relations

$$\mathbf{P}(A) = \mathbf{P}(AB) + \mathbf{P}(A\overline{B}) = \mathbf{P}(B) - \mathbf{P}(B\overline{A}) + \mathbf{P}(A\overline{B}).$$

Theorem 3.4.4 (The approximation theorem) *Let* $\langle \Omega, \mathfrak{F}, \mathbf{P}\rangle$ *be a probability space and* $\mathfrak{A}$ *the* σ*-algebra generated by an algebra* $\mathcal{A}$ *of events from* $\mathfrak{F}$. *Then, for any* $A \in \mathfrak{A}$, *there exists a sequence* $A_n \in \mathcal{A}$ *such that*

$$\lim_{n\to\infty} d(A, A_n) = 0. \tag{3.4.3}$$

By the last inequality from (3.4.2), the assertion of the theorem means that $\mathbf{P}(A) = \lim_{n\to\infty} \mathbf{P}(A_n)$ and that each event $A \in \mathfrak{A}$ can be represented, up to a set of zero probability, as a limit of a sequence of events from the generating algebra $\mathcal{A}$ (see also Appendix 1).

Proof[7] We will call an event $A \in \mathfrak{F}$ *approximable* if there exists a sequence $A_n \in \mathcal{A}$ possessing property (3.4.3), i.e. $d(A_n, A) \to 0$.

Since $d(A, A) = 0$, the class of approximable events $\mathfrak{A}^*$ contains $\mathcal{A}$. Therefore to prove the theorem it suffices to verify that $\mathfrak{A}^*$ is a σ-algebra.

The fact that $\mathfrak{A}^*$ is an algebra is obvious, for the relations $A \in \mathfrak{A}^*$ and $B \in \mathfrak{A}^*$ imply that $\overline{A}$, $A \cup B$, $A \cap B \in \mathfrak{A}$. (For instance, if $d(A, A_n) \to 0$ and $d(B, B_n) \to 0$, then by the third inequality in (3.4.2) one has $d(AB, A_nB_n) \le d(A, A_n) + d(B, B_n) \to 0$, so that $AB \in \mathfrak{A}^*$.)

Now let $C = \bigcap_{k=1}^{\infty} C_k$ where $C_k \in \mathfrak{A}^*$. Since $\mathfrak{A}^*$ is an algebra, we have $D_n = \bigcup_{k=1}^{n} C_k \in \mathfrak{A}^*$; moreover,

$$d(D_n, C) = \mathbf{P}(C - D_n) = \mathbf{P}(C) - \mathbf{P}(D_n) \to 0.$$

Therefore one can choose $A_n \in \mathcal{A}$ so that $d(D_n, A_n) < 1/n$, and consequently by virtue of (3.4.2) we have

$$d(C, A_n) \le d(C, D_n) + d(D_n, A_n) \to 0.$$

Thus $C \in \mathfrak{A}^*$ and hence $\mathfrak{A}^*$ forms a σ-algebra. The theorem is proved. □

Proof of Theorem 3.4.3 is now easy. If $A_1 \in \mathfrak{A}_1$ and $A_2 \in \mathfrak{A}_2$, then by Theorem 3.4.4 there exist sequences $A_{1n} \in \mathcal{A}_1$ and $A_{2n} \in \mathcal{A}_2$ such that $d(A_i, A_{in}) \to 0$ as $n \to \infty$, $i = 1, 2$. Putting $B = A_1A_2$ and $B_n = A_{1n}A_{2n}$, we obtain that

$$d(B, B_n) \le d(A_1, A_{1n}) + d(A_2, A_{2n}) \to 0$$

as $n \to \infty$ and

$$\mathbf{P}(A_1A_2) = \lim_{n\to\infty} \mathbf{P}(B_n) = \lim_{n\to\infty} \mathbf{P}(A_{1n})\mathbf{P}(A_{2n}) = \mathbf{P}(A_1)\mathbf{P}(A_2).$$ □

3.4.3 Relations Between the Introduced Notions

We will need one more definition. Let ξ be a random variable (or vector) given on a probability space $\langle \Omega, \mathfrak{F}, \mathbf{P}\rangle$.

---

[7] The theorem is also a direct consequence of the lemma from Appendix 1.

Definition 3.4.4 The class $\mathfrak{F}_\xi$ of events from $\mathfrak{F}$ of the form $A = \xi^{-1}(B) = \{\omega : \xi(\omega) \in B\}$, where B are Borel sets, is called the *σ-algebra generated by the random variable* ξ.

It is evident that $\mathfrak{F}_\xi$ is a σ-algebra since to each operation on sets A there corresponds the same operation on the sets $B = \xi(A)$ forming a σ-algebra.

The σ-algebra $\mathfrak{F}_\xi$ generated by the random variable ξ will also be denoted by $\sigma(\xi)$.

Consider, for instance, a probability space $\langle \Omega, \mathfrak{B}, \mathbf{P} \rangle$, where $\Omega = \mathbb{R}$ is the real line and $\mathfrak{B}$ is the σ-algebra of Borel sets. If

$$\xi = \xi(\omega) = \begin{cases} 0, & \omega < 0, \\ 1, & \omega \ge 0, \end{cases}$$

then $\mathfrak{F}_\xi$ clearly consists of four sets: $\mathbb{R}$, $\varnothing$, $\{\omega < 0\}$ and $\{\omega \ge 0\}$. Such a random variable ξ cannot distinguish "finer" sets from $\mathfrak{B}$. On the other hand, it is obvious that ξ will be measurable ($\{\xi \in B\} \in \mathfrak{B}_1$) with respect to any other "richer" sub-σ-algebra $\mathfrak{B}_1$, such that $\sigma(\xi) \subset \mathfrak{B}_1 \subset \mathfrak{B}$.

If $\xi = \xi(\omega) = \lfloor \omega \rfloor$ is the integral part of ω, then $\mathfrak{F}_\xi$ will be the σ-algebra of sets composed of the events $\{k \le \omega < k+1\}$, $k = \dots, -1, 0, 1, \dots$

Finally, if $\xi(\omega) = \varphi(\omega)$ where φ is continuous and monotone, $\varphi(\infty) = \infty$ and $\varphi(-\infty) = -\infty$, then $\mathfrak{F}_\xi$ coincides with the σ-algebra of Borel sets $\mathfrak{B}$.

Lemma 3.4.1 *Let ξ and η be two random variables given on $\langle \Omega, \mathfrak{F}, \mathbf{P} \rangle$, the variable ξ being measurable with respect to $\sigma(\eta)$. Then ξ and η are functionally related, i.e. there exists a Borel function g such that $\xi = g(\eta)$.*

Proof By assumption,

$$A_{k,n} = \left\{ \xi \in \left[\frac{k}{2^n}, \frac{k+1}{2^n} \right) \right\} \in \sigma(\eta).$$

Denote by $B_{k,n} = \{\eta(\omega) : \omega \in A_{k,n}\}$ the images of the sets $A_{k,n}$ on the line $\mathbb{R}$ under the mapping $\eta(\omega)$ and put $g_n(x) = k/2^n$ for $x \in B_{k,n}$. Then $g_n(\eta) = [2^n \varepsilon]/2^n$ and because $A_{k,n} \in \sigma(\eta)$, $B_{k,n} \in \mathfrak{B}$ and g_n is a Borel function. Since $g_n(x) \uparrow$ for any x, the limit $\lim_{n\to\infty} g_n(x) = g(x)$ exists and is also a Borel function. It remains to observe that $\varepsilon = \lim_{n\to\infty} g_n(\eta) = g(\eta)$ by the very construction. □

Now we formulate an evident proposition relating independence of random variables and σ-algebras.

Random variables $\xi_1, \dots, \xi_n$ are independent if and only if the σ-algebras $\sigma(\xi_1), \dots, \sigma(\xi_n)$ are independent.

This is a direct consequence of the definitions of independence of random variables and σ-algebras.

Now we can prove Theorem 3.4.1. First note that finite unions of semi-intervals $[\cdot,\cdot)$ (perhaps with infinite end points) form a σ-algebra generating the Borel σ-algebra on the line: $\mathfrak{B} = \sigma(\mathcal{A})$.

Proof of Theorem 3.4.1 Since in one direction the assertion of the theorem is obvious, it suffices to verify that the equality $F(x_1, \dots, x_n) = F_{\xi_1}(x_1) \cdots F_{\xi_n}(x_n)$ for the joint distribution function implies the independence of $\sigma(\xi_1), \dots, \sigma(\xi_n)$. Put for simplicity $n = 2$ and denote by Δ and Λ the semi-intervals $[x_1, x_2)$ and $[y_1, y_2)$, respectively. The following equalities hold:

$$\begin{aligned}\mathbf{P}(\xi_1 \in \Delta, \xi_2 \in \Lambda) &= \mathbf{P}\big(\xi_1 \in [x_1, x_2), \xi_2 \in [y_1, y_2)\big)\\ &= F(x_2, y_2)F(x_1, y_2) - F(x_2, y_1) + F(x_1, y_1)\\ &= \big(F_{\xi_1}(x_2) - F_{\xi_1}(x_1)\big)\big(F_{\xi_2}(y_2) - F_{\xi_2}(y_1)\big)\\ &= \mathbf{P}\{\xi_1 \in \Delta\}\mathbf{P}\{\xi_2 \in \Lambda\}.\end{aligned}$$

Consequently, if Δ_i, $i = 1, \dots, n$, and Λ_j, $j = 1, \dots, m$, are two systems of disjoint semi-intervals, then

$$\begin{aligned}\mathbf{P}\left(\xi_1 \in \bigcup_{i=1}^{n} \Delta_i, \xi_2 \in \bigcup_{j=1}^{m} \Lambda_j\right) &= \sum_{i,j} \mathbf{P}(\xi_1 \in \Delta_i, \xi_2 \in \Lambda_j)\\ &= \sum_{i,j} \mathbf{P}(\xi_1 \in \Delta_i)\mathbf{P}(\xi_2 \in \Lambda_j)\\ &= \mathbf{P}\left(\xi_1 \in \bigcup_{i=1}^{n} \Delta_i\right)\mathbf{P}\left(\xi_2 \in \bigcup_{j=1}^{m} \lambda_j\right). \qquad (3.4.4)\end{aligned}$$

But the class of events $\{\omega : \xi(\omega) \in A\} = \xi^{-1}(A)$, where $A \in \mathcal{A}$, forms, along with $\mathcal{A}$, an algebra (we will denote it by $\alpha(\xi)$), and one has $\sigma(\alpha(\xi)) = \sigma(\xi)$. In (3.4.4) we proved that $\alpha(\xi_1)$ and $\alpha(\xi_2)$ are independent. Therefore by Theorem 3.4.3 the σ-algebras $\sigma(\xi_1) = \sigma(\alpha(\xi_1))$ and $\sigma(\xi_2) = \sigma(\alpha(\xi_1))$ are also independent. The theorem is proved. □

It is convenient to state the following fact as a theorem.

Theorem 3.4.5 *Let φ_1 and φ_2 be Borel functions and ξ_1 and ξ_2 be independent random variables. Then $\eta_1 = \varphi_1(\xi_1)$ and $\eta_2 = \varphi_2(\xi_2)$ are also independent random variables.*

Proof We have to verify that, for any Borel sets B_1 and B_2,

$$\mathbf{P}\big(\varphi_1(\xi_1) \in B_1, \varphi_2(\xi_2) \in B_2\big) = \mathbf{P}\big(\varphi_1(\xi_1) \in B_1\big)\mathbf{P}\big(\varphi_2(\xi_2) \in B_2\big). \qquad (3.4.5)$$

But the sets $\{x : \varphi_i(x) \in B_i\} = \varphi^{-1}(B_i) = B_i^*$, $i = 1, 2$, are again Borel sets. Therefore

$$\big\{\omega : \varphi_i(\xi_i) \in B_i\big\} = \big\{\omega : \xi_i \in B_i^*\big\},$$

and the required multiplicativity of probability (3.4.5) follows from the independence of ξ_i. The theorem is proved. □

Let $\{\xi_n\}_{n=1}^{\infty}$ be a sequence of independent random variables. Consider the random variables $\xi_k, \xi_{k+1}, \dots, \xi_m$ where $k < m \le \infty$. Denote by $\sigma(\xi_k, \dots, \xi_m)$ (for $m = \infty$ we will write $\sigma(\xi_k, \xi_{k+1}, \dots)$) the σ-algebra generated by the events $\bigcap_{i=k}^{m} A_i$, where $A_i \in \sigma(\xi_i)$.

Definition 3.4.5 The σ-algebra $\sigma(\xi_k, \dots, \xi_m)$ is said to be *generated by the random variables* $\xi_k, \dots, \xi_m$.

In the sequel we will need the following proposition.

Theorem 3.4.6 *For any* $k \ge 1$, *the* σ*-algebra* $\sigma(\xi_{n+k})$ *is independent of* $\sigma(\xi_1, \dots, \xi_n)$.

Proof To prove the assertion, we make use of Theorem 3.4.3. To this end we have to verify that the algebra $\mathcal{A}$ generated by sets of the form $B = \bigcap_{i=1}^{n} A_i$, where $A_i \in \sigma(\xi_i)$, is independent of $\sigma(\xi_{n+k})$. Let $A \in \sigma(\xi_{n+k})$, then it follows from the independence of the σ-algebras $\sigma(\xi_1), \sigma(\xi_2), \dots, \sigma(\xi_n), \sigma(\xi_{n+k})$ that

$$\mathbf{P}(AB) = \mathbf{P}(A)\mathbf{P}(A_1)\cdots\mathbf{P}(A_n) = \mathbf{P}(A)\cdot\mathbf{P}(B).$$

In a similar way we verify that

$$\mathbf{P}\left(\bigcup_{i=1}^{n} A_i A\right) = \mathbf{P}\left(\bigcup_{i=1}^{n} A_i\right)\mathbf{P}(A)$$

(one just has to represent $\bigcup_{i=1}^{n} A_i$ as a union of disjoint events from $\mathcal{A}$). Thus the algebra $\mathcal{A}$ is independent of $\sigma(\xi_{n+k})$. Hence $\sigma(\xi_1, \dots, \xi_n)$ and $\sigma(\xi_{n+k})$ are independent. The theorem is proved. □

It is not hard to see that similar conclusions can be made about vector-valued random variables $\xi_1, \xi_2, \dots$ defining their independence using the relation

$$\mathbf{P}(\xi_1 \in B_1, \dots, \xi_n \in B_n) = \prod \mathbf{P}(\xi_j \in B_j),$$

where B_j are Borel sets in spaces of respective dimensions.

In conclusion of this section note that one can always construct a probability space $\langle \Omega, \mathfrak{F}, \mathbf{P}\rangle$ ($\langle \mathbb{R}^n, \mathfrak{B}^n, \mathbf{P}_\xi\rangle$) on which independent random variables $\xi_1, \dots, \xi_n$ with prescribed distribution functions F_{ξ_j} are given whenever these distributions F_{ξ_j} are known. This follows immediately from Sect. 3.3, since in our case the joint distribution function $F_\xi(x_1, \dots, x_n)$ of the vector $\xi = (\xi_1, \dots, \xi_n)$ is uniquely determined by the distribution functions $F_{\xi_j}(x)$ of the variables ξ_j:

$$F_\xi(x_1, \dots, x_n) = \prod_{1}^{n} F_{\xi_j}(x_j).$$

3.5 * On Infinite Sequences of Random Variables

We have already mentioned infinite sequences of random variables. Such sequences will repeatedly be objects of our studies below. However, there arises the question of whether one can define an infinite sequence on a probability space in such a way that its components possess certain prescribed properties (for instance, that they will be independent and identically distributed).

As we saw, one can always define a *finite* sequence of independent random variables by choosing for the "compound" random variable $(\xi_1, \dots, \xi_n)$ the sample space $\mathbb{R}_1 \times \mathbb{R}_2 \times \cdots \times \mathbb{R}_n = \mathbb{R}^n$ and σ-algebra $\mathfrak{B}_1 \times \mathfrak{B}_1 \times \cdots \times \mathfrak{B}_n = \mathfrak{B}^n$ generated by sets of the form $B_1 \times B_2 \times \cdots \times B_n \subset \mathbb{R}^n$, B_i being Borel sets. It suffices to define probability on the algebra of these sets. In the infinite-dimensional case, however, the situation is more complicated. Theorem 3.2.1 and its extensions to the multivariate case are insufficient here. One should define probability on an algebra of events from $\mathbb{R}^\infty = \prod_{k=1}^\infty \mathbb{R}_k$ so that its closure under countably many operations $\cup$ and $\cap$ form the σ-algebra $\mathfrak{B}^\infty$ generated by the products $\bigcap B_{j_k}$, $B_{j_k} \in \mathfrak{B}_{j_k}$.

Let N be a subset of integers. Denote by $\mathbb{R}^N = \prod_{k\in N} \mathbb{R}_k$ the direct product of the spaces $\mathbb{R}_k$ over $k \in N$, $\mathfrak{B}^N = \prod_{k\in N} \mathfrak{B}_k$. We say that distributions $\mathbf{P}_{N'}$ and $\mathbf{P}_{N''}$ on $\langle \mathbb{R}^{N'}, \mathfrak{B}^{N'} \rangle$ and $\langle \mathbb{R}^{N''}, \mathfrak{B}^{N''} \rangle$, respectively, are *consistent* if the measures induced by $\mathbf{P}_{N'}$ and $\mathbf{P}_{N''}$ on the intersection $\mathbb{R}^N = \mathbb{R}^{N'} \cap \mathbb{R}^{N''}$ (here $N = N' \cap N''$) coincide with each other. The measures on $\mathbb{R}^N$ are said to be the *projections* of $\mathbf{P}_{N'}$ and $\mathbf{P}_{N''}$, respectively, on $\mathbb{R}^N$. An answer to the above question about the existence of an infinite sequence of random variables is given by the following theorem (the proof of which is given in Appendix 2).

Theorem 3.5.1 (Kolmogorov) *Specifying a family of consistent distributions* $\mathbf{P}_N$ *on finite-dimensional spaces* $\mathbb{R}^N$ *defines a unique probability measure* $\mathbf{P}_\infty$ *on* $\langle \mathbb{R}^\infty, \mathfrak{B}^\infty \rangle$ *such that each probability* $\mathbf{P}_N$ *is the projection of* $\mathbf{P}_\infty$ *onto* $\mathbb{R}^N$.

It follows from this theorem, in particular, that one can always define on an appropriate space an infinite sequence of arbitrary independent random variables. Indeed, direct products of measures given on $\mathbb{R}_1, \mathbb{R}_2, \dots$ for different products $\mathbb{R}^{N'}$ and $\mathbb{R}^{N''}$ are always consistent.

3.6 Integrals

3.6.1 Integral with Respect to Measure

As we have already noted, defining a probability space includes specifying a finite countably additive measure. This enables one to consider integrals with respect to the measure,

$$\int g\big(\xi(\omega)\big)\mathbf{P}(d\omega) \tag{3.6.1}$$

over the set Ω for a Borel function g and any random variable ξ on $\langle\Omega, \mathfrak{F}, \mathbf{P}\rangle$ (recall that $g(x)$ is said to be Borel if, for any t, $\{x : g(x) < t\}$ is a Borel set on the real line).

The definition, construction and basic properties of the integral with respect to a measure are assumed to be familiar to the reader. If the reader feels his or her background is insufficient in this aspect, we recommend Appendix 3 which contains all the necessary information. However, the reader could skip this material if he/she is willing to restrict him/herself to considering only discrete or absolutely continuous distributions for which integrals with respect to a measure become sums or conventional Riemann integrals. It would also be useful for the sequel to know the Stieltjes integral; see the comments in the next subsection.

We already know that a random variable $\xi(\omega)$ induces a measure $\mathbf{F}_\xi$ on the real line which is specified by the equality

$$\mathbf{F}_\xi\bigl([x, y)\bigr) = \mathbf{P}(x \le \xi \le y) = F_\xi(y) - F_\xi(x).$$

Using this measure, one can write the integral (3.6.1) as

$$\int g\bigl(\xi(\omega)\bigr)\mathbf{P}(d\omega) = \int g(x)\mathbf{F}_\xi(dx).$$

This is just the result of the substitution $x = \xi(\omega)$. It can be proved simply by writing down the definitions of both integrals. The integral on the right hand side is called the *Lebesgue–Stieltjes integral* of the function $g(x)$ with respect to the measure $\mathbf{P}_\xi$ and can also be written as

$$\int g(x)\,dF_\xi(x). \tag{3.6.2}$$

3.6.2 The Stieltjes Integral

The integral (3.6.2) is often just called the Stieltjes integral, or the Riemann–Stieltjes integral which is defined in a somewhat different way and for a narrower class of functions.

If $g(x)$ is a continuous function, then the Lebesgue–Stieltjes integral coincides with the Riemann–Stieltjes integral which is equal by definition to

$$\int g(x)\,dF(x) = \lim_{\substack{b\to\infty\\ a\to-\infty}} \lim_{N\to\infty} \sum_{k=0}^{N} g(\widetilde{x}_k)\bigl[F(x_{k+1}) - F(x_k)\bigr], \tag{3.6.3}$$

where the limit on the right-hand side does not depend on the choice of partitions $x_0, x_1, \dots, x_N$ of the semi-intervals $[a, b)$ and points $\widetilde{x}_k \in \Delta_k = [x_k, x_{k+1})$. Partitions $x_0, x_1, \dots, x_N$ are different for different N's and have the property that $\max_k(x_{k+1} - x_k) \to 0$ as $N \to \infty$.

Indeed, as we know (see Appendix 3), the Lebesgue–Stieltjes integral is

$$\int g(x)\,dF(x) = \lim_{\substack{b\to\infty\\ a\to-\infty}} \lim_{N\to\infty} \int_a^b g_N(x)\mathbf{F}_\xi(dx), \tag{3.6.4}$$

where g_N is any sequence of simple functions (assuming finitely many values) converging monotonically to $g(x)$. We see from these definitions that it suffices to show that the integrals $\int_a^b$ with finite integration limits coincide. Since the Lebesgue–Stieltjes integral $\int_a^b g\,dF$ of a continuous function g always exists, we could obtain its value by taking the sequence g_N to be any of the two sequences of simple functions g_N^* and g_N^{**} which are constant on the semi-intervals Δ_k and equal on them to

$$g_N^*(x_k) = \sup_{x\in\Delta_k} g(x) \quad \text{and} \quad g_N^{**}(x_k) = \inf_{x\in\Delta_k} g(x),$$

respectively. Both sequences in (3.6.4) constructed from g_N^* and g_N^{**} will clearly converge monotonically from different sides to the same limit equal to the Lebesgue–Stieltjes integral

$$\int_a^b g(x)\,dF(x).$$

But for any $\widetilde{x}_k \in \Delta_k$, one has

$$g_N^{**}(x_k) \le g(\widetilde{x}_k) \le g_N^*(x_k),$$

and therefore the integral sum in (3.6.3) will be between the bounds

$$\int_a^b g_N^{**}\,dF(x) \le \sum_{k=0}^{N} g(\widetilde{x}_k)\big[F(x_{k+1}) - F(x_k)\big] \le \int_a^b g_N^*\,dF(x).$$

These inequalities prove the required assertion about the coincidence of the integrals.

It is not hard to verify that (3.6.3) and (3.6.4) will also coincide when $F(x)$ is continuous and $g(x)$ is a function of bounded variation. In that case,

$$\int_a^b g(x)\,dF(x) = g(x)F(x)\big|_a^b - \int_a^b F(x)\,dg(x).$$

Making use of this fact, we can extend the definition of the Riemann–Stieltjes integral to the case when $g(x)$ is a function of bounded variation and $F(x)$ is an arbitrary distribution function. Indeed, let $F(x) = F_c(x) + F_d(x)$ be a representation of $F(x)$ as a sum of its continuous and discrete components, and $y_1, y_2, \ldots$ be the jump points of $F_d(x)$:

$$p_k = F_d(y_k + 0) - F_d(y_k) > 0.$$

Then one has to put by definition

$$\int g(x)\,dF(x) = \sum p_k g(y_k) + \int g(x)\,dF_c(x),$$

where the Riemann–Stieltjes integral $\int g\,dF_c(x)$ can be understood, as we have already noted, in the sense of definition (3.6.3).

We will say, as is generally accepted, that $\int g\,dF$ exists if the integral $\int |g|\,dF$ is finite. It is easy to see from the definition of the Stieltjes integral that, for step functions $F(x)$ (the distribution is discrete), the integral becomes the sum

$$\int g(x)\,dF(x) = \sum_k g(x_k)\big(F(x_k+0) - F(x_k)\big) = \sum_k g(x_k)\mathbf{P}(\xi = x_k),$$

where $x_1, x_2, \ldots$ are jump points of $F(x)$. If

$$F(x) = \int_{-\infty}^{x} p(x)\,dx$$

is absolutely continuous and $p(x)$ and $g(x)$ are Riemann integrable, then the Stieltjes integral

$$\int g(x)\,dF(x) = \int g(x)p(x)\,dx$$

becomes a conventional Riemann integral.

We again note that for a *reader who is not familiar with Stieltjes integral techniques and integration with respect to measures, it is possible to continue reading the book keeping in mind only the last two interpretations of the integral.* This would be quite sufficient for an understanding of the exposition. Moreover, most of the distributions which are important from the practical point of view are just of one of these types: either discrete or absolutely continuous.

We recall some other properties of the Stieltjes integral (following immediately from definitions (3.6.4) or (3.6.3) and (3.6.5)):

$$\int_a^b dF = F(b) - F(a);$$

$$\int_a^b g\,dF = \int_a^c g\,dF + \int_c^b g\,dF \quad \text{if } g \text{ or } F \text{ is continuous at the point } c;$$

$$\int (g_1 + g_2)\,dF = \int g_1\,dF + \int g_2\,dF;$$

$$\int cg\,dF = c\int g\,dF \quad \text{for } c = \text{const};$$

$$\int_a^b g\,dF = gF\big|_a^b - \int_a^b F\,dg$$

if g is a function of bounded variation.

3.6.3 Integrals of Multivariate Random Variables. The Distribution of the Sum of Independent Random Variables

Integrals with respect to measure (3.6.1) make sense for multivariate variables $\xi(\omega) = (\xi_1(\omega), \ldots, \xi_n(\omega))$ as well (one cannot say the same about Riemann–

Stieltjes integrals (3.6.3)). We mean here the integral

$$\int_{\Omega} g(\xi_1(\omega), \dots, \xi_n(\omega))\mathbf{P}(d\omega), \tag{3.6.5}$$

where g is a measurable function mapping $\mathbb{R}^n$ into $\mathbb{R}$, so that $g(\xi_1(\omega), \dots, \xi_n(\omega))$ is a measurable mapping of Ω into $\mathbb{R}$.

If $\langle \mathbb{R}^n, \mathfrak{B}^n, \mathbf{F}_\xi \rangle$ is a sample probability space for ξ, then the integral (3.6.5) can be written as

$$\int_{\mathbb{R}^n} g(x)\mathbf{F}_\xi(dx), \quad x = (x_1, \dots, x_n) \in \mathbb{R}^n.$$

Now turn to the case when the components $\xi_1, \dots, \xi_n$ of the vector ξ are independent and assume first that $n = 2$. For sets

$$B = B_1 \times B_2 = \{(x_1, x_2) : x_1 \in B_1,\ x_2 \in B_2\} \subset \mathbb{R}^2,$$

where B_1 and B_2 are measurable subsets of $\mathbb{R}$, one has the equality

$$\mathbf{P}(\xi \in B) = \mathbf{P}(\xi_1 \in B_1,\ \xi_2 \in B_2) = \mathbf{P}(\xi_1 \in B_1)\mathbf{P}(\xi_2 \in B_2). \tag{3.6.6}$$

In that case one says that the measure $\mathbf{F}_{\xi_1,\xi_2}(dx_1, dx_2) = \mathbf{P}(\xi_1 \in dx_1, \xi_2 \in dx_2)$ on $\mathbb{R}^2$, corresponding to (ξ_1, ξ_2), is a *direct product of the measures*

$$\mathbf{F}_{\xi_1}(dx_1) = \mathbf{P}(\xi_1 \in dx_1) \quad \text{and} \quad \mathbf{F}_{\xi_2}(dx_2) = \mathbf{P}(\xi_2 \in dx_2).$$

As we already know, equality (3.6.6) uniquely specifies a measure on $\langle \mathbb{R}^2, \mathfrak{B}^2 \rangle$ from the given distributions of ξ_1 and ξ_2 on $\langle \mathbb{R}, \mathfrak{B} \rangle$. It turns out that the integral

$$\int g(x_1, x_2)\mathbf{F}_{\xi_1\xi_2}(dx_1, dx_2) \tag{3.6.7}$$

with respect to the measure $\mathbf{F}_{\xi_1,\xi_2}$ can be expressed in terms of integrals with respect to the measures $\mathbf{F}_{\xi_1}$ and $\mathbf{F}_{\xi_2}$. Namely, Fubini's theorem holds true (for the proof see Appendix 3 or property 5A in Sect. 4.8).

Theorem 3.6.1 (Theorem on iterated integration) *For a Borel function $g(x, y) \ge 0$ and independent ξ_1 and ξ_2,*

$$\int g(x_1, x_2)\mathbf{F}_{\xi_1\xi_2}(dx_1, dx_2) = \int \left[\int g(x_1, x_2)\mathbf{F}_{\xi_2}(dx_2) \right] \mathbf{F}_{\xi_1}(dx_1). \tag{3.6.8}$$

If $g(x, y)$ can assume values of different signs, then the existence of the integral on the left-hand side of (3.6.8) *is required for the equality* (3.6.8). *The order of integration on the right-hand side of* (3.6.8) *may be changed.*

It is shown in Appendix 3 that the measurability of $g(x, y)$ implies that of the integrands on the right-hand side of (3.6.8).

Corollary 3.6.1 *Let $g(x_1, x_2) = g_1(x_1)g_2(x_2)$. Then, if at least one of the following three conditions is met*:

(1) $g_1 \ge 0$, $g_2 \ge 0$,

(2) $\int g_1(x_1)g_2(x_2)\mathbf{F}_{\xi_1\xi_2}(dx_1, dx_2)$ *exists,*
(3) $\int g_j(x_j)\mathbf{F}_{\xi_j}(dx_j)$, $j = 1, 2$, *exist,*

then

$$\int g_1(x_1)g_2(x_2)\mathbf{F}_{\xi_1\xi_2}(dx_1, dx_2) = \int g_1(x_1)\mathbf{F}_{\xi_1}(dx_1)\int g_2(x_2)\mathbf{F}_{\xi_2}(dx_2). \quad (3.6.9)$$

To avoid trivial complications, we assume that $\mathbf{P}(g_j(\xi_j) = 0) \neq 1$, $j = 1, 2$.

Proof Under any of the first two conditions, the assertion of the corollary follows immediately from Fubini's theorem. For arbitrary g_1, g_2, put $g_j = g_j^+ - g_j^-$, $g_j^\pm \geq 0$, $j = 1, 2$. If $\int g_j^\pm d\mathbf{F}_\xi < \infty$ (we will use here the abridged notation for integrals), then

$$\begin{aligned}\int g_1 g_2\, d\mathbf{F}_{\xi_1}\, d\mathbf{F}_{\xi_2} &= \int g_1^+ g_2^+ d\mathbf{F}_{\xi_1}\, d\mathbf{F}_{\xi_2} - \int g_1^+ g_2^-\, d\mathbf{F}_{\xi_1}\, d\mathbf{F}_{\xi_2} \\ &\quad - \int g_1^- g_2^+\, d\mathbf{F}_{\xi_1}\, d\mathbf{F}_{\xi_2} + \int g_1^- g_2^-\, d\mathbf{F}_{\xi_1}\, d\mathbf{F}_{\xi_2} \\ &= \int g_1^+\, d\mathbf{F}_{\xi_1} \int g_2^+\, d\mathbf{F}_{\xi_2} - \int g_1^+\, d\mathbf{F}_{\xi_1} \int g_2^+\, d\mathbf{F}_{\xi_2} \\ &\quad - \int g_1^-\, d\mathbf{F}_{\xi_1} \int g_2^+\, d\mathbf{F}_{\xi_2} + \int g_1^-\, d\mathbf{F}_{\xi_1} \int g_2^-\, d\mathbf{F}_{\xi_2} \\ &= \int g_1\, d\mathbf{F}_{\xi_1} \int g_2\, d\mathbf{F}_{\xi_2}.\end{aligned}$$

□

Corollary 3.6.2 *In the special case when* $g(x_1, x_2) = I_B(x_1, x_2)$ *is the indicator of a set* $B \in \mathfrak{B}^2$, *we obtain the formula for sequential computation of the measure of* B:

$$\mathbf{P}\big((\xi_1, \xi_2) \in B\big) = \int \mathbf{P}\big((x_1, \xi_2) \in B\big)\mathbf{F}_{\xi_1}(dx_1).$$

The probability of the event $\{(x_1, \xi_2) \in B\}$ could also be written as $\mathbf{P}(\xi_2 \in B_{x_1}) = \mathbf{P}_{\xi_2}(B_{x_1})$ where $B_{x_1} = \{x_2 : (x_1, x_2) \in B\}$ is the "section" of the set B at the point x_1.

If $B = \{(x_1, x_2) : x_1 + x_2 < x\}$, we get

$$\begin{aligned}\mathbf{P}\big((\xi_1, \xi_2) \in B\big) &= \mathbf{P}(\xi_1 + \xi_2 < x) \equiv F_{\xi_1+\xi_2}(x) \\ &= \int \mathbf{P}(x_1 + \xi_2 < x)\mathbf{F}_{\xi_1}(dx_1) \\ &= \int F_{\xi_2}(x - x_1)\, dF_{\xi_1}(x_1). \qquad (3.6.10)\end{aligned}$$

We have obtained a formula for the distribution function of the sum of independent random variables expressing $F_{\xi_1+\xi_2}$ in terms of F_{ξ_1} and F_{ξ_2}. The integral on the right-hand side of (3.6.10) is called the *convolution* of the distribution functions

$F_{\xi_1}(x)$ and $F_{\xi_2}(x)$ and is denoted by $F_{\xi_1} * F_{\xi_2}(x)$. In the same way one can obtain the equality

$$\mathbf{P}(\xi_1 + \xi_2 < x) = \int_{-\infty}^{\infty} F_{\xi_1}(x-t)\,dF_{\xi_2}(t).$$

Observe that the right-hand side here could also be considered as a result of integrating

$$\int dF_{\xi_1}(t)F_{\xi_2}(x-t)$$

by parts.

If at least one of the distribution functions has a density, the convolution also has a density. This follows immediately from the formulas for convolution. Let, for instance,

$$F_{\xi_2}(x) = \int_{-\infty}^{x} f_{\xi_2}(u)\,du.$$

Then

$$\begin{aligned} F_{\xi_1+\xi_2}(x) &= \int_{-\infty}^{\infty} \mathbf{F}_{\xi_1}(dt) \int_{-\infty}^{x} f_{\xi_2}(u-t)\,du \\ &= \int_{-\infty}^{x} \left(\int_{-\infty}^{\infty} \mathbf{F}_{\xi_1}(dt) f_{\xi_2}(u-t) \right) du, \end{aligned}$$

so that the density of the sum $\xi_1 + \xi_2$ equals

$$f_{\xi_1+\xi_2}(x) = \int_{-\infty}^{\infty} \mathbf{F}_{\xi_1}(dt) f_{\xi_2}(x-t) = \int_{-\infty}^{\infty} f_{\xi_2}(x-t)\,dF_{\xi_1}(t).$$

Example 3.6.1 Let $\xi_1, \xi_2, \dots$ be independent random variables uniformly distributed over $[0, 1]$, i.e. $\xi_1, \xi_2, \dots$ have the same distribution function with density

$$f(x) = \begin{cases} 1, & x \in [0,1], \\ 0, & x \notin [0,1]. \end{cases} \tag{3.6.11}$$

Then the density of the sum $\xi_1 + \xi_2$ is

$$f_{\xi_1+\xi_2}(x) = \int_0^1 f(x-t)\,dt = \begin{cases} 0, & x \notin [0,2], \\ x, & x \in [0,1], \\ 2-x, & x \in [1,2]. \end{cases} \tag{3.6.12}$$

The integral present here is clearly the length of the intersection of the segments $[0, 1]$ and $[x - 1, x]$. The graph of the density of the sum $\xi_1 + \xi_2 + \xi_3$ will consist of three pieces of parabolas:

$$f_{\xi_1+\xi_2+\xi_3}(x) = \int_0^1 f_{\xi_1+\xi_2}(x-t)\,dt = \begin{cases} 0, & x \notin [0,3], \\ \frac{x^2}{2}, & x \in [0,1], \\ 1 - \frac{(2-x)^2}{2} - \frac{(x-1)^2}{2}, & x \in [1,2], \\ \frac{(3-x)^2}{2}, & x \in [2,3]. \end{cases}$$

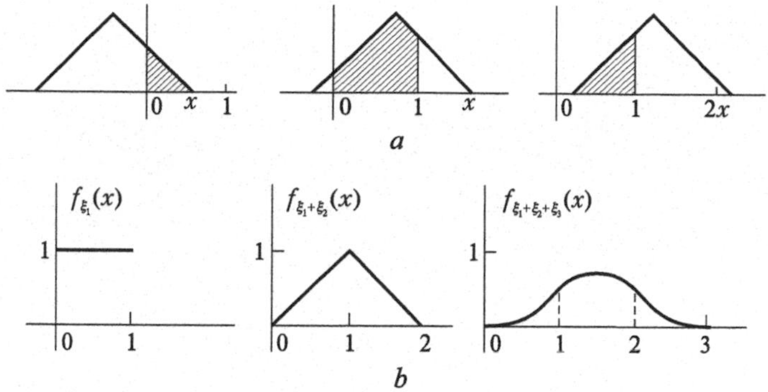

Fig. 3.2 Illustration to Example 3.6.1. The *upper row* visualizes the computation of the convolution integral for the density of $\xi_1 + \xi_2 + \xi_3$. The *lower row* displays the densities of ξ_1, $\xi_1 + \xi_2$, and $\xi_1 + \xi_2 + \xi_3$, respectively

The computation of this integral is visualised in Fig. 3.2, where the shaded areas correspond to the values of $f_{\xi_1+\xi_2+\xi_3}(x)$ for different x. The shape of the densities of ξ_1, $\xi_1 + \xi_2$ and $\xi_1 + \xi_2 + \xi_3$ is shown in Fig. 3.2b. The graph of the density of the sum $\xi_1 + \xi_2 + \xi_3 + \xi_4$ will consist of four pieces of cubic parabolas and so on. If we shift the origin to the point $n/2$, then, as n increases, the shape (up to a scaling transformation) of the density of the sum $\xi_1 + \cdots + \xi_n$ will be closer and closer to that of the function e^{-x^2}. We will see below that this is not due to chance.

In connection with this example we could note that if ξ and η are two independent random variables, ξ having the distribution function $F(x)$ and η being uniformly distributed over [0, 1], then the density of the sum $\xi + \eta$ at the point x is equal to

$$f_{\xi+\eta}(x) = \int dF(t) f_\eta(x-t) = \int_{x-1}^{x} dF(t) = F(x) - F(x-1).$$

Chapter 4
Numerical Characteristics of Random Variables

Abstract This chapter opens with Sect. 4.1 introducing the concept of the expectation of random variable as the respective Lebesgue integral and deriving its key properties, illustrated by a number of examples. Then the concepts of conditional distribution functions and conditional expectations given an event are presented and discussed in detail in Sect. 4.2, one of the illustrations introducing the ruin problem for the simple random walk. In the Sects. 4.3 and 4.4, expectations of independent random variables and those of sums of random numbers of random variables are considered. In Sect. 4.5, Kolmogorov–Prokhorov's theorem is proved for the case when the number of random terms in the sum is independent of the future, followed by the derivation of Wald's identity. After that, moments of higher orders are introduced and discussed, starting with the variance in Sect. 4.5 and proceeding to covariance and correlation coefficient and their key properties in Sect. 4.6. Section 4.7 is devoted to the fundamental moment inequalities: Cauchy–Bunjakovsky's inequality (a.k.a. Cauchy–Schwarz inequality), Hölder's and Jensen's inequalities, followed by inequalities for probabilities (Markov's and Chebyshev's inequalities). Section 4.8 extends the concept of conditional expectation (given a random variable or sigma-algebra), starting with the discrete case, then turning to square-integrable random variables and using projections, and finally considering the general case basing on the Radon–Nykodim theorem (proved in Appendix 3). The properties of the conditional expectation are studied, following by introducing the concept of conditional distribution given a random variable and illustrating it by several examples in Sect. 4.9.

4.1 Expectation

Definition 4.1.1 The (*mathematical*) *expectation*, or *mean value*, of a random variable ξ given on a probability space $\langle \Omega, \mathfrak{F}, \mathbf{P} \rangle$ is defined as the quantity

$$\mathbf{E}\xi = \int_{\Omega} \xi(\omega)\mathbf{P}(d\omega).$$

Let $\xi^{\pm} = \max(0, \pm\xi)$. The values $\mathbf{E}\xi^{\pm} \geq 0$ are always well defined (see Appendix 3). We will say that $\mathbf{E}\xi$ *exists* if $\max(\mathbf{E}\xi^{+}, \mathbf{E}\xi^{-}) < \infty$.

A.A. Borovkov, *Probability Theory*, Universitext,
DOI 10.1007/978-1-4471-5201-9_4,

We will say that $\mathbf{E}\xi$ is *well defined* if $\min(\mathbf{E}\xi^+, \mathbf{E}\xi^-) < \infty$. In this case the difference $\mathbf{E}\xi^+ - \mathbf{E}\xi^-$ is always well defined, but $\mathbf{E}\xi = \mathbf{E}\xi^+ - \mathbf{E}\xi^-$ may be $\pm\infty$.

By virtue of the above remarks (see Sect. 3.6) one can also define $\mathbf{E}\xi$ as

$$\mathbf{E}\xi := \int x\mathbf{F}_\xi(dx) = \int x\,dF(x), \tag{4.1.1}$$

where $F(x)$ is the distribution function of ξ. It follows from the definition that $\mathbf{E}\xi$ exists if $\mathbf{E}|\xi| < \infty$. It is not hard to see that $\mathbf{E}\xi$ does not exist if, for instance, $1 - F(x) > 1/x$ for all sufficiently large x.

We already know that if $F(x)$ is a step function then the Stieltjes integral (4.1.1) becomes the sum

$$\mathbf{E}\xi := \sum_k x_k \mathbf{P}(\xi = x_k).$$

If $F(x)$ has a density $f(x)$, then

$$\mathbf{E}\xi := \int x f(x)\,dx,$$

so that $\mathbf{E}\xi$ is the point of the "centre of gravity" of the distribution F of the unit mass on the real line and corresponds to the natural interpretation of the mean value of the distribution.

If $g(x)$ is a Borel function, then $\eta = g(\xi)$ is again a random variable and

$$\mathbf{E}g(\xi) = \int g\big(\xi(\omega)\big)\mathbf{P}(d\omega) = \int g(x)\,dF(x) = \int x\,dF_{g(\xi)}(x).$$

The last equality follows from definition (4.1.1).

The basic properties of expectations coincide with those of the integral:

E1. *If a and b are constants, then* $\mathbf{E}(a + b\xi) = a + b\mathbf{E}\xi$.
E2. $\mathbf{E}(\xi_1 + \xi_2) = \mathbf{E}(\xi_1) + \mathbf{E}(\xi_2)$, *if any two of the expectations appearing in the formula exist.*
E3. *If $a \le \xi \le b$, then $a \le \mathbf{E}\xi \le b$. The inequality $\mathbf{E}\xi \le \mathbf{E}|\xi|$ always holds.*
E4. *If $\xi \ge 0$ and $\mathbf{E}\xi = 0$, then $\xi = 0$ with probability* 1.
E5. *The probability of an event A can be expressed in terms of expectations as*

$$\mathbf{P}(A) = \mathbf{E}\mathrm{I}(A),$$

where $\mathrm{I}(A)$ is the random variable equal to the indicator of the event A: $\mathrm{I}(A) = 1$ if $\omega \in A$ and $\mathrm{I}(A) = 0$ otherwise.

For further properties of expectations, see Appendix 3.

We consider several examples.

Example 4.1.1 Expectations related to the Bernoulli scheme. Let $\xi \in \mathbf{B}_p$, i.e. ξ assumes two values: 0 with probability q and 1 with probability p, where $p + q = 1$. Then

$$\mathbf{E}\xi = 0 \times \mathbf{P}(\xi = 0) + 1 \times \mathbf{P}(\xi = 1) = p.$$

Now consider a sequence of trials in the Bernoulli scheme until the time of the first "success". In other words, consider a sequence of independent variables $\xi_1, \xi_2, \dots$ distributed as ξ until the time

$$\eta := \min\{k \ge 1 : \xi_k = 1\}.$$

It is evident that η is a random variable,

$$\mathbf{P}(\eta = k) = q^{k-1} p, \quad k \ge 1,$$

so that $\eta - 1$ has the geometric distribution. Consequently,

$$\mathbf{E}\eta = \sum_{k=1}^{\infty} k q^{k-1} p = \frac{p}{(1-q)^2} = \frac{1}{p}.$$

If we put $S_n := \sum_1^n \xi_k$, then clearly $\mathbf{E}S_n = np$. Now define, for an integer $N \ge 1$, the random variable $\eta = \min\{k \ge 1 : S_k = N\}$ as the "first passage time" of level N by the sequence S_n. One has

$$\mathbf{P}(\eta = k) = \mathbf{P}(S_{k-1} = N - 1)p,$$

$$\mathbf{E}\eta = p \sum_{N}^{\infty} k \binom{k-1}{N-1} p^{N-1} q^{k-N} = \frac{p^N}{(N-1)!} \sum_{k=N}^{\infty} k(k-1)\cdots(k-N+1) q^{k-N}.$$

The sum here is equal to the N-th derivative of the function $\psi(z) = \sum_0^\infty z^k = 1/(1-z)$ at the point $z = q$, i.e. it equals $N!/p^{N+1}$. Thus $\mathbf{E}\eta = N/p$. As we will see below, this equality could be obtained as an obvious consequence of the results of Sect. 4.4.

Example 4.1.2 If $\xi \in \mathbf{\Phi}_{a,\sigma^2}$ then

$$\begin{aligned}
\mathbf{E}\xi &= \int t\phi_{a,\sigma^2}(t)\,dt = \int t \frac{1}{\sigma\sqrt{2\pi}} e^{-\frac{(t-a)^2}{2\sigma^2}}\,dt \\
&= \frac{1}{\sigma\sqrt{2\pi}} \int (t-a) e^{-\frac{(t-a)^2}{2\sigma^2}}\,dt + \frac{a}{\sigma\sqrt{2\pi}} \int e^{-\frac{(t-a)^2}{2\sigma^2}}\,dt \\
&= \frac{1}{\sigma\sqrt{2\pi}} \int z e^{-\frac{z^2}{2\sigma^2}}\,dz + a = a.
\end{aligned}$$

Thus the parameter a of the normal law is equal to the expectation of the latter.

Example 4.1.3 If $\xi \in \mathbf{\Pi}_\mu$, then $\mathbf{E}\xi = \mu$. Indeed,

$$\mathbf{E}\xi = \sum_{k=0}^{\infty} k \frac{\mu^k}{k!} e^{-\mu} = \mu e^{-\mu} \sum_{k=1}^{\infty} \frac{\mu^{k-1}}{(k-1)!} = \mu.$$

Example 4.1.4 If $\xi \in \mathbf{U}_{0,1}$, then

$$\mathbf{E}\xi = \int_0^1 x\,dx = \frac{1}{2}.$$

It follows from property E1 that, for $\xi \in \mathbf{U}_{a,b}$, one has

$$\mathbf{E}\xi = a + \frac{b-a}{2} = \frac{a+b}{2}.$$

If $\xi \in \mathbf{K}_{0,1}$ then the expectation $\mathbf{E}\xi$ does not exist. That follows from the fact that the integral $\int \frac{x\,dx}{1+x^2}$ diverges.

Example 4.1.5 We now consider an example that is, in a sense, close to Example 4.1.1 on the computation of $\mathbf{E}\eta$, but which is more complex and corresponds to computing the mean value of the duration of a chess tournament in a real-life situation. In Sect. 3.4 we described a simple probabilistic model of a chess tournament. The first player wins in a given game, independently of the outcomes of the previous games, with probability p, loses with probability q, $p+q<1$, and makes a tie with probability $1-p-q$. Of course, this is a rather rough first approximation since in a real-life tournament there is apparently no independence. On the other hand, it is rather unlikely that, for balanced high level players, the above probabilities would substantially vary from game to game or depend on the outcomes of their previous results. A more complex model incorporating dependence of p and q of the outcomes of previous games will be considered in Example 13.4.2.

Assume that the tournament continues until one of the two participants wins N games (then this player will be declared the winner). For instance, the 1984 individual World Championship match between A. Karpov and G. Kasparov was organised just according to this scheme with $N=6$. What can one say about the expectation $\mathbf{E}\eta$ of the duration η of the tournament?

As was shown in Example 3.3.1,

$$\mathbf{P}(\eta = n) = p\sum_{i=0}^{N-1} p(n-1; N-1, i) + q\sum_{i=0}^{N-1} p(n-1; i, N-1),$$

where

$$p(n; i, j) = \frac{n!}{i!j!(n-i-j)!}p^i q^j (1-p-q)^{n-i-j}.$$

Therefore, under obvious conventions on the summation indices,

$$\mathbf{E}\eta = \frac{1}{(N-1)!}\sum_{i=0}^{N-1}\frac{p^N q^i + p^i q^N}{i!}\sum_{n=0}^{N-1} n(n-1)\times\cdots$$
$$\times (n-i-N+1)(1-p-q)^{n-i-N}.$$

The sum over n was calculated in Example 4.1.1 to be $(N+i)!/(p+q)^{N+i+1}$. Consequently,

$$\begin{aligned}\mathbf{E}\eta &= \frac{N}{p+q}\sum_{i=0}^{N-1}\frac{(p^N q^i + p^i q^N)(N+i)!}{i!N!(p+q)^{i+N}}\\ &= \frac{N}{p+q}\sum_{n=0}^{N-1}\binom{N+i}{i}\bigl[r^N(1-r)^i + r^i(1-r)^N\bigr],\end{aligned}$$

where $r = p/(p+q)$.

In his interview of 3 March 1985 to the newspaper "Izvestija", Karpov said that in qualifying tournaments he would lose, on average, 1 game out of 20, and that Kasparov's results were similar. If we put in our simple model $p = q = 1/20$ (strictly speaking, one cannot make such a conclusion from the given data; the relation $p = q = 1/20$ should be considered rather as one of many possible hypotheses) then, for $N = 6$, direct calculations show that $(r = 1/2)$

$$\mathbf{E}\eta = \frac{15}{8}\left[1 + 21\left(1 + \frac{5}{8} + \frac{11}{16}\right)\right] \approx 93.$$

Thus, provided that our simplest model is adequate, the expected duration of a tournament turns out to be very large. The fact that the match between Karpov and Kasparov was interrupted by the decision of the chairman of the World Chess Federation after 48 games because the match had dragged on, might serve as a confirmation of the correctness of the assumptions we made.

Taking into account the results of the match and consequent games between Karpov and Kasparov could lead to estimates (approximate values) for the quantities p and q that would differ from $1/20$.

For our model, one also has the following simple inequality:

$$\frac{N}{p+q} < \mathbf{E}\eta < \frac{2N-1}{p+q}.$$

It follows from the relation $\eta_N \le \eta \le \eta_{2N-1}$, where η_N is the number of games until the time when the total of the points gained by both players reaches N. By virtue of Example 4.1.1, $\mathbf{E}\eta_N = N/(p+q)$.

Example 4.1.6 In the problem on cells in Sects. 1.3 and 1.4, we considered the probability that at least one of the n cells in which r particles are placed at random is empty. Find the expectation of the number $S_{n,r}$ of empty cells after r particles have been placed. If A_k denotes the event that the k-th cell is empty and $\mathrm{I}(A_k)$ is the indicator of this event then

$$S_{n,r} = \sum_1^n \mathrm{I}(A_k), \qquad \mathbf{E}S_{n,r} = \sum_1^n \mathbf{P}(A_k) = n\left(1 - \frac{1}{n}\right)^r.$$

Note now that $\mathbf{E}S_{n,r}$ is close to 0 if $(1 - 1/n)^r$ is small compared with $1/n$, i.e. when $-r\ln(1 - 1/n) - \ln n$ is large. For large n,

$$\ln\left(1 - \frac{1}{n}\right) = -\frac{1}{n} + o\left(\frac{1}{n^2}\right),$$

and the required relation will hold if $(r - n\ln n)/n$ is large. In our case (cf. property E4), the smallness of $\mathbf{E}S_{n,r}$ will clearly imply that of $\mathbf{P}(A) = \mathbf{P}(S_{n,r} > 0)$.

4.2 Conditional Distribution Functions and Conditional Expectations

Let $\langle\Omega,\mathfrak{F},\mathbf{P}\rangle$ be a probability space and $B\in\mathfrak{F}$ be an event with $\mathbf{P}(B)>0$. Form a new probability space $\langle\Omega,\mathfrak{F},\mathbf{P}_B\rangle$, where $\mathbf{P}_B$ is defined for $A\in\mathfrak{F}$ by the equality

$$\mathbf{P}_B(A):=\mathbf{P}(A|B).$$

It is easy to verify that the probability properties P1, P2 and P3 hold for $\mathbf{P}_B$. Let ξ be a random variable on $\langle\Omega,\mathfrak{F},\mathbf{P}\rangle$. It is clearly a random variable on the space $\langle\Omega,\mathfrak{F},\mathbf{P}_B\rangle$ as well.

Definition 4.2.1 The expectation of ξ in the space $\langle\Omega,\mathfrak{F},\mathbf{P}_B\rangle$ is called the *conditional expectation of* ξ *given* B and is denoted by $\mathbf{E}(\xi\,|\,B)$:

$$\mathbf{E}(\xi\,|\,B)=\int_\Omega \xi(\omega)\mathbf{P}_B(d\omega).$$

By the definition of the measure $\mathbf{P}_B$,

$$\mathbf{E}(\xi\,|\,B)=\int_\Omega \xi(\omega)\mathbf{P}(d\omega|B)=\frac{1}{\mathbf{P}(B)}\int_\Omega \xi(\omega)\mathbf{P}(d\omega\cap B)=\frac{1}{\mathbf{P}(B)}\int_B \xi(\omega)\mathbf{P}(d\omega).$$

The last integral differs from $\mathbf{E}\xi$ in that the integration in it is carried over the set B only. We will denote this integral by

$$\mathbf{E}(\xi;\,B):=\int_B \xi(\omega)\mathbf{P}(d\omega),$$

so that

$$\mathbf{E}(\xi\,|\,B)=\frac{1}{\mathbf{P}(B)}\mathbf{E}(\xi;\,B).$$

It is not hard to see that the function

$$F(x\,|\,B):=\mathbf{P}_B(\xi<x)=\mathbf{P}(\xi<x\,|\,B)$$

is the distribution function of the random variable ξ on $\langle\Omega,\mathfrak{F},\mathbf{P}_B\rangle$.

Definition 4.2.2 The function $F(x\,|\,B)$ is called the *conditional distribution function* of ξ (in the "conditional" space $\langle\Omega,\mathfrak{F},\mathbf{P}_B\rangle$) *given* B.

The quantity $\mathbf{E}(\xi\,|\,B)$ can evidently be rewritten as

$$\int x\,dF(x\,|\,B).$$

If the σ-algebra $\sigma(\xi)$ generated by the random variable ξ does not depend on the event B, then $\mathbf{P}_B(A)=\mathbf{P}(A)$ for any $A\in\sigma(\xi)$. Therefore, in that case

$$F(x\,|\,B)=F(x),\qquad \mathbf{E}(\xi\,|\,B)=\mathbf{E}\xi,\qquad \mathbf{E}(\xi;\,B)=\mathbf{P}(B)\mathbf{E}\xi. \tag{4.2.1}$$

Let $\{B_n\}$ be a (possibly finite) sequence of disjoint events such that $\bigcup B_n = \Omega$ and $\mathbf{P}(B_n) > 0$ for any n. Then

$$\begin{aligned}\mathbf{E}\xi &= \int_\Omega \xi(\omega)\mathbf{P}(d\omega) = \sum_n \int_{B_n} \xi(\omega)\mathbf{P}(d\omega)\\ &= \sum_n \mathbf{E}(\xi; B_n) = \sum_n \mathbf{P}(B_n)\mathbf{E}(\xi \,|\, B_n).\end{aligned} \tag{4.2.2}$$

We have obtained the *total probability formula for expectations*. This formula can be rather useful.

Example 4.2.1 Let the lifetime of a device be a random variable ξ with a distribution function $F(x)$. We know that the device has already worked for a units of time. What is the distribution of the residual lifetime? What is the expectation of the latter?

Clearly, in this problem we have to find $\mathbf{P}(\xi - a \ge x | \xi \ge a)$ and $\mathbf{E}(\xi - a | \xi \ge a)$. Of course, it is assumed that

$$P(a) := \mathbf{P}(\xi \ge a) > 0.$$

By the above formulae,

$$\mathbf{P}(\xi - a \ge x | \xi \ge a) = \frac{P(x+a)}{P(a)}, \qquad \mathbf{E}(\xi - a | \xi \ge a) = \frac{1}{P(a)} \int_0^\infty x\, dF(x+a).$$

It is interesting to note the following. In many applied problems, especially when one deals with the operation of complex devices consisting of a large number of reliable parts, the distribution of ξ can be assumed to be *exponential*:

$$P(x) = \mathbf{P}(\xi \ge x) = e^{-\mu x}, \quad \mu > 0.$$

(The reason for this will become clear later, when considering the Poisson theorem and Poisson process. Computers could serve as examples of such devices.) But, for the exponential distribution, it turns out that the residual lifetime distribution

$$\mathbf{P}(\xi - a \ge x | \xi \ge a) = \frac{P(x+a)}{P(a)} = e^{-\mu x} = P(x) \tag{4.2.3}$$

coincides with the lifetime distribution of a new device. In other words, a new device and a device which has already worked without malfunction for some time a are, from the viewpoint of their future failure-free operation time distributions, equivalent.

It is not hard to understand that the exponential distribution (along with its discrete analogue $\mathbf{P}(\xi = k) = q^k(1-q)$, $k = 0, 1, \ldots$) is the only distribution possessing the above remarkable property. One can see that, from equality (4.2.3), we necessarily have

$$P(x+a) = P(x)P(a).$$

Example 4.2.2 Assume that n machines are positioned so that the distance between the i-th and j-th machines is $a_{i,j}$, $1 \le i, j \le n$. Each machine requires service from time to time (tuning, repair, etc.). Assume that the service is to be done by a single worker and that the probability that a given new call for service comes from the j-th machine is p_j ($\sum_{j=1}^n p_j = 1$). If, for instance, the worker has just completed servicing the i-th machine, then with probability p_j (not depending on i) the next machine to be served will be the j-th machine; the worker will then need to go to it and cover a distance of a_{ij} units. What is the mean length of such a passage?

Let B_i denote the event that the i-th machine was serviced immediately before a given passage. Then $\mathbf{P}(B_i) = p_i$, and the probability that the worker will move from the i-th machine to the j-th machine, $j = 1, \dots, n$, is equal to p_j. The length ξ of the passage is a_{ij}. Hence

$$\mathbf{E}(\xi | B_i) = \sum_{j=1}^{n} p_j a_{i,j},$$

and by the total probability formula

$$\mathbf{E}\xi = \sum_{i=1}^{n} \mathbf{P}(B_i)\mathbf{E}(\xi | B_i) = \sum_{i,j=1}^{n} p_j p_i a_{ij}.$$

The obtained expression enables one to compare different variants of positioning machines from the point of view of minimisation of the quantity $\mathbf{E}\xi$ under given restrictions on a_{ij}. For instance, if $a_{ij} \ge 1$ and all the machines are of the same type ($p_j = 1/n$) then, provided they are positioned along a straight line (with unit steps between them), one gets $a_{ij} = |j - i|$ and[1]

$$\mathbf{E}\xi = \frac{1}{n^2} \sum_{i,j=1}^{n} |j - i| = \frac{2}{n^2} \sum_{k=1}^{n-1} k(n-k) = \frac{n-1}{3}\left(1 + \frac{1}{n}\right),$$

so that, for large n, the value of $\mathbf{E}\xi$ is close to $n/3$. Thus, if there are s calls a day then the average total distance covered daily by the worker is approximately $sn/3$. It is easy to show that positioning machines around a circle would be better but still not optimal.

Example 4.2.3 As was already noticed, not all random variables (distributions) have expectations. The respective examples are by no means pathological: for instance, the Cauchy distribution $\mathrm{K}_{\alpha,\sigma}$ has this property. Now we will consider a problem on random walks in which there also arise random variables having no expectations. This is the problem on the so-called *fair* game. Two gamblers take part in the game. The initial capital of the first gambler is z units. This gambler wins or loses each

[1] To compute the sum, it suffices to note that

$$\sum_{k=1}^{n-1} k(k-1) = \frac{1}{3}(n-2)(n-1)n$$

(compare the initial values and increments of the both sides).

play of the game with probability $1/2$ independently of the outcomes of the previous plays, his capital increasing or decreasing by one unit, respectively. Let $z + S_k$ be the capital of the first gambler after the k-th play, $\eta(z)$ is the number of steps until his ruin in the game versus an infinitely rich adversary, i.e.

$$\eta(z) = \min\{k : z + S_k = 0\}, \quad \eta(0) = 0.$$

If $\inf_k S_k > -z$ (i.e. the first gambler is never ruined), we put $\eta(z) = \infty$.

First we show that $\eta(z)$ is a *proper* random variable, i.e. a random variable assuming finite values with probability 1. For the first gambler, this will mean that he goes bankrupt with probability 1 whatever his initial capital z is. Here one could take Ω to be the "sample" space consisting of all possible sequences made up of 1 and -1. Each such sequence ω would describe a "trajectory" of the game. (For example, -1 in the k-th place means that the first gambler lost the k-th play.) We leave it to the reader as an exercise to complete the construction of the probability space $\langle \Omega, \mathfrak{F}, \mathbf{P} \rangle$. Clearly, one has to do this so that the probability of any first n outcomes of the game (the first n components of ω are fixed) is equal to 2^{-n}.

Put

$$u(z) := \mathbf{P}\big(\eta(z) < \infty\big), \quad u(0) := 1,$$

and denote by B_1 the event that the first component of ω is 1 (the gambler won in the first play) and B_2 that this component is -1 (the gambler lost). Noticing that $\mathbf{P}(\eta(z) < \infty | B_1) = u(z+1)$ (if the first play is won, the capital becomes $z+1$), we obtain by the total probability formula that, for $z \ge 1$,

$$\begin{aligned} u(z) &= \mathbf{P}(B_1)\mathbf{P}\big(\eta(z) < \infty | B_1\big) + \mathbf{P}(B_2)\mathbf{P}\big(\eta(z) < \infty | B_2\big) \\ &= \frac{1}{2}u(z+1) + \frac{1}{2}u(z-1). \end{aligned}$$

Putting $\delta(z) := u(z+1) - u(z)$, $z \ge 0$, we conclude from here that $\delta(z) - \delta(z-1) = 0$, and hence $\delta(z) = \delta = \text{const}$. Since

$$u(z+1) = u(0) + \sum_{k=1}^{z} \delta(k) = u(0) + z\delta,$$

it is evident that δ can be nothing but 0, so that $u(z) \equiv 1$ for all z.

Thus, in a game against an infinitely rich adversary, the gambler will be ruined with probability 1. This explains, to some extent, the fact that all reckless gamblers (not stopping "at the right time"; choosing this "right time" is a separate problem) go bankrupt sooner or later. Even if the game is fair.

We show now that although $\eta(z)$ is a proper random variable, $\mathbf{E}\eta(z) = \infty$. Assume the contrary:

$$v(z) := \mathbf{E}\eta(z) < \infty.$$

Similarly to the previous argument, we notice that $\mathbf{E}(\eta(z)|B_1) = 1 + v(z+1)$ (the capital became $z+1$, one play has already been played). Therefore by the total probability formula we find for $z \ge 1$ that

$$v(z) = \frac{1}{2}\big(1 + v(z+1)\big) + \frac{1}{2}\big(1 + v(z-1)\big), \quad v(0) = 0.$$

It can be seen from this formula that if $v(z) < \infty$, then $v(k) < \infty$ for all k. Set $\Delta(z) = v(z+1) - v(z)$. Then the last equality can be written down for $z \geq 1$ as

$$-1 = \frac{1}{2}\Delta(z) - \frac{1}{2}\Delta(z-1),$$

or

$$\Delta(z) = \Delta(z-1) - 2.$$

From this equality we find that $\Delta(z) = \Delta(0) - 2z$. Therefore

$$v(z) = \sum_{k=0}^{z-1} \Delta(k) = z\Delta(0) - z(z-1) = zv(1) - z(z-1).$$

It follows that $\mathbf{E}\eta(z) < 0$ for sufficiently large z. But $\eta(z)$ is a positive random variable and hence $\mathbf{E}\eta(z) \geq 0$. The contradiction shows that the assumption on the finiteness of the expectation of $\eta(z)$ is wrong.

4.3 Expectations of Functions of Independent Random Variables

Theorem 4.3.1

1. *Let ξ and η be independent random variables and $g(x, y)$ be a Borel function. Then if $g \geq 0$ or $\mathbf{E}g(\xi, \eta)$ is finite, then*

$$\mathbf{E}g(\xi, \eta) = \mathbf{E}\big[\mathbf{E}g(x, \eta)|_{x=\xi}\big]. \tag{4.3.1}$$

2. *Let $g(x, y) = g_1(x)g_2(y)$. If $g_1(\xi) \geq 0$ and $g_2(\eta) \geq 0$, or both $\mathbf{E}g_1(\xi)$ and $\mathbf{E}g_2(\eta)$ exist, then*

$$\mathbf{E}g(\xi, \eta) = \mathbf{E}g_1(\xi)\mathbf{E}g_2(\eta). \tag{4.3.2}$$

The expectation $\mathbf{E}g(\xi, \eta)$ exists if and only if both $\mathbf{E}g_1(\xi)$ and $\mathbf{E}g_2(\eta)$ exist. (We exclude here the trivial cases $\mathbf{P}(g_1(\xi) = 0) = 1$ and $\mathbf{P}(g_2(\eta) = 0) = 1$ to avoid trivial complications.)

Proof The first assertion of the theorem is a paraphrase of Fubini's theorem in terms of expectations. The first part of the second assertion follows immediately from Corollary 3.6.1 of Fubini's theorem. Since $|g_1(\xi)| \geq 0$ and $|g_2(\eta)| \geq 0$ and these random variables are independent, one has

$$\mathbf{E}\big|g_1(\xi)g_2(\eta)\big| = \mathbf{E}\big|g_1(\xi)\big|\mathbf{E}\big|g_2(\eta)\big|.$$

Now the last assertion of the theorem follows immediately, for one clearly has $\mathbf{E}|g_1(\xi)| \neq 0$, $\mathbf{E}|g_2(\eta)| \neq 0$. □

Remark 4.3.1 Formula (4.3.1) could be considered as the total probability formula for computing the expectation $\mathbf{E}g(\xi, \eta)$. Assertion (4.3.2) could be written down without loss of generality in the form

$$\mathbf{E}\xi\eta = \mathbf{E}\xi\mathbf{E}\eta. \tag{4.3.3}$$

To get (4.3.2) from this, one has to take $g_1(\xi)$ instead of ξ and $g_2(\eta)$ instead of η—these will again be independent random variables.

Examples of the use of Theorem 4.3.1 were given in Sect. 3.6 and will be appearing in the sequel.

The converse to (4.3.2) or (4.3.3) does not hold. There exist dependent random variables ξ and η such that

$$\mathbf{E}\xi\eta = \mathbf{E}\xi\,\mathbf{E}\eta.$$

Let, for instance, ζ and ξ be independent and $\mathbf{E}\xi = \mathbf{E}\zeta = 0$. Put $\eta = \xi\zeta$. Then ξ and η are clearly dependent (excluding some trivial cases when, say, $\xi = \text{const}$), but

$$\mathbf{E}\xi\eta = \mathbf{E}\xi^2\zeta = \mathbf{E}\xi^2\mathbf{E}\zeta = 0 = \mathbf{E}\xi\,\mathbf{E}\eta.$$

4.4 Expectations of Sums of a Random Number of Random Variables

Assume that a sequence $\{\xi_n\}_{n=1}^{\infty}$ of independent random variables (or random vectors) and an integer-valued random variable $\nu \ge 0$ are given on a probability space $\langle\Omega, \mathfrak{F}, \mathbf{P}\rangle$.

Property E2 of expectations implies that, for sums $S_n = \sum_{i=1}^{n} \xi_i$, the following equality holds:

$$\mathbf{E}S_n = \sum_{i=1}^{n} \mathbf{E}\xi_i.$$

In particular, if $a_k = \mathbf{E}\xi_k = a$ do not depend on k then $\mathbf{E}S_n = an$.

What can be said about the expectation of the sum s_ν of the *random number* ν of random variables $\xi_1, \xi_2, \ldots$? To answer this question we need to introduce some new notions.

Let $\mathfrak{F}_{k,n} := \sigma(\xi_k, \ldots, \xi_n)$ be the σ-algebra generated by the $n - k + 1$ random variables $\xi_k, \ldots, \xi_n$.

Definition 4.4.1 A random variable ν is said to be *independent of the future* if the event $\{\nu \le n\}$ does not depend on $\mathfrak{F}_{n+1,\infty}$.

Let, further, a family of embedded σ-algebras $\mathfrak{F}_n : \mathfrak{F}_n \subset \mathfrak{F}_{n+1}$ be given, such that $\mathfrak{F}_{1,n} = \sigma(\xi_1, \ldots, \xi_n) \subset \mathfrak{F}_n$.

Definition 4.4.2 A random variable ν is said to be a *Markov* (or *stopping*) *time* with respect to the family $\{\mathfrak{F}_n\}$, if $\{\nu \le n\} \in \mathfrak{F}_n$.

Often $\mathfrak{F}_n$ is taken to be $\mathfrak{F}_{1,n} = \sigma(\xi_1, \ldots, \xi_n)$. We will call a stopping time with respect to $\mathfrak{F}_{1,n}$ simply a *stopping (or Markov) time* without indicating the corresponding family of σ-algebras. In this case, knowing the values of $\xi_1, \ldots, \xi_n$ enables us to say whether the event $\{\nu \le n\}$ has occurred or not.

If the ξ_n are independent (the σ-algebras $\mathfrak{F}_{1,n}$ and $\mathfrak{F}_{n+1,\infty}$ are independent) then the requirement of independence of the future is wider than the Markov property, because if ν is a stopping time with respect to $\{\mathfrak{F}_{1,n}\}$ then, evidently, the random variable ν does not depend on the future.

As for a converse statement, one can only assert the following. If ν does not depend on the future and the ξ_k are independent then one can construct a family of embedded σ-algebras $\{\mathfrak{F}_n\}$, $\mathfrak{F}_n \supset \mathfrak{F}_{1,n}$, such that ν is a stopping time with respect to $\mathfrak{F}_n$ ($\{\nu \le n\} \subset \mathfrak{F}_n$) and $\mathfrak{F}_n$ does not depend on $\mathfrak{F}_{n+1,\infty}$. As $\mathfrak{F}_n$, we can take the σ-algebra generated by $\mathfrak{F}_{1,n}$ and the events $\{\nu = k\}$ for $k \le n$. For instance, a random variable ν independent of $\{\xi_i\}$ clearly does not depend on the future, but is not a stopping time. Such ν will be a stopping time only with respect to the family $\{\mathfrak{F}_n\}$ constructed above.

It should be noted that, formally, any random variable can be made a stopping time using the above construction (but, generally speaking, there will be no independence of $\mathfrak{F}_n$ and $\mathfrak{F}_{n_1,\infty}$). However, such a construction is unsubstantial and not particularly useful. In all the examples below the variables ν not depending on the future are stopping times defined in a rather natural way.

Example 4.4.1 Let ν be the number of the first random variable in the sequence $\{\xi_n\}_{n=1}^{\infty}$ which is greater than or equal to N, i.e. $\nu = \inf\{k : \xi_k \ge N\}$. Clearly, ν is a stopping time, since

$$\{\nu \le n\} = \bigcup_{k=1}^{n} \{\xi_k \ge N\} \in \mathfrak{F}_{1,n}.$$

If ξ_k are independent, then evidently ν is independent of the future.

The same can be said about the random variable

$$\eta(t) := \min\{k : S_k \ge N\}, \quad S_k = \sum_{j=1}^{k} \xi_j.$$

Note that the random variables ν and $\eta(t)$ may be improper (e.g., $\eta(t)$ is not defined on the event $\{S := \sup S_k < N\}$). The random variable $\theta := \min\{k : S_k = S\}$ is not a stopping time and depends on the future.

The term "Markov" random variable (or Markov time) will become clearer after introducing the notion of Markovity in Chap. 13. The term "stopping time" is related to the nature of a large number of applied problems in which such random variables arise. As a typical example, the following procedure could be considered. Let ξ_k be the number of defective items in the k-th lot produced by a factory. Statistical quality control is carried out as follows. The whole production is rejected if, in sequential testing of the lots, it turns out that, for some n, the value of the sum

$$S_n = \sum_{k=1}^{n} \xi_k$$

exceeds a given admissible level $a + bn$. The lot number ν for which this happens,

$$\nu := \min\{n : S_n \ge a + bn\},$$

is a stopping time for the whole testing procedure. To avoid a lengthy testing, one also introduces a (literal) stopping time

$$\nu^* := \min\{n : S_n \le -A + bn\},$$

where $A > 0$ is chosen so large as to guarantee, with a high probability, a sufficient quality level for the whole production (assuming, say, that the ξ_k are identically distributed). It is clear that ν and ν^* both satisfy the definition of a Markov or stopping time.

Consider the sum $S_\nu = \xi_1 + \cdots + \xi_\nu$ of a random number of random variables. This sum is also called a *stopped sum* in the case when ν is a stopping time.

Theorem 4.4.1 (Kolmogorov–Prokhorov) *Let an integer-valued random variable* ν *be independent of the future. If*

$$\sum_{k=1}^{\infty} \mathbf{P}(\nu \ge k)\mathbf{E}|\xi_k| < \infty \tag{4.4.1}$$

then

$$\mathbf{E}S_\nu = \sum_{k=1}^{\infty} \mathbf{P}(\nu \ge k)\mathbf{E}\xi_k. \tag{4.4.2}$$

If $\xi_k \ge 0$ *then condition* (4.4.1) *is superfluous.*

Proof The summand ξ_k is present in the sum S_ν if and only if the event $\{\nu \ge k\}$ occurs. Thus the following representation holds for the sum S_ν:

$$S_\nu = \sum_{k=1}^{\infty} \xi_k \mathrm{I}(\nu \ge k),$$

where $\mathrm{I}(B)$ is the indicator of the event B. Put $S_{\nu,n} := \sum_{k=1}^{n} \xi_k \mathrm{I}(\nu \ge k)$. If $\xi_k \ge n$ then $S_{\nu,n} \uparrow S_\nu$ for each ω as $n \to \infty$, and hence, by the monotone convergence theorem (see Theorem A3.3.1 in Appendix 3),

$$\mathbf{E}S_\nu = \lim_{n\to\infty} \mathbf{E}S_{\nu,n} = \lim_{n\to\infty} \sum_{k=1}^{n} \mathbf{E}\xi_k \mathrm{I}(\nu \ge k).$$

But the event $\{\nu \ge k\}$ complements the event $\{\nu \le k-1\}$ and therefore does not depend on $\sigma(\xi_k, \xi_{k+1}, \ldots)$ and, in particular, on $\sigma(\xi_k)$. Hence, putting $a_k := \mathbf{E}\xi_k$ we get $\mathbf{E}\xi_k \mathrm{I}(\nu \ge k) = a_k \mathbf{P}(\nu \ge k)$, and

$$\mathbf{E}S_\nu = \lim_{k\to\infty} \sum_{k=1}^{n} a_k \mathbf{P}(\nu_k \ge k) = \sum_{k=1}^{\infty} a_k \mathbf{P}(\nu \ge k). \tag{4.4.3}$$

This proves (4.4.2) for $\xi_k \ge 0$.

Now assume ξ_k can take values of both signs. Put

$$\xi_k^* := |\xi_k|, \qquad a_k^* := \mathbf{E}\xi_k^*, \qquad Z_n := \sum_{k=1}^{n} \xi_k^*, \qquad Z_{\nu,n} := \sum_{k=1}^{n} \xi_k^* \mathrm{I}(\nu \ge k).$$

Applying (4.4.3), we obtain by virtue of (4.4.1) that

$$\mathbf{E}Z_\nu = \sum_{k=1}^{\infty} a_k^* \mathbf{P}(\nu \ge k) < \infty.$$

Since $|S_{\nu,n}| \le Z_{\nu,n} \le Z_\nu$, by the monotone convergence theorem (see Corollary 6.1.3 or (the Fatou–Lebesgue) Theorem A3.3.2 in Appendix 3) we have

$$\mathbf{E}S_\nu = \lim_{n\to\infty} \mathbf{E}S_{\nu,n} = \sum a_k \mathbf{P}(\nu \ge k),$$

where the series on the right-hand side absolutely converges by virtue of (4.4.1).

The theorem is proved. □

Put

$$a^* := \max a_k, \qquad a_* := \min a_k,$$

where, as above, $a_k = \mathbf{E}\xi_k$.

Theorem 4.4.2 *Let* $\sup_k \mathbf{E}|\xi_k| < \infty$ *and* ν *be a random variable which does not depend on the future. Then the following assertions hold true.*

(a) *If* $\mathbf{E}\nu < \infty$ *(or* $\mathbf{E}Z_\nu < \infty$*, where* $Z_n = \sum_{k=1}^n |\xi_k|$*) then* $\mathbf{E}S_\nu$ *exists and*

$$a_*\mathbf{E}\nu \le \mathbf{E}S_\nu \le a^*\mathbf{E}\nu. \tag{4.4.4}$$

(b) *If* $\mathbf{E}S_\nu$ *is well defined (and may be* $\pm\infty$*),* $a_* > 0$ *and, for any* $N \ge 1$,

$$\mathbf{E}(S_N - a_*N;\, \nu > N) \le c,$$

where c *does not depend on* N*, then* (4.4.4) *holds true.*

(c) *If* $\xi_k \ge 0$ *then* (4.4.4) *is always valid.*

If $S_\nu \ge \text{const}$ *a.s. then condition* (c) *clearly implies* (b).

The case $a^* < 0$ in assertions (b)–(c) can be treated in exactly the same way.

If ν does not depend on $\{\xi_k\}$, $a_* = a^* = a > 0$, then $\mathbf{E}(S_N;\, \nu > N) = aN\mathbf{P}(\nu > N)$ and the condition in (b) holds. But the assumption $a_* = a^*$ is inessential here, and, for independent ν and $\{\xi\}$, (4.4.4) is always true, since in this case

$$\mathbf{E}S_\nu = \sum \mathbf{P}(\nu = k)\mathbf{E}S_k \le a^* \sum k\mathbf{P}(\nu = k) = a^*\mathbf{E}\nu.$$

The reverse inequality $\mathbf{E}S_\nu \ge a_*\mathbf{E}\nu$ is verified in the same way.

Proof of Theorem 4.4.2

(a) First note that

$$\sum_{k=1}^{\infty} \mathbf{P}(\nu \ge k) = \sum_{k=1}^{\infty}\sum_{i=k}^{\infty} \mathbf{P}(\nu = i) = \sum_{i=1}^{\infty} i\mathbf{P}(\nu = i) = \mathbf{E}\nu.$$

Note also that, for $\mathbf{E}|\xi_k| \le c < \infty$, the condition $\mathbf{E}\nu < \infty$ (or $\mathbf{E}Z_\nu < \infty$) turns into condition (4.4.1), and assertion (4.4.4) follows from (4.4.2). Therefore, if $\mathbf{E}\nu < \infty$

then Theorem 4.4.2 is a direct consequence of Theorem 4.4.1. The same is true in case (d).

Consider now assertions (b) and (c).

For a fixed $N > 0$, introduce the random variable

$$\nu_N := \min(\nu, N),$$

which, together with ν, does not depend on the future. Indeed, if $n \le N$ then the event $\{\nu_N \le n\} = \{\nu \le n\}$ does not depend on $\mathfrak{F}_{n+1,\infty}$. If $n > N$ then the event $\{\nu_N \le N\}$ is certain and hence it too does not depend on $\mathfrak{F}_{n+1,\infty}$.

(b) If $\mathbf{E}\nu < \infty$ then (4.4.4) is proved. Now let $\mathbf{E}\nu = \infty$. We have to prove that $\mathbf{E}S_\nu = \infty$. Since $\mathbf{E}\nu_N < \infty$, the relations

$$\mathbf{E}S_{\nu_N} = \mathbf{E}(S_\nu;\, \nu \le N) + \mathbf{E}(S_N;\, \nu > N) \ge a_*\bigl(\mathbf{E}(\nu;\, \nu \le N) + N\mathbf{P}(\nu > N)\bigr) \quad (4.4.5)$$

are valid by (a). Together with the conditions in (b) this implies that

$$\mathbf{E}(S_\nu;\, \nu \le N) \ge a_*\mathbf{E}(\nu;\, \nu \le N) - c \to \infty$$

as $N \to \infty$. Since S_ν is well defined, we have

$$\mathbf{E}(S_\nu;\, \nu \le N) \to \mathbf{E}S_\nu$$

as $N \to \infty$ (see Corollary A3.2.1 in Appendix 3). Therefore necessarily $\mathbf{E}S_\nu = \infty$.

(c) Here it is again sufficient to show that $\mathbf{E}S_\nu = \infty$ in the case when $\mathbf{E}\nu = \infty$. It follows from (4.4.5) that

$$\begin{aligned}\mathbf{E}S_\nu &= \mathbf{E}(S_\nu;\, \nu \le N) + \mathbf{E}(S_\nu;\, \nu > N)\\ &\ge \mathbf{E}\bigl[S_\nu - (S_N - a_*N);\, \nu > N\bigr] + a_*\mathbf{E}(\nu;\, \nu \le N) \ge a_*\mathbf{E}(\nu;\, \nu \le N) - c \to \infty\end{aligned}$$

as $N \to \infty$, and thus $\mathbf{E}S_\nu = \infty$.

The theorem is proved. □

Theorem 4.4.2 implies the following famous result.

Theorem 4.4.3 (Wald's identity) *Assume $a = \mathbf{E}\xi_k$ does not depend on k, $\sup_k \mathbf{E}|\xi_k| < \infty$, and a random variable ν is independent of the future. Then, under at least one of the conditions* (a)–(d) *of Theorem* 4.4.2 (*with a_* replaced by a*),

$$\mathbf{E}S_\nu = a\mathbf{E}\nu. \quad (4.4.6)$$

If $a = 0$ and $\mathbf{E}\nu = \infty$ then identity (4.4.6) can hold, since there would be an ambiguity of type $0 \cdot \infty$ on the right-hand side of (4.4.6).

Remark 4.4.1 If there is no independence of the future then equality (4.4.6) is, generally speaking, untrue. Let, for instance, $a = \mathbf{E}\xi_k < 0$, $\theta := \min\{k : S_k = S\}$ and $S := \sup_k S_k$ (see Example 4.4.1; see Chaps. 10–12 for conditions of finiteness of $\mathbf{E}S$ and $\mathbf{E}\theta$). Then $S_\theta = S > 0$ and $\mathbf{E}S > 0$, while $a\mathbf{E}\theta < 0$. Hence, (4.4.6) cannot hold true for $\nu = \theta$.

We saw that if there is no assumption on the finiteness of $\mathbf{E}\nu$ then, even in the case $a > 0$, in order for (4.4.6) to hold, additional conditions are needed, e.g., conditions (b)–(d). Without these conditions identity (4.4.6) is, generally speaking, not valid, as shown by the following example.

Example 4.4.2 Let the random variables ζ_k be independent and identically distributed, and

$$\mathbf{E}\zeta_k = 0, \qquad \mathbf{E}\zeta_k^2 = 1, \qquad \mathbf{E}|\zeta_k|^3 = \mu < \infty,$$
$$\xi_k := 1 + \sqrt{2k}\zeta_k, \qquad \nu := \min\{k : S_k < 0\}.$$

We will show below in Example 20.2.1 that ν is a proper random variable, i.e.

$$\mathbf{P}(\nu < \infty) = \mathbf{P}\left(\bigcup_{n=1}^{\infty}\{S_n < 0\}\right) = 1.$$

It is also clear that ν is a Markov time independent of the future and $\mathbf{E}\xi_k = a = 1$. But one has $\mathbf{E}S_\nu < 0$, while $a\mathbf{E}\nu > 0$, and hence equality (4.4.6) cannot be valid. (Here necessarily $\mathbf{E}\nu = \infty$, since otherwise condition (a) would be satisfied and (4.4.6) would hold.)

However, if the ξ_k are independent and identically distributed and ν is a *stopping time* then statement (4.4.6) is always valid whenever its right-hand side is well defined. We will show this below in Theorem 11.3.2 by virtue of the laws of large numbers.

Conditions (b) and (c) in Theorem 4.4.2 were used in the case $\mathbf{E}\nu = \infty$. However, in some problems these conditions can be used to prove the finiteness of $\mathbf{E}\nu$. The following example confirms this observation.

Example 4.4.3 Let $\xi_1, \xi_2, \ldots$ be independent and identically distributed and $a = \mathbf{E}\xi_1 > 0$. For a fixed $t \ge 0$, consider, as a stopping time (and a variable independent of the future), the random variable

$$\nu = \eta(t) := \min\{k : S_k \ge t\}.$$

Clearly, $S_N < t$ on the set $\eta(t) > N$ and $S_{\eta(t)} \ge t$. Therefore conditions (b) and (c) are satisfied, and hence

$$\mathbf{E}S_{\eta(t)} = a\mathbf{E}\eta(t).$$

We now show that $\mathbf{E}\eta(t) < \infty$. In order to do this, we consider the "trimmed" random variables $\xi_k^{(N)} := \min(N, \xi_k)$ and choose N large enough for the inequality $a^{(N)} := \mathbf{E}\xi^{(N)} > 0$ to hold true. Let $S_K^{(N)}$ and $\eta^{(N)}(t)$ be defined similarly to S_k and $\eta(t)$, but for the sequence $\{\xi_j^{(N)}\}$. Then evidently $S_{\eta^{(N)}(t)}^{(N)} \le t + N$, $\eta(t) \le \eta^{(N)}(t)$,

$$a^{(N)}\mathbf{E}\eta^{(N)}(t) \le t + N, \qquad \mathbf{E}\eta(t) \le \frac{t+N}{a^{(N)}} < \infty.$$

If $a = 0$ then $\mathbf{E}\eta(t) = \infty$. This can be seen from the fair game example ($\xi_k = \pm 1$ with probability $1/2$; see Example 4.2.3). In the general case, this will be shown below in Chap. 12. As was noted above, in this case the right-hand side of (4.4.6) turns into the indeterminacy $0 \cdot \infty$, but the left-hand side may equal any finite number, as in the case of the fair game where $S_{\eta(t)} = t$.

If we take ν to be the Markov time

$$\nu = \mu(t) := \min\{k : |S_k| \ge t\},$$

where ξ_k may assume values of both signs, then, to prove (4.4.6), it is apparently easier to verify the condition of assertion (a) in Theorem 4.4.2. Let us show that $\mathbf{E}\mu(t) < \infty$. It is clear that, for any given $t > 0$, there exists an N such that

$$q := \min\bigl[\mathbf{P}(S_N > 2t), \mathbf{P}(S_N < -2t)\bigr] > 0.$$

($N = 1$ if the ξ_k are bounded from below.) For such N,

$$\inf_{|v| \le t} \mathbf{P}\bigl(|v + S_N| > t\bigr) > 2q.$$

Hence, in each N steps, the random walk $\{S_k\}$ has a chance to leave the strip $|v| \le t$ with probability greater than $2q$, whatever point v, $|v| \le t$, it starts from. Therefore,

$$\mathbf{P}\bigl(\mu(t) > kN\bigr) = \mathbf{P}\Bigl(\max_{j \le kN} |S_j| < t\Bigr) < \mathbf{P}\Biggl(\bigcap_{j=1}^{k} \bigl\{|S_{jN}| < t\bigr\}\Biggr) < (1 - 2q)^k.$$

This implies that $\mathbf{P}(\mu(t) > kN)$ decreases exponentially as k grows and that $\mathbf{E}\mu(t)$ is finite.

Example 4.4.4 A chain reaction scheme. Suppose we have a single initial particle which either disappears with probability q or turns into m similar particles with probability $p = 1 - q$. Each particle from the new generation behaves in the same way independently of the fortunes of the other particles. What is the expectation of the number ζ_n of particles in the n-th generation?

Consider the "double sequence" $\{\xi_k^{(n)}\}_{k=1,n=1}^{\infty\ \ \infty}$ of independent identically distributed random variables assuming the values m and 0 with probabilities p and q, respectively. The sequences $\{\xi_k^{(1)}\}_{k=1}^{\infty}$, $\{\xi_k^{(2)}\}_{k=1}^{\infty}$, … will clearly be mutually independent. Using these sequences, one can represent the variables ζ_n ($\zeta_0 = 1$) as

$$\zeta_1 = \xi_{\zeta_0}^{(1)} = \xi_1^{(1)},$$

$$\zeta_2 = \xi_1^{(2)} + \cdots + \xi_{\zeta_1}^{(2)}$$

$$\cdots \quad \cdots \quad \cdots \quad \cdots$$

$$\zeta_n = \xi_1^{(n)} + \cdots + \xi_{\zeta_{n-1}}^{(n)},$$

where the number of summands in the equation for ζ_n is ζ_{n-1}, the number of "parent particles". Since the sequence $\{\xi_k^{(n)}\}$ is independent of ζ_{n-1}, $\xi_k^{(n)} \ge 0$, and $\mathbf{E}\xi_k^{(n)} = pm$, by virtue of Wald's identity we have

$$\mathbf{E}\zeta_n = \mathbf{E}\xi_1^{(n)} \mathbf{E}\zeta_{n-1} = pm\mathbf{E}\zeta_{n-1} = (pm)^n.$$

Example 4.4.5 We return to the fair game of two gamblers described in Example 4.2.3, but now assume that the respective capitals $z_1 > 0$ and $z_2 > 0$ of the gamblers are finite. Introduce random variables ξ_k representing the gains of the first gambler in the respective (k-th) play. The variables ξ_k are obviously independent, and

$$\xi_k = \begin{cases} 1 & \text{with probability } 1/2, \\ -1 & \text{with probability } 1/2. \end{cases}$$

The quantity $z_1 + S_k = z_1 + \sum_{j=1}^{k} \xi_j$ will be the capital of the first gambler and $z_2 - S_k$ the capital of the second gambler after k plays. The quantity

$$\eta := \min\{k : z_1 + S_k = 0 \text{ or } z_2 - S_k = 0\}$$

is the time until the end of the game, i.e. until the ruin of one of the gamblers. The question is what is the probability P_i that the i-th gambler wins (for $i = 1, 2$)?

Clearly, η is a Markov time, $S_\eta = -z_1$ with probability P_2 and $S_\eta = z_2$ with probability $P_1 = 1 - P_2$. Therefore,

$$\mathbf{E}S_\eta = P_1 z_2 - P_2 z_1.$$

If $\mathbf{E}\eta < \infty$, then by Wald's identity we have

$$P_1 z_2 - P_2 z_1 = \mathbf{E}\eta\mathbf{E}\xi_1 = 0.$$

From this we find that $P_i = z_i/(z_1 + z_2)$.

It remains to verify that $\mathbf{E}\eta$ is finite. Let, for the sake of simplicity, $z_1 + z_2 = 2z$ be even. With probability $2^{-\min(z_1,z_2)} \ge 2^{-z}$, the game can be completed in $\min(z_1, z_2) \le z$ plays. Since the total capital of both players remains unchanged during the game,

$$\mathbf{P}(\eta > z) \le 1 - 2^{-z}, \quad \dots, \quad \mathbf{P}(\eta > Nz) \le \left(1 - 2^{-z}\right)^N.$$

This evidently implies the finiteness of

$$\mathbf{E}\eta = \sum_{k=0}^{\infty} \mathbf{P}(\eta > k).$$

We will now give a less trivial example of a random variable ν which is independent of the future, but is not a stopping time.

Example 4.4.6 Consider two mutually independent sequences of independent positive random variables $\xi_1, \xi_2, \dots$ and $\zeta_1, \zeta_2, \dots$, such that $\xi_j \in F$ and $\zeta_j \in G$. Further, consider a system consisting of two devices. After starting the system, the first device operates for a random time ξ_1 after which it breaks down. Then the second device replaces the first one and works for ξ_2 time units (over the time interval $(\xi_1, \xi_1 + \xi_2)$). Immediately after the first device's breakdown, one starts repairing it, and the repair time is ζ_2. If $\zeta_2 > \xi_2$, then at the time $\xi_1 + \xi_2$ of the second device's failure both devices are faulty and the system fails. If $\zeta_2 \le \xi_2$, then at the time $\xi_1 + \xi_2$ the first device starts working again and works for ξ_3 time units, while the second device will be under repair for ζ_3 time units. If $\zeta_3 > \xi_3$, the system fails. If $\zeta_3 \le \xi_3$, the second device will start working, etc. What is the expectation of the failure-free operation time τ of the system?

Let $\nu := \min\{k \ge 2 : \zeta_k > \xi_k\}$. Then clearly $\tau = \xi_1 + \cdots + \xi_\nu$, where the ξ_j are independent and identically distributed and $\{\nu \le n\} \in \sigma(\xi_1, \dots, \xi_n; \zeta_1, \dots, \zeta_\nu)$. This means that ν is independent of the future. At the same time, if $\zeta_j \ne \text{const}$, then $\{\nu \le n\} \notin \mathfrak{F}_{1,n} = \sigma(\xi_1, \dots, \xi_n)$ and ν is not a Markov time with respect to $\mathfrak{F}_{1,n}$. Since $\xi_k \ge 0$, by Wald's identity $\mathbf{E}\tau = \mathbf{E}\nu\, \mathbf{E}\xi_1$. Since

$$\{\nu = k\} = \bigcap_{j=2}^{k-1} \{\eta_j \le \zeta_j\} \cap \{\eta_k > \zeta_k\}, \quad k \ge 2,$$

one has $\mathbf{P}(\nu = k) = q^{k-2}(1-q)$, $k \ge 2$, where

$$q = \mathbf{P}(\eta_k \le \zeta_k) = \int dF(t)\, G(t+0).$$

Consequently,

$$\mathbf{E}\nu = \sum_{k=2}^{\infty} kq^{k-2}(1-q) = 1 + \sum_{k=1}^{\infty} kq^{k-1}(1-q) = 1 + \frac{1}{1-q},$$

$$\mathbf{E}\tau = \mathbf{E}\xi_1 \frac{2-q}{1-q}.$$

Wald's identity has a number of extensions (we will discuss these in more detail in Sects. 10.3 and 15.2).

4.5 Variance

We introduce one more numerical characteristic for random variables.

Definition 4.5.1 The *variance* $\mathrm{Var}(\xi)$ of a random variable ξ is the quantity

$$\mathrm{Var}(\xi) := \mathbf{E}(\xi - \mathbf{E}\xi)^2.$$

It is a measure of the "dispersion" or "spread" of the distribution of ξ. The variance is equal to the inertia moment of the distribution of unit mass along the line. We have

$$\mathrm{Var}(\xi) = \mathbf{E}\xi^2 - 2\mathbf{E}\xi\,\mathbf{E}\xi + (\mathbf{E}\xi)^2 = \mathbf{E}\xi^2 - (\mathbf{E}\xi)^2. \tag{4.5.1}$$

The variance could also be defined as $\min_a \mathbf{E}(\xi - a)^2$. Indeed, by that definition

$$\mathrm{Var}(\xi) = \mathbf{E}\xi^2 + \min_a\bigl(a^2 - 2a\mathbf{E}\xi\bigr) = \mathbf{E}\xi^2 - (\mathbf{E}\xi)^2,$$

since the minimum of $a^2 - 2a\mathbf{E}\xi$ is attained at the point $a = \mathbf{E}\xi$. This remark shows that the quantity $a = \mathbf{E}\xi$ is the best mean square estimate (approximation) of the random variable ξ.

The quantity $\sqrt{\mathrm{Var}(\xi)}$ is called the *standard deviation* of ξ.

Example 4.5.1 Let $\xi \in \mathbf{\Phi}_{a,\sigma^2}$. As we saw in Example 4.1.2, $a = \mathbf{E}\xi$. Therefore,

$$\operatorname{Var}(\xi) = \int (x-a)^2 \frac{1}{\sigma\sqrt{2\pi}} e^{-(x-a)^2/2\sigma^2}\,dx = \frac{\sigma^2}{\sqrt{2\pi}} \int t^2 e^{-t^2/2}\,dt.$$

The last equality was obtained by the variable change $(x-a)/\sigma = t$. Integrating by parts, one gets

$$\operatorname{Var}(\xi) = -\frac{\sigma^2}{\sqrt{2\pi}} t e^{-t^2/2}\bigg|_{-\infty}^{\infty} + \frac{\sigma^2}{\sqrt{2\pi}} \int e^{-t^2/2}\,dt = \sigma^2.$$

Example 4.5.2 Let $\xi \in \mathbf{\Pi}_\mu$. In Example 4.1.3 we computed the expectation $\mathbf{E}\xi = \mu$. Hence

$$\operatorname{Var}(\xi) = \mathbf{E}\xi^2 - (\mathbf{E}\xi)^2 = \sum_{k=0}^{\infty} k^2 \frac{\mu^k e^{-\mu}}{k!} - \mu^2$$

$$= \sum_{k=2}^{\infty} \frac{k(k-1)\mu^k}{k!} e^{-\mu} + \sum_{k=0}^{\infty} \frac{k\mu^k}{k!} e^{-\mu} - \mu^2 = \mu^2 + \mu - \mu^2 = \mu.$$

Example 4.5.3 For $\xi \in \mathbf{U}_{0,1}$, one has

$$\mathbf{E}\xi^2 = \int_0^1 x^2\,dx = \frac{1}{3}, \qquad \mathbf{E}\xi = \frac{1}{2}.$$

By (4.5.1) we obtain $\operatorname{Var}(\xi) = \frac{1}{12}$.

Example 4.5.4 For $\xi \in \mathbf{B}_p$, by virtue of the relations $\xi^2 = \xi$ and $\mathbf{E}\xi^2 = \mathbf{E}\xi = p$ we obtain $\operatorname{Var}(\xi) = p - p^2 = p(1-p)$.

Consider now some properties of the variance.

D1. $\operatorname{Var}(\xi) \ge 0$, *with* $\operatorname{Var}(\xi) = 0$ *if and only if* $\mathbf{P}(\xi = c) = 1$, *where c is a constant* (not depending on ω).

The first assertion is obvious, for $\operatorname{Var}(\xi) = \mathbf{E}(\xi - \mathbf{E}\xi)^2 \ge 0$. Let $\mathbf{P}(\xi = c) = 1$, then $(\mathbf{E}\xi)^2 = \mathbf{E}\xi^2 = c^2$ and hence

$$\operatorname{Var}(\xi) = c^2 - c^2 = 0.$$

If $\operatorname{Var}(\xi) = \mathbf{E}(\xi - \mathbf{E}\xi)^2 = 0$ then (since $(\xi - \mathbf{E}\xi)^2 \ge 0$) $\mathbf{P}(\xi - \mathbf{E}\xi = 0) = 1$, or $\mathbf{P}(\xi = \mathbf{E}\xi) = 1$ (see property E4).

D2. *If a and b are constants then*

$$\operatorname{Var}(a + b\xi) = b^2 \operatorname{Var}(\xi).$$

This property follows immediately from the definition of $\operatorname{Var}(\xi)$.

D3. *If random variables ξ and η are independent then*

$$\operatorname{Var}(\xi + \eta) = \operatorname{Var}(\xi) + \operatorname{Var}(\eta).$$

Indeed,

$$\begin{aligned}\mathrm{Var}(\xi+\eta) &= \mathbf{E}(\xi+\eta)^2-(\mathbf{E}\xi+\mathbf{E}\eta)^2\\ &= \mathbf{E}\xi^2+2\mathbf{E}\xi\mathbf{E}\eta+\mathbf{E}\eta^2-(\mathbf{E}\xi)^2-(\mathbf{E}\eta)^2-2\mathbf{E}\xi\mathbf{E}\eta\\ &= \mathbf{E}\xi^2-(\mathbf{E}\xi)^2+\mathbf{E}\eta^2-(\mathbf{E}\eta)^2=\mathrm{Var}(\xi)+\mathrm{Var}(\eta).\end{aligned}$$

It is seen from the computations that the variance will be additive not only for independent ξ and η, but also whenever

$$\mathbf{E}\xi\eta=\mathbf{E}\xi\mathbf{E}\eta.$$

Example 4.5.5 Let $\nu\ge 0$ be an integer-valued random variable independent of a sequence $\{\xi_j\}$ of independent identically distributed random variables, $\mathbf{E}\nu<\infty$ and $\mathbf{E}\xi_j=a$. Then, as we know, $\mathbf{E}S_\nu=a\mathbf{E}\nu$. What is the variance of S_ν?

By the total probability formula,

$$\begin{aligned}\mathrm{Var}(S_\nu) &= \mathbf{E}(S_\nu-\mathbf{E}S_\nu)^2=\sum\mathbf{P}(\nu=k)\mathbf{E}(S_k-\mathbf{E}S_\nu)^2\\ &= \sum\mathbf{P}(\nu=k)\big[\mathbf{E}(S_k-ak)^2+(ak-a\mathbf{E}\nu)^2\big]\\ &= \sum\mathbf{P}(\nu=k)k\,\mathrm{Var}(\xi_1)+a^2\mathbf{E}(\nu-\mathbf{E}\nu)^2=\mathrm{Var}(\xi_1)\mathbf{E}\nu+a^2\,\mathrm{Var}(\nu).\end{aligned}$$

This equality is equivalent to the relation

$$\mathbf{E}(S_\nu-\nu a)^2=\mathbf{E}\nu\cdot\mathrm{Var}(\xi_1).$$

In this form, the relation remains valid for any stopping time ν (see Chap. 15). Making use of it, one can find in Example 4.4.5 the expectation of the time η until the end of the fair game, when the initial capitals z_1 and z_2 of the players are finite. Indeed, in that case $a=0$, $\mathrm{Var}(\xi_1)=1$ and

$$\mathbf{E}S_\eta^2=\mathrm{Var}(\xi_1)\,\mathbf{E}\eta=z_1^2P_2+z_2^2P_1.$$

We find from this that $\mathbf{E}\eta=z_1z_2$.

4.6 The Correlation Coefficient and Other Numerical Characteristics

Two random variables ξ and η could be functionally (deterministically) dependent: $\xi=g(\eta)$; they could be dependent, but not in a deterministic way; finally, they could be independent. The correlation coefficient of random variables is a quantity which can be used to quantify the degree of dependence of the variables on each other.

All the random variables to appear in the present section are assumed to have finite non-zero variances.

A random variable ξ is said to be *standardised* if $\mathbf{E}\xi=0$ and $\mathrm{Var}(\xi)=1$. Any random variable ξ can be reduced by a linear transformation to a standardised one by putting $\xi_1:=(\xi-\mathbf{E}\xi)/\sqrt{\mathrm{Var}(\xi)}$. Let ξ and η be two random variables and ξ_1 and η_1 the respective standardised random variables.

Definition 4.6.1 The *correlation coefficient* of the random variables ξ and η is the quantity $\rho(\xi,\eta)=\mathbf{E}\xi_1\eta_1$.

Properties of the correlation coefficient.
1. $|\rho(\xi,\eta)|\le 1$.

Proof Indeed,

$$0\le \operatorname{Var}(\xi_1\pm\eta_1)=\mathbf{E}(\xi_1\pm\eta_1)^2=2\pm 2\rho(\xi,\eta).$$

It follows that $|\rho(\xi,\eta)|\le 1$. □

2. *If ξ and η are independent then* $\rho(\xi,\eta)=0$.
This follows from the fact that ξ_1 and η_1 are also independent in this case. □

The converse assertion is not true, of course. In Sect. 4.3 we gave an example of dependent random variables ξ and η such that $\mathbf{E}\xi=0$, $\mathbf{E}\eta=0$ and $\mathbf{E}\xi\eta=0$. The correlation coefficient of these variables is equal to 0, yet they are dependent. However, as we will see in Chap. 7, for a normally distributed vector (ξ,η) the equality $\rho(\xi,\eta)=0$ is necessary and sufficient for the independence of its components.

Another example where the non-correlation of random variables implies their independence is given by the Bernoulli scheme. Let $\mathbf{P}(\xi=1)=p$, $\mathbf{P}(\xi=0)=1-p$, $\mathbf{P}(\eta=1)=q$ and $\mathbf{P}(\eta=0)=1-q$. Then

$$\mathbf{E}\xi=p,\qquad \mathbf{E}\eta=p,\qquad \operatorname{Var}(\xi)=p(1-p),\qquad \operatorname{Var}(\eta)=q(1-q),$$

$$\rho(\xi,\eta)=\frac{\mathbf{E}(\xi-p)(\eta-q)}{\sqrt{pq(1-p)(1-q)}}.$$

The equality $\rho(\xi,\eta)=0$ means that $\mathbf{E}\xi\eta=pq$, or, which is the same,

$$\begin{aligned}
&\mathbf{P}(\xi=1,\eta=1)=\mathbf{P}(\xi=1)\mathbf{P}(\eta=1),\\
&\mathbf{P}(\xi=1,\eta=0)=\mathbf{P}(\xi=1)-\mathbf{P}(\xi=1,\eta=1)=p-pq=\mathbf{P}(\xi=1)\mathbf{P}(\eta=0),
\end{aligned}$$

and so on.

One can easily obtain from this that, in the general case, ξ and η are independent if

$$\rho\big(f(\xi),g(\eta)\big)=0$$

for any bounded measurable functions f and g. It suffices to take $f=I_{(-\infty,x)}$, $g=I_{(-\infty,y)}$, then derive that $\mathbf{P}(\xi<x,\eta<y)=\mathbf{P}(\xi<x)\mathbf{P}(\eta<y)$, and make use of the results of the previous chapter.

3. $|\rho(\xi,\eta)|=1$ *if and only if there exist numbers a and $b\ne 0$ such that* $\mathbf{P}(\eta=a+b\xi)=1$.

Proof Let $\mathbf{P}(\eta=a+b\xi)=1$. Set $\mathbf{E}\xi=\alpha$ and $\sqrt{\operatorname{Var}(\xi)}=\sigma$; then

$$\rho(\xi,\eta)=\mathbf{E}\frac{\xi-\alpha}{\sigma}\cdot\frac{a+b\xi-a-b\alpha}{|b|\sigma}=\operatorname{sign} b.$$

Assume now that $|\rho(\xi,\eta)|=1$. Let, for instance, $\rho(\xi,\eta)=1$. Then

$$\mathrm{Var}(\xi_1-\eta_1)=2\big(1-\rho(\xi,\eta)\big)=0.$$

By property D1 of the variance, this can be the case if and only if

$$\mathbf{P}(\xi_1-\eta_1=c)=1.$$

If $\rho(\xi,\eta)=-1$ then we get $\mathrm{Var}(\xi_1+\eta_1)=0$, and hence

$$\mathbf{P}(\xi_1+\eta_1=c)=1.$$

□

If $\rho>0$ then the random variables ξ and η are said to be *positively correlated*; if $\rho<0$ then ξ and η are said to be *negatively correlated*.

Example 4.6.1 Consider a transmitting device. A random variable ξ denotes the magnitude of the transmitted signal. Because of interference, a receiver gets the variable $\eta=\alpha\xi+\Delta$ (α is the amplification coefficient, Δ is the noise). Assume that the random variables Δ and ξ are independent. Let $\mathbf{E}\xi=a$, $\mathrm{Var}(\xi)=1$, $\mathbf{E}\Delta=0$ and $\mathrm{Var}(\Delta)=\sigma^2$. Compute the correlation coefficient of ξ and η:

$$\rho(\xi,\eta)=\mathbf{E}\left((\xi-a)\frac{\alpha\xi+\Delta-a\alpha}{\sqrt{\alpha^2+\sigma^2}}\right)=\frac{\alpha}{\sqrt{\alpha^2+\sigma^2}}.$$

If σ is a large number compared to the amplification α, then ρ is close to 0 and η essentially does not depend on ξ. If σ is small compared to α, then ρ is close to 1, and one can easily reconstruct ξ from η.

We consider some further numerical characteristics of random variables. One often uses the so-called higher order moments.

Definition 4.6.2 The *k-th order moment* of a random variable ξ is the quantity $\mathbf{E}\xi^k$. The quantity $\mathbf{E}(\xi-\mathbf{E}\xi)^k$ is called the *k-th order central moment*, so the variance is the second central moment of ξ.

Given a random vector $(\xi_1,\dots,\xi_n)$, the quantity $\mathbf{E}\xi_1^{k_1}\cdots\xi_n^{k_n}$ is called the *mixed moment of order* $k_1+\cdots+k_n$. Similarly, $\mathbf{E}(\xi_1-\mathbf{E}\xi_1)^{k_1}\cdots(\xi_n-\mathbf{E}\xi_n)^{k_n}$ is said to be the *central mixed moment* of the same order.

For independent random variables, mixed moments are evidently equal to the products of the respective usual moments.

4.7 Inequalities

4.7.1 Moment Inequalities

Theorem 4.7.1 (Cauchy–Bunjakovsky's inequality) *If* ξ_1 *and* ξ_2 *are arbitrary random variables, then*

$$\mathbf{E}|\xi_1\xi_2|\le\left[\mathbf{E}\xi_1^2\mathbf{E}\xi_2^2\right]^{1/2}.$$

This inequality is also sometimes called the *Schwarz inequality*.

Proof The required relation follows from the inequality $2|ab| \le a^2 + b^2$ if one puts $a^2 = \xi_1^2/\mathbf{E}\xi_1^2$, $b^2 = \xi_2^2/\mathbf{E}\xi_2^2$ and takes the expectations of the both sides. □

The Cauchy–Bunjakovsky inequality is a special case of more general inequalities.

Theorem 4.7.2 *For $r > 1$, $\frac{1}{r} + \frac{1}{s} = 1$, one has Hölder's inequality*:

$$\mathbf{E}|\xi_1\xi_2| \le \left[\mathbf{E}|\xi_1|^r\right]^{1/r}\left[\mathbf{E}|\xi_2|^s\right]^{1/s},$$

and Minkowski's inequality:

$$\left[\mathbf{E}|\xi_1 + \xi_2|^r\right]^{1/r} \le \left[\mathbf{E}|\xi_1|^r\right]^{1/r} + \left[\mathbf{E}|\xi_2|^r\right]^{1/r}.$$

Proof Since x^r is, for $r > 1$, a convex function in the domain $x > 0$, which at the point $x = 1$ is equal to 1 and has derivative equal to r, one has $r(x - 1) \le x^r - 1$ for all $x > 0$. Putting $x = (a/b)^{1/r}$ ($a > 0$, $b > 0$), we obtain

$$a^{1/r}b^{1-1/r} - b \le \frac{a}{r} - \frac{b}{r},$$

or, which is the same, $a^{1/r}b^{1/s} \le a/r + b/r$. If one puts

$$a := \frac{|\xi_1|^r}{\mathbf{E}|\xi_1|^r}, \qquad b := \frac{|\xi_2|^s}{\mathbf{E}|\xi_2|^s}$$

and takes the expectations, one gets Hölder's inequality.

To prove Minkowski's inequality, note that, by the inequality $|\xi_1 + \xi_2| \le |\xi_1| + |\xi_2|$, one has

$$\mathbf{E}|\xi_1 + \xi_2|^r \le \mathbf{E}|\xi_1||\xi_1 + \xi_2|^{r-1} + \mathbf{E}|\xi_2||\xi_1 + \xi_2|^{r-1}.$$

Applying Hölder's inequality to the terms on the right-hand side, we obtain

$$\mathbf{E}|\xi_1 + \xi_2|^r \le \left\{\left[\mathbf{E}|\xi_1|^r\right]^{1/r} + \left[\mathbf{E}|\xi_2|^r\right]^{1/r}\right\}\left[\mathbf{E}|\xi_1 + \xi_2|^{(r-1)s}\right]^{1/s}.$$

Since $(r - 1)s = r$, $1 - 1/s = 1/r$, and Minkowski's inequality follows. □

It is obvious that, for $r = s = 2$, Hölder's inequality becomes the Schwarz inequality.

Theorem 4.7.3 (Jensen's inequality) *If* $\mathbf{E}\xi$ *exists and* $g(x)$ *is a convex function, then* $g(\mathbf{E}\xi) \le \mathbf{E}g(\xi)$.

Proof If $g(x)$ is convex then for any y there exists a number $g^1(y)$ such that, for all x,

$$g(x) \ge g(y) + (x - y)g^1(y).$$

Putting $x = \xi$, $y = \mathbf{E}\xi$, and taking the expectations of the both sides of this inequality, we obtain

$$\mathbf{E}g(\xi) \geq g(\mathbf{E}\xi). \qquad \square$$

The following corollary is also often useful.

Corollary 4.7.1 *For any* $0 < v < u$,

$$\left[\mathbf{E}|\xi|^v\right]^{1/v} \leq \left[\mathbf{E}|\xi|^u\right]^{1/u}. \tag{4.7.1}$$

This inequality shows, in particular, that if the u-th order moment exists, then the moments of any order $v < u$ also exist.

Inequality (4.7.1) follows from Hölder's inequality, if one puts $\xi_1 := |\xi|^v$, $\xi_2 := 1$, $r := u/v$, or from Jensen's inequality with $g(x) = |x|^{u/v}$ and $|\xi|^v$ in place of ξ.

4.7.2 Inequalities for Probabilities

Theorem 4.7.4 *Let* $\xi \geq 0$ *with probability* 1. *Then, for any* $x > 0$,

$$\mathbf{P}(\xi \geq x) \leq \frac{\mathbf{E}(\xi; \xi \geq x)}{x} \leq \frac{\mathbf{E}\xi}{x}. \tag{4.7.2}$$

If $\mathbf{E}\xi < \infty$ *then* $\mathbf{P}(\xi \geq x) = o(1/x)$ *as* $x \to \infty$.

Proof The inequality is proved by the following relations:

$$\mathbf{E}\xi \geq \mathbf{E}(\xi; \xi \geq x) \geq x\mathbf{E}(1; \xi \geq x) = x\mathbf{P}(\xi \geq x).$$

If $\mathbf{E}\xi < \infty$ then $\mathbf{E}(\xi; \xi \geq x) \to 0$ as $x \to \infty$. This proves the second statement of the theorem. $\square$

If a function $g(x) \geq 0$ is monotonically increasing, then clearly $\{\xi : g(\xi) \geq g(\varepsilon)\} = \{\xi : \xi \geq \varepsilon\}$ and, applying Theorem 4.7.4 to the random variable $\eta = g(\xi)$, one gets

Corollary 4.7.2 *If* $g(x) \uparrow$, $g(x) \geq 0$, *then*

$$\mathbf{P}(\xi \geq x) \leq \frac{\mathbf{E}(g(\xi); \xi \geq x)}{g(x)} \leq \frac{\mathbf{E}g(\xi)}{g(x)}.$$

In particular, for $g(x) = e^{\lambda x}$,

$$\mathbf{P}(\xi \geq x) \leq e^{-\lambda x}\mathbf{E}e^{\lambda\xi}, \quad \lambda > 0.$$

Corollary 4.7.3 (Chebyshev's inequality) *For an arbitrary random variable* ξ *with a finite variance,*

$$\mathbf{P}\big(|\xi - \mathbf{E}\xi| \geq x\big) \leq \frac{\mathrm{Var}(\xi)}{x^2}. \tag{4.7.3}$$

To prove (4.7.3), it suffices to apply Theorem 4.7.4 to the random variable $\eta = (\xi - \mathbf{E}\xi)^2 \ge 0$. □

The assertions of Theorem 4.7.4 and Corollary 4.7.2 are also often called Chebyshev's inequalities (or *Chebyshev type inequalities*), since in regard to their proofs, they are unessential generalisations of (4.7.3).

Using Chebyshev's inequality, we can bound probabilities of various deviations of ξ knowing only $\mathbf{E}\xi$ and $\mathrm{Var}(\xi)$. As one of the first applications of this inequality, we will derive the so-called *law of large numbers in Chebyshev's form* (the law of large numbers in a more general form will be obtained in Chap. 8).

Theorem 4.7.5 *Let $\xi_1, \xi_2, \ldots$ be independent identically distributed random variables with expectation $\mathbf{E}\xi_j = a$ and finite variance σ^2 and let $S_n = \sum_{j=1}^n \xi_j$. Then, for any $\varepsilon > 0$,*

$$\mathbf{P}\left(\left|\frac{S_n}{n} - a\right| > \varepsilon\right) \le \frac{\sigma^2}{n\varepsilon^2} \to 0$$

as $n \to \infty$.

We will discuss this assertion in Chaps. 5, 6 and 8.

Proof of Theorem 4.7.5 follows from Chebyshev's inequality, for

$$\mathbf{E}\frac{S_n}{n} = a, \qquad \mathrm{Var}\left(\frac{S_n}{n}\right) = \frac{n\sigma^2}{n^2} = \frac{\sigma^2}{n}.$$ □

Now we will give a computational example of the use of Chebyshev's inequality.

Example 4.7.1 Assume we decided to measure the diameter of the lunar disk using photographs made with a telescope. Due to atmospheric interference, measurements of pictures made at different times give different results. Let $\xi - a$ denote the deviation of the result of a measurement from the true value a, $\mathbf{E}\xi = a$ and $\sigma = \sqrt{\mathrm{Var}(\xi)} = 1$ on a certain scale. Carry out a series of n independent measurements and put $\zeta_n := \frac{1}{n}(\xi_1 + \cdots + \xi_n)$. Then, as we saw, $\mathbf{E}\zeta_n = a$, $\mathrm{Var}(\zeta_n) = \sigma^2/n$. Since the variance of the average of the measurements decreases as the number of observations increases, it is natural to estimate the quantity a by ζ_n.

How many observations should be made to ensure $|\zeta_n - a| \le 0.1$ with a probability greater than 0.99? That is, we must have $\mathbf{P}(|\zeta_n - a| \le 0.1) > 0.99$, or $\mathbf{P}(|\zeta_n - a| > 0.1) \le 0.01$. By Chebyshev's inequality, $\mathbf{P}(|\zeta_n - a| > 0.1) \le \sigma^2/(n \cdot 0.01)$. Therefore, if n is chosen so that $\sigma^2/(n \cdot 0.01) \le 0.01$ then the required inequality will be satisfied. Hence we get $n \ge 10^4$.

The above example illustrates the possibility of using Chebyshev's inequality to bound the probabilities of the deviations of random variables. However, this example is an even better illustration of how crude Chebyshev's inequality is for practical purposes. If the reader returns to Example 4.7.1 after meeting with the central limit theorem in Chap. 8, he/she will easily calculate that, to achieve the required accuracy, one actually needs to conduct not 10^4, but only 670 observations.

4.8 Extension of the Notion of Conditional Expectation

In conclusion to the present chapter, we will introduce a notion which, along with those we have already discussed, is a useful and important tool in probability theory. Giving the reader the option to skip this section in the first reading of the book, we avoid direct use of this notion until Chaps. 13 and 15.

4.8.1 Definition of Conditional Expectation

In Sect. 4.2 we introduced the notion of conditional expectation given an arbitrary event B with $\mathbf{P}(B) > 0$ that was defined by the equality

$$\mathbf{E}(\xi \mid B) := \frac{\mathbf{E}(\xi;\, B)}{\mathbf{P}(B)}, \qquad (4.8.1)$$

where

$$\mathbf{E}(\xi;\, B) = \int_B \xi \, d\mathbf{P} = \mathbf{E}\xi \mathrm{I}_B,$$

$\mathrm{I}_B = \mathrm{I}_B(\omega)$ being the indicator of the set B. We have already seen and will see many times in what follows that this is a very useful notion. Definition 4.8.1 introducing this notion has, however, the deficiency that it makes no sense when $\mathbf{P}(B) = 0$. How could one overcome this deficiency?

The fact that the condition $\mathbf{P}(B) > 0$ should not play any substantial role could be illustrated by the following considerations. Assume that ξ and η are independent, $B = \{\eta = x\}$ and $\mathbf{P}(B) > 0$. Then, for any measurable function $\varphi(x, y)$, one has according to (4.8.1) that

$$\mathbf{E}\big[\varphi(\xi, \eta) \big| \eta = x\big] = \frac{\mathbf{E}\varphi(\xi, \eta)\mathrm{I}_{\{\eta = x\}}}{\mathbf{P}(\eta = x)} = \frac{\mathbf{E}\varphi(\xi, x)\mathrm{I}_{\{\eta = x\}}}{\mathbf{P}(\eta = x)} = \mathbf{E}\varphi(\xi, x). \qquad (4.8.2)$$

The last equality holds because the random variables $\varphi(\xi, x)$ and $\mathrm{I}_{\{\eta = x\}}$ are independent, being functions of ξ and η respectively, and consequently

$$\mathbf{E}\varphi(\xi, \eta)\mathrm{I}_{\{\eta = x\}} = \mathbf{E}\varphi(\xi, x)\mathbf{P}(\eta = x).$$

Relations (4.8.2) show that the notion of conditional expectation could also retain its meaning in the case when the probability of the condition is 0, for the equality

$$\mathbf{E}\big[\varphi(\xi, \eta) \big| \eta = x\big] = \mathbf{E}\varphi(\xi, x)$$

itself looks quite natural for independent ξ and η and is by no means related to the assumption that $\mathbf{P}(\eta = x) > 0$.

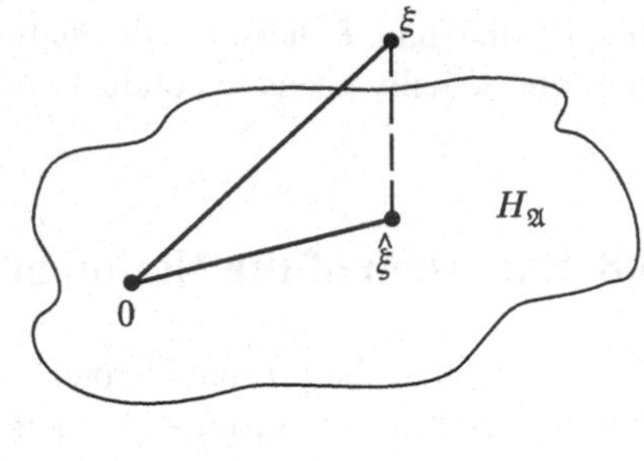

Fig. 4.1 Conditional expectation as the projection of ξ onto $H_{\mathfrak{A}}$

Let $\mathfrak{A}$ be a sub-σ-algebra of $\mathfrak{F}$. We will now define the notion of the conditional expectation of a random variable ξ given $\mathfrak{A}$, which will be denoted by $\mathbf{E}(\xi|\mathfrak{A})$. First we will give the definition in the "discrete" case, but in such a way that it can easily be extended.

Recall that we call discrete the case when the σ-algebra $\mathfrak{A}$ is formed (generated) by an at most countable sequence of disjoint events $A_1, A_2, \dots, \bigcup_j A_j = \Omega$, $\mathbf{P}(A_j) > 0$. We will write this as $\mathfrak{A} = \sigma(A_1, A_2, \dots)$, which means that the elements of $\mathfrak{A}$ are all possible unions of the sets $A_1, A_2, \dots$.

Let L_2 be the collection of all random variables (all the measurable functions $\xi(\omega)$ defined on $\langle \Omega, \mathfrak{F}, \mathbf{P} \rangle$) for which $\mathbf{E}\xi^2 < \infty$. In the linear space L_2 one can introduce the inner product $(\xi, \eta) = \mathbf{E}(\xi\eta)$ (whereby L_2 becomes a Hilbert space with the norm $\|\xi\| = (\mathbf{E}\xi^2)^{1/2}$; we identify two random variables ξ_1 and ξ_2 if $\|\xi_1 - \xi_2\| = 0$, see also Remark 6.1.1).

Now consider the linear space $H_{\mathfrak{A}}$ of all functions of the form

$$\xi(\omega) = \sum_k c_k \mathrm{I}_{A_k}(\omega),$$

where $\mathrm{I}_{A_k}(\omega)$ are indicators of the sets A_k. The space $H_{\mathfrak{A}}$ is clearly the space of all $\mathfrak{A}$-measurable functions, and one could think of it as the space spanned by the orthogonal system $\{\mathrm{I}_{A_k}(\omega)\}$ in L_2.

We now turn to the definition of conditional expectation. We know that the conventional expectation $a = \mathbf{E}\xi$ of $\xi \in L_2$ can be defined as the unique point at which the minimum value of the function $\varphi(a) = \mathbf{E}(\xi - a)^2$ is attained (see Sect. 4.5). Consider now the problem of minimising the functional $\varphi(a) = \mathbf{E}(\xi - a(\omega))^2$, $\xi \in L_2$, over all $\mathfrak{A}$-measurable functions $a(\omega)$ from $H_{\mathfrak{A}}$.

Definition 4.8.1 Let $\xi \in L_2$. The $\mathfrak{A}$-measurable function $a(\omega)$ on which the minimum $\min_{a \in H_{\mathfrak{A}}} \varphi(a)$ is attained is said to be the *conditional expectation of* ξ *given* $\mathfrak{A}$ and is denoted by $\mathbf{E}(\xi|\mathfrak{A})$.

Thus, unlike the conventional expectations, the conditional expectation $\mathbf{E}(\xi|\mathfrak{A})$ is a *random variable*. Let us consider it in more detail. It is evident that the minimum of $\varphi(a)$ is attained when $a(\omega)$ is the projection $\widehat{\xi}$ of the element ξ in the space L_2 onto $H_{\mathfrak{A}}$, i.e. the element $\widehat{\xi} \in H_{\mathfrak{A}}$ for which $\xi - \widehat{\xi} \perp H_{\mathfrak{A}}$ (see Fig. 4.1). In that case, for any $a \in H_{\mathfrak{A}}$,

$$\widehat{\xi} - a \in H_{\mathfrak{A}}, \qquad \xi - \widehat{\xi} \perp \widehat{\xi} - a,$$
$$\varphi(a) = \mathbf{E}(\xi - \widehat{\xi} + \widehat{\xi} - a)^2 = \mathbf{E}(\xi - \widehat{\xi})^2 + \mathbf{E}(\widehat{\xi} - a)^2,$$
$$\varphi(a) \ge \varphi(\widehat{\xi}),$$

and $\varphi(a) = \varphi(\widehat{\xi})$ if $a = \widehat{\xi}$ a.s.

Thus, in L_2 the conditional expectation operation is just an *orthoprojector* onto $H_{\mathfrak{A}}$ ($\hat{\xi} = \mathbf{E}(\xi \,|\, \mathfrak{A})$ is the projection of ξ onto $H_{\mathfrak{A}}$).

Since, for a discrete σ-algebra $\mathfrak{A}$, the element $\hat{\xi}$, being an element of $H_{\mathfrak{A}}$, has the form $\hat{\xi} = \sum c_k \mathrm{I}_{A_k}$, the orthogonality condition $\xi - \hat{\xi} \perp H_{\mathfrak{A}}$ (or, which is the same, $\mathbf{E}(\xi - \hat{\xi})\,\mathrm{I}_{A_k} = 0$) determines uniquely the coefficients c_k:

$$\mathbf{E}(\xi \mathrm{I}_{A_k}) = c_k \mathbf{P}(A_k), \qquad c_k = \frac{\mathbf{E}(\xi;\, A_k)}{\mathbf{P}(A_k)} = \mathbf{E}(\xi \,|\, A_k),$$

so that

$$\mathbf{E}(\xi \,|\, \mathfrak{A}) = \widehat{\xi} = \sum_k \mathbf{E}(\xi \,|\, A_k)\mathrm{I}_{A_k}.$$

Thus the *random variable* $\mathbf{E}(\xi \,|\, \mathfrak{A})$ *is constant on* A_k *and, on these sets, is equal to the average value of* ξ *on* A_k.

If ξ and $\mathfrak{A}$ are independent (i.e. $\mathbf{P}(\xi \in B;\, A_k) = \mathbf{P}(\xi \in B)\mathbf{P}(A_k)$) then clearly $\mathbf{E}(\xi;\, A_k) = \mathbf{E}\xi\, \mathbf{P}(A_k)$ and $\widehat{\xi} = \mathbf{E}\xi$. If $\mathfrak{A} = \mathfrak{F}$ then $\mathfrak{F}$ is also discrete, ξ is constant on the sets A_k and hence $\widehat{\xi} = \xi$.

Now note the following basic properties of conditional expectation which allow one to get rid of the two special assumptions (that $\xi \in L_2$ and $\mathfrak{A}$ is discrete), which were introduced at first to gain a better understanding of the nature of conditional expectation:

(1) $\widehat{\xi}$ *is* $\mathfrak{A}$*-measurable.*
(2) *For any event* $A \in \mathfrak{A}$,

$$\mathbf{E}(\widehat{\xi};\, A) = \mathbf{E}(\xi;\, A).$$

The former property is obvious. The latter follows from the fact that any event $A \in \mathfrak{A}$ can be represented as $A \in \bigcup_k A_{j_k}$, and hence

$$\mathbf{E}(\widehat{\xi};\, A) = \sum_k \mathbf{E}(\widehat{\xi};\, A_{j_k}) = \sum_k c_{j_k}\mathbf{P}(A_{j_k}) = \sum_k \mathbf{E}(\xi;\, A_{j_k}) = \mathbf{E}(\xi;\, A).$$

The meaning of this property is rather clear: averaging the variable ξ over the set A gives the same result as averaging the variable $\widehat{\xi}$ which has already been averaged over A_{j_k}.

Lemma 4.8.1 *Properties* (1) *and* (2) *uniquely determine the conditional expectation and are equivalent to Definition* 4.8.1.

Proof In one direction the assertion of the lemma has already been proved. Assume now that conditions (1) and (2) hold. $\mathfrak{A}$-measurability of $\widehat{\xi}$ means that $\widehat{\xi}$ is constant on each set A_k. Denote by c_k the value of $\widehat{\xi}$ on A_k. Since $A_k \in \mathfrak{A}$, it follows from property (2) that

$$\mathbf{E}(\widehat{\xi}; A_k) = c_k \mathbf{P}(A_k) = \mathbf{E}(\xi; A_k),$$

and hence, for $\omega \in A_k$,

$$\widehat{\xi} = c_k = \frac{\mathbf{E}(\xi; A_k)}{\mathbf{P}(A_k)}.$$

The lemma is proved. □

Now we can give the general definition of conditional expectation.

Definition 4.8.2 Let ξ be a random variable on a probability space $\langle \Omega, \mathfrak{F}, \mathbf{P} \rangle$ and $\mathfrak{A} \subset \mathfrak{F}$ an arbitrary sub-σ-algebra of $\mathfrak{F}$. *The conditional expectation of* ξ *given* $\mathfrak{A}$ is a random variable $\hat{\xi}$ which is denoted by $\mathbf{E}(\xi | \mathfrak{A})$ and has the following two properties:

(1) $\hat{\xi}$ *is* $\mathfrak{A}$*-measurable.*
(2) *For any* $A \in \mathfrak{A}$, *one has* $\mathbf{E}(\hat{\xi}; A) = \mathbf{E}(\xi; A)$.

In this definition, the random variable ξ can be both scalar and vector-valued.

There immediately arises the question of whether such a random variable exists and is unique. In the discrete case we saw that the answer to this question is positive. In the general case, the following assertion holds true.

Theorem 4.8.1 *If* $\mathbf{E}|\xi|$ *is finite, then the function* $\widehat{\xi} = \mathbf{E}(\xi | \mathfrak{A})$ *in Definition* 4.8.2 *always exists and is unique up to its values on a set of probability* 0.

Proof First assume that ξ is scalar and $\xi \geq 0$. Then the set function

$$\mathbf{Q}(A) = \int_A \xi \, d\mathbf{P} = \mathbf{E}(\xi; A), \qquad A \in \mathfrak{A}$$

will be a measure on $\langle \Omega, \mathfrak{A} \rangle$ which is absolutely continuous with respect to $\mathbf{P}$, for $\mathbf{P}(A) = 0$ implies $\mathbf{Q}(A) = 0$. Therefore, by the Radon–Nykodim theorem (see Appendix 3), there exists an $\mathfrak{A}$-measurable function $\widehat{\xi} = \mathbf{E}(\xi | \mathfrak{A})$ which is unique up to its values on a set of measure zero and such that

$$\mathbf{Q}(A) = \int_A \widehat{\xi} \, d\mathbf{P} = \mathbf{E}(\widehat{\xi}; A).$$

In the general case we put $\xi = \xi^+ - \xi^-$, where $\xi^+ := \max(0, \xi) \geq 0$, $\xi^- := \max(0, -\xi) \geq 0$, $\widehat{\xi} := \widehat{\xi}^+ - \widehat{\xi}^-$ and $\widehat{\xi}^\pm$ are conditional expectations of $\xi^\pm$. This proves the existence of the conditional expectation, since $\widehat{\xi}$ satisfies conditions (1) and (2) of Definition 4.8.2. This will also imply uniqueness, for the assumption on non-uniqueness of $\widehat{\xi}$ would imply non-uniqueness of $\widehat{\xi}^+$ or $\widehat{\xi}^-$. The proof for vector-valued ξ reduces to the one-dimensional case, since the components of $\widehat{\xi}$ will possess properties (1) and (2) and, for the components, the existence and uniqueness have already been proved. □

The essence of the above proof is quite transparent: by condition (2), for any $A \in \mathfrak{A}$ we are given the value

$$\mathbf{E}(\widehat{\xi}; A) = \int_A \widehat{\xi}\, d\mathbf{P},$$

i.e. the values of the integrals of $\hat{\xi}$ over all sets $A \in \mathfrak{A}$ are given. This clearly should define an $\mathfrak{A}$-measurable function $\hat{\xi}$ uniquely up to its values on a set of measure zero.

The meaning of $\mathbf{E}(\xi \,|\, \mathfrak{A})$ remains the same: roughly speaking, this is the result of averaging of ξ over "indivisible" elements of $\mathfrak{A}$.

If $\mathfrak{A} = \mathfrak{F}$ then evidently $\widehat{\xi} = \xi$ satisfies properties (1) and (2) and therefore $\mathbf{E}(\xi \,|\, \mathfrak{F}) = \xi$.

Definition 4.8.3 Let ξ and η be random variables on $\langle \Omega, \mathfrak{F}, \mathbf{P} \rangle$ and $\mathfrak{A} = \sigma(\eta)$ be the σ-algebra generated by the random variable η. Then $\mathbf{E}(\xi \,|\, \mathfrak{A})$ is also called the *conditional expectation of* ξ *given* η.

To simplify the notation, we will sometimes write $\mathbf{E}(\xi \,|\, \eta)$ instead of $\mathbf{E}(\xi \,|\, \sigma(\eta))$. This does not lead to confusion.

Since $\mathbf{E}(\xi \,|\, \eta)$ is, by definition, a $\sigma(\eta)$-measurable random variable, this means (see Sect. 3.5) that there exists a measurable function $g(x)$ for which $\mathbf{E}(\xi \,|\, \eta) = g(\eta)$. By analogy with the discrete case, one can interpret the quantity $g(x)$ as the result of averaging ξ over the set $\{\eta = x\}$. (Recall that in the discrete case $g(x) = \mathbf{E}(\xi \,|\, \eta = x)$.)

Definition 4.8.4 If $\xi = \mathrm{I}_C$ is the indicator of a set $C \in \mathfrak{F}$, then $\mathbf{E}(\mathrm{I}_C \,|\, \mathfrak{A})$ is called the *conditional probability* $\mathbf{P}(C \,|\, \mathfrak{A})$ *of the event* C *given* $\mathfrak{A}$. If $\mathfrak{A} = \sigma(\eta)$, we speak of the conditional probability $\mathbf{P}(C \,|\, \eta)$ of the event C given η.

4.8.2 Properties of Conditional Expectations

1. *Conditional expectations have the properties of conventional expectations, the only difference being that they hold almost surely* (*with probability* 1):

(a) $\mathbf{E}(a + b\xi \,|\, \mathfrak{A}) = a + b\mathbf{E}(\xi \,|\, \mathfrak{A})$.
(b) $\mathbf{E}(\xi_1 + \xi_2 \,|\, \mathfrak{A}) = \mathbf{E}(\xi_1 \,|\, \mathfrak{A}) + \mathbf{E}(\xi_2 \,|\, \mathfrak{A})$.
(c) *If* $\xi_1 \le \xi_2$ *a.s., then* $\mathbf{E}(\xi_1 \,|\, \mathfrak{A}) \le \mathbf{E}(\xi_2 \,|\, \mathfrak{A})$ *a.s.*

To prove, for instance, property (a), one needs to verify, according to Definition 4.8.2, that

(1) $a + b\mathbf{E}(\xi \,|\, \mathfrak{A})$ is an $\mathfrak{A}$-measurable function;
(2) $\mathbf{E}(a + b\xi;\, A) = \mathbf{E}(a + b\mathbf{E}(\xi \,|\, \mathfrak{A});\, A)$ for any $A \in \mathfrak{A}$.

Here (1) is evident; (2) follows from the linearity of conventional expectation (or integral).

Property (b) is proved in the same way.

To prove (c), put, for brevity, $\widehat{\xi_i} := \mathbf{E}(\xi_i \,|\, \mathfrak{A})$. Then, for any $A \in \mathfrak{A}$,

$$\int_A \widehat{\xi_1}\, d\mathbf{P} = \mathbf{E}(\widehat{\xi_1};\, A) = \mathbf{E}(\xi_1;\, A) \le \mathbf{E}(\xi_2;\, A) = \int_A \widehat{\xi_2}\, d\mathbf{P}, \qquad \int_A (\widehat{\xi_2} - \widehat{\xi_1})\, d\mathbf{P} \ge 0.$$

This implies that $\widehat{\xi_2} - \widehat{\xi_1} \ge 0$ a.s.

2. *Chebyshev's inequality. If* $\xi \ge 0$, $x > 0$, *then* $\mathbf{P}(\xi \ge x \,|\, \mathfrak{A}) \le \mathbf{E}(\xi \,|\, \mathfrak{A})/x$.

This property follows from 1(c), since $\mathbf{P}(\xi \ge x \,|\, \mathfrak{A}) = \mathbf{E}(\mathrm{I}_{\{\xi \ge x\}} \,|\, \mathfrak{A})$, where I_A is the indicator of the event A, and one has the inequality $\mathrm{I}_{\{\xi \ge x\}} \le \xi/x$.

3. *If* $\mathfrak{A}$ *and* $\sigma(\eta)$ *are independent, then* $\mathbf{E}(\xi \,|\, \mathfrak{A}) = \mathbf{E}\xi$. Since $\widehat{\xi} = \mathbf{E}\xi$ is an $\mathfrak{A}$-measurable function, it remains to verify the second condition from Definition 4.8.2: for any $A \in \mathfrak{A}$,

$$\mathbf{E}(\hat{\xi};\, A) = \mathbf{E}(\xi;\, A).$$

This equality follows from the independence of the random variables I_A and ξ and the relations $\mathbf{E}(\xi;\, A) = \mathbf{E}(\xi \mathrm{I}_A) = \mathbf{E}\xi\mathbf{E}\mathrm{I}_A = \mathbf{E}(\widehat{\xi};\, A)$.

It follows, in particular, that if ξ and η are independent, then $\mathbf{E}(\xi \,|\, \eta) = \mathbf{E}\xi$. If the σ-algebra $\mathfrak{A}$ is trivial, then clearly one also has $\mathbf{E}(\xi \,|\, \mathfrak{A}) = \mathbf{E}\xi$.

4. *Convergence theorems that are true for conventional expectations hold for conditional expectations as well.* For instance, the following assertion is true.

Theorem 4.8.2 (Monotone convergence theorem) *If* $0 \le \xi_n \uparrow \xi$ *a.s. then*

$$\mathbf{E}(\xi_n \,|\, \mathfrak{A}) \uparrow \mathbf{E}(\xi \,|\, \mathfrak{A}) \quad \textit{a.s.}$$

Indeed, it follows from $\xi_{n+1} \ge \xi_n$ a.s. that $\widehat{\xi_{n+1}} \ge \widehat{\xi_n}$ a.s., where $\widehat{\xi_n} = \mathbf{E}(\xi_n \,|\, \mathfrak{A})$. Therefore there exists an $\mathfrak{A}$-measurable random variable $\widehat{\xi}$ such that $\widehat{\xi_n} \uparrow \widehat{\xi}$ a.s. By the conventional monotone convergence theorem, for any $A \in \mathfrak{A}$,

$$\int_A \widehat{\xi_n}\, d\mathbf{P} \to \int_A \widehat{\xi}\, d\mathbf{P}, \qquad \int_A \xi_n\, d\mathbf{P} \to \int_A \xi\, d\mathbf{P}.$$

Since the left-hand sides of these relations coincide, the same holds for the right-hand sides. This means that $\widehat{\xi} = \mathbf{E}(\xi \,|\, \mathfrak{A})$.

5. *If* η *is an* $\mathfrak{A}$*-measurable scalar random variable,* $\mathbf{E}|\xi| < \infty$, *and* $\mathbf{E}|\xi\eta| < \infty$, *then*

$$\mathbf{E}(\eta\xi \,|\, \mathfrak{A}) = \eta\mathbf{E}(\xi \,|\, \mathfrak{A}). \tag{4.8.3}$$

If $\xi \ge 0$ *and* $\eta \ge 0$ *then the moment conditions are superfluous.*

In other words, in regard to the conditional expectation operation, $\mathfrak{A}$-measurable random variables behave as constants in conventional expectations (cf. property 1(a)).

In order to prove (4.8.3), note that if $\eta = \mathrm{I}_B$ (the indicator of a set $B \in \mathfrak{A}$) then the assertion holds since, for any $A \in \mathfrak{A}$,

$$\int_A \mathbf{E}(\mathrm{I}_B\xi \,|\, \mathfrak{A})\, d\mathbf{P} = \int_A \mathrm{I}_B\xi\, d\mathbf{P} = \int_{AB} \xi\, d\mathbf{P} = \int_{AB} \mathbf{E}(\xi \,|\, \mathfrak{A})\, d\mathbf{P} = \int_A \mathrm{I}_B\mathbf{E}(\xi \,|\, \mathfrak{A})\, d\mathbf{P}.$$

This together with the linearity of conditional expectations implies that the assertion holds for all simple functions η.

If $\xi \ge 0$ and $\eta \ge 0$ then, taking a sequence of simple functions $0 \le \eta_n \uparrow \eta$ and applying the monotone convergence theorem to the equality

$$\mathbf{E}(\eta_n \xi \,|\, \mathfrak{A}) = \eta_n \mathbf{E}(\xi \,|\, \mathfrak{A}),$$

we obtain (4.8.3). Transition to the case of arbitrary ξ and η is carried out in the standard way—by considering positive and negative parts of the random variables ξ and η. In addition, to ensure that the arising differences and sums make sense, we require the existence of the expectations $\mathbf{E}|\xi|$ and $\mathbf{E}|\xi\eta|$.

6. *All the basic inequalities for conventional expectations remain true for conditional expectations as well, in particular, Cauchy–Bunjakovsky's inequality*

$$\mathbf{E}\big(|\xi_1 \xi_2| \,|\, \mathfrak{A}\big) \le \big[\mathbf{E}\big(\xi_1^2 \,|\, \mathfrak{A}\big)\mathbf{E}\big(\xi_2^2 \,|\, \mathfrak{A}\big)\big]^{1/2}$$

and Jensen's inequality: if $\mathbf{E}|\xi| < \infty$ *then, for any convex function* g,

$$g\big(\mathbf{E}(\xi \,|\, \mathfrak{A})\big) \le \mathbf{E}\big(g(\xi) \,|\, \mathfrak{A}\big). \tag{4.8.4}$$

Cauchy–Bunjakovsky's inequality can be proved in exactly the same way as for conventional expectations, for its proof requires no properties of expectations other than linearity.

Jensen's inequality is a consequence of the following relation. By convexity of $g(x)$, for any y, there exists a number $g^*(y)$ such that $g(x) \ge g(y) + (x - y)g^*(y)$ ($g^*(y) = g'(y)$ if g is differentiable at the point y). Put $x = \xi$, $y = \widehat{\xi} = \mathbf{E}(\xi \,|\, \mathfrak{A})$, and take conditional expectations of the both sides of the inequality. Then, assuming for the moment that

$$\mathbf{E}\big(\big|(\xi - \widehat{\xi})g^*(\widehat{\xi})\big|\big) < \infty \tag{4.8.5}$$

(this can be proved if $\mathbf{E}|g(\xi)| < \infty$), we get

$$\mathbf{E}\big[(\xi - \widehat{\xi})g^*(\widehat{\xi}) \,|\, \mathfrak{A})\big] = g^*(\widehat{\xi})\mathbf{E}(\xi - \widehat{\xi} \,|\, \mathfrak{A}) = 0$$

by virtue of property 5. Thus we obtain (4.8.4). In the general case note that the function $g^*(y)$ is nondecreasing. Let (y_{-N}, y_N) be the maximal interval on which $|g^*(y)| < N$. Put

$$g_N(y) := \begin{cases} g(y) & \text{if } y \in [y_{-N}, y_N], \\ g(y_{\pm N}) \pm (y - y_{\pm N})N & \text{if } y \gtrless y_{\pm N}. \end{cases}$$

($y_{\pm N}$ can take infinite values if $\pm g^*(y)$ are bounded as $y \to \infty$. Note that the values of $g^*(y)$ are always bounded from below as $y \to \infty$ and from above as $y \to -\infty$, hence $g^*(y_{\pm N}) \gtrless 0$ for N large enough.) The support function $g_N^*(y)$ corresponding to $g_N(y)$ has the form

$$g_N^*(y) = \max\big[-N, \min\big(N, g^*(y)\big)\big]$$

and, consequently, is bounded for each N. Therefore, condition (4.8.5) is satisfied for $g_N^*(y)$ (recall that $\mathbf{E}|\xi| < \infty$) and hence

$$g_N(\widehat{\xi}) \le \mathbf{E}\big(g_N(\xi) \,|\, \mathfrak{A}\big).$$

Further, we have $g_N(y) \uparrow g(y)$ as $N \to \infty$ for all y. Therefore the left-hand side of this inequality converges everywhere to $g(\widehat{\xi})$ as $N \to \infty$, but the right-hand side converges to $\mathbf{E}(g(\xi)|\mathfrak{A})$ by Theorem 4.8.2. Property 6 is proved. □

7. *The total probability formula*

$$\mathbf{E}\xi = \mathbf{E}\mathbf{E}(\xi\,|\mathfrak{A})$$

follows immediately from property 2 of Definition 4.8.2 with $A = \Omega$.

8. *Iterated averaging* (an extension of property 7): if $\mathfrak{A} \subset \mathfrak{A}_1 \subset \mathfrak{F}$ then

$$\mathbf{E}(\xi\,|\mathfrak{A}) = \mathbf{E}\big[\mathbf{E}(\xi\,|\mathfrak{A}_1)|\mathfrak{A}\big].$$

Indeed, for any $A \in \mathfrak{A}$, since $A \in \mathfrak{A}_1$ one has

$$\int_A \mathbf{E}\big[\mathbf{E}(\xi\,|\mathfrak{A}_1)|\mathfrak{A}\big]\,d\mathbf{P} = \int_A \mathbf{E}(\xi\,|\mathfrak{A}_1)\,d\mathbf{P} = \int_A \xi\,d\mathbf{P} = \int_A \mathbf{E}(\xi\,|\mathfrak{A})\,d\mathbf{P}.$$

The properties 1, 3–5, 7 and 8 clearly hold for both scalar- and vector-valued random variables ξ. The next property we will single out.

9. *For $\xi \in L_2$, the minimum of $\mathbf{E}(\xi - a(\omega))^2$ over all $\mathfrak{A}$-measurable functions $a(\omega)$ is attained at $a(\omega) = \mathbf{E}(\xi\,|\mathfrak{A})$.*

Indeed, $\mathbf{E}(\xi - a(\omega))^2 = \mathbf{E}\mathbf{E}((\xi - a(\omega))^2|\mathfrak{A})$, but $a(\omega)$ behaves as a constant in what concerns the operation $\mathbf{E}(\cdot|\mathfrak{A})$ (see property 5), so that

$$\mathbf{E}\big((\xi - a(\omega))^2|\mathfrak{A}\big) = \mathbf{E}\big((\xi - \mathbf{E}(\xi\,|\mathfrak{A}))^2|\mathfrak{A}\big) + \big(\mathbf{E}(\xi\,|\mathfrak{A}) - a(\omega)\big)^2$$

and the minimum of this expression is attained at $a(\omega) = \mathbf{E}(\xi\,|\mathfrak{A})$.

This property proves the *equivalence of Definitions* 4.8.1 *and* 4.8.2 *in the case when* $\xi \in L_2$ (in both definitions, conditional expectation is defined up to its values on a set of measure 0). In this connection note once again that, in L_2, the operation of taking conditional expectations is the projection onto $H_{\mathfrak{A}}$ (see our comments to Definition 4.8.1).

Property 9 can be extended to the multivariate case in the following form: *for any nonnegative definite matrix V, the minimum* $\min(\xi - a(\omega))V(\xi - a(\omega))^T$ *over all $\mathfrak{A}$-measurable functions $a(\omega)$ is attained at $a(\omega) = \mathbf{E}(\xi\,|\mathfrak{A})$.*

The assertions proved above in the case where $\xi \in L_2$ and the σ-algebra $\mathfrak{A}$ is countably generated will surely hold true for an arbitrary σ-algebra $\mathfrak{A}$, but the substantiation of this fact requires additional work.

In conclusion we note that property 5 admits, under wide assumptions, the following generalisation:

5A. *If η is $\mathfrak{A}$-measurable and $g(\omega, \eta)$ is a measurable function of its arguments $\omega \in \Omega$ and $\eta \in \mathbb{R}^k$ such that $\mathbf{E}|g(\omega, \eta)|\mathfrak{A})| < \infty$, then*

$$\mathbf{E}\big(g(\omega, \eta)|\mathfrak{A}\big) = \mathbf{E}\big(g(\omega, y)|\mathfrak{A}\big)\big|_{y=\eta}. \tag{4.8.6}$$

This implies the double expectation (or total probability) formula.

$$\mathbf{E}g(\omega, \eta) = \mathbf{E}\big[\mathbf{E}\big(g(\omega, y)|\mathfrak{A}\big)\big|_{y=\eta}\big],$$

which can be considered as an extension of Fubini's theorem (see Sects. 4.6 and 3.6). Indeed, if $g(\omega, y)$ is independent of $\mathfrak{A}$, then

$$\mathbf{E}\big(g(\omega, y)\big|\mathfrak{A}\big) = \mathbf{E}g(\omega, y), \qquad \mathbf{E}\big(g(\omega, \eta)\big|\mathfrak{A}\big) = \mathbf{E}g(\omega, y)\big|_{y=\eta},$$
$$\mathbf{E}g(\omega, \eta) = \mathbf{E}\big[\mathbf{E}g(\omega, y)\big|_{y=\eta}\big].$$

In regard to its form, this is Fubini's theorem, but here η is a vector-valued random variable, while ω can be of an arbitrary nature.

We will prove property 5A under the simplifying assumption that there exists a sequence of simple functions η_n such that $g(\omega, \eta_n) \uparrow g(\omega, \eta)$ and $h(\omega, \eta_n) \uparrow h(\omega, \eta)$ a.s., where $h(\omega, y) = \mathbf{E}(g(\omega, y)|\mathfrak{A}))$. Indeed, let $\eta_n = y_k$ for $\omega \in A_k \subset \mathfrak{A}$. Then

$$g(\omega, \eta_n) = \sum g(\omega, y_k)\mathrm{I}_{A_k}.$$

By property 5 it follows that (4.8.6) holds for the functions η_n. It remains to make use of the monotone convergence theorem (property 4) in the equality $\mathbf{E}(g(\omega, \eta_n)|\mathfrak{A})) = h(\omega, \eta_n)$.

4.9 Conditional Distributions

Along with conditional expectations, one can consider *conditional distributions* given sub-σ-algebras and random variables. In the present section, we turn our attention to the latter.

Let ξ and η be two random variables on $\langle \Omega, \mathfrak{F}, \mathbf{P}\rangle$ taking values in $\mathbb{R}^s$ and $\mathbb{R}^k$, respectively, and let $\mathfrak{B}^s$ be the σ-algebra of Borel sets in $\mathbb{R}^s$.

Definition 4.9.1 A function $\mathbf{F}(B|y)$ of two variables $y \in \mathbb{R}^k$ and $B \in \mathfrak{B}^s$ is called the *conditional distribution of ξ given $\eta = y$* if:

1. For any B, $\mathbf{F}(B|\eta)$ is the conditional probability $\mathbf{P}(\xi \in B|\eta)$ of the event $\{\xi \in B\}$ given η, i.e. $\mathbf{F}(B|y)$ is a Borel function of y such that, for any $A \in \mathfrak{B}^k$,

$$\mathbf{E}\big(\mathbf{F}(B|\eta); \eta \in A\big) \equiv \int_A \mathbf{F}(B|y)\mathbf{P}(\eta \in dy) = \mathbf{P}(\xi \in B, \eta \in A).$$

2. For any y, $\mathbf{F}(B|y)$ is a probability distribution in B.

Sometimes we will write the function $\mathbf{F}(B|y)$ in a more "deciphered" form as $\mathbf{F}(B|y) = \mathbf{P}(\xi \in B|\eta = y)$.

We know that, for each $B \in \mathfrak{B}^s$, there exists a Borel function $g_B(y)$ such that $g_B(\eta) = \mathbf{P}(\xi \in B|\eta)$. Thus, putting $P(B|y) := g_B(y)$, we will satisfy condition 1 of Definition 4.9.1. Condition 2, however, does not follow from the properties of conditional expectations and by no means needs to hold: indeed, since conditional probability $\mathbf{P}(\xi \in B|\eta)$ is defined for each B up to its values on a set N_B of zero measure (so that there exist many variants of conditional expectation), and this set can be different for each B. Therefore, if the union

$$N=\bigcup_{B\in\mathfrak{B}^s} N_B$$

has a non-zero probability, it could turn out that, for instance, the equalities

$$\mathbf{P}(\xi\in B_1\cup B_2|\eta)=\mathbf{P}(\xi\in B_1|\eta)+\mathbf{P}(\xi\in B_2|\eta)$$

(additivity of probability) for all disjoint B_1 and B_2 from $\mathfrak{B}^s$ hold for no $\omega\in N$, i.e. on an ω-set N of positive probability, the function $g_B(y)$ will not be a distribution as a function of B.

However, in the case *when ξ is a random variable taking values in $\mathbb{R}^s$ with the σ-algebra $\mathfrak{B}^s$ of Borel sets, one can always choose $g_B(\eta)=\mathbf{P}(\xi\in B|\eta)$ such that $g_B(y)$ will be a conditional distribution.*[2]

As one might expect, conditional probabilities possess the natural property that conditional expectations can be expressed as integrals with respect to conditional distributions.

Theorem 4.9.1 *For any measurable function $g(x)$ mapping $\mathbb{R}^s$ into $\mathbb{R}$ such that $\mathbf{E}|g(\xi)|<\infty$, one has*

$$\mathbf{E}\big(g(\xi)|\eta\big)=\int g(x)\mathbf{F}(dx|\eta). \tag{4.9.1}$$

Proof It suffices to consider the case $g(x)\ge 0$. If $g(x)=\mathrm{I}_A(x)$ is the indicator of a set A, then formula (4.9.1) clearly holds. Therefore it holds for any simple (i.e. assuming only finitely many values) function $g_n(x)$. It remains to take a sequence $g_n\uparrow g$ and make use of the monotonicity of both sides of (4.9.1) and property 4 from Sect. 4.8. □

In real-life problems, to compute conditional distributions one can often use the following simple rule which we will write in the form

$$\mathbf{P}(\xi\in B|\eta=y)=\frac{\mathbf{P}(\xi\in B,\eta\in dy)}{\mathbf{P}(\eta\in dy)}. \tag{4.9.2}$$

Both conditions of Definition 4.9.1 will clearly be formally satisfied.

If ξ and η have a joint density, this equality will have a precise meaning.

Definition 4.9.2 Assume that, for each y, the conditional distribution $\mathbf{F}(B|y)$ is absolutely continuous with respect to some measure μ in $\mathbb{R}^s$:

$$\mathbf{P}(\xi\in B|\eta=y)=\int_B f(x|y)\mu(dx).$$

Then the density $f(x|y)$ is said to be the *conditional density of ξ (with respect to the measure μ) given $\eta=y$.*

[2] For more details, see e.g. [12, 14, 26].

In other words, a measurable function $f(x|y)$ of two variables x and y is the conditional density of ξ given $\eta = y$ if:

(1) For any Borel sets $A \subset \mathbb{R}^k$ and $B \subset \mathbb{R}^s$,

$$\int_{y\in A}\int_{x\in B} f(x|y)\mu(dx)\mathbf{P}(\eta \in dy) = \mathbf{P}(\xi \in B, \eta \in A). \tag{4.9.3}$$

(2) For any y, the function $f(x|y)$ is a probability density.

It follows from Theorem 4.9.1 that if there exists a conditional density, then

$$\mathbf{E}\big(g(\xi)\big|\eta\big) = \int g(x)f(x|\eta)\mu(dx).$$

If we additionally assume that the distribution of η has a density $q(y)$ with respect to some measure λ in $\mathbb{R}^k$, then we can re-write (4.9.3) in the form

$$\int_{y\in A}\int_{x\in B} f(x|y)q(y)\,\mu(dx)\,\lambda(dy) = \mathbf{P}(\xi \in B, \eta \in A). \tag{4.9.4}$$

Consider now the direct product of spaces $\mathbb{R}^s$ and $\mathbb{R}^k$ and the direct product of measures $\mu \times \lambda$ on it (if $C = B \times A$, $B \subset \mathbb{R}^s$, $A \subset \mathbb{R}^k$ then $\mu \times \lambda(C) = \mu(B)\lambda(A)$). In the product space, relation (4.9.4) evidently means that the joint distribution of ξ and η in $\mathbb{R}^s \times \mathbb{R}^k$ has a density with respect to $\mu \times \lambda$ which is equal to

$$f(x, y) = f(x|y)q(y).$$

The converse assertion is also true.

Theorem 4.9.2 *If the joint distribution of ξ and η in $\mathbb{R}^s \times \mathbb{R}^k$ has a density $f(x, y)$ with respect to $\mu \times \lambda$, then the function*

$$f(x|y) = \frac{f(x, y)}{q(y)}, \quad \textit{where } q(y) = \int f(x, y)\,\mu(dx),$$

is the conditional density of ξ given $\eta = y$, and the function $q(y)$ is the density of η with respect to the measure λ.

Proof The assertion on $q(y)$ is obvious, since

$$\int_A q(y)\,\lambda(dy) = \mathbf{P}(\eta \in A).$$

It remains to observe that $f(x|y) = f(x, y)/q(y)$ satisfies all the conditions from Definition 4.9.2 of conditional density (equality (4.9.4), which is equivalent to (4.9.3), clearly holds here). □

Theorem 4.9.2 gives a precise meaning to (4.9.2) when ξ and η have densities.

Example 4.9.1 Let ξ_1 and ξ_2 be independent random variables, $\xi_1 \Subset \mathbf{\Pi}_{\lambda_1}$, $\xi_2 \Subset \mathbf{\Pi}_{\lambda_2}$. What is the distribution of ξ_1 given $\xi_1 + \xi_2 = n$? We could easily compute the desired conditional probability $\mathbf{P}(\xi_1 = k|\xi_1 + \xi_2 = n)$, $k \le n$, without using Theorem 4.9.2, for $\xi_1 + \xi_2 \Subset \mathbf{\Pi}_{\lambda_1+\lambda_2}$ and the probability of the event $\{\xi_1 + \xi_2 = n\}$ is positive. Retaining this possibility for comparison, we will still make formal use of Theorem 4.9.2. Here ξ_1 and $\eta = \xi_1 + \xi_2$ have densities (equal to the corresponding probabilities) with respect to the counting measure, so that

$$f(k,n) = \mathbf{P}(\xi_1 = k, \eta = n) = \mathbf{P}(\xi_1 = k, \xi_2 = n-k) = e^{-\lambda_1-\lambda_2}\frac{\lambda_1^k\lambda_2^{n-k}}{k!(n-k)!},$$

$$q(n) = \mathbf{P}(\eta = n) = e^{-\lambda_1-\lambda_2}\frac{(\lambda_1+\lambda_2)^n}{n!}.$$

Therefore the required density (probability) is equal to

$$f(k|n) = \mathbf{P}(\xi_1 = k \,|\, \eta = n) = \frac{f(k,n)}{q(n)} = \frac{n!}{k!(n-k)!}p^k(1-p)^{n-k},$$

where $p = \lambda/(\lambda_1 + \lambda_2)$. Thus the conditional distribution of ξ_1 given the fixed sum $\xi_1 + \xi_2 = n$ is a binomial distribution. In particular, if $\xi_1, \ldots, \xi_r$ are independent, $\xi_i \in \mathbf{\Pi}_\lambda$, then the conditional distribution of ξ_1 given the fixed sum $\xi_1 + \cdots + \xi_r = n$ will be $\mathbf{B}^n_{1/r}$, which does not depend on λ.

The next example answers the same question as in Example 4.9.1 but for normally distributed random variables.

Example 4.9.2 Let $\mathbf{\Phi}_{a,\sigma^2}$ be the two-dimensional joint normal distribution of random variables ξ_1 and ξ_2, where $a = (a_1, a_2)$, $a_i = \mathbf{E}\xi_i$, and $\sigma^2 = \|\sigma_{i,j}\|$ is the covariance matrix, $\sigma_{ij} = \mathbf{E}(\xi_i - a_i)(\xi_j - a_j)$, $i, j = 1, 2$. The determinant of σ^2 is

$$|\sigma^2| = \sigma_{11}\sigma_{22} - \sigma_{12}^2 = \sigma_{11}\sigma_{22}(1-\rho^2),$$

where ρ is the correlation coefficient of ξ_1 and ξ_2. Thus, if $|\rho| \neq 1$ then the covariance matrix is non-degenerate and has the inverse

$$A = (\sigma^2)^{-1} = \frac{1}{|\sigma^2|}\begin{Vmatrix} \sigma_{22} & -\sigma_{12} \\ -\sigma_{12} & \sigma_{11} \end{Vmatrix} = \frac{1}{1-\rho^2}\begin{Vmatrix} \frac{1}{\sigma_{11}} & -\frac{\rho}{\sqrt{\sigma_{11}\sigma_{22}}} \\ -\frac{\rho}{\sqrt{\sigma_{11}\sigma_{12}}} & \frac{1}{\sigma_{22}} \end{Vmatrix}.$$

Therefore the joint density of ξ_1 and ξ_2 (with respect to Lebesgue measure) is (see Sect. 3.3)

$$f(x,y) = \frac{1}{2\pi\sqrt{\sigma_{11}\sigma_{22}(1-\rho^2)}} \times \exp\left\{-\frac{1}{2(1-\rho^2)}\left[\frac{(x-a_1)^2}{\sigma_{11}} - \frac{2\rho(x-a_1)(y-a_2)}{\sqrt{\sigma_{11}\sigma_{22}}} + \frac{(y-a_2)^2}{\sigma_{22}}\right]\right\}. \tag{4.9.5}$$

The one-dimensional densities of ξ_1 and ξ_2 are, respectively,

$$f(x) = \frac{1}{\sqrt{2\pi\sigma_{11}}}e^{-(x-a_1)^2/(2\sigma_{11})}, \qquad q(y) = \frac{1}{\sqrt{2\pi\sigma_{22}}}e^{-(y-a_2)^2/(2\sigma_{22})}. \tag{4.9.6}$$

Hence the conditional density of ξ_1 given $\xi_2 = y$ is

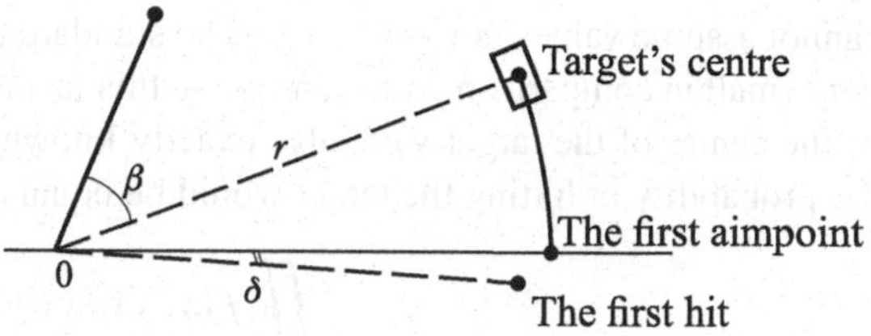

Fig. 4.2 Illustration to Example 4.9.4. Positions of the target's centre, the first aimpoint, and the first hit

$$f(x|y) = \frac{f(x,y)}{q(y)}$$
$$= \frac{1}{\sqrt{2\pi\sigma_{11}(1-\rho^2)}} \exp\left\{-\frac{1}{2\sigma_{11}(1-\rho^2)}\left(x - a_1 - \rho\sqrt{\frac{\sigma_{11}}{\sigma_{22}}}(y - a_2)\right)^2\right\},$$

which is the density of the normal distribution with mean $a_1 + \rho\sqrt{\frac{\sigma_{11}}{\sigma_{22}}}(y - a_2)$ and variance $\sigma_{11}(1-\rho^2)$.

This implies that $f(x|y)$ coincides with the unconditional density of $f(x)$ in the case $\rho = 0$ (and hence ξ_1 and ξ_2 are independent), and that the conditional expectation of ξ_1 given ξ_2 is

$$\mathbf{E}(\xi_1|\xi_2) = a_1 + \rho\sqrt{\sigma_{11}/\sigma_{22}}(\xi_2 - a_2).$$

The straight line $x = a_1 + \rho\sqrt{\sigma_{11}/\sigma_{22}}(y - a_2)$ is called the *regression line* of ξ_1 on ξ_2. It gives the best mean-square approximation for ξ_1 given $\xi_2 = y$.

Example 4.9.3 Consider the problem of computing the density of the random variable $\xi = \varphi(\zeta, \eta)$ when ζ and η are independent. It follows from formula (4.9.3) with $A = \mathbb{R}^k$ that the density of the distribution of ξ can be expressed in terms of the conditional density $f(x|y)$ as

$$f(x) = \int f(x|y)\mathbf{P}(\eta \in dy).$$

In our problem, by $f(x|y)$ one should understand the density of the random variable $\varphi(\zeta, y)$, since $\mathbf{P}(\xi \in B|\eta = y) = \mathbf{P}(\varphi(\zeta, y) \in B)$.

Example 4.9.4 Target shooting with adjustment. A gun fires at a target of a known geometric form. Introduce the polar system of coordinates, of which the origin is the position of the gun. The distance r (see Fig. 4.2) from the gun to a certain point which is assumed to be the centre of the target is precisely known to the crew of the gun, while the azimuth is not. However, there is a spotter who communicates to the crew after the first trial shoot what the azimuth deviation of the hitting point from the centre of the target is.

Suppose the scatter of the shells fired by the gun (the deviation (ξ, η) of the hitting point from the aimpoint) is described, in the polar system of coordinates, by the two-dimensional normal distribution with density (4.9.5) with $\alpha = 0$. In Sect. 8.4 we will see why the deviation is normally distributed. Here we will neglect the circumstance that the azimuth deviation ξ cannot exceed π while the distance deviation ξ

cannot assume values in $(-\infty, -r)$. (The standard deviations σ_1 and σ_2 are usually very small in comparison with r and π, so this fact is insignificant.) If the azimuth β of the centre of the target were also exactly known along with the distance r, then the probability of hitting the target would be equal to

$$\iint_{B(r,\beta)} f(x, y)\,dx\,dy,$$

where $B(r, \beta) = \{(x, y) : (r + x, \beta + y) \in B\}$ and the set B represents the target. However, the azimuth is communicated to the crew of the gun by the spotter based on the result of the trial shot, i.e. the spotter reports it with an error δ distributed according to the normal law with the density $q(y)$ (see (4.9.6)). What is the probability of the event A that, in these circumstances, the gun will hit the target from the second shot? If $\delta = z$, then the azimuth is communicated with the error z and

$$\mathbf{P}(A|\delta = z) = \iint_{B(r,\beta)} f(x, y - z)\,dx\,dy =: \varphi(z).$$

Therefore,

$$\mathbf{P}(A) = \mathbf{E}\big[\mathbf{P}(A\,|\,\delta)\big] = \mathbf{E}\varphi(\delta) = \frac{1}{\sigma_2\sqrt{2\pi}} \int_{-\infty}^{\infty} e^{-z^2/(2\sigma_2^2)} \varphi(z)\,dz.$$

Example 4.9.5 The segment $[0, 1]$ is broken "at random" (i.e. with the uniform distribution of the breaking point) into two parts. Then the larger part is also broken "at random" into two parts. What is the probability that one can form a triangle from the three fragments?

The triangle can be formed if there occurs the event B that all the three fragments have lengths smaller than $1/2$. Let ω_1 and ω_2 be the distances from the points of the first and second breaks to the origin. Use the complete probability formula

$$\mathbf{P}(B) = \mathbf{E}\mathbf{P}(B|\omega_1).$$

Since ω_1 is distributed uniformly over $[0, 1]$, one only has to calculate the conditional probability $\mathbf{P}(B|\omega_1)$. If $\omega_1 < 1/2$ then ω_2 is distributed uniformly over $[\omega_1, 1]$. One can construct a triangle provided that $1/2 < \omega_2 < 1/2 + \omega_1$. Therefore $\mathbf{P}(B|\omega_1) = \omega_1/(1 - \omega_1)$ on the set $\{\omega_1 < 1/2\}$. We easily find from symmetry that, for $\omega_1 > 1/2$,

$$\mathbf{P}(B|\omega_1) = \frac{1 - \omega_1}{\omega_1}.$$

Hence

$$\mathbf{P}(B) = 2\int_0^{1/2} \frac{x}{1 - x}\,dx = -1 + 2\int_0^{1/2} \frac{dx}{1 - x}\,dx = -1 + 2\ln 2.$$

One could also solve this problem using a direct "geometric" method. The density $f(x, y)$ of the joint distribution of (ω_1, ω_2) is

$$f(x, y) = \begin{cases} \frac{1}{1-x} & \text{if } x < 1/2,\ y \in [x, 1], \\ \frac{1}{x} & \text{if } x \geq 1/2,\ y \in [0, x], \\ 0 & \text{otherwise.} \end{cases}$$

It remains to compute the integral of this function over the domain corresponding to B.

All the above examples were on conditional expectations given *random variables* (not σ-algebras).

The need for conditional expectations given σ-algebras arises where it is difficult to manage working just with conditional expectations given random variables. Assume, for instance, that a certain process is described by a sequence of random variables $\{\xi_j\}_{j=-\infty}^{\infty}$ which are not independent. Then the most convenient way to describe the distribution of ξ_1 given the whole "history" (i.e. the values ξ_0, ξ_{-1}, $\xi_{-2}, \ldots$) is to take the conditional distribution of ξ_1 given $\sigma(\xi_0, \xi_{-1}, \ldots)$. It would be difficult to confine oneself here to conditional distributions given random variables only. Respective examples are given in Chaps. 13, 15–22.

Chapter 5
Sequences of Independent Trials with Two Outcomes

Abstract The weak and strong laws of large numbers are established for the Bernoulli scheme in Sect. 5.1. Then the local limit theorem on approximation of the binomial probabilities is proved in Sect. 5.2 using Stirling's formula (covering both the normal approximation zone and the large deviations zone). The same section also contains a refinement of that result, including a bound for the relative error of the approximation, and an extension of the local limit theorem to polynomial distributions. This is followed by the derivation of the de Moivre–Laplace theorem and its refinements in Sect. 5.3. In Sect. 5.4, the coupling method is used to prove the Poisson theorem for sums of non-identically distributed independent random indicators, together with sharp approximation error bounds for the total variation distance. The chapter ends with derivation of large deviation inequalities for the Bernoulli scheme in Sect. 5.5.

5.1 Laws of Large Numbers

Suppose we have a sequence of trials in each of which a certain event A can occur with probability p independently of the outcomes of other trials. Form a sequence of random variables as follows. Put $\xi_k = 1$ if the event A has occurred in the k-th trial, and $\xi_k = 0$ otherwise. Then $(\xi_k)_{k=1}^{\infty}$ will be a sequence of independent random variables which are identically distributed according to the Bernoulli law: $\mathbf{P}(\xi_k = 1) = p$, $\mathbf{P}(\xi_k = 0) = q = 1 - p$, $\mathbf{E}\xi_k = p$, $\operatorname{Var}(\xi_k) = pq$. The sum $S_n = \xi_1 + \cdots + \xi_n \in \mathbf{B}_p^n$ is simply the number of occurrences of the event A in the first n trials. Clearly $\mathbf{E}S_n = np$ and $\operatorname{Var}(S_n) = npq$.

The following assertion is called the *law of large numbers* for the Bernoulli scheme.

Theorem 5.1.1 *For any* $\varepsilon > 0$

$$\mathbf{P}\left(\left|\frac{S_n}{n} - p\right| > \varepsilon\right) \to 0 \quad \text{as } n \to \infty.$$

This assertion is a direct consequence of Theorem 4.7.5. One can also obtain the following stronger result:

A.A. Borovkov, *Probability Theory*, Universitext,
DOI 10.1007/978-1-4471-5201-9_5, © Springer-Verlag London 2013

Theorem 5.1.2 (The Strong Law of Large Numbers for the Bernoulli Scheme) *For any $\varepsilon > 0$, as $n \to \infty$,*

$$\mathbf{P}\left(\sup_{k\ge n}\left|\frac{S_k}{k} - p\right| > \varepsilon\right) \to 0.$$

The interpretation of this result is that the notion of probability which we introduced in Chaps. 1 and 2 corresponds to the intuitive interpretation of probability as the limiting value of the relative frequency of the occurrence of the event. Indeed, S_n/n could be considered as the relative frequency of the event A for which $\mathbf{P}(A) = p$. It turned out that, in a certain sense, S_n/n converges to p.

Proof of Theorem 5.1.2 One has

$$\begin{aligned}\mathbf{P}\left(\sup_{k\ge n}\left|\frac{S_k}{k} - p\right| > \varepsilon\right) &= \mathbf{P}\left(\bigcup_{k=n}^{\infty}\left\{\left|\frac{S_k}{k} - p\right| > \varepsilon\right\}\right) \\ &\le \sum_{k=n}^{\infty}\mathbf{P}\left(\left|\frac{S_k}{k} - p\right| > \varepsilon\right) \le \sum_{k=n}^{\infty}\frac{\mathbf{E}(S_k - kp)^4}{k^4\varepsilon^4}.\end{aligned} \tag{5.1.1}$$

Here we again made use of Chebyshev's inequality but this time for the fourth moments. Expanding we find that

$$\begin{aligned}\mathbf{E}(S_k - kp)^4 &= \mathbf{E}\left(\sum_{j=1}^{k}(\xi_j - p)\right)^4 = \sum_{j=1}^{k}\mathbf{E}(\xi_j - p)^4 + 6\sum_{i<j}(\xi_i - p)^2(\xi_j - p)^2 \\ &= k\left(pq^4 + qp^4\right) + 3k(k-1)(pq)^2 \le k + k(k-1) = k^2.\end{aligned} \tag{5.1.2}$$

Thus the probability we want to estimate does not exceed the sum

$$\varepsilon^{-4}\sum_{k=n}^{\infty}k^{-2} \to 0 \quad \text{as } n \to \infty.$$

□

It is not hard to see that we would not have found the required bound if we used Chebyshev's inequality with second moments in (5.1.1).

We could also note that one actually has much stronger bounds for $\mathbf{P}(|S_k - kp| > \varepsilon k)$ than those that we made use of above. These will be derived in Sect. 5.5.

Corollary 5.1.1 *If $f(x)$ is a continuous function on $[0, 1]$ then, as $n \to \infty$,*

$$\mathbf{E}f\left(\frac{S_n}{n}\right) \to f(p) \tag{5.1.3}$$

uniformly in p.

Proof For any $\varepsilon > 0$,

$$\begin{aligned}\mathbf{E}\left|f\left(\frac{S_n}{n}\right) - f(p)\right| &\le \mathbf{E}\left(\left|f\left(\frac{S_n}{n}\right) - f(p)\right|; \left|\frac{S_n}{n} - p\right| \le \varepsilon\right) \\ &\quad + \mathbf{E}\left(\left|f\left(\frac{S_n}{n}\right) - f(p)\right|; \left|\frac{S_n}{n} - p\right| > \varepsilon\right) \\ &\le \sup_{|x|\le\varepsilon} \left|f(p+x) - f(p)\right| + \delta_n(\varepsilon),\end{aligned}$$

where the quantity $\delta(\varepsilon)$ is independent of p by virtue of (5.1.1), (5.1.2), and since $\delta_n(\varepsilon) \to 0$ as $n \to \infty$. □

Corollary 5.1.2 *If $f(x)$ is continuous on $[0, 1]$, then, as $n \to \infty$,*

$$\sum_{k=0}^{n} f\left(\frac{k}{n}\right)\binom{n}{k} x^k (1-x)^{n-k} \to f(x)$$

uniformly in $x \in [0, 1]$.

This relation is just a different form of (5.1.3) since

$$\mathbf{P}(S_n = k) = \binom{n}{k} p^k (1-p)^{n-k}$$

(see Chap. 1). This relation implies the well-known *Weierstrass theorem* on approximation of continuous functions by polynomials. Moreover, the required polynomials are given here explicitly—they are *Bernstein polynomials*.

5.2 The Local Limit Theorem and Its Refinements

5.2.1 The Local Limit Theorem

We know that $\mathbf{P}(S_n = k) = \binom{n}{k} p^k q^{n-k}$, $q = 1 - p$. However, this formula becomes very inconvenient for computations with large n and k, which raises the question about the asymptotic behaviour of the probability $\mathbf{P}(S_n = k)$ as $n \to \infty$.

In the sequel, we will write $a_n \sim b_n$ for two number sequences $\{a_n\}$ and $\{b_n\}$ if $a_n/b_n \to 1$ as $n \to \infty$. Such sequences $\{a_n\}$ and $\{b_n\}$ will be said to be *equivalent*.

Set

$$H(x) = x \ln\frac{x}{p} + (1-x)\ln\frac{1-x}{1-p}, \qquad p^* = \frac{k}{n}. \tag{5.2.1}$$

Theorem 5.2.1 *As $k \to \infty$ and $n - k \to \infty$,*

$$\mathbf{P}(S_n = k) = \mathbf{P}\left(\frac{S_n}{n} = p^*\right) \sim \frac{1}{\sqrt{2\pi n p^*(1-p^*)}} \exp\bigl\{-nH\bigl(p^*\bigr)\bigr\}. \tag{5.2.2}$$

Proof We will make use of Stirling's formula according to which $n! \sim \sqrt{2\pi n} n^n e^{-n}$ as $n \to \infty$. One has

$$\begin{aligned}\mathbf{P}(S_n = k) &= \binom{n}{k} p^k q^{n-k} \sim \sqrt{\frac{n}{2\pi k(n-k)}} \frac{n^n}{k^k (n-k)^{n-k}} p^k (1-p)^{n-k} \\ &= \frac{1}{\sqrt{2\pi n p^*(1-p^*)}} \\ &\quad \times \exp\left\{ -k \ln \frac{k}{n} - (n-k) \ln \frac{n-k}{n} + k \ln p + (n-k) \ln(1-p) \right\} \\ &= \frac{1}{\sqrt{2\pi n p^*(1-p^*)}} \exp\left\{ -n\left[p^* \ln p^* + (1-p^*) \ln(1-p^*) \right.\right. \\ &\quad \left.\left. - p^* \ln p - (1-p^*) \ln(1-p) \right] \right\} \\ &= \frac{1}{\sqrt{2\pi n p^*(1-p^*)}} \exp\left\{ nH(p^*) \right\}.\end{aligned}$$

□

If $p^* = k/n$ is close to p, then one can find another form for the right-hand side of (5.2.2) which is of significant interest. Note that the function $H(x)$ is analytic on the interval (0, 1). Since

$$H'(x) = \ln \frac{x}{p} - \ln \frac{1-x}{1-p}, \qquad H''(x) = \frac{1}{p} + \frac{1}{1-x}, \tag{5.2.3}$$

one has $H(p) = H'(p) = 0$ and, as $p^* - p \to 0$,[1]

$$H(p^*) = \frac{1}{2}\left(\frac{1}{p} + \frac{1}{q} \right) (p^* - p)^2 + O\left(|p^* - p|^3 \right).$$

Therefore if $p^* \sim p$ and $n(p^* - p)^3 \to 0$ then

$$\mathbf{P}(S_n = k) \sim \frac{1}{\sqrt{2\pi pq}} \exp\left\{ -\frac{n}{2pq} (p^* - p)^2 \right\}.$$

Putting

$$\Delta = \frac{1}{\sqrt{npq}}, \qquad \varphi(x) = \frac{1}{\sqrt{2\pi}} e^{-x^2/2},$$

one obtains the following assertion.

Corollary 5.2.1 *If $z = n(p^* - p) = k - np = o(n^{2/3})$ then*

$$\mathbf{P}(S_n = k) = \mathbf{P}(S_n - np = z) \sim \varphi(z\Delta)\Delta, \tag{5.2.4}$$

where $\varphi = \varphi_{0,1}(x)$ is evidently the density of the normal distribution with parameters (0, 1).

---

[1] According to standard conventions, we will write $a(z) = o(b(z))$ as $z \to z_0$ if $b(z) > 0$ and $\lim_{z \to z_0} \frac{a(z)}{b(z)} = 0$, and $a(z) = O(b(z))$ as $z \to z_0$ if $b(z) > 0$ and $\limsup_{z \to z_0} \frac{|a(z)|}{b(z)} < \infty$.

This formula also enables one to estimate the probabilities of the events of the form $\{S_n < k\}$.

If p^* differs substantially from p, then one could estimate the probabilities of such events using the results of Sect. 1.3.

Example 5.2.1 In a jury consisting of an odd number $n = 2m + 1$ of persons, each member makes a correct decision with probability $p = 0.7$ independently of the other members. What is the minimum number of members for which the verdict rendered by the majority of jury members will be correct with a probability of at least 0.99?

Put $\xi_k = 1$ if the k-th jury member made a correct decision and $\xi_k = 0$ otherwise. We are looking for odd numbers n for which $\mathbf{P}(S_n \le m) \le 0.01$. It is evident that such a trustworthy decision can be achieved only for large values of n. In that case, as we established in Sect. 1.3, the probability $\mathbf{P}(S_n \le m)$ is approximately equal to

$$\frac{(n+1-m)p}{(n+1)p-m}\mathbf{P}(S_n = m) \approx \frac{p}{2p-1}\mathbf{P}(S_n = m).$$

Using Theorem 5.2.1 and the fact that in our problem

$$p^* \approx \frac{1}{2}, \qquad H\left(\frac{1}{2}\right) = -\frac{1}{2}\ln 4p(1-p), \qquad H'\left(\frac{1}{2}\right) = \ln\left(\frac{1-p}{p}\right),$$

we get

$$\begin{aligned}
\mathbf{P}(S_n \le m) &\approx \frac{p}{2p-1}\sqrt{\frac{2}{\pi n}}\exp\left\{-nH\left(\frac{1}{2}-\frac{1}{2n}\right)\right\} \\
&\approx \frac{p}{2p-1}\sqrt{\frac{2}{\pi n}}\exp\left\{-nH\left(\frac{1}{2}\right)+\frac{1}{2}H'\left(\frac{1}{2}\right)\right\} \\
&\approx \frac{\sqrt{2\pi(1-p)}}{(2p-1)\sqrt{\pi n}}\left(\sqrt{4p(1-p)}\right)^n \approx 0.915\frac{1}{\sqrt{n}}(0.84)^{n/2}.
\end{aligned}$$

On the right-hand side there is a monotonically decreasing function $a(n)$. Solving the equation $a(n) = 0.01$ we get the answer $n = 33$. The same result will be obtained if one makes use of the explicit formulae.

5.2.2 Refinements of the Local Theorem

It is not hard to bound the error of approximation (5.2.2). If, in Stirling's formula $n! = \sqrt{2\pi n}n^n e^{-n+\theta(n)}$, we make use of the well-known inequalities[2]

$$\frac{1}{12n+1} < \theta(n) < \frac{1}{12n},$$

then the same argument will give the following refinement of Theorem 5.2.1.

[2] See, e.g., [12], Sect. 2.9.

Theorem 5.2.2

$$\mathbf{P}(S_n = k) = \frac{1}{\sqrt{2\pi np^*(1-p^*)}} \exp\bigl\{nH\bigl(p^*\bigr) + \theta(k,n)\bigr\}, \tag{5.2.5}$$

where

$$\bigl|\theta(k,n)\bigr| = \bigl|\theta(n) - \theta(k)\theta(n-k)\bigr| < \frac{1}{12k} + \frac{1}{12(n-k)} = \frac{1}{12np^*(1-p^*)}. \tag{5.2.6}$$

Relation (5.2.4) could also be refined as follows.

Theorem 5.2.3 *For all k such that $|p^* - p| \le \frac{1}{2}\min(p,q)$ one has*

$$\mathbf{P}(S_n = k) = \varphi(z\Delta)\Delta\bigl(1 + \varepsilon(k,n)\bigr),$$

where

$$1 + \varepsilon(k,n) = \exp\left\{\vartheta\left(\frac{|z|^3\Delta^4}{3} + \left(|z| + \frac{1}{6}\right)\Delta^2\right)\right\}, \quad |\vartheta| < 1.$$

As one can easily see from the properties of the Taylor expansion of the function e^x, the order of magnitude of the term $\varepsilon(k,n)$ in the above formulae coincides with that of the argument of the exponential. Hence it follows from Theorem 5.2.3 that for $z = k - np = o(\Delta^{-4/3})$ or, which is the same, $z = o(n^{2/3})$, one still has (5.2.4).

Proof We will make use of Theorem 5.2.2. In addition to formulae (5.2.3) one can write:

$$H^{(k)} = \frac{(-1)^k(k-2)!}{x^{k-1}} + \frac{(k-2)!}{(1-x)^{k-1}}, \quad k \ge 2,$$

$$H\bigl(p^*\bigr) = \frac{1}{2pq}\bigl(p^* - p\bigr)^2 + R_1,$$

where we can estimate the residual $R_1 = \sum_{k=3}^{\infty} \frac{H^{(k)}(p)}{k!}(p^* - p)$. Taking into account that

$$\bigl|H^{(k)}(p)\bigr| \le (k-2)!\left(\frac{1}{p^{k-1}} + \frac{1}{q^{k-1}}\right), \quad k \ge 2,$$

and letting for brevity $|p^* - p| = \delta$, we get for $\delta \le \frac{1}{2}\min(p,q)$ the bounds

$$|R_1| \le \sum_{k=3}^{\infty} \frac{(k-2)!}{k!}\left(\frac{1}{p^{k-1}} + \frac{1}{q^{k-1}}\right) \le \frac{\delta^3}{6}\left(\frac{1}{p^2}\frac{1}{1-\frac{\delta}{p}} + \frac{1}{q^2}\frac{1}{1-\frac{\delta}{q}}\right)$$
$$\le \frac{\delta}{6}\left(\frac{2}{p^2} + \frac{2}{q^2}\right) < \frac{\delta^3}{3(pq)^2}.$$

From this it follows that

$$-nH(p^*) = -\frac{(k-np)^2}{2npq} + \frac{\vartheta_1|k-np|^3}{3(npq)^2} = -\frac{z^2\Delta^2}{2} + \frac{\vartheta_1|z|^3\Delta^4}{3}, \quad |\vartheta_1| < 1. \tag{5.2.7}$$

We now turn to the other factors in equality (5.2.5) and consider the product $p^*(1-p^*)$. Since $-p \le 1-p-p^* \le 1-p$, we have

$$|p^*(1-p^*) - p(1-p)| = |(p-p^*)(1-p-p^*)| \le |p^*-p|\max(p,q).$$

This implies in particular that, for $|p^*-p| < \frac{1}{2}\min(p,q)$, one has

$$|p^*(1-p^*) - pq| < \frac{1}{2}pq, \qquad p^*(1-p^*) > \frac{1}{2}pq.$$

Therefore one can write along with (5.2.6) that, for the values of k indicated in Theorem 5.2.3,

$$|\theta(k,n)| < \frac{1}{6npq} = \frac{\Delta^2}{6}. \tag{5.2.8}$$

It remains to consider the factor $[p^*(1-p^*)]^{-1/2}$. Since for $|\gamma| < 1/2$

$$|\ln(1+\gamma)| = \left|\int_1^{1+\gamma} \frac{1}{x}dx\right| < 2|\gamma|,$$

one has for $\delta = |p^*-p| < (1/2)\min(p,q)$ the relations

$$\begin{aligned} \ln(p^*(1-p^*)) &= \ln pq + \ln\left(1 + \frac{p^*(1-p^*)-pq}{pq}\right) \\ &= \ln(pq) + \ln\left(1 - \frac{\vartheta^*\delta}{pq}\right), \quad |\vartheta^*| < \max(p,q); \\ \ln\left(1 - \frac{\vartheta^*\delta}{pq}\right) &= -\frac{2\vartheta_2\delta}{pq}, \quad |\vartheta_2| < \max(p,q), \\ [p^*(1-p^*)]^{-1/2} &= [pq]^{-1/2}\exp\left\{\frac{\vartheta_2\delta}{pq}\right\}. \end{aligned} \tag{5.2.9}$$

Using representations (5.2.7)–(5.2.9) and the assertion of Theorem 5.2.2 completes the proof. □

One can see from the above estimates that the bounds for ϑ in the statement of Theorem 5.2.3 can be narrowed if we consider smaller deviations $|p^*-p|$—if they, say, do not exceed the value $\alpha\min(p,q)$ where $\alpha < 1/2$.

The relations for $\mathbf{P}(S_n = k)$ that we found are the so-called *local limit theorems for the Bernoulli scheme* and their refinements.

5.2.3 The Local Limit Theorem for the Polynomial Distributions

The basic asymptotic formula given in Theorem 5.2.1 admits a natural extension to the polynomial distribution $\mathbf{B}_p^n$, $p = (p_1, \dots, p_r)$, when, in a sequence of independent trials, in each of the trials one has not two but $r \ge 2$ possible outcomes $A_1, \dots, A_r$ of which the probabilities are equal to $p_1, \dots, p_r$, respectively. Let $S_n^{(j)}$ be the number of occurrences of the event A_j in n trials,

$$S_n = \left(S_n^{(1)}, \dots, S_n^{(r)}\right), \qquad k = (k_1, \dots, k_r), \qquad p^* = \frac{k}{n},$$

and put $H(x) = \sum x_i \ln(x_i/p_i)$, $x = (x_1, \dots, x_r)$. Clearly, $S_n \in \mathbf{B}_p^n$. The following assertion is a direct extension of Theorem 5.2.1.

Theorem 5.2.4 *If each of the r variables $k_1, \dots, k_r$ is either zero or tends to ∞ as $n \to \infty$ then*

$$\mathbf{P}(S_n = k) \sim (2\pi n)^{(1-r_0)/2} \Bigg(\prod_{\substack{j=1 \\ p_j^* \ne 0}}^{r} p_j^* \Bigg)^{-1/2} \exp\bigl\{-nH\bigl(p^*\bigr)\bigr\},$$

where r_0 is the number of variables $k_1, \dots, k_r$ which are not equal to zero.

Proof As in the proof of Theorem 5.2.1, we will use Stirling's formula

$$n! \sim \sqrt{2\pi n} e^{-n} n^n$$

as $n \to \infty$. Assuming without loss of generality that all $k_j \to \infty$, $j = 1, \dots, r$, we get

$$\begin{aligned} \mathbf{P}(S_n = k) &\sim (2\pi)^{(1-r)/2} \left(\frac{n}{\prod_{j=1}^r k_j} \right)^{1/2} \prod_{j=1}^r \left(\frac{np_j}{k_j} \right)^{k_j} \\ &= (2\pi n)^{(1-r)/2} \Bigg(\prod_{j=1}^r p_j^* \Bigg)^{-1/2} \exp\Bigg\{ n \sum_{j=1}^r \frac{k_j}{n} \ln \frac{p_j n}{k_j} \Bigg\}. \end{aligned}$$ □

5.3 The de Moivre–Laplace Theorem and Its Refinements

Let a and b be two fixed numbers and $\zeta_n = (S_n - np)/\sqrt{npq}$. Then

$$\mathbf{P}(a < \zeta_n < b) = \sum_{a\sqrt{npq} < z < b\sqrt{npq}} \mathbf{P}(S_n - np = z).$$

If, instead of $\mathbf{P}(S_n - np = z)$, we substitute here the values $\varphi(z\Delta)\Delta$ (see Corollary 5.2.1), we will get an integral sum $\sum_{a<z\Delta<b} \varphi(z\Delta)\Delta$ corresponding to the integral $\int_a^b \varphi(x)\,dx$.

Thus relations (5.2.4) make the equality

$$\lim_{n\to\infty} \mathbf{P}(a < \zeta_n < b) = \int_a^b \varphi(x)\,dx = \Phi(b) - \Phi(a) \qquad (5.3.1)$$

plausible, where $\Phi(x)$ is the normal distribution function with parameters (0, 1):

$$\Phi(x) = \frac{1}{\sqrt{2\pi}} \int_{-\infty}^{x} e^{-t^2/2}\,dt.$$

This is the *de Moivre–Laplace theorem*, which is one of the so-called *integral limit theorems* that describe probabilities of the form $\mathbf{P}(S_n < x)$. In Chap. 8 we will derive more general integral theorems from which (5.3.1) will follow as a special case.

Theorem 5.2.3 makes it possible to obtain (5.3.1) together with an error bound or, in other words, *with a bound for the convergence rate.*

Let A and B be integers,

$$a = \frac{A - np}{\sqrt{npq}}, \qquad b = \frac{B - np}{\sqrt{npq}}. \qquad (5.3.2)$$

Theorem 5.3.1 *Let* $b > a$, $c = \max(|a|, |b|)$, *and*

$$\rho = \frac{c^3 + 3c}{3}\Delta + \frac{\Delta^2}{6}.$$

If $\Delta = 1/\sqrt{npq} \le 1/2$ *and* $\rho \le 1/2$ *then*

$$\mathbf{P}(A \le S_n < B) = \mathbf{P}(a \le \zeta_n < b) = \int_a^b \varphi(t)\,dt(1 + \vartheta_1 \Delta c)(1 + 2\vartheta_2 \rho), \qquad (5.3.3)$$

where $|\vartheta_i| \le 1$, $i = 1, 2$.

This theorem shows that the left-hand side in (5.3.3) can be equivalent to $\Phi(b) - \Phi(a)$ for growing a and b as well. In that case, $\Phi(b) - \Phi(a)$ can converge to 0, and knowing the *relative* error in (5.3.1) is more convenient since its smallness enables one to establish that of the absolute error as well, but not vice versa.

Proof First we note that, for all k such that $|z| = |k - np| < c\sqrt{npq}$, the conditions of Theorem 5.2.3 will hold. Indeed, to have the inequality $|p^* - p| < (1/2)\min(p, q)$ it suffices that $|k - np| < npq/2 = 1/(2\Delta^2)$. This inequality will hold if $c < 1/(2\Delta)$. But since $\rho \le 1/2$, one has

$$\frac{c(c^2 + 3)\Delta}{3} < 1/2, \qquad c\Delta < 1/2.$$

Thus, for each k such that $a\sqrt{npq} \le z < b\sqrt{npq}$, we can make use of Theorem 5.2.3 to conclude that

$$\mathbf{P}(A \le S_n < B) = \sum_{a\sqrt{npq} \le z < b\sqrt{npq}} \mathbf{P}(S_n = k)$$
$$= \sum_{a \le z\Delta < b} \varphi(z\Delta)\Delta\left[1 + \left(\exp\left\{\vartheta\left(\frac{|z|^3\Delta^4}{3} + \left(|z| + \frac{1}{6}\right)\Delta^2\right)\right\} - 1\right)\right], \tag{5.3.4}$$

where $|\vartheta| < 1$. Since, for $\rho \le 1$,

$$\left|\frac{e^p - 1}{\rho}\right| < e - 1 < 2,$$

the absolute value of the correction term in (5.3.4) does not exceed (substituting there $z\Delta = c$)

$$\left|\exp\left\{\vartheta\left(\frac{c^3\Delta}{3} + c\Delta + \frac{\Delta^2}{6}\right)\right\} - 1\right| \le 2\vartheta\left(\frac{c^3\Delta}{3} + c\Delta + \frac{\Delta^2}{6}\right) = 2\vartheta p.$$

Therefore

$$\mathbf{P}(A \le S_n < B) = \sum_{a \le z\Delta < b} \varphi(z\Delta)\Delta[1 + 2\vartheta_1\rho], \tag{5.3.5}$$

where $|\vartheta_1| < 1$.

Now we transform the sum on the right-hand side of the last equality. To this end, note that, for any smooth function $\varphi(x)$,

$$\left|\Delta\varphi(x) - \int_x^{x+\Delta} \varphi(t)\,dt\right| = \frac{\Delta^2}{2} \max_{x \le t \le x+\Delta} |\varphi'(t)|. \tag{5.3.6}$$

But for the function $\varphi(x) = (2\pi)^{-1/2}e^{-x^2/2}$ one has $\varphi'(x) = -x\varphi(x)$ and the maximum value of $\varphi(t)$ on the segment $[x, x + \Delta]$, $|x| \le c$, differs from the minimum value by not more than the factor $\exp\{c\Delta + \Delta^2/2\}$. Therefore, for $|x| \le c$, one has by virtue of (5.3.6)

$$\left|\Delta\varphi(x) - \int_x^{x+\Delta} \varphi(t)\,dt\right|$$
$$\le \frac{\Delta^2 c}{2} e^{c\Delta + \Delta^2/2} \min_{x \le t \le x+\Delta} \varphi(t) \le \frac{\Delta c}{2} e^{c\Delta + \Delta^2/2} \int_x^{x+\Delta} \varphi(t)\,dt.$$

Since $c\Delta + \Delta^2/2 < 1/2 + 1/8$, $e^{c\Delta + \Delta^2/2} \le 2$, we have the representation

$$\Delta\varphi(x) = \int_x^{x+\Delta} \varphi(t)\,dt\,(1 + \vartheta_1\Delta c), \quad |\vartheta_1| < 1.$$

Substituting this into (5.3.5) we obtain the assertion of the theorem. □

Thus by Theorem 5.3.1 the difference

$$\left|\mathbf{P}(x \le \zeta_n < y) - \big(\Phi(y) - \Phi(x)\big)\right| \tag{5.3.7}$$

can be effectively, yet rather roughly, bounded from above by a quantity of the order $1/\sqrt{npq}$ if $x = a$, $y = b$ (assuming that a and b are values which can be represented in the form $(k - np)\Delta$, see (5.3.2)). If x and y do not belong to the mentioned lattice with the span Δ then the error (5.3.7) will still be of the same order since, for instance, when y varies, $\mathbf{P}(x \le \zeta_n < y)$ remains constant on the semi-intervals of the form $(a + k\Delta, a + (k + 1)\Delta]$, while the function $\Phi(y) - \Phi(x)$ increases monotonically with a bounded derivative. A similar argument holds for the left end point x. It is important to note that the error order $1/\sqrt{npq}$ *cannot be improved*, for the jumps of the distribution function of ζ_n are just of this order of magnitude by Theorem 5.2.2.

Theorem 5.3.1 enables one to use the normal approximation for $\mathbf{P}(x \le \zeta_n < y)$ in the so-called *large deviations range* as well, when both x and y grow in absolute value and are of the same sign. In that case, both $\Phi(y) - \Phi(x)$ and the probability to be approximated tend to zero. Therefore the approximation can be considered satisfactory *only if*

$$\frac{\mathbf{P}(x \le \zeta_n < y)}{(\Phi(y) - \Phi(x))} \to 1. \tag{5.3.8}$$

As Theorem 5.3.1 shows, this convergence will take place if

$$c = \max\bigl(|x|, |y|\bigr) = o\bigl(\Delta^{-1/3}\bigr)$$

or, which is the same, $c = o(n^{1/6})$. For more details about large deviation probabilities, see Chap. 9.

For larger values of c, as one could verify using Theorem 5.2.1, relation (5.3.8) will, generally speaking, not hold.

In conclusion we note that since

$$\mathbf{P}\bigl(|\zeta_n| > b\bigr) \to 0$$

as $b \to \infty$, it follows immediately from Theorem 5.3.1 that, for any fixed y,

$$\lim_{n\to\infty} \mathbf{P}(\zeta_n < y) = \Phi(y).$$

Later we will show that this assertion remains true under much wider assumptions, when ζ_n is a scaled sum of arbitrary distributed random variables having finite variances.

5.4 The Poisson Theorem and Its Refinements

5.4.1 *Quantifying the Closeness of Poisson Distributions to Those of the Sums S_n*

As we saw from the bounds in the last section, the de Moivre–Laplace theorem gives a good approximation to the probabilities of interest if the number npq (the variance of S_n) is large. This number will grow together with n if p and q are fixed positive numbers. But what will happen in a problem where, say, $p = 0.001$ and $n = 1000$ so that $np = 1$? Although n is large here, applying the de Moivre–Laplace theorem in such a problem would be meaningless. It turns out that in this case the distribution $\mathbf{P}(S_n = k)$ can be well approximated by the Poisson distribution $\mathbf{\Pi}_\mu$ with an appropriate parameter value μ (see Sect. 5.4.2). Recall that

$$\mathbf{\Pi}_\mu(B) = \sum_{0 \le k \in B} e^{-\mu} \frac{\mu^k}{k!}.$$

Put $np = \mu$.

Theorem 5.4.1 *For all sets B,*

$$\left|\mathbf{P}(S_n \in B) - \mathbf{\Pi}_\mu(B)\right| \le \frac{\mu^2}{n}.$$

We could prove this assertion in the same way as the local theorem, making use of the explicit formula for $\mathbf{P}(S_n = k)$. However, we can prove it in a simpler and nicer way which could be called the *common probability space method*, or *coupling method*. The method is often used in research in probability theory and consists, in our case, of constructing on a common probability space random variables S_n and S_n^*, the latter being as close to S_n as possible and distributed according to the Poisson distribution.

It is also important that the common probability space method admits, without any complications, extension to the case of *non-identically* distributed random variables, when the probability of getting 1 in a particular trial depends on the number of the trial. Thus we will now prove a more general assertion of which Theorem 5.4.1 is a special case.

Assume that we are given a sequence of independent random variables $\xi_1, \ldots, \xi_n$, such that $\xi_j \Subset \mathbf{B}_{p_j}$. Put, as above, $S_n = \sum_{j=1}^n \xi_j$. The theorem we state below is intended for approximating the probability $\mathbf{P}(S_n = k)$ when p_j are small and the number $\mu = \sum_{j=1}^n p_j$ is "comparable" with 1.

Theorem 5.4.2 *For all sets B,*

$$\left|\mathbf{P}(S_n \in B) - \mathbf{\Pi}_\mu(B)\right| \le \sum_{j=1}^n p_j^2.$$

To prove this theorem we will need an important "stability" property of the Poisson distribution.

Lemma 5.4.1 *If η_1 and η_2 are independent, $\eta_2 \Subset \mathbf{\Pi}_{\mu_1}$ and $\eta_2 \Subset \mathbf{\Pi}_{\mu_2}$, then*[3]

$$\eta_1 + \eta_2 \Subset \mathbf{\Pi}_{\mu_1+\mu_2}.$$

[3]This fact will also easily follow from the properties of characteristic functions dealt with in Chap. 7.

Proof By the total probability formula,

$$\begin{aligned}\mathbf{P}(\eta_1+\eta_2=k) &= \sum_{j=0}^{k}\mathbf{P}(\eta_1=j)\mathbf{P}(\eta_2=k-j)\\ &= \sum_{j=0}^{k}\frac{\mu_1^j e^{-\mu_1}}{j!}\cdot\frac{\mu_2^{k-j}e^{-\mu_2}}{(k-j)!}=\frac{1}{k!}e^{-(\mu_1+\mu_2)}\sum_{j=0}^{k}\binom{k}{j}\mu_1^j\mu_2^{k-j}\\ &= \frac{(\mu_1+\mu_2)^k e^{-(\mu_1+\mu_2)}}{k!}.\end{aligned}$$

□

Proof of Theorem 5.4.2 Let $\omega_1,\dots,\omega_n$ be independent random variables, each being the identity function ($\xi(\omega_k)=\omega_k$) on the unit interval with the uniform distribution. We can assume that the vector $\omega=(\omega_1,\dots,\omega_n)$ is given as the identity function on the unit n-dimensional cube Ω with the uniform distribution.

Now construct the random variables ξ_j and ξ_j^* on Ω as follows:

$$\xi_j(\omega)=\begin{cases}0 & \text{if } \omega_j<1-p_j,\\ 1 & \text{if } \omega_j\ge 1-p_j,\end{cases}\qquad \xi_j^*(\omega)=\begin{cases}0 & \text{if } \omega_j<e^{-p_j},\\ k\ge 1 & \text{if } \omega_j\in[\pi_{k-1},\pi_k),\end{cases}$$

where $\pi_k=\sum_{m\le k}e^{-p_j}\frac{(p_j)^m}{m!}$, $k=0,1,\dots.$

It is evident that the $\xi_j(\omega)$ are independent and $\xi_j(\omega)\in\mathbf{B}_{p_j}$; $\xi_j^*(\omega)$ are also jointly independent with $\xi_j^*(\omega)\in\boldsymbol{\Pi}_{p_j}$. Now note that since $1-p_j\le e^{-p_j}$ one has $\xi_j(\omega)\ne\xi_j^*(\omega)$ only if $\omega_j\in[1-p_j,e^{-p_j})$ or $\omega_j\in[e^{-p_j}+p_je^{-p_j},1]$. Hence

$$\mathbf{P}(\xi_j\ne\xi_j^j)=(e^{-p_j}-1+p_j)+(1-e^{-p_j}-p_je^{-p_j})=p_j(1-e^{-p_j})\le p_j^2$$

and

$$\mathbf{P}(S_n\ne S_n^*)\le\mathbf{P}\Big(\bigcup_j|\xi_j\ne\xi_j^*|\Big)\le\sum p_j^2,$$

where $S_n^*=\sum_{j=1}^n\xi_j^*\in\boldsymbol{\Pi}_\mu$.

Now we can write

$$\begin{aligned}\mathbf{P}(S_n\in B) &= \mathbf{P}(S_n\in B,\,S_n=S_n^*)+\mathbf{P}(S_n\in B,\,S_n\ne S_n^*)\\ &= \mathbf{P}(S_n^*\in B)-\mathbf{P}(S_n^*\in B,\,S_n\ne S_n^*)+\mathbf{P}(S_n\in B,\,S_n\ne S_n^*),\end{aligned}$$

so that

$$\begin{aligned}&\big|\mathbf{P}(S_n\in B)-\mathbf{P}(S_n^*\in B)\big|\\ &\quad\le\big|\mathbf{P}(S_n^*\in B,\,S_n\ne S_n^*)-\mathbf{P}(S_n\in B,\,S_n\ne S_n^*)\big|\le\mathbf{P}(S_n\ne S_n^*).\end{aligned}\tag{5.4.1}$$

The assertion of the theorem follows from this in an obvious way. □

Remark 5.4.1 One can give other common probability space constructions as well. One of them will be used now to show that there exists a better Poisson approximation to the distribution of S_n.

Namely, let $\xi_j^*(\omega)$ be independent random variables distributed according to the Poisson laws with parameters $r_j = -\ln(1-p_j) \ge p_j$, so that $\mathbf{P}(\xi_j^* = 0) = e^{-r_j} = 1 - p_j$. Then $\xi_j(\omega) = \min\{1, \xi_j^*(\omega)\} \in \mathbf{B}_{p_j}$ and, moreover,

$$\mathbf{P}\Biggl(\bigcup_{j=1}^{n}\{\xi_j(\omega) \ne \xi_j^*(\omega)\}\Biggr) \le \sum_{j=1}^{n}\mathbf{P}(\xi_j^*(\omega) \ge 2) = \sum_{j=1}^{n}\bigl(1 - e^{-r_j} - r_j e^{-r_j}\bigr).$$

But for $r = -\ln(1-p)$ one has the inequality

$$1 - e^{-r} - re^{-r} = p + (1-p)\ln(1-p) \le p + (1-p)\biggl(-p - \frac{p^2}{2}\biggr)$$
$$= \frac{p^2}{2}(1+p).$$

Hence for the new Poisson approximation we have

$$\mathbf{P}\bigl(S_n^* \ne S_n\bigr) \le \frac{1}{2}\sum_{j=1}^{n} p_j^2(1+p_j).$$

Putting $\lambda = -\sum_{j=1}^{n}\ln(1-p_j) \ge \sum_{j=1}^{n} p_j$, the same argument as above will lead to the bound

$$\sup_B\bigl|\mathbf{P}(S_n \in B) - \mathbf{\Pi}_\lambda(B)\bigr| \le \frac{1}{2}\sum_{j=1}^{n} p_j^2(1+p_j).$$

This bound of the rate of approximation given by the Poisson distribution with a "slightly shifted" parameter is better than that obtained in Theorem 5.4.2. Moreover, one could note that, in the new construction, $\xi_j \le \xi_j^*$, $S_n \le S_n^*$, and consequently

$$\mathbf{P}(S_n \ge k) \le \mathbf{P}\bigl(S_n^* \ge k\bigr) = \mathbf{\Pi}_\lambda\bigl([k,\infty)\bigr).$$

5.4.2 The Triangular Array Scheme. The Poisson Theorem

Now we will return back to the case of identically distributed ξ_k. To obtain from Theorem 5.4.2 a limit theorem of the type similar to that of the de Moivre–Laplace theorem (see (5.3.1)), one needs a somewhat different setup. In fact, to ensure that np remains bounded as n increases, $p = \mathbf{P}(\xi_k = 1)$ needs to converge to zero which cannot be the case when we consider a fixed sequence of random variables $\xi_1, \xi_2, \ldots$.

We introduce a sequence of rows (of growing length) of random variables:

$$\begin{array}{lllll}
\xi_1^{(1)}; & & & & \\
\xi_1^{(2)}, & \xi_2^{(2)}; & & & \\
\xi_1^{(3)}, & \xi_2^{(3)}, & \xi_1^{(1)}; & & \\
\cdots & \cdots & \cdots & \cdots & \cdots \\
\xi_1^{(n)}, & \xi_2^{(n)}, & \xi_3^{(n)}, & \ldots, & \xi_n^{(n)}.
\end{array}$$

This is the so-called *triangular array scheme*. The superscript denotes the row number, while the subscript denotes the number of the variable in the row.

Assume that the variables $\xi_k^{(n)}$ in the n-th row are independent and $\xi_k^{(n)} \in \mathbf{B}_{p_n}$, $k = 1, \dots, n$.

Corollary 5.4.1 (The Poisson theorem) *If $np_n \to \mu > 0$ as $n \to \infty$ then, for each fixed k,*

$$\mathbf{P}(S_n = k) \to \mathbf{\Pi}_\mu(\{k\}), \tag{5.4.2}$$

where $S_n = \xi_1^{(n)} + \dots + \xi_n^{(n)}$.

Proof This assertion is an immediate corollary of Theorem 5.4.1. It can also be obtained directly, by noting that it follows from the equality

$$\mathbf{P}(S_n = k) = \binom{n}{k} p^k (1-p)^{n-k}$$

that

$$\mathbf{P}(S_n = 0) = e^{n \ln(1-p)} \sim e^{-\mu}, \qquad \frac{\mathbf{P}(S_n = k+1)}{\mathbf{P}(S_n = k)} = \frac{n-k}{k+1} \frac{p}{1-p} \sim \frac{\mu}{k+1}. \quad \square$$

Theorem 5.4.2 implies an analogue of the Poisson theorem in a more general case as well, when the $\xi_j^{(n)}$ are not necessarily identically distributed[4] and can take values different from 0 and 1.

Corollary 5.4.2 *Assume that $p_{jn} = \mathbf{P}(\xi_j^{(n)} = 1)$ depend on n and j so that*

$$\max_j p_{jn} \to 0, \qquad \sum_{j=1}^n p_{jn} \to \mu > 0, \qquad \mathbf{P}(\xi_j^{(n)} = 0) = 1 - p_{jn} + o(p_{jn}).$$

Then (5.4.2) *holds.*

Proof To prove the corollary, one has to use Theorem 5.4.2 and the fact that

$$\mathbf{P}\left(\bigcup_{j=1}^n \{\xi_j^{(n)} \neq 0, \xi_j^{(n)} \neq 1\}\right) \le \sum_{j=1}^n o(p_{jn}) = o(1),$$

which means that, with probability tending to 1, all the variables $\xi_j^{(n)}$ assume the values 0 and 1 only. $\square$

One can clearly obtain from Theorems 5.4.1 and 5.4.2 somewhat stronger assertions than the above. In particular,

$$\sup_B \left| \mathbf{P}(S_n \in B) - \mathbf{\Pi}_\mu(B) \right| \to 0 \quad \text{as } n \to \infty.$$

[4] An extension of the de Moivre–Laplace theorem to the case of non-identically distributed random variables is contained in the central limit theorem from Sect. 8.4.

Note that under the assumptions of Theorem 5.4.1 this convergence will also take place in the case where $np \to \infty$ but only if $np^2 \to 0$. At the same time, the refinement of the de Moivre–Laplace theorem from Sect. 5.3 shows that the normal approximation for the distribution of S_n holds if $np \to \infty$ (for simplicity we assume that $p \le q$ so that $npq \ge \frac{1}{2}np \to \infty$).

Thus there exist sequences $p \in \{p : np \to \infty,\ np^2 \to 0\}$ such that *both the normal and the Poisson approximations* are valid. In other words, the domains of applicability of the normal and Poisson approximations *overlap*.

We see further from Theorem 5.4.1 that the convergence rate in Corollary 5.4.1 is determined by a quantity of the order of n^{-1}. Since, as $n \to \infty$,

$$\mathbf{P}(S_n = 0) - \mathbf{\Pi}_\mu\big(\{0\}\big) = e^{n\ln(1-p)} - e^{-\mu} \sim \frac{\mu^2}{2\pi} e^{-\mu},$$

this estimate cannot be substantially improved. However, for large k (in the large deviations range, say) such an estimate for the difference

$$\mathbf{P}(S_n = k) - \mathbf{\Pi}_\mu\big(\{k\}\big)$$

becomes rough. (This is because, in (5.4.1), we neglected not only the different signs of the correction terms but also the rare events $\{S_n = k\}$ and $\{S_n^* = k\}$ that appear in the arguments of the probabilities.) Hence we see, as in Sect. 5.4, the necessity for having approximations of which both *absolute* and *relative* errors are small.

Now we will show that the *asymptotic equivalence relations*

$$\mathbf{P}(S_n = k) \sim \mathbf{\Pi}_\mu\big(\{k\}\big)$$

remain valid when k and μ grow (along with n) in such a way that

$$k = o\big(n^{2/3}\big), \qquad \mu = o\big(n^{2/3}\big), \qquad |k - \mu| = o(\sqrt{n}\,).$$

Proof Indeed,

$$\begin{aligned}\mathbf{P}(S_n = k) &= \binom{n}{k} p^k (1-p)^{n-k} = \frac{n(n-1)\cdots(n-k+1)}{k!} p^k (1-p)^{n-k} \\ &= \frac{(nk)^k}{k!} e^{-pn} \left(1 - \frac{1}{n}\right) \cdots \left(1 - \frac{k-1}{n}\right)(1-p)^{n-k} e^{pn} \\ &= \mathbf{\Pi}_\mu\big(\{k\}\big) e^{\varepsilon(k,n)}.\end{aligned}$$

Thus we have to prove that, for values of k and μ from the indicated range,

$$\varepsilon(k,n) := \ln\left[\left(1 - \frac{1}{n}\right) \cdots \left(1 - \frac{k-1}{n}\right)(1-p)^{n-k} e^{pn}\right] = o(1). \tag{5.4.3}$$

We will obtain this relation together with the form of the correction term. Namely, we will show that

$$\varepsilon(k,n) = \frac{k - (k-\mu)^2}{2n} + O\left(\frac{k^3 + \mu^3}{n^2}\right), \tag{5.4.4}$$

and hence

$$\mathbf{P}(S_n = k) = \left(1 + \frac{k - (k-\mu)^2}{2n} + O\left(\frac{k^3 + \mu^3}{n^2}\right)\right)\Pi_\mu(\{k\}).$$

We make use of the fact that, as $\alpha \to 0$,

$$\ln(1-\alpha) = -\alpha - \frac{\alpha^2}{2} + O(\alpha^3).$$

Then relations (5.4.3) and (5.4.4) will follow from the equalities

$$\sum_{j=1}^{k-1} \ln\left(1 - \frac{j}{n}\right) = -\sum_{j=1}^{k-1} \frac{j}{n} + O\left(\frac{k^3}{n^2}\right) = -\frac{k(k-1)}{2n} + O\left(\frac{k^3}{n^2}\right),$$

$$(n-k)\ln(1-p) + pn = (n-k)\left(-p - \frac{p^2}{2} + O\left(p^3\right)\right) + pn$$

$$= -\frac{\mu^2}{2n} + \frac{k\mu}{n} + O\left(\frac{\mu^3}{n^2}\right). \qquad \square$$

In conclusion we note that the approximate Poisson formula

$$\mathbf{P}(S_n = k) \approx \frac{\mu^k}{k!} e^{-\mu}$$

is widely used in various applications and has, as experience and the above estimates show, a rather high accuracy even for moderate values of n.

Now we consider several examples of the use of the de Moivre–Laplace and Poisson theorems for approximate computations.

Example 5.4.1 Suppose we are given 10^4 packets of grain. It is known that there are 5000 tagged grains in the packets. What is the probability that, in a particular fixed packet, there is at least one tagged grain? We can assume that the tagged grains are distributed to packets at random. Then the probability that a particular tagged grain will be in the chosen packet is $p = 10^{-4}$. Since there are 5000 such grains, this will be the number of trials, i.e. $n = 5000$. Define a random variable ξ_k as follows: $\xi_k = 1$ if the k-th grain is in the chosen packet, and $\xi_k = 0$ otherwise. Then

$$S_{5000} = \sum_{k=1}^{5000} \xi_k$$

will be the number of tagged grains in our packet. By Theorem 5.4.1, $\mathbf{P}(S_{5000} = 0) \approx e^{-np} = e^{-0.5}$ so that the desired probability is approximately equal to $1 - e^{-0.5}$. The accuracy of this relation turns out to be rather high (by Theorem 5.4.1, the error does not exceed $2^{-1} \times 10^{-4}$). If we used the Poisson theorem instead of Theorem 5.4.1, we would have to imagine a triangular array of Bernoulli random variables, our ξ_k constituting the 5000-th row of the array. Moreover, we would assume that, for the n-th row, one has $np_n = 0.5$. Thus the conditions of the Poisson theorem would be met and we could make use of the limit theorem to find the approximate equality we have already obtained.

Example 5.4.2 A similar argument can be used in the following problem. There are n dangerous bacteria in a reservoir of capacity V from which we take a sample of volume $v \ll V$. What is the probability that we will find the bacteria in the test sample?

One usually assumes that the probability p that any given bacterium will be in the test sample is equal to the ratio v/V. Moreover, it is also assumed that the presence of a given bacterium in the sample does not depend on whether the remaining $n-1$ bacteria are in the test sample or not. In other words, one usually postulates that the mechanism of bacterial transfer into the test sample is equivalent to a sequence of n independent trials with "success" probability equal to $p = v/V$ in each trial.

Introducing random variables ξ_k as above, we obtain a description of the number of bacteria in the test sample by the sum $S_n = \sum_{k=1}^{n} \xi_k$ in the Bernoulli scheme. If nv is comparable in magnitude with V then by the Poisson theorem the desired probability will be equal to

$$\mathbf{P}(S_n > 0) \approx 1 - e^{-nv/V}.$$

Similar models are also used to describe the number of visible stars in a certain part of the sky far away from the Milky Way. Namely, it is assumed that if there are n visible stars in a region R then the probability that there are k visible stars in a subregion $r \subset R$ is

$$\binom{n}{k} p^k (1-p)^k,$$

where p is equal to the ratio $S(r)/S(R)$ of the areas of the regions r and R respectively.

Example 5.4.3 Suppose that the probability that a newborn baby is a boy is constant and equals 0.512 (see Sect. 3.4.1).

Consider a group of 10^4 newborn babies and assume that it corresponds to a series of 10^4 independent trials of which the outcomes are the events that either a boy or girl is born. What is the probability that the number of boys among these newborn babies will be greater than the number of girls by at least 200?

Define random variables as follows: $\xi_k = 1$ if the k-th baby is a boy and $\xi_k = 0$ otherwise. Then $S_n = \sum_{k=1}^{10^4} \xi_k$ is the number of boys in the group. The quantity $npq \sim 2.5 \times 10^3$ is rather large here, hence applying the integral limit (de Moivre–Laplace) theorem we obtain for the desired probability the value

$$\begin{aligned}\mathbf{P}(S_n \ge 5100) &= 1 - \mathbf{P}\left(\frac{S_n - np}{\sqrt{npq}} < \frac{5100 - 5120}{\sqrt{2500}}\right) \\ &\approx 1 - \Phi(-20/50) = 1 - \Phi(-0.4) \approx 0.66.\end{aligned}$$

To find the numerical values of $\Phi(x)$ one usually makes use of suitable statistical computer packages or calculators.

In our example, $\Delta = 1/\sqrt{npq} \approx 1/50$, and a satisfactory approximation by the de Moivre–Laplace formula will certainly be ensured (see Theorem 5.3.1) for $c \le 2.5$.

If, however, we have to estimate the probability that the proportion of boys exceeds 0.55, we will be dealing with large deviation probabilities when to estimate $\mathbf{P}(S_n > 5500)$ one would rather use the approximate relation obtained in Sect. 1.3 by virtue of which ($k = 0.45n$, $q = 0.488$) one has

$$\mathbf{P}(S_n > 5500) \approx \frac{(n+1-k)q}{(n+1)q-k}\mathbf{P}(S_n = 5500).$$

Applying Theorem 5.2.1 we find that

$$\mathbf{P}(S_n > 5500) \approx \frac{0.55q}{q-0.45}\frac{1}{\sqrt{2\pi n 0.25}}e^{-nH(0.55)} \le \frac{1}{5}e^{-25} < 10^{-11}.$$

Thus if we assume for a moment that 100 million babies are born on this planet each year and group them into batches of 10 thousand, then, to observe a group in which the proportion of boys exceeds the mean value by just 3.8 % we will have to wait, on average, 10 million years (see Example 4.1.1 in Sect. 4.1).

It is clear that the normal approximation can be used for numerical evaluation of probabilities for the problems from Example 5.4.3 provided that the values of np are large.

5.5 Inequalities for Large Deviation Probabilities in the Bernoulli Scheme

In conclusion of the present chapter we will derive several useful inequalities for the Bernoulli scheme. In Sect. 5.2 we introduced the function

$$H(x) = x\ln\frac{x}{p} + (1-x)\ln\frac{1-x}{1-p},$$

which plays an important role in Theorems 5.2.1 and 5.2.2 on the asymptotic behaviour of the probability $\mathbf{P}(S_n = k)$. We also considered there the basic properties of this function.

Theorem 5.5.1 *For* $z \ge 0$,

$$\begin{aligned} \mathbf{P}(S_n - np \ge z) &\le \exp\bigl\{-nH(p+z/n)\bigr\}, \\ \mathbf{P}(S_n - np \le -z) &\le \exp\bigl\{-nH(p-z/n)\bigr\}. \end{aligned} \tag{5.5.1}$$

Moreover, for all p,

$$H(p+x) \ge 2x^2, \tag{5.5.2}$$

so that each of the probabilities in (5.5.1) *does not exceed* $\exp\{-2z^2/n\}$ *for any* p.

To compare it with assertion (5.2.2) of Theorem 5.2.1, the first inequality from Theorem 5.5.1 can be re-written in the form

$$\mathbf{P}\left(\frac{S_n}{n} \geq p^*\right) \leq \exp\{-nH(p^*)\}.$$

The inequalities (5.5.1) are close, to some extent, to the de Moivre–Laplace theorem since, for $z = o(n^{2/3})$,

$$-nH\left(p + \frac{z}{n}\right) = -\frac{z^2}{2npq} + o(1).$$

The last assertion, together with (5.5.2), can be interpreted as follows: deviating by z or more from the mean value np has the maximum probability when $p = 1/2$.

If $z/\sqrt{n} \to \infty$, then both probabilities in (5.5.1) converge to zero as $n \to \infty$ for they correspond to large deviations of the sum S_n from the mean np. As we have already said, they are called *large deviation probabilities.*

Proof of Theorem 5.5.1 In Corollary 4.7.2 of the previous chapter we established the inequality

$$\mathbf{P}(\xi \geq x) \leq e^{-\lambda x}\mathbf{E}e^{\lambda\xi}.$$

Applying it to the sum S_n we get

$$\mathbf{P}(S_n \geq np + z) \leq e^{-\lambda(np+z)}\mathbf{E}e^{\lambda S_n}.$$

Since $\mathbf{E}e^{\lambda S_n} = \prod_{k=1}^n \mathbf{E}e^{\lambda\xi_k}$ and the random variables $e^{\lambda\xi_k}$ are independent,

$$\mathbf{E}e^{\lambda S_n} = \prod_{k=1}^n \mathbf{E}e^{\lambda\xi_k} = (pe^\lambda + q)^n = \big(1 + p(e^\lambda - 1)\big)^n,$$

$$\mathbf{P}(S_n \geq np + z) \leq \big[\big(1 + p(e^\lambda - 1)\big)e^{-\lambda(p+\alpha)}\big]^n, \qquad \alpha = z/n.$$

The expression in brackets is equal to

$$\mathbf{E}e^{-\lambda[\xi_k-(p+\alpha)]} = pe^{\lambda(1-p-\alpha)} + (1-p)e^{-\lambda(p+\alpha)}.$$

Therefore, being the sum of two convex functions, it is a convex function of λ. The equation for the minimum point $\lambda(\alpha)$ of the function has the form

$$-(p-\alpha)\big(1 + p(e^\lambda - 1)\big) + pe^\lambda = 0,$$

from which we find that

$$e^{\lambda(\alpha)} = \frac{(p+\alpha)q}{p(q-\alpha)},$$

$$\begin{aligned}
\big(1 + p(e^{\lambda(\alpha)} - 1)\big)e^{-\lambda(\alpha)(p+\alpha)} &= \frac{q}{q-\alpha}\left[\frac{p(q-\alpha)}{(p+\alpha)q}\right]^{p+\alpha} \\
&= \frac{p^{p+\alpha}q^{q-\alpha}}{(p+\alpha)^{p+\alpha}(q-\alpha)^{q-\alpha}} \\
&= \exp\left\{-(p+\alpha)\ln\frac{p+\alpha}{p} - (q-\alpha)\ln\frac{q-\alpha}{q}\right\} \\
&= \exp\{-H(p+\alpha)\}.
\end{aligned}$$

The first of the inequalities (5.5.1) is proved. The second inequality follows from the first if we consider the latter as the inequality for the number of zeros.

It follows further from (5.2.1) that $H(p) = H'(p) = 0$ and $H''(x) = 1/x(1-x)$. Since the function $x(1-x)$ attains its maximum value on the interval [0, 1] at the point $x = 1/2$, one has $H''(x) \geq 4$ and hence

$$H(p+\alpha) \geq \frac{\alpha^2}{2} \cdot 4 = 2\alpha^2. \qquad \square$$

For analogues of Theorem 5.5.1 for sums of arbitrary random variables, see Chap. 9 and Appendix 8. Example 9.1.2 shows that the function $H(\alpha)$ is the so-called *deviation function* for the Bernoulli scheme. This function is important in describing large deviation probabilities.

Chapter 6
On Convergence of Random Variables and Distributions

Abstract In this chapter, several different types of convergence used in Probability Theory are defined and relationships between them are elucidated. Section 6.1 deals with convergence in probability and convergence with probability one (the almost sure convergence), presenting some criteria for them and, in particular, discussing the concept of Cauchy sequences (in probability and almost surely). Then the continuity theorem is established (convergence of functions of random variables) and the concept of uniform integrability is introduced and discussed, together with its consequences (in particular, for convergence in mean of suitable orders). Section 6.2 contains an extensive discussion of weak convergence of distributions. The chapter ends with Sect. 6.3 presenting criteria for weak convergence of distributions, including the concept of distribution determining classes of functions and that of tightness.

6.1 Convergence of Random Variables

In previous chapters we have already encountered several assertions which dealt with convergence, in some sense, of the distributions of random variables or of the random variables themselves. Now we will give definitions of different types of convergence and elucidate the relationships between them.

6.1.1 Types of Convergence

Let a sequence of random variables $\{\xi_n\}$ and a random variable ξ be given on a probability space $\langle \Omega, \mathfrak{F}, \mathbf{P} \rangle$.

Definition 6.1.1 The sequence $\{\xi_n\}$ *converges in probability*[1] to ξ if, for any $\varepsilon > 0$,

$$\mathbf{P}\big(|\xi_n - \xi| > \varepsilon\big) \to 0 \quad \text{as } n \to \infty.$$

[1] In the set-theoretic terminology, convergence in probability means convergence in measure.

A.A. Borovkov, *Probability Theory*, Universitext,
DOI 10.1007/978-1-4471-5201-9_6, © Springer-Verlag London 2013

One writes this as

$$S_n \xrightarrow{p} \xi \quad \text{as } n \to \infty.$$

In this notation, the assertion of the law of large numbers for the Bernoulli scheme could be written as

$$\frac{S_n}{n} \xrightarrow{p} p,$$

since S_n/n can be considered as a sequence of random variables given on a common probability space.

Definition 6.1.2 We will say that the *sequence ξ_n converges to ξ with probability* 1 (or *almost surely*: $\xi_n \to \xi$ a.s., $\xi_n \xrightarrow{a.s.} \xi$), if $\xi_n(\omega) \to \xi(\omega)$ as $n \to \infty$ for all $\omega \in \Omega$ except for ω from a set $N \subset \Omega$ of null probability: $\mathbf{P}(N) = 0$. This convergence can also be called convergence *almost everywhere* (a.e.) with respect to the measure $\mathbf{P}$.

Convergence $\xi_n \xrightarrow{a.s.} \xi$ implies convergence $\xi_n \xrightarrow{p} \xi$. Indeed, if we assume that the convergence in probability does not take place then there exist $\varepsilon > 0$, $\delta > 0$, and a sequence n_k such that, for the sequence of events $A_k = \{|\xi_{n_k} - \xi| > \varepsilon\}$, we have $\mathbf{P}(A_k) \ge \delta$ for all k. Let B consist of all elementary events belonging to infinitely many A_k, i.e. $B = \bigcap_{m=1}^{\infty} \bigcup_{k=m}^{\infty} A_k$. Then, clearly for $\omega \in B$, the convergence $\xi_n(\omega) \to \xi(\omega)$ is impossible. But $B = \bigcap_{m=1}^{\infty} B_m$, where $B_m = \bigcup_{k \ge m} A_k$ are decreasing events ($B_{m+1} \subset B_m$), $\mathbf{P}(B_m) \ge \mathbf{P}(A_{n_m}) \ge \delta$ and, by the continuity axiom, $\mathbf{P}(B_m) \to \mathbf{P}(B)$ as $m \to \infty$. Therefore $\mathbf{P}(B) \ge \delta$ and a.s. convergence is impossible. The obtained contradiction proves the desired statement. □

The converse assertion, that convergence in probability implies a.s. convergence, is, generally speaking, not true, as we will see below. However in one important special case such a converse holds true.

Theorem 6.1.1 *If ξ_n is monotonically increasing or decreasing then convergence $\xi_n \xrightarrow{p} \xi$ implies that $\xi_n \xrightarrow{a.s.} \xi$.*

Proof Assume, without loss of generality, that $\xi \equiv 0$, $\xi_n \ge 0$, $\xi_n \downarrow$ and $\xi_n \xrightarrow{p} \xi$. If convergence $\xi_n \xrightarrow{a.s.} \xi$ did not hold, there would exist an $\varepsilon > 0$ and a set A with $\mathbf{P}(A) > \delta > 0$ such that $\sup_{k \ge n} \xi_k > \varepsilon$ for $\omega \in A$ and all n. But $\sup_{k \ge n} \xi_k = \xi_n$ and hence we have

$$\mathbf{P}(\xi_n > \varepsilon) \ge \mathbf{P}(A) > \delta > 0$$

for all n, which contradicts the assumed convergence $\xi_n \xrightarrow{p} 0$. □

Thus convergence in probability is determined by the behaviour of the numerical sequence $\mathbf{P}(|\xi_n - \xi| > \varepsilon)$. Is it possible to characterise convergence with probability 1 in a similar way? Set $\zeta_n := \sup_{k \ge n} |\xi_n - \xi|$.

Corollary 6.1.1 *$\xi_n \xrightarrow{a.s.} \xi$ if and only if $\zeta_n \xrightarrow{p} 0$, or, which is the same, when, for any $\varepsilon > 0$,*

$$\mathbf{P}\Bigl(\sup_{k\ge n} |\xi_k - \xi| > \varepsilon\Bigr) \to 0 \quad \text{as } n \to \infty. \tag{6.1.1}$$

Proof Clearly $\xi_n \to \xi$ a.s. if and only if $\zeta_n \to 0$ a.s. But the sequence ζ_n decreases monotonically and it remains to make use of Theorem 6.1.1, which implies that $\zeta_n \xrightarrow{p} 0$ if and only if $\zeta_n \xrightarrow{a.s.} 0$. The corollary is proved. □

In the above argument, the random variables ξ_n and ξ could be improper, where the random variables ξ_n and ξ are only defined on a set B and $\mathbf{P}(B) \in (0, 1)$. (These random variables can take infinite values on $\Omega \setminus B$.) In this case, all the considerations concerning convergence are carried out on the set $B \subset \Omega$ only.

In the introduced terminology, the assertion of the strong law of large numbers for the Bernoulli scheme (Theorem 5.1.2) can be stated, by virtue of (6.1.1), as convergence $S_n/n \to p$ with probability 1.

We have already noted that convergence almost surely implies convergence in probability. Now we will give an example showing that the converse assertion is, generally speaking, not true. Let $\langle \Omega, \mathfrak{F}, \mathbf{P}\rangle$ be the unit circle with the σ-algebra of Borel sets and uniform distribution. Put $\xi(\omega) \equiv 1$, $\xi_n(\omega) = 2$ on the arc $[r(n), r(n)+1/n]$ and $\xi_n(\omega) = 1$ outside the arc. Here $r(n) = \sum_{k=1}^{n} \frac{1}{k}$. It is obvious that $\xi_n \xrightarrow{p} \xi$. At the same time, $r(n) \to \infty$ as $n \to \infty$, and the set on which ξ_n converges to ξ is empty (we can find no ω for which $\xi_n(\omega) \to \xi(\omega)$).

However, if $\mathbf{P}(|\xi_n - \xi| > \varepsilon)$ decreases as $n \to \infty$ sufficiently fast, then convergence in probability will also become a.s. convergence. In particular, relation (6.1.1) gives the following sufficient condition for convergence with probability 1.

Theorem 6.1.2 *If the series $\sum_{k=1}^{\infty} \mathbf{P}(|\xi_n - \xi| > \varepsilon)$ converges for any $\varepsilon > 0$, then $\xi_n \to \xi$ a.s.*

Proof This assertion is obvious, for

$$\mathbf{P}\Bigl(\bigcup_{k\ge n}\bigl\{|\xi_k - \xi| > \varepsilon\bigr\}\Bigr) \le \sum_{k=n}^{\infty} \mathbf{P}\bigl(|\xi_n - \xi| > \varepsilon\bigr).$$ □

It is this criterion that has actually been used in proving the strong law of large numbers for the Bernoulli scheme.

One cannot deduce a converse assertion about the convergence rate to zero of the probability $\mathbf{P}(|\xi_n - \xi| > \varepsilon)$ from the a.s. convergence. The reader can easily construct an example where $\xi_n \to \xi$ a.s., while $\mathbf{P}(|\xi_n - \xi| > \varepsilon)$ converges to zero arbitrarily slowly.

Theorem 6.1.2 implies the following result.

Corollary 6.1.2 *If $\xi_n \xrightarrow{p} \xi$, then there exists a subsequence $\{n_k\}$ such that $\xi_{n_k} \to \xi$ a.s. as $k \to \infty$.*

Proof This assertion is also obvious since it suffices to take n_k such that $\mathbf{P}(|\xi_{n_k} - \xi| > \varepsilon) \le 1/k^2$ and then make use of Theorem 6.1.2. □

There is one more important special case where convergence in probability $\xi_n \xrightarrow{p} \xi$ implies convergence $\xi_n \to \xi$ a.s. This is the case when the ξ_n are sums of independent random variables. Namely, the following assertion is true. *If* $\xi_n = \sum_{k=1}^n \eta_k$, η_k *are independent, then convergence of* ξ_n *in probability implies convergence with probability* 1. This assertion will be proved in Sect. 11.2.

Finally we consider a third type of convergence of random variables.

Definition 6.1.3 We will say that ξ_n *converges to* ξ *in the* r*-th order mean* (*in mean* if $r = 1$; *in mean square* if $r = 2$) if, as $n \to \infty$,

$$\mathbf{E}|\xi_n - \xi|^r \to 0.$$

This convergence will be denoted by $\xi_n \xrightarrow{(r)} \xi$.

Clearly, by Chebyshev's inequality $\xi_n \xrightarrow{(r)} \xi$ implies that $\xi_n \xrightarrow{p} \xi$. On the other hand, convergence $\xrightarrow{(r)}$ does not follow from a.s. convergence (and all the more from convergence in probability). Thus convergence in probability is the weakest of the three types of convergence we have introduced.

Note that, under additional conditions, convergence $\xi_n \xrightarrow{p} \xi$ can imply that $\xi_n \xrightarrow{(r)} \xi$ (see Theorem 6.1.7 below). For example, it will be shown in Corollary 6.1.4 that if $\xi_n \xrightarrow{p} \xi$ and $\mathbf{E}|\xi_n|^{r+\alpha} < c$ for some $\alpha > 0$, $c < \infty$ and all n, then $\xi_n \xrightarrow{(r)} \xi$.

Definition 6.1.4 A sequence ξ_n is said to be a *Cauchy sequence in probability* (a.s., *in mean*) if, for any $\varepsilon > 0$,

$$\mathbf{P}\big(|\xi_n - \xi_m| > \varepsilon\big) \to 0$$

$$\Big(\mathbf{P}\Big(\sup_{n\ge m} |\xi_n - \xi_m| > \varepsilon\Big) \to 0,\ \mathbf{E}|\xi_n - \xi_m|^r \to 0\Big)$$

as $n \to \infty$ and $m \to \infty$.

Theorem 6.1.3 (Cauchy convergence test) $\xi_n \to \xi$ *in one of the senses* $\xrightarrow{p}$, $\xrightarrow{a.s.}$ *or* $\xrightarrow{(r)}$ *if and only if* ξ_n *is a Cauchy sequence in the respective sense.*

Proof That ξ_n is a Cauchy sequence follows from convergence by virtue of the inequalities

$$|\xi_n - \xi_m| \le |\xi_n - \xi| + |\xi_m - \xi|,$$

$$\sup_{n\ge m} |\xi_n - \xi_m| \le \sup_{n\ge m} |\xi_n - \xi| + |\xi_m - \xi| \le 2 \sup_{n\ge m} |\xi_n - \xi|,$$

$$|\xi_n - \xi_m|^r \le C_r\big(|\xi_n - \xi|^r + |\xi_m - \xi|^r\big)$$

for some C_r.

Now assume that ξ_n is a Cauchy sequence in probability. Choose a sequence $\{n_k\}$ such that

$$\mathbf{P}\big(|\xi_n - \xi_m| > 2^{-k}\big) < 2^{-k}$$

for $n \ge n_k$, $m \ge n_k$. Put

$$\xi'_k := \xi_{n_k}, \qquad A_k := \big\{|\xi'_k - \xi'_{k+1}| > 2^{-k}\big\}, \qquad \eta = \sum_{k=1}^{\infty} I(A_k).$$

Then $\mathbf{P}(A_k) \le 2^{-k}$ and $\mathbf{E}\eta = \sum_{k=1}^{\infty} \mathbf{P}(A_k) \le 1$. This means, of course, that the number of occurrences of the events A_k is a proper random variable: $\mathbf{P}(\eta < \infty) = 1$, and hence with probability 1 finitely many events A_k occur. This means that, for any ω for which $\eta(\omega) < \infty$, there exists a $k_0(\omega)$ such that $|\xi'_k(\omega) - \xi'_{k+1}(\omega)| \le 2^{-k}$ for all $k \ge k_0(\omega)$. Therefore one has the inequality $|\xi'_k(\omega) - \xi'_l(\omega)| \le 2^{-k+1}$ for all $k \ge k_0(\omega)$ and $l \ge k_0(\omega)$, which means that $\xi'_n(\omega)$ is a numerical Cauchy sequence and hence there exists a value $\xi(\omega)$ such that $|\xi'_k(\omega) - \xi(\omega)| \to 0$ as $k \to \infty$. This means, in turn, that $\xi'_k \xrightarrow{a.s.} \xi$ and hence

$$\mathbf{P}\big(|\xi_n - \xi| \ge \varepsilon\big) \le \mathbf{P}\left(|\xi_n - \xi_{n_k}| \ge \frac{\varepsilon}{2}\right) + \mathbf{P}\left(|\xi_{n_k} - \xi| > \frac{\varepsilon}{2}\right) \to 0$$

as $n \to \infty$ and $k \to \infty$.

Now assume that ξ_n is a Cauchy sequence in mean. Then, by Chebyshev's inequality, it will be a Cauchy sequence in probability and hence, by Corollary 6.1.2, there will exist a random variable ξ and a subsequence $\{n_k\}$ such that $\xi_{n_k} \xrightarrow{a.s.} \xi$. Now we will show that $\mathbf{E}|\xi_n - \xi|^r \to 0$. For a given $\varepsilon > 0$, choose an n such that $\mathbf{E}|\xi_k - \xi_l|^r < \varepsilon$ for $k \ge n$ and $l \ge n$. Then, by Fatou's lemma (see Appendix 3),

$$\begin{aligned}\mathbf{E}|\xi_n - \xi|^r &= \mathbf{E} \lim_{n_k \to \infty} |\xi_n - \xi_{n_k}|^r \\ &= \mathbf{E} \liminf_{n_k \to \infty} |\xi_n - \xi_{n_k}|^r \le \liminf_{n_k \to \infty} \mathbf{E}|\xi_n - \xi_{n_k}|^r \le \varepsilon.\end{aligned}$$

This means that $\mathbf{E}|\xi_n - \xi|^r \to 0$ as $n \to \infty$.

It remains to verify the assertion of the theorem related to a.s. convergence. We already know that if ξ_n is a Cauchy sequence in probability (or a.s.) then there exist a ξ and a subsequence ξ_{n_k} such that $\xi_{n_k} \xrightarrow{a.s.} \xi$. Therefore, if we put $n_{k(n)} := \min\{n_k : n_k \ge n\}$, then

$$\mathbf{P}\Big(\sup_{k \ge n} |\xi_k - \xi| \ge \varepsilon\Big) \le \mathbf{P}\Big(\sup_{k \ge n} |\xi_k - \xi_{n_{k(n)}}| \ge \varepsilon/2\Big) + \mathbf{P}\big(|\xi_{n_{k(n)}} - \xi| \ge \varepsilon/2\big) \to 0$$

as $n \to \infty$. The theorem is proved. □

Remark 6.1.1 If we introduce the space L_r of all random variables ξ on $\langle \Omega, \mathfrak{F}, \mathbf{P}\rangle$ for which $\mathbf{E}|\xi|^r < \infty$ and the norm $\|\xi\| = (\mathbf{E}|\xi|^r)^{1/r}$ on it (the triangle inequality $\|\xi_1 + \xi_2\| \le \|\xi_1\| + \|\xi_2\|$ is then nothing else but Minkowski's inequality, see Theorem 4.7.2), then the assertion of Theorem 6.1.3 on convergence $\xrightarrow{(r)}$ (which is convergence in the norm of L_r, for we identify random variables ξ_1 and ξ_2 if $\|\xi_1 - \xi_2\| = 0$) means that L_r is complete and hence is a Banach space.

The space of *all* random variables on $\langle \Omega, \mathfrak{F}, \mathbf{P} \rangle$ can be metrised so that convergence in the metric will be equivalent to convergence in probability. For instance, one could put

$$\rho(\xi_1, \xi_2) := \mathbf{E}\frac{|\xi_1 - \xi_2|}{1 + |\xi_1 - \xi_2|}.$$

Since

$$\frac{|x+y|}{1+|x+y|} \le \frac{|x|}{1+|x|} + \frac{|y|}{1+|y|}$$

always holds, $\rho(\xi_1, \xi_2)$ satisfies all the axioms of a metric. It is not difficult to see that relations $\rho(\xi_1, \xi_2) \to 0$ and $\xi_n \xrightarrow{p} 0$ are equivalent. The assertion of Theorem 6.1.3 related to convergence $\xrightarrow{p}$ means that the metric space we introduced is complete.

6.1.2 The Continuity Theorem

Now we will derive the following "continuity theorem".

Theorem 6.1.4 *Let $\xi_n \xrightarrow{a.s.} \xi$ ($\xi_n \xrightarrow{p} \xi$) and $H(s)$ be a function continuous everywhere with respect to the distribution of the random variable ξ (i.e. $H(s)$ is continuous at each point of a set S such that $\mathbf{P}(\xi \in S) = 1$). Then*

$$H(\xi_n) \xrightarrow{a.s.} H(\xi) \qquad \bigl(H(\xi_n) \xrightarrow{p} H(\xi)\bigr).$$

Proof Let $\xi_n \xrightarrow{a.s.} \xi$. Since the sets $A = \{\omega : \xi_n(\omega) \to \xi(\omega)\}$ and $B = \{\omega : \xi(\omega) \in S\}$ are both of probability 1, $\mathbf{P}(AB) = \mathbf{P}(A) + \mathbf{P}(B) - \mathbf{P}(A \cup B) = 1$. But one has $H(\xi_n) \to H(\xi)$ on the set AB. Convergence with probability 1 is proved.

Now let $\xi_n \xrightarrow{p} \xi$. If we assume that convergence $H(\xi_n) \xrightarrow{p} H(\xi)$ does not take place then there will exist $\varepsilon > 0$, $\delta > 0$ and a subsequence $\{n'\}$ such that

$$\mathbf{P}\bigl(\bigl|H(\xi_{n'}) - H(\xi)\bigr| > \varepsilon\bigr) > \delta.$$

But $\xi_{n'} \xrightarrow{p} \xi$ and hence there exists a subsequence $\{n''\}$ such that $\xi_{n''} \xrightarrow{a.s.} \xi$ and $H(\xi_{n''}) \xrightarrow{a.s.} H(\xi)$. This contradicts the assumption we made, for the latter implies that

$$\mathbf{P}\bigl(\bigl|H(\xi_{n''}) - H(\xi)\bigr| > \varepsilon\bigr) > \delta.$$

The theorem is proved. □

6.1.3 Uniform Integrability and Its Consequences

Now we will consider this question: in what cases does convergence in probability imply convergence in mean?

The main condition that ensures the transition from convergence in probability to convergence in mean is associated with the notion of uniform integrability.

Definition 6.1.5 A sequence $\{\xi_n\}$ is said to be *uniformly integrable* if

$$\sup_n \mathbf{E}\big(|\xi_n|;\ |\xi_n| > N\big) \to 0 \quad \text{as } N \to \infty.$$

A sequence of independent identically distributed random variables with finite mean is, clearly, uniformly integrable.

If $\{\xi_n\}$ is uniformly integrable then so are $\{c\xi_n\}$ and $\{\xi_n + c\}$, where $c = \text{const}$.

Let us present some further, less evident, properties of uniform integrability.

U1. *If the sequences* $\{\xi_n'\}$ *and* $\{\xi_n''\}$ *are uniformly integrable then the sequences defined by* $\zeta_n = \max(|\xi_n'|, |\xi_n''|)$ *and* $\zeta_n = \xi_n' + \xi_n''$ *are also uniformly integrable.*

Proof Indeed, for $\zeta_n = \max(|\xi_n'|, |\xi_n''|)$ we have

$$\begin{aligned}\mathbf{E}(\zeta_n;\ \zeta_n > N) &= \mathbf{E}\big(\zeta_n;\ \zeta_n > N, |\xi_n'| > |\xi_n''|\big) + \mathbf{E}\big(\zeta_n;\ \zeta_n > N, |\xi_n'| \le |\xi_n''|\big)\\ &\le \mathbf{E}\big(|\xi_n'|;\ |\xi_n'| > N\big) + \mathbf{E}\big(|\xi_n''|;\ |\xi_n''| \ge N\big) \to 0\end{aligned}$$

as $N \to \infty$.

Since

$$|\xi_n' + \xi_n''| \le |\xi_n'| + |\xi_n''| \le 2\max\big(|\xi_n'|, |\xi_n''|\big),$$

from the above it follows that the sequence defined by the sum $\zeta_n = \xi_n' + \xi_n''$ is also uniformly integrable. □

U2. *If* $\{\xi_n\}$ *is uniformly integrable then* $\sup_n \mathbf{E}|\xi_n| \le c < \infty$.

Proof Indeed, choose N so that

$$\sup_n \mathbf{E}\big(|\xi_n|;\ |\xi_n| > N\big) \le 1.$$

Then

$$\sup_n \mathbf{E}|\xi_n| = \sup_n \big[\mathbf{E}\big(|\xi_n|;\ |\xi_n| \le N\big) + \mathbf{E}\big(|\xi_n|;\ |\xi_n| > N\big)\big] \le N + 1. \qquad \square$$

The converse assertion is not true. For example, for a sequence

$$\xi_n : \mathbf{P}(\xi_n = n) = 1/n = 1 - \mathbf{P}(\xi_n = 0)$$

one has $\mathbf{E}|\xi_n| = 1$, but the sequence is not uniformly integrable.

If we somewhat strengthen the above statement U2, it becomes "characteristic" for uniform integrability.

Theorem 6.1.5 *For a sequence* $\{\xi_n\}$ *to be uniformly integrable, it is necessary and sufficient that there exists a function* $\psi(x)$ *such that*

$$\frac{\psi(x)}{x} \uparrow \infty \quad \text{as } x \uparrow \infty, \qquad \sup_n \mathbf{E}\psi\big(|\xi_n|\big) < c < \infty. \tag{6.1.2}$$

In the necessity assertion one can choose a convex function ψ.

Proof Without loss of generality we can assume that $\xi_i \ge 0$.

The *sufficiency* is evident, since, putting $v(x) := \frac{\psi(x)}{x}$, we get

$$\mathbf{E}(\xi_n;\ \xi_n \ge N) \le \frac{1}{v(N)} \mathbf{E}\big(\xi_n v(\xi_n);\ \xi_n \ge N\big) \le \frac{c}{v(N)}.$$

To prove the *necessity*, put

$$\varepsilon(N) := \sup_n \mathbf{E}(\xi_n;\ \xi_n \ge N).$$

Then, by virtue of uniform integrability, $\varepsilon(N) \downarrow 0$ as $N \uparrow \infty$. Choose a sequence $N_k \uparrow \infty$ as $k \uparrow \infty$ such that

$$\sum_{k=1}^{\infty} \sqrt{\varepsilon(N_k)} < c_1 < \infty,$$

and put

$$g(x) = x\big(\varepsilon(N_k)\big)^{-1/2} \quad \text{for } x \in [N_k, N_{k+1}).$$

Since

$$\frac{g(N_k - 0)}{N_k} = \big(\varepsilon(N_{k-1})\big)^{-1/2} \le \big(\varepsilon(N_k)\big)^{-1/2} = \frac{g(N_k)}{N_k},$$

we have $\frac{g(x)}{x} \uparrow \infty$ as $x \to \infty$. Further,

$$\begin{aligned}
\mathbf{E}g(\xi_n) &= \sum_k \mathbf{E}\big[g(\xi_n);\ \xi_n \in [N_k, N_{k+1})\big] \\
&= \sum_k \mathbf{E}\big[\xi_n \big(\varepsilon(N_k)\big)^{-1/2};\ \xi_n \in [N_k, N_{k+1})\big] \\
&\le \sum_k \big(\varepsilon(N_k)\big)^{-1/2} \varepsilon(N_k) = \sum_k \sqrt{\varepsilon(N_k)} < c_1,
\end{aligned}$$

where the right-hand side does not depend on n. Therefore, to prove the theorem it is sufficient to construct a function $\psi \le g$ which is convex and such that $\frac{\psi(x)}{x} \uparrow \infty$ as $x \uparrow \infty$.

Define the function $\psi(x)$ as the continuous polygon with nodes $(N_k, g(N_k - 0))$. Since

$$\frac{g(N_k - 0)}{N_k} = \varepsilon(N_{k-1})^{-1/2}$$

monotonically increases as k grows, ψ is a lower envelope curve for the discontinuous function $g(x) \ge \psi(x)$. The monotonicity of $\frac{\psi(x)}{x}$ follows from the fact that, on the interval $[N_k, N_{k+1})$, this function can be represented as

$$\frac{\psi(x)}{x} = a_{k,\psi} - \frac{b_k}{x},$$

where $b_k > 0$, because the values $\psi(N_{k+1}-0)$ and $g(N_{k+1}-0)$ coincide, while the angular incline $a_{k,\psi}$ of the function ψ on the interval $[N_k, N_{k+1})$ is greater than the "radial" incline $a_{k,g}$ of the function g:

$$a_{k,g} = \frac{g(N_{k+1}-0) - g(N_k)}{N_{k+1} - N_k} < \frac{g(N_{k+1}-0) - g(N_k - 0)}{N_{k+1} - N_k} = a_{k,\psi}.$$

It is clear that $\frac{\psi(x)}{x}$ increases unboundedly, for

$$\frac{\psi(N_k)}{N_k} = \frac{g(N_k - 0)}{N_k} = \varepsilon(N_{k-1})^{-1/2} \uparrow \infty$$

as $k \to \infty$. The theorem is proved. □

In studying the mean values of sums of random variables, the following theorem on *uniform integrability of average values*, following from Theorem 6.1.5, plays an important role.

Theorem 6.1.6 *Let $\xi_1, \xi_2, \ldots$ be an arbitrary uniformly integrable sequence of random variables,*

$$p_{i,n} \ge 0, \qquad \sum_{i=1}^{n} p_{i,n} = 1, \qquad \zeta_n = \sum_{k=1}^{n} |\xi_i| p_{i,n}.$$

Then the sequence $\{\zeta_n\}$ is uniformly integrable as well.

Proof Let $\psi(x)$ be the convex function from Theorem 6.1.5 satisfying properties (6.1.2). Then, by that theorem,

$$\mathbf{E}\psi(\zeta_n) = \mathbf{E}\psi\left(\sum_{i=1}^{n} p_{i,n}|\xi_i|\right) \le \mathbf{E}\sum_{i=1}^{n} p_{i,n}\psi(|\xi_i|) \le c.$$

It remains to make use of Theorem 6.1.5 again. □

Now we will show that convergence in probability together with uniform integrability imply convergence in mean.

Theorem 6.1.7 *Let $\xi_n \xrightarrow{p} \xi$ and $\{\xi_n\}$ be uniformly integrable. Then $\mathbf{E}|\xi|$ exists and, as $n \to \infty$,*

$$\mathbf{E}|\xi_n - \xi| \to 0.$$

If, moreover, $\{|\xi_n^r|\}$ is uniformly integrable then $\xi_n \xrightarrow{(r)} \xi$.

Conversely, if, for an $r \ge 1$, $\xi_n \xrightarrow{(r)} \xi$ and $\mathbf{E}|\xi|^r < \infty$, then $\{|\xi_n|^r\}$ is uniformly integrable.

In the law of large numbers for the Bernoulli scheme (see Theorem 5.1.1) we proved that the normed sum S_n/n converges to p in probability. Since $0 \le S_n/n \le 1$,

S_n/n is clearly uniformly integrable and the convergence in mean $\mathbf{E}|S_n/n - p|^r \to 0$ holds for any r. This fact can also be established directly. For a more substantiative example of application of Theorems 6.1.6 and 6.1.7, see Sect. 8.1.

Proof We show that $\mathbf{E}\xi$ exists. By the properties of integrals (see Lemma A3.2.3 in Appendix 3), if $\mathbf{E}|\zeta| < \infty$ then $\mathbf{E}(\zeta; A_n) \to 0$ as $\mathbf{P}(A_n) \to 0$. Since $\mathbf{E}\xi_n < \infty$, for any N and ε one has

$$\begin{aligned} \mathbf{E}\min(|\xi|, N) &= \lim_{n\to\infty} \left[\mathbf{E}\min(|\xi|, N);\ |\xi_n - \xi| < \varepsilon\right] \\ &\le \lim_{n\to\infty} \mathbf{E}\min(|\xi| + \varepsilon, N) \le c + \varepsilon. \end{aligned}$$

It follows that $\mathbf{E}|\xi| \le c$.

Further, for brevity, put $\eta_n = |\xi_n - \xi|$. Then $\eta_n \xrightarrow{p} 0$ and η_n are uniformly integrable together with ξ_n. For any N and ε, one has

$$\begin{aligned} \mathbf{E}\eta_n &= \mathbf{E}(\eta_n;\ \eta_n \le \varepsilon) + \mathbf{E}(\eta_n;\ N \ge \eta_n > \varepsilon) + \mathbf{E}(\eta_n;\ \eta_n \ge N) \\ &\le \varepsilon + N\mathbf{P}(\eta_n \ge \varepsilon) + \mathbf{E}(\eta_n;\ \eta_n > N). \end{aligned} \tag{6.1.3}$$

Choose N so that $\sup_n \mathbf{E}(\eta_n;\ \eta_n > N) \le \varepsilon$. Then, for such an N,

$$\limsup_{n\to\infty} \mathbf{E}\eta_n \le 2\varepsilon.$$

Since ε is arbitrary, $\mathbf{E}\eta_n \to 0$ as $n \to \infty$.

The relation $\mathbf{E}|\xi_n - \xi|^r \to 0$ can be proved in the same way as (6.1.3), since $\eta_n^r = |\xi_n - \xi|^r \xrightarrow{p} 0$ and η_n^r are uniformly integrable together with $|\xi_n|^r$.

Now we will prove the converse assertion. Let, for simplicity, $r = 1$. One has

$$\begin{aligned} \mathbf{E}\big(|\xi_n|;\ |\xi_n| > N\big) &\le \mathbf{E}\big(|\xi_n - \xi|;\ |\xi_n| > N\big) + \mathbf{E}\big(|\xi|;\ |\xi_n| > N\big) \\ &\le \mathbf{E}|\xi_n - \xi| + \mathbf{E}\big(|\xi|;\ |\xi_n| > N\big) \\ &\le \mathbf{E}|\xi_n - \xi| + \mathbf{E}\big(|\xi|;\ |\xi_n - \xi| > 1\big) + \mathbf{E}\big(|\xi|;\ |\xi| > N - 1\big). \end{aligned}$$

The first term on the right-hand side tends to zero by the assumption, and the second term, by Lemma A3.2.3 from Appendix 3, which we have just mentioned, and the fact that $\mathbf{P}(|\xi_n - \xi| > 1) \to 0$. The last term does not depend on n and can be made arbitrarily small by choosing N. Theorem 6.1.7 is proved. □

Now we can derive yet another *continuity theorem* which has the following form.

Theorem 6.1.8 *If* $\xi_n \xrightarrow{p} \xi$, $H(s)$ *satisfies the conditions of Theorem* 6.1.4, *and* $H(\xi_n)$ *is uniformly integrable, then, as* $n \to \infty$,

$$\mathbf{E}\big|H(\xi_n) - H(\xi)\big| \to 0$$

and, in particular, $\mathbf{E}H(\xi_n) \to \mathbf{E}H(\xi)$.

This assertion follows from Theorems 6.1.4 and 6.1.7, for $H(\xi_n) \xrightarrow{p} H(\xi)$ by Theorem 6.1.4.

Sometimes it is convenient to distinguish between *left* and *right* uniform integrability. We will say that a sequence $\{\xi_n\}$ is *right (left) uniformly integrable* if

$$\sup \mathbf{E}(\xi_n;\ \xi_n \ge N) \to 0 \qquad \bigl(\sup \mathbf{E}\bigl(|\xi_n|;\ \xi_n \le -N\bigr) \to 0\bigr)$$

as $N \to \infty$. It is evident that a sequence $\{\xi_n\}$ is uniformly integrable if and only if it is both right and left uniformly integrable.

Lemma 6.1.1 *A sequence $\{\xi_n\}$ is right uniform integrable if at least one of the following conditions is met:*

1. *For any sequence $N(n) \to \infty$ as $n \to \infty$, one has*

$$\mathbf{E}\bigl(\xi_n;\ \xi_n > N(n)\bigr) \to 0.$$

   (This condition is clearly also necessary for uniform integrability.)
2. $\xi_n \le \eta$, *where* $\mathbf{E}\eta < \infty$.
3. $\mathbf{E}(\xi_n^+)^{1+\alpha} < c < \infty$ *for some* $\alpha > 0$ *(here* $x^+ = \max(0, x)$*).*
4. ξ_n *is left uniformly integrable,* $\xi_n \xrightarrow{p} \xi$*, and* $\mathbf{E}\xi_n \to \mathbf{E}\xi < \infty$.

Proof

1. If the sequence $\{\xi_n\}$ were not right uniformly integrable, there would exist an $\varepsilon > 0$ and subsequences $n' \to \infty$ and $N' = N'(n') \to \infty$ such that $\mathbf{E}(\xi_{n'};\ \xi_{n'} > N') > \varepsilon$. But this contradicts condition 1.
2. $\mathbf{E}(\xi_n;\ \xi_n > N) \le \mathbf{E}(\eta;\ \eta > N) \to 0$ as $N \to \infty$.
3. $\mathbf{E}(\xi_n;\ \xi_n > N) \le \mathbf{E}(\xi_n^{1+\alpha} N^{-\alpha};\ \xi_n > N) \le N^{-\alpha} c \to 0$ as $N \to \infty$.
4. Without loss of generality, put $\xi := 0$. Then

$$\mathbf{E}(\xi_n;\ \xi_n > N) = \mathbf{E}\xi_n - \mathbf{E}(\xi_n;\ \xi_n < -N) - \mathbf{E}\bigl(\xi_n;\ |\xi_n| \le N\bigr).$$

The first two terms on the right-hand side vanish as $n \to \infty$ for any $N = N(n) \to \infty$. For the last term, for any $\varepsilon > 0$, one has

$$\begin{aligned}\bigl|\mathbf{E}\bigl(\xi_n;\ |\xi_n| \le N\bigr)\bigr| &\le \bigl|\mathbf{E}\bigl(\xi_n;\ |\xi_n| \le \varepsilon\bigr)\bigr| + \bigl|\mathbf{E}\bigl(\xi_n;\ \varepsilon < |\xi_n| \le N\bigr)\bigr| \\ &\le \varepsilon + N\mathbf{P}\bigl(|\xi_n| > \varepsilon\bigr).\end{aligned}$$

For any given $\varepsilon > 0$, choose an $n(\varepsilon)$ such that, for all $n \ge n(\varepsilon)$, we would have $\mathbf{P}(|\xi_n| > \varepsilon) < \varepsilon$, and put $N(\varepsilon) := \lfloor 1/\sqrt{\varepsilon} \rfloor$. This will mean that, for all $n \ge n(\varepsilon)$ and $N \le N(\varepsilon)$, one has $\mathbf{E}(\xi_n;\ |\xi_n| \le N) < \varepsilon + \sqrt{\varepsilon}$, and therefore condition 1 of the lemma holds for $\mathbf{E}(\xi_n;\ \xi_n > N)$. The lemma is proved. □

Now, based on the above, we can state three useful corollaries.

Corollary 6.1.3 (The dominated convergence theorem) *If* $\xi_n \xrightarrow{p} \xi$, $|\xi_n| < \eta$, *and* $\mathbf{E}\eta < \infty$ *then* $\mathbf{E}\xi$ *exists and* $\mathbf{E}\xi_n \to \mathbf{E}\xi$.

Corollary 6.1.4 *If* $\xi_n \xrightarrow{p} \xi$ *and* $\mathbf{E}|\xi_n|^{r+\alpha} < c < \infty$ *for some* $\alpha > 0$ *then* $\xi_n \xrightarrow{(r)} \xi$.

Corollary 6.1.5 *If* $\xi_n \xrightarrow{p} \xi$ *and* $H(x)$ *is a continuous bounded function, then* $\mathbf{E}|H(\xi_n) - \mathbf{E}H(\xi)| \to 0$ *as* $n \to \infty$.

In conclusion of the present section, we will derive one more auxiliary proposition that can be useful.

Lemma 6.1.2 (On integrals over sets of small probability) *If* $\{\xi_n\}$ *is a uniformly integrable sequence and* $\{A_n\}$ *is an arbitrary sequence of events such that* $\mathbf{P}(A_n) \to 0$, *then* $\mathbf{E}(|\xi_n|;\ A_n) \to 0$ *as* $n \to \infty$.

Proof Put $B_n := \{|\xi_n| \le N\}$. Then

$$\begin{aligned}\mathbf{E}\big(|\xi_n|;\ A_n\big) &= \mathbf{E}\big(|\xi_n|;\ A_n B_n\big) + \mathbf{E}\big(|\xi_n|;\ A_n \overline{B}_n\big) \\ &\le N\mathbf{P}(A_n) + \mathbf{E}\big(|\xi_n|;\ |\xi_n| > N\big).\end{aligned}$$

For a given $\varepsilon > 0$, first choose N so that the second summand on the right-hand side does not exceed $\varepsilon/2$ and then an n such that the first summand does not exceed $\varepsilon/2$. We obtain that, by choosing n large enough, we can make $\mathbf{E}(|\xi_n|;\ A_n)$ less than ε. The lemma is proved. □

6.2 Convergence of Distributions

In Sect. 6.1 we introduced three types of convergence which can be used to characterise the closeness of random variables given on a common probability space. But what can one do if random variables are given on different probability spaces (or if it is not known where they are given) which nevertheless have similar distributions? (Recall, for instance, the Poisson or de Moivre–Laplace theorems.) In such cases one should be able to characterise the closeness of the distributions themselves. Having found an apt definition for such a closeness, in many problems we will be able to approximate the required but hard to come by distributions by known and, as a rule, simpler distributions.

Now what distributions should be considered as close? We are clearly looking for a definition of convergence of a sequence of distribution functions $F_n(x)$ to a distribution function $F(x)$. It would be natural, for instance, that the distributions of the variables $\xi_n = \xi + 1/n$ should converge to that of ξ as $n \to \infty$. Therefore requiring in the definition of convergence that $\sup_x |F_n(x) - F(x)|$ is small would be unreasonable since this condition is not satisfied for the distributions of $\xi + 1/n$ and ξ if $F(x) = \mathbf{P}(\xi < x)$ has at least one point of discontinuity.

We will define the convergence of F_n to F as that which arises when one considers convergence in probability.

Definition 6.2.1 We will say that distribution functions F_n *converge weakly* to a distribution function F as $n \to \infty$, and denote this by $F_n \Rightarrow F$ if, for any continuous bounded function $f(x)$,

$$\int f(x)\,dF_n(x) \to \int f(x)\,dF(x). \tag{6.2.1}$$

Considering the distributions $\mathbf{F}_n(B)$ and $\mathbf{F}(B)$ (B are Borel sets) corresponding to F_n and F, we say that $\mathbf{F}_n$ converges weakly to $\mathbf{F}$ and write $\mathbf{F}_n \Rightarrow \mathbf{F}$. One can clearly re-write (6.2.1) as

$$\int f(x)\,\mathbf{F}_n(dx) \to \int f(x)\,\mathbf{F}(dx) \quad \text{or} \quad \mathbf{E}f(\xi_n) \to \mathbf{E}f(\xi) \tag{6.2.2}$$

(cf. Corollary 6.1.5), where $\xi_n \in \mathbf{F}_n$ and $\xi \in \mathbf{F}$.

Another possible definition of weak convergence follows from the next assertion.

Theorem 6.2.1 [2] $\mathbf{F}_n \Rightarrow \mathbf{F}$ *if and only if* $F_n(x) \to F(x)$ *at each point of continuity* x *of* F.

Proof Let (6.2.1) hold. Consider an $\varepsilon > 0$ and a continuous function $f_\varepsilon(t)$ which is equal to 1 for $t < x$ and to 0 for $t \ge x + \varepsilon$, and varies linearly on $[x, x+\varepsilon]$. Since

$$F_n(x) = \int_{-\infty}^{x} f_\varepsilon(t)\,dF_n(t) \le \int f_\varepsilon(t)\,dF_n(t),$$

by virtue of (6.2.1) one has

$$\limsup_{n\to\infty} F_n(x) \le \int f_\varepsilon(t)\,dF(t) \le F(x+\varepsilon).$$

If x is a point of continuity of F then

$$\limsup_{n\to\infty} F_n(x) \le F(x)$$

since ε is arbitrary.

In the same way, using the function $f_\varepsilon^*(t) = f_\varepsilon(t+\varepsilon)$, we obtain the inequality

$$\liminf_{n\to\infty} F_n(x) \ge F(x).$$

We now prove the converse assertion. Let $-M$ and N be points of continuity of F such that $F(-M) < \varepsilon/5$ and $1 - F(N) < \varepsilon/5$. Then $F_n(-M) < \varepsilon/4$ and $1 - F_n(N) < \varepsilon/4$ for all sufficiently large n. Therefore, assuming for simplicity that $|f| \le 1$, we obtain that

$$\int f\,dF_n \quad \text{and} \quad \int f\,dF \tag{6.2.3}$$

[2] In many texts on probability theory the condition of the theorem is given as the definition of weak convergence. However, the definition in terms of the relation (6.2.2) is apparently more appropriate for it continues to remain valid for distributions on arbitrary topological spaces (see, e.g. [1, 25]).

will differ from

$$\int_{-M}^{N} f\,dF_n \quad \text{and} \quad \int_{-M}^{N} f\,dF,$$

respectively, by less than $\varepsilon/2$. Construct on the semi-interval $(-M, N]$ a step function f_ε with jumps at the points of continuity of F which differs from f by less than $\varepsilon/2$. Outside $(-M, N]$ we set $f_\varepsilon := 0$. We can put, for instance,

$$f_\varepsilon(x) := \sum_{j=1}^{k} f(x_j)\delta_j(x),$$

where $x_0 = -M < x_1 < \cdots < x_k = N$ are appropriately chosen points of continuity of F, and $\delta_j(x)$ is the indicator function of the semi-interval $(x_{j-1}, x_j]$. Then $\int f_\varepsilon\,dF_n$ and $\int f_\varepsilon\,dF$ will differ from the respective integrals in (6.2.3), for sufficiently large n, by less than ε. At the same time,

$$\int f_\varepsilon\,dF_n = \sum_{j=1}^{k} f(x_j)\bigl[F_n(x_j) - F_n(x_{j-1})\bigr] \to \int f_\varepsilon\,dF.$$

Since $\varepsilon > 0$ is arbitrary, the last relation implies (6.2.1). (Indeed, one just has to make use of the inequality

$$\limsup \int f\,dF_n \le \varepsilon + \limsup \int f_\varepsilon\,dF_n = \varepsilon + \int f_\varepsilon\,dF \le 2\varepsilon + \int f\,dF$$

and a similar inequality for $\liminf \int f\,dF_n$.) The theorem is proved. □

For remarks on different and, in a certain sense, simpler proofs of the second assertion of Theorem 6.2.1, see the end of Sect. 6.3 and Sect. 7.4.

Remark 6.2.1 Repeating with obvious modifications the above-presented proof, we can get a somewhat different equivalent of convergence (4): *convergence of differences $F_n(y) - F_n(x) \to F(y) - F(x)$ for any points of continuity x and y of F.*

Remark 6.2.2 If $F(x)$ is continuous then convergence $\mathbf{F}_n \Rightarrow \mathbf{F}$ is equivalent to the *uniform* convergence $\sup_x |F_n(x) - F(x)| \to 0$.

We leave the proof of the last assertion to the reader. It follows from the fact that convergence $F_n(x) \to F(x)$ at any x implies, by virtue of the continuity of F, uniform convergence on any finite interval. The uniform smallness of $F_n(x) - F(x)$ on the "tails" is ensured by the smallness of $F(x)$ and $1 - F(x)$.

Remark 6.2.3 If distributions $\mathbf{F}_n$ and $\mathbf{F}$ are discrete and have jumps at the same points $x_1, x_2, \ldots$ then $\mathbf{F}_n \Rightarrow \mathbf{F}$ will clearly be equivalent to the convergence of the probabilities of the values $x_1, x_2, \ldots$ $(F_n(x_k + 0) - F_n(x_k) \to F(x_k + 0) - F(x_k))$.

We introduce some notation which will be convenient for the sequel. Let ξ_n and ξ be some random variables (given, generally speaking, on different probability spaces) such that $\xi_n \in \mathbf{F}_n$ and $\xi \in \mathbf{F}$.

Definition 6.2.2 If $\mathbf{F}_n \Rightarrow \mathbf{F}$ we will say that ξ_n *converges to* ξ *in distribution* and write $\xi_n \Rightarrow \xi$.

We used here the same symbol $\Rightarrow$ as for the weak convergence, but this leads to no confusion.

It is clear that $\xi_n \xrightarrow{p} \xi$ implies $\xi_n \Rightarrow \xi$, but not vice versa.

At the same time the following assertion holds true.

Lemma 6.2.1 *If* $\xi_n \Rightarrow \xi$ $(\mathbf{F}_n \Rightarrow \mathbf{F})$ *then one can construct random variables* ξ_n' *and* ξ' *on a common probability space so that* $\mathbf{P}(\xi_n' < x) = \mathbf{P}(\xi_n < x) = F_n(x)$, $\mathbf{P}(\xi' < x) = \mathbf{P}(\xi < x) = F(x)$, *and*

$$\xi_n' \xrightarrow{a.s.} \xi'.$$

Proof Define the quantile transforms (see Definition 3.2.6) by

$$F_n^{-1}(t) := \sup\{x : F_n(x) \le t\}, \qquad F^{-1}(t) := \sup\{x : F(x) \le t\}.$$

(If $F(x)$ is continuous and strictly increasing then $F^{-1}(t)$ coincides with the solution to the equation $F(v) = t$.) Let $\eta \in \mathbf{U}_{0,1}$. Put

$$\xi_n' := F_n^{-1}(\eta) \in F_n, \quad \xi' := F^{-1}(\eta) \in F$$

(cf. Theorem 3.2.2), and show that $\xi_n' \xrightarrow{a.s.} \xi'$. In order to do that, it suffices to prove that $F_n^{-1}(y) \to F^{-1}(y)$ for almost all $y \in [0, 1]$.

The functions F and F^{-1} are monotone and hence each of them has at most a countable set of discontinuity points. This means that, for all $y \in [0, 1]$ with the possible exclusion of the points from a countable set T, the function $F^{-1}(y)$ will be continuous.

So let y be a point of continuity of $F^{(-1)}$, and $F^{(-1)}(y) = x$.

For $t \le y$, choose a continuous strictly increasing function $G^{(-1)}(t)$ such that

$$G^{(-1)}(y) = F^{(-1)}(y), \qquad G^{(-1)}(t) \le F^{(-1)}(t) \quad \text{for } t \le y.$$

Denote by $G(v)$, $v \le x$, the function inverse to $G^{(-1)}(t)$. Clearly, $G(v)$ dominates the function $F(v)$ in the domain $v \le x$. By virtue of the continuity and strict monotonicity of the functions $G^{(-1)}$ and G (in the domain under consideration), for $\varepsilon > 0$ we have

$$G(x - \varepsilon) = y - \delta(\varepsilon),$$

where $\delta(\varepsilon) > 0$, $\delta(\varepsilon) \to 0$ as $\varepsilon \to 0$. Choose an ε such that $x - \varepsilon$ is a point of continuity of F. Then, for all n large enough,

$$F_n(x - \varepsilon) \le F(x - \varepsilon) + \frac{\delta(\varepsilon)}{2} \le G(x - \varepsilon) + \frac{\delta(\varepsilon)}{2} = y - \frac{\delta(\varepsilon)}{2}.$$

The opposite inequality can be proved in a similar way. Since ε can be arbitrarily small, we obtain that, for almost all y,

$$F_n^{-1}(y) \to F^{(-1)}(y) \quad \text{as } n \to \infty.$$

Hence $F_n^{(-1)}(\eta) \to F^{(-1)}(\eta)$ with probability 1 with respect to the distribution of η. The lemma is proved. □

Lemma 6.2.1 remains true for vector-valued random variables as well.

Sometimes it is also convenient to have a simple symbol for the relation "the distribution of ξ_n converges weakly to $\mathbf{F}$". We will write this relation as

$$\xi_n \Subset\!\!\Rightarrow \mathbf{F}, \tag{6.2.4}$$

so that the symbol $\Subset\!\!\Rightarrow$ expresses the same fact as $\Rightarrow$ but relates objects of a different nature in the same way as the symbol $\Subset$ in the relation $\xi \Subset \mathbf{P}$ (on the left-hand side in (6.2.4) we have random variables, while on the right hand side there is a distribution).

In these terms, the assertion of the Poisson theorem could be written as $S_n \Subset\!\!\Rightarrow \mathbf{\Pi}_\mu$, while the statement of the law of large numbers for the Bernoulli scheme takes the form $S_n/n \Subset\!\!\Rightarrow \mathbf{I}_p$.

The coincidence of the distributions of ξ and η will be denoted by $\xi \stackrel{d}{=} \eta$.

Lemma 6.2.2 *If $\xi_n \Rightarrow \xi$ and $\varepsilon_n \xrightarrow{p} 0$ then $\xi_n + \varepsilon_n \Rightarrow \xi$.*

If $\xi_n \Rightarrow \xi$ and $\gamma_n \xrightarrow{p} 1$ then $\xi_n \gamma_n \Rightarrow \xi$.

Proof Let us prove the first assertion. For any t and $\delta > 0$ such that t and $t \pm \delta$ are points of continuity of $\mathbf{P}(\xi < t)$, one has

$$\begin{aligned}\limsup_{n\to\infty} \mathbf{P}(\xi_n + \varepsilon_n < t) &= \limsup_{n\to\infty} \mathbf{P}(\xi_n + \varepsilon_n < t,\ \varepsilon_n > -\delta) \\ &\le \limsup_{n\to\infty} \mathbf{P}(\xi_n < t + \delta) = \mathbf{P}(\xi < t + \delta).\end{aligned}$$

Similarly,

$$\liminf_{n\to\infty} \mathbf{P}(\xi_n + \varepsilon_n < t) \ge \mathbf{P}(\xi < t - \delta).$$

Since $\mathbf{P}(\xi < t \pm \delta)$ can be chosen arbitrary close to $\mathbf{P}(\xi < t)$ by taking a sufficiently small δ, the required convergence follows.

The second assertion can be proved in the same way. The lemma is proved. □

Now we will give analogues of Theorems 6.1.4 and 6.1.7 in terms of distributions.

Theorem 6.2.2 *If $\xi_n \Rightarrow \xi$ and a function $H(s)$ satisfies the conditions of Theorem* 6.1.4 *then $H(\xi_n) \Rightarrow H(\xi)$.*

Theorem 6.2.3 *If $\xi_n \Rightarrow \xi$ and the sequence $\{\xi_n\}$ is uniformly integrable then* $\mathbf{E}\xi$ *exists and* $\mathbf{E}\xi_n \to \mathbf{E}\xi$.

Proof *There are two ways of proving these theorems.* One of them consists of reducing them to Theorems 6.1.4 and 6.1.7. To this end, one has to construct random variables $\xi'_n = F_n^{(-1)}(\eta)$ and $\xi' = F^{(-1)}(\eta)$, where $\eta \in \mathbf{U}_{0,1}$ and $F_n^{(-1)}$ and $F^{(-1)}$ are the quantile transforms of F_n and F, respectively, and prove that $\xi'_n \xrightarrow{p} \xi'$ (we already know that $F^{(-1)}(\eta) \in \mathbf{F}$; if F is discontinuous or not strictly increasing, then $F^{(-1)}$ should be defined as in Lemma 6.2.1).

Another approach is to prove the theorems anew using the language of distributions. Under inessential additional assumptions, such proofs are sometimes even simpler. To illustrate this, assume, for instance, in Theorem 6.2.3 that the function H is continuous. One has to prove that $\mathbf{E}g(H(\xi_n)) \to \mathbf{E}g(H(\xi))$ for any continuous bounded function g. But this is an immediate consequence of (6.2.1) and (6.2.2), for $f = g \circ H$ (f is the composition of the functions g and H).

In Theorem 6.2.3 assume that $\xi_n \ge 0$ (this does not restrict the generality). Then, integrating by parts, we get

$$\mathbf{E}\xi_n = -\int_0^\infty x\,d\mathbf{P}(\xi_n \ge x) = \int_0^\infty \mathbf{P}(\xi_n \ge x)\,dx. \tag{6.2.5}$$

Since by virtue of uniform integrability

$$\sup_n \int_N^\infty \mathbf{P}(\xi_n \ge x)\,dx \le \sup_n \mathbf{E}(\xi_n;\ \xi_n \ge N) \to 0$$

as $N \to \infty$, the integral in (6.2.5) is uniformly convergent. Moreover, $\mathbf{P}(\xi_n \ge x) \to \mathbf{P}(\xi \ge x)$ a.s., and therefore

$$\lim_{n\to\infty} \mathbf{E}\xi_n = \lim_{n\to\infty} \int_0^\infty \mathbf{P}(\xi_n \ge x)\,dx = \int_0^\infty \mathbf{P}(\xi \ge x)\,dx = \mathbf{E}\xi. \qquad \square$$

Conditions ensuring uniform integrability are contained in Lemma 6.1.1. Now we will give a modification of assertion 4 of this lemma for the case of weak convergence.

Lemma 6.2.3 *If $\{\xi_n\}$ is left uniformly integrable, $\xi_n \Rightarrow \xi$ and $\mathbf{E}\xi_n \to \mathbf{E}\xi$ then $\{\xi_n\}$ is uniformly integrable.*

We suggest to the reader to construct examples showing that all three conditions of the lemma are essential.

Lemma 6.2.3 implies, in particular, that *if $\xi_n \ge 0$, $\xi_n \Rightarrow \xi$ and $\mathbf{E}\xi_n \to \mathbf{E}\xi$ then $\{\xi_n\}$ is uniformly integrable.*

As for Theorems 6.2.2 and 6.2.3, two alternative ways to prove the result are possible here. One of them consists of using Lemma 6.1.1. We will present here a different, somewhat simpler, proof.

Proof of Lemma 6.2.3 For simplicity assume that $\xi_n \geq 0$. Suppose that the lemma is not valid. Then there exist an $\varepsilon > 0$ and subsequences $n' \to \infty$ and $N(n') \to \infty$ such that

$$\mathbf{E}\big(\xi_{n'};\ \xi_{n'} > N(n')\big) > \varepsilon.$$

Since

$$\mathbf{E}\xi_{n'} = \mathbf{E}(\xi_{n'};\ \xi_{n'} \leq N) + \mathbf{E}(\xi_{n'};\ \xi_{n'} > N),$$

for any N that is a point of continuity of the distribution of ξ, one has

$$\mathbf{E}\xi = \lim_{n\to\infty} \xi_{n'} \geq \mathbf{E}(\xi;\ \xi \leq N) + \varepsilon.$$

Choose an N such that the first summand on the right-hand side exceeds $\mathbf{E}\xi - \varepsilon/2$. Then we obtain the contradiction $\mathbf{E}\xi \geq \mathbf{E}\xi + \varepsilon/2$, which proves the lemma.

We leave it to the reader to extend the proof to the case of arbitrary left uniformly integrable $\{\xi_n\}$. □

The following theorem can also be useful.

Theorem 6.2.4 *Suppose that $\xi_n \Rightarrow \xi$, $H(s)$ is differentiable at a point a, and $b_n \to 0$ as $n \to \infty$. Then*

$$\frac{1}{b_n}\big(H(a + b_n\xi_n) - H(a)\big) \Rightarrow \xi H'(a).$$

If $H'(a) = 0$ and $H''(a)$ exists then

$$\frac{1}{b_n^2}\big(H(a + b_n\xi_n) - H(a)\big) \Rightarrow \frac{\xi^2}{2} H''(a).$$

Proof Consider the function

$$h(x) = \begin{cases} \frac{H(a+x)-H(a)}{x} & \text{if } x \neq 0, \\ H'(a) & \text{if } x = 0, \end{cases}$$

which is continuous at the point $x = 0$. Since $b_n\xi_n \Rightarrow 0$, by Theorem 6.2.2 one has $h(b_n\xi_n) \Rightarrow h(0) = H'(a)$. Using the theorem again (this time for two-dimensional distributions), we get

$$\frac{H(a + b_n\xi_n) - H(a)}{b_n} = h(b_n\xi_n)\xi_n \Rightarrow H'(a)\xi.$$

The second assertion is proved in the same way. □

A multivariate analogue of this theorem will look somewhat more complicated. The reader could obtain it himself, following the lines of the argument proving Theorem 6.2.4.

6.3 Conditions for Weak Convergence

Now we will return to the concept of weak convergence. We have two criteria for this convergence: relation (6.2.1) and Theorem 6.2.1. However, from the point of view of their possible applications (their verification in concrete problems) both these criteria are inconvenient. For instance, proving, say, convergence $\mathbf{E}f(\xi_n) \to \mathbf{E}f(\xi)$ not for all continuous bounded functions f but just for elements f of a certain rather narrow class of functions that has a simple and clear nature would be much easier. It is obvious, however, that such a class cannot be very narrow.

Before stating the basic assertions, we will introduce a few concepts.

Extend the class $\mathcal{F}$ of all distribution functions to the class $\mathcal{G}$ of all functions G satisfying conditions F1 and F2 from Sect. 3.2 and conditions $G(-\infty) \ge 0$, $G(\infty) \le 1$. Functions G from $\mathcal{G}$ could be called generalised distribution functions. One can think of them as distribution functions of improper random variables assuming infinite values with positive probabilities, so that $G(-\infty) = \mathbf{P}(\xi = -\infty)$ and $1 - G(\infty) = \mathbf{P}(\xi = \infty)$. We will write $G_n \Rightarrow G$ for $G_n \in \mathcal{G}$ and $G \in \mathcal{G}$ if $G_n(x) \to G(x)$ at all points of continuity of $G(x)$.

Theorem 6.3.1 (Helly) *The class $\mathcal{G}$ is compact with respect to convergence $\Rightarrow$, i.e. from any sequence $\{G_n\}$, $G_n \in \mathcal{G}$, one can choose a convergent subsequence $G_{n_k} \Rightarrow G \in \mathcal{G}$.*

For the proof of Theorem 6.3.1, see Appendix 4.

Corollary 6.3.1 *If each convergent subsequence $\{G_{n_k}\}$ of $\{G_n\}$ with $G_n \in \mathcal{G}$ converges to G then $G_n \Rightarrow G$.*

Proof If $G_n \not\Rightarrow G$ then there exists a point of continuity x_0 of G such that $G_n(x_0) \not\to G(x_0)$. Since $G_n(x_0) \in [0, 1]$, there exists a convergent subsequence G_{n_k} such that $G_{n_k}(x_0) \to g \ne G(x_0)$. This, however, is impossible by our assumption, for $G_{n_k}(x_0) \to G(x_0)$. □

The reason for extending the class $\mathcal{F}$ of all distribution functions is that it is not compact (in the sense of Theorem 6.3.1) and convergence $F_n \Rightarrow G$, $F_n \in \mathcal{F}$, does not imply that $G \in \mathcal{F}$. For example, the sequence

$$F_n(x) = \begin{cases} 0 & \text{if } x \le -n, \\ 1/2 & \text{if } -n < x \le n, \\ 1 & \text{if } x > n \end{cases} \tag{6.3.1}$$

converges everywhere to the function $G(x) \equiv 1/2 \notin \mathcal{F}$ corresponding to an improper random variable taking the values $\pm\infty$ with probabilities 1/2.

However, dealing with the class $\mathcal{G}$ is also not very convenient. The fact is that convergence at points of continuity $G_n \Rightarrow G$ in the class $\mathcal{G}$ is not equivalent to convergence

$$\int f\,dG_n \to \int f\,dG$$

(see example (6.3.1) for $f \equiv 1$), and the integrals $\int f\,dG$ do not specify G uniquely (they specify the *increments* of G, but not the values $G(-\infty)$ and $G(\infty)$). Now we will introduce two concepts that will help to avoid the above-mentioned inconvenience.

Definition 6.3.1 A sequence of distributions $\{\mathbf{F}_n\}$ (or distribution functions $\{F_n\}$) is said to be *tight* if, for any $\varepsilon > 0$, there exists an N such that

$$\inf_n \mathbf{F}_n\big([-N, N]\big) > 1 - \varepsilon. \tag{6.3.2}$$

Definition 6.3.2 A class $\mathfrak{L}$ of continuous bounded functions is said to be *distribution determining* if the equality

$$\int f(x)\,dF(x) = \int f(x)\,dG(x), \quad F \in \mathfrak{F},\ G \in \mathfrak{G},$$

for all $f \in \mathfrak{L}$ implies that $F = G$ (or, which is the same, if the relation $\mathbf{E} f(\xi) = \mathbf{E} f(\eta)$ for all $f \in \mathfrak{L}$, where one of the random variables ξ and η is proper, implies that $\xi \stackrel{d}{=} \eta$).

The next theorem is the main result of the present section.

Theorem 6.3.2 *Let $\mathfrak{L}$ be a distribution determining class and $\{F_n\}$ a sequence of distributions. For the existence of a distribution $F \in \mathfrak{F}$ such that $F_n \Rightarrow F$ it is necessary and sufficient that:*[3]

(1) *the sequence $\{F_n\}$ is tight; and*
(2) $\lim_{n\to\infty} \int f\,dF_n$ *exists for all $f \in \mathfrak{L}$.*

Proof The *necessary* part is obvious.

Sufficiency. By Theorem 6.3.1 there exists a subsequence $F_{n_k} \Rightarrow F \in \mathfrak{G}$. But by condition (1) one has $F \in \mathfrak{F}$. Indeed, if $x \ge N$ is a point of continuity of F then, by Definition 6.3.1, $F(x) = \lim F_{n_k}(x) \ge 1 - \varepsilon$. In a similar way we establish that for $x \le -N$ one has $F(x) < \varepsilon$. Since ε is arbitrary, we have $F(-\infty) = 0$ and $F(\infty) = 1$.

Further, take another convergent subsequence $F_{n'_k} \Rightarrow G \in \mathfrak{F}$. Then, for any $f \in \mathfrak{L}$, one has

$$\lim \int f\,dF_{n_k} = \int f\,dF, \qquad \lim \int f\,dF_{n'_k} = \int f\,dG. \tag{6.3.3}$$

But, by condition (2),

$$\int f\,dF = \int f\,dG, \tag{6.3.4}$$

and hence $F = G$. The theorem is proved by virtue of Corollary 6.3.1. □

---

[3]In this form the theorem persists for spaces of a more general nature. The role of the segments $[-N, N]$ in (6.3.2) is played in that case by compact sets (cf. [1, 14, 25, 31]).

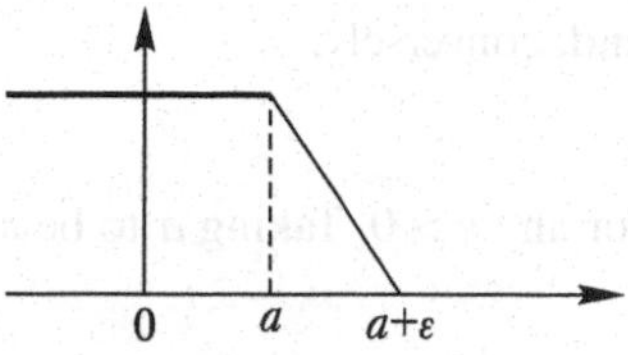

Fig. 6.1 The plot of the function $f_{a,\varepsilon}(x)$ from Example 6.3.1

If one needs to prove convergence to a "known" distribution $F \in \mathcal{F}$, the tightness condition in Theorem 6.3.2 becomes redundant.

Corollary 6.3.2 *Let $\mathcal{L}$ be a distribution determining class and*

$$\int f\,dF_n \to \int f\,dF, \quad F \in \mathcal{G}, \tag{6.3.5}$$

for any $f \in \mathcal{L}$. Moreover, assume that at least one of the following three conditions is met:

(1) *the sequence $\{F_n\}$ is tight;*
(2) *$F \in \mathcal{F}$;*
(3) *$f \equiv 1 \in \mathcal{L}$ (i.e. (6.3.5) holds for $f \equiv 1$).*

Then $F \in \mathcal{F}$ and $F_n \Rightarrow F$.

The proof of the corollary is almost next to obvious. Under condition (1) the assertion follows immediately from Theorem 6.3.2. Condition (3) and convergence (6.3.5) imply condition (2). If (2) holds, then $F \in \mathcal{F}$ in relations (6.3.3) and (6.3.4), and therefore $G = F$. □

Since, as a rule, at least one of conditions (1)–(3) is satisfied (as we will see below), the basic task is to verify convergence (6.3.5) for the class $\mathcal{L}$.

Note also that, in the case where one proves *convergence to a distribution $F \in \mathcal{F}$ "known" in advance, the whole arrangement of the argument can be different and simpler*. One such alternative approach is presented in Sect. 7.4.

Now we will give several examples of distributions determining classes $\mathcal{L}$.

Example 6.3.1 The class $\mathcal{L}_0$ of functions having the form

$$f_{a,\varepsilon}(x) = \begin{cases} 1 & \text{if } x \le a, \\ 0 & \text{if } x \ge a + \varepsilon. \end{cases}$$

On the segment $[a, a+\varepsilon]$ the functions $f_{a,\varepsilon}$ are defined to be linear and continuous (a plot of $f_{a,\varepsilon}(x)$ is given in Fig. 6.1). It is a two-parameter family of functions.

We show that $\mathcal{L}_0$ is a distribution determining class. Let

$$\int f\,dF = \int f\,dG$$

for all $f \in \mathcal{L}_0$. Then

$$F(a) \le \int f_{a,\varepsilon}\,dF = \int f_{a,\varepsilon}\,dG \le G(a+\varepsilon),$$

and, conversely,

$$G(a) \le F(a+\varepsilon)$$

for any $\varepsilon > 0$. Taking a to be a point of continuity of both F and G, we obtain that

$$F(a) = G(a).$$

Since this is valid for all points of continuity, we get $F = G$.

One can easily verify in a similar way that the class $\widehat{\mathcal{L}}_0$ of "trapezium-shaped" functions $f(x) = \min(f_{a,\varepsilon}, 1 - f_{b,\varepsilon})$, $a < b$, is also distribution determining.

Example 6.3.2 The class $\mathcal{L}_1$ of continuous bounded functions such that, for each $f \in \mathcal{L}_0$ (or $f \in \widehat{\mathcal{L}}_0$), there exists a sequence $f_n \in \mathcal{L}_1$, $\sup_x |f(x)| < M < \infty$, for which $\lim_{n\to\infty} f_n(x) = f(x)$ for each $x \in \mathbb{R}$.

Let

$$\int f\,dF = \int f\,dG$$

for all $f \in \mathcal{L}_1$. By the dominated convergence theorem,

$$\lim \int f_n\,dF = \int f\,dF, \qquad \lim \int f_n\,dG = \int f\,dG, \qquad f \in \mathcal{L}_0.$$

Therefore

$$\int f\,dF = \int f\,dG, \quad f \in \mathcal{L}_0, \qquad \mathfrak{F} = \mathfrak{G}$$

and hence $\mathcal{L}_1$ is a distribution determining class.

Example 6.3.3 The class C_k of all bounded functions $f(x)$ having bounded uniformly continuous k-th derivatives $f^{(k)}(x)$ ($\sup_x |f^{(k)}(x)| < \infty$), $k \ge 1$.

It is evident that C_k is a distribution determining class for it is a special case of an $\mathcal{L}_1$ class.

In the same way one can see that the subclass $C_k^0 \subset C_k$ of functions having finite support (vanishing outside a finite interval) is also distribution determining. This follows from the fact that C_k^0 is an $\mathcal{L}_1$-class with respect to the class $\widehat{\mathcal{L}}_0$ of trapezium-shaped (and therefore having compact support) functions.

It is clear that the class C_k satisfies condition (3) from Corollary 6.3.2 ($f \equiv 1 \in C_k$). Therefore, to prove convergence $F_n \Rightarrow F \in \mathfrak{F}$ it suffices to verify convergence (6.3.5) for $f \in C_k$ only.

If one takes $\mathcal{L}$ to be the class C_k^0 of differentiable functions with finite support then relation (6.3.5) together with condition (2) of Corollary 6.3.2 could be re-written as

$$\int F_n f'\,dx \to \int F f'\,dx, \qquad F \in \mathfrak{F}. \tag{6.3.6}$$

(One has to integrate (6.3.5) by parts and use the fact that f' also has a finite support.) The convergence criterion (6.3.6) is sometimes useful. It can be used to show, for example, that (6.3.5) follows from convergence $F_n(x) \to F(x)$ at all points of continuity of F (i.e. almost everywhere), since that convergence and the dominated convergence theorem imply (6.3.6) which is equivalent to (6.3.5).

Example 6.3.4 One of the most important distribution determining classes is the one-parameter family of complex-valued functions $\{e^{itx}\}$, $t \in \mathbb{R}$.

The next chapter will be devoted to studying the properties of $\int e^{itx} dF(x)$.

After obvious changes, all the material in the present chapter can be extended to the multivariate case.

Chapter 7
Characteristic Functions

Abstract Section 7.1 begins with formal definitions and contains an extensive discussion of the basic properties of characteristic functions, including those related to the nature of the underlying distributions. Section 7.2 presents the proofs of the inversion formulas for both densities and distribution functions, and also in the space of square integrable functions. Then the fundamental continuity theorem relating pointwise convergence of characteristic functions to weak convergence of the respective distributions is proved in Sect. 7.3. The result is illustrated by proving the Poisson theorem, with a bound for the convergence rate, in Sect. 7.4. After that, the previously presented theory is extended in Sect. 7.5 to the multivariate case. Some applications of characteristic functions are discussed in Sect. 7.6, including the stability properties of the normal and Cauchy distributions and an in-depth discussion of the gamma distribution and its properties. Section 7.7 introduces the concept of generating functions and uses it to analyse the asymptotic behaviour of a simple Markov discrete time branching process. The obtained results include the formula for the eventual extinction probability, the asymptotic behaviour of the non-extinction probabilities in the critical case, and convergence in that case of the conditional distributions of the scaled population size given non-extinction to the exponential law.

7.1 Definition and Properties of Characteristic Functions

As a preliminary remark, note that together with real-valued random variables $\xi(\omega)$ we could also consider complex-valued random variables, by which we mean functions of the form $\xi_1(\omega) + i\xi_2(\omega)$, (ξ_1, ξ_2) being a random vector. It is natural to put $\mathbf{E}(\xi_1 + i\xi_2) = \mathbf{E}\xi_1 + i\mathbf{E}\xi_2$. Complex-valued random variables $\xi = \xi_1 + i\xi_2$ and $\eta = \eta_1 + i\eta_2$ are *independent* if the σ-algebras $\sigma(\xi_1, \xi_2)$ and $\sigma(\eta_1, \eta_2)$ generated by the vectors (ξ_1, ξ_2) and (η_1, η_2), respectively, are independent. It is not hard to verify that, for such random variables,

$$\mathbf{E}\xi\eta = \mathbf{E}\xi\mathbf{E}\eta.$$

Definition 7.1.1 The *characteristic function* (ch.f.) of a real-valued random variable ξ is the complex-valued function

A.A. Borovkov, *Probability Theory*, Universitext,
DOI 10.1007/978-1-4471-5201-9_7, © Springer-Verlag London 2013

$$\varphi_\xi(t) := \mathbf{E}e^{it\xi} = \int e^{itx}\, dF(x),$$

where t is real.

If the distribution function $F(x)$ has a density $f(x)$ then the ch.f. is equal to

$$\varphi_\xi(t) = \int e^{itx} f(x)\, dx$$

and is just the Fourier transform of the function $f(x)$.[1] In the general case, the ch.f. is the *Fourier–Stieltjes transform* of the function $F(x)$.

The ch.f. exists for any random variable ξ. This follows immediately from the relation

$$\left|\varphi_\xi(t)\right| \le \int \left|e^{itx}\right| dF(x) \le \int 1\, dF(x) = 1.$$

Ch.f.s are a powerful tool for studying properties of the sums of independent random variables.

7.1.1 Properties of Characteristic Functions

1. *For any random variable* ξ,

$$\varphi_\xi(0) = 1 \quad \text{and} \quad \left|\varphi_\xi(t)\right| \le 1 \quad \text{for all } t.$$

This property is obvious.

2. *For any random variable* ξ,

$$\varphi_{a\xi+b}(t) = e^{itb}\varphi_\xi(ta).$$

Indeed,

$$\varphi_{a\xi+b}(t) = \mathbf{E}e^{it(a\xi+b)} = e^{itb}\mathbf{E}e^{iat\xi} = e^{itb}\varphi_\xi(ta).$$ □

[1]More precisely, in classical mathematical analysis, the Fourier transform $\varphi(t)$ of a function $f(t)$ from the space L_1 of integrable functions is defined by the equation

$$\varphi(t) = \frac{1}{\sqrt{2\pi}} \int e^{itx} f(t)\, dt$$

(the difference from ch.f. consists in the factor $1/\sqrt{2\pi}$). Under this definition the inversion formula has a "symmetric" form: if $\varphi \in L_1$ then

$$f(x) = \frac{1}{\sqrt{2\pi}} \int e^{-itx}\varphi(t)\, dt.$$

This representation is more symmetric than the inversion formula for ch.f. (7.2.1) in Sect. 7.2 below.

3. *If $\xi_1, \dots, \xi_n$ are independent random variables then the ch.f. of the sum $S_n = \xi_1 + \dots + \xi_n$ is equal to*

$$\varphi_{S_n}(t) = \varphi_{\xi_1}(t) \cdots \varphi_{\xi_n}(t).$$

Proof This follows from the properties of the expectation of the product of independent random variables. Indeed,

$$\begin{aligned}\varphi_{S_n}(t) = \mathbf{E}e^{it(\xi_1+\dots+\xi_n)} &= \mathbf{E}e^{it\xi_1}e^{it\xi_2}\dots e^{it\xi_n} \\ &= \mathbf{E}e^{it\xi_1}\mathbf{E}e^{it\xi_2}\dots\mathbf{E}e^{it\xi_n} = \varphi_{\xi_1}(t)\varphi_{\xi_2}(t)\cdots\varphi_{\xi_n}(t).\end{aligned}$$ □

Thus to the convolution $F_{\xi_1} * F_{\xi_2}$ there corresponds the product $\varphi_{\xi_1}\varphi_{\xi_2}$.

4. *The ch.f. $\varphi_\xi(t)$ is a uniformly continuous function.*
Indeed, as $h \to 0$,

$$\left|\varphi(t+h) - \varphi(t)\right| = \left|\mathbf{E}\left(e^{i(t+h)\xi} - e^{it\xi}\right)\right| \le \mathbf{E}\left|e^{ih\xi} - 1\right| \to 0$$

by the dominated convergence theorem (see Corollary 6.1.2) since $|e^{ih\xi} - 1| \xrightarrow{p} 0$ as $h \to 0$, and $|e^{ih\xi} - 1| \le 2$. □

5. *If the k-th moment exists*: $\mathbf{E}|\xi|^k < \infty$, $k \ge 1$, *then there exists a continuous k-th derivative of the function $\varphi_\xi(t)$, and $\varphi^{(k)}(0) = i^k\mathbf{E}\xi^k$.*

Proof Indeed, since

$$\left|\int ixe^{itx}\,dF(x)\right| \le \int |x|\,dF(x) = \mathbf{E}|\xi| < \infty,$$

the integral $\int ixe^{itx}\,dF(x)$ converges uniformly in t. Therefore one can differentiate under the integral sign:

$$\varphi'(t) = i\int xe^{itx}\,dF(x), \qquad \varphi'(0) = i\mathbf{E}\xi.$$

Further, one can argue by induction. If, for $l < k$,

$$\varphi^{(l)}(t) = i^l\int x^l e^{itx}\,dF(x),$$

then

$$\varphi^{(l+1)}(t) = i^{l+1}\int x^{l+1}e^{itx}\,dF(x)$$

by the uniform convergence of the integral on the right-hand side. Therefore

$$\varphi^{(l+1)}(0) = i^{l+1}\mathbf{E}\xi^{l+1}.$$ □

Property 5 implies that if $\mathbf{E}|\xi|^k < \infty$ then, in a neighbourhood of the point $t = 0$, one has the expansion

$$\varphi(t) = 1 + \sum_{j=1}^{k} \frac{(it)^j}{j!} \mathbf{E}\xi^j + o(|t^k|). \tag{7.1.1}$$

The converse assertion is only partially true:

If a derivative of an even order $\varphi^{(2k)}$ *exists then*

$$\mathbf{E}|\xi|^{2k} < \infty, \qquad \varphi^{(2k)}(0) = (-1)^k \mathbf{E}\xi^{2k}.$$

We will prove the property for $k = 1$ (for $k > 1$ one can employ induction). It suffices to verify that $\mathbf{E}|\xi|^2$ is finite. One has

$$-\frac{2\varphi(0) - \varphi(2h) - \varphi(-2h)}{4h^2} = \mathbf{E}\left(\frac{e^{ih\xi} - e^{-ih\xi}}{2h}\right)^2 = \mathbf{E}\frac{\sin^2 h\xi}{h^2}.$$

Since $h^{-2}\sin^2 h\xi \to \xi^2$ as $h \to 0$, by Fatou's lemma

$$\begin{aligned} -\varphi''(0) &= \lim_{h\to 0}\left(\frac{2\varphi(0) - \varphi(2h) - \varphi(-2h)}{4h^2}\right) = \lim_{h\to 0} \mathbf{E}\frac{\sin^2 h\xi}{h^2} \\ &\geq \mathbf{E} \lim_{h\to 0} \frac{\sin^2 h\xi}{h^2} = \mathbf{E}\xi^2. \end{aligned}$$

□

6. *If* $\xi \geq 0$ *then* $\varphi_\xi(\lambda)$ *is defined in the complex plane for* $\operatorname{Im}\lambda \geq 0$. *Moreover,* $|\varphi_\xi(\lambda)| \leq 1$ *for such* λ, *and in the domain* $\operatorname{Im}\lambda > 0$, $\varphi_\xi(\lambda)$ *is analytic and continuous including on the boundary* $\operatorname{Im}\lambda = 0$.

Proof That $\varphi(\lambda)$ is analytic follows from the fact that, for $\operatorname{Im}\lambda > 0$, one can differentiate under the integral sign the right-hand side of

$$\varphi_\xi(\lambda) = \int_0^\infty e^{i\lambda x}\, dF(x).$$

(For $\operatorname{Im}\lambda > 0$ the integrand decreases exponentially fast as $x \to \infty$.) □

Continuity is proved in the same way as in property 4. This means that for non-negative ξ the ch.f. $\varphi_\xi(\lambda)$ uniquely determines the function

$$\psi(s) = \varphi_\xi(is) = \mathbf{E}e^{-s\xi}$$

of real variable $s \geq 0$, which is called the *Laplace* (or *Laplace–Stieltjes*) *transform* of the distribution of ξ.

The converse assertion also follows from properties of analytic functions: the *Laplace transform* $\psi(s)$ *on the half-line* $s \geq 0$ *uniquely determines the ch.f.* $\varphi_\xi(\lambda)$.

7. $\overline{\varphi}_\xi(t) = \varphi_\xi(-t) = \varphi_{-\xi}(t)$, *where the bar denotes the complex conjugate.*

Proof The relations follow from the equalities

$$\overline{\varphi}_\xi(t) = \overline{\mathbf{E}e^{it\xi}} = \mathbf{E}\overline{e^{it\xi}} = \mathbf{E}e^{-it\xi}. \qquad \square$$

This implies the following property.

7A. *If ξ is symmetric (has the same distribution as $-\xi$) then its ch.f. is real* ($\varphi_\xi(t) = \varphi_\xi(-t)$).

One can show that the converse is also true; to this end one has to make use of the uniqueness theorem to be discussed below.

Now we will find the ch.f.s of the basic probability laws.

Example 7.1.1 If $\xi = a$ with probability 1, i.e. $\xi \in \mathbf{I}_a$, then $\varphi_\xi(t) = e^{ita}$.

Example 7.1.2 If $\xi \in \mathbf{B}_p$ then $\varphi_\xi(t) = pe^{it} + (1-p) = 1 + p(e^{it} - 1)$.

Example 7.1.3 If $\xi \in \mathbf{\Phi}_{0,1}$ then $\varphi_\xi(t) = e^{-t^2/2}$.

Indeed,

$$\varphi(t) = \varphi_\xi(t) = \frac{1}{\sqrt{2\pi}} \int_{-\infty}^{\infty} e^{itx - x^2/2}\,dx.$$

Differentiating with respect to t and integrating by parts ($xe^{-x^2/2}\,dx = -de^{-x^2/2}$), we get

$$\varphi'(t) = \frac{1}{\sqrt{2\pi}} \int ixe^{itx-x^2/2}\,dx = -\frac{1}{\sqrt{2\pi}} \int te^{itx-x^2/2}\,dx = -t\varphi(t),$$

$$\bigl(\ln\varphi(t)\bigr)' = -t, \qquad \ln\varphi(t) = -\frac{t^2}{2} + c.$$

Since $\varphi(0) = 1$, one has $c = 0$ and $\varphi(t) = e^{-t^2/2}$. $\square$

Now let η be a normal random variable with parameters (a, σ). Then it can be represented as $\eta = \sigma\xi + a$, where ξ is normally distributed with parameters $(0, 1)$. The ch.f. of η can be found using Property 2:

$$\varphi_\eta(t) = e^{ita} e^{-(t\sigma)^2/2} = e^{ita - t^2\sigma^2/2}.$$

Differentiating $\varphi_\eta(t)$ for $\eta \in \mathbf{\Phi}_{0,\sigma^2}$, we will obtain that $\mathbf{E}\eta^k = 0$ for odd k, and $\mathbf{E}\eta^k = \sigma^k(k-1)(k-3)\cdots 1$ for $k = 2, 4, \ldots$.

Example 7.1.4 If $\xi \in \mathbf{\Pi}_\mu$ then

$$\varphi_\xi(t) = \mathbf{E}e^{it\xi} = \sum_k e^{itk}\frac{\mu^k}{k!}e^{-\mu} = e^{-\mu}\sum_k \frac{(\mu e^{it})^k}{k!} = e^{-\mu}e^{\mu e^{it}} = \exp\bigl[\mu\bigl(e^{it} - 1\bigr)\bigr].$$

Example 7.1.5 If ξ has the exponential distribution $\mathbf{\Gamma}_\alpha$ with density $\alpha e^{-\alpha x}$ for $x \ge 0$, then

$$\varphi_\xi(t) = \alpha \int_0^\infty e^{itx-\alpha x}\,dx = \frac{\alpha}{\alpha - it}.$$

Therefore, if ξ has the "double" exponential distribution with density $\frac{1}{2}e^{-|x|}$, $-\infty < x < \infty$, then

$$\varphi_\xi(t) = \frac{1}{2}\left(\frac{1}{1-it} + \frac{1}{1+it}\right) = \frac{1}{1+t^2}.$$

If ξ has the geometric distribution $\mathbf{P}(\xi = k) = (1-p)p^k$, $k = 0, 1, \ldots$, then

$$\varphi_\xi(t) = \frac{1-p}{1-pe^{it}}.$$

Example 7.1.6 If $\xi \in \mathbf{K}_{0,1}$ (has the density $[\pi(1+x^2)]^{-1}$) then $\varphi_\xi(t) = e^{-|t|}$. The reader will easily be able to prove this somewhat later, using the inversion formula and Example 7.1.5.

Example 7.1.7 If $\xi \in \mathbf{U}_{0,1}$, then

$$\varphi_\xi(t) = \int_0^1 e^{itx}\,dx = \frac{e^{it}-1}{it}.$$

By virtue of Property 3, the ch.f.s of the sums $\xi_1 + \xi_2$, $\xi_1 + \xi_2 + \xi_3, \ldots$ that we considered in Example 3.6.1 will be equal to

$$\varphi_{\xi_1+\xi_2}(t) = -\frac{(e^{it}-1)^2}{t^2}, \qquad \varphi_{\xi_1+\xi_2+\xi_3}(t) = -\frac{(e^{it}-1)^3}{t^3}, \qquad \ldots.$$

We return to the general case. How can one verify whether one or another function φ is characteristic or not? Sometimes one can do this using the above properties. We suggest the reader to determine whether the functions $(1+t)^{-1}$, $1+t$, $\sin t$, $\cos t$ are characteristic, and if so, to which distributions they correspond.

In the general case the posed question is a difficult one. We state without proof one of the known results.

Bochner–Khinchin's Theorem *A necessary and sufficient condition for a continuous function $\varphi(t)$ with $\varphi(0) = 1$ to be characteristic is that it is nonnegatively defined, i.e., for any real $t_1, \ldots, t_n$ and complex $\lambda_1, \ldots, \lambda_n$, one has*

$$\sum_{k,j=1}^{n} \varphi(t_k - t_j)\lambda_k \overline{\lambda}_j \ge 0$$

($\overline{\lambda}$ is the complex conjugate of λ).

Note that the necessity of this condition is almost obvious, for if $\varphi(t) = \mathbf{E}e^{it\xi}$ then

$$\sum_{k,j=1}^{n} \varphi(t_k - t_j)\lambda_k\overline{\lambda}_j = \mathbf{E}\sum_{k,j=1}^{n} e^{i(t_k-t_j)\xi}\lambda_k\overline{\lambda}_j = \mathbf{E}\left|\sum_{k=1}^{n} \lambda_k e^{it_k\xi}\right|^2 \ge 0.$$

7.1.2 The Properties of Ch.F.s Related to the Structure of the Distribution of ξ

8. *If the distribution of ξ has a density then $\varphi_\xi(t) \to 0$ as $|t| \to \infty$.*

This is a direct consequence of the Lebesgue theorem on Fourier transforms. The converse assertion is false.

In general, the smoother $F(x)$ is the faster $\varphi_\xi(t)$ vanishes as $|t| \to \infty$. The formulas in Example 7.1.7 are typical in this respect. If the density $f(x)$ has an integrable k-th derivative then, by integrating by parts, we get

$$\varphi_\xi(t) = \int e^{itx} f(x)\,dx = \frac{1}{it}\int e^{itx} f'(x)\,dx = \cdots = \frac{1}{(it)^k}\int e^{itx} f^{(k)}(x)\,dx,$$

which implies that

$$\varphi_\xi(t) \le \frac{c}{|t|^k}.$$

8A. *If the distribution of ξ has a density of bounded variation then*

$$\bigl|\varphi_\xi(t)\bigr| \le \frac{c}{|t|}.$$

This property is also validated by integration by parts:

$$\bigl|\varphi_\xi(t)\bigr| = \left|\frac{1}{it}\int e^{itx}\,df(x)\right| \le \frac{1}{|t|}\int \bigl|df(x)\bigr|.$$

9. *A random variable ξ has a lattice distribution with span $h > 0$* (see Definition 3.2.3) *if and only if*

$$\left|\varphi_\xi\left(\frac{2\pi}{h}\right)\right| = 1, \qquad \left|\varphi_\xi\left(\frac{v}{h}\right)\right| < 1 \tag{7.1.2}$$

if v is not a multiple of 2π.

Clearly, without loss of generality we can assume $h = 1$. Moreover, since

$$\bigl|\varphi_{\xi-a}(t)\bigr| = \bigl|e^{-ita}\varphi_\xi(t)\bigr| = \bigl|\varphi_\xi(t)\bigr|,$$

the properties (7.1.2) are invariant with respect to the shift by a. Thus we can assume the shift a is equal to zero and thus change the lattice distribution condition in Property 9 to the arithmeticity condition (see Definition 3.2.3). Since $\varphi_\xi(t)$ is a periodic function, Property 9 can be rewritten in the following equivalent form:

The distribution of a random variable ξ is arithmetic if and only if

$$\varphi_\xi(2\pi) = 1, \qquad \left|\varphi_\xi(t)\right| < 1 \quad \text{for all } t \in (0, 2\pi). \tag{7.1.3}$$

Proof If ξ has an arithmetic distribution then

$$\varphi_\xi(t) = \sum_k \mathbf{P}(\xi = k)e^{itk} = 1$$

for $t = 2\pi$. Now let us prove the second relation in (7.1.3). Assume the contrary: for some $v \in (0, 2\pi)$, we have $|\varphi_\xi(v)| = 1$ or, which is the same,

$$\varphi_\xi(v) = e^{ibv}$$

for some real b. The last relation implies that

$$\varphi_{\xi-b}(v) = 1 = \mathbf{E}\cos v(\xi - b) + i\mathbf{E}\sin v(\xi - b), \qquad \mathbf{E}\left[1 - \cos v(\xi - b)\right] = 0.$$

Hence, by Property E4 in Sect. 4.1, $\cos v(\xi - b) = 1$ and $v(\xi - b) = 2\pi k(\omega)$ with probability 1, where $k(\omega)$ is an integer. Thus $\xi - b$ is a multiple of $2\pi/v > 1$. This contradicts the assumption that the span of the lattice equals 1, and hence proves (7.1.3).

Conversely, let (7.1.3) hold. As we saw, the first relation in (7.1.3) implies that ξ takes only integer values. If we assume that the lattice span equals $h > 1$ then, by the first part of the proof and the first relation in (7.1.2), we get $|\varphi(2\pi/h)| = 1$, which contradicts the first relation in (7.1.3). Property 9 is proved. □

The next definition looks like a tautology.

Definition 7.1.2 The distribution of ξ is called *non-lattice* if it is not a lattice distribution.

10. *If the distribution of ξ is non-lattice then*

$$\left|\varphi_\xi(t)\right| < 1 \quad \text{for all } t \neq 0.$$

Proof Indeed, if we assume the contrary, i.e. that $|\varphi(u)| = 1$ for some $u \neq 0$, then, by Property 9, we conclude that the distribution of ξ is a lattice with span $h = 2\pi/u$ or with a lesser span. □

11. *If the distribution of ξ has an absolutely continuous component of a positive mass $p > 0$, then it is clearly non-lattice and, moreover,*

$$\limsup_{|t|\to\infty} \left|\varphi_\xi(t)\right| \le 1 - p.$$

This assertion follows from Property 8.

Arithmetic distributions occupy an important place in the class of lattice distributions.

For arithmetic distributions, the ch.f. $\varphi_\xi(t)$ is a function of the variable $z = e^{it}$ and is periodic in t with period 2π. Hence, in this case it is sufficient to know the behaviour of the ch.f. on the interval $[-\pi, \pi]$ or, which is the same, to know the behaviour of the function

$$p_\xi(z) := \mathbf{E}z^\xi = \sum z^k \mathbf{P}(\xi = k)$$

on the unit circle $|z| = 1$.

Definition 7.1.3 The function $p_\xi(z)$ is called the *generating function of the random variable* ξ (or of the distribution of ξ).

Since $p_\xi(e^{it}) = \varphi_\xi(t)$ is a ch.f., all the properties of ch.f.s remain valid for generating functions, with the only changes corresponding to the change of variable. For more on applications of generating functions, see Sect. 7.7.

7.2 Inversion Formulas

Thus for any random variable there exists a corresponding ch.f. We will now show that the set $\mathcal{L}$ of functions e^{itx} is a distribution determining class, i.e. that the distribution can be uniquely reconstructed from its ch.f. This is proved using inversion formulas.

7.2.1 The Inversion Formula for Densities

Theorem 7.2.1 *If the ch.f. $\varphi(t)$ of a random variable ξ is integrable then the distribution of ξ has the bounded density*

$$f(x) = \frac{1}{2\pi} \int e^{-itx} \varphi(t)\, dt. \tag{7.2.1}$$

This fact is known from classical Fourier analysis, but we shall give a proof of a probabilistic character.

Proof First we will establish the following (Parseval's) identity: for any fixed $\varepsilon > 0$,

$$\begin{aligned} p_\varepsilon(t) &:= \frac{1}{2\pi} \int e^{-itu} \varphi(u) e^{-\varepsilon^2 u^2/2}\, du \\ &\equiv \frac{1}{\sqrt{2\pi}\,\varepsilon} \int \exp\left\{-\frac{(u-t)^2}{2\varepsilon^2}\right\} \mathbf{F}(du), \end{aligned} \tag{7.2.2}$$

where $\mathbf{F}$ is the distribution of ξ. We begin with the equality

$$\frac{1}{\sqrt{2\pi}} \int \exp\left\{ix\frac{\xi - t}{\varepsilon} - \frac{x^2}{2}\right\} dx = \exp\left\{-\frac{(\xi - t)^2}{2\varepsilon^2}\right\}, \tag{7.2.3}$$

both sides of which being the value of the ch.f. of the normal distribution with parameters $(0, 1)$ at the point $(\xi - t)/\varepsilon$. After changing the variable $x = \varepsilon u$, the left-hand side of this equality can be rewritten as

$$\frac{\varepsilon}{\sqrt{2\pi}} \int \exp\left\{iu(\xi - t) - \frac{\varepsilon^2 u^2}{2}\right\} du.$$

If we take expectations of both sides of (7.2.3), we obtain

$$\frac{\varepsilon}{\sqrt{2\pi}} \int e^{-iut} \varphi(u) e^{-\frac{\varepsilon^2 u^2}{2}} du = \int \exp\left\{-\frac{(u-t)^2}{2\varepsilon^2}\right\} \mathbf{F}(du).$$

This proves (7.2.2).

To prove the theorem first consider the left-hand side of the equality (7.2.2). Since $e^{-\varepsilon^2 u^2/2} \to 1$ as $\varepsilon \to 0$, $|e^{-\frac{\varepsilon^2 u^2}{2}}| \le 1$ and $\varphi(u)$ is integrable, as $\varepsilon \to 0$ one has

$$p_\varepsilon(t) \to \frac{1}{2\pi} \int e^{-itu} \varphi(u)\, du = p_0(t) \tag{7.2.4}$$

uniformly in t, because the integral on the left-hand side of (7.2.2) is uniformly continuous in t. This implies, in particular, that

$$\int_a^b p_\varepsilon(t)\, dt \to \int_a^b p_0(t). \tag{7.2.5}$$

Now consider the right-hand side of (7.2.2). It represents the density of the sum $\xi + \varepsilon\eta$, where ξ and η are independent and $\eta \in \mathbf{\Phi}_{0,1}$. Therefore

$$\int_a^b p_\varepsilon(t)\, dt = \mathbf{P}(a < \xi + \varepsilon\eta \le b). \tag{7.2.6}$$

Since $\xi + \varepsilon\eta \xrightarrow{p} \xi$ as $\varepsilon \to 0$ and the limit $\int_a^b p_\varepsilon(t)\, dt$ exists for any fixed a and b by virtue of (7.2.5), this limit (see (7.2.6)) cannot be anything other than $\mathbf{F}([a, b))$.

Thus, from (7.2.5) and (7.2.6) we get

$$\int_a^b p_0(t)\, dt = \mathbf{F}\big([a, b)\big).$$

This means that the distribution $\mathbf{F}$ has the density $p_0(t)$, which is defined by relation (7.2.4). The boundedness of $p_0(t)$ evidently follows from the integrability of φ:

$$p_0(t) \le \frac{1}{2\pi} \int \big|\varphi(t)\big|\, dt < \infty.$$

The theorem is proved. □

7.2.2 The Inversion Formula for Distributions

Theorem 7.2.2 *If $F(x)$ is the distribution function of a random variable ξ and $\varphi(t)$ is its ch.f., then, for any points of continuity x and y of the function $F(x)$*,[2]

$$F(y) - F(x) = \frac{1}{2\pi} \lim_{\sigma \to 0} \int \frac{e^{-itx} - e^{-ity}}{it} \varphi(t) e^{-t^2\sigma^2}\, dt. \tag{7.2.7}$$

If the function $\varphi(t)/t$ is integrable at infinity then the passage to the limit under the integral sign is justified and one can write

$$F(y) - F(x) = \frac{1}{2\pi} \int \frac{e^{-itx} - e^{-ity}}{it} \varphi(t)\, dt. \tag{7.2.8}$$

Proof Suppose first that the ch.f. $\varphi(t)$ is integrable. Then $F(x)$ has a density $f(x)$ and the assertion of the theorem in the form (7.2.8) follows if we integrate both sides of Eq. (7.2.1) over the interval with the end points x and y and change the order of integration (which is valid because of the absolute convergence).[3]

Now let $\varphi(t)$ be the characteristic function of a random variable ξ with an arbitrary distribution $\mathbf{F}$. On a common probability space with ξ, consider a random variable η which is independent of ξ and has the normal distribution with parameters $(0, 2\sigma^2)$. As we have already pointed out, the ch.f. of η is $e^{-t^2\sigma^2}$.

This means that the ch.f. of $\xi + \eta$, being equal to $\varphi(t)e^{-t^2\sigma^2}$, is integrable. Therefore by (7.2.8) one will have

$$F_{\xi+\eta}(y) - F_{\xi+\eta}(x) = \frac{1}{2\pi} \int_{-\infty}^{\infty} \frac{e^{-itx} - e^{-ity}}{it} \varphi(t) e^{-t^2\sigma^2}\, dt. \tag{7.2.9}$$

Since $\eta \xrightarrow{p} 0$ as $\sigma \to 0$, we have $\mathbf{F}_{\xi+\eta} \Rightarrow \mathbf{F}$ (see Chap. 6). Therefore, if x and y are points of continuity of $\mathbf{F}$, then $F(y) - F(x) = \lim_{\sigma \to 0}(F_{\xi+\eta}(y) - F_{\xi+\eta}(x))$. This, together with (7.2.9), proves the assertion of the theorem. □

In the proof of Theorem 7.2.2 we used a method which might be called the "smoothing" of distributions. It is often employed to overcome technical difficulties related to the inversion formula.

Corollary 7.2.1 (Uniqueness Theorem) *The ch.f. of a random variable uniquely determines its distribution function.*

---

[2] In the literature, the inversion formula is often given in the form

$$F(y) - F(x) = \frac{1}{2\pi} \lim_{A \to \infty} \int_{-A}^{A} \frac{e^{-itx} - e^{-ity}}{it} \varphi(t)\, dt$$

which is equivalent to (7.2.7).

[3] Formula (7.2.8) can also be obtained from (7.2.1) without integration by noting that $(F(x) - F(y))/(y - x)$ is the value at zero of the convolution of two densities: $f(x)$ and the uniform density over the interval $[-y, -x]$ (see also the remark at the end of Sect. 3.6). The ch.f. of the convolution is equal to $\frac{e^{-itx} - e^{-ity}}{(y-x)it} \varphi(t)$.

The proof follows from the inversion formula and the fact that $\mathbf{F}$ is uniquely determined by the differences $F(y) - F(x)$.

For *lattice* random variables the inversion formula becomes simpler. Let, for the sake of simplicity, ξ be an integer-valued random variable.

Theorem 7.2.3 *If $p_\xi(z) := \mathbf{E}z^\xi$ is the generating function of an arithmetic random variable then*

$$\mathbf{P}(\xi = k) = \frac{1}{2\pi i}\int_{|z|=1} p_\xi(z)z^{-k-1}\,dz. \tag{7.2.10}$$

Proof Turning to the ch.f. $\varphi_\xi(t) = \sum_j e^{itj}\mathbf{P}(\xi = j)$ and changing the variables $z = it$ in (7.2.10) we see that the right-hand side of (7.2.10) equals

$$\frac{1}{2\pi}\int_{-\pi}^{\pi} e^{-itk}\varphi_\xi(t)\,dt = \frac{1}{2\pi}\sum_j \mathbf{P}(\xi = j)\int_{-\pi}^{\pi} e^{it(j-k)}\,dt.$$

Here all the integrals on the right-hand side are equal to zero, except for the integral with $j = k$ which is equal to 2π. Thus the right-hand side itself equals $\mathbf{P}(\xi = k)$. The theorem is proved. □

Formula (7.2.10) is nothing else but the formula for Fourier coefficients and has a simple geometric interpretation. The functions $\{\mathrm{e}_k = e^{itk}\}$ form an orthonormal basis in the Hilbert space $L_2(-\pi, \pi)$ of square integrable complex-valued functions with the inner product

$$(f, g) = \frac{1}{2\pi}\int_{-\pi}^{\pi} f(t)\overline{g}(t)\,dt$$

($\overline{g}$ is the complex conjugate of g). If $\varphi_\xi = \sum \mathrm{e}_k\mathbf{P}(\xi = k)$ then it immediately follows from the equality $\varphi_\xi = \sum \mathrm{e}_k(_\xi, \mathrm{e}_k)$ that

$$\mathbf{P}(\xi = k) = (\varphi_\xi, \mathrm{e}_k) = \frac{1}{2\pi}\int_{-\pi}^{\pi} e^{-itk}\varphi_\xi(t)\,dt.$$

7.2.3 The Inversion Formula in L_2. The Class of Functions that Are Both Densities and Ch.F.s

First consider some properties of ch.f.s related to the inversion formula. As a preliminary, note that, in classical Fourier analysis, one also considers the Fourier transforms of functions f from the space L_2 of square-integrable functions. Since in this case a function f is not necessarily integrable, the Fourier transform is defined as the integral in the principal value sense:[4]

$$\varphi(t) := \lim_{N\to\infty} \varphi_{(N)}(t), \qquad \varphi_{(N)}(t) := \int_{-N}^{N} e^{itx} f(x)\,dx, \qquad (7.2.11)$$

where the limit is taken in the sense of convergence in L_2:

$$\int \left|\varphi(t) - \varphi_{(N)}(t)\right|^2 dx \to 0 \quad \text{as } N \to \infty.$$

Since by Parseval's equality

$$\|f\|_{L_2} = \frac{1}{2\pi}\|\varphi\|_{L_2}, \quad \text{where } \|g\|_{L_2} = \left[\int |g|^2(t)\,dt\right]^{1/2},$$

the Fourier transform maps the space L_2 into itself (there is no such isometricity for Fourier transforms in L_1). *Here the inversion formula* (7.2.1) *holds true but the integral in* (7.2.1) *is understood in the principal value sense.*

Denote by $\mathcal{F}$ and $\mathcal{H}$ the class of all densities and the class of all ch.f.s, respectively, and by $\mathcal{H}_{1,+} \subset L_1$ the class of *nonnegative real-valued integrable* ch.f.s, so that the elements of $\mathcal{H}_{1,+}$ are in $\mathcal{F}$ up to the normalising factors. Further, let $(\mathcal{H}_{1,+})^{(-1)}$ be the inverse image of the class $\mathcal{H}_{1,+}$ in $\mathcal{F}$ for the mapping $f \to \varphi$, i.e. the class of densities whose ch.f.s lie in $\mathcal{H}_{1,+}$. It is clear that functions f from $(\mathcal{H}_{1,+})^{(-1)}$ and φ from $\mathcal{H}_{1,+}$ are necessarily symmetric (see Property 7A in Sect. 7.1) and that $f(0) \in (0, \infty)$. The last relation follows from the fact that, by the inversion formula for $\varphi \in \mathcal{H}_{1,+}$, we have

$$\|\varphi\| := \|\varphi\|_{L_1} = \int \varphi(t)\,dt = 2\pi f(0).$$

Further, denote by $(\mathcal{H}_{1,+})_{\|\cdot\|}$ the class of *normalised* functions $\frac{\varphi}{\|\varphi\|}$, $\varphi \in \mathcal{H}_{1,+}$, so that $(\mathcal{H}_{1,+})_{\|\cdot\|} \subset \mathcal{F}$, and denote by $\mathcal{F}^{(2,*)}$ the class of *convolutions of symmetric densities* from L_2:

$$\mathcal{F}^{(2,*)} := \left\{ f^{(2)*}(x) : f \in L_2,\ f \text{ is symmetric} \right\},$$

where

$$f^{(2)*}(x) = \int f(t) f(x-t)\,dt.$$

Theorem 7.2.4 *The following relations hold true*:

$$(\mathcal{H}_{1,+})^{(-1)} = (\mathcal{H}_{1,+})_{\|\cdot\|}, \qquad \mathcal{F}^{(2,*)} \subset (\mathcal{H}_{1,+})_{\|\cdot\|}.$$

The class $(\mathcal{H}_{1,+})_{\|\cdot\|}$ may be called the class of *densities conjugate to* $f \in (\mathcal{H}_{1,+})^{(-1)}$. It turns out that this class coincides with the inverse image $(\mathcal{H}_{1,+})^{(-1)}$. The second statement of the theorem shows that this inverse image is a very rich

---

[4]Here we again omit the factor $\frac{1}{\sqrt{2\pi}}$ (cf. the footnote on page 154).

class and provides sufficient conditions for the density f to have a conjugate. We will need these conditions in Sect. 8.7.

Proof of Theorem 7.2.4 Let $f \in (\mathcal{H}_{1,+})^{(-1)}$. Then the corresponding ch.f. φ is in $\mathcal{H}_{1+}$ and the inversion formula (7.2.1) is applicable. Multiplying its right-hand side by $\frac{2\pi}{\|\varphi\|}$, we obtain an expression for the ch.f. (at the point $-t$) of the density $\frac{\varphi}{\|\varphi\|}$ (recall that $\varphi \geq 0$ is symmetric if $\varphi \in \mathcal{H}_{1,+}$). This means that $\frac{2\pi f}{\|\varphi\|}$ is a ch.f. and, moreover, that $f \in (\mathcal{H}_{1,+})_{\|\cdot\|}$.

Conversely, suppose that $f^* := \frac{\varphi}{\|\varphi\|} \in (\mathcal{H}_{1,+})_{\|\cdot\|}$. Then $f^* \in \mathcal{F}$ is symmetric, and the inversion formula can be applied to φ:

$$f(x) = \frac{1}{2\pi}\int e^{-itx}\varphi(t)\,dt = \frac{1}{2\pi}\int e^{itx}\varphi(t)\,dt, \qquad \frac{2\pi f(t)}{\|\varphi\|} = \int e^{itx} f^*(x)\,dx.$$

Since the ch.f. $\varphi^*(t) := \frac{2\pi f(t)}{\|\varphi\|}$ belongs to $\mathcal{H}_{1,+}$, one has $f^* \in (\mathcal{H}_{1,+})^{(-1)}$.

We now prove the second assertion. Suppose that $f \in L_2$. Then $\varphi \in L_2$ and $\varphi^2 \in L_1$. Moreover, by virtue of the symmetry of f and Property 7A in Sect. 7.1, the function φ is real-valued, so $\varphi^2 \geq 0$. This implies that $\varphi^2 \in \mathcal{H}_{1,+}$. Since φ^2 is the ch.f. of the density $f^{(2)*}$, we have $f^{(2)*} \in (\mathcal{H}_{1,+})^{(-1)}$. The theorem is proved. □

Note that *any bounded density f belongs to L_2*. Indeed, since the Lebesgue measure of $\{x : f(x) \geq 1\}$ is always less than 1, for $f(\cdot) \leq N$ we have

$$\|f\|^2_{L_2} = \int f^2(x)\,dx \leq \int_{f(x)<1} f(x)\,dx + N^2 \int_{f(x)\geq 1} dx \leq 1 + N^2. \qquad \square$$

Thus we have obtained the following result.

Corollary 7.2.2 *For any bounded symmetric density f, the convolution $f^{(2)*}$ is, up to a constant factor, the ch.f. of a random variable.*

Example 7.2.1 The "triangle" density

$$g(x) = \begin{cases} 1 - |x| & \text{if } |x| \leq 1, \\ 0 & \text{if } |x| > 1, \end{cases}$$

being the convolution of the two uniform distributions on $[-1/2, 1/2]$ (cf. Example 3.6.1) is also a ch.f. We suggest the reader to verify that the preimage of this ch.f. is the density

$$f(x) = \frac{1}{2\pi}\frac{\sin^2 x/2}{x^2}$$

(the density conjugate to g). Conversely, the density g is conjugate to f, and the functions $8\pi f(t)$ and $g(t)$ will be ch.f.s for g and f, respectively.

These assertions will be useful in Sect. 8.7.

7.3 The Continuity (Convergence) Theorem

Let $\{\varphi_n(t)\}_{n=1}^{\infty}$ be a sequence of ch.f.s and $\{F_n\}_{n=1}^{\infty}$ the sequence of the respective distribution functions. Recall that the symbol $\Rightarrow$ denotes the weak convergence of distributions introduced in Chap. 6.

Theorem 7.3.1 (The Continuity Theorem) *A necessary and sufficient condition for the convergence $F_n \Rightarrow F$ as $n \to \infty$ is that $\varphi_n(t) \to \varphi(t)$ for any t, $\varphi(t)$ being the ch.f. corresponding to F.*

The theorem follows in an obvious way from Corollary 6.3.2 (here two of the three sufficient conditions from Corollary 6.3.2 are satisfied: conditions (2) and (3)). The proof of the theorem can be obtained in a simpler way as well. This way is presented in Sect. 7.4 of the previous editions of this book.

In Sect. 7.1, for nonnegative random variables ξ we introduced the notion of the Laplace transform $\psi(s) := \mathbf{E}e^{-s\xi}$. Let $\psi_n(s)$ and $\psi(s)$ be Laplace transforms corresponding to F_n and F. The following analogue of Theorem 7.3.1 holds for Laplace transforms:

In order that $F_n \Rightarrow F$ as $n \to \infty$ it is necessary and sufficient that $\psi_n(s) \to \psi(s)$ for each $s \ge 0$.

Just as in Theorem 7.3.1, this assertion follows from Corollary 6.3.2, since the class $\{f(x) = e^{-sx},\ s \ge 0\}$ is (like $\{e^{itx}\}$) a distribution determining class (see Property 6 in Sect. 7.1) and, moreover, the sufficient conditions (2) and (3) of Corollary 6.3.2 are satisfied.

Theorem 7.3.1 has a deficiency: one needs to know in advance that the function $\varphi(t)$ to which the ch.f.s converge is a ch.f. itself. However, one could have no such prior information (see e.g. Sect. 8.8). In this connection there arises a natural question under what conditions the limiting function $\varphi(t)$ will be characteristic.

The answer to this question is given by the following theorem.

Theorem 7.3.2 *Let*

$$\varphi_n(t) = \int e^{itx}\, dF_n(x)$$

be a sequence of ch.f.s and $\varphi_n(t) \to \varphi(t)$ as $n \to \infty$ for any t.

Then the following three conditions are equivalent:

(a) *$\varphi(t)$ is a ch.f.;*
(b) *$\varphi(t)$ is continuous at $t = 0$;*
(c) *the sequence $\{F_n\}$ is tight.*

Thus if we establish that $\varphi_n(t) \to \varphi(t)$ and one of the above three conditions is met, then we can assert that there exists a distribution F such that φ is the ch.f. of F and $F_n \Rightarrow F$.

Proof The equivalence of conditions (a) and (c) follows from Theorem 6.3.2. That (a) implies (b) is known. It remains to establish that (c) follows from (b). First we will show that the following lemma is true. □

Lemma 7.3.1 *If φ is the ch.f. of ξ then, for any $u > 0$,*

$$\mathbf{P}\left(|\xi| > \frac{2}{u}\right) \le \frac{1}{u} \int_{-u}^{u} \left[1 - \varphi(t)\right] dt.$$

Proof The right-hand side of this inequality is equal to

$$\frac{1}{u} \int_{-u}^{u} \int_{-\infty}^{\infty} \left(1 - e^{-itx}\right) dF(x)\, dt,$$

where F is the distribution function of ξ. Changing the order of integration and noting that

$$\int_{-u}^{u} \left(1 - e^{-itx}\right) dt = \left(t + \frac{e^{-itx}}{ix}\right)\Bigg|_{-u}^{u} = 2u\left(1 - \frac{\sin ux}{ux}\right),$$

we obtain that

$$\begin{aligned} \frac{1}{u} \int_{-u}^{u} \left[1 - \varphi(t)\right] dt &= 2 \int_{-\infty}^{\infty} \left(1 - \frac{\sin ux}{ux}\right) dF(x) \\ &\ge 2 \int_{|x|>2/u} \left(1 - \left|\frac{\sin ux}{ux}\right|\right) dF(x) \\ &\ge 2 \int_{|x|>2/u} \left(1 - \frac{1}{|ux|}\right) dF(x) \ge \int_{|x|>2/u} dF(x). \end{aligned}$$

The lemma is proved. □

Now suppose that condition (b) is met. By Lemma 7.3.1

$$\limsup_{n\to\infty} \int_{|x|>2/u} dF_n(x) \le \limsup_{n\to\infty} \frac{1}{u} \int_{-u}^{u} \left[1 - \varphi_n(t)\right] dt = \frac{1}{u} \int_{-u}^{u} \left[1 - \varphi(t)\right] dt.$$

Since $\varphi(t)$ is continuous at 0 and $\varphi(0) = 1$, the mean value on the right-hand side can clearly be made arbitrarily small by choosing sufficiently small u. This obviously means that condition (c) is satisfied. The theorem is proved. □

Using ch.f.s one can not only establish convergence of distribution functions but also estimate the rate of this convergence in the cases when one can estimate how fast $\varphi_n - \varphi$ vanishes. We will encounter respective examples in Sect. 7.5.

We will mostly use the machinery of ch.f.s in Chaps. 8, 12 and 17. In the present chapter we will also touch upon some applications of ch.f.s, but they will only serve as illustrations.

7.4 The Application of Characteristic Functions in the Proof of the Poisson Theorem

Let $\xi_1, \dots, \xi_n$ be independent integer-valued random variables,

$$S_n = \sum_1^k \xi_k, \qquad \mathbf{P}(\xi_k = 1) = p_k, \qquad \mathbf{P}(\xi_k = 0) = 1 - p_k - q_k.$$

The theorem below is a generalisation of the theorems established in Sect. 5.4.[5]

Theorem 7.4.1 *One has*

$$\big|\mathbf{P}(S_n = k) - \boldsymbol{\Pi}_\mu(\{k\})\big| \le \sum_{k=1}^n p_k^2 + 2\sum_{k=1}^n q_k, \quad \textit{where } \mu = \sum_{k=1}^n p_k.$$

Thus, if one is given a triangle array $\xi_{1n}, \xi_{2n}, \dots, \xi_{nn}$, $n = 1, 2, \dots$, of independent integer-valued random variables,

$$S_n = \sum_{k=1}^n \xi_{kn}, \qquad \mathbf{P}(\xi_{kn} = 1) = p_{kn}, \qquad \mathbf{P}(\xi_{kn} = 0) = 1 - p_{kn} - q_{kn},$$

$$\mu = \sum_{k=1}^n p_{kn},$$

then a sufficient condition for convergence of the difference $\mathbf{P}(S_n = k) - \boldsymbol{\Pi}_\mu(\{k\})$ to zero is that

$$\sum_{k=1}^n q_{kn} \to 0, \qquad \sum_{k=1}^n p_{kn}^2 \to 0.$$

Since

$$\sum_{k=1}^n p_{kn}^2 \le \mu \max_{k \le n} p_{kn},$$

the last condition is always met if

$$\max_{k \le n} p_{kn} \to 0, \qquad \mu \le \mu_0 = \text{const}.$$

---

[5]This extension is not really substantial since close results could be established using Theorem 5.2.2 in which ξ_k can only take the values 0 and 1. It suffices to observe that the probability of the event $A = \bigcup_k \{\xi_k \ne 0,\ \xi_k \ne 1\}$ is bounded by the sum $\sum q_k$ and therefore

$$\mathbf{P}(S_n = k) = \theta_1 \sum q_k + \Big(1 - \theta_2 \sum q_k\Big)\mathbf{P}(S_n = k | \overline{A}), \quad \theta_i \le 1,\ i = 1, 2,$$

where $\mathbf{P}(S_n = k|\overline{A}) = \mathbf{P}(S_n^* = k)$ and S_n^* are sums of independent random variables ξ_k^* with

$$\mathbf{P}(\xi_k^* = 1) = p_k^* = \frac{p_k}{1 - q_k}, \qquad \mathbf{P}(\xi_k^* = 0) = 1 - p_k^*.$$

To prove the theorem we will need two auxiliary assertions.

Lemma 7.4.1 *If* $\operatorname{Re}\beta \le 0$ *then*

$$|e^\beta - 1| \le |\beta|, \qquad |e^\beta - 1 - \beta| \le |\beta|^2/2, \qquad |e^\beta - 1 - \beta - \beta^2/2| \le |\beta|^3/6.$$

Proof The first two inequalities follow from the relations (we use here the change of variables $t = \beta v$ and the fact that $|e^s| \le 1$ for $\operatorname{Re} s \le 0$)

$$|e^\beta - 1| = \left| \int_0^\beta e^t\,dt \right| = \left| \beta \int_0^1 e^{\beta v}\,dv \right| \le |\beta|,$$

$$|e^\beta - 1 - \beta| = \left| \int_0^\beta (e^t - 1)\,dt \right| = \left| \beta \int_0^1 (e^{\beta v} - 1)\,dv \right| \le |\beta|^2 \int_0^1 v\,dv = |\beta^2|/2.$$

The last inequality is proved in the same way. □

Lemma 7.4.2 *If* $|a_k| \le 1$, $|b_k| \le 1$, $k = 1, \ldots, n$, *then*

$$\left| \prod_{k=1}^n a_k - \prod_{k=1}^n b_k \right| \le \sum_{k=1}^n |a_k - b_k|.$$

Thus if $\varphi_k(t)$ *and* $\theta_k(t)$ *are ch.f.s then, for any* t,

$$\left| \prod_{k=1}^n \varphi_k(t) - \prod_{k=1}^n \theta_k(t) \right| \le \sum_{k=1}^n \bigl|\varphi_k(t) - \theta_k(t)\bigr|.$$

Proof Put $A_n = \prod_{k=1}^n a_k$ and $B_n = \prod_{k=1}^n b_k$. Then $|A_n| \le 1$, $|B_n| \le 1$, and

$$\begin{aligned}
|A_n - B_n| &= |A_{n-1}a_n - B_{n-1}b_n| \\
&= \bigl|(A_{n-1} - B_{n-1})a_n + (a_n - b_n)B_{n-1}\bigr| \le |A_{n-1} - B_{n-1}| + |a_n - b_n|.
\end{aligned}$$

Applying this inequality n times, we obtain the required relation. □

Proof of Theorem 7.4.1 One has

$$\varphi_k(t) := \mathbf{E}e^{it\xi_k} = 1 + p_k\bigl(e^{it} - 1\bigr) + q_k\bigl(\gamma_k(t) - 1\bigr),$$

where $\gamma_k(t)$ is the ch.f. of some integer-valued random variable. By independence of the random variables ξ_k,

$$\varphi_{S_n}(t) = \prod_{k=1}^n \varphi_k(t).$$

Let further $\zeta \in \mathbf{\Pi}_\mu$. Then

$$\varphi_\zeta(t) = \mathbf{E}e^{it\zeta} = e^{\mu(e^{it}-1)} = \prod_{k=1}^n \theta_k(t),$$

where $\theta_k(t) = e^{p_k(e^{it}-1)}$. Therefore the difference between the ch.f.s φ_{S_n} and φ_ζ can be bounded by Lemma 7.4.2 as follows:

$$|\varphi_{S_n}(t) - \varphi_\zeta(t)| = \left|\prod_{k=1}^n \varphi_k - \prod_{k=1}^n \theta_k\right| \le \sum_{k=1}^n |\varphi_k - \theta_k|,$$

where by Lemma 7.4.1 (note that $\mathrm{Re}(e^{it} - 1) \le 0$)

$$\begin{aligned}\left|\theta_k(t) - 1 - p_k\left(e^{it} - 1\right)\right| &\le \frac{p_k^2|e^{it} - 1|^2}{2} = \frac{p_k^2}{2}\left(\sin^2 t + (1 - \cos t)^2\right) \\ &= p_k^2\left(\frac{\sin^2 t}{2} + 2\sin^4 \frac{t}{2}\right),\end{aligned} \tag{7.4.1}$$

$$\sum_{k=1}^n |\varphi_k - \theta_k| \le 2\sum_{k=1}^n q_k + \sum_{k=1}^n p_k^2\left(\frac{\sin^2 t}{2} + 2\sin^4 \frac{t}{2}\right).$$

It remains to make use of the inversion formula (7.2.10):

$$\begin{aligned}\left|\mathbf{P}(S_n = k) - \mathbf{\Pi}_\mu\left(\{k\}\right)\right| &\le \left|\frac{1}{2\pi}\int_{-\pi}^{\pi} e^{-ikt}\left(\varphi_{S_n}(t) - \varphi_\zeta(t)\right) dt\right| \\ &\le \frac{1}{\pi}\int_0^\pi \left[2\sum_{k=1}^n q_k + \sum_{k=1}^n p_k^2\left(\frac{\sin^2 t}{2} + 2\sin^4 \frac{t}{2}\right)\right] dt \\ &= 2\sum_{k=1}^n q_k + \sum_{k=1}^n p_k^2,\end{aligned}$$

for

$$\frac{1}{2\pi}\int_0^\pi \sin^2 t\, dt = \frac{1}{4}, \qquad \frac{2}{\pi}\int_0^\pi \sin^4 \frac{t}{2}\, dt = \frac{3}{4}.$$

The theorem is proved. □

If one makes use of the inequality $|e^{it} - 1| \le 2$ in (7.4.1), the computations will be simplified, there will be no need to calculate the last two integrals, but the bounds will be somewhat worse:

$$\sum |\varphi_k - \theta_k| \le 2\left(\sum q_k + \sum p_k^2\right),$$

$$\left|\mathbf{P}(S_n = k) - \mathbf{\Pi}_\mu\left(\{k\}\right)\right| \le 2\left(\sum q_k + \sum p_k^2\right).$$

7.5 Characteristic Functions of Multivariate Distributions. The Multivariate Normal Distribution

Definition 7.5.1 Given a random vector $\xi = (\xi_1, \xi_2, \ldots, \xi_d)$, its ch.f. (the ch.f. of its distribution) is defined as the function of the vector variable $t = (t_1, \ldots, t_d)$ equal to

$$\varphi_\xi(t) := \mathbf{E}e^{it\xi^T} = \mathbf{E}e^{i(t,\xi)} = \mathbf{E}\exp\left\{i\sum_{k=1}^d t_k\xi_k\right\}$$
$$= \int \exp\left\{i\sum_{k=1}^d t_k x_k\right\}\mathbf{F}_{\xi_1,\dots,\xi_d}(dx_1,\dots,dx_d),$$

where ξ^T is the transpose of ξ (a column vector), and (t,ξ) is the inner product.

The ch.f.s of multivariate distributions possess all the properties (with obvious amendments of their statements) listed in Sects. 7.1–7.3.

It is clear that $\varphi_\xi(0)=1$ and that $|\varphi_\xi(t)|\le 1$ and $\varphi_\xi(-t)=\overline{\varphi_\xi(t)}$ always hold. Further, $\varphi_\xi(t)$ is everywhere continuous. If there exists a mixed moment $\mathbf{E}\xi_1^{k_1}\cdots\xi_d^{k_d}$ then φ_ξ has the respective derivative of order $k_1+\dots+k_d$:

$$\left.\frac{\partial\varphi_\xi^{k_1+\dots+k_d}(t)}{\partial t_1^{k_1}\dots\partial t_d^{k_d}}\right|_{t=0} = i^{k_1+\dots+k_d}\mathbf{E}\xi_1^{k_1}\cdots\xi_d^{k_d}.$$

If all the moments of some order exist, then an expansion of the function $\varphi_\xi(t)$ similar to (7.1.1) is valid in a neighbourhood of the point $t=0$.

If $\varphi_\xi(t)$ is known, then the ch.f. of any subcollection of the random variables $(\xi_{k_1},\dots,\xi_{k_j})$ can obviously be obtained by setting all t_k except $t_{k_1},\dots,t_{k_j}$ to be equal to 0.

The following theorems are simple extensions of their univariate analogues.

Theorem 7.5.1 (The Inversion Formula) *If Δ is a parallelepiped defined by the inequalities $a_k<x<b_k$, $k=1,\dots,d$, and the probability $\mathbf{P}(\xi\in\Delta)$ is continuous on the faces of the parallelepiped, then*

$$\mathbf{P}(\xi\in\Delta) = \lim_{\sigma\to 0}\frac{1}{(2\pi)^d}\int\cdots\int\left(\prod_{k=1}^d\frac{e^{-it_ka_k}-e^{-it_kb_k}}{it_k}e^{-t_k^2\sigma^2}\right)\varphi_\xi(t)\,dt_1\cdots dt_d.$$

If the random vector ξ has a density $f(x)$ and its ch.f. $\varphi_\xi(t)$ is integrable, then the inversion formula can be written in the form

$$f(x) = \frac{1}{(2\pi)^d}\int e^{-i(t,x)}\varphi_\xi(t)\,dt.$$

If a function $g(x)$ is such that its Fourier transform

$$\widetilde{g}(t) = \int e^{i(t,x)}g(x)\,dx$$

is integrable (and this is always the case for sufficiently smooth $g(x)$) then the Parseval equality holds:

$$\mathbf{E}g(\xi) = \mathbf{E}\frac{1}{(2\pi)^d}\int e^{-i(t,\xi)}\widetilde{g}(t)\,dt = \frac{1}{(2\pi)^d}\int\varphi_\xi(-t)\widetilde{g}(t)\,dt.$$

As before, the inversion formula implies the theorem on one-to-one correspondence between ch.f.s and distribution functions and together with it the fact that $\{e^{i(t,x)}\}$ is a distribution determining class (cf. Definition 6.3.2).

The weak convergence of distributions $\mathbf{F}_n(B)$ in the d-dimensional space to a distribution $\mathbf{F}(B)$ is defined in the same way as in the univariate case: $\mathbf{F}_{(n)} \Rightarrow \mathbf{F}$ if

$$\int f(x)\, d\mathbf{F}_{(n)}(dx) \to \int f(x)\, d\mathbf{F}(dx)$$

for any continuous and bounded function $f(x)$.

Denote by $\varphi_n(t)$ and $\varphi(t)$ the ch.f.s of distributions $\mathbf{F}_n$ and $\mathbf{F}$, respectively.

Theorem 7.5.2 (Continuity Theorem) *A necessary and sufficient condition for the weak convergence* $\mathbf{F}_{(n)} \Rightarrow \mathbf{F}$ *is that, for any* t, $\varphi_n(t) \to \varphi(t)$ *as* $n \to \infty$.

In the case where one can establish convergence of $\varphi_n(t)$ to some function $\varphi(t)$, there arises the question of whether $\varphi(t)$ will be the ch.f. of some distribution, or, which is the same, whether the sequence $\mathbf{F}_{(n)}$ will converge weakly to some distribution $\mathbf{F}$. Answers to these questions are given by the following assertion. Let Δ_N be the cube defined by the inequality $\max_k |x_k| < N$.

Theorem 7.5.3 (Continuity Theorem) *Suppose a sequence* $\varphi_n(t)$ *of ch.f.s converges as* $n \to \infty$ *to a function* $\varphi(t)$ *for each* t. *Then the following three conditions are equivalent*:

(a) $\varphi(t)$ *is a ch.f.*;
(b) $\varphi(t)$ *is continuous at the point* $t = 0$;
(c) $\limsup_{n\to\infty} \int_{x \notin \Delta_N} \mathbf{F}_{(n)}(dx) \to 0$ *as* $N \to \infty$.

All three theorems from this section can be proved in the same way as in the univariate case.

Example 7.5.1 The multivariate normal distribution is defined as a distribution with density (see Sect. 3.3)

$$f_\xi(x) = \frac{|A|^{1/2}}{(2\pi)^{d/2}} e^{-Q(x)/2},$$

where

$$Q(x) = xAx^T = \sum_{i,j=1}^{d} a_{ij} x_i x_j,$$

and $|A|$ is the determinant of a positive definite matrix $A = \|a_{ij}\|$.

This is a *centred* normal distribution for which $\mathbf{E}\xi = 0$. The distribution of the vector $\xi + a$ for any constant vector a is also called normal.

Find the ch.f. of ξ. Show that

$$\varphi_\xi(t) = \exp\left\{-\frac{t\sigma^2 t^T}{2}\right\}, \tag{7.5.1}$$

where $\sigma^2 = A^{-1}$ is the matrix inverse to A and coinciding with the covariance matrix $\|\sigma_{ij}\|$ of ξ:

$$\sigma_{ij} = \mathbf{E}\xi_i\xi_j.$$

Indeed,

$$\varphi_\xi(t) = \frac{\sqrt{|A|}}{(2\pi)^{d/2}} \int_{-\infty}^{\infty} \cdots \int_{-\infty}^{\infty} \exp\left\{itx^T - \frac{xAx^T}{2}\right\} dx_1 \cdots dx_d. \qquad (7.5.2)$$

Choose an orthogonal matrix C such that $CAC^T = D$ is a diagonal matrix, and denote by $\mu_1, \ldots, \mu_n$ the values of its diagonal elements. Change the variables by putting $x = yC$ and $t = vC$. Then

$$|A| = |D| = \prod_{k=1}^{d} \mu_k,$$

$$itx^T - \frac{1}{2}xAx^T = ivy^T - \frac{1}{2}yDy^T = i\sum_{k=1}^{d} v_k y_k - \frac{1}{2}\sum_{k=1}^{n} \mu_k y_k^2,$$

and, by Property 2 of ch.f.s of the univariate normal distributions,

$$\varphi_\xi(t) = \frac{\sqrt{|A|}}{(2\pi)^{d/2}} \prod_{k=1}^{d} \int_{-\infty}^{\infty} \exp\left\{iv_k y_k - \frac{\mu_k y_k^2}{2}\right\} dy_k = \sqrt{|A|} \prod_{k=1}^{d} \frac{1}{\sqrt{\mu_k}} \exp\left\{-\frac{v_k^2}{2\mu_k}\right\}$$
$$= \exp\left\{-\frac{vD^{-1}v^T}{2}\right\} = \exp\left\{-\frac{tC^TD^{-1}Ct^T}{2}\right\} = \exp\left\{-\frac{tA^{-1}t^T}{2}\right\}.$$

On the other hand, since all the moments of ξ exist, in a neighbourhood of the point $t = 0$ one has

$$\varphi_\xi(t) = 1 - \frac{1}{2}tA^{-1}t^T + o\left(\sum t_k^2\right) = 1 + it\mathbf{E}\xi^T + \frac{1}{2}t\sigma^2 t^T + o\left(\sum t_k^2\right).$$

From this it follows that $\mathbf{E}\xi = 0$, $A^{-1} = \sigma^2$.

Formula (7.5.1) that we have just proved implies the following property of normal distributions: *the components of the vector* $(\xi_1, \ldots, \xi_d)$ *are independent if and only if the correlation coefficients* $\rho(\xi_i, \xi_j)$ *are zero for all* $i \neq j$. Indeed, if σ^2 is a diagonal matrix, then $A = \sigma^{-2}$ is also diagonal and $f_\xi(x)$ is equal to the product of densities. Conversely, if $(\xi_1, \ldots, \xi_d)$ are independent, then A is a diagonal matrix, and hence σ^2 is also diagonal.

7.6 Other Applications of Characteristic Functions. The Properties of the Gamma Distribution

7.6.1 Stability of the Distributions Φ_{α,σ^2} and $\mathbf{K}_{\alpha,\sigma}$

The stability property means, roughly speaking, that the distribution type is preserved under summation of random variables (this description of stability is not exact, for more detail see Sect. 8.8).

The sum of independent normally distributed random variables is also normally distributed. Indeed, let ξ_1 and ξ_2 be independent and normally distributed with parameters (a_1, σ_1^2) and (a_2, σ_2^2), respectively. Then the ch.f. of $\xi_1 + \xi_2$ is equal to

$$\begin{aligned}\varphi_{\xi_1+\xi_2}(t) = \varphi_{\xi_1}(t)\varphi_{\xi_2}(t) &= \exp\left\{ita_1 - \frac{t^2\sigma_1^2}{2}\right\}\exp\left\{ita_2 - \frac{t^2\sigma_2^2}{2}\right\} \\ &= \exp\left\{it(a_1+a_2) - \frac{t^2}{2}\left(\sigma_1^2+\sigma_2^2\right)\right\}.\end{aligned}$$

Thus the sum $\xi_1 + \xi_2$ is again a normal random variable, with parameters $(a_1 + a_2, \sigma_1^2 + \sigma_2^2)$.

Normality is also preserved when taking sums of *dependent* random variables (components of an arbitrary normally distributed random vector). This immediately follows from the form of the ch.f. of the multivariate normal law found in Sect. 7.5. One just has to note that to get the ch.f. of the sum $\xi_1 + \cdots + \xi_n$ it suffices to put $t_1 = \cdots = t_n = t$ in the expression

$$\varphi_{(\xi_1,\dots,\xi_n)}(t_1,\dots,t_n) = \mathbf{E}\exp\{it_1\xi_1 + \cdots + it_n\xi_n\}.$$

The sum of independent random variables distributed according to the Poisson law also has a Poisson distribution. Indeed, consider two independent random variables $\xi_1 \in \mathbf{\Pi}_{\lambda_1}$ and $\xi_2 \in \mathbf{\Pi}_{\lambda_2}$. The ch.f. of their sum is equal to

$$\varphi_{\xi_1+\xi_2}(t) = \exp\left\{\lambda_1\left(e^{it}-1\right)\right\}\exp\left\{\lambda_2\left(e^{it}-1\right)\right\} = \exp\left\{(\lambda_1+\lambda_2)\left(e^{it}-1\right)\right\}.$$

Therefore $\xi_1 + \xi_2 \in \mathbf{\Pi}_{\lambda_1+\lambda_2}$.

The sum of independent random variables distributed according to the Cauchy law also has a Cauchy distribution. Indeed, if $\xi_1 \in \mathbf{K}_{\alpha_1,\sigma_1}$ and $\xi_2 \in \mathbf{K}_{\alpha_2,\sigma_2}$, then

$$\begin{aligned}\varphi_{\xi_1+\xi_2}(t) &= \exp\left\{i\alpha_1 t - \sigma_1|t|\right\}\exp\left\{i\alpha_2 t - \sigma_2|t|\right\} \\ &= \exp\left\{i(\alpha_1+\alpha_2)t - (\sigma_1+\sigma_2)|t|\right\}; \\ \xi_1 + \xi_2 &\in \mathbf{K}_{\alpha_1+\alpha_2,\sigma_1+\sigma_2}.\end{aligned}$$

The above assertions are closely related to the fact that the normal and Poisson laws are, as we saw, limiting laws for sums of independent random variables (the Cauchy distribution has the same property, see Sect. 8.8). Indeed, if $S_{2n}/\sqrt{2n}$ converges in distribution to a normal law (where $S_k = \sum_{j=1}^k \xi_j$, ξ_j are independent and identically distributed) then it is clear that $S_n/\sqrt{n}$ and $(S_{2n} - S_n)/\sqrt{n}$ will also converge to a normal law so that the sum of two asymptotically normal random variables also has to be asymptotically normal.

Note, however, that due to its arithmetic structure the random variable $\xi \in \mathbf{\Pi}_\lambda$ (as opposed to $\xi \in \mathbf{\Phi}_{a,\sigma^2}$ or $\xi \in \mathbf{K}_{\alpha,\sigma}$) cannot be transformed by any normalisation (linear transformation) into a random variable again having the Poisson distribution but with another parameter. For this reason the Poisson distribution cannot be stable in the sense of Definition 8.8.2.

It is not hard to see that the other distributions we have met do not possess the above-mentioned property of preservation of the distribution type under summation of random variables. If, for instance, ξ_1 and ξ_2 are uniformly distributed over $[0, 1]$ and independent then F_{ξ_1} and $F_{\xi_1+\xi_2}$ are substantially different functions (see Example 3.6.1).

7.6.2 The Γ-distribution and its properties

In this subsection we will consider one more rather wide-spread type of distribution closely related to the normal distribution and frequently used in applications. This is the so-called *Pearson gamma distribution* $\mathbf{\Gamma}_{\alpha,\lambda}$. We will write $\xi \in \mathbf{\Gamma}_{\alpha,\lambda}$ if ξ has density

$$f(x;\alpha,\lambda) = \begin{cases} \frac{\alpha^\lambda}{\Gamma(\lambda)} x^{\lambda-1} e^{-\alpha x}, & x \ge 0, \\ 0, & x < 0, \end{cases}$$

depending on two parameters $\alpha > 0$ and $\lambda > 0$, where $\Gamma(\lambda)$ is the gamma function

$$\Gamma(\lambda) = \int_0^\infty x^{\lambda-1} e^{-x}\,dx, \quad \lambda > 0.$$

It follows from this equality that $\int f(x;\alpha,\lambda)\,dx = 1$ (one needs to make the variable change $\alpha x = y$). If one differentiates the ch.f.

$$\varphi(t) = \varphi(t;\alpha,\lambda) = \frac{\alpha^\lambda}{\Gamma(\lambda)} \int_0^\infty x^{\lambda-1} e^{itx-\alpha x}\,dx$$

with respect to t and then integrates by parts, the result will be

$$\begin{aligned} \varphi'(t) &= \frac{\alpha^\lambda}{\Gamma(\lambda)} \int_0^\infty i x^\lambda e^{itx-\alpha x}\,dx = \frac{\alpha^\lambda}{\Gamma(\lambda)} \frac{i\lambda}{\alpha - it} \int_0^\infty x^{\lambda-1} e^{itx-\alpha x}\,dx \\ &= \frac{i\lambda}{\alpha - it}\varphi(t); \\ &\big(\ln\varphi(t)\big)' = \big(-\lambda \ln(\alpha - it)\big)', \qquad \varphi(t) = c(\alpha - it)^{-\lambda}. \end{aligned}$$

Since $\varphi(0) = 1$ one has $c = \alpha^\lambda$ and $\varphi(t) = (1 - it/\alpha)^{-\lambda}$.

It follows from the form of the ch.f. that the subfamily of distributions $\mathbf{\Gamma}_{\alpha,\lambda}$ for a fixed α also has a certain stability property: if $\xi_1 \in \mathbf{\Gamma}_{\alpha,\lambda_1}$ and $\xi_2 \in \mathbf{\Gamma}_{\alpha,\lambda_2}$ are independent, then $\xi_1 + \xi_2 \in \mathbf{\Gamma}_{\alpha,\lambda_1+\lambda_2}$.

An example of a particular gamma distribution is given, for instance, by the distribution of the random variable

$$\chi_n^2 = \sum_{i=1}^{n} \xi_i^2,$$

where ξ_i are independent and normally distributed with parameters (0, 1). This is the so-called *chi-squared distribution with n degrees of freedom* playing an important role in statistics.

To find the distribution of χ_n^2 it suffices to note that, by virtue of the equality

$$\mathbf{P}(\chi_1^2 < x) = \mathbf{P}(|\xi_1| < \sqrt{x}) = \frac{2}{\sqrt{2\pi}} \int_0^{\sqrt{x}} e^{-u^2/2}\,du,$$

the density of χ_1^2 is equal to

$$\frac{1}{\sqrt{2\pi}} e^{-x/2} x^{-1/2} = f(x; 1/2, 1/2), \qquad \chi_1^2 \in \mathbf{\Gamma}_{1/2,1/2}.$$

This means that the ch.f. of χ_n^2 is

$$\varphi^n(t; 1/2, 1/2) = (1 - 2it)^{-n/2} = \varphi(t; 1/2, n/2)$$

and corresponds to the density $f(t; 1/2, n/2)$.

Another special case of the gamma distribution is the *exponential distribution* $\mathbf{\Gamma}_\alpha = \mathbf{\Gamma}_{\alpha,1}$ with density

$$f(x; \alpha, 1) = \alpha e^{-\alpha x}, \quad x \ge 0,$$

and characteristic function

$$\varphi(x; \alpha, 1) = \left(1 - \frac{it}{\alpha}\right)^{-1}.$$

We leave it to the reader to verify with the help of ch.f.s that if $\xi_j \in \mathbf{\Gamma}_{\alpha_j}$ and are independent, $\alpha_j \ne \alpha_l$ for $j \ne l$, then

$$\mathbf{P}\left(\sum_{j=1}^{n} \xi_j > x\right) = \sum_{j=1}^{n} e^{-\alpha_j x} \prod_{\substack{l=1 \\ l \ne j}}^{n} \left(1 - \frac{\alpha_j}{\alpha_l}\right)^{-1}.$$

In various applications (in particular, in queueing theory, cf. Sect. 12.4), the so-called *Erlang distribution* is also of importance. This is a distribution with density $f(x; \alpha, \lambda)$ for integer λ. The Erlang distribution is clearly a λ-fold convolution of the exponential distribution with itself.

We find the expectation and variance of a random variable ξ that has the gamma distribution with parameters (α, λ):

$$\mathbf{E}\xi = -i\varphi'(0; \alpha, \lambda) = \frac{\lambda}{\alpha}, \qquad \mathbf{E}\xi^2 = -i\varphi''(0; \alpha, \lambda) = \frac{\lambda(\lambda+1)}{\alpha^2},$$

$$\mathrm{Var}(\xi) = \frac{\lambda(\lambda+1)}{\alpha^2} - \left(\frac{\lambda}{\alpha}\right)^2 = \frac{\lambda}{\alpha^2}.$$

Distributions from the gamma family, and especially the exponential ones, are often (and justifiably) used to approximate distributions in various applied problems. We will present three relevant examples.

Example 7.6.1 Consider a complex device. The failure of at least one of n parts comprising the device means the breakdown of the whole device. The lifetime distribution of any of the parts is usually well described by the exponential law. (The reasons for this could be understood with the help of the Poisson theorem on rare events. See also Example 2.4.1 and Chap. 19.)

Thus if the lifetimes ξ_j of the parts are independent, and for the part number j one has

$$\mathbf{P}(\xi_j > x) = e^{-\alpha_j x}, \quad x > 0,$$

then the lifetime of the whole device will be equal to $\eta_n = \min(\xi_1, \ldots, \xi_n)$ and we will get

$$\mathbf{P}(\eta_n > x) = \mathbf{P}\left(\bigcap_{j=1}^{n} \{\xi_j > x\}\right) = \prod_{j=1}^{n} \mathbf{P}(\xi_j > x) = \exp\left\{-x \sum_{i=1}^{n} \alpha_i\right\}.$$

This means that η_n will also have the exponential distribution, and since

$$\mathbf{E}\xi_j = 1/\alpha_j,$$

the mean failure-free operation time of the device will be equal to

$$\mathbf{E}\eta_n = \left(\sum_{i=1}^{n} \frac{1}{\mathbf{E}\xi_i}\right)^{-1}.$$

Example 7.6.2 Now turn to the distribution of $\zeta_n = \max(\xi_1, \ldots, \xi_n)$, where ξ_i are independent and all have the $\boldsymbol{\Gamma}$-distribution with parameters (α, λ). We could consider, for instance, a queueing system with n channels. (That could be, say, a multiprocessor computer solving a problem using the complete enumeration algorithm, each of the processors of the machine checking a separate variant.) Channel number i is busy for a random time ξ_i. After what time will the whole system be free? This random time will clearly have the same distribution as ζ_n.

Since the ξ_i are independent, we have

$$\mathbf{P}(\zeta_n < x) = \mathbf{P}\left(\bigcap_{j=1}^{n} \{\xi_j < x\}\right) = \left[\mathbf{P}(\xi_1 < x)\right]^n. \tag{7.6.1}$$

If n is large, then for approximate calculations we could find the limiting distribution of ζ_n as $n \to \infty$. Note that, for any fixed x, $\mathbf{P}(\zeta_n < x) \to 0$ as $n \to \infty$.

Assuming for simplicity that $\alpha = 1$ (the general case can be reduced to this one by changing the scale), we apply L'Hospital's rule to see that, as $x \to \infty$,

$$\mathbf{P}(\xi_j < x) = \int_x^\infty \frac{1}{\Gamma(\lambda)} y^{\lambda-1} e^{-y}\, dy \sim \frac{x^{\lambda-1}}{\Gamma(\lambda)} e^{-x}.$$

Letting $n \to \infty$ and

$$x = x(n) = \ln\bigl[n(\ln n)^{\lambda-1}/\Gamma(\lambda)\bigr] + u, \quad u = \text{const},$$

we get

$$\mathbf{P}(\xi_j > x) \sim \frac{(\ln n)^{\lambda-1}}{\Gamma(\lambda)} \frac{\Gamma(\lambda)}{n(\ln n)^{\lambda-1}} e^{-u} = \frac{e^{-u}}{n}.$$

Therefore for such x and $n \to \infty$ we obtain by (7.6.1) that

$$\mathbf{P}(\zeta_n < x) = \left(1 - \frac{e^{-u}}{n}\bigl(1 + o(1)\bigr)\right)^n \to e^{-e^{-u}}.$$

Thus we have established the existence of the limit

$$\lim_{n\to\infty} \mathbf{P}\left(\zeta_n - \ln\left[\frac{n(\ln n)^{\lambda-1}}{\Gamma(\lambda)}\right] < u\right) = e^{-e^{-u}},$$

or, which is the same, that

$$\zeta_n - \ln\left[\frac{n(\ln n)^{\lambda-1}}{\Gamma(\lambda)}\right] \Subset F_0, \qquad F_0(u) = e^{-e^{-u}}.$$

In other words, for large n the variable ζ_n admits the representation

$$\zeta_n \approx \ln\left[\frac{n(\ln n)^{\lambda-1}}{\Gamma(\lambda)}\right] + \zeta^0, \quad \text{where } \zeta^0 \Subset F_0.$$

Example 7.6.3 Let ξ_1 and ξ_2 be independent with $\xi_1 \Subset \boldsymbol{\Gamma}_{\alpha,\lambda_1}$ and $\xi_2 \Subset \boldsymbol{\Gamma}_{\alpha,\lambda_2}$. What is the distribution of $\xi_1/(\xi_1 + \xi_2)$? We will make use of Theorem 4.9.2. Since the joint density $f(x, y)$ of ξ_1 and $\eta = \xi_1 + \xi_2$ is equal to

$$f(x, y) = f(x; \alpha, \lambda_1) f(y - x; \alpha, \lambda_2),$$

the density of η is

$$q(y) = f(y; \alpha, \lambda_1 + \lambda_2),$$

and the conditional density $f(x \mid y)$ of ξ_1 given $\eta = y$ is equal to

$$f(x \mid y) = \frac{f(x, y)}{q(y)} = \frac{\Gamma(\lambda_1 + \lambda_2)}{\Gamma(\lambda_1)\Gamma(\lambda_2)} \frac{x^{\lambda_1-1}(y - x)^{\lambda_2-1}}{y^{\lambda_1+\lambda_2-1}}, \qquad x \in [0, y].$$

By the formulas from Sect. 3.2 the conditional density of $\xi_1/y = \xi_1/(\xi_1+\xi_2)$ (given the same condition $\xi_1 + \xi_2 = y$) is equal to

$$y f(yx \mid y) = \frac{\Gamma(\lambda_1 + \lambda_2)}{\Gamma(\lambda_1)\Gamma(\lambda_2)} x^{\lambda_1-1}(1 - x)^{\lambda_2-1}, \quad x \in [0, 1].$$

This distribution does not depend on y (nor on α). Hence the conditional density of $\xi_1/(\xi_1 + \xi_2)$ will have the same property, too. We obtain the so-called *beta distribution* $\mathbf{B}_{\lambda_1,\lambda_2}$ with parameters λ_1 and λ_2 defined on the interval [0, 1]. In particular, for $\lambda_1 = \lambda_2 = 1$, the distribution is uniform: $\mathbf{B}_{1,1} = \mathbf{U}_{0,1}$.

7.7 Generating Functions. Application to Branching Processes. A Problem on Extinction

7.7.1 Generating Functions

We already know that if a random variable ξ is integer-valued, i.e.

$$\mathbf{P}\Big(\bigcup_k \{\xi = k\}\Big) = 1,$$

then the ch.f. $\varphi_\xi(t)$ will actually be a function of $z = e^{it}$, and, along with its ch.f., the distribution of ξ can be specified by its generating function

$$p_\xi(z) := \mathbf{E}z^\xi = \sum_k z^k \mathbf{P}(\xi = k).$$

The inversion formula can be written here as

$$\mathbf{P}(\xi = k) = \frac{1}{2\pi}\int_{-\pi}^{\pi} e^{-itk}\varphi_\xi(t)\,dt = \frac{1}{2\pi i}\int_{|z|=1} z^{-k-1} p_\xi(z)\,dz. \tag{7.7.1}$$

As was already noted (see Sect. 7.2), relation (7.7.1) is simply the formula for Fourier coefficients (since $e^{itk} = \cos tk + i \sin tk$).

If ξ and η are independent random variables, then the distribution of $\xi + \eta$ will be given by the convolution of the sequences $\mathbf{P}(\xi = k)$ and $\mathbf{P}(\eta = k)$:

$$\mathbf{P}(\xi + \eta = n) = \sum_{k=-\infty}^{\infty} \mathbf{P}(\xi = k)\mathbf{P}(\eta = n - k)$$

(the total probability formula). To this convolution there corresponds the product of the generating functions:

$$p_{\xi+\eta}(z) = p_\xi(z)p_\eta(z).$$

It is clear from the examples considered in Sect. 7.1 that the generating functions of random variables distributed according to the Bernoulli and Poisson laws are

$$p_\xi(z) = 1 + p(z - 1), \qquad p_\xi(z) = \exp\{\mu(z - 1)\},$$

respectively.

One can see from the definition of the generating function that, for a nonnegative random variable $\xi \ge 0$, the function $p_\xi(z)$ is defined for $|z| \le 1$ and is analytic in the domain $|z| < 1$.

7.7.2 The Simplest Branching Processes

Now we turn to sequences of random variables which describe the so-called *branching processes*. We have already encountered a simple example of such a process when describing a chain reaction scheme in Example 4.4.4. Consider a more general scheme of a branching process. Imagine particles that can produce other particles of the same type; these could be neutrons in chain reactions, bacteria reproducing according to certain laws etc. Assume that initially there is a single particle (the "null generation") that, as a result of a "division" act, transforms with probabilities f_k, $k = 0, 1, 2, \dots$, into k particles of the same type,

$$\sum_{k=0}^{\infty} f_k = 1.$$

The new particles form the "first generation". Each of the particles from that generation behaves itself in the same way as the initial particle, independently of what happened before and of the other particles from that generation. Thus we obtain the "second generation", and so on. Denote by ζ_n the number of particles in the n-th generation. To describe the sequence ζ_n, introduce, as we did in Example 4.4.4, independent sequences of independent identically distributed random variables

$$\bigl\{\xi_j^{(1)}\bigr\}_{j=1}^{\infty}, \quad \bigl\{\xi_j^{(2)}\bigr\}_{j=1}^{\infty}, \quad \dots,$$

where $\xi_j^{(n)}$ have the distribution

$$\mathbf{P}\bigl(\xi_j^{(n)} = k\bigr) = f_k, \quad k = 0, 1, \dots.$$

Then the sequence ζ_n can be represented as

$$\begin{aligned}
\zeta_0 &= 1, \\
\zeta_1 &= \xi_1^{(1)}, \\
\zeta_2 &= \xi_1^{(2)} + \dots + \xi_{\zeta_1}^{(2)}, \\
&\dots\dots\dots\dots\dots\dots \\
\zeta_n &= \xi_1^{(n)} + \dots + \xi_{\zeta_{n-1}}^{(n)}.
\end{aligned}$$

These are sums of random numbers of random variables. Since $\xi_1^{(n)}, \xi_2^{(n)}, \dots$ do not depend on ζ_{n-1}, for the generating function $f_{(n)}(z) = \mathbf{E}z^{\zeta_n}$ we obtain by the total probability formula that

$$\begin{aligned}
f_{(n)}(z) &= \sum_{k=0}^{\infty} \mathbf{P}(\zeta_{n-1} = k)\mathbf{E}z^{\xi_1^{(n)} + \dots + \xi_k^{(n)}} \\
&= \sum_{k=0}^{\infty} \mathbf{P}(\zeta_{n-1} = k) f^k(z) = f_{(n-1)}\bigl(f(z)\bigr),
\end{aligned} \tag{7.7.2}$$

where

$$f(z) := f_{(1)}(z) = \mathbf{E}z^{\xi_1^{(n)}} = \sum_{k=0}^{\infty} f_k z^k.$$

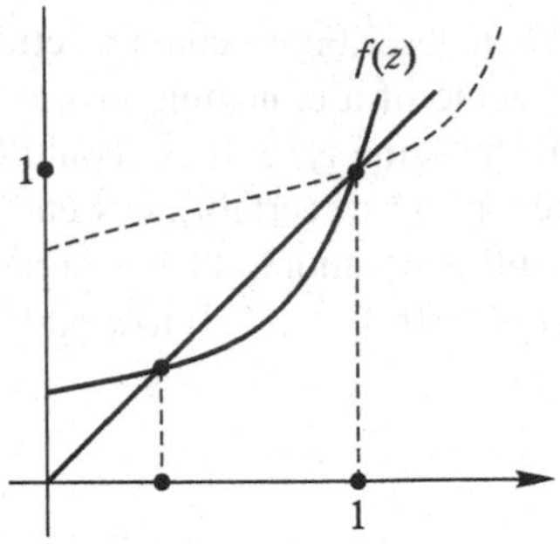

Fig. 7.1 Finding the extinction probability of a branching process: it is given by the smaller of the two solutions to the equation $z = f(z)$

Denote by $f_n(z)$ the n-th iterate of the function $f(z)$, i.e. $f_1(z) = f(z)$, $f_2(z) = f(f(z))$, $f_3(z) = f(f_2(z))$ and so on. Then we conclude from (7.7.2) by induction that the generating function of ζ_n is equal to the n-th iterate of $f(z)$:

$$\mathbf{E}z^{\zeta_n} = f_{(n)}(z).$$

From this one can easily obtain, by differentiating at the point $z = 1$, recursive relations for the moments of ζ_n.

How can one find the extinction probability of the process? By extinction we will understand the event that all ζ_n starting from some n will be equal to 0. (If $\zeta_n = 0$ then clearly $\zeta_{n+1} = \zeta_{n+2} = \cdots = 0$, because $\mathbf{P}(\zeta_{n+1} = 0 | \zeta_n = 0) = 1$.) Set $A_k = \{\zeta_k = 0\}$. Then extinction is the event $\bigcup_{k=1}^{\infty} A_k$. Since $A_n \subset A_{n+1}$, the extinction probability q is equal to $q = \lim_{n\to\infty} \mathbf{P}(A_n)$.

Theorem 7.7.1 *The extinction probability q is equal to the smallest nonnegative solution of the equation $q = f(q)$.*

Proof One has $\mathbf{P}(A_n) = f_n(0) \le 1$, and this sequence is non-increasing. Passing in the equality

$$f_{n+1}(0) = f\big(f_n(0)\big) \tag{7.7.3}$$

to the limit as $n \to \infty$, we obtain

$$q = f(q), \quad q \le 1.$$

This is an equation for the extinction probability. Let us analyse its solutions. The function $f(z)$ is convex (as $f''(z) \ge 0$) and non-decreasing in the domain $z \ge 0$ and $f'(1) = m$ is the mean number of offspring of a single particle. First assume that $\mathbf{P}(\xi_1^{(1)} = 1) < 1$. If $m \le 1$ then $f(z) > z$ for $z < 1$ and hence $q = 1$. If $m > 1$ then by convexity of f the equation $q = f(q)$ has exactly two solutions on the interval $[0, 1]$: $q_1 < 1$ and $q_2 = 1$ (see Fig. 7.1). Assume that $q = q_2 = 1$. Then the sequence $\delta_n = 1 - f_n(0)$ will monotonically converge to 0, and $f(1 - \delta_n) < 1 - \delta_n$ for sufficiently large n. Therefore, for such n,

$$\delta_{n+1} = 1 - f(1 - \delta_n) > \delta_n,$$

which is a contradiction as δ_n is a decreasing sequence. This means that $q = q_1 < 1$. Finally, in the case $\mathbf{P}(\xi_1^{(1)} = 1) = f_1 = 1$ one clearly has $f(z) \equiv z$ and $q = 0$. The theorem is proved. □

Now consider in more detail the case $m = 1$, which is called *critical*. We know that in this case the extinction probability q equals 1. Let $q_n = \mathbf{P}(A_n) = f_n(0)$ be the probability of extinction by time n. How fast does q_n converge to 1? By (7.7.3) one has $q_{n+1} = f(q_n)$. Therefore the probability $p_n = 1 - q_n$ of non-extinction of the process by time n satisfies the relation

$$p_{n+1} = g(p_n), \qquad g(x) = 1 - f(1 - x).$$

It is also clear that $\gamma_n = p_n - p_{n+1}$ is the probability that extinction will occur on step n.

Theorem 7.7.2 *If $m = f'(1) = 1$ and $0 < b := f''(1) < \infty$ then $\gamma_n \sim \frac{2}{bn^2}$ and $p_n \sim \frac{2}{bn}$ as $n \to \infty$.*

Proof If the second moment of the number of offspring of a single particle is finite ($b < \infty$) then the derivative $g''(0) = -b$ exists and therefore, since $g(0) = 0$ and $g'(0) = f'(1) = 1$, one has

$$g(x) = x - \frac{b}{2}x^2 + o(x^2), \quad x \to \infty.$$

Putting $x = p_n \to 0$, we find for the sequence $a_n = 1/p_n$ that

$$a_{n+1} - a_n = \frac{p_n - p_{n+1}}{p_n p_{n+1}} = \frac{bp_n^2(1 + o(1))}{2p_n^2(1 - bp_n/2 + o(p_n))} \to \frac{b}{2},$$

$$a_n = a_1 + \sum_{k=1}^{n-1}(a_{k+1} - a_k) \sim \frac{bn}{2}, \qquad p_n \sim \frac{2}{bn}.$$

The theorem is proved. □

Now consider the problem on the distribution of the number ζ_n of particles given $\zeta_n > 0$.

Theorem 7.7.3 *Under the assumptions of Theorem 7.7.2, the conditional distribution of $p_n\xi_n$ (or $2\zeta_n/(bn)$) given $\zeta_n > 0$ converges as $n \to \infty$ to the exponential distribution*:

$$\mathbf{P}(p_n\zeta_n > x | \zeta_n > 0) \to e^{-x}, \quad x > 0.$$

The above statement means, in particular, that given $\zeta_n > 0$, the number of particles ζ_n is of order n as $n \to \infty$.

Proof Consider the Laplace transform (see Property 6 in Sect. 7.1) of the conditional distribution of $p_n\zeta_n$ (given $\zeta_n > 0$):

$$\mathbf{E}\big(e^{-sp_n\zeta_n} | \zeta_n > 0\big) = \frac{1}{p_n}\sum_{k=1}^{\infty} e^{-skp_n}\mathbf{P}(\zeta_n = k). \tag{7.7.4}$$

We will make use of the fact that, if we could find an N such that $e^{-sp_n} = 1 - p_N$, which is the probability of extinction by time N, then the right-hand side of (7.7.4) will give, by the total probability formula, the conditional probability of the extinction of the process by time $n+N$ given its non-extinction at time n. We can evaluate this probability using Theorem 7.7.2.

Since $p_n \to 0$, for any fixed $s > 0$ one has

$$e^{-sp_n} - 1 \sim -sp_n \sim -\frac{2s}{bn}.$$

Clearly, one can always choose $N \sim n/s$, $s_n \sim s$, $s_n \downarrow s$ such that $e^{-s_n p_n} - 1 = -p_N$. Therefore $e^{-s_n p_n k} = (1 - p_N)^k$ and the right-hand side of (7.7.4) can be rewritten for $s = s_n$ as

$$\begin{aligned}\frac{1}{p_n}\sum_{k=1}^{\infty}\mathbf{P}(\zeta_n = k)(1 - p_N)^k &= \frac{1}{p_n}\mathbf{P}(\zeta_n > 0,\ \zeta_{n+N} = 0)\\ &= \frac{p_n - p_{n+N}}{p_n}\\ &= 1 - \frac{p_{n+N}}{p_n} \sim 1 - \frac{n}{n+N} = \frac{N}{n+N} \to \frac{1}{1+s}.\end{aligned}$$

Now note that

$$\mathbf{E}\big(e^{-sp_n\zeta_n}\big|\zeta_n > 0\big) - \mathbf{E}\big(e^{-s_n p_n\zeta_n}\big|\zeta_n > 0\big) = \mathbf{E}\big[e^{-sp_n\zeta_n}\big(1 - e^{-(s_n - s)p_n\zeta_n}\big|\zeta_n > 0\big)\big].$$

Since $e^{-\alpha} \le 1$ and $1 - e^{-\alpha} \le \alpha$ for $\alpha \ge 0$, and $\mathbf{E}\zeta_n = 1$, $\mathbf{E}(\zeta_n | \zeta_n > 0) = 1/p_n$, it is easily seen that the positive (since $s_n > s$) difference of the expectations in the last formula does not exceed

$$(s_n - s)p_n\mathbf{E}(\zeta_n | \zeta_n > 0) = s_n - s \to 0.$$

Therefore the Laplace transform (7.7.4) converges, as $n \to \infty$, to $1/(1+s)$. Since $1/(1+s)$ is the Laplace transform of the exponential distribution:

$$\int_0^{\infty} e^{-sx - x}\,dx = \frac{1}{1+s},$$

we conclude by the continuity theorem (see the remark after Theorem 7.3.1 in Sect. 7.3) that the conditional distribution of interest converges to the exponential law.[6]

In Sect. 15.4 (Example 15.4.1) we will obtain, as consequences of martingale convergence theorems, assertions about the behaviour of ζ_n as $n \to \infty$ for branching processes in the case $\mu > 1$ (the so-called *supercritical* processes). □

[6] The simple proof of Theorem 7.7.3 that we presented here is due to K.A. Borovkov.

Chapter 8
Sequences of Independent Random Variables. Limit Theorems

Abstract The chapter opens with proofs of Khintchin's (weak) Law of Large Numbers (Sect. 8.1) and the Central Limit Theorem (Sect. 8.2) the case of independent identically distributed summands, both using the apparatus of characteristic functions. Section 8.3 establishes general conditions for the Weak Law of Large Numbers for general sequences of independent random variables and also conditions for the respective convergence in mean. Section 8.4 presents the Central Limit Theorem in the triangular array scheme (the Lindeberg–Feller theorem) and its corollaries, illustrated by several insightful examples. After that, in Sect. 8.5 an alternative method of compositions is introduced and used to prove the Central Limit Theorem in the same situation, establishing an upper bound for the convergence rate for the uniform distance between the distribution functions in the case of finite third moments. This is followed by an extension of the above results to the multivariate case in Sect. 8.6. Section 8.7 presents important material not to be found in other textbooks: the so-called integro-local limit theorems on convergence to the normal distribution (the Stone–Shepp and Gnedenko theorems), including versions for sums of random variables depending on a parameter. These results will be of crucial importance in Chap. 9, when proving theorems on exact asymptotic behaviour of large deviation probabilities. The chapter concludes with Sect. 8.8 establishing integral, integro-local and local theorems on convergence of the distributions of scaled sums on independent identically distributed random variables to non-normal stable laws.

8.1 The Law of Large Numbers

Theorem 8.1.1 (Khintchin's Law of Large Numbers) *Let* $\{\xi_n\}_{n=1}^{\infty}$ *be a sequence of independent identically distributed random variables having a finite expectation* $\mathbf{E}\xi_n = a$ *and let* $S_n := \xi_1 + \cdots + \xi_n$. *Then*

$$\frac{S_n}{n} \xrightarrow{p} a \quad \text{as } n \to \infty.$$

The above assertion together with Theorems 6.1.6 and 6.1.7 imply the following.

A.A. Borovkov, *Probability Theory*, Universitext,
DOI 10.1007/978-1-4471-5201-9_8, © Springer-Verlag London 2013

Corollary 8.1.1 *Under the conditions of Theorem* 8.1.1, *as well as convergence of* S_n/n *in probability, convergence in mean also takes place*:

$$\mathbf{E}\left|\frac{S_n}{n}-a\right|\to 0 \quad \text{as } n\to\infty.$$

Note that the condition of independence of ξ_k and the very assertion of the theorem assume that all the random variables ξ_k are given on a common probability space.

From the physical point of view, the stated law of large numbers is the simplest ergodic theorem which means, roughly speaking, that for random variables their "time averages" and "space averages" coincide. This applies to an even greater extent to the strong law of large numbers, by virtue of which $S_n/n\to a$ with probability 1.

Under more strict assumptions (existence of variance) Theorem 8.1.1 was obtained in Sect. 4.7 as a consequence of Chebyshev's inequality.

Proof of Theorem 8.1.1 We have to prove that, for any $\varepsilon>0$,

$$\mathbf{P}\left(\left|\frac{S_n}{n}-a\right|>\varepsilon\right)\to 0$$

as $n\to\infty$. The above relation is equivalent to the weak convergence of distributions $S_n/n \Leftrightarrow \mathbf{I}_a$. Therefore, by the continuity theorem and Example 7.1.1 it suffices to show that, for any fixed t,

$$\varphi_{S_n/n}(t)\to e^{iat}.$$

The ch.f. $\varphi(t)$ of the random variable ξ_k has, in a certain neighbourhood of 0, the property $|\varphi(t)-1|<1/2$. Therefore for such t one can define the function $l(t)=\ln\varphi(t)$ (we take the principal value of the logarithm). Since ξ_n has finite expectation, the derivative

$$l'(0)=\frac{\varphi'(0)}{\varphi(0)}=ia$$

exists. For each fixed t and sufficiently large n, the value of $l(t/n)$ is defined and

$$\varphi_{S_n/n}(t)=\varphi^n(t/n)=e^{l(t/n)n}.$$

Since $l(0)=0$, one has

$$e^{l(t/n)n}=\exp\left\{t\frac{l(t/n)-l(0)}{t/n}\right\}\to e^{l'(0)t}=e^{iat}$$

as $n\to\infty$. The theorem is proved. □

8.2 The Central Limit Theorem for Identically Distributed Random Variables

Let, as before, $\{\xi_n\}$ be a sequence of independent identically distributed random variables. But now we assume, along with the expectation $\mathbf{E}\xi_n = a$, the existence of the variance $\operatorname{Var}\xi_n = \sigma^2$. We retain the notation $S_n = \xi_1 + \cdots + \xi_n$ for sums of our random variables and $\Phi(x)$ for the normal distribution function with parameters (0, 1). Introduce the sequence of random variables

$$\zeta_n = \frac{S_n - an}{\sigma\sqrt{n}}.$$

Theorem 8.2.1 *If* $0 < \sigma^2 < \infty$, *then* $\mathbf{P}(\zeta_n < x) \to \Phi(x)$ *uniformly in* x $(-\infty < x < \infty)$ *as* $n \to \infty$.

In such a case, the sequence $\{\zeta_n\}$ is said to be *asymptotically normal.*

It follows from $\zeta_n \Rightarrow \zeta \in \mathbf{\Phi}_{0,1}$, $\zeta_n^2 \ge 0$, $\mathbf{E}\zeta_n^2 = \mathbf{E}\zeta^2 = 1$ and from Lemma 6.2.3 that the sequence $\{\zeta_n^2\}$ is uniformly integrable. Therefore, as well as the weak convergence $\zeta_n \Rightarrow \zeta$, $\zeta \in \mathbf{\Phi}_{0,1}$ $(\mathbf{E}f(\zeta_n) \to \mathbf{E}f(\zeta)$ for any bounded continuous $f)$, one also has convergence $\mathbf{E}f(\zeta_n) \to \mathbf{E}f(\zeta)$ for any continuous f such that $|f(x)| < c(1 + x^2)$ (see Theorem 6.2.3).

Proof of Theorem 8.2.1 The uniform convergence is a consequence of the weak convergence and continuity of $\Phi(x)$. Further, we may assume without loss of generality that $a = 0$, for otherwise we could consider the sequence $\{\xi'_n = \xi_n - a\}_{n=1}^{\infty}$ without changing the sequence $\{\zeta_n\}$. Therefore, to prove the required convergence, it suffices to show that $\varphi_{\zeta_n}(t) \to e^{-t^2/2}$ when $a = 0$. We have

$$\varphi_{\zeta_n}(t) = \varphi^n\left(\frac{t}{\sigma\sqrt{n}}\right), \qquad \text{where } \varphi(t) = \varphi_{\xi_k}(t).$$

Since $\mathbf{E}\xi_n^2$ exists, $\varphi''(t)$ also exists and, as $t \to 0$, one has

$$\varphi(t) = \varphi(0) + t\varphi'(0) + \frac{t^2}{2}\varphi''(0) + o(t^2) = 1 - \frac{t^2\sigma^2}{2} + o(t^2). \tag{8.2.1}$$

Therefore, as $n \to \infty$,

$$\begin{aligned} \ln\varphi_{\zeta_n}(t) &= n\ln\left[1 - \frac{\sigma^2}{2}\left(\frac{t}{\sigma\sqrt{n}}\right)^2 + o\left(\frac{t^2}{n}\right)\right] \\ &= n\left[-\frac{t^2}{2n} + o\left(\frac{t^2}{n}\right)\right] = -\frac{t^2}{2} + o(1) \to -\frac{t^2}{2}. \end{aligned}$$

The theorem is proved. □

8.3 The Law of Large Numbers for Arbitrary Independent Random Variables

Now we proceed to elucidating conditions under which the law of large numbers and the central limit theorem will hold in the case when ξ_k are independent but not necessarily identically distributed. The problem will not become more complicated if, from the very beginning, we consider a more general situation where one is given an arbitrary series $\xi_{1,n}, \dots, \xi_{n,n}$, $n = 1, 2, \dots$ of independent random variables, where the distributions of $\xi_{k,n}$ may depend on n. This is the so-called *triangular array scheme*.

Put

$$\zeta_n := \sum_{k=1}^{n} \xi_{k,n}.$$

From the viewpoint of the results to follow, we can assume without loss of generality that

$$\mathbf{E}\xi_{k,n} = 0. \tag{8.3.1}$$

Assume that the following condition is met: as $n \to \infty$,

$$D_1 := \sum_{k=1}^{n} \mathbf{E} \min\left(|\xi_{k,n}|, |\xi_{k,n}|^2\right) \to 0. \tag{$\mathbf{D}_1$}$$

Theorem 8.3.1 (The Law of Large Numbers) *If conditions* (8.3.1) *and* $[\mathbf{D}_1]$ *are satisfied, then* $\zeta_n \Leftrightarrow \mathbf{I}_0$ *or, which is the same,* $\zeta_n \xrightarrow{p} 0$ *as* $n \to \infty$.

Example 8.3.1 Assume $\xi_k = \xi_{k,n}$ do not depend on n, $\mathbf{E}\xi_k = 0$ and $\mathbf{E}|\xi_k|^s \le m_s < \infty$ for $1 < s \le 2$. For such s, there exists a sequence $b(n) = o(n)$ such that $n = o(b^s(n))$. Since, for $\xi_{k,n} = \xi_k / b(n)$,

$$\begin{aligned}
\mathbf{E}\min\left(|\xi_{k,n}|, \xi_{k,n}^2\right) &= \mathbf{E}\left[\left|\frac{\xi_k}{b(n)}\right|^2; |\xi_k| \le b(n)\right] + \mathbf{E}\left[\frac{|\xi_k|}{b(n)}; |\xi_k| > b(n)\right] \\
&\le \mathbf{E}\left[\left|\frac{\xi_k}{b(n)}\right|^s; |\xi_k| \le b(n)\right] + \mathbf{E}\left[\left|\frac{\xi_k}{b(n)}\right|^s; |\xi_k| > b(n)\right] \\
&= m_s b^{-s}(n),
\end{aligned}$$

we have

$$D_1 \le n m_s b^{-s}(n) \to 0,$$

and hence $S_n / b(n) \xrightarrow{p} 0$.

A more general sufficient condition (compared to $m_s < \infty$) for the law of large numbers is contained in Theorem 8.3.3 below. Theorem 8.1.1 is an evident corollary of that theorem.

Now consider condition $[\mathbf{D}_1]$ in more detail. It can clearly also be written in the form

$$D_1 = \sum_{k=1}^{n} \mathbf{E}\big(|\xi_{k,n}|;\ |\xi_{k,n}| > 1\big) + \sum_{k=1}^{n} \mathbf{E}\big(|\xi_{k,n}|^2;\ |\xi_{k,n}| \le 1\big) \to 0.$$

Next introduce the condition

$$M_1 := \sum_{k=1}^{n} \mathbf{E}|\xi_{k,n}| \le c < \infty \qquad (8.3.2)$$

and the condition

$$M_1(\tau) := \sum_{k=1}^{n} \mathbf{E}\big(|\xi_{k,n}|;\ |\xi_{k,n}| > \tau\big) \to 0 \qquad [\mathbf{M1}]$$

for any $\tau > 0$ as $n \to \infty$. Condition $[\mathbf{M}_1]$ could be called a Lindeberg type condition (the Lindeberg condition $[\mathbf{M}_2]$ will be introduced in Sect. 8.4).

The following lemma explains the relationship between the introduced conditions.

Lemma 8.3.1 1. $\{[\mathbf{M}_1] \cap (3.2)\} \subset [\mathbf{D}_1]$. 2. $[\mathbf{D}_1] \subset [\mathbf{M}_1]$.

That is, conditions $[\mathbf{M}_1]$ and (8.3.2) imply $[\mathbf{D}_1]$, and condition $[\mathbf{D}_1]$ implies $[\mathbf{M}_1]$.

It follows from Lemma 8.3.1 that under condition (8.3.2), conditions $[\mathbf{D}_1]$ and $[\mathbf{M}_1]$ are equivalent.

Proof of Lemma 8.3.1 1. Let conditions (8.3.2) and $[\mathbf{M}_1]$ be met. Then, for

$$\tau \le 1, \qquad g_1(x) = \min\big(|x|, |x|^2\big),$$

one has

$$\begin{aligned} D_1 &= \sum_{k=1}^{n} \mathbf{E} g_1(\xi_{k,n}) \le \sum_{k=1}^{n} \mathbf{E}\big(|\xi_{k,n}|;\ |\xi_{k,n}| > \tau\big) + \sum_{k=1}^{n} \mathbf{E}\big(|\xi_{k,n}|^2;\ |\xi_{k,n}| \le \tau\big) \\ &\le M_1(\tau) + \tau \sum_{k=1}^{n} \mathbf{E}\big(|\xi_{k,n}|;\ |\xi_{k,n}| \le \tau\big) \le M_1(\tau) + \tau M_1(0). \end{aligned} \qquad (8.3.3)$$

Since $M_1(0) = M_1 \le c$ and τ can be arbitrary small, we have $D_1 \to 0$ as $n \to \infty$.

2. Conversely, let condition $[\mathbf{D}_1]$ be met. Then, for $\tau \le 1$,

$$M_1(\tau) \le \sum_{k=1}^{n} \mathbf{E}\big(|\xi_{k,n}|;\ |\xi_{k,n}| > 1\big)$$

$$+\tau^{-1}\sum_{k=1}^{n}\mathbf{E}\big(|\xi_{k,n}|^2;\ \tau<|\xi_{k,n}|\le 1\big)\le\tau^{-1}D_1\to 0 \qquad (8.3.4)$$

as $n\to\infty$ for any $\tau>0$. The lemma is proved. □

Let us show that condition [$\mathbf{M}_1$] (as well as [$\mathbf{D}_1$]) is essential for the law of large numbers to hold.

Consider the random variables

$$\xi_{k,n}=\begin{cases}1-\frac{1}{n} & \text{with probability } \frac{1}{n},\\ -\frac{1}{n} & \text{with probability } 1-\frac{1}{n}.\end{cases}$$

For them, $\mathbf{E}\xi_{k,n}=0$, $\mathbf{E}|\xi_{k,n}|=\frac{2(n-1)}{n^2}\sim\frac{2}{n}$, $M_1\le 2$, condition (8.3.2) is met, but $M_1(\tau)=\frac{n-1}{n}>\frac{1}{2}$ for $n>2$, $\tau<1/2$, and thus condition [$\mathbf{M}_1$] is not satisfied. Here the number ν_n of positive $\xi_{k,n}$, $1\le k\le n$, converges in distribution to a random variable ν having the Poisson distribution with parameter $\lambda=1$. The sum of the remaining $\xi_{k,n}$s is equal to $-\frac{(n-\nu_n)}{n}\xrightarrow{p}-1$. Therefore, $\zeta_n+1\Leftrightarrow\boldsymbol{\Pi}_1$ and the law of large numbers does not hold.

Each of the conditions [$\mathbf{D}_1$] and [$\mathbf{M}_1$] imply the uniform smallness of $\mathbf{E}|\xi_{k,n}|$:

$$\max_{1\le k\le n}\mathbf{E}|\xi_{k,n}|\to 0 \quad \text{as } n\to\infty. \qquad (8.3.5)$$

Indeed, equation [$\mathbf{M}_1$] means that there exists a sufficiently slowly decreasing sequence $\tau_n\to 0$ such that $M_1(\tau_n)\to 0$. Therefore

$$\max_{k\le n}\mathbf{E}|\xi_{k,n}|\le\max_{k\le n}\big[\tau_n+\mathbf{E}\big(|\xi_{k,n}|;\ |\xi_{k,n}|>\tau_n\big)\big]\le\tau_n+M_1(\tau_n)\to 0. \qquad (8.3.6)$$

In particular, (8.3.5) implies the negligibility of the summands $\xi_{k,n}$.

We will say that $\xi_{k,n}$ are *negligible*, or, equivalently, have property [$\mathbf{S}$], if, for any $\varepsilon>0$,

$$\max_{k\le n}\mathbf{P}\big(|\xi_{k,n}|>\varepsilon\big)\to 0 \quad \text{as } n\to\infty. \qquad [\mathbf{S}]$$

Property [$\mathbf{S}$] could also be called *uniform convergence of* $\xi_{k,n}$ *in probability to zero*. Property [$\mathbf{S}$] follows immediately from (8.3.5) and Chebyshev's inequality. It also follows from stronger relations implied by [$\mathbf{M}_1$]:

$$\begin{aligned}\mathbf{P}\Big(\max_{k\le n}|\xi_{k,n}|>\varepsilon\Big)&=\mathbf{P}\Big(\bigcup_{k\le n}\big\{|\xi_{k,n}|>\varepsilon\big\}\Big)\\ &\le\sum_{k\le n}\mathbf{P}\big(|\xi_{k,n}|>\varepsilon\big)\le\varepsilon^{-1}\sum_{k\le n}\mathbf{E}\big(|\xi_{k,n}|;\ |\xi_{k,n}|>\varepsilon\big)\to 0. \qquad [\mathbf{S}_1]\end{aligned}$$

We now turn to proving the law of large numbers. We will give two versions of the proof. The first one illustrates the classical method of characteristic functions.

The second version is based on elementary inequalities and leads to a stronger assertion about convergence in mean.[1]

Here is the first version.

Proof of Theorem 8.3.1[2] Put

$$\varphi_{k,n}(t) := \mathbf{E}e^{it\xi_{k,n}}, \qquad \Delta_k(t) := \varphi_{k,n}(t) - 1.$$

One has to prove that, for each t,

$$\varphi_{\zeta_n}(t) = \mathbf{E}e^{it\zeta_n} = \prod_{k=1}^{n} \varphi_{k,n}(t) \to 1,$$

as $n \to \infty$. By Lemma 7.4.2

$$\begin{aligned}\bigl|\varphi_{\zeta_n}(t) - 1\bigr| &= \left|\prod_{k=1}^{n} \varphi_{k,n}(t) - \prod_{k=1}^{n} 1\right| \le \sum_{k=1}^{n} \bigl|\Delta_k(t)\bigr| \\ &= \sum_{k=1}^{n} \bigl|\mathbf{E}e^{it\xi_{k,n}} - 1\bigr| = \sum_{k=1}^{n} \bigl|\mathbf{E}\bigl(e^{it\xi_{k,n}} - 1 - it\xi_{k,n}\bigr)\bigr|.\end{aligned}$$

By Lemma 7.4.1 we have (for $g_1(x) = \min(|x|, x^2)$)

$$\bigl|e^{itx} - 1 - itx\bigr| \le \min\bigl(2|tx|, t^2x^2/2\bigr) \le 2g_1(tx) \le 2h(t)g_1(t),$$

where $h(t) = \max(|t|, |t|^2)$. Therefore

$$\bigl|\varphi_{\zeta_n}(t) - 1\bigr| \le 2h(t) \sum_{k=1}^{n} \mathbf{E}g_1(\xi_{k,n}) = 2h(t)D_1 \to 0.$$

The theorem is proved. □

The last inequality shows that $|\varphi_{\zeta_n}(t) - 1|$ admits a bound in terms of D_1. It turns out that $\mathbf{E}|\zeta_n|$ also admits a bound in terms of D_1. Now we will give the second version of the proof that actually leads to a stronger variant of the law of large numbers.

Theorem 8.3.2 *Under conditions* (8.3.1) *and* $[\mathbf{D}_1]$ *one has* $\mathbf{E}|\zeta_n| \to 0$ *(i.e.* $\zeta_n \xrightarrow{(1)} 0$*).*

---

[1] The second version was communicated to us by A.I. Sakhanenko.

[2] There exists an alternative "direct" proof of Theorem 8.3.1 using not ch.f.s but the so-called truncated random variables and estimates of their variances. However, because of what follows, it is more convenient for us to use here the machinery of ch.f.s.

The assertion of Theorem 8.3.2 clearly means *the uniform integrability of* $\{\zeta_n\}$; it implies Theorem 8.3.1, for

$$\mathbf{P}\big(|\zeta_n| > \varepsilon\big) \le \mathbf{E}|\zeta_n|/\varepsilon \to 0 \quad \text{as } n \to \infty.$$

Proof of Theorem 8.3.2 Put

$$\xi'_{k,n} := \begin{cases} \xi_{k,n} & \text{if } |\xi_{k,n}| \le 1, \\ 0 & \text{otherwise,} \end{cases}$$

and $\xi''_{k,n} := \xi_{k,n} - \xi'_{k,n}$. Then $\xi_{k,n} = \xi'_{k,n} + \xi''_{k,n}$ and $\zeta_n = \zeta'_n + \zeta''_n$ with an obvious convention for the notations ζ'_n, ζ''_n. By the Cauchy–Bunjakovsky inequality,

$$\begin{aligned} \mathbf{E}|\zeta_n| &\le \mathbf{E}\big|\zeta'_n - \mathbf{E}\zeta'_n\big| + \mathbf{E}\big|\zeta''_n - \mathbf{E}\zeta''_n\big| \le \sqrt{\mathbf{E}\big(\zeta'_n - \mathbf{E}\zeta'_n\big)^2} + \mathbf{E}\big|\zeta''_n\big| + \big|\mathbf{E}\zeta''_n\big| \\ &\le \sqrt{\sum \operatorname{Var}\big(\xi'_{k,n}\big)} + 2\sum \mathbf{E}\big|\xi''_{k,n}\big| \le \sqrt{\sum \mathbf{E}\big(\xi'_{k,n}\big)^2} + 2\sum \mathbf{E}\big|\xi''_{k,n}\big| \\ &= \Big[\sum \mathbf{E}\big(\xi^2_{k,n};\, |\xi_{k,n}| \le 1\big)\Big]^{1/2} \\ &\quad + 2\sum \mathbf{E}\big(|\xi_{k,n}|;\, |\xi_{k,n}| > 1\big) \le \sqrt{D_1} + 2D_1 \to 0, \end{aligned}$$

if $D_1 \to 0$. The theorem is proved. □

Remark 8.3.1 It can be seen from the proof of Theorem 8.3.2 that the argument will remain valid if we replace the independence of $\xi_{k,n}$ by the weaker condition that $\xi'_{k,n}$ are non-correlated. It will also be valid if $\xi'_{k,n}$ are only weakly correlated so that

$$\mathbf{E}\big(\zeta'_n - \mathbf{E}\zeta'_n\big)^2 \le c \sum \operatorname{Var}\big(\xi'_{k,n}\big), \quad c < \infty.$$

If $\{\xi_k\}$ is a given *fixed* (not dependent on n) sequence of independent random variables, $S_n = \sum_{k=1}^n \xi_k$ and $\mathbf{E}\xi_k = a_k$, then one looks at the applicability of the law of large numbers to the sequences

$$\xi_{k,n} = \frac{\xi_k - a_k}{b(n)}, \qquad \zeta_n = \sum \xi_{k,n} = \frac{1}{b(n)}\left(S_n - \sum_{k=1}^n a_k\right), \tag{8.3.7}$$

where $\xi_{k,n}$ satisfy (8.3.1), and $b(n)$ is an unboundedly increasing sequence. In some cases it is natural to take $b(n) = \sum_{k=1}^n \mathbf{E}|\xi_k|$ if this sum increases unboundedly. Without loss of generality we can set $a_k = 0$. The next assertion follows from Theorem 8.3.2.

Corollary 8.3.1 *If, as* $n \to \infty$,

$$D_1 := \frac{1}{b(n)} \sum \mathbf{E}\min\big(|\xi_k|,\, \xi_k^2/b(n)\big) \to 0$$

or, for any $\tau > 0$,

$$M_1(\tau) = \frac{1}{b(n)} \sum \mathbf{E}\big(|\xi_k|;\, |\xi_k| > \tau b(n)\big) \to 0, \quad b(n) = \sum_{k=1}^{n} \mathbf{E}|\xi_k| \to \infty, \tag{8.3.8}$$

then $\zeta_n \xrightarrow{(1)} 0$.

Now we will present an important sufficient condition for the law of large numbers that is very close to condition (8.3.8) and which explains to some extent its essence. In addition, in many cases this condition is easier to check. Let $b_k = \mathbf{E}|\xi_k|$, $\overline{b}_n = \max_{k \le n} b_k$, and, as before,

$$S_n = \sum_{k=1}^{n} \xi_k, \qquad b(n) = \sum_{k=1}^{n} b_k.$$

The following assertion is a direct generalisation of Theorem 8.1.1 and Corollary 8.1.1.

Theorem 8.3.3 *Let* $\mathbf{E}\xi_k = 0$, *the sequence of normalised random variables* ξ_k / b_k *be uniformly integrable and* $\overline{b}_n = o(b(n))$ *as* $n \to \infty$. *Then*

$$\frac{S_n}{b(n)} \xrightarrow{(1)} 0.$$

If $\overline{b}_n \le b < \infty$ *then* $b(n) \le bn$ *and* $\frac{S_n}{n} \xrightarrow{(1)} 0$.

Proof Since

$$\mathbf{E}\big(|\xi_k|;\, |\xi_k| > \tau b(n)\big) \le b_k \mathbf{E}\left(\left|\frac{\xi_k}{b_k}\right|;\, \left|\frac{\xi_k}{b_k}\right| > \tau \frac{b(n)}{\overline{b}_n}\right) \tag{8.3.9}$$

and $\frac{b(n)}{\overline{b}_n} \to \infty$, the uniform integrability of $\{\frac{\xi_k}{b_k}\}$ implies that the right-hand side of (8.3.9) is $o(b_k)$ uniformly in k (i.e. it admits a bound $\varepsilon(n) b_k$, where $\varepsilon(n) \to 0$ as $n \to \infty$ and does not depend on k). Therefore

$$M_1(\tau) = \frac{1}{b(n)} \sum_{k=1}^{n} \mathbf{E}\big(|\xi_k|;\, |\xi_k| > \tau b(n)\big) \to 0$$

as $n \to \infty$, and condition (8.3.8) is met. The theorem is proved. □

Remark 8.3.2 If, in the context of the law of large numbers, we are interested in convergence in probability, only then can we generalise Theorem 8.3.3. In particular, *convergence*

$$\frac{S_n}{b(n)} \xrightarrow{p} 0$$

will still hold if a finite number of the summands ξ_k (e.g., for $k \le l$, l being fixed) *are completely arbitrary* (they can even fail to have expectations) *and the sequence* $\xi_k^* = \xi_{k+l}$, $k \ge 1$, *satisfies the conditions of Theorem* 8.3.3, *where* $b(n)$ *is defined for the variables* ξ_k^* *and has the property* $\frac{b(n-1)}{b(n)} \to 1$ *as* $n \to \infty$.

This assertion follows from the fact that

$$\frac{S_n}{b(n)} = \frac{S_l}{b(n)} + \frac{S_n - S_l}{b(n-l)} \cdot \frac{b(n-l)}{b(n)}, \qquad \frac{S_l}{b(n)} \xrightarrow{p} 0, \qquad \frac{b(n-l)}{b(n)} \to 1,$$

and by Theorem 8.3.3

$$\frac{S_n - S_l}{b(n-l)} \xrightarrow{p} 0 \quad \text{as } n \to \infty.$$

Now we will show that the uniform integrability condition in Theorem 8.3.3 (as well as condition $M_1(\tau) \to 0$) is essential for convergence $\zeta_n \xrightarrow{p} 0$. Consider a sequence of random variables

$$\xi_j = \begin{cases} 2^s - 1 & \text{with probability } 2^{-s}, \\ -1 & \text{with probability } 1 - 2^{-s} \end{cases}$$

for $j \in I_s := (2^{s-1}, 2^s]$, $s = 1, 2, \ldots$; $\xi_1 = 0$. Then $\mathbf{E}\xi_j = 0$, $\mathbf{E}|\xi_j| = 2(1 - 2^{-s})$ for $j \in I_s$, and, for $n = 2^k$, one has

$$b(n) = \sum_{s=1}^{k} 2\big(1 - 2^{-s}\big)|I_s|,$$

where $|I_s| = 2^s - 2^{s-1} = 2^{s-1}$ is the number of points in I_k. Hence, as $k \to \infty$,

$$\begin{aligned} b(n) &\sim 2\big[\big(1 - 2^{-k}\big)2^{k-1} + \big(1 - 2^{-k+1}\big)2^{k-2} + \cdots\big] \\ &\sim 2^k + 2^{k-1} + \ldots \sim 2^{k+1} = 2n. \end{aligned}$$

Observe that the uniform integrability condition is clearly not met here. The distribution of the number $\nu^{(s)}$ of jumps of magnitude $2^s - 1$ on the interval I_s converges, as $s \to \infty$, to the Poisson distribution with parameter $1/2 = \lim_{s\to\infty} 2^{-s}|I_s|$, while the distribution of $2^{-s}(S_{2^s} - S_{2^{s-1}})$ converges to the distribution of $\nu - 1/2$, where $\nu \Subset \mathbf{\Pi}_{1/2}$. Hence, assuming that $n = 2^k$, and partitioning the segment $[2, n]$ into the intervals $(2^{s-1}, 2^s]$, $s = 1, \ldots, k$, we obtain that the distribution of S_n/n converges, as $k \to \infty$, to the distribution of

$$\frac{S_n}{n} = 2^{-k} \sum_{s=1}^{k} \frac{S_{2^s} - S_{2^{s-1}}}{2^s} 2^s \Rightarrow \sum_{l=0}^{\infty} (\nu_l - 1/2)2^{-l} =: \zeta,$$

where ν_l, $l = 0, 1, \ldots$, are independent copies of ν. Clearly, $\zeta \not\equiv 0$, and so convergence $\frac{S_n}{n} \xrightarrow{p} 0$ fails to take place.

Let us return to arbitrary $\xi_{k,n}$. In order for $[\mathbf{D}_1]$ to hold it suffices that the following condition is met: *for some* s, $2 \ge s > 1$,

$$\sum_{k=1}^{n} \mathbf{E}|\xi_{k,n}|^s \to 0. \qquad [\mathbf{L}_s]$$

This assertion is evident, since $g_1(x) \le |x|^s$ for $2 \ge s > 1$. Conditions $[\mathbf{L}_s]$ could be called the modified Lyapunov conditions (cf. the Lyapunov condition $[\mathbf{L}_s]$ in Sect. 8.4).

To prove Theorem 8.3.2, we used the so-called "truncated versions" $\xi'_{k,n}$ of the random variables $\xi_{k,n}$. Now we will consider yet another variant of the law of large numbers, in which *conditions* are expressed in terms of truncated random variables.

Denote by $\xi^{(N)}$ the result of truncation of the random variable ξ at level N:

$$\xi^{(N)} = \max\bigl[-N, \min(N, \xi)\bigr].$$

Theorem 8.3.4 *Let the sequence of random variables* $\{\xi_k\}$ *in* (8.3.7) *satisfy the following condition: for any given* $\varepsilon > 0$, *there exist* N_k *such that*

$$\frac{1}{b(n)} \sum_{k=1}^{n} \mathbf{E}\bigl|\xi_k - \xi_k^{(N_k)}\bigr| < \varepsilon, \qquad \frac{1}{b(n)} \sum_{k=1}^{n} N_k < N < \infty.$$

Then the sequence $\{\zeta_n\}$ *converges to zero in mean*: $\zeta_n \xrightarrow{(1)} 0$.

Proof Clearly $a_k^{(N_k)} := \mathbf{E}\xi_k^{(N_k)} \to a_k$ as $N_k \to \infty$ and $|a_k^{(N_k)}| \le N_k$. Further, we have

$$\mathbf{E}|\zeta_n| = \frac{1}{b(n)}\mathbf{E}\Bigl|\sum(\xi_k - a_k)\Bigr| \le \frac{1}{b(n)} \sum \mathbf{E}\bigl|\xi_k - \xi_k^{(N_k)}\bigr| + \mathbf{E}\biggl|\sum \frac{\xi_k^{(N_k)} - a_k^{(N_k)}}{b(n)}\biggr| + \frac{1}{b(n)} \sum \bigl|a_k^{(N_k)} - a_k\bigr|.$$

Here the second term on the right-hand side converges to zero, since the sum under the expectation satisfies the conditions of Theorem 8.3.1 and is bounded. But the first and the last terms do not exceed ε. Since the left-hand side does not depend on ε, we have $\mathbf{E}|\zeta_n| \to 0$ as $n \to \infty$. □

Corollary 8.3.2 *If* $b(n) = n$ *and, for sufficiently large* N *and all* $k \le n$,

$$\mathbf{E}\bigl|\xi_k - \xi_k^{(N)}\bigr| < \varepsilon,$$

then $\zeta_n \xrightarrow{(1)} 0$.

The corollary follows from Theorem 8.3.4, since the conditions of the corollary clearly imply the conditions of Theorem 8.3.4.

It is obvious that, for identically distributed ξ_k, the conditions of Corollary 8.3.2 are always met, and we again obtain a generalisation of Theorem 8.1.1 and Corollary 8.1.1.

If $\mathbf{E}|\xi_k|^r < \infty$ for $r \ge 1$, then we can also establish in a similar way that

$$\frac{S_n}{n} \xrightarrow{(r)} a.$$

Remark 8.3.3 Condition $[\mathbf{D}_1]$ (or $[\mathbf{M}_1]$) is not necessary for convergence $\zeta_n \xrightarrow{p} 0$ even when (8.3.2) and (8.3.5) hold, as the following example demonstrates. Let $\xi_{k,n}$ assume the values $-n$, 0, and n with probabilities $1/n^2$, $1-2/n^2$, and $1/n^2$, respectively. Here $\zeta_n \xrightarrow{p} 0$, since $\mathbf{P}(\zeta_n \ne 0) \le \mathbf{P}(\bigcup\{\xi_{k,n} \ne 0\}) \le 2/n \to 0$, $\mathbf{E}|\xi_{k,n}| = 2/n \to 0$ and $M_1 = \sum \mathbf{E}|\xi_{k,n}| = 2 < \infty$. At the same time, $\sum \mathbf{E}(|\xi_{k,n}|; |\xi_{k,n}| \ge 1) = 2 \not\to \infty$, so that conditions $[\mathbf{D}_1]$ and $[\mathbf{M}_1]$ are not satisfied.

However, if we require that

$$\begin{aligned} &\xi_{k,n} \ge -\varepsilon_{k,n}, \qquad \varepsilon_{k,n} \ge 0, \\ &\max_{k \le n} \varepsilon_{k,n} \to 0, \qquad \sum_{k=1}^{n} \varepsilon_{k,n} \le c < \infty, \end{aligned} \tag{8.3.10}$$

then condition $[\mathbf{D}_1]$ will become necessary for convergence $\zeta_n \xrightarrow{p} 0$.

Before proving that assertion we will establish several auxiliary relations that will be useful in the sequel. As above, put $\Delta_k(t) := \varphi_{k,n}(t) - 1$.

Lemma 8.3.2 *One has*

$$\sum_{k=1}^{n} \bigl|\Delta_k(t)\bigr| \le |t| M_1.$$

If condition $[\mathbf{S}]$ *holds, then for each* t, *as* $n \to \infty$,

$$\max_{k \le n} \bigl|\Delta_k(t)\bigr| \to 0.$$

If a random variable ξ *with* $\mathbf{E}\xi = 0$ *is bounded from the left*: $\xi > -c$, $c > 0$, *then* $\mathbf{E}|\xi| \le 2c$.

Proof By Lemma 7.4.1,

$$\bigl|\Delta_k(t)\bigr| \le \mathbf{E}\bigl|e^{it\xi_{k,n}} - 1\bigr| \le |t|\,\mathbf{E}|\xi_{k,n}|, \qquad \sum \bigl|\Delta_k(t)\bigr| \le |t| M_1.$$

Further,

$$\begin{aligned}\left|\Delta_k(t)\right| &\le \mathbf{E}\left(\left|e^{it\xi_{k,n}}-1\right|;\,|\xi_{k,n}|\le\varepsilon\right)+\mathbf{E}\left(\left|e^{it\xi_{k,n}}-1\right|;\,|\xi_{k,n}|>\varepsilon\right)\\ &\le |t|\varepsilon+2\mathbf{P}\left(|\xi_{k,n}|>\varepsilon\right).\end{aligned}$$

Since ε is arbitrary here, the second assertion of the lemma now follows from condition [**S**].

Put

$$\xi^+ := \max(0;\xi)\ge 0,\qquad \xi^- := -\left(\xi-\xi^+\right)\ge 0.$$

Then $\mathbf{E}\xi=\mathbf{E}\xi^+-\mathbf{E}\xi^-=0$ and $\mathbf{E}|\xi|=\mathbf{E}\xi^++\mathbf{E}\xi^-=2\mathbf{E}\xi^-\le 2c$. The lemma is proved. □

From the last assertion of the lemma it follows that (8.3.10) implies (8.3.2) and (8.3.5).

Lemma 8.3.3 *Let conditions* [**S**] *and* (8.3.2) *be satisfied. A necessary and sufficient condition for convergence $\varphi_{\zeta_n}(t)\to\varphi(t)$ is that*

$$\sum_{k=1}^n \Delta_k(t)\to\ln\varphi(t).$$

Proof Observe that

$$\operatorname{Re}\Delta_k(t)=\operatorname{Re}\left(\varphi_{k,n}(t)-1\right)\le 0,\qquad \left|e^{\Delta_k(t)}\right|\le 1,$$

and therefore, by Lemma 7.4.2,

$$\begin{aligned}\left|\varphi_{z_n}(t)-e^{\sum\Delta_k(t)}\right| &= \left|\prod_{k=1}^n\varphi_{k,n}(t)-\prod_{k=1}^n e^{\Delta_k(t)}\right|\\ &\le \sum_{k=1}^n\left|\varphi_{k,n}(t)-e^{\Delta_k(t)}\right|=\sum_{k=1}^n\left|e^{\Delta_k(t)}-1-\Delta_k(t)\right|\\ &\le \frac{1}{2}\sum_{k=1}^n\Delta_k^2(t)\le\frac{1}{2}\max_k\left|\Delta_k(t)\right|\sum_{k=1}^n\left|\Delta_k(t)\right|.\end{aligned}$$

By Lemma 8.3.2 and conditions [**S**] and (8.3.2), the expression on the left-hand side converges to 0 as $n\to\infty$. Therefore, if $\varphi_{z_n}(t)\to\varphi(t)$ then $\exp\{\sum\Delta_k(t)\}\to\varphi(t)$, and vice versa. The lemma is proved. □

The next assertion complements Theorem 8.3.1.

Theorem 8.3.5 *Assume that relations* (8.3.1) *and* (8.3.10) *hold. Then condition* [$\mathbf{D}_1$] (*or condition* [$\mathbf{M}_1$]) *is necessary for the law of large numbers.*

Proof If the law of large numbers holds then $\varphi_{z_n}(t) \to 1$ and, hence by Lemma 8.3.3 (recall that (8.3.10) implies (8.3.2), (8.3.5) and [**S**])

$$\sum_{k=1}^{n} \Delta_k(t) = \sum_{k=1}^{n} \mathbf{E}\big(e^{it\xi_{k,n}} - 1 - it\xi_{k,n}\big) \to 0.$$

Moreover, by Lemma 7.4.1

$$\sum_{k=1}^{n} \mathbf{E}\big(\big|e^{it\xi_{k,n}} - 1 - it\xi_{k,n}\big|; |\xi_{k,n}| \le \varepsilon_{k,n}\big)$$

$$\le \frac{1}{2}\sum_{k=1}^{n} \mathbf{E}\big(|xi_{k,n}|^2; |\xi_{k,n}| \le \varepsilon_{k,n}\big) \le \sum_{k=1}^{n} \varepsilon_{k,n}^2 \le \max_k \varepsilon_{k,n} \sum_{k=1}^{n} \varepsilon_{k,n} \to 0.$$

Therefore, if the law of large numbers holds, then by virtue of (8.3.10)

$$\sum_{k=1}^{n} \mathbf{E}\big(e^{it\xi_{k,n}} - 1 - it\xi_{k,n}; \xi_{k,n} > \varepsilon_{k,n}\big) \to 0.$$

Consider the function $\alpha(x) = (e^{ix} - 1)/ix$. It is not hard to see that the inequality $|\alpha(x)| \le 1$ proved in Lemma 7.4.1 is strict for $x > \varepsilon > 0$, and hence there exists a $\delta(\tau) > 0$ for $\tau > 0$ such that $\mathrm{Re}(1 - \alpha(x)) \ge \delta(\tau)$ for $x > \tau$. This is equivalent to $\mathrm{Im}(1 + ix - e^{ix}) \ge \delta(\tau)x$, so that

$$x \le \frac{1}{\delta(\tau)} \mathrm{Im}\big(1 + ix - e^{ix}\big) \quad \text{for } x > \tau.$$

From this we find that

$$\mathbf{E}_1(\tau) = \sum_{k=1}^{n} \mathbf{E}\big(|\xi_{k,n}|; |\xi_{k,n}| > \tau\big) = \sum_{k=1}^{n} \mathbf{E}(\xi_{k,n}; \xi_{k,n} > \tau)$$

$$\le \frac{1}{\delta(\tau)} \mathrm{Im} \sum_{k=1}^{n} \mathbf{E}\big(1 + i\xi_{k,n} - e^{i\xi_{k,n}}; \xi_{k,n} > \varepsilon_{k,n}\big) \to 0.$$

Thus condition [$\mathbf{M}_1$] holds. Together with relation (8.3.2), that follows from (8.3.10), this condition implies [$\mathbf{D}_1$]. The theorem is proved. □

There seem to exist some conditions that are wider than (8.3.10) and under which condition $[\mathbf{D}_1]$ is necessary for convergence $\zeta_n \xrightarrow{(1)} 0$ in mean (condition (8.3.10) is too restrictive).

8.4 The Central Limit Theorem for Sums of Arbitrary Independent Random Variables

As in Sect. 8.3, we consider here a triangular array of random variables $\xi_{1,n}, \dots, \xi_{n,n}$ and their sums

$$\zeta_n = \sum_{k=1}^{n} \xi_{k,n}. \tag{8.4.1}$$

We will assume that $\xi_{k,n}$ have finite second moments:

$$\sigma_{k,n}^2 := \operatorname{Var}(\xi_{k,n}) < \infty,$$

and suppose, without loss of generality, that

$$\mathbf{E}\xi_{k,n} = 0, \qquad \sum_{k=1}^{n} \sigma_{k,n}^2 = \operatorname{Var}(\zeta_n) = 1. \tag{8.4.2}$$

We introduce the following condition: *for some* $s > 2$,

$$D_2 := \sum_{k=1}^{n} \mathbf{E} \min\big(\xi_{k,n}^2, |\xi_{k,n}|^s\big) \to 0 \quad \textit{as } n \to \infty, \tag{$[\mathbf{D}_2]$}$$

which is to play an important role in what follows. Our arguments related to condition $[\mathbf{D}_2]$ and also to conditions $[\mathbf{M}_2]$ and $[\mathbf{L}_s]$ to be introduced below will be quite similar to the ones from Sect. 8.3 that were related to conditions $[\mathbf{D}_1]$, $[\mathbf{M}_1]$ and $[\mathbf{L}_s]$.

We also introduce *the Lindeberg condition*: for any $\tau > 0$, as $n \to \infty$,

$$M_2(\tau) := \sum_{k=1}^{n} \mathbf{E}\big(|\xi_{k,n}|^2; |\xi_{k,n}| > \tau\big) \to 0. \tag{$[\mathbf{M}_2]$}$$

The following assertion is an analogue of Lemma 8.3.1.

Lemma 8.4.1 1. $\{[\mathbf{M}_2] \cap (4.2)\} \subset [\mathbf{D}_2]$. 2. $[\mathbf{D}_2] \subset [\mathbf{M}_2]$.

That is, conditions $[\mathbf{M}_2]$ and (8.4.2) imply $[\mathbf{D}_2]$, and condition $[\mathbf{D}_2]$ implies $[\mathbf{M}_2]$.

From Lemma 8.4.1 it follows that, under condition (8.4.2), conditions $[\mathbf{D}_2]$ and $[\mathbf{M}_2]$ are equivalent.

Proof of Lemma 8.4.1 1. Let conditions $[\mathbf{M}_2]$ and (8.4.2) be met. Put

$$g_2(x) := \min(x^2, |x|^s), \quad s > 2.$$

Then (cf. (8.3.3), (8.3.4); $\tau \le 1$)

$$D_2 = \sum_{k=1}^{n} \mathbf{E} g_2(\xi_{k,n}) \le \sum_{k=1}^{n} \mathbf{E}\big(\xi_{k,n}^2;\ |\xi_{k,n}| > \tau\big) + \sum_{k=1}^{n} \mathbf{E}\big(|\xi_{k,n}|^s;\ |\xi_{k,n}| \le \tau\big)$$
$$\le M_2(\tau) + \tau^{s-2} M_2(0) = M_2(\tau) + \tau^{s-2}.$$

Since τ is arbitrary, we have $D_2 \to 0$ as $n \to \infty$.

2. Conversely, suppose that $[\mathbf{D}_2]$ holds. Then

$$M_2(\tau) \le \sum_{k=1}^{n} \mathbf{E}\big(\xi_{k,n}^2;\ |\xi_{k,n}| > 1\big) + \frac{1}{\tau^{s-2}} \sum_{k=1}^{n} \big(|\xi_{k,n}|^s;\ \tau < |\xi_{k,n}| \le 1\big) \le \frac{1}{\tau^{s-2}} D_2 \to 0$$

for any $\tau > 0$, as $n \to \infty$. The lemma is proved. □

Lemma 8.4.1 also implies that *if* (8.4.2) *holds, then condition* $[\mathbf{D}_2]$ *is "invariant" with respect to* $s > 2$.

Condition $[\mathbf{D}_2]$ can be stated in a more general form:

$$\sum_{k=1}^{n} \mathbf{E}\xi_{k,n}^2 h\big(|\xi_{k,n}|\big) \to 0,$$

where $h(x)$ is any function for which $h(x) > 0$ for $x > 0$, $h(x) \uparrow$, $h(x) \to 0$ as $x \to 0$, and $h(x) \to c < \infty$ as $x \to \infty$. All the key properties of condition $[\mathbf{D}_2]$ will then be preserved. The Lindeberg condition clarifies the meaning of condition $[\mathbf{D}_2]$ from a somewhat different point of view. In Lindeberg's condition, $h(x) = I_{(\tau,\infty)}$, $\tau \in (0, 1)$. A similar remark may be made with regard to conditions $[\mathbf{D}_1]$ and $[\mathbf{M}_1]$ in Sect. 8.3.

In a way similar to what we did in Sect. 8.3 when discussing condition $[\mathbf{M}_1]$, one can easily verify that condition $[\mathbf{M}_2]$ implies convergence (see (8.3.6))

$$\max_{k \le n} \operatorname{Var}(\xi_{k,n}) \to 0 \tag{8.4.3}$$

and the negligibility of $\xi_{k,n}$ (property $[\mathbf{S}]$). Moreover, one obviously has the inequality

$$M_1(\tau) \le \frac{1}{\tau} M_2(\tau).$$

For a given fixed (independent of n) sequence $\{\xi_k\}$ of independent random variables,

$$S_n = \sum_{k=1}^{\infty} \xi_k, \qquad \mathbf{E}\xi_k = a_k, \qquad \operatorname{Var}(\xi_k) = \sigma_k^2, \tag{8.4.4}$$

one considers the asymptotic behaviour of the normed sums

$$\zeta_n = \frac{1}{B_n}\left(S_n - \sum_{k=1}^{\infty} a_k\right), \qquad B_n^2 = \sum_{k=1}^{\infty} \sigma_k^2, \tag{8.4.5}$$

that are clearly also of the form (8.4.1) with $\xi_{k,n} = (\xi_k - a_k)/B_n$.

Conditions $[\mathbf{D}_1]$ and $[\mathbf{M}_2]$ for ξ_k will take the form

$$\begin{aligned} D_2 &= \frac{1}{B_n^2} \sum_{k=1}^{\infty} \mathbf{E}\min\left((\xi_k - a_k)^2, \frac{|\xi_k - a_k|^s}{B_n^{s-2}}\right) \to 0, \quad s > 2; \\ M_2(\tau) &= \frac{1}{B_n^2} \sum_{k=1}^{\infty} \mathbf{E}\big((\xi_k - a_k)^2; |\xi_k - a_k| > \tau B_n\big) \to 0, \quad \tau > 0. \end{aligned} \tag{8.4.6}$$

Theorem 8.4.1 (The Central Limit Theorem) *If the sequences of random variables* $\{\xi_{k,n}\}_{k=1}^{\infty}$, $n = 1, 2, \ldots$, *satisfy conditions* (8.4.2) *and* $[\mathbf{D}_2]$ (*or* $[\mathbf{M}_2]$) *then, as* $n \to \infty$, $\mathbf{P}(\zeta_n < x) \to \Phi(x)$ *uniformly in* x.

Proof It suffices to verify that

$$\varphi_{\zeta_n}(t) = \prod_{k=1}^{\infty} \varphi_{k,n}(t) \to e^{-t^2/2}.$$

By Lemma 7.4.2,

$$\begin{aligned} \left|\varphi_{\zeta_n}(t) - e^{-t^2/2}\right| &= \left|\prod_{k=1}^{n} \varphi_{k,n}(t) - \prod_{k=1}^{n} e^{-t^2\sigma_{k,n}^2/2}\right| \\ &\le \sum_{k=1}^{n} \left|\varphi_{k,n}(t) - e^{-t^2\sigma_{k,n}^2/2}\right| \\ &\le \sum_{k=1}^{n} \left|\varphi_{k,n}(t) - 1 + \frac{1}{2}t^2\sigma_{k,n}^2\right| \\ &\quad + \sum_{k=1}^{n} \left|e^{-t^2\sigma_{k,n}^2/2} - 1 + \frac{1}{2}t^2\sigma_{k,n}^2\right|. \end{aligned} \tag{8.4.7}$$

Since by Lemma 7.4.1, for $s \le 3$,

$$\left|e^{ix} - 1 - ix + \frac{x^2}{2}\right| \le \min\left(x^2, \frac{|x^3|}{6}\right) \le g_2(x)$$

(see the definition of the function g_2 in the beginning of the proof of Lemma 8.4.1), the first sum on the right-hand side of (8.4.7) does not exceed

$$\sum_{k=1}^{\infty} \left| \mathbf{E}\left(e^{it\xi_{k,n}} - 1 - it\xi_{k,n} + \frac{1}{2} t^2 \xi_{k,n}^2 \right) \right|$$

$$\leq \sum_{k=1}^{\infty} \mathbf{E} g_2\big(|t\xi_{k,n}|\big) \leq h(t) \sum_{k=1}^{\infty} \mathbf{E} g_2\big(|\xi_{k,n}|\big) \leq h(t) D_2 \to 0,$$

where $h(t) = \max(t^2, |t|^3)$. The last sum in (8.4.7) (again by Lemma 7.4.1) does not exceed (see (8.4.2) and (8.4.3))

$$\frac{t^4}{8} \sum_{k=1}^{n} \sigma_{k,n}^4 \leq \frac{t^4}{8} \max_k \sigma_{k,n}^2 \sum_{k=1}^{n} \sigma_{k,n}^2 \leq \frac{t^4}{8} \max_k \sigma_{k,n}^2 \to 0 \quad \text{as } n \to \infty.$$

The theorem is proved. □

If we change the second relation in (8.4.2) to $\mathbf{E}\zeta_n \to \sigma^2 > 0$, then, introducing the new random variables $\xi'_{k,n} = \xi_{k,n} / \sqrt{\operatorname{Var} \zeta_n}$ and using continuity theorems, it is not hard to obtain from Theorem 8.4.1 (see e.g. Lemma 6.2.2), the following assertion, which sometimes proves to be more useful in applications than Theorem 8.4.1.

Corollary 8.4.1 *Assume that* $\mathbf{E}\xi_{k,n} = 0$, $\operatorname{Var}(\zeta_n) \to \sigma^2 > 0$, *and condition* $[\mathbf{D}_2]$ (*or* $[\mathbf{M}_2]$) *is satisfied. Then* $\zeta_n \Leftrightarrow \boldsymbol{\Phi}_{0,\sigma_2}$.

Remark 8.4.1 A sufficient condition for $[\mathbf{D}_2]$ and $[\mathbf{M}_2]$ is provided by the more restrictive *Lyapunov condition*, the verification of which is sometimes easier. Assume that (8.4.2) holds. For $s > 2$, the quantity

$$L_s := \sum_{k=1}^{n} \mathbf{E}|\xi_{k,n}|^2$$

is called the *Lyapunov fraction of the s-th order*. The condition

$$L_s \to 0 \quad \text{as } n \to \infty \qquad [\mathbf{L}_s]$$

is called the *Lyapunov condition*.

The quantity L_s is called a *fraction* since for $\xi_{k,n} = (\xi_k - a)/B_n$ (where $a_k = \mathbf{E}\xi_k$, $B_n^2 = \sum_{k=1}^{n} \operatorname{Var}(\xi_k)$ and ξ_k do not depend on n), it has the form

$$L_s = \frac{1}{B_n^s} \sum_{k=1}^{n} \mathbf{E}|\xi_k - a_k|^s.$$

If the ξ_k are identically distributed, $a_k = a$, $\operatorname{Var}(\xi_k) = \sigma^2$, and $\mathbf{E}|\xi_k - a|^s = \mu < \infty$, then

$$L_s = \frac{\mu}{\sigma^s n^{(s-2)/2}} \to 0.$$

The sufficiency of the Lyapunov condition follows from the obvious inequalities $g_2(x) \le |x|^s$ for any s, $D_2 \le L_s$.

In the case of (8.4.4) and (8.4.5) we can give a sufficient condition for the integral limit theorem that is very close to the Lindeberg condition $[\mathbf{M}_2]$; the former condition elucidates to some extent the essence of the latter (cf. Theorem 8.3.3), and in many cases it is easier to verify. Put $\overline{\sigma}_n = \max_{k\le n} \sigma_k$. Theorem 8.4.1 implies the following assertion which is a direct extension of Theorem 8.2.1

Theorem 8.4.2 *Let conditions* (8.4.4) *and* (8.4.5) *be satisfied, the sequence of normalised random variables* ξ_k^2/σ_k^2 *be uniformly integrable and* $\overline{\sigma}_n = o(B_n)$ *as* $n \to \infty$. *Then* $\zeta_n \Leftrightarrow \mathbf{\Phi}_{0,1}$.

Proof of Theorem 8.4.2 repeats, to some extent, the proof of Theorem 8.3.3. For simplicity assume that $a_k = 0$. Then

$$\mathbf{E}\bigl(\xi_k^2;\, |\xi_k| > \tau B_n\bigr) \le \sigma_k^2 \mathbf{E}\left(\frac{\xi_k^2}{\sigma_k^2};\, \left|\frac{\xi_k}{\sigma_k}\right| > \tau \frac{B_n}{\overline{\sigma}_n}\right), \tag{8.4.8}$$

where $B_n/\overline{\sigma}_n \to \infty$. Hence, it follows from the uniform integrability of $\{\frac{\xi_k^2}{\sigma_k^2}\}$ that the right-hand side of (8.4.8) is $o(\sigma_k^2)$ uniformly in k. This means that

$$M_2(\tau) = \frac{1}{B_n^2} \sum_{k=1}^{n} \mathbf{E}\bigl(\xi_k^2;\, |\xi_k| > \tau B_n\bigr) \to 0$$

as $n \to \infty$ and condition (8.4.6) (or condition $[\mathbf{M}_2]$) is satisfied. The theorem is proved. □

Remark 8.4.2 We can generalise the assertion of Theorem 8.4.2 (cf. Remark 8.3.3). In particular, *convergence* $\zeta_n \Leftrightarrow \mathbf{\Phi}_{0,1}$ *still takes place if a finite number of summands* ξ_k (*e.g., for* $k \le l$, l *being fixed*) *are completely arbitrary, and the sequence* $\xi_k^* := \xi_{k+l}$, $k \ge 1$, *satisfies the conditions of Theorem* 8.4.2, *in which we put* $\sigma_k^2 = \operatorname{Var}(\xi_k^*)$, $B_n^2 = \sum_{k=1}^n \sigma_k^2$, *and it is also assumed that* $\frac{B_{n-1}}{B_n} \to 1$ as $n \to \infty$.

This assertion follows from the fact that

$$\frac{S_n}{B_n} = \frac{S_l}{B_n} + \frac{S_n - S_l}{B_{n-l}} \cdot \frac{B_{n-l}}{B_n},$$

where $\frac{S_l}{B_n} \xrightarrow{p} 0$, $\frac{B_{n-l}}{B_n} \to 1$ and, by Theorem 8.4.2, $\frac{S_n - S_l}{B_{n-l}} \Leftrightarrow \mathbf{\Phi}_{0,1}$ as $n \to \infty$.

Remark 8.4.3 The uniform integrability condition that was used in Theorem 8.4.2 can be used for the triangular array scheme as well. In this more general case the uniform integrability should mean the following: *the sequences* $\eta_{1,n},\dots,\eta_{n,n}$, $n=1,2,\dots$, *in the triangular array scheme are uniformly integrable if there exists a function* $\varepsilon(N)\downarrow 0$ *as* $N\uparrow\infty$ *such that, for all* n,

$$\max_{j\le n}\mathbf{E}\bigl(|\eta_{j,n}|;\,|\eta_{j,n}|>N\bigr)\le\varepsilon(N).$$

It is not hard to see that, with such an interpretation of uniform integrability, the assertion of Theorem 8.4.2 holds true for the triangular array scheme as well provided that the sequence $\{\frac{\xi_{j,n}^2}{\sigma_{j,n}^2}\}$ is uniformly integrable and $\max_{j\le n}\sigma_{j,n}=o(1)$ as $n\to\infty$.

Example 8.4.1 We will clarify the difference between the Lindeberg condition and uniform integrability of $\{\frac{\xi_k^2}{\sigma_k^2}\}$ in the following example. Let η_k be independent bounded identically distributed random variables, $\mathbf{E}\,\eta_k=0$, $\mathbf{D}\eta_k=1$ and $g(k)>\sqrt{2}$ be an arbitrary function. Put

$$\xi_k:=\begin{cases}\eta_k & \text{with probability } 1-2g^{-2}(k),\\ \pm g(k) & \text{with probability } g^{-2}(k).\end{cases}$$

Then clearly $\mathbf{E}\xi_k=0$, $\sigma_k^2:=\mathbf{D}\xi_k=3-2g^{-2}(k)\in(2,3)$ and $B_n^2\in(2n,3n)$. The uniform integrability of $\{\frac{\xi_k^2}{\sigma_k^2}\}$, or the uniform integrability of $\{\xi_k^2\}$ which means the same in our case, excludes the case where $g(k)\to\infty$ as $k\to\infty$. The Lindeberg condition is wider and allows the growth of $g(k)$, except for the case where $g(k)>c\sqrt{k}$. If $g(k)=o(\sqrt{k})$, then the Lindeberg condition is satisfied because, for any fixed $\tau>0$,

$$\mathbf{E}\bigl(\xi^2;\,|\xi_k|>\tau\sqrt{k}\bigr)=0$$

for all large enough k.

Remark 8.4.4 Let us show that condition $[\mathbf{M}_2]$ (or $[\mathbf{D}_2]$) is essential for the central limit theorem. Consider random variables

$$\xi_{k,n}=\begin{cases}\pm\frac{1}{\sqrt{2}} & \text{with probability } \frac{1}{n},\\ 0 & \text{with probability } 1-\frac{2}{n}.\end{cases}$$

They satisfy conditions (8.4.2), $[\mathbf{S}]$, but not the Lindeberg condition as $M_2(\tau)=1$ for $\tau<\frac{1}{\sqrt{2}}$. The number ν_k of non-zero summands converges in distribution to a random variable ν having the Poisson distribution with parameter 2. Therefore, ζ_n will clearly converge in distribution not to the normal law, but to $\sum_{j=1}^{\nu}\gamma_j$, where γ_j are independent and take values ± 1 with probability $1/2$.

Note also that conditions $[\mathbf{D}_2]$ or $[\mathbf{M}_2]$ are not necessary for convergence of the distributions of ζ_n to the normal distribution. Indeed, consider the following example: $\xi_{1,n} \in \mathbf{\Phi}_{0,1}$, $\xi_{2,n} = \cdots = \xi_{n,n} = 0$. Conditions (8.4.2) are clearly met, $\mathbf{P}(\zeta_n < x) = \Phi(x)$, but the variables $\xi_{k,n}$ are not negligible and therefore do not satisfy conditions $[\mathbf{D}_2]$ and $[\mathbf{M}_2]$.

If, however, as well as convergence $\zeta_n \Leftrightarrow \mathbf{\Phi}_{0,1}$ we require that the $\xi_{k,n}$ are negligible, then conditions $[\mathbf{D}_2]$ and $[\mathbf{M}_2]$ become necessary.

Theorem 8.4.3 *Suppose that the sequences of independent random variables* $\{\xi_{k,n}\}_{k=1}^n$ *satisfy conditions* (8.4.2) *and* [**S**]. *Then condition* $[\mathbf{D}_1]$ (*or* $[\mathbf{M}_2]$) *is necessary and sufficient for convergence* $\zeta_n \Leftrightarrow \mathbf{\Phi}_{0,1}$.

First note that the assertions of Lemmas 8.3.2 and 8.3.3 remain true, up to some inessential modifications, if we substitute conditions (8.3.2) and [**S**] with (8.4.2) and [**S**].

Lemma 8.4.2 *Let conditions* (8.4.2) *and* [**S**] *hold. Then* ($\Delta_k(t) = \varphi_{k,n}(t) - 1$)

$$\max_{k\le n}\big|\Delta_k(t)\big| \to 0, \quad \sum\big|\Delta_k(t)\big| \le \frac{t^2}{2},$$

and the assertion of Lemma 8.3.3, *that the convergence* (8.3.10) *is necessary and sufficient for convergence* $\varphi_{\zeta_n}(t) \to \varphi(t)$, *remain completely true.*

Proof We can retain all the arguments in the proofs of Lemmas 8.3.2 and 8.3.3 except for one place where $\sum|\Delta_k(t)|$ is bounded. Under the new conditions, by Lemma 7.4.1, we have

$$\big|\Delta_k(t)\big| = \big|\varphi_{k,n}(t) - 1 - it\mathbf{E}\xi_{k,n}\big| \le \mathbf{E}\big|e^{it\xi_{k,n}} - 1 - it\xi_{k,n}\big| \le \frac{t^2}{2}\mathbf{E}\xi_{k,n}^2,$$

so that

$$\sum\big|\Delta_k(t)\big| \le \frac{t^2}{2}.$$

No other changes in the proofs of Lemmas 8.3.2 and 8.3.3 are needed. □

Proof of Theorem 8.4.3 Sufficiency is already proved. To prove necessity, we make use of Lemma 8.4.1. If $\varphi_{\zeta_n}(t) \to e^{-t^2/2}$, then by virtue of that lemma, for $\Delta_k(t) = \varphi_{k,n}(t) - 1$, one has

$$\sum_{k=1}^n \Delta_k(t) \to \ln\varphi(t) = -\frac{t^2}{2}.$$

For $t = 1$ the above relation can be written in the form

$$R_n := \sum_{k=1}^n \mathbf{E}\biggl(e^{i\xi_{k,n}} - 1 - i\xi_{k,n} + \frac{1}{2}\xi_{k,n}^2\biggr) \to 0. \tag{8.4.9}$$

Put $\alpha(x) := (e^{ix} - 1 - ix)/x^2$. It is not hard to see that the inequality $|\alpha(x)| \le 1/2$ proved in Lemma 7.4.1 is strict for $x \neq 0$, and

$$\sup_{|x|\ge\tau} |\alpha(x)| < \frac{1}{2} - \delta(\tau),$$

where $\delta(\tau) > 0$ for $\tau > 0$. This means that, for $|x| \ge \tau > 0$,

$$\mathrm{Re}\left[\alpha(x) + \frac{1}{2}\right] \ge \delta(\tau) > 0, \qquad x^2 \le \frac{1}{\delta(\tau)} \mathrm{Re}\left(e^{ix} - 1 - ix + \frac{x^2}{2}\right),$$

$$\mathbf{E}\big(\xi_{k,n}^2; |\xi_{k,n}| > \tau\big) \le \frac{1}{\delta(\tau)} \mathrm{Re}\,\mathbf{E}\left(e^{i\xi_{k,n}} - 1 - i\xi_{k,n} + \frac{\xi_{k,n}^2}{2}\right),$$

and hence by virtue of (8.4.9), for any $\tau > 0$,

$$M_2(\tau) \le \frac{1}{\delta(\tau)} |R_n| \to 0$$

as $n \to \infty$. The theorem is proved. □

Corollary 8.4.2 *Assume that* (8.4.2) *holds and*

$$\max_{k\le n} \mathrm{Var}(\xi_{k,n}) \to 0. \tag{8.4.10}$$

Then a necessary and sufficient condition for convergence $\zeta_n \Leftrightarrow \mathbf{\Phi}_{0,1}$ *is that*

$$\eta_n := \sum_{k=1}^n \xi_{k,n}^2 \Leftrightarrow \mathbf{I}_1$$

(or that $\eta_n \xrightarrow{p} 1$*).*

Proof Let $\eta_n \Leftrightarrow \mathbf{I}_1$. The random variables $\xi'_{k,n} = \xi_{k,n}^2 - \sigma_{k,n}^2$ satisfy, by virtue of (8.4.10), condition (8.3.10) and satisfy the law of large numbers:

$$\sum_{k=1}^n \xi'_{k,n} = \eta_n - 1 \xrightarrow{p} 0.$$

Therefore, by Theorem 8.3.5, the $\xi'_{k,n}$ satisfy condition $[\mathbf{M}_1]$: for any $\tau > 0$,

$$\sum_{k=1}^n \mathbf{E}\big(|\xi_{k,n}^2 - \sigma_{k,n}^2|; |\xi_{k,n}^2 - \sigma_{k,n}^2| > \tau\big) \to 0. \tag{8.4.11}$$

But by (8.4.10) this condition is clearly equivalent to condition $[\mathbf{M}_2]$ for $\xi_{k,n}$, and hence $\zeta_n \Leftrightarrow \mathbf{\Phi}_{0,1}$.

Conversely, if $\zeta_n \Leftrightarrow \Phi_{0,1}$, then [$\mathbf{M}_2$] holds for $\xi_{k,n}$ which implies (8.4.11). Since, moreover,

$$\sum_{k=1}^{n} \mathbf{E}|\xi'_{k,n}| \le 2 \sum_{k=1}^{n} \operatorname{Var}(\xi_{k,n}) = 2,$$

relation (8.3.2) holds for $\xi'_{k,n}$, and by Theorem 8.3.1

$$\sum_{k=1}^{n} \xi'_{k,n} = \eta_n - 1 \xrightarrow{p} 0.$$

The corollary is proved. □

Example 8.4.2 Let ξ_k, $k = 1, 2, \ldots$, be independent random variables with distributions

$$\mathbf{P}(\xi_k = k^\alpha) = \mathbf{P}(\xi_k = -k^\alpha) = \frac{1}{2}.$$

Evidently, ξ_k can be represented as $\xi_k = k^\alpha \eta_k$, where $\eta_k \stackrel{d}{=} \eta$ are independent,

$$\mathbf{P}(\eta = 1) = \mathbf{P}(\eta = -1) = \frac{1}{2}, \qquad \operatorname{Var}(\eta) = 1, \qquad \sigma_k^2 = \operatorname{Var}(\xi_k) = k^{2\alpha}.$$

Let us show that, for all $\alpha \ge -1/2$, the random variables S_n/B_n are asymptotically normal. Since

$$\frac{\xi_k^2}{\sigma_k^2} \stackrel{d}{=} \eta^2$$

are uniformly integrable, by Theorem 8.4.2 it suffices to verify the condition

$$\overline{\sigma}_n = \max_{k \le n} \sigma_k = o(B_n).$$

In our case $\overline{\sigma}_n = \max(1, n^{2\alpha})$ and, for $\alpha > -1/2$,

$$B_n^2 = \sum_{k=1}^{n} k^{2\alpha} \sim \int_0^n x^{2\alpha}\, d\alpha = \frac{n^{2\alpha+1}}{2\alpha+1}.$$

For $\alpha = -1/2$, one has

$$B_n^2 = \sum_{k=1}^{n} k^{-1} \sim \ln n.$$

Clearly, in these cases $\overline{\sigma}_n = o(B_n)$ and the asymptotical normality of S_n/n holds.

If $\alpha < -1/2$ then the sequence B_n converges, condition $\overline{\sigma}_n = 1 = o(B_n)$ is not satisfied and the asymptotical normality of S_n/B_n fails to take place.

Note that, for $\alpha = -1/2$, the random variable

$$S_n = \sum_{k=1}^{n} \frac{\eta_k}{\sqrt{k}}$$

will be "comparable" with $\sqrt{\ln n}$ with a high probability, while the sums

$$\sum_{k=1}^{n} \frac{(-1)^k}{\sqrt{k}}$$

converge to a constant.

A rather graphical and well-known illustration of the above theorems is the scattering of shells when shooting at a target. The fact is that the trajectory of a shell is influenced by a large number of independent factors of which the individual effects are small. These are deviations in the amount of gun powder, in the weight and size of a shell, variations in the humidity and temperature of the air, wind direction and velocities at different altitudes and so on. As a result, the deviation of a shell from the aiming point is described by the normal law with an amazing accuracy.

Similar observations could be made about errors in measurements when their accuracy is affected by many "small" factors. (There even exists a theory of errors of which the crucial element is the central limit theorem.)

On the whole, the central limit theorem has a lot of applications in various areas. This is due to its *universality and robustness* under small deviations from the assumptions of the theorem, and its relatively *high accuracy* even for moderate values of n. The first two noted qualities mean that:

(1) the theorem is applicable to variables $\xi_{k,n}$ with *any distributions* so long as the variances of $\xi_{k,n}$ exist and are "negligible";

(2) the presence of a "moderate" dependence[3] between $\xi_{k,n}$ does not change the normality of the limiting distribution.

To illustrate the accuracy of the normal approximation, consider the following example. Let $F_n(x) = \mathbf{P}(S_n/\sqrt{n} < x)$ be the distribution function of the normalised sum S_n of independent variables ξ_k uniformly distributed over $[-\sqrt{3}, \sqrt{3}]$, so that $\operatorname{Var}(\xi_k) = 1$. Then it turns out that already for $n = 5$ (!) the maximum of $|F_n(x) - \Phi(x)|$ over the whole axis of x-values does not exceed 0.006 (the maximum is attained near the points $x = \pm 0.7$).

And still, despite the above circumstances, one has to be careful when applying the central limit theorem. For instance, one cannot expect high accuracy from the normal approximation when estimating probabilities of rare events, say when studying large deviation probabilities (this issue has already been discussed in Sect. 5.3).

---

[3]There exist several conditions characterising admissible dependence of $\xi_{k,n}$. Such considerations are beyond the scope of the present book, but can be found in the special literature. See e.g. [20].

After all, the theorem only ensures the smallness of the difference

$$\left|\Phi(x) - \mathbf{P}(\zeta < x)\right| \tag{8.4.12}$$

for large n. Suppose we want to use the normal approximation to find an x_0 such that the event $\{\zeta_n > x_0\}$ would occur on average once in 1000 trials (a problem of this sort could be encountered by an experimenter who wants to ensure that, in a single experiment, such an event will not occur). Even if the difference (8.4.12) does not exceed 0.02 (which can be a good approximation) then, using the normal approximation, we risk making a serious error. It can turn out, say, that $1 - \Phi(x_0) = 10^{-3}$ while $\mathbf{P}(\zeta < x) \approx 0.02$, and then the event $\{\zeta_n > x_0\}$ will occur much more often (on average, once in each 50 trials).

In Chap. 9 we will consider the problem of large deviation probabilities that enables one to handle such situations. In that case one looks for a function $P(n, x)$ such that $\mathbf{P}(\zeta < x)/P(n, x) \to 1$ as $n \to \infty$, $x \to \infty$. The function $P(n, x)$ turns out to be, generally speaking, different from $1 - \Phi(x)$. We should note however that using the approximation $P(n, x)$ requires more restrictive conditions on $\{\xi_{k,n}\}$.

In Sect. 8.7 we will consider the so-called *integro-local* and *local limit theorems* that establish convergence of the density of ζ_n to that of the normal law and enables one to estimate probabilities of rare events of another sort—say, of the form $\{a < \zeta_n < b\}$ where a and b are close to each other.

8.5* Another Approach to Proving Limit Theorems. Estimating Approximation Rates

The approach to proving the principal limit theorems for the distributions of sums of random variables that we considered in Sects. 8.1–8.4 was based on the use of ch.f.s. However, this is by far not the only method of proof of such assertions. Nowadays there exist several rather simple proofs of both the laws of large numbers and the central limit theorem that do not use the apparatus of ch.f.s. (This, however, does not belittle that powerful, well-developed, and rather universal tool.) Moreover, these proofs sometimes enable one to obtain more general results. As an illustration, we will give below a proof of the central limit theorem that extends, in a certain sense, Theorems 8.4.1 and 8.4.3 and provides an estimate of the convergence rate (although not the best one).

Along with the random variables $\xi_{k,n}$ in the triangular array scheme under assumption (8.4.2), consider mutually independent and independent of the sequence $\{\xi_{k,n}\}_{k=1}^n$ random variables $\eta_{k,n} \in \mathbf{\Phi}_{0,\sigma^2_{k,n}}$, $\sigma_{k,n} := \operatorname{Var}(\xi_{k,n})$, so that

$$\eta_n := \sum_{k=1}^n \eta_{k,n} \in \mathbf{\Phi}_{0,1}.$$

Set[4]

$$\mu_{k,n} := \mathbf{E}|\xi_{k,n}|^3, \qquad \nu_{k,n} := \mathbf{E}|\eta_{k,n}|^3 = c_3\sigma_{k,n}^3 \le c_3\mu_{k,n},$$

$$\mu_{k,n}^0 := \int |x|^3 \left| d\bigl(F_{k,n}(x) - \Phi_{k,n}(x)\bigr)\right| \le \mu_{k,n} + \nu_{k,n},$$

$$L_3 := \sum_{k=1}^n \mu_{k,n}, \qquad N_3 := \sum_{k=1}^n \nu_{k,n}, \qquad L_3^0 := \sum_{k=1}^n \mu_{k,n}^0 \le L_3 + N_3 \le (1+c_3)L_3.$$

Here $F_{k,n}$ and $\Phi_{k,n}$ are the distribution functions of $\xi_{k,n}$ and $\eta_{k,n}$, respectively. The quantities L_3 and N_3 are the third order Lyapunov fractions for the sequences $\{\xi_{k,n}\}$ and $\{\eta_{k,n}\}$. The quantities $\mu_{k,n}^0$ are called the third order *pseudomoments* and L_s^0 the Lyapunov fractions for pseudomoments. Clearly, $N_3 \le c_3 L_3 \to 0$, provided that the Lyapunov condition holds. As we have already noted, for $\xi_{k,n} = (\xi_k - a_k)/B_n$, where $a_k = \mathbf{E}\xi_k$, $B_n^2 = \sum_1^n \operatorname{Var}(\xi_k)$, and ξ_k do not depend on n, one has

$$L_3 = \frac{1}{B_n^3}\sum_{k=1}^n \mu_k, \qquad \mu_k = \mathbf{E}|\xi_k - a_k|^3.$$

If, moreover, ξ_k are identically distributed, then

$$L_3 = \frac{\mu_1}{\sigma^3\sqrt{n}}.$$

Our first task here is to estimate the closeness of $\mathbf{E}f(\zeta_n)$ to $\mathbf{E}f(\eta_n)$ for sufficiently smooth f. This problem could be of independent interest. Assume that f belongs to the class C_3 of all bounded functions with uniformly continuous and bounded third derivatives: $\sup_x |f^{(3)}(x)| \le f_3$.

Theorem 8.5.1 *If $f \in C_3$ then*

$$\bigl|\mathbf{E}f(\zeta_n) - \mathbf{E}f(\eta_n)\bigr| \le \frac{f_3 L_3^0}{6} \le \frac{f_3}{6}(L_3 + N_3). \tag{8.5.1}$$

Proof Put, for $1 < l \le n$,

$$X_l := \xi_{1,n} + \cdots + \xi_{l-1,n} + \eta_{l,n} + \cdots + \eta_{n,n},$$
$$Z_l := \xi_{1,n} + \cdots + \xi_{l-1,n} + \eta_{l+1,n} + \cdots + \eta_{n,n},$$
$$X_1 := \eta_n, \qquad X_{n+1} = \zeta_n.$$

Then

$$X_{l+1} = Z_l + \xi_{l,n}, \qquad X_l = Z_l + \eta_{l,n}, \tag{8.5.2}$$

[4] If $\eta \in \mathbf{\Phi}_{0,1}$ then $c_3 = \mathbf{E}|\eta|^3 = \frac{2}{\sqrt{2\pi}}\int_0^\infty x^3 e^{-x^2/2}dx = \frac{4}{\sqrt{2\pi}}\int_0^\infty te^{-t}dt = \frac{4}{\sqrt{2\pi}}$.

$$f(\zeta_n) - f(\eta_n) = \sum_{l=1}^{n} \bigl[f(X_{l+1}) - f(X_l)\bigr]. \tag{8.5.3}$$

Now we will make use of the following lemma.

Lemma 8.5.1 *Let $f \in C_3$ and Z, ξ and η be independent random variables with*

$$\mathbf{E}\xi = \mathbf{E}\eta = a, \qquad \mathbf{E}\xi^2 = \mathbf{E}\eta^2 = \sigma^2, \qquad \mu^0 = \int |x^3| \bigl|d\bigl(F_\xi(x) - F_\eta(x)\bigr)\bigr| < \infty.$$

Then

$$\bigl|\mathbf{E} f(Z+\xi) - \mathbf{E} f(Z+\eta)\bigr| \le \frac{f_3 \mu^0}{6}. \tag{8.5.4}$$

Applying this lemma to (8.5.3), we get

$$\bigl|\mathbf{E}\bigl[f(X_{l+1}) - f(X_1)\bigr]\bigr| \le \frac{f_3 \mu^0}{6}$$

which after summation gives (8.5.1). The theorem is proved. □

Thus to complete the argument proving Theorem 8.5.1 it remains to prove Lemma 8.5.1.

Proof of Lemma 8.5.1 Set $g(x) := \mathbf{E} f(Z+x)$. It is evident that g, being the result of the averaging of f, has all the smoothness properties of f and, in particular, $|g'''(x)| \le f_3$. By virtue of the independence of Z, ξ and η, we have

$$\mathbf{E} f(Z+\xi) - \mathbf{E} f(Z+\eta) = \int g(x)\, d\bigl(F_\xi(x) - F_\eta(x)\bigr). \tag{8.5.5}$$

For the integrand, we make use of the expansion

$$g(x) = g(0) + x g'(0) + \frac{x^2}{2} g''(0) + \frac{x^3}{6} g'''(\theta_x), \quad \theta_x \in [0, x].$$

Since the first and second moments of ξ coincide with those of η, we obtain for the right-hand side of (8.5.5) the bound

$$\left|\frac{1}{6} x^3 g'''(\theta_x)\, d\bigl(F_\xi(x) - F_\eta(x)\bigr)\right| \le \frac{f_3 \mu^0}{6}.$$

The lemma is proved. □

Remark 8.5.1 In exactly the same way one can establish the representation

$$\bigl|\mathbf{E} f(\zeta_n) - \mathbf{E} f(\eta_n)\bigr| \le \frac{g'''(0)}{6} \sum_{k=1}^{n} \mathbf{E}\bigl(\xi_{k,n}^3 - \eta_{k,n}^3\bigr) + \frac{f_4 L_4^0}{24}, \tag{8.5.6}$$

under obvious conventions for the notations f_4 and L_4^0. This bound can improve upon (8.5.1) if the differences $\mathbf{E}(\xi_{k,n}^3 - \eta_{k,n}^3)$ are small. If, for instance, $\xi_{k,n} = (\xi_k - a)/(\sigma\sqrt{n})$, ξ_k are identically distributed, and the third moments of $\xi_{k,n}$ and $\eta_{k,n}$ coincide, then on the right-hand side of (8.5.6) we will have a quantity of the order $1/n$.

Theorem 8.5.1 extends Theorem 8.4.1 in the case when $s = 3$. The extension is that, to establish convergence $\zeta_n \Leftrightarrow \mathbf{\Phi}_{0,1}$, one no longer needs the negligibility of $\xi_{k,n}$. If, for example, $\xi_{1,n} \in \mathbf{\Phi}_{0,1/2}$ (in that case $\mu_{1,n}^0 = 0$) and $L_3^0 \to 0$, then $\mathbf{E}f(\zeta_n) \to \mathbf{E}f(\eta)$, $\eta \in \mathbf{\Phi}_{0,1}$, for any f from the class C_3. Since C_3 is a distribution determining class (see Chap. 6), it remains to make use of Corollary 6.3.2.

We can strengthen the above assertion.

Theorem 8.5.2 *For any* $x \in \mathbb{R}$,

$$\bigl|\mathbf{P}(\zeta_n < x) - \Phi(x)\bigr| \le c\bigl(L_3^0\bigr)^{1/4}, \tag{8.5.7}$$

where c is an absolute constant.

Proof Take an arbitrary function $h \in C_3$, $0 \le h \le 1$, such that $h(x) = 1$ for $x \le 0$ and $h(x) = 0$ for $x \ge 1$, and put $h_3 = \sup_x |h'''(x)|$. Then, for the function $f(x) = h((x-t)/\varepsilon)$, we will have $f_3 = \sup_x |f'''(x)| \le h_3/\varepsilon^3$, and by Theorem 8.5.1

$$\begin{aligned}\mathbf{P}(\zeta_n < t) &\le \mathbf{E}f(\zeta_n) \le \mathbf{E}f(\eta) + \frac{f_3 L_3^0}{6} \\ &\le \mathbf{P}(\eta < t + \varepsilon) + \frac{h_3 L_3^0}{6\varepsilon^3} \le \mathbf{P}(\eta < t) + \frac{\varepsilon}{\sqrt{2\pi}} + \frac{h_3 L_3^0}{6\varepsilon^3}.\end{aligned}$$

The last inequality holds since the maximum of the derivative of the normal distribution function $\Phi(t) = \mathbf{P}(\eta < t)$ is equal to $1/\sqrt{2\pi}$. Establishing in the same way the converse inequality and putting $\varepsilon = (L_3^0)^{1/4}$, we arrive at (8.5.7). The theorem is proved. □

The bound in Theorem 8.5.2 is, of course, not the best one. And yet inequality (8.5.7) shows that we will have a good normal approximation for $\mathbf{P}(\zeta_n < x)$ in the large deviations range (i.e. for $|x| \to \infty$) as well—at least for those x for which

$$\bigl(1 - \Phi\bigl(|x|\bigr)\bigr)\bigl(L_3^0\bigr)^{-1/4} \to \infty \tag{8.5.8}$$

as $n \to \infty$. Indeed, in that case, say, for $x = |x| > 0$,

$$\left|\frac{\mathbf{P}(\zeta_n \ge x)}{1 - \Phi(x)} - 1\right| \le \frac{c(L_3^0)^{1/4}}{1 - \Phi(x)} \to 0.$$

Since by L'Hospital's rule

$$1 - \Phi(x) = \frac{1}{\sqrt{2\pi}} \int_x^\infty e^{-t^2/2}\, dt \sim \frac{1}{\sqrt{2\pi}\, x} e^{-x^2/2} \quad \text{as } x \to \infty,$$

(8.5.8) holds for $|x| < c_1\sqrt{-\ln L_3^0}$ with an appropriately chosen constant c_1.

In Chap. 20 we will obtain an extension of Theorems 8.5.1 and 8.5.2.

The problem of refinements and approximation rate bounds in the central limit theorem and other limit theorems is one of the most important in probability theory, because solving it will tell us how precise and efficient the applications of these theorems to practical problems will be. First of all, one has to find the true order of the decay of

$$\Delta_n = \sup_x \bigl|\mathbf{P}(\zeta_n < x) - \Phi(x)\bigr|$$

in n (or, say, in L_3 in the case of non-identically distributed variables). There exist at least two approaches to finding sharp bounds for Δ_n. The first one, the so-called *method of characteristic functions*, is based on the unimprovable bound for the closeness of the ch.f.s

$$\left|\ln \varphi_{\zeta_n}(t) + \frac{t^2}{2}\right| < cL_3$$

that the reader can obtain by him/herself, using Lemma 7.4.1 and somewhat modifying the argument in the proof of Theorem 8.4.1. The principal technical difficulties here are in deriving, using the inversion formula, the same order of smallness for Δ_n.

The second approach, the so-called *method of compositions*, has been illustrated in the present section in Theorem 8.5.1 (the idea of the method is expressed, to a certain extent, by relation (8.5.3)). It will be using just that method that we will prove in Appendix 5 the following general result (Cramér–Berry–Esseen):

Theorem 8.5.3 *If $\xi_{k,n} = (\xi_k - a_k)/B_n$, where ξ_k do not depend on n, then*

$$\sup_x \bigl|\mathbf{P}(\zeta_n < x) - \Phi(x)\bigr| \le cL_3,$$

where c is an absolute constant.

In the case of identically distributed ξ_k the right-hand side of the above inequality becomes $c\mu_1/(\sigma^3\sqrt{n})$. It was established that in this case $(2\pi)^{-1/2} < c < 0.4774$, while in the case of non-identically distributed summands $c < 0.5591$.[5]

One should keep in mind that the above theorems and the bounds for the constant c are universal and therefore hold under the most unfavourable conditions (from the point of view of the approximation). In real problems, the convergence rate is usually much better.

---

[5] See [33].

8.6 The Law of Large Numbers and the Central Limit Theorem in the Multivariate Case

In this section we assume that $\xi_{1,n}, \ldots, \xi_{n,n}$ are random vectors in the triangular array scheme,

$$\mathbf{E}\xi_{k,n} = 0, \qquad \zeta_n = \sum_{k=1}^{n} \xi_{k,n}.$$

The law of large numbers $\zeta_n \xrightarrow{p} 0$ follows immediately from Theorem 8.3.1, if we assume that the components of $\xi_{k,n}$ satisfy the conditions of that theorem. Thus we can assume that Theorem 8.3.1 was formulated and proved for vectors.

Dealing with the central limit theorem is somewhat more complicated. Here we will assume that $\mathbf{E}|\xi_{k,n}|^2 < \infty$, where $|x|^2 = (x,x)$ is square of the norm of x. Let

$$\sigma_{k,n}^2 := \mathbf{E}\,\xi_{k,n}^T \xi_{k,n}, \qquad \sigma_n^2 := \sum_{k=1}^{n} \sigma_{k,n}^2$$

(the superscript T denotes transposition, so that $\xi_{k,n}^T$ is a column vector).

Introduce the condition

$$\sum_{k=1}^{n} \mathbf{E}\min\bigl(|\xi_{k,n}|^2, |\xi_{k,n}|^s\bigr) \to 0, \quad s > 2, \qquad [\mathbf{D}_2]$$

and the Lindeberg condition

$$\sum_{k=1}^{n} \mathbf{E}\bigl(|\xi_{k,n}|^2; |\xi_{k,n}| > \tau\bigr) \to 0 \qquad [\mathbf{M}_2]$$

as $n \to \infty$ for any $\tau > 0$. As in the univariate case, we can easily verify that conditions $[\mathbf{D}_2]$ and $[\mathbf{M}_2]$ are equivalent provided that $\operatorname{tr}\sigma_n^2 := \sum_{j=1}^{d} (\sigma_n^2)_{jj} < c < \infty$.

Theorem 8.6.1 *If $\sigma_n^2 \to \sigma^2$, where σ^2 is a positive definite matrix, and condition $[\mathbf{D}_2]$ (or $[\mathbf{M}_2]$) is met, then*

$$\zeta_n \Leftrightarrow \mathbf{\Phi}_{0,\sigma^2}.$$

Corollary 8.6.1 ("The conventional" central limit theorem) *If $\xi_1, \xi_2, \ldots$ is a sequence of independent identically distributed random vectors, $\mathbf{E}\xi_k = 0$, $\sigma^2 = \mathbf{E}\,\xi_k^T \xi_k$ and $S_n = \sum_{k=1}^{n} \xi_k$ then, as $n \to \infty$,*

$$\frac{S_n}{\sqrt{n}} \Leftrightarrow \mathbf{\Phi}_{0,\sigma^2}.$$

This assertion is a consequence of Theorem 8.6.1, since the random variables $\xi_{k,n} = \xi_k/\sqrt{n}$ satisfy its conditions.

Proof of Theorem 8.6.1 Consider the characteristic functions

$$\varphi_{k,n}(t) := \mathbf{E}e^{i(t,\xi_{k,n})}, \qquad \varphi_n(t) := \mathbf{E}e^{i(t,\zeta_n)} = \prod_{k=1}^{n} \varphi_{k,n}(t).$$

In order to prove the theorem we have to verify that, for any t, as $n \to \infty$,

$$\varphi_n(t) \to \exp\left\{-\frac{1}{2} t\sigma^2 t^T\right\}.$$

We make use of Theorem 8.4.1. We can interpret $\varphi_{k,n}(t)$ and $\varphi_n(t)$ as the ch.f.s

$$\varphi_{k,n}^{\theta}(v) = \mathbf{E}\exp(iv\xi_{k,n}^{\theta}), \qquad \varphi_n^{\theta}(v) = \mathbf{E}\exp(iv\zeta_n^{\theta})$$

of the random variables $\xi_{k,n}^{\theta} = (\xi_{k,n},\theta)$, $\zeta_n^{\theta} = (\zeta_n,\theta)$, where $\theta = t/|t|$, $v = |t|$. Let us show that the scalar random variables $\xi_{k,n}^{\theta}$ satisfy the conditions of Theorem 8.4.1 (or Corollary 8.4.1) for the univariate case. Clearly,

$$\mathbf{E}\xi_{k,n}^{\theta} = 0, \qquad \sum_{k=1}^{n} \mathbf{E}\left(\xi_{k,n}^{\theta}\right)^2 = \sum_{k=1}^{n} \mathbf{E}(\xi_{k,n},\theta)^2 = \theta\sigma_n^2\theta^T \to \theta\sigma^2\theta^T > 0.$$

That condition $[\mathbf{D}_2]$ is satisfied follows from the obvious inequalities

$$(\xi_{k,n},\theta)^2 = \left|\xi_{k,n}^{\theta}\right|^2 \le |\xi_{k,n}|^2, \qquad \sum_{k=1}^{n} \mathbf{E}g_2(\xi_{k,n}^{\theta}) \le \sum_{k=1}^{n} \mathbf{E}g_2(|\xi_{k,n}|),$$

where $g_2(x) = \min(x^2, |x|^s)$, $s > 2$. Thus, for any v and θ (i.e., for any t), by Corollary 8.4.1 of Theorem 8.4.1

$$\varphi_n(t) = \mathbf{E}\exp\{iv\zeta_n^{\theta}\} \to \exp\left\{-\frac{1}{2}v^2\theta\sigma^2\theta^T\right\} = \exp\left\{-\frac{1}{2}t\sigma^2 t^T\right\}.$$

The theorem is proved. □

Theorem 8.6.1 does not cover the case where the entries of the matrix σ_n^2 grow unboundedly or behave in such away that the rank of the limiting matrix σ^2 becomes less than the dimension of the vectors $\xi_{k,n}$. This can happen when the variances of different components of $\xi_{k,n}$ have different orders of decay (or growth). In such a case, one should consider the transformed sums $\zeta_n' = \zeta_n\sigma_n^{-1}$ instead of ζ_n. Theorem 8.6.1 is actually a consequence of the following more general assertion which, in turn, follows from Theorem 8.6.1.

Theorem 8.6.2 *If the random variables* $\xi_{k,n}' = \xi_{k,n}\sigma_n^{-1}$ *satisfy condition* $[\mathbf{D}_2]$ *(or* $[\mathbf{M}_2]$*) then* $\zeta_n' \Leftrightarrow \mathbf{\Phi}_{0,E}$, *where* E *is the identity matrix.*

8.7 Integro-Local and Local Limit Theorems for Sums of Identically Distributed Random Variables with Finite Variance

Theorem 8.2.1 from Sect. 8.2 is called the *integral* limit theorem. To understand the reasons for using such a name, one should compare this assertion with (more accurate) limit theorems of another type, that describe the asymptotic behaviour of the *densities* of the distributions of S_n (if any) or the asymptotics of the probabilities of sums S_n hitting a fixed interval. It is natural to call the theorems for densities *local* theorems. Theorems similar to Theorem 8.2.1 can be obtained from the local ones (if the densities exist) by integrating, and it is natural to call them *integral theorems*. Assertions about the asymptotics of the probabilities of S_n hitting an interval are "intermediate" between the local and integral theorems, and it is natural to call them *integro-local theorems*. In the literature, such statements are often also referred to as *local*, apparently because they describe the probability of the localisation of the sum S_n in a given interval.

8.7.1 Integro-Local Theorems

Integro-local theorems describe the asymptotics of

$$\mathbf{P}\big(S_n \in [x, x+\Delta)\big)$$

as $n \to \infty$ for a fixed $\Delta > 0$. Probabilities of this type for increasing Δ (or for $\Delta = \infty$) can clearly be obtained by summing the corresponding probabilities for fixed Δ.

We will derive integro-local and local theorems with the inversion formulas from Sect. 8.7.2.

For the sake of brevity, put

$$\Delta[x) = [x, x+\Delta)$$

and denote by $\phi(x) = \phi_{0,1}(x)$ the density of the standard normal distribution. Below we will restrict ourselves to the investigation of the sums $S_n = \xi_1 + \cdots + \xi_n$ of independent identically distributed random variables $\xi_k \stackrel{d}{=} \xi$.

Theorem 8.7.1 (The Stone–Shepp integro-local theorem) *Let ξ be a non-lattice random variable, $\mathbf{E}\xi = 0$ and $\mathbf{E}\xi^2 = \sigma^2 < \infty$. Then, for any fixed $\Delta > 0$, as $n \to \infty$,*

$$\mathbf{P}\big(S_n \in \Delta[x)\big) = \frac{\Delta}{\sigma\sqrt{n}}\phi\left(\frac{x}{\sigma\sqrt{n}}\right) + o\left(\frac{1}{\sqrt{n}}\right), \tag{8.7.1}$$

where the remainder term $o(1/\sqrt{n})$ is uniform in x.

Remark 8.7.1 Since relation (8.7.1) is valid for any fixed Δ, it will also be valid when $\Delta = \Delta_n \to 0$ slowly enough as $n \to \infty$. If $\Delta = \Delta_n$ grows then the asymptotics of $\mathbf{P}(S_n \in \Delta[x))$ can be obtained by summing the right-hand sides of (8.7.1) for, say, $\Delta = 1$ (if $\Delta_n \to \infty$ is integer-valued). Thus the integral theorem follows from the integro-local one but not vice versa.

Remark 8.7.2 By virtue of the properties of densities (see Sect. 3.2), the right-hand side of representation (8.7.1) has the same form as if the random variable $\zeta_n = S_n/(\sigma\sqrt{n})$ had the density $\phi(v) + o(1)$, although the existence of the density of S_n (or ζ_n) is not assumed in the theorem.

Proof of Theorem 8.7.1 First prove the theorem under the simplifying assumption that condition

$$\limsup_{|t|\to\infty} \bigl|\varphi(t)\bigr| < 1 \tag{8.7.2}$$

is satisfied (the Cramér condition on the ch.f.). Property 11 of ch.f.s (see Sect. 8.7.1) implies that this condition is always met if the distribution of the sum S_m, for some $m \ge 1$, has a positive absolutely continuous component. The proof of Theorem 8.7.1 in its general form is more complicated and will be given at the end of this section, in Sect. 8.7.3.

In order to use the inversion formula (7.2.8), we employ the "smoothing method" and consider, along with S_n, the sums

$$Z_n = S_n + \eta_\delta, \tag{8.7.3}$$

where $\eta_\delta \in \mathbf{U}_{-\delta,0}$. Since the ch.f. $\varphi_{\eta_\delta}(t)$ of the random variable η_δ, being equal to

$$\varphi_{\eta_\delta}(t) = \frac{1 - e^{-it\delta}}{it\delta}, \tag{8.7.4}$$

possesses the property that the function $\varphi_{\eta_\delta}(t)/t$ is integrable at infinity, for the increments of the distribution function $G_n(x)$ of the random variable Z_n (its ch.f. divided by t is integrable, too) we can use formula (7.2.8):

$$\begin{aligned} G_n(x+\Delta) - G_n(x) = \mathbf{P}\bigl(Z_n \in \Delta[x)\bigr) &= \frac{1}{2\pi}\int e^{-itx}\frac{1-e^{-it\Delta}}{it}\,\varphi^n(t)\varphi_{\eta_\delta}(t)\,dt \\ &= \frac{\Delta}{2\pi}\int e^{-itx}\varphi^n(t)\widehat{\varphi}(t)\,dt, \end{aligned} \tag{8.7.5}$$

where $\widehat{\varphi}(t) = \varphi_{\eta_\delta}(t)\varphi_{\eta_\Delta}(t)$ (cf. (7.2.8)) is the ch.f. of the sum of independent random variables η_δ and η_Δ. We obtain that the difference $G_n(x+\Delta) - G_n(x)$, up to the factor Δ, is nothing else but the value of the density of the random variable $S_n + \eta_\delta + \eta_\Delta$ at the point x.

Split the integral on the right-hand side of (8.7.5) into the two subintegrals: one over the domain $|t| < \gamma$ for some $\gamma < 1$, and the other—over the complementary domain. Put $x = v\sqrt{n}$ and consider first

$$I_1 := \int_{|t|<\gamma} e^{-itv\sqrt{n}}\varphi^n(t)\widehat{\varphi}(t)\,dt = \frac{1}{\sqrt{n}}\int_{|u|<\gamma\sqrt{n}} e^{-iuv}\varphi^n\left(\frac{u}{\sqrt{n}}\right)\widehat{\varphi}\left(\frac{u}{\sqrt{n}}\right)du.$$

Without loss of generality we can assume $\sigma = 1$, and by (8.2.1) obtain that

$$1-\varphi(t) = \frac{t^2}{2} + o(t^2),$$

$$\ln\varphi(t) = \ln\bigl[1-\bigl(1-\varphi(t)\bigr)\bigr] = -\frac{t^2}{2} + o\bigl(t^2\bigr) \quad \text{as } t \to 0. \tag{8.7.6}$$

Hence

$$n\ln\varphi\left(\frac{u}{\sqrt{n}}\right) = -\frac{u^2}{2} + h_n(u), \tag{8.7.7}$$

where $h_n(u) \to 0$ for any fixed u as $n \to \infty$. Moreover, for γ small enough, in the domain $|u| < \gamma\sqrt{n}$ we have

$$\bigl|h_n(u)\bigr| \le \frac{u^2}{6},$$

so the right-hand side of (8.7.7) does not exceed $-u^2/3$. Now we can rewrite I_1 in the form

$$I_1 = \frac{1}{\sqrt{n}}\int_{|u|<\gamma\sqrt{n}} \exp\left\{-iuv - \frac{u^2}{2} + h_n(u)\right\}\widehat{\varphi}\left(\frac{u}{\sqrt{n}}\right)du, \tag{8.7.8}$$

where $|\widehat{\varphi}(u/\sqrt{n})| \le 1$ and $\widehat{\varphi}(u/\sqrt{n}) \to 1$ for any fixed u as $n \to \infty$. Therefore, by virtue of the dominated convergence theorem,

$$\sqrt{n}\,I_1 \to \int \exp\left\{-iuv - \frac{u^2}{2}\right\}du \tag{8.7.9}$$

uniformly in v, since the integral on the right-hand side of (8.7.8) is uniformly continuous in v. But the integral on the right-hand side of (8.7.9) is simply (up to the factor $1/(2\pi)$) the result of applying the inversion formula to the ch.f. of the normal distribution, so that

$$\lim_{n\to\infty}\sqrt{n}\,I_1 = \sqrt{2\pi}\,e^{-v^2/2}. \tag{8.7.10}$$

It remains to consider the integral

$$I_2 := \int_{|t|\ge\gamma} e^{-itv\sqrt{n}}\varphi^n(t)\widehat{\varphi}(t)\,dt.$$

By virtue of (8.7.2) and non-latticeness of the distribution of ξ,

$$q := \sup_{|t|\ge\gamma} |\varphi(t)| < 1 \tag{8.7.11}$$

and therefore

$$|I_2| \le q^n \int_{|t|\ge\gamma} |\widehat{\varphi}(t)|\, dt \le q^n c(\Delta,\delta), \qquad \lim_{n\to\infty} \sqrt{n} I_2 = 0 \tag{8.7.12}$$

uniformly in v, where $c(\Delta,\delta)$ depends on Δ and δ only. We have established that, for $x = v\sqrt{n}$, as $n \to \infty$, the relations

$$\begin{aligned} I_1 + I_2 &= \sqrt{\frac{2\pi}{n}} e^{-v^2/2} + o\left(\frac{1}{\sqrt{n}}\right), \\ \mathbf{P}\big(Z_n \in \Delta[x)\big) &= \frac{\Delta}{\sqrt{2\pi n}} e^{-x^2/(2n)} + o\left(\frac{1}{\sqrt{n}}\right) \end{aligned} \tag{8.7.13}$$

hold uniformly in v (see (8.7.5)). This means that representation (8.7.13) holds uniformly for all x.

Further, by (8.7.3),

$$\big\{Z_n \in [x, x+\Delta-\delta)\big\} \subset \big\{S_n \in \Delta[x)\big\} \subset \big\{Z_n \in [x-\delta, x+\Delta)\big\} \tag{8.7.14}$$

and, so, in particular,

$$\mathbf{P}\big(S_n \in \Delta[x)\big) \le \frac{\Delta+\delta}{\sqrt{2\pi n}} e^{-(x-\delta)^2/(2n)} + o\left(\frac{1}{\sqrt{n}}\right) = \frac{\Delta+\delta}{\sqrt{2\pi n}} e^{-x^2/(2n)} + o\left(\frac{1}{\sqrt{n}}\right).$$

By (8.7.14) an analogous converse inequality also holds. Since δ is arbitrary, this is possible only if

$$\mathbf{P}\big(S_n \in \Delta[x)\big) = \frac{\Delta}{\sqrt{2\pi n}} e^{-x^2/(2n)} + o\left(\frac{1}{\sqrt{n}}\right). \tag{8.7.15}$$

The theorem is proved. □

8.7.2 Local Theorems

If the distribution of S_n has a density than we can obtain local theorems on the asymptotics of this density.

Theorem 8.7.2 *Let* $\mathbf{E}\xi = 0$, $\mathbf{E}\xi^2 = \sigma^2 < \infty$ *and suppose there exists an* $m \ge 1$ *such that at least one of the following three conditions is met*:

(a) *the distribution of* S_m *has a bounded density*;

(b) *the distribution of* S_m *has a density from* L_2;
(c) *the ch.f.* $\varphi^m(t)$ *of the sum* S_m *is integrable.*

Then, for $n \ge m$, *the distribution of the sum* S_n *has density* $f_{S_n}(x)$ *for which the representation*

$$f_{S_n}(x) = \frac{1}{\sqrt{2\pi n}\sigma} \exp\left\{-\frac{x^2}{2n\sigma^2}\right\} + o\left(\frac{1}{\sqrt{n}}\right) \qquad (8.7.16)$$

holds uniformly in x *as* $n \to \infty$.

Conditions (a)–(c) *are equivalent to each other (possibly with different values of* m).

Proof We first establish the equivalence of (a)–(c). The fact that a bounded density belongs to L_2 was proved in Sect. 7.2.3. Conversely, if $f \in L_2$ then

$$\begin{aligned}\left|f^{(2)*}(t)\right| &= \left|\int f(u) f(t-u)\,du\right| \\ &\le \left[\int f^2(u)\,du \times \int f^2(t-u)\,du\right]^{1/2} = \int f^2(u)\,du < \infty.\end{aligned}$$

Hence the relationship $f_{S_m} \in L_2$ implies the boundedness of $f_{S_{2m}}$, and thus (a) and (b) are equivalent.

If φ^m is integrable then by Theorem 7.2.2 the density f_{S_m} exists and is bounded. Conversely, if f_{S_m} is bounded then $f_{S_m} \in L_2$, $\varphi_{S_m} \in L_2$ and $\varphi_{S_{2m}} \in L_1$ (see Sect. 8.7.2). This proves the equivalence of (a) and (c).

We will now prove (8.7.16). By the inversion formula (7.2.1),

$$f_{S_n}(x) = \frac{1}{2\pi} \int e^{-itx} \varphi^n(t)\,dt.$$

Here the integral on the right-hand side does not "qualitatively" differ from the integral on the right-hand side of (8.7.5), we only have to put $\widehat{\varphi}(t) \equiv 1$ in the part I_1 of the integral (8.7.5) (the integral over the set $|t| < \gamma$), and, in the part I_2 (over the set $|t| \ge \gamma$), to replace the integrable function $\widehat{\varphi}(t)$ with the integrable function $\varphi^m(t)$ and to replace the function $\varphi^n(t)$ with $\varphi^{n-m}(t)$. After these changes the whole argument in the proof of relation (8.7.13) remains valid, and therefore the same relation (up to the factor Δ) will hold for

$$f_{S_n}(x) = \frac{1}{\sqrt{2\pi n}\,\sigma} \exp\left\{-\frac{x^2}{2n\sigma^2}\right\} + o\left(\frac{1}{\sqrt{n}}\right).$$

The theorem is proved. □

Theorem 8.7.2 implies that the density f_{ζ_n} of the random variable $\zeta_n = \frac{S_n}{\sigma\sqrt{n}}$ converges to the density ϕ of the standard normal law:

$$f_{\zeta_n}(v) \to \phi(v)$$

uniformly in v as $n \to \infty$.

For instance, the density of the uniform distribution over $[-1, 1]$ satisfies the conditions of this theorem, and hence the density of S_n at the point $x = v\sigma\sqrt{n}$ ($\sigma^2 = 1/3$) will behave as $\frac{1}{\sigma\sqrt{2\pi n}} e^{-v^2/(2\sigma^2)}$ (cf. the remark to Example 3.6.1).

In the *arithmetic case*, where the random variable ξ is integer-valued and the greatest common divisor of all possible values of ξ equals 1 (see Sect. 7.1), it is the asymptotics of the probabilities $\mathbf{P}(S_n = x)$ for integer x that become the subject of interest for local theorems. In this case we cannot assume without loss of generality that $\mathbf{E}\xi = 0$.

Theorem 8.7.3 (Gnedenko) *Let* $\mathbf{E}\,\xi = a$, $\mathbf{E}\,\xi^2 = \sigma^2 < \infty$ *and* ξ *have an arithmetic distribution. Then, uniformly over all integers* x, *as* $n \to \infty$,

$$\mathbf{P}(S_n = x) = \frac{1}{\sqrt{2\pi n}\sigma} \exp\left\{\frac{(x - an)^2}{2n\sigma^2}\right\} + o\left(\frac{1}{\sqrt{n}}\right). \qquad (8.7.17)$$

Proof When proving limit theorems for arithmetic ξ, it is more convenient to use the generating functions (see Sects. 7.1, 7.7)

$$p(z) \equiv p_\xi(z) := \mathbf{E}\, z^\xi, \quad |z| = 1,$$

so that $p(e^{it}) = \varphi(t)$, where φ is the ch.f. of ξ.

In this case the inversion formulas take the following form (see (7.2.10)): for integer x,

$$\mathbf{P}(\xi = x) = \frac{1}{2\pi i} \int_{|z|=1} z^{-x-1} p(z)\, dz,$$

$$\mathbf{P}(S_n = x) = \frac{1}{2\pi i} \int_{|z|=1} z^{-x-1} p^n(z)\, dz = \frac{1}{2\pi} \int_{-\pi}^{\pi} e^{-itx} \varphi^n(t)\, dt.$$

As in the proof of Theorem 8.7.1, here we split the integral on the right-hand side into two subintegrals: over the domain $|t| < \gamma$ and over the complementary set. The treatment of the first subintegral

$$I_1 := \int_{|t|<\gamma} e^{-itx} \varphi^n(t)\, dt = \int_{|t|<\gamma} e^{-ity} \left[e^{-ita} \varphi(t)\right]^n dt$$

for $y = x - an$ differs from the considerations for I_1 in Theorem 8.7.1 only in that it is simpler and yields (see (8.7.10))

$$I_1 = \frac{\sqrt{2\pi}}{\sigma\sqrt{n}} \exp\left\{-\frac{y^2}{2\pi\sigma^2}\right\} + o\left(\frac{1}{\sqrt{n}}\right).$$

Similarly, the treatment of the second subintegral differs from that of I_2 in Theorem 8.7.1 in that it becomes simpler, since the range of integration here is compact and on that one has

$$|\varphi(t)| \le q(\gamma) < 1. \tag{8.7.18}$$

Therefore, as in Theorem 8.7.1,

$$I_2 = o\left(\frac{1}{\sqrt{n}}\right), \qquad \mathbf{P}(S_n = x) = \frac{1}{\sqrt{2\pi n}\sigma} \exp\left\{-\frac{y^2}{2n\sigma^2}\right\} + o\left(\frac{1}{\sqrt{n}}\right).$$

The theorem is proved. □

Evidently, for the values of y of order $\sqrt{n}$ Theorem 8.7.3 is a generalisation of the local limit theorem for the Bernoulli scheme (see Corollary 5.2.1).

8.7.3 The Proof of Theorem 8.7.1 in the General Case

To prove Theorem 8.7.1 in the general case we will use the same approach as in Sect. 7.1. We will again employ the smoothing method, but now, when specifying the random variable Z_n in (8.7.3), we will take $\theta\eta$ instead of η_δ, where $\theta = \text{const}$, η is a random variable with the ch.f. from Example 7.2.1 (see the end of Sect. 7.2) equal to

$$\varphi_\eta(t) = \begin{cases} 1 - |t|, & |t| \le 1; \\ 0, & |t| > 1, \end{cases}$$

so that for $Z_n = S_n + \theta\eta$, similarly to (8.7.5), we have

$$\mathbf{P}\big(Z_n \in \Delta[x)\big) = \frac{\Delta}{2\pi} \int_{|t| \le \frac{1}{\theta}} e^{-itx} \varphi^n(t) \varphi_{\eta_\Delta}(t) \varphi_{\theta\eta}(t)\, dt, \tag{8.7.19}$$

where $\varphi_{\theta\eta}(t) = \max(0, 1 - \theta|t|)$. As in Sect. 8.7.1, split the integral on the right-hand side of (8.7.19) into two subintegrals: I_1 over the domain $|t| < \gamma$ and I_2 over the domain $\gamma \le |t| \le 1/\theta$. The asymptotic behaviour of these integrals is investigated in almost the same way as in Sect. 8.7.1, but is somewhat simpler, since the domain of integration in I_2 is compact, and so, by the non-latticeness of ξ, one has on it the upper bound

$$q := \sup_{\gamma \le |t| \le 1/\theta} |\varphi(t)| < 1. \tag{8.7.20}$$

Therefore, to bound I_2 we no longer need condition (8.7.2).

Thus we have established, as above, relation (8.7.13).

To derive from this fact the required relation (8.7.15) we will need the following.

Lemma 8.7.1 *Let $f(y)$ be a bounded uniformly continuous function, η an arbitrary proper random variable independent of S_n and $b(n) \to \infty$ as $n \to \infty$. If, for any* *fixed* $\Delta > 0$ *and* $\theta > 0$, *as* $n \to \infty$, *we have*

$$\mathbf{P}\big(S_n + \theta\eta \in \Delta[x)\big) = \frac{\Delta}{b(n)}\left[f\left(\frac{x}{b(n)}\right) + o(1)\right], \tag{8.7.21}$$

then

$$\mathbf{P}\big(S_n \in \Delta[x)\big) = \frac{\Delta}{b(n)}\left[f\left(\frac{x}{b(n)}\right) + o(1)\right]. \tag{8.7.22}$$

In this assertion we can take S_n to be any sequence of random variables satisfying (8.7.21). In this section we will set $b(n)$ to be equal to $\sqrt{n}$, but later (see the proof of Theorem A7.2.1 in Appendix 7) we will need some other sequences as well.

Proof Put $\theta := \delta^2\Delta$, where $\delta > 0$ will be chosen later, $\Delta_\pm := (1 \pm 2\delta)\Delta$, $\Delta_\pm[x) := [x, x + \Delta_\pm)$ and $f_0 := \max f(y)$. We first obtain an upper bound for $\mathbf{P}(S_n \in \Delta[x))$. We have

$$\mathbf{P}\big(Z_n \in \Delta_+[x - \Delta\delta)\big) \ge \mathbf{P}\big(Z_n \in \Delta_+[x - \Delta\delta); |\eta| < 1/\delta\big).$$

On the event $|\eta| < 1/\delta$ one has $-\delta\Delta < \theta\eta < \delta\Delta$, and hence on this event

$$\big\{Z_n \in \Delta_+[x - \Delta\delta)\big\} \supset \big\{S_n \in \Delta[x)\big\}.$$

Thus, by independence of η and S_n,

$$\mathbf{P}\big(Z_n \in \Delta_+[x - \Delta\delta)\big) \ge \mathbf{P}\big(S_n \in \Delta[x); |\eta| < 1/\delta\big) = \mathbf{P}\big(S_n \in \Delta[x)\big)\big(1 - h(\delta)\big),$$

where $h(\delta) := \mathbf{P}(|\eta| \ge 1/\delta) \to 0$ as $\delta \to 0$. By condition (8.7.21) and the uniform integrability of f we obtain

$$\begin{aligned}\mathbf{P}\big(S_n \in \Delta[x)\big) &\le \mathbf{P}\big(Z_n \in \Delta_+[x - \Delta\delta)\big)\big(1 - h(\delta)\big)^{-1} \\ &\le \left[\frac{\Delta}{b(n)} f\left(\frac{x}{b(n)}\right) + \frac{2\delta\Delta f_0}{b(n)} + o\left(\frac{1}{b(n)}\right)\right]\big(1 - h(\delta)\big)^{-1}.\end{aligned} \tag{8.7.23}$$

If, for a given $\varepsilon > 0$, we choose $\delta > 0$ such that

$$\big(1 - h(\delta)\big)^{-1} \le 1 + \frac{\varepsilon\Delta}{3}, \qquad 2\delta f_0 \le \frac{\varepsilon}{3},$$

then we derive from (8.7.23) that, for all n large enough and ε small enough,

$$\mathbf{P}\big(S_n \in \Delta[x)\big) \le \frac{\Delta}{b(n)}\left(f\left(\frac{x}{b(n)}\right) + \varepsilon\right). \tag{8.7.24}$$

This implies, in particular, that for all x,

$$\mathbf{P}\big(S_n \in \Delta[x)\big) \le \frac{\Delta}{b(n)}(f_0 + \varepsilon). \tag{8.7.25}$$

Now we will obtain a lower bound for $\mathbf{P}(S_n \in \Delta[x))$. For the event

$$A := \big\{Z_n \in \Delta_-[x + \Delta\delta)\big\}$$

we have

$$\mathbf{P}(A) = \mathbf{P}\big(A;\ |\eta| < 1/\delta\big) + \mathbf{P}\big(A;\ |\eta| \ge 1/\delta\big). \tag{8.7.26}$$

On the event $|\eta| < 1/\delta$ we have

$$\big\{Z_n \in \Delta_-[x + \Delta\delta)\big\} \subset \big\{S_n \in \Delta[x)\big\},$$

and hence

$$\mathbf{P}\big(A;\ |\eta| < 1/\delta\big) \le \mathbf{P}\big(S_n \in \Delta[x)\big). \tag{8.7.27}$$

Further, by independence of η and S_n and inequality (8.7.25),

$$\begin{aligned}
\mathbf{P}\big(A;\ |\eta| \ge 1/\delta\big) &= \mathbf{E}\big[\mathbf{P}(A \mid \eta);\ |\eta| \ge 1/\delta\big] \\
&= \mathbf{E}\big[\mathbf{P}\big(S_n \in \Delta_-[x + \theta\eta + \Delta\delta) \mid \eta\big);\ |\eta| \ge 1/\delta\big] \\
&\le \frac{\Delta}{b(n)}(f_0 + \varepsilon)h(\delta).
\end{aligned}$$

Therefore, combining (8.7.26), (8.7.27) and (8.7.21), we get

$$\mathbf{P}\big(S_n \in \Delta[x)\big) \ge \frac{\Delta}{b(n)} f\left(\frac{x}{b(n)}\right) - \frac{2\delta\Delta f_0}{b(n)} + o\left(\frac{1}{b(n)}\right) - \frac{\Delta}{b(n)}(f_0 + \varepsilon)h(\delta).$$

In addition, choosing δ such that

$$f_0 h(\delta) < \frac{\varepsilon}{3}, \qquad 2\delta f_0 < \frac{\varepsilon}{3},$$

we obtain that, for all n large enough and ε small enough,

$$\mathbf{P}\big(S_n \in \Delta[x)\big) \ge \frac{\Delta}{b(n)}\left(f\left(\frac{x}{b(n)}\right) - \varepsilon\right). \tag{8.7.28}$$

Since ε is arbitrarily small, inequalities (8.7.24) and (8.7.28) prove the required relation (8.7.22). The lemma is proved. □

To prove the theorem it remains to apply Lemma 8.7.1 in the case (see (8.7.13)) where $f = \phi$ and $b(n) = \sqrt{n}$. Theorem 8.7.1 is proved. □

8.7.4 Uniform Versions of Theorems 8.7.1–8.7.3 for Random Variables Depending on a Parameter

In the next chapter, we will need uniform versions of Theorems 8.7.1–8.7.3, where the summands ξ_k depend on a parameter λ. Denote such summands by $\xi_{(\lambda)k}$, the corresponding distributions by $\mathbf{F}_{(\lambda)}$, and put

$$S_{(\lambda)n} := \sum_{k=1}^{n} \xi_{(\lambda)k},$$

where $\xi_{(\lambda)k}$ are independent copies of $\xi_{(\lambda)} \in \mathbf{F}_{(\lambda)}$. If λ is only determined by the number of summands n then we will be dealing with the triangular array scheme considered in Sects. 8.3–8.6 (the summands there were denoted by $\xi_{k,n}$). In the general case we will take the segment $[0, \lambda_1]$ for some $\lambda_1 > 0$ as the parametric set, keeping in mind that $\lambda \in [0, \lambda_1]$ may depend on n (in the triangular array scheme one can put $\lambda = 1/n$).

We will be interested in what conditions must be imposed on a family of distributions $\mathbf{F}_{(\lambda)}$ for the assertions of Theorems 8.7.1–8.7.3 to hold uniformly in $\lambda \in [0, \lambda_1]$. We introduce the following notation:

$$a(\lambda) = \mathbf{E}\xi_{(\lambda)}, \qquad \sigma^2(\lambda) = \mathrm{Var}(\xi_{(\lambda)}), \qquad \varphi_{(\lambda)}(t) = \mathbf{E}e^{it\xi_{(\lambda)}}.$$

The next assertion is an analogue of Theorem 8.7.1.

Theorem 8.7.1A *Let the distributions $\mathbf{F}_{(\lambda)}$ satisfy the following properties: $0 < \sigma_1 < \sigma(\lambda) < \sigma_2 < \infty$, where σ_1 and σ_2 do not depend on λ:*

(a) *the relation*

$$\varphi_{(\lambda)}(t) - 1 - ia(\lambda)t + \frac{t^2 m_2(\lambda)}{2} = o(t^2), \quad m_2(\lambda) := \mathbf{E}\xi^2_{(\lambda)}, \tag{8.7.29}$$

holds uniformly in $\lambda \in [0, \lambda_1]$ as $t \to 0$, i.e. there exist a $t_0 > 0$ and a function $\varepsilon(t) \to 0$ as $t \to 0$, independent of λ, such that, for all $|t| \le t_0$, the absolute value of the left-hand side of (8.7.29) *does not exceed $\varepsilon(t)t^2$;*

(b) *for any fixed $0 < \theta_1 < \theta_2 < \infty$,*

$$q_{(\lambda)} := \sup_{\theta_1 \le |t| \le \theta_2} \left|\varphi_{(\lambda)}(t)\right| \le q < 1, \tag{8.7.30}$$

where q does not depend on λ.

Then, for each fixed $\Delta > 0$,

$$\mathbf{P}\big(S_{(\lambda)n} - na(\lambda) \in \Delta[x)\big) = \frac{\Delta}{\sigma(\lambda)\sqrt{n}}\phi\left(\frac{x}{\sigma(\lambda)\sqrt{n}}\right) + o\left(\frac{1}{\sqrt{n}}\right), \tag{8.7.31}$$

where the remainder term $o(1/\sqrt{n})$ is uniform in x and $\lambda \in [0, \lambda_1]$.

Proof Going through the proof of Theorem 8.7.1 in its general form (see Sect. 7.3), we see that, to ensure the validity of all the proofs of the intermediate assertions in their uniform forms, it suffices to have uniformity in the following two places:

(a) the uniformity in λ of the estimate $o(t^2)$ as $t \to 0$ in relation (8.7.6) for the expansion of the ch.f. of the random variable $\xi = \frac{\xi_{(\lambda)} - a(\lambda)}{\sigma(\lambda)}$;

(b) the uniformity in relation (8.7.20) for the same ch.f.

We verify the uniformity in (8.7.6). For $\varphi(t) = \mathbf{E}\, e^{it\xi}$, we have by (8.7.29)

$$\ln \varphi(t) = -\frac{ita(\lambda)}{\sigma(\lambda)} + \ln \varphi_{(\lambda)}\left(\frac{t}{\sigma(\lambda)}\right)$$
$$= -\frac{t^2(m_2(\lambda) - a^2(\lambda))}{2\sigma^2(\lambda)} + o(t^2) = -\frac{t^2}{2} + o(t^2),$$

where the remainder term is uniform in λ.

The uniformity in relation (8.7.20) clearly follows from condition b), since $\sigma(\lambda)$ is uniformly separated from both 0 and ∞. The theorem is proved. □

Remark 8.7.3 Conditions (a) and (b) of Theorem 8.7.1A are essential for (8.7.31) to hold. To see this, consider random variables ξ and η with fixed distributions, $\mathbf{E}\xi = \mathbf{E}\eta = 0$ and $\mathbf{E}\xi^2 = \mathbf{E}\eta^2 = 1$. Let $\lambda \in [0, 1]$ and the random variable $\xi_{(\lambda)}$ be defined by

$$\xi_{(\lambda)} := \begin{cases} \xi & \text{with probability } 1 - \lambda, \\ \frac{\eta}{\sqrt{\lambda}} & \text{with probability } \lambda, \end{cases} \tag{8.7.32}$$

so that $\mathbf{E}\,\xi_{(\lambda)} = 0$ and $\mathrm{Var}(\xi_{(\lambda)}) = 2 - \lambda$ (in the case of the triangular array scheme one can put $\lambda = 1/n$). Then, under the obvious notational conventions, for $\lambda = t^2$, $t \to 0$, we have

$$\varphi_{(\lambda)}(t) = (1 - \lambda)\varphi_\xi(t) + \lambda\varphi_\eta\left(\frac{t}{\sqrt{\lambda}}\right) = 1 - \frac{3t^2}{2} + o(t^2) + t^2\varphi_\eta(1).$$

This implies that (8.7.29) does not hold and hence condition a) is not met for the values of λ in the vicinity of zero. At the same time, the uniform versions of relation (8.7.31) and the central limit theorem will fail to hold. Indeed, putting $\lambda = 1/n$, we obtain the triangular array scheme, in which the number ν_n of the summands of the form $\eta_i/\sqrt{\lambda}$ in the sum $S_{(\lambda)n} = \sum_{i=1}^n \xi_{(\lambda)i}$ converges in distribution to $\nu \in \mathbf{\Pi}_1$ and

$$\frac{1}{\sqrt{n(2-\lambda)}} S_{(\lambda)n} \stackrel{d}{=} \frac{S_{n-\nu_n}}{\sqrt{2n-1}} + \frac{H_{\nu_n}}{\sqrt{2 - 1/n}}, \qquad \text{where } H_k = \sum_{i=1}^k \eta_i.$$

The first term on the right-hand side weakly converges in distribution to $\zeta \in \mathbf{\Phi}_{0,1/2}$, while the second term converges to $H_\nu/\sqrt{2}$. Clearly, the sum of these independent summands is, generally speaking, not distributed normally with parameters (0, 1).

To see that condition (b) is also essential, consider an arithmetic random variable ξ with $\mathbf{E}\xi = 0$ and $\operatorname{Var}(\xi) = 1$, take η to be a random variable with the uniform distribution $\mathbf{U}_{-1,1}$, and put

$$\xi_{(\lambda)} := \begin{cases} \xi & \text{with probability } 1-\lambda, \\ \eta & \text{with probability } \lambda. \end{cases}$$

Here the random variable $\xi_{(\lambda)}$ is non-lattice (its distribution has an absolutely continuous component), but

$$\varphi_{(\lambda)}(2\pi) = (1-\lambda) + \lambda\varphi_\eta(2\pi), \qquad q_{(\lambda)} \ge 1 - 2\lambda.$$

Again putting $\lambda = 1/n$, we get the triangular array scheme for which condition (b) is not met. Relation (8.7.31) does not hold either, since, in the previous notation, the sum $S_{(\lambda)n}$ is integer-valued with probability $\mathbf{P}(\nu_n = 0) = e^{-1}$, so that its distribution will have atoms at integer points with probabilities comparable, by Theorem 8.7.3, with the right-hand side of (8.7.31). This clearly contradicts (8.7.31).

If we put $\lambda = 1/n^2$ then the sum $S_{(\lambda)n}$ will be integer-valued with probability $(1 - 1/n^2)^n \to 1$, and the failure of relation (8.7.31) becomes even more evident.

Uniform versions of the local Theorems 8.7.2 and 8.7.3 are established in a completely analogous way.

Theorem 8.7.2A *Let the distributions* $\mathbf{F}_{(\lambda)}$ *satisfy the conditions of Theorem* 8.7.1A *with* $\theta_2 = \infty$ *and the conditions of Theorem* 8.7.2, *in which conditions* (a)–(c) *are understood in the uniform sense (i.e.,* $\max_x f_{S_{(\lambda)m}}(x)$ *or the norm of* $f_{S_{(\lambda)m}}$ *in* L_2 *or* $\int |\varphi^m_{(\lambda)}(t)|\,dt$ *are bounded uniformly in* $\lambda \in [0, \lambda_1]$).

Then representation (8.7.16) *holds for* $f_{S_{(\lambda)n}}(x)$ *uniformly in* x *and* λ, *provided that on its right-hand side we replace* σ *by* $\sigma(\lambda)$.

Proof The conditions of Theorem 8.7.2A are such that they enable one to obtain the proof of the uniform version without any noticeable changes in the arguments proving Theorems 8.7.1A and 8.7.2. □

The following assertion is established in the same way.

Theorem 8.7.3A *Let the arithmetic distributions* $\mathbf{F}_{(\lambda)}$ *satisfy the conditions of Theorem* 8.7.1A *for* $\theta_2 = \pi$. *Then representation* (8.7.17) *holds uniformly in* x *and* λ, *provided that* a *and* σ *on its right-hand side are replaced with* $a(\lambda)$ *and* $\sigma(\lambda)$, *respectively.*

Remark 8.7.3 applies to Theorems 8.7.2A and 8.7.3A as well.

8.8 Convergence to Other Limiting Laws

As we saw in previous sections, the normal law occupies a special place among all distributions—it is the limiting law for normed sums of arbitrary distributed random variables. There arises the natural question of whether there exist any other limiting laws for sums of independent random variables.

It is clear from the proof of Theorem 8.2.1 for identically distributed random variables that the character of the limiting law is determined by the behaviour of the ch.f. of the summands in the vicinity of 0. If $\mathbf{E}\xi = 0$ and $\mathbf{E}\xi^2 = \sigma^2 = -\varphi''(0)$ exist, then

$$\varphi\left(\frac{1}{\sqrt{n}}\right) = 1 + \frac{\varphi''(0)t^2}{2n} + o\left(\frac{1}{n}\right),$$

and this determines the asymptotic behaviour of the ch.f. of $S_n/\sqrt{n}$, equal to $\varphi^n(t\sqrt{n})$, which leads to the normal limiting law. Therefore, if one is looking for different limiting laws for the sums $S_n = \xi_1 + \cdots + \xi_n$, it is necessary to renounce the condition that the variance is finite or, which is the same, that $\varphi''(0)$ exists. In this case, however, we will have to impose some conditions on the regular variation of the functions $F_+(x) = \mathbf{P}(\xi \ge x)$ and/or $F_-(x) = \mathbf{P}(\xi < -x)$ as $x \to \infty$, which we will call the *right* and the *left tail* of the distribution of ξ, respectively. We will need the following concepts.

Definition 8.8.1 A positive (Lebesgue) measurable function $L(t)$ is called a *slowly varying function* (s.v.f.) as $t \to \infty$, if, for any fixed $v > 0$,

$$\frac{L(vt)}{L(t)} \to 1 \quad \text{as } t \to \infty. \tag{8.8.1}$$

A function $V(t)$ is called a *regularly varying function* (r.v.f.) (of index $-\beta$) as $t \to \infty$ if it can be represented as

$$V(t) = t^{-\beta} L(t), \tag{8.8.2}$$

where $L(t)$ is an s.v.f. as $t \to \infty$.

One can easily see that, similarly to (8.8.1), the characteristic property of regularly varying functions is the convergence

$$\frac{V(vt)}{V(t)} \to v^{-\beta} \quad \text{as } t \to \infty \tag{8.8.3}$$

for any fixed $v > 0$. Thus an s.v.f. is an r.v.f. of index zero.

Among typical representatives the class of s.v.f.s are the logarithmic function and its powers $\ln^\gamma t$, $\gamma \in \mathbb{R}$, linear combinations thereof, multiple logarithms, functions with the property that $L(t) \to L = \text{const} \ne 0$ as $t \to \infty$ etc. As an example of *a bounded oscillating* s.v.f. we mention

$$L_0(t) = 2 + \sin(\ln\ln t), \quad t > 1.$$

The main properties of r.v.f.s are given in Appendix 6.

As has already been noted, for $S_n/b(n)$ to converge to a "nondegenerate" limiting law under a suitable normalisation $b(n)$, we will have to impose conditions on the regular variation of the distribution tails of ξ. More precisely, we will need a regular variation of the "two-sided tail"

$$F_0(t) = F_-(t) + F_+(t) = \mathbf{P}\big(\xi \notin [-t, t)\big).$$

We will assume that the following condition is satisfied for some $\beta \in (0, 2]$, $\rho \in [-1, 1]$:

$[\mathbf{R}_{\beta,\rho}]$ The two-sided tail $F_0(x) = F_-(x) + F_+(x)$ is an r.v.f. as $x \to \infty$, i.e. it can be represented as

$$F_0(x) = t^{-\beta} L_{F_0}(x), \quad \beta \in (0, 2], \tag{8.8.4}$$

where $L_{F_0}(x)$ is an s.v.f., and the following limit exists

$$\rho_+ := \lim_{x\to\infty} \frac{F_+(x)}{F_0(x)} \in [0, 1], \qquad \rho := 2\rho_+ - 1. \tag{8.8.5}$$

If $\rho_+ > 0$, then clearly the right tail $F_+(x)$ is an r.v.f. like $F_0(x)$, i.e. it can be represented as

$$F_+(x) = V(x) := x^{-\beta} L(x), \quad \beta \in (0, 2],\ L(x) \sim \rho_+ L_{F_0}(x).$$

(Here, and likewise in Appendix 6, we use the symbol V to denote an r.v.f.) If $\rho_+ = 0$, then the right tail $F_+(x) = o(F_0(x))$ is not assumed to be regularly varying.

Relation (8.8.5) implies that the following limit also exists

$$\rho_- := \lim_{x\to\infty} \frac{F_-(x)}{F_0(x)} = 1 - \rho_+.$$

If $\rho_- > 0$, then, similarly to the case of the right tail, the left tail $F_-(x)$ can be represented as

$$F_-(x) = W(x) := x^{-\beta} L_W(x), \quad \beta \in (0, 2],\ L_W(x) \sim \rho_- L_{F_0}(x).$$

If $\rho_- = 0$, then the left tail $F_-(x) = o(F_0(x))$ is not assumed to be regularly varying.

The parameters $\rho_\pm$ are related to the parameter ρ in the notation $[\mathbf{R}_{\beta,\rho}]$ through the equalities

$$\rho = \rho_+ - \rho_- = 2\rho_+ - 1 \in [-1, 1].$$

Clearly, in the case $\beta < 2$ we have $\mathbf{E}\xi^2 = \infty$, so that the representation

$$\varphi(t) = 1 - \frac{t^2\sigma^2}{2} + o\big(t^2\big) \quad \text{as } t \to 0$$

no longer holds, and the central limit theorem is not applicable. If $\mathbf{E}\xi$ exists and is finite then everywhere in what follows it will be assumed without loss of generality that

$$\mathbf{E}\xi = 0.$$

Since $F_0(x)$ is non-increasing, there always exists the "generalised" inverse function $F_0^{(-1)}(u)$ understood as

$$F_0^{(-1)}(u) := \inf\{x : F_0(x) < u\}.$$

If the function F_0 is strictly monotone and continuous then $b = F_0^{(-1)}(u)$ is the unique solution to the equation

$$F_0(b) = u, \quad u \in (0, 1).$$

Set

$$\zeta_n := \frac{S_n}{b(n)},$$

wherein the case $\beta > 2$ we define the normalising factor $b(n)$ by

$$b(n) := F_0^{(-1)}(1/n). \tag{8.8.6}$$

For $\beta = 2$ put

$$b(n) := Y^{(-1)}(1/n), \tag{8.8.7}$$

where

$$\begin{aligned} Y(x) &:= 2x^{-2} \int_0^x yF_0(y)\,dy = 2x^{-2}\left[\int_0^x yF_+(y)\,dy + \int_0^x yF_-(y)\,dy\right] \\ &= x^{-2}\mathbf{E}\left(\xi^2;\ -x \le \xi < x\right) = x^{-2}L_Y(x), \end{aligned} \tag{8.8.8}$$

L_Y is an s.v.f. (see Theorem A6.2.1(iv) in Appendix 6). It follows from Theorem A6.2.1(v) in Appendix 6 that, under condition (8.8.4), we have

$$b(n) = n^{1/\beta}L_b(n), \quad \beta \le 2,$$

where L_b is an s.v.f.

We introduce the functions

$$V_I(x) = \int_0^x V(y)\,dy, \qquad V^I(x) = \int_x^\infty V(y)\,dy.$$

8.8.1 The Integral Theorem

Theorem 8.8.1 *Let condition $[\mathbf{R}_{\beta,\rho}]$ be satisfied. Then the following assertions hold true.*

(i) *For $\beta \in (0,2)$, $\beta \neq 1$ and the normalising factor* (8.8.6), *as $n \to \infty$,*

$$\zeta_n \Rightarrow \zeta^{(\beta,\rho)}. \tag{8.8.9}$$

The distribution $\mathbf{F}_{\beta,\rho}$ of the random variable $\zeta^{(\beta,\rho)}$ depends on parameters β and ρ only and has a ch.f. $\varphi^{(\beta,\rho)}(t)$, given by

$$\varphi^{(\beta,\rho)}(t) := \mathbf{E}e^{it\zeta^{(\beta,\rho)}} = \exp\bigl\{|t|^\beta B(\beta,\rho,\vartheta)\bigr\}, \tag{8.8.10}$$

where $\vartheta = \operatorname{sign} t$,

$$B(\beta,\rho,\vartheta) = \Gamma(1-\beta)\left[i\rho\vartheta \sin\frac{\beta\pi}{2} - \cos\frac{\beta\pi}{2}\right] \tag{8.8.11}$$

and, for $\beta \in (1,2)$, we put $\Gamma(1-\beta) = \Gamma(2-\beta)/(1-\beta)$.

(ii) *When $\beta = 1$, for the sequence ζ_n with the normalising factor* (8.8.6) *to converge to a limiting law, the former, generally speaking, needs to be centred. More precisely, as $n \to \infty$, the following convergence takes place*:

$$\zeta_n - A_n \Rightarrow \zeta^{(1,\rho)}, \tag{8.8.12}$$

where

$$A_n = \frac{n}{b(n)}\bigl[V_I\bigl(b(n)\bigr) - W_I\bigl(b(n)\bigr)\bigr] - \rho C, \tag{8.8.13}$$

$C \approx 0.5772$ is the Euler constant, and

$$\varphi^{(1,\rho)}(t) = \mathbf{E}e^{it\zeta^{(1,\rho)}} = \exp\left\{-\frac{\pi|t|}{2} - i\rho t \ln|t|\right\}. \tag{8.8.14}$$

If $n[V_I(b(n)) - W_I(b(n))] = o(b(n))$, then $\rho = 0$ and we can put $A_n = 0$. If $\mathbf{E}\xi$ exists and equals zero then

$$A_n = \frac{n}{b(n)}\bigl[W^I\bigl(b(n)\bigr) - V^I\bigl(b(n)\bigr)\bigr] - \rho C.$$

If $\mathbf{E}\xi = 0$ and $\rho \neq 0$ then $\rho A_n \to -\infty$ as $n \to \infty$.

(iii) *For $\beta = 2$ and the normalising factor* (8.8.7), *as $n \to \infty$,*

$$\zeta_n \Rightarrow \zeta^{(2,\rho)}, \qquad \varphi^{(2,\rho)}(t) := \mathbf{E}e^{it\zeta^{(2,\rho)}} = e^{-t^2/2},$$

so that $\zeta^{(2,\rho)}$ has the standard normal distribution that is independent of ρ.

The Proof of Theorem 8.8.1 is based on the same considerations as the proof of Theorem 8.2.1, i.e. on using the asymptotic behaviour of the ch.f. $\varphi(t)$ in the vicinity of zero. But here it will be somewhat more difficult from the technical viewpoint. This is why the proof of Theorem 8.8.1 appears in Appendix 7. □

Remark 8.8.1 The last assertion of the theorem (for $\beta = 2$) shows that the limiting distribution may be normal even in the case of infinite variance of ξ.

Besides with the normal distribution, we also note "extreme" limit distributions, corresponding to the $\rho = \pm 1$ where the ch.f. $\varphi^{(\beta,\rho)}$ (or the respective Laplace transform) takes a very simple form. Let, for example, $\rho = -1$. Since $e^{i\pi\vartheta/2} = \vartheta i$, then, for $\beta \neq 1, 2$,

$$\begin{aligned} B(\beta, -1, \vartheta) &= -\Gamma(1-\beta)\left[i \sin \frac{\beta\pi\vartheta}{2} + \cos \frac{\beta\pi\vartheta}{2}\right] \\ &= -\Gamma(1-\beta)e^{i\beta\pi\vartheta/2} = -\Gamma(1-\beta)(i\vartheta)^{\beta}, \\ \varphi^{(\beta,-1)}(t) &= \exp\bigl\{-\Gamma(1-\beta)(it)^{\beta}\bigr\}, \\ \mathbf{E}\, e^{\lambda\zeta^{(\beta,-1)}} &= \exp\bigl\{-\Gamma(1-\beta)\lambda^{\beta}\bigr\}, \quad \operatorname{Re}\lambda \ge 0. \end{aligned}$$

Similarly, for $\beta = 1$, by (8.8.14) and the equalities $-\frac{\pi\vartheta}{2} = i\,\frac{i\pi\vartheta}{2} = i \ln i\vartheta$ we have

$$\begin{aligned} \ln \varphi^{(1,-1)}(t) &= -\frac{\pi\vartheta t}{2} + it \ln|t| = it \ln i\vartheta + it \ln|t| = it \ln it, \\ \mathbf{E}\, e^{\lambda\zeta^{(1,-1)}} &= \exp\{\lambda \ln \lambda\}, \quad \operatorname{Re}\lambda \ge 0. \end{aligned}$$

A similar formula is valid for $\rho = 1$.

Remark 8.8.2 If $\beta < 2$, then by virtue of the properties of s.v.f.s (see Theorem A6.2.1(iv) in Appendix 6), as $x \to \infty$,

$$\int_0^x yF_0(y)\,dy = \int_0^x y^{1-\beta}L_{F_0}(y)\,dy \sim \frac{1}{2-\beta}x^{2-\beta}L_{F_0}(x) = \frac{1}{2-\beta}x^2F_0(x).$$

Therefore, for $\beta < 2$, we have $Y(x) \sim 2(2-\beta)^{-1}F_0(x)$,

$$Y^{(-1)}(1/n) \sim F_0^{(-1)}\left(\frac{2-\beta}{2n}\right) \sim \left(\frac{2}{2-\beta}\right)^{1/\beta} F_0^{(-1)}(1/n)$$

(cf. (8.8.6)). On the other hand, for $\beta = 2$ and $\sigma^2 := \mathbf{E}\xi^2 < \infty$ one has

$$Y(x) \sim x^{-2}\sigma^2, \qquad b(n) = Y^{(-1)}(1/n) \sim \sqrt{\sigma n}.$$

Thus normalisation (8.8.7) is "transitional" from normalisation (8.8.6) (up to the constant factor $(2/(2-\beta))^{1/\beta}$) to the standard normalisation $\sigma\sqrt{n}$ in the central limit theorem in the case where $\mathbf{E}\xi^2 < \infty$. This also means that normalisation (8.8.7) is "universal" and can be used for all $\beta \le 2$ (as it is done in many textbooks on probability theory). However, as we will see below, in the case $\beta < 2$ normalisation (8.8.6) is easier and simpler to deal with, and therefore we will use that scaling.

Recall that $\mathbf{F}_{\beta,\rho}$ denotes the distribution of the random variable $\zeta^{(\beta,\rho)}$. The parameter β takes values in the interval $(0, 2]$, the parameter $\rho = \rho_+ - \rho_-$ can assume any value from $[-1, 1]$. The role of the parameters β and ρ will be clarified below.

Theorem 8.8.1 implies that each of the laws $\mathbf{F}_{\beta,\rho}$, $0 < \beta \le 2$ and $-1 \le \rho \le 1$ is *limiting* for the distributions of suitably normalised sums of independent identically distributed random variables. It follows from the law of large numbers that the degenerate distribution $\mathbf{I}_a$ concentrated at the point a is also a limiting one. Denote the set of all such distributions by $\mathfrak{S}_0$. Furthermore, it is not hard to see that if $\mathbf{F}$ is a distribution from the class $\mathfrak{S}_0$ then the law that differs from $\mathbf{F}$ by scaling and shifting, i.e. the distribution $\mathbf{F}_{\{a,b\}}$ defined, for some fixed $b > 0$ and a, by the relation

$$\mathbf{F}_{\{a,b\}}(B) := \mathbf{F}\left(\frac{B-a}{b}\right), \quad \text{where } \frac{B-a}{b} = \{u \in \mathbb{R} : ub + a \in B\},$$

is also limiting for the distributions of sums of random variables $(S_n - a_n)/b_n$ as $n \to \infty$ for appropriate $\{a_n\}$ and $\{b_n\}$.

It turns out that the *class of distributions $\mathfrak{S}$ obtained by the above extension from $\mathfrak{S}_0$ exhausts all the limiting laws for sums of identically distributed independent random variables.*

Another characterisation of the class of limiting laws $\mathfrak{S}$ is also possible.

Definition 8.8.2 We call a distribution $\mathbf{F}$ *stable* if, for any a_1, a_2, $b_1 > 0$, $b_2 > 0$, there exist a and $b > 0$ such that

$$\mathbf{F}_{\{a_1,b_1\}} * \mathbf{F}_{\{a_2,b_2\}} = \mathbf{F}_{\{a,b\}}.$$

This definition means that the convolution of a stable distribution $\mathbf{F}$ with itself again yields the same distribution $\mathbf{F}$, up to a scaling and shift (or, which is the same, for independent random variables $\xi_i \in \mathbf{F}$ we have $(\xi_1 + \xi_2 - a)/b \in \mathbf{F}$ for appropriate a and b).

In terms of the ch.f. φ, the stability property has the following form: for any $b_1 > 0$ and $b_2 > 0$, there exist a and $b > 0$ such that

$$\varphi(tb_1)\varphi(tb_2) = e^{ita}\varphi(tb), \quad t \in \mathbb{R}. \tag{8.8.15}$$

Denote the class of all stable laws by $\mathfrak{S}^S$. The remarkable fact is that *the class of all limiting laws $\mathfrak{S}$ (for $(S_n - a_n)/b_n$* for *some a_n and b_n) and the class of all stable laws $\mathfrak{S}^S$ coincide.*

If, under a suitable normalisation, as $n \to \infty$,

$$\zeta_n \Rightarrow \zeta^{(\beta,\rho)},$$

then one says that *the distribution* $\mathbf{F}$ *of the summands ξ belongs to the domain of attraction of the stable law* $\mathbf{F}_{\beta,\rho}$.

Theorem 8.8.1 means that, if $\mathbf{F}$ satisfies condition $[\mathbf{R}_{\beta,\rho}]$, then $\mathbf{F}$ belongs to the domain of attraction of the stable law $\mathbf{F}_{\beta,\rho}$.

One can prove the converse assertion (see e.g. Chap. XVII, § 5 in [30]): if $\mathbf{F}$ belongs to the domain of attraction of a stable law $\mathbf{F}_{\beta,\rho}$ for $\beta < 2$, then $[\mathbf{R}_{\beta,\rho}]$ is satisfied.

As for the role of the parameters β and ρ, note the following. The parameter β *characterises the rate of convergence to zero as* $x \to \infty$ for the functions

$$F_{\beta,\rho,-}(x) := \mathbf{F}_{\beta,\rho}\big((-\infty, -x)\big) \quad \text{and} \quad F_{\beta,\rho,+}(x) := \mathbf{F}_{\beta,\rho}\big([x, \infty)\big).$$

One can prove that, for $\rho_+ > 0$, as $t \to \infty$,

$$F_{\beta,\rho,+}(t) \sim \rho_+ t^{-\beta}, \tag{8.8.16}$$

and, for $\rho_- > 0$, as $t \to \infty$,

$$F_{\beta,\rho,-}(t) \sim \rho_- t^{-\beta}. \tag{8.8.17}$$

Note that, for $\xi \in \mathbf{F}_{\beta,\rho}$, the asymptotic relations in Theorem 8.8.1 turn into precise equalities provided that we replace in them $b(n)$ with $b_n := n^{1/\beta}$. In particular,

$$\mathbf{P}\left(\frac{S_n}{b_n} \ge t\right) = F_{\beta,\rho,+}(t). \tag{8.8.18}$$

This follows from the fact that $[\varphi^{(\beta,\rho)}(t/b_n)]^n$ coincides with $\varphi^{(\beta,\rho)}(t)$ (see (8.8.10)) and hence the distribution of the normalised sum S_n/b_n coincides with the distribution of the random variable ξ.

The parameter ρ taking values in $[-1, 1]$ is the *measure of asymmetry* of the distribution $\mathbf{F}_{\beta,\rho}$. If, for instance, $\rho = 1$ ($\rho_- = 0$), then, for $\beta < 1$, the distribution $\mathbf{F}_{\beta,1}$ is concentrated entirely on the positive half-line. This is evident from the fact that in this case $\mathbf{F}_{\beta,1}$ can be considered as the limiting distribution for the normalised sums of independent identically distributed random variables $\xi_k \ge 0$ (with $F_-(0) = 0$). Since all the prelimit distributions are concentrated on the positive half-line, so is the limiting distribution.

Similarly, for $\rho = -1$ and $\beta < 1$, the distribution $\mathbf{F}_{\beta,-1}$ is entirely concentrated on the negative half-line. For $\rho = 0$ ($\rho_+ = \rho_- = 1/2$) the ch.f. of the distribution $\mathbf{F}_{\beta,0}$ will be real, and the distribution $\mathbf{F}_{\beta,0}$ itself is symmetric.

As we saw above, the ch.f.s $\varphi^{(\beta,\rho)}(t)$ of stable laws $\mathbf{F}_{\beta,\rho}$ admit closed-form representations. They are clearly integrable over $\mathbb{R}$, and the same is true for the functions $t^k \varphi^{(\beta,\rho)}(t)$ for any $k \ge 1$. Therefore all the stable distributions have densities that are differentiable arbitrarily many times (see e.g. the inversion formula (7.2.1)). As for explicit forms of these densities, they are only known for a few laws. Among them are:

1. The normal law $\mathbf{F}_{2,\rho}$ (which does not depend on ρ).

2. The Cauchy distribution $\mathbf{F}_{1,0}$ with density $2/(\pi^2+4x^2)$, $-\infty < x < \infty$. Scaling the x-axis with a factor of $\pi/2$ transforms this density into the form $1/\pi(1+x^2)$ corresponding to $\mathbf{K}_{0,1}$.

3. The Lévy distribution. This law can be obtained from the explicit form for the distribution of the maximum of the Wiener process. This will be the distribution $\mathbf{F}_{1/2,1}$ with parameters 1/2, 1 and density (up to scaling; cf. (8.8.16))

$$f^{(1/2,1)}(x) = \frac{1}{\sqrt{2\pi}x^{3/2}} e^{-1/(2x)}, \quad x > 0$$

(this density has a first hitting time of level 1 by the standard Wiener process, see Theorem 19.2.2).

8.8.2 The Integro-Local and Local Theorems

Under the conditions of this section we can also obtain integro-local and local theorems in the same way as in Sect. 8.7 in the case of convergence to the normal law. As in Sect. 8.7, integro-local theorems deal here with the asymptotics of

$$\mathbf{P}\big(S_n \in \Delta[x)\big), \qquad \Delta[x) = [x, x+\Delta)$$

as $n \to \infty$ for a fixed $\Delta > 0$.

As we can see from Theorem 8.8.1, the ch.f. $\varphi^{(\beta,\rho)}(t)$ of the stable law $\mathbf{F}_{\beta,\rho}$ is integrable, and hence, by the inversion formula, there exists a uniformly continuous density $f^{(\beta,\rho)}$ of the distribution $\mathbf{F}_{\beta,\rho}$. (As has already been noted, it is not difficult to show that $f^{(\beta,\rho)}$ is differentiable arbitrarily many times, see Sect. 7.2.)

Theorem 8.8.2 (The Stone integro-local theorem) *Let ξ be a non-lattice random variable and the conditions of Theorem* 8.8.1 *be met. Then, for any fixed $\Delta > 0$, as $n \to \infty$,*

$$\mathbf{P}\big(S_n \in \Delta[x)\big) = \frac{\Delta}{b(n)} f^{(\beta,\rho)}\left(\frac{x}{b(n)}\right) + o\left(\frac{1}{b(n)}\right), \tag{8.8.19}$$

where the remainder term $o(\frac{1}{b(n)})$ is uniform over x.

If $\beta = 1$ and $\mathbf{E}|\xi|$ does not exist then, on the right-hand side of (8.8.20), *we must replace $f^{(\beta,\rho)}(\frac{x}{b(n)})$ with $f^{(\beta,\rho)}(\frac{x}{b(n)} - A_n)$, where A_n is defined in* (8.8.13).

All the remarks to the integro-local Theorem 8.7.1 hold true here as well, with evident changes.

If the distribution of S_n has a density then we can find the asymptotics of that density.

Theorem 8.8.3 *Let there exist an $m \ge 1$ such that at least one of conditions (a)–(c) of Theorem 8.7.2 is satisfied. Moreover, let the conditions of Theorem 8.8.1 be met. Then for the density $f_{S_n}(x)$ of the distribution of S_n one has the representation*

$$f_{S_n}(x) = \frac{1}{b(n)} f^{(\beta,\rho)}\left(\frac{x}{b(n)}\right) + o\left(\frac{1}{b(n)}\right) \tag{8.8.20}$$

which holds uniformly in x as $n \to \infty$.

If $\beta = 1$ and $\mathbf{E}|\xi|$ does not exist then, on the right-hand side of (8.8.20), we must replace $f^{(\beta,\rho)}(\frac{x}{b(n)})$ with $f^{(\beta,\rho)}(\frac{x}{b(n)} - A_n)$, where A_n is defined in (8.8.13).

The assertion of Theorem 8.8.3 can be rewritten for $\zeta_n = \frac{S_n}{b(n)} - A_n$ as

$$f_{\zeta_n}(v) \to f^{(\beta,\rho)}(v)$$

for any v as $n \to \infty$.

For integer-valued ξ_k the following theorem holds true.

Theorem 8.8.4 *Let the distribution of ξ be arithmetic and the conditions of Theorem 8.8.1 be met. Then, uniformly for all integers x, as $n \to \infty$,*

$$\mathbf{P}(S_n = x) = \frac{1}{b(n)} f^{(\beta,\rho)}\left(\frac{x - an}{b(n)}\right) + o\left(\frac{1}{\sqrt{n}}\right), \tag{8.8.21}$$

where $a = \mathbf{E}\xi$ if $\mathbf{E}|\xi|$ exists and $a = 0$ if $\mathbf{E}|\xi|$ does not exist, $\beta \ne 1$. If $\beta = 1$ and $\mathbf{E}|\xi|$ does not exist then, on the right-hand side of (8.8.21), we must replace $f^{(\beta,\rho)}(\frac{x-an}{b(n)})$ with $f^{(\beta,\rho)}(\frac{x}{b(n)} - A_n)$.

The proofs of Theorems 8.8.2–8.8.4 mostly repeat those of Theorems 8.7.1–8.7.3 and can be found in Appendix 7.

8.8.3 An Example

In conclusion we will consider an example.

In Sect. 12.8 we will see that in the fair game considered in Example 4.2.3 the ruin time $\eta(z)$ of a gambler with an initial capital of z units satisfies the relation $\mathbf{P}(\eta(z) \ge n) \sim z\sqrt{2/\pi n}$ as $n \to \infty$. In particular, for $z = 1$,

$$\mathbf{P}\big(\eta(1) \ge n\big) \sim \sqrt{2/\pi n}. \tag{8.8.22}$$

It is not hard to see (for more detail, see also Chap. 12) that $\eta(z)$ has the same distribution as $\eta_1 + \eta_2 + \cdots + \eta_z$, where η_j are independent and distributed as $\eta(1)$.

Thus for studying the distribution of $\eta(z)$ when z is large, by virtue of (8.8.22), one can make use of Theorem 8.8.4 (with $\beta = 1/2$, $b(n) = 2n^2/\pi$), by which

$$\lim_{z\to\infty} \mathbf{P}\left(\frac{2\pi\eta(x)}{z^2} < x\right) = F_{1/2,1}(x) \tag{8.8.23}$$

is the Lévy stable law with parameters $\beta = 1/2$ and $\rho = 1$. Moreover, for integer x and $z \to \infty$,

$$\mathbf{P}\big(\eta(z) = x\big) = \frac{\pi}{2z^2} f^{(1/2,1)}\left(\frac{x\pi}{2z^2}\right) + o\left(\frac{1}{z^2}\right).$$

These assertions enable one to obtain the limiting distribution for the number of crossings of an arbitrary strip $[u, v]$ by the trajectory $S_1, \ldots, S_n$ in the case where

$$\mathbf{P}(\xi_k = -1) = \mathbf{P}(\xi_k = -1) = 1/2.$$

Indeed, let for simplicity $u = 0$. By the first *positive crossing* of the strip $[0, v]$ we will mean the Markov time

$$\eta_+ := \min\{k : S_k = v\}.$$

The first *negative crossing* of the strip is then defined as the time $\eta_+ + \eta_-$, where

$$\eta_- := \min\{k : S_{\eta_+ + k} = 0\}.$$

The time $\eta_1 = \eta_+ + \eta_-$ will also be the time of the "double crossing" of $[0, v]$. The variables $\eta_\pm$ are distributed as $\eta(v)$ and are independent, so that η_1 has the same distribution as $\eta(2v)$. The variable $H_k = \eta_1(2v) + \cdots + \eta_k(2v)$, where $\eta_i(2v)$ have the same distribution as $\eta(2v)$ and are independent, is the time of the k-th double crossing. Therefore

$$\nu(n) := \max\{k : H_k \le n\} = \min\{k : H_k > n\} - 1$$

is the number of double crossings of the strip $[0, v]$ by time n. Now we can prove the following assertion:

$$\lim_{n\to\infty} \mathbf{P}\left(\frac{\nu(n)}{\sqrt{n}} \ge x\right) = F_{1/2,1}\left(\frac{\pi}{2v^2x^2}\right). \tag{8.8.24}$$

To prove it, we will make use of the following relation (which will play, in its more general form, an important role in Chap. 10):

$$\big\{\nu(n) \ge k\big\} = \{H_k \le n\},$$

where H_k is distributed as $\eta(2vk)$. If $n/k^2 \to s^2$ as $n \to \infty$, then by virtue of (8.8.23)

$$\mathbf{P}(H_k \le n) = \mathbf{P}\left(\frac{2\pi H_k}{(2vk)^2} \le \frac{2\pi n}{(2vk)^2}\right) \to F_{1/2,1}\left(\frac{\pi s^2}{2v^2}\right),$$

and therefore

$$\mathbf{P}\left(\frac{\nu(n)}{\sqrt{n}} \geq x\right) = \mathbf{P}\big(\nu(n) \geq x\sqrt{n}\big) = \mathbf{P}(H_{\lfloor x\sqrt{n}\rfloor} \leq n) \to F_{1/2,1}\left(\frac{\pi}{2v^2x^2}\right).$$

(Here for $k = \lfloor x\sqrt{n}\rfloor$ one has $n/k^2 \to s^2 = 1/x^2$.) Relation (8.8.24) is proved. □

Assertion (8.8.24) will clearly remain true for the number of crossings of the strip $[u, v]$, $u \neq 0$; one just has to replace v with $v - u$ on the right-hand side of (8.8.24). It is also clear that (8.8.24) enables one to find the limiting distribution of the number of "simple" (not double) crossings of $[u, v]$ since the latter is equal to $2\nu(n)$ or $2\nu(n)+1$.

Chapter 9
Large Deviation Probabilities for Sums of Independent Random Variables

Abstract The material presented in this chapter is unique to the present text. After an introductory discussion of the concept and importance of large deviation probabilities, Cramér's condition is introduced and the main properties of the Cramér and Laplace transforms are discussed in Sect. 9.1. A separate subsection is devoted to an in-depth analysis of the key properties of the large deviation rate function, followed by Sect. 9.2 establishing the fundamental relationship between large deviation probabilities for sums of random variables and those for sums of their Cramér transforms, and discussing the probabilistic meaning of the rate function. Then the logarithmic Large Deviations Principle is established. Section 9.3 presents integro-local, integral and local theorems on the exact asymptotic behaviour of the large deviation probabilities in the so-called Cramér range of deviations. Section 9.4 is devoted to analysing various types of the asymptotic behaviours of the large deviation probabilities for deviations at the boundary of the Cramér range that emerge under different assumptions on the distributions of the random summands. In Sect. 9.5, the behaviour of the large deviation probabilities is found in the case of heavy-tailed distributions, namely, when the distributions tails are regularly varying at infinity. These results are used in Sect. 9.6 to find the asymptotics of the large deviation probabilities beyond the Cramér range of deviations, under special assumptions on the distribution tails of the summands.

Let $\xi, \xi_1, \xi_2, \ldots$ be a sequence of independent identically distributed random variables,

$$\mathbf{E}\xi_k = 0, \qquad \mathbf{E}\xi_k^2 = \sigma^2 < \infty, \qquad S_n = \sum_{k=1}^{n} \xi_k.$$

Suppose that we have to evaluate the probability $\mathbf{P}(S_n \ge x)$. If $x \sim v\sqrt{n}$ as $n \to \infty$, $v = \text{const}$, then by the integral limit theorem

$$\mathbf{P}(S_n \ge x) \sim 1 - \Phi\left(\frac{v}{\sigma}\right) \tag{9.0.1}$$

as $n \to \infty$. But if $x \gg \sqrt{n}$, then the integral limit theorem enables one only to conclude that $\mathbf{P}(S_n \ge x) \to 0$ as $n \to \infty$, which in fact contains no quantitative

A.A. Borovkov, *Probability Theory*, Universitext,
DOI 10.1007/978-1-4471-5201-9_9, © Springer-Verlag London 2013

information on the probability we are after. Essentially the same can happen for fixed but "relatively" large values of v/σ. For example, for $v/\sigma \ge 3$ and the values of n around 100, the relative accuracy of the approximation in (9.0.1) becomes, generally speaking, bad (the true value of the left-hand side can be several times greater or smaller than that of the right-hand side). Studying the asymptotic behaviour of $\mathbf{P}(S_n \ge x)$ for $x \gg \sqrt{n}$ as $n \to \infty$, which is not known to us yet, could fill these gaps. This problem is highly relevant since questions of just this kind arise in many problems of mathematical statistics, insurance theory, the theory of queueing systems, etc. For instance, in mathematical statistics, finding *small* probabilities of errors of the first and second kind of statistical tests when the sample size n is large leads to such problems (e.g. see [7]). In these problems, we have to find explicit functions $P(n,x)$ such that

$$\mathbf{P}(S_n \ge x) = P(n,x)\bigl(1 + o(1)\bigr) \tag{9.0.2}$$

as $n \to \infty$. Thus, unlike the case of normal approximation (9.0.1), here we are looking for approximations $P(n,x)$ with a *relatively* small error rather than an *absolutely* small error. If $P(n,x) \to 0$ in (9.0.2) as $n \to \infty$, then we will speak of the probabilities of *rare events*, or of the *probabilities of large deviations* of sums S_n. Deviations of the order $\sqrt{n}$ are called *normal deviations*.

In order to study large deviation probabilities, we will need some notions and assertions.

9.1 Laplace's and Cramér's Transforms. The Rate Function

9.1.1 The Cramér Condition. Laplace's and Cramér's Transforms

In all the sections of this chapter, except for Sect. 9.5, the following *Cramér condition* will play an important role.

[C] *There exists a* $\lambda \ne 0$ *such that*

$$\mathbf{E}e^{\lambda\xi} = \int e^{\lambda y}\mathbf{F}(dy) < \infty. \tag{9.1.1}$$

We will say that the *right-side* (*left-side*) Cramér condition holds if $\lambda > 0$ ($\lambda < 0$) in (9.1.1). If (9.1.1) is valid for some negative and positive λ (i.e. in a neighbourhood of the point $\lambda = 0$), then we will say that the *two-sided Cramér's condition* is satisfied.

The Cramér condition can be interpreted as characterising a fast (at least exponentially fast) rate of decay of the tails $F_\pm(t)$ of the distribution $\mathbf{F}$. If, for instance, we have (9.1.1) for $\lambda > 0$, then by Chebyshev's inequality, for $t > 0$,

$$F_+(t) := \mathbf{P}(\xi \ge t) \le e^{-\lambda t}\mathbf{E}e^{\lambda\xi},$$

i.e. $F_+(t)$ decreases at least exponentially fast. Conversely, if, for some $\mu > 0$, one has $F_+(t) \le ce^{-\mu t}$, $t > 0$, then, for $\lambda \in (0, \mu)$,

$$\int_0^\infty e^{\lambda y}\mathbf{F}(dy) = -\int_0^\infty e^{\lambda y}\, dF_+(y) = F_+(0) + \lambda \int_0^\infty e^{\lambda y} F_+(y)\, dy$$
$$\le F_+(0) + c\lambda \int_0^\infty e^{(\lambda-\mu)y} dy = F_+(0) + \frac{c\lambda}{\mu - \lambda} < \infty.$$

Since the integral $\int_{-\infty}^0 e^{\lambda y}\mathbf{F}(dy)$ is finite for any $\lambda > 0$, we have $\mathbf{E}e^{\lambda\xi} < \infty$ for $\lambda \in (0, \mu)$.

The situation is similar for the left tail $F_-(t) := \mathbf{P}(\xi < -t)$ provided that (9.1.1) holds for some $\lambda < 0$.

Set

$$\lambda_+ := \sup\{\lambda : \mathbf{E}e^{\lambda\xi} < \infty\}, \qquad \lambda_- := \inf\{\lambda : \mathbf{E}e^{\lambda\xi} < \infty\}.$$

Condition **[C]** is equivalent to $\lambda_+ > \lambda_-$. The right-side Cramér condition means that $\lambda_+ > 0$; the two-sided condition means that $\lambda_+ > 0 > \lambda_-$. Clearly, the ch.f. $\varphi(t) = \mathbf{E}e^{it\xi}$ is analytic in the complex plane in the strip $-\lambda_+ < \operatorname{Im} t < -\lambda_-$. This follows from the differentiability of $\varphi(t)$ in this region of the complex plane, since the integral $\int |ye^{ity}|\mathbf{F}(dy)$ for the said values of $\operatorname{Im} t$ converges uniformly in $\operatorname{Re} t$.

Here and henceforth by the *Laplace transform* (Laplace–Stieltjes or Laplace–Lebesgue) of the distribution $\mathbf{F}$ of the random variable ξ we shall mean the function

$$\psi(\lambda) := \mathbf{E}e^{\lambda\xi} = \varphi(-i\lambda),$$

which conflicts with Sect. 7.1.1 (and the terminology of mathematical analysis), according to which the term Laplace's transform refers to the function $\mathbf{E}e^{-\lambda\xi} = \varphi(i\lambda)$. The reason for such a slight inconsistency in terminology (only the sign of the argument differs, this changes almost nothing) is our reluctance to introduce new notation or to complicate the old notation. Nowhere below will it cause confusion.[1]

As well as condition **[C]**, we will also assume that the random variable ξ is nondegenerate, i.e. $\xi \not\equiv \text{const}$ or, which is the same, $\operatorname{Var}\xi > 0$.

The main properties of Laplace's transform.

As was already noted in Sect. 7.1.1, Laplace's transform, like the ch.f., uniquely characterises the distribution $\mathbf{F}$. Moreover, it has the following properties, which are similar to the corresponding properties of ch.f.s (see Sect. 7.1). Under obvious conventions of notation,

(Ψ1) $\psi_{a+b\xi}(\lambda) = e^{\lambda a}\psi_\xi(b\lambda)$, *if a and b are constant.*

---

[1] In the literature, the function $\mathbf{E}e^{\lambda\xi}$ is sometimes called the "moment generating function".

(Ψ2) *If $\xi_1, \dots, \xi_n$ are independent and $S_n = \sum_{j=1}^n \xi_j$, then*

$$\psi_{S_n}(\lambda) = \prod_{j=1}^n \psi_{\xi_j}(\lambda).$$

(Ψ3) *If $\mathbf{E}|\xi|^k < \infty$ and the right-side Cramér condition is satisfied then the function ψ_ξ is k-times right differentiable at the point $\lambda = 0$,*

$$\psi_\xi^{(k)}(0) = \mathbf{E}\xi^k =: m_k$$

and, as $\lambda \downarrow 0$,

$$\psi_\xi(\lambda) = 1 + \sum_{j=1}^k \frac{\lambda^j}{j!} m_j + o(\lambda^k).$$

This also implies that, as $\lambda \downarrow 0$, the representation

$$\ln \psi_\xi(\lambda) = \sum_{j=1}^k \frac{\gamma_j \lambda^j}{j!} + o(\lambda^k), \tag{9.1.2}$$

holds, where γ_j are the so-called *semi-invariants* (or *cumulants*) of order j of the random variable ξ. One can easily verify that

$$\gamma_1 = m_1, \quad \gamma_2 = m_2^0 = \sigma^2, \quad \gamma_3 = m_3^0, \quad \dots, \tag{9.1.3}$$

where $m_k^0 = \mathbf{E}(\xi - m_1)^k$ is the central moment of order k.

Definition 9.1.1 Let condition **[C]** be met. *The Cramér transform at the point λ of the distribution* $\mathbf{F}$ is the distribution[2]

$$\mathbf{F}_{(\lambda)}(dy) = \frac{e^{\lambda y}\mathbf{F}(dy)}{\psi(\lambda)}. \tag{9.1.4}$$

[2]In some publications the transform (9.1.4) is also called the *Esscher transform*. However, the systematic use of transform (9.1.4) for the study of large deviations was first done by Cramér.

If we study the probabilities of large deviations of sums of random variables using the inversion formula, similarly to what was done for normal deviations in Chap. 8, then we will necessarily come to employ the so-called *saddle-point method*, which consists of moving the contour of integration so that it passes through the so-called *saddle point*, at which the exponent in the integrand function, as we move along the imaginary axis, attains its minimum (and, along the real axis, attains its maximum; this explains the name "saddle point"). Cramér's transform does essentially the same, making such a translation of the contour of integration even before applying the inversion formula, and reduces the large deviation problem to the normal deviation problem, where the inversion formula is not needed if we use the results of Chap. 8. It is this technique that we will follow in the present chapter.

Clearly, the distributions $\mathbf{F}$ and $\mathbf{F}_{(\lambda)}$ are mutually absolutely continuous (see Sect. 3.5 of Appendix 3) with density

$$\frac{\mathbf{F}_{(\lambda)}(dy)}{\mathbf{F}(dy)} = \frac{e^{\lambda y}}{\psi(\lambda)}.$$

Denote a random variable with distribution $\mathbf{F}_{(\lambda)}$ by $\xi_{(\lambda)}$.

The Laplace transform of the distribution $\mathbf{F}_{(\lambda)}$ is obviously equal to

$$\mathbf{E}e^{\mu\xi_{(\lambda)}} = \frac{\psi(\lambda+\mu)}{\psi(\lambda)}. \tag{9.1.5}$$

Clearly,

$$\mathbf{E}\xi_{(\lambda)} = \frac{\psi'(\lambda)}{\psi(\lambda)} = \big(\ln\psi(\lambda)\big)', \qquad \mathbf{E}\xi^2_{(\lambda)} = \frac{\psi''(\lambda)}{\psi(\lambda)},$$

$$\mathrm{Var}(\xi_{(\lambda)}) = \frac{\psi''(\lambda)}{\psi(\lambda)} - \left(\frac{\psi'(\lambda)}{\psi(\lambda}\right)^2 = \big(\ln\psi(\lambda)\big)''.$$

Since $\psi''(\lambda) > 0$ and $\mathrm{Var}(\xi_{(\lambda)}) > 0$, the foregoing implies one more important property of the Laplace transform.

(Ψ4) *The functions* $\psi(\lambda)$ *and* $\ln\psi(\lambda)$ *are strictly convex, and*

$$\mathbf{E}\xi_{(\lambda)} = \frac{\psi'(\lambda)}{\psi(\lambda)}$$

strictly increases on (λ_-, λ_+).

The analyticity of $\psi(\lambda)$ in the strip $\mathrm{Re}\,\lambda \in (\lambda_-, \lambda_+)$ can be supplemented by the following "extended" continuity property on the segment $[\lambda_-, \lambda_+]$ (in the strip $\mathrm{Re}\,\lambda \in [\lambda_-, \lambda_+]$).

(Ψ5) *The function* $\psi(\lambda)$ *is continuous "inside"* $[\lambda_-, \lambda_+]$, *i.e.* $\psi(\lambda_\pm \mp 0) = \psi(\lambda_\pm)$ (where the cases $\psi(\lambda_\pm) = \infty$ are not excluded).

Outside the segment $[\lambda_-, \lambda_+]$ such continuity, generally speaking, does not hold as, for example, is the case when $\psi(\lambda_+) < \infty$ and $\psi(\lambda_+ + 0) = \infty$, which takes place, say, for the distribution $\mathbf{F}$ with density $f(x) = cx^{-3}e^{-\lambda_+ x}$ for $x \ge 1$, $c = \mathrm{const}$.

9.1.2 The Large Deviation Rate Function

Under condition **[C]**, the *large deviation rate function* will play the determining role in the description of asymptotics of probabilities $\mathbf{P}(S_n \ge x)$.

Definition 9.1.2 *The large deviation rate function* (or, for brevity, simply the *rate function*) Λ *of a random variable* ξ is defined by

$$\Lambda(\alpha) := \sup_{\lambda} \big(\alpha\lambda - \ln \psi(\lambda)\big). \tag{9.1.6}$$

The meaning of the name will become clear later. In classical analysis, the right-hand side of (9.1.6) is known as the *Legendre transform* of the function $\ln \psi(\lambda)$.

Consider the function $A(\alpha, \lambda) = \alpha\lambda - \ln \psi(\lambda)$ of the supremum appearing in (9.1.6). The function $-\ln \psi(\lambda)$ is strictly concave (see property (Ψ4)), and hence so is the function $A(\alpha, \lambda)$ (note also that $A(\alpha, \lambda) = -\ln \psi_\alpha(\lambda)$, where $\psi_\alpha(\lambda) = e^{-\lambda\alpha}\psi(\lambda)$ is the Laplace transform of the distribution of the random variable $\xi - \alpha$ and, therefore, from the "qualitative point of view", $A(\alpha, \lambda)$ possesses all the properties of the function $-\ln \psi(\lambda)$). The foregoing implies that there always exists a *unique* point $\lambda = \lambda(\alpha)$ (on the "extended" real line $[-\infty, \infty]$) at which the supremum in (9.1.6) is attained. As α grows, the value of $A(\alpha, \lambda)$ for $\lambda > 0$ increases (proportionally to λ), and for $\lambda < 0$ it decreases. Therefore, the graph of $A(\alpha, \lambda)$ as the function of λ will, roughly speaking, "roll over" to the right as α grows. This means that the maximum point $\lambda(\alpha)$ will also move to the right (or stay at the same place if $\lambda(\alpha) = \lambda_+$).

We now turn to more precise formulations. On the interval $[\lambda_-, \lambda_+]$, there exists the derivative (respectively, the right and the left derivative at the endpoints $\lambda_\pm$)

$$A'_\lambda(\alpha, \lambda) = \alpha - \frac{\psi'(\lambda)}{\psi(\lambda)}. \tag{9.1.7}$$

The parameters

$$\alpha_\pm = \frac{\psi'(\lambda_\pm \mp 0)}{\psi(\lambda_\pm \mp 0)}, \qquad \alpha_- < \alpha_+, \tag{9.1.8}$$

will play an important role in what follows. The value of α_+ determines the angle at which the curve $\ln \psi(\lambda)$ "sticks" into the point $(\lambda_+, \ln \psi(\lambda_+))$. The quantity α_- has a similar meaning. If $\alpha \in [\alpha_-, \alpha_+]$ then the equation $A'_\lambda(\alpha, \lambda)=0$, or (see (9.1.7))

$$\frac{\psi'(\lambda)}{\psi(\lambda)} = \alpha, \tag{9.1.9}$$

always has a unique solution $\lambda(\alpha)$ on the segment $[\lambda_-, \lambda_+]$ ($\lambda_\pm$ can be infinite). This solution $\lambda(\alpha)$, being the inverse of an analytical and strictly increasing function $\frac{\psi'(\lambda)}{\psi(\lambda)}$ on (λ_-, λ_+) (see (9.1.9)), is also analytical and strictly increasing on (α_-, α_+),

$$\lambda(\alpha) \uparrow \lambda_+ \quad \text{as } \alpha \uparrow \alpha_+; \qquad \lambda(\alpha) \downarrow \lambda_- \quad \text{as } \alpha \downarrow \alpha_-. \tag{9.1.10}$$

The equalities

$$\Lambda(\alpha) = \alpha\lambda(\alpha) - \ln \psi\big(\lambda(\alpha)\big), \qquad \frac{\psi'(\lambda(\alpha))}{\psi(\lambda(\alpha))} = \alpha \tag{9.1.11}$$

yield

$$\Lambda'(\alpha) = \lambda(\alpha) + \alpha\lambda'(\alpha) - \frac{\psi'(\lambda(\alpha))}{\psi(\lambda(\alpha))}\lambda'(\alpha) = \lambda(\alpha).$$

Recalling that

$$\frac{\psi'(0)}{\psi(0)} = m_1 = \mathbf{E}\xi, \quad 0 \in [\lambda_-, \lambda_+],\ m_1 \in [\alpha_-, \alpha_+],$$

we obtain the following representation for the function Λ:

(Λ1) *If* $\alpha_0 \in [\alpha_-, \alpha_+]$, $\alpha \in [\alpha_-, \alpha_+]$ *then*

$$\Lambda(\alpha) = \Lambda(\alpha_0) + \int_{\alpha_0}^{\alpha} \lambda(v)\,dv. \qquad (9.1.12)$$

Since $\lambda(m_1) = \Lambda(m_1) = 0$ (*this follows from* (9.1.9) *and* (9.1.11)), *we obtain, in particular, for* $\alpha_0 = m_1$, *that*

$$\Lambda(\alpha) = \int_{m_1}^{\alpha} \lambda(v)\,dv. \qquad (9.1.13)$$

The functions $\lambda(\alpha)$ *and* $\Lambda(\alpha)$ *are analytic on* (α_-, α_+).

Now consider what happens outside the segment $[\alpha_-, \alpha_+]$. Assume for definiteness that $\lambda_+ > 0$. We will study the behaviour of the functions $\lambda(\alpha)$ and $\Lambda(\alpha)$ near the point α_+ and for $\alpha > \alpha_+$. Similar results hold true in the vicinity of the point α_- in the case $\lambda_- < 0$.

First let $\lambda_+ = \infty$, i.e. the function $\ln\psi(\lambda)$ is analytic on the whole semiaxis $\lambda > 0$, and the tail $F_+(t)$ decays as $t \to \infty$ faster than any exponential function. Denote by

$$s_\pm = \pm \sup\{t : F_\pm(t) > 0\}$$

the boundaries of the support of $\mathbf{F}$. Without loss of generality, we will assume that

$$s_+ > 0, \quad s_- < 0. \qquad (9.1.14)$$

This can always be achieved by shifting the random variable, similarly to our assuming, without loss of generality, $\mathbf{E}\xi = 0$ in many theorems of Chap. 8, where we used the fact that the problem of studying the distribution of S_n is "invariant" with respect to a shift. (We can also note that $\Lambda_{\xi-a}(\alpha - a) = \Lambda_\xi(\alpha)$, see property ($\Lambda$4) below, and that (9.1.14) always holds provided that $\mathbf{E}\xi = 0$.)

(Λ2) (i) *If* $\lambda_+ = \infty$ *then* $\alpha_+ = s_+$.

Hence, for $s_+ = \infty$, we always have $\alpha_+ = \infty$ and so for any $\alpha \geq \alpha_-$ we are dealing with the already considered "regular" case, where (9.1.12) and (9.1.13) hold true.

(ii) *If* $s_+ < \infty$ *then* $\lambda_+ = \infty$, $\alpha_+ = s_+$,

$$\Lambda(\alpha_+) = -\ln \mathbf{P}(\xi = s_+), \qquad \Lambda(\alpha) = \infty \quad \text{for } \alpha > \alpha_+.$$

Similar assertions hold true for s_-, α_-, λ_-.

Proof (i) First let $s_+ < \infty$. Then the asymptotics of $\psi(\lambda)$ and $\psi'(\lambda)$ as $\lambda \to \infty$ is determined by the integrals in a neighbourhood of the point s_+: for any fixed $\varepsilon > 0$,

$$\psi(\lambda) \sim \mathbf{E}\big(e^{\lambda\xi}; \xi > s_+ - \varepsilon\big), \qquad \psi'(\lambda) \sim \mathbf{E}\big(\xi e^{\lambda\xi}; \xi > s_+ - \varepsilon\big)$$

as $\lambda \to \infty$. Hence

$$\alpha_+ = \lim_{\lambda\to\infty} \frac{\psi'(\lambda)}{\psi(\lambda)} = \lim_{\lambda\to\infty} \frac{\mathbf{E}(\xi e^{\lambda\xi}; \xi > s_+ - \varepsilon)}{\mathbf{E}(e^{\lambda\xi}; \xi > s_+ - \varepsilon)} = s_+.$$

If $s_+ = \infty$, then $\ln \psi(\lambda)$ grows as $\lambda \to \infty$ faster than any linear function and therefore the derivative $(\ln \psi(\lambda))'$ increases unboundedly, $\alpha_+ = \infty$.

(ii) The first two assertions are obvious. Further, let $p_+ = \mathbf{P}(\xi = s_+) > 0$. Then

$$\psi(\lambda) \sim p_+ e^{\lambda s_+},$$
$$\alpha\lambda - \ln \psi(\lambda) = \alpha\lambda - \ln p_+ - \lambda s_+ + o(1) = (\alpha - \alpha_+)\lambda - \ln p_+ + o(1)$$

as $\lambda \to \infty$. This and (9.1.11) imply that

$$\Lambda(\alpha) = \begin{cases} -\ln p_+ & \text{for } \alpha = \alpha_+, \\ \infty & \text{for } \alpha > \alpha_+. \end{cases}$$

If $p_+ = 0$, then the relation $\psi(\lambda) = o(e^{\lambda s_+})$ as $\lambda \to \infty$ similarly implies $\Lambda(\alpha_+) = \infty$. Property ($\Lambda$2) is proved. □

Now let $0 < \lambda_+ < \infty$. If $\alpha_+ < \infty$, then necessarily $\psi(\lambda_+) < \infty$, $\psi(\lambda_+ + 0) = \infty$ and $\psi'(\lambda_+) < \infty$ (here we mean the left derivative). If we assume that $\psi(\lambda_+) = \infty$, then $\ln \psi(\lambda_+) = \infty$, $(\ln \psi(\lambda))' \to \infty$ as $\lambda \uparrow \lambda_+$ and $\alpha_+ = \infty$, which contradicts the assumption $\alpha_+ < \infty$. Since $\psi(\lambda) = \infty$ for $\lambda > \lambda_+$, the point $\lambda(\alpha)$, having reached the value λ_+ as α grows, will stop at that point. So, for $\alpha \ge \alpha_+$, we have

$$\lambda(\alpha) = \lambda_+, \qquad \Lambda(\alpha) = \alpha\lambda_+ - \ln \psi(\lambda_+) = \Lambda(\alpha_+) + \lambda_+(\alpha - \alpha_+). \tag{9.1.15}$$

Thus, in this case, for $\alpha \ge \alpha_+$ the function $\lambda(\alpha)$ remains constant, while $\Lambda(\alpha)$ grows linearly. Relations (9.1.12) and (9.1.13) remain true.

If $\alpha_+ = \infty$, then $\alpha < \alpha_+$ for all finite $\alpha \ge \alpha_-$, and we again deal with the "regular" case that we considered earlier (see (9.1.12) and (9.1.13)). Since $\lambda(\alpha)$ does not decrease, these relations imply the convexity of $\Lambda(\alpha)$.

In summary, we can formulate the following property.

($\Lambda 3$) *The functions* $\lambda(\alpha)$ *and* $\Lambda(\alpha)$ *can only be discontinuous at the points* $s_\pm$ *and under the condition* $\mathbf{P}(\xi = s_\pm) > 0$. *These points separate the domain* (s_-, s_+) where the function Λ is finite and continuous (*in the extended sense*) *from the domain* $\alpha \notin [s_-, s_+]$ *where* $\Lambda(\alpha) = \infty$. *In the domain* $[s_-, s_+]$ *the function* Λ *is convex. (If we define convexity in the "extended" sense, i.e. including infinite values as well, then* Λ *is convex on the entire real line.) The function* Λ *is analytic in the interval* (α_-, α_+). *If* $\lambda_+ < \infty$ *and* $\alpha_+ < \infty$, *then on the half-line* (α_+, ∞) *the function* $\Lambda(\alpha)$ *is linear with slope* λ_+; *at the boundary point* α_+ *the continuity of the first derivatives persists. If* $\lambda_+ = \infty$, *then* $\Lambda(\alpha) = \infty$ *on* (α_+, ∞). *The function* $\Lambda(\alpha)$ *possesses a similar property on* $(-\infty, \alpha_-)$.

If $\lambda_- = 0$, then $\alpha_- = m_1$ and $\lambda(\alpha) = \Lambda(\alpha) = 0$ for $\alpha \le m_1$.

Indeed, since $\lambda(m_1) = 0$ and $\psi(\lambda) = \infty$ for $\lambda < \lambda_- = 0 = \lambda(m_1)$, as the value of α decreases to $\alpha_- = m_1$, the point $\lambda(\alpha)$, having reached the value 0, will stop, and $\lambda(\alpha) = 0$ for $\alpha \le \alpha_- = m_1$. This and the first identity in (9.1.11) also imply that $\Lambda(\alpha) = 0$ for $\alpha \le m_1$.

If $\lambda_- = \lambda_+ = 0$ (condition **[C]** is not met), then $\lambda(\alpha) = \Lambda(\alpha) \equiv 0$ for all α. This is obvious, since the value of the function under the sup sign in (9.1.6) equals $-\infty$ for all $\lambda \ne 0$. In this case the limit theorems presented in the forthcoming sections will be of little substance.

We will also need the following properties of the function Λ.

($\Lambda 4$) *Under obvious notational conventions, for independent random variables* ξ *and* η, *we have*

$$\Lambda_{\xi+\eta}(\alpha) = \sup_\lambda\bigl(\alpha\lambda - \ln\psi_\xi(\lambda) - \ln\psi_\eta(\lambda)\bigr) = \inf_\gamma\bigl(\Lambda_\xi(\gamma) + \Lambda_\eta(\alpha-\gamma)\bigr),$$

$$\Lambda_{c\xi+b}(\alpha) = \sup_\lambda\bigl(\alpha\lambda - \lambda b - \ln\psi_\xi(\lambda c)\bigr) = \Lambda_\xi\left(\frac{\alpha-b}{c}\right).$$

Clearly, $\inf_\gamma$ in the former relation is attained at the point γ at which $\lambda_\xi(\gamma) = \lambda_\eta(\alpha-\gamma)$. If ξ and η are identically distributed then $\gamma = \alpha/2$ and therefore

$$\Lambda_{\xi+\eta}(\alpha) = \Lambda_\xi\left(\frac{\alpha}{2}\right) + \Lambda_\eta\left(\frac{\alpha}{2}\right) = 2\Lambda_\xi\left(\frac{\alpha}{2}\right).$$

($\Lambda 5$) *The function* $\Lambda(\alpha)$ *attains its minimal value* 0 *at the point* $\alpha = \mathbf{E}\xi = m_1$. *For definiteness, assume that* $\alpha_+ > 0$. *If* $m_1 = 0$ *and* $\mathbf{E}|\xi^k| < \infty$, *then*

$$\lambda(0) = \Lambda(0) = \Lambda'(0) = 0, \quad \Lambda''(0) = \frac{1}{\gamma_2}, \quad \Lambda'''(0) = -\frac{\gamma_3}{\gamma_2^2}, \quad \dots \tag{9.1.16}$$

(In the case $\alpha_- = 0$ *the right derivatives are intended.) As* $\alpha \downarrow 0$, *one has the representation*

$$\Lambda(\alpha) = \sum_{j=2}^{k} \frac{\Lambda^{(j)}(0)}{j!}\alpha^j + o(\alpha^k). \tag{9.1.17}$$

The semi-invariants γ_j were defined in (9.1.2) and (9.1.3).

If the two-sided Cramér condition is satisfied then the series expansion (9.1.17) of the function $\Lambda(\alpha)$ holds for $k = \infty$. This series is called the *Cramér series*.

Verifying properties (Λ4) and (Λ5) is not difficult, and is left to the reader.

(Λ6) *The following inversion formula is valid*: *for* $\lambda \in (\lambda_-, \lambda_+)$,

$$\ln \psi(\lambda) = \sup_{\alpha}\big(\alpha\lambda - \Lambda(\alpha)\big). \tag{9.1.18}$$

This means that the rate function uniquely determines the Laplace transform $\psi(\lambda)$ and hence the distribution $\mathbf{F}$ as well. Formula (9.1.18) also means that subsequent double applications of the Legendre transform to the convex function $\ln \psi(\lambda)$ leads to the same original function.

Proof We denote by $T(\lambda)$ the right-hand side of (9.1.18) and show that $T(\lambda) = \ln \psi(\lambda)$ for $\lambda \in (\lambda_-, \lambda_+)$. If, in order to find the supremum in (9.1.18), we equate to zero the derivative in α of the function under the sup sign, then we will get the equation

$$\lambda = \Lambda'(\alpha) = \lambda(\alpha). \tag{9.1.19}$$

Since $\lambda(\alpha)$, $\alpha \in (\alpha_-, \alpha_+)$, is the function inverse to $(\ln \psi(\lambda))'$ (see (9.1.9)), for $\lambda \in (\lambda_-, \lambda_+)$ Eq. (9.1.19) clearly has the solution

$$\alpha = a(\lambda) := \big(\ln \psi(\lambda)\big)'. \tag{9.1.20}$$

Taking into account the fact that $\lambda(a(\lambda)) \equiv \lambda$, we obtain

$$T(\lambda) = \lambda a(\lambda) - \Lambda\big(a(\lambda)\big),$$
$$T'(\lambda) = a(\lambda) + \lambda a'(\lambda) - \lambda\big(a(\lambda)\big)a'(\lambda) = a(\lambda).$$

Since $a(0) = m_1$ and $T(0) = -\Lambda(m_1) = 0$, we have

$$T(\lambda) = \int_0^{\lambda} a(u)\,du = \ln \psi(\lambda). \tag{9.1.21}$$

The assertion is proved, and so is yet another inversion formula (the last equality in (9.1.21), which expresses $\ln \psi(\lambda)$ as the integral of the function $a(\lambda)$ inverse to $\lambda(\alpha)$). □

(Λ7) *The exponential Chebyshev inequality. For* $\alpha \ge m_1$, *we have*

$$\mathbf{P}(S_n \ge \alpha n) \le e^{-n\Lambda(\alpha)}.$$

Proof If $\alpha \ge m_1$, then $\lambda(\alpha) \ge 0$. For $\lambda = \lambda(\alpha) \ge 0$, we have

$$\psi^n(\lambda) \ge \mathbf{E}\big(e^{\lambda S_n};\, S_n \ge \alpha n\big) \ge e^{\lambda\alpha n}\mathbf{P}(S_n \ge \alpha n);$$
$$\mathbf{P}(S_n \ge \alpha n) \le e^{-\alpha n\lambda(\alpha)+n\ln\psi(\lambda(\alpha))} = e^{-n\Lambda(\alpha)}. \qquad \square$$

We now consider a few examples, where the values of $\lambda_\pm$, $\alpha_\pm$, and the functions $\psi(\lambda)$, $\lambda(\alpha)$, $\Lambda(\alpha)$ can be calculated in an explicit form.

Example 9.1.1 If $\xi \in \mathbf{\Phi}_{0,1}$, then

$$\psi(\lambda) = e^{\lambda^2/2}, \qquad |\lambda_\pm| = |\alpha_\pm| = \infty, \qquad \lambda(\alpha) = \alpha, \qquad \Lambda(\alpha) = \frac{\alpha^2}{2}.$$

Example 9.1.2 For the Bernoulli scheme $\xi \in \mathbf{B}_p$, we have

$$\psi(\lambda) = pe^\lambda + q, \qquad |\lambda_\pm| = \infty, \quad \alpha_+ = 1, \quad \alpha_- = 0, \quad m_1 = \mathbf{E}\xi = p,$$
$$\lambda(\alpha) = \ln\frac{\alpha(1-p)}{p(1-\alpha)}, \quad \Lambda(\alpha) = \alpha\ln\frac{\alpha}{p} + (1-\alpha)\ln\frac{1-\alpha}{1-p} \quad \text{for } \alpha \in (0,1),$$
$$\Lambda(0) = -\ln(1-p), \quad \Lambda(1) = -\ln p, \quad \Lambda(\alpha) = \infty \quad \text{for } \alpha \notin [0,1].$$

Thus the function $H(\alpha) = \Lambda(\alpha)$, which described large deviation probabilities for S_n in the local Theorem 5.2.1 for the Bernoulli scheme, is nothing else but the rate function. Below, in Sect. 9.3, we will obtain generalisations of Theorem 5.2.1 for arbitrary arithmetic distributions.

Example 9.1.3 For the exponential distribution $\mathbf{\Gamma}_\beta$, we have

$$\psi(\lambda) = \frac{\beta}{\beta-\lambda}, \quad \lambda_+ = \beta, \quad \lambda_- = -\infty, \quad \alpha_+ = \infty, \quad \alpha_- = 0, \quad m_1 = \frac{1}{\beta},$$
$$\lambda(\alpha) = \beta - \frac{1}{\alpha}, \quad \Lambda(\alpha) = \alpha\beta - 1 - \ln\alpha\beta \quad \text{for } \alpha > 0.$$

Example 9.1.4 For the centred Poisson distribution with parameter β, we have

$$\psi(\lambda) = \exp\big\{\beta\big[e^\lambda - 1 - \lambda\big]\big\}, \quad |\lambda_\pm| = \infty, \quad \alpha_- = -\beta, \quad \alpha_+ = \infty, \quad m_1 = 0,$$
$$\lambda(\alpha) = \ln\frac{\beta+\alpha}{\beta}, \quad \Lambda(\alpha) = (\alpha+\beta)\ln\frac{\alpha+\beta}{\beta} - \alpha \quad \text{for } \alpha > -\beta.$$

9.2 A Relationship Between Large Deviation Probabilities for Sums of Random Variables and Those for Sums of Their Cramér Transforms. The Probabilistic Meaning of the Rate Function

9.2.1 A Relationship Between Large Deviation Probabilities for Sums of Random Variables and Those for Sums of Their Cramér Transforms

Consider the Cramér transform of $\mathbf{F}$ at the point $\lambda = \lambda(\alpha)$ for $\alpha \in [\alpha_-, \alpha_+]$ and introduce the notation $\xi^{(\alpha)} := \xi_{(\lambda(\alpha))}$,

$$S_n^{(\alpha)} := \sum_{i=1}^{n} \xi_i^{(\alpha)},$$

where $\xi_i^{(\alpha)}$ are independent copies of $\xi^{(\alpha)}$. The distribution $\mathbf{F}^{(\alpha)} := \mathbf{F}_{(\lambda(\alpha))}$ of the random variable $\xi^{(\alpha)}$ is called *the Cramér transform of* $\mathbf{F}$ *with parameter* α. The random variables $\xi^{(\alpha)}$ are also called Cramér transforms, but of the original random variable ξ. The relationship between the distributions of S_n and $S_n^{(\alpha)}$ is established in the following assertion.

Theorem 9.2.1 *For $x = n\alpha$, $\alpha \in (\alpha_-, \alpha_+)$, and any $t > 0$, one has*

$$\mathbf{P}\big(S_n \in [x, x+t)\big) = e^{-n\Lambda(\alpha)} \int_0^t e^{-\lambda(\alpha)z}\mathbf{P}\big(S_n^{(\alpha)} - \alpha n \in dz\big). \tag{9.2.1}$$

Proof The Laplace transform of the distribution of the sum $S_n^{(\alpha)}$ is clearly equal to

$$\mathbf{E}e^{\mu S_n^{(\alpha)}} = \left[\frac{\psi(\mu + \lambda(\alpha))}{\psi(\lambda(\alpha))}\right]^n \tag{9.2.2}$$

(see (9.1.5)). On the other hand, consider the Cramér transform $(S_n)_{(\lambda(\alpha))}$ of S_n at the point $\lambda(\alpha)$. Applying (9.1.5) to the distribution of S_n, we obtain

$$\mathbf{E}e^{\mu (S_n)_{(\lambda(\alpha))}} = \frac{\psi^n(\mu + \lambda(\alpha))}{\psi^n(\lambda(\alpha))}.$$

Since this expression coincides with (9.2.2), *the Cramér transform of S_n at the point $\lambda(\alpha)$ coincides in distribution with the sum $S_n^{(\alpha)}$ of the transforms $\xi_i^{(\alpha)}$*. In other words,

$$\frac{\mathbf{P}(S_n \in dv)e^{\lambda(\alpha)v}}{\psi^n(\lambda(\alpha))} = \mathbf{P}\big(S_n^{(\alpha)} \in dv\big) \tag{9.2.3}$$

or, which is the same,

$$\mathbf{P}(S_n \in dv) = e^{-\lambda(\alpha)v + n\ln\psi(\lambda(\alpha))}\mathbf{P}\bigl(S_n^{(\alpha)} \in dv\bigr) = e^{-n\Lambda(\alpha)+\lambda(\alpha)(n\alpha - v)}\mathbf{P}\bigl(S_n^{(\alpha)} \in dv\bigr).$$

Integrating this equality in v from x to $x+t$, letting $x := n\alpha$ and making the change of variables $v - n\alpha = z$, we get

$$\begin{aligned}\mathbf{P}\bigl(S_n \in [x, x+t)\bigr) &= e^{-n\Lambda(\alpha)} \int_x^{x+t} e^{\lambda(\alpha)(n\alpha - v)}\mathbf{P}\bigl(S_n^{(\alpha)} \in dv\bigr)\\ &= e^{-n\Lambda(\alpha)} \int_0^t e^{-\lambda(\alpha)z}\mathbf{P}\bigl(S_n^{(\alpha)} - \alpha n \in dz\bigr).\end{aligned}$$

The theorem is proved. □

Since for $\alpha \in [\alpha_-, \alpha_+]$ we have

$$\mathbf{E}\xi^{(\alpha)} = \frac{\psi'(\lambda(\alpha))}{\psi(\lambda(\alpha))} = \alpha$$

(see (9.1.11)), one has $\mathbf{E}(S_n^{(\alpha)} - \alpha n) = 0$ and so for $t \le c\sqrt{n}$ we have probabilities of *normal* deviations of $S_n^{(\alpha)} - \alpha n$ on the right-hand side of (9.2.1). This allows us to *reduce the problem on large deviations of S_n to the problem on normal deviations of $S_n^{(\alpha)}$*. If $\alpha > \alpha_+$, then formula (9.2.1) is still rather useful, as will be shown in Sects. 9.4 and 9.5.

9.2.2 The Probabilistic Meaning of the Rate Function

In this section we will prove the following assertion, which clarifies the probabilistic meaning of the function $\Lambda(\alpha)$.

Denote by $\Delta[\alpha) := [\alpha, \alpha + \Delta)$ the interval of length Δ with the left end at the point α. The notation $\Delta_n[\alpha)$, where Δ_n depends on n, will have a similar meaning.

Theorem 9.2.2 *For each fixed α and all sequences Δ_n converging to 0 as $n \to \infty$ slowly enough, one has*

$$\Lambda(\alpha) = -\lim_{n\to\infty} \frac{1}{n} \ln \mathbf{P}\biggl(\frac{S_n}{n} \in \Delta_n[\alpha)\biggr). \tag{9.2.4}$$

This relation can also be written as

$$\mathbf{P}\biggl(\frac{S_n}{n} \in \Delta_n[\alpha)\biggr) = e^{-n\Lambda(\alpha)+o(n)}.$$

Proof of Theorem 9.2.2 First let $\alpha \in (\alpha_-, \alpha_+)$. Then

$$\mathbf{E}\xi^{(\alpha)} = \alpha, \qquad \operatorname{Var}\xi^{(\alpha)} = \big(\ln \psi(\lambda)\big)''_{\lambda=\lambda(\alpha)} < \infty$$

and hence, as $n \to \infty$ and $\Delta_n \to 0$ slowly enough (e.g., for $\Delta_n \ge n^{-1/3}$), by the central limit theorem we have

$$\mathbf{P}\big(S_n^{(\alpha)} - \alpha n \in [0, \Delta_n n)\big) \to 1/2.$$

Therefore, by Theorem 9.2.1 for $t = \Delta_n n$, $x = \alpha n$ and by the mean value theorem,

$$\mathbf{P}\big(S_n \in [x, x+t)\big) = \left(\frac{1}{2} + o(1)\right) e^{-n\Lambda(\alpha) - \lambda(\alpha)\Delta_n n\theta}, \quad \theta \in (0,1);$$

$$\frac{1}{n} \ln \mathbf{P}\big(S_n \in [x, x+t)\big) = -\Lambda(\alpha) - \lambda(\alpha)\theta\Delta_n + o(1) = -\Lambda(\alpha) + o(1)$$

as $n \to \infty$. This proves (9.2.4) for $\alpha \in (\alpha_-, \alpha_+)$.

The further proof is divided into three stages.

(1) *The upper bound in the general case.* Now let α be arbitrary and $|\lambda(\alpha)| < \infty$. By Theorem 9.2.1 for $t = n\Delta_n$, we have

$$\mathbf{P}\left(\frac{S_n}{n} \in \Delta_n[\alpha)\right) \le \exp\big\{-n\Lambda(\alpha) + \max\big(|\lambda(0)|, |\lambda(\alpha)|\big) n\Delta_n\big\}.$$

If $\Delta_n \to 0$ then

$$\limsup_{n\to\infty} \frac{1}{n} \ln \mathbf{P}\left(\frac{S_n}{n} \in \Delta_n[\alpha)\right) \le -\Lambda(\alpha). \tag{9.2.5}$$

(This inequality can also be obtained from the exponential Chebyshev's inequality (Λ7).)

(2) *The lower bound in the general case.* Let $|\lambda(\alpha)| < \infty$ and $|s_\pm| = \infty$. Introduce "truncated" random variables ${}^{(N)}\xi$ with the distribution

$$\mathbf{P}\big({}^{(N)}\xi \in B\big) = \frac{\mathbf{P}(\xi \in B;\ |\xi| < N)}{\mathbf{P}(|\xi| < N)} = \mathbf{P}\big(\xi \in B \mid |\xi| < N\big)$$

and endow all the symbols that correspond to ${}^{(N)}\xi$ with the left superscript (N). Then clearly, for each λ,

$$\mathbf{E}\big(e^{\lambda\xi};\ |\xi| < N\big) \uparrow \psi(\lambda), \qquad \mathbf{P}\big(|\xi| < N\big) \uparrow 1$$

as $N \to \infty$, so that

$${}^{(N)}\psi(\lambda) = \frac{\mathbf{E}(e^{\lambda\xi};\ |\xi| < N)}{\mathbf{P}(|\xi| < N)} \to \psi(\lambda).$$

The functions ${}^{(N)}\Lambda(\alpha)$ and $\Lambda(\alpha)$ are the upper bounds for the concave functions $\alpha\lambda - \ln {}^{(N)}\psi(\lambda)$ and $\alpha\lambda - \ln \psi(\lambda)$, respectively. Therefore for each α we also have convergence ${}^{(N)}\Lambda(\alpha) \to \Lambda(\alpha)$ as $N \to \infty$.

Further,

$$\mathbf{P}\biggl(\frac{S_n}{n} \in \Delta_n[\alpha)\biggr) \ge \mathbf{P}\biggl(\frac{S_n}{n} \in \Delta_n[\alpha); |\xi_j| < N, j = 1, \dots, N\biggr)$$
$$= \mathbf{P}^n\bigl(|\xi| < N\bigr)\mathbf{P}\biggl(\frac{{}^{(N)}S_n}{n} \in \Delta_n[\alpha)\biggr).$$

Since $s_\pm = \pm\infty$, one has ${}^{(N)}\alpha_\pm = \pm N$ and, for N large enough, we have $\alpha \in ({}^{(N)}\alpha_-, {}^{(N)}\alpha_+)$. Hence we can apply the first part of the proof of the theorem by virtue of which, as $\Delta_n \to 0$,

$$\frac{1}{n} \ln \mathbf{P}\biggl(\frac{{}^{(N)}S_n}{n} \in \Delta_n[\alpha)\biggr) = -{}^{(N)}\Lambda(\alpha) + o(1),$$
$$\frac{1}{n} \ln \mathbf{P}\biggl(\frac{S_n}{n} \in \Delta_n[\alpha)\biggr) \ge -{}^{(N)}\Lambda(\alpha) + o(1) + \ln \mathbf{P}\bigl(|\xi| < N\bigr).$$

The right-hand side of the last inequality can be made arbitrarily close to $-\Lambda(\alpha)$ by choosing a suitable N. Since the left-hand side of this inequality does not depend on N, we have

$$\liminf_{n\to\infty} \frac{1}{n} \ln \mathbf{P}\biggl(\frac{S_n}{n} \in \Delta_n[\alpha)\biggr) \ge -\Lambda(\alpha). \tag{9.2.6}$$

Together with (9.2.5), this proves (9.2.4).

(3) It remains to remove the restrictions stated at the beginning of stages (1) and (2) of the proof, i.e. to consider the cases $|\lambda(\alpha)| = \infty$ and $\min|s_\pm| < \infty$. These two relations are connected with each other since, for instance, the equality $\lambda(\alpha) = \lambda_+ = \infty$ can only hold if $\alpha \ge \alpha_+ = s_+ < \infty$ (see property $(\Lambda 2)$). For $\alpha > s_+$, relation (9.2.4) is evident, since $\mathbf{P}(S_n/n \in \Delta_n[\alpha)) = 0$ and $\Lambda(\alpha) = \infty$. For $\alpha = \alpha_+ = s_+$ and $p_+ = \mathbf{P}(\xi = s_+)$, we have, for any $\Delta > 0$,

$$\mathbf{P}\biggl(\frac{S_n}{n} \in \Delta[\alpha_+)\biggr) = \mathbf{P}(S_n = n\alpha_+) = p_+^n. \tag{9.2.7}$$

Since in this case $\Lambda(\alpha_+) = -\ln p_+$ (see $(\Lambda 2)$), the equality (9.2.4) holds true.

The case $\lambda(\alpha) = \lambda_- = -\infty$ with $s_- > -\infty$ is considered in a similar way. However, due to the asymmetry of the interval $\Delta[\alpha)$ with respect to the point α, there are small differences. Instead of an equality in (9.2.7) we only have the inequality

$$\mathbf{P}\biggl(\frac{S_n}{n} \in \Delta_n[\alpha_-)\biggr) \ge \mathbf{P}(S_n = n\alpha_-) = p_-^n, \qquad p_- = \mathbf{P}(\xi = \alpha_-). \tag{9.2.8}$$

Therefore we also have to use the exponential Chebyshev's inequality (see (Λ7)) applying it to $-S_n$ for $s_- = \alpha_- < 0$:

$$\mathbf{P}\left(\frac{S_n}{n} \in \Delta_n[\alpha_-)\right) \le \mathbf{P}\left(\frac{S_n}{n} < \alpha_- + \Delta_n\right) \le e^{-n\Lambda(\alpha_- + \Delta_n)}. \tag{9.2.9}$$

Relations (9.2.8), (9.2.9), the equality $\Lambda(\alpha_-) = -\ln p_-$, and the right continuity of $\Lambda(\alpha)$ at the point α_- imply (9.2.4) for $\alpha = \alpha_-$. The theorem is proved. □

9.2.3 The Large Deviations Principle

It is not hard to derive from Theorem 9.2.2 a corollary on the asymptotics of the probabilities of S_n/n hitting an arbitrary Borel set. Denote by (B) and $[B]$ the interior and the closure of B, respectively ((B) is the union of all open intervals contained in B). Put

$$\Lambda(B) := \inf_{\alpha \in B} \Lambda(\alpha).$$

Theorem 9.2.3 *For any Borel set B, the following inequalities hold*:

$$\liminf_{n\to\infty} \frac{1}{n} \ln \mathbf{P}\left(\frac{S_n}{n} \in B\right) \ge -\Lambda\big((B)\big), \tag{9.2.10}$$

$$\limsup_{n\to\infty} \frac{1}{n} \ln \mathbf{P}\left(\frac{S_n}{n} \in B\right) \le -\Lambda\big([B]\big). \tag{9.2.11}$$

If $\Lambda((B)) = \Lambda([B])$, then the following limit exists:

$$\lim_{n\to\infty} \frac{1}{n} \ln \mathbf{P}\left(\frac{S_n}{n} \in B\right) = -\Lambda(B). \tag{9.2.12}$$

This assertion is called the *large deviation principle*. It is one of the so-called "rough" ("logarithmic") limit theorems that describe the asymptotic behaviour of $\ln \mathbf{P}(S_n/n \in B)$. It is usually impossible to derive from this assertion the asymptotics of the probability $\mathbf{P}(S_n/n \in B)$ itself. (In the equality $\mathbf{P}(S_n/n \in B) = \exp\{-n\Lambda(B) + o(n)\}$, the term $o(n)$ may grow in absolute value.)

Proof Without losing generality, we can assume that $B \subset [s_-, s_+]$ (since $\Lambda(\alpha) = \infty$ outside that domain).

We first prove (9.2.10). Let $\alpha_{(B)}$ be such that

$$\Lambda\big((B)\big) \equiv \inf_{\alpha \in (B)} \Lambda(\alpha) = \Lambda(\alpha_{(B)})$$

(recall that $\Lambda(\alpha)$ is continuous on $[s_-, s_+]$). Then there exist a sequence of points α_k and a sequence of intervals $(\alpha_k - \delta_k, \alpha_k + \delta_k)$, where $\delta_k \to 0$, lying in (B) and converging to the point $\alpha_{(B)}$, such that

$$\Lambda\big((B)\big) = \inf_k \Lambda\big((\alpha_k - \delta_k, \alpha_k + \delta_k)\big).$$

Here clearly

$$\inf_k \Lambda\big((\alpha_k - \delta_k, \alpha_k + \delta_k)\big) = \inf_k \Lambda(\alpha_k),$$

and for a given $\varepsilon > 0$, there exists a $k = K$ such that $\Lambda(\alpha_K) < \Lambda((B)) + \varepsilon$. Since $\Delta_n[\alpha_k) \subset (\alpha_k - \delta_k, \alpha_k + \delta_k)$ for large enough n (here $\Delta_n[\alpha_k)$ is from Theorem 9.2.2), we have by Theorem 9.2.2 that, as $n \to \infty$,

$$\begin{aligned}
\frac{1}{n} \ln \mathbf{P}\bigg(\frac{S_n}{n} \in B\bigg) &\geq \frac{1}{n} \ln \mathbf{P}\bigg(\frac{S_n}{n} \in (B)\bigg) \\
&\geq \frac{1}{n} \ln \mathbf{P}\bigg(\frac{S_n}{n} \in (\alpha_K - \delta_K, \alpha_K + \delta_K)\bigg) \\
&\geq \frac{1}{n} \ln \mathbf{P}\bigg(\frac{S_n}{n} \in \Delta_n[\alpha_K)\bigg) \geq -\Lambda(\alpha_K) + o(1) \\
&\geq -\Lambda\big((B)\big) - \varepsilon + o(1).
\end{aligned}$$

As the left-hand side of this inequality does not depend on ε, inequality (9.2.10) is proved.

We now prove inequality (9.2.11). Denote by $\alpha_{[B]}$ the point at which $\inf_{\alpha \in [B]} \Lambda(\alpha) = \Lambda(\alpha_{[B]})$ is attained (this point always belongs to $[B]$ since $[B]$ is closed). If $\Lambda(\alpha_{[B]}) = 0$, then the inequality is evident. Now let $\Lambda(\alpha_{[B]}) > 0$. By convexity of Λ the equation $\Lambda(\alpha) = \Lambda(\alpha_{[B]})$ can have a second solution $\alpha'_{[B]}$. Assume it exists and, for definiteness, $\alpha'_{[B]} < \alpha_{[B]}$. The relation $\Lambda([B]) = \Lambda(\alpha_{[B]})$ means that the set $[B]$ does not intersect with $(\alpha'_{[B]}, \alpha_{[B]})$ and

$$\mathbf{P}\bigg(\frac{S_n}{n} \in B\bigg) \leq \mathbf{P}\bigg(\frac{S_n}{n} \in [B]\bigg) \leq \mathbf{P}\bigg(\frac{S_n}{n} \leq \alpha'_{[B]}\bigg) + \mathbf{P}\bigg(\frac{S_n}{n} \geq \alpha_{[B]}\bigg). \tag{9.2.13}$$

Moreover, in this case $m_1 \in (\alpha'_{[B]}, \alpha_{[B]})$ and each of the probabilities on the right-hand side of (9.2.13) can be bounded using the exponential Chebyshev's inequality (see (Λ7)) by the value $e^{-n\Lambda(\alpha_{[B]})}$. This implies (9.2.11).

If the second solution $\alpha'_{[B]}$ does not exist, then one of the summands on the right-hand side of (9.2.13) equals zero, and we obtain the same result.

The second assertion of the theorem (Eq. (9.2.12)) is evident.

The theorem is proved. □

Using Theorem 9.2.3, we can complement Theorem 9.2.2 with the following assertion.

Corollary 9.2.1 *The following limit always exists*

$$\lim_{\Delta\to 0}\lim_{n\to\infty}\frac{1}{n}\ln \mathbf{P}\left(\frac{S_n}{n}\in\Delta[\alpha)\right)=-\Lambda(\alpha). \qquad (9.2.14)$$

Proof Take the set B in Theorem 9.2.3 to be the interval $B=\Delta[\alpha)$. If $\alpha\notin[s_-,s_+]$ then the assertion is obvious (since both sides of (9.2.14) are equal to $-\infty$). If $\alpha=s_\pm$ then (9.2.14) is already proved in (9.2.7), (9.2.8) and (9.2.9).

It remains to consider points $\alpha\in(s_-,s_+)$. For such α, the function $\Lambda(\alpha)$ is continuous and $\alpha+\Delta$ is also a point of continuity of Λ for Δ small enough, and hence

$$\Lambda\big((B)\big)=\Lambda\big([B]\big)\to\Lambda(\alpha)$$

as $\Delta\to 0$. Therefore by Theorem 9.2.3 the inner limit in (9.2.14) exists and converges to $-\Lambda(\alpha)$ as $\Delta\to 0$.

The corollary is proved. □

Note that the assertions of Theorems 9.2.2 and 9.2.3 and their corollaries are "universal"—they contain no restrictions on the distribution $\mathbf{F}$.

9.3 Integro-Local, Integral and Local Theorems on Large Deviation Probabilities in the Cramér Range

9.3.1 Integro-Local and Integral Theorems

In this subsection, under the assumption that the Cramér condition $\lambda_+>0$ is met, we will find the asymptotics of probabilities $\mathbf{P}(S_n\in\Delta[x))$ for scaled deviations $\alpha=x/n$ from the so-called *Cramér* (or *regular*) *range*, i.e. for the range $\alpha\in(\alpha_-,\alpha_+)$ in which the rate function $\Lambda(\alpha)$ is analytic.

In the non-lattice case, in addition to the condition $\lambda_+>0$, we will assume without loss of generality that $\mathbf{E}\xi=0$. In this case necessarily

$$\alpha_-\le 0,\qquad \alpha_+=\frac{\psi'(\lambda_+)}{\psi(\lambda_+)}>0,\qquad \lambda(0)=0.$$

The length Δ of the interval may depend on n in some cases. In such cases, we will write Δ_n instead of Δ, as we did earlier. The value

$$\sigma_\alpha^2=\frac{\psi''(\lambda(\alpha))}{\psi(\lambda(\alpha))}-\alpha^2 \qquad (9.3.1)$$

is clearly equal to $\mathrm{Var}(\xi^{(\alpha)})$ (see (9.1.5) and the definition of $\xi^{(\alpha)}$ in Sect. 9.2).

Theorem 9.3.1 *Let $\lambda_+ > 0$, $\alpha \in [0, \alpha_+)$, ξ be a non-lattice random variable, $\mathbf{E}\xi = 0$ and $\mathbf{E}\xi^2 < \infty$. If $\Delta_n \to 0$ slowly enough as $n \to \infty$, then*

$$\mathbf{P}\big(S_n \in \Delta_n[x)\big) = \frac{\Delta_n}{\sigma_\alpha \sqrt{2\pi n}} e^{-n\Lambda(\alpha)}\big(1 + o(1)\big), \tag{9.3.2}$$

where $\alpha = x/n$, and, for each fixed $\alpha_1 \in (0, \alpha_+)$, the remainder term $o(1)$ is uniform in $\alpha \in [0, \alpha_1]$ for any fixed $\alpha_1 \in (0, \alpha_+)$.

A similar assertion is valid in the case when $\lambda_- < 0$ and $\alpha \in (\alpha_-, 0]$.

Proof The proof is based on Theorems 9.2.1 and 8.7.1A. Since the conditions of Theorem 9.2.1 are satisfied, we have

$$\mathbf{P}\big(S_n \in \Delta_n[x)\big) = e^{-n\Lambda(\alpha)} \int_0^{\Delta_n} e^{-\lambda(\alpha)z} \mathbf{P}\big(S_n^{(\alpha)} - \alpha n \in dz\big).$$

As $\lambda(\alpha) \le \lambda(\alpha_+ - \varepsilon) < \infty$ and $\Delta_n \to 0$, one has $e^{-\lambda(\alpha)z} \to 1$ uniformly in $z \in \Delta_n[0)$ and hence, as $n \to \infty$,

$$\mathbf{P}\big(S_n \in \Delta_n[x)\big) = e^{-n\Lambda(\alpha)} \mathbf{P}\big(S_n^{(\alpha)} - \alpha n \in \Delta_n[0)\big)\big(1 + o(1)\big) \tag{9.3.3}$$

uniformly in $\alpha \in [0, \alpha_+ - \varepsilon]$.

We now show that Theorem 8.7.1A is applicable to the random variables $\xi^{(\alpha)} = \xi_{(\lambda(\alpha))}$. That $\sigma_\alpha = \sigma(\lambda(\alpha))$ is bounded away from 0 and from ∞ for $\alpha \in [0, \alpha_1]$ is evident. (The same is true of all the theorems in this section.) Therefore, it remains to verify whether conditions (a) and (b) of Theorem 8.7.1A are met for $\lambda = \lambda(\alpha) \in [0, \lambda_1]$, $\lambda_1 := \lambda(\alpha_1) < \lambda_+$ and $\varphi_{(\lambda)}(t) = \frac{\psi(\lambda+it)}{\psi(\lambda)}$ (see (9.1.5)). We have

$$\psi(\lambda + it) = \psi(\lambda) + it\psi'(\lambda) - \frac{t^2}{2}\psi''(\lambda) + o\big(t^2\big)$$

as $t \to 0$, where the remainder term is uniform in λ if the function $\psi''(\lambda + iu)$ is uniformly continuous in u. The required uniform continuity can easily be proved by imitating the corresponding result for ch.f.s (see property 4 in Sect. 7.1). This proves condition (a) in Theorem 8.7.1A with

$$a(\lambda) = \frac{\psi'(\lambda)}{\psi(\lambda)}, \qquad m_2(\lambda) = \frac{\psi''(\lambda)}{\psi(\lambda)}.$$

Now we will verify condition (b) in Theorem 8.7.1A. Assume the contrary: there exists a sequence $\lambda_k \in [0, \lambda_1]$ such that

$$q_{\lambda_k} := \sup_{\theta_1 \le |t| \le \theta_2} \frac{|\psi(\lambda_k + it)|}{\psi(\lambda_k)} \to 1$$

as $k \to \infty$. By the uniform continuity of ψ in that domain, there exist points $t_k \in [\theta_1, \theta_2]$ such that, as $k \to \infty$,

$$\frac{\psi(\lambda_k + it_k)}{\psi(\lambda_k)} \to 1.$$

Since the region $\lambda \in [0, \lambda_1]$, $|t| \in [\theta_1, \theta_2]$ is compact, there exists a subsequence $(\lambda_{k'}, t_{k'}) \to (\lambda_0, t_0)$ as $k' \to \infty$. Again using the continuity of ψ, we obtain the equality

$$\frac{|\psi(\lambda_0 + it_0)|}{\psi(\lambda_0)} = 1, \tag{9.3.4}$$

which contradicts the non-latticeness of $\xi_{(\lambda_0)}$. Property (b) is proved.

Thus we can now apply Theorem 8.7.1A to the probability on the right-hand side of (9.3.3). Since $\mathbf{E}\xi^{(\alpha)} = \alpha$ and $\mathbf{E}(\xi^{(\alpha)})^2 = \frac{\psi''(\lambda(\alpha))}{\psi(\lambda(\alpha))}$, this yields

$$\begin{aligned} \mathbf{P}\big(S_n \in \Delta_n[x)\big) &= e^{-n\Lambda(\alpha)}\left(\frac{\Delta_n}{\sigma_\alpha\sqrt{n}}\phi(0) + o\left(\frac{1}{\sqrt{n}}\right)\right) \\ &= \frac{\Delta_n}{\sigma_\alpha\sqrt{2\pi n}} e^{-n\Lambda(\alpha)}\big(1 + o(1)\big) \end{aligned} \tag{9.3.5}$$

uniformly in $\alpha \in [0, \alpha_1]$ (or in $x \in [0, \alpha_1 n]$), where the values of

$$\sigma_\alpha^2 = \mathbf{E}\big(\xi^{(\alpha)} - \alpha\big)^2 = \frac{\psi''(\lambda(\alpha))}{\psi(\lambda(\alpha))} - \alpha^2$$

are bounded away from 0 and from ∞. The theorem is proved. □

From Theorem 9.3.1 we can now derive integro-local theorems and integral theorems for fixed or growing Δ. Since in the normal deviation range (when x is comparable with $\sqrt{n}$) we have already obtained such results, to simplify the exposition we will consider here large deviations only, when $x \gg \sqrt{n}$ or, which is the same, $\alpha = x/n \gg 1/\sqrt{n}$. To be more precise, we will assume that there exists a function $N(n) \to \infty$, $N(n) = o(\sqrt{n})$ as $n \to \infty$, such that $x \ge N(n)\sqrt{n}$ ($\alpha \ge N(n)/\sqrt{n}$).

Theorem 9.3.2 *Let $\lambda_+ > 0$, $\alpha \in [0, \alpha_+)$, ξ be non-lattice, $\mathbf{E}\xi = 0$ and $\mathbf{E}\xi^2 < \infty$. Then, for any $\Delta \ge \Delta_0 > 0$, $x \ge N(n) = o(\sqrt{n})$, $N(n) \to \infty$ as $n \to \infty$, one has*

$$\mathbf{P}\big(S_n \in \Delta[x)\big) = \frac{e^{-n\Lambda(\alpha)}}{\sigma_\alpha \lambda(\alpha)\sqrt{2\pi n}}\big(1 - e^{-\lambda(\alpha)\Delta}\big)\big(1 + o(1)\big), \tag{9.3.6}$$

$o(1)$ being uniform in $\alpha = x/n \in [N(n)/\sqrt{n}, \alpha_1]$ and $\Delta \ge \Delta_0$ for each fixed $\alpha_1 \in (0, \alpha_+)$.

In particular (for $\Delta = \infty$),

$$\mathbf{P}(S_n \ge x) = \frac{e^{-n\Lambda(\alpha)}}{\sigma_\alpha \lambda(\alpha)\sqrt{2\pi n}}\big(1 + o(1)\big). \tag{9.3.7}$$

Proof Partition the interval $\Delta[x)$ into subintervals $\Delta_n[x + k\Delta_n)$, $k = 0, \dots, \Delta/\Delta_n - 1$, where $\Delta_n \to 0$ and, for simplicity, we assume that $M = \Delta/\Delta_n$ is an integer. Then, by Theorem 9.2.1, as $\Delta_n \to 0$,

$$\begin{aligned}
&\mathbf{P}\big(S_n \in \Delta_n[x + k\Delta_n)\big) \\
&\quad = \mathbf{P}\big(S_n \in \big[x, x + (k+1)\Delta_n\big)\big) - \mathbf{P}\big(S_n \in [x, x + k\Delta_n)\big) \\
&\quad = e^{-n\Lambda(\alpha)} \int_{k\Delta_n}^{(k+1)\Delta_n} e^{-\lambda(\alpha)z} \mathbf{P}\big(S_n^{(\alpha)} - \alpha n \in dz\big) \\
&\quad = e^{-n\Lambda(\alpha) - \lambda(\alpha)k\Delta_n} \mathbf{P}\big(S_n^{(\alpha)} - \alpha n \in \Delta_n[k\Delta_n)\big)\big(1 + o(1)\big)
\end{aligned} \tag{9.3.8}$$

uniformly in $\alpha \in [0, \alpha_1]$. Here, similarly to (9.3.5), by Theorem 8.7.1A we have

$$\mathbf{P}\big(S_n^{(\alpha)} - \alpha n \in \Delta_n[k\Delta_n)\big) = \frac{\Delta_n}{\sigma_\alpha\sqrt{n}} \phi\left(\frac{k\Delta_n}{\sigma_\alpha\sqrt{n}}\right) + o\left(\frac{1}{\sqrt{n}}\right) \tag{9.3.9}$$

uniformly in k and α. Since

$$\mathbf{P}\big(S_n \in \Delta[x)\big) = \sum_{k=0}^{M-1} \mathbf{P}\big(S_n \in \Delta_n[x + k\Delta)\big),$$

substituting the values (9.3.8) and (9.3.9) into the right-hand side of the last equality, we obtain

$$\begin{aligned}
\mathbf{P}\big(S_n \in \Delta[x)\big) &= \frac{e^{-n\Lambda(\alpha)}}{\sigma_\alpha\sqrt{n}} \sum_{k=0}^{M-1} \Delta_n e^{-\lambda(\alpha)k\Delta_n} \left(\phi\left(\frac{k\Delta_n}{\sigma_\alpha\sqrt{n}}\right) + o(1)\right) \\
&= \frac{e^{-n\Lambda(\alpha)}}{\sigma_\alpha\sqrt{n}} \int_0^{\Delta - \Delta_n} e^{-\lambda(\alpha)z} \left(\phi\left(\frac{z}{\sigma_\alpha\sqrt{n}}\right) + o(1)\right) dz.
\end{aligned} \tag{9.3.10}$$

After the variable change $\lambda(\alpha)z = u$, the right-hand side can be rewritten as

$$\frac{e^{-n\lambda(\alpha)}}{\sigma_\alpha\lambda(\alpha)\sqrt{n}} \int_0^{(\Delta - \Delta_n)\lambda(\alpha)} e^{-u} \left(\phi\left(\frac{u}{\sigma_\alpha\lambda(\alpha)\sqrt{n}}\right) + o(1)\right) du, \tag{9.3.11}$$

where the remainder term $o(1)$ is uniform in $\alpha \in [0, \alpha_1]$, $\Delta \ge \Delta_0$, and u from the integration range. Since $\lambda(\alpha) \sim \alpha/\sigma^2$ for small α (see (9.1.12) and (9.1.16)), for $\alpha \ge N(n)/\sqrt{n}$ we have

$$\lambda(\alpha) > \frac{N(n)}{\sigma^2\sqrt{n}}\big(1 + o(1)\big), \qquad \sigma_\alpha\lambda(\alpha)\sqrt{n} > \frac{\sigma_\alpha N(n)}{\sigma^2} \to \infty.$$

Therefore, for any fixed u, one has

$$\phi\left(\frac{u}{\sigma_\alpha\lambda(\alpha)\sqrt{n}}\right) \to \phi(0) = \frac{1}{\sqrt{2\pi}}.$$

Moreover, $\phi(v) \le 1/\sqrt{2\pi}$ for all v. Hence, by (9.3.10) and (9.3.11),

$$\mathbf{P}\big(S_n \in \Delta[x)\big) = \frac{e^{-n\Lambda(\alpha)}}{\sigma_\alpha \lambda(\alpha)\sqrt{2\pi n}} \int_0^{\lambda(\alpha)\Delta} e^{-u} du \big(1+o(1)\big)$$

$$= \frac{e^{-n\Lambda(\alpha)}}{\sigma_\alpha \lambda(\alpha)\sqrt{2\pi n}} \big(1 - e^{-\lambda(\alpha)\Delta}\big)\big(1+o(1)\big)$$

uniformly in $\alpha \in [0, \alpha_1]$ and $\Delta \ge \Delta_0$. Relation (9.3.7) clearly follows from (9.3.6) with $\Delta = \infty$. The theorem is proved. □

Note that if $\mathbf{E}|\xi|^k < \infty$ (for $\lambda_+ > 0$ this is a restriction on the rate of decay of the left tails $\mathbf{P}(\xi < -t)$, $t > 0$), then expansion (9.1.17) is valid and, for deviations $x = o(n)$ ($\alpha = o(1)$) such that $n\alpha^k = x^k/n^{k-1} \le c = \text{const}$, we can change the exponent $n\Lambda(\alpha)$ in (9.3.6) and (9.3.7) to

$$n\Lambda(\alpha) = n \sum_{j=2}^{k} \frac{\Lambda^{(j)}(0)}{j!} \alpha^j + o\big(n\alpha^k\big), \tag{9.3.12}$$

where $\Lambda^{(j)}(0)$ are found in (9.1.16). For $k = 3$, the foregoing implies the following.

Corollary 9.3.1 *Let $\lambda_+ > 0$, $\mathbf{E}|\xi|^3 < \infty$, ξ be non-lattice, $\mathbf{E}\xi = 0$, $\mathbf{E}\xi^2 = \sigma^2$, $x \gg \sqrt{n}$ and $x = o(n^{2/3})$ as $n \to \infty$. Then*

$$\mathbf{P}(S_n \ge x) \sim \frac{\sigma\sqrt{n}}{x\sqrt{2\pi}} \exp\left\{-\frac{x^2}{2n\sigma^2}\right\} \sim \Phi\left(-\frac{x}{\sigma\sqrt{n}}\right). \tag{9.3.13}$$

In the last relation we used the symmetry of the standard normal law, i.e. the equality $1 - \Phi(t) = \Phi(-t)$. Assertion (9.3.13) shows that in the case $\lambda_+ > 0$ and $\mathbf{E}|\xi|^3 < \infty$ the asymptotic equivalence

$$\mathbf{P}(S_n \ge x) \sim \Phi\left(-\frac{x}{\sigma\sqrt{n}}\right)$$

persists outside the range of normal deviations as well, up to the values $x = o(n^{2/3})$. If $\mathbf{E}\xi^3 = 0$ and $\mathbf{E}\xi^4 < \infty$, then this equivalence holds true up to the values $x = o(n^{3/4})$. For larger x this equivalence, generally speaking, no longer holds.

Proof of Corollary 9.3.1 The first relation in (9.3.13) follows from Theorem 9.3.2 and (9.3.12). The second follows from the asymptotic equivalence

$$\int_x^\infty e^{-\frac{u^2}{2}} du \sim \frac{e^{-x^2/2}}{x},$$

which is easy to establish, using, for example, l'Hospital's rule. □

9.3.2 Local Theorems

In this subsection we will obtain analogues of the local Theorems 8.7.2 and 8.7.3 for large deviations in the Cramér range. To simplify the exposition, we will formulate the theorem for densities, assuming that the following condition is satisfied:

[D] *The distribution* **F** *has a bounded density* $f(x)$ *such that*

$$f(x) = e^{-\lambda_+ x + o(x)} \quad \text{as } x \to \infty, \text{ if } \lambda_+ < \infty; \tag{9.3.14}$$

$$f(x) \le ce^{-\lambda x} \quad \text{for any fixed } \lambda > 0,\ c = c(\lambda),\ \text{if } \lambda_+ = \infty. \tag{9.3.15}$$

Since inequalities of the form (9.3.14) and (9.3.15) always hold, by the exponential Chebyshev inequality, for the right tails

$$F_+(x) = \int_x^\infty f(u)\,du,$$

condition **[D]** is not too restrictive. It only eliminates sharp "bursts" of $f(x)$ as $x \to \infty$.

Denote by $f_n(x)$ the density of the distribution of S_n.

Theorem 9.3.3 *Let*

$$\mathbf{E}\xi = 0, \qquad \mathbf{E}\xi^2 < \infty, \qquad \lambda_+ > 0, \qquad \alpha = \frac{x}{n} \in [0, \alpha_+),$$

and condition **[D]** *be met. Then*

$$f_n(x) = \frac{e^{-n\Lambda(\alpha)}}{\sigma_\alpha \sqrt{2\pi n}} \bigl(1 + o(1)\bigr),$$

where the remainder term $o(1)$ *is uniform in* $\alpha \in [0, \alpha_1]$ *for any fixed* $\alpha_1 \in (0, \alpha_+)$.

Proof The proof is based on Theorems 9.2.1 and 8.7.2A. Denote by $f_n^{(\alpha)}(x)$ the density of the distribution of $S_n^{(\alpha)}$. Relation (9.2.3) implies that, for $x = \alpha n$, $\alpha \in [\alpha_-, \alpha_+]$, we have

$$f_n(x) = e^{-\lambda(\alpha)x} \psi^n\bigl(\lambda(\alpha)\bigr) f_n^{(\alpha)}(x) = e^{-n\Lambda(\alpha)} f_n^{(\alpha)}(x). \tag{9.3.16}$$

Since $\mathbf{E}\xi^{(\alpha)} = \alpha$, we see that $\mathbf{E}(S_n^{(\alpha)} - x) = 0$ and the density value $f_n^{(\alpha)}(x)$ coincides with the density of the distribution of the sum $S_n^{(\alpha)} - \alpha n$ at the point 0. In order to use Theorems 8.7.1A and 8.7.2A, we have to verify conditions (a) and (b) for $\theta_2 = \infty$ in these theorems and also the uniform boundedness in $\alpha \in [0, \alpha_1]$ of

$$\int \bigl|\varphi_{(\lambda(\alpha))}(t)\bigr|^m dt \tag{9.3.17}$$

for some integer $m \geq 1$, where $\varphi_{(\lambda(\alpha))}$ is the ch.f. of $\xi^{(\alpha)}$ (the uniform version of condition (c) in Theorem 8.7.2). By condition **[D]** the density

$$f^{(\alpha)}(v) = \frac{e^{\lambda(\alpha)v} f(v)}{\psi(\lambda(\alpha))}$$

in bounded uniformly in $\alpha \in [0, \alpha_1]$ (for such α one has $\lambda(\alpha) \in [0, \lambda_1]$, $\lambda_1 = \lambda(\alpha_1) < \lambda_+$). Hence the integral

$$\int \bigl(f^{(\alpha)}(v)\bigr)^2 dv$$

is also uniformly bounded, and so, by virtue of Parseval's identity (see Sect. 7.2), is the integral

$$\int \bigl|\varphi_{(\lambda(\alpha))}(t)\bigr|^2 dt.$$

This means that the required uniform boundedness of integral (9.3.17) is proved for $m = 2$.

Conditions (a) and (b) for $\theta_2 < \infty$ were verified in the proof of Theorem 9.3.1. It remains to extend the verification of condition (b) to the case $\theta_2 = \infty$. This can be done by following an argument very similar to the one used in the proof of Theorem 9.3.1 in the case of finite θ_2. Let $\theta_2 = \infty$. If we assume that there exist sequences $\lambda_k \in [0, \lambda_{+,\varepsilon}]$ and $|t_k| \geq \theta_1$ such that

$$\frac{|\psi(\lambda_k + it_k)|}{\psi(\lambda_k)} \to 1,$$

then, by compactness of $[0, \lambda_{+,\varepsilon}]$, there will exist sequences $\lambda'_k \to \lambda_0 \in [0, \lambda_{+,\varepsilon}]$ and t'_k such that

$$\frac{|\psi(\lambda'_k + it'_k)|}{\psi(\lambda_0)} \to 1. \tag{9.3.18}$$

But by virtue of condition **[D]** the family of functions $\psi(\lambda + it)$, $t \in \mathbb{R}$, is equicontinuous in $\lambda \in [0, \lambda_{+,\varepsilon}]$. Therefore, along with (9.3.18), we also have convergence

$$\frac{|\psi(\lambda_0 + it'_k)|}{\psi(\lambda_0)} \to 1, \qquad |t_k| \geq \theta_1 > 0,$$

which contradicts the inequality

$$\sup_{|t| \geq \theta_1} \frac{|\psi(\lambda_0 + it)|}{\psi(\lambda_0)} < 1$$

that follows from the existence of density.

Thus property (b) is proved for $\theta_2 = \infty$, and we can use Theorem 8.7.2A, which implies that

$$f_n^{(\alpha)}(x) = \frac{1}{\sigma(\lambda(\alpha))\sqrt{2\pi n}}\big(1 + o(1)\big).$$

This, together with (9.3.16), proves Theorem 9.3.3. □

Remark 9.3.1 We can see from the proof that, in Theorem 9.3.3, as a more general condition instead of condition **[D]** one could also consider the integrability of $\psi^m(\lambda + it)$ for any fixed $\lambda \in [0, \lambda_1]$, $\lambda_1 < \lambda_+$, or condition **[D]** imposed on S_m for some $m \geq 1$.

For arithmetic distributions we cannot assume without loss of generality that $m_1 = \mathbf{E}\xi = 0$, but that does not change much in the formulations of the assertions. If $\lambda_+ > 0$, then $\alpha_+ = \psi'(\lambda_+)/\psi(\lambda_+) > m_1$ and the scaled deviations $\alpha = x/n$ for the Cramér range must lie in the region $[m_1, \alpha_+)$.

Theorem 9.3.4 *Let $\lambda_+ > 0$, $\mathbf{E}\xi^2 < \infty$ and the distribution of ξ be arithmetic. Then, for integer x,*

$$\mathbf{P}(S_n = x) = \frac{e^{-n\Lambda(\alpha)}}{\sigma_\alpha\sqrt{2\pi n}}\big(1 + o(1)\big),$$

where the remainder term $o(1)$ is uniform in $\alpha = x/n \in [m_1, \alpha_1]$ for any fixed $\alpha_1 \in (m_1, \alpha_+)$.

A similar assertion is valid in the case when $\lambda_- < 0$ and $\alpha \in (\alpha_-, m_1]$.

Proof The proof does not differ much from that of Theorem 9.3.1. By (9.2.3),

$$\mathbf{P}(S_n = x) = e^{-\lambda(\alpha)x}\psi^{-n}\big(\lambda(\alpha)\big)\mathbf{P}\big(S_n^{(\alpha)} = x\big) = e^{-n\Lambda(\alpha)}\mathbf{P}\big(S_n^{(\alpha)} = x\big),$$

where $\mathbf{E}\xi^{(\alpha)} = \alpha$ for $\alpha \in [m_1, \alpha_+)$. In order to compute $\mathbf{P}(S_n^{(\alpha)} = x)$ we have to use Theorem 8.7.3A. The verification of conditions (a) and (b) of Theorem 8.7.1A, which are assumed to hold in Theorem 8.7.3A, is done in the same way as in the proof of Theorem 9.3.1, the only difference being that relation (9.3.4) for $t_0 \in [\theta_1, \pi]$ will contradict the arithmeticity of the distribution of ξ. Since $a(\lambda(\alpha)) = \mathbf{E}\xi^{(\alpha)} = \alpha$, by Theorem 8.7.3A we have

$$\mathbf{P}\big(S_n^{(\alpha)} = x\big) = \frac{1}{\sigma_\alpha\sqrt{2\pi n}}\big(1 + o(1)\big)$$

uniformly in $\alpha = x/n \in [m_1, \alpha_1]$. The theorem is proved. □

9.4 Integro-Local Theorems at the Boundary of the Cramér Range

9.4.1 Introduction

In this section we again assume that Cramér's condition $\lambda_+ > 0$ is met. If $\alpha_+ = \infty$ then the theorems of Sect. 9.3 describe the large deviation probabilities for any $\alpha = x/n$. But if $\alpha_+ < \infty$ then the approaches of Sect. 9.3 do not enable one to find the asymptotics of probabilities of large deviations of S_n for scaled deviations $\alpha = x/n$ in the vicinity of the point α_+.

In this section we consider the case $\alpha_+ < \infty$. If in this case $\lambda_+ = \infty$, then, by property $(\Lambda 2)$(i), we have $\alpha_+ = s_+ = \sup\{t : F_+(t) > 0\}$, and therefore the random variables ξ_k are bounded from above by the value α_+, $\mathbf{P}(S_n \geq x) = 0$ for $\alpha = x/n > \alpha_+$. We will not consider this case in what follows. Thus we will study the case $\alpha_+ < \infty$, $\lambda_+ < \infty$.

In the present and the next sections, we will confine ourselves to considering integro-local theorems in the non-lattice case with $\Delta = \Delta_n \to 0$ since, as we saw in the previous section, local theorems differ from the integro-local theorems only in that they are simpler. As in Sect. 9.3, the integral theorems can be easily obtained from the integro-local theorems.

9.4.2 The Probabilities of Large Deviations of S_n in an $o(n)$-Vicinity of the Point $\alpha_+ n$; the Case $\psi''(\lambda_+) < \infty$

In this subsection we will study the asymptotics of $\mathbf{P}(S_n \in \Delta[x))$, $x = \alpha n$, when α lies in the vicinity of the point $\alpha_+ < \infty$ and, moreover, $\psi''(\lambda_+) < \infty$. (The case of distributions $\mathbf{F}$, for which $\lambda_+ < \infty$, $\alpha_+ < \infty$ and $\psi''(\lambda_+) < \infty$, will be illustrated later, in Lemma 9.4.1.) Under the above-mentioned conditions, the Cramér transform $\mathbf{F}_{(\lambda_+)}$ is well defined at the point λ_+, and the random variable $\xi^{(\alpha_+)}$ with the distribution $\mathbf{F}_{(\lambda_+)}$ has mean α_+ and a finite variance:

$$\mathbf{E}\xi^{(\alpha_+)} = \frac{\psi'(\lambda_+)}{\psi(\lambda_+)} = \alpha_+, \qquad \operatorname{Var}(\xi^{(\alpha_+)}) = \sigma_{\alpha_+}^2 = \frac{\psi''(\lambda_+)}{\psi(\lambda_+)} - \alpha_+^2 \tag{9.4.1}$$

(cf. (9.3.1)).

Theorem 9.4.1 *Let ξ be a non-lattice random variable,*

$$\lambda_+ \in (0, \infty), \qquad \psi''(\lambda_+) < \infty, \qquad y = x - \alpha_+ n = o(n).$$

If $\Delta_n \to 0$ slowly enough as $n \to \infty$ then

$$\mathbf{P}\big(S_n \in \Delta_n[x)\big) = \frac{\Delta_n}{\sigma_{\alpha_+}\sqrt{2\pi n}} e^{-n\Lambda(\alpha_+) - \lambda_+ y} \left(\exp\left\{-\frac{y^2}{\sigma_{\alpha_+}^2 n}\right\} + o(1)\right),$$

where

$$\alpha = \frac{x}{n}, \qquad \sigma_{\alpha_+}^2 = \frac{\psi''(\lambda_+)}{\psi(\lambda_+)} - \alpha_+^2,$$

and the remainder term $o(1)$ *is uniform in* y.

Proof As in the proof of Theorem 9.3.1, we use the Cramér transform, but now at the fixed point λ_+, so there will be no triangular array scheme when analysing the sums $S_n^{(\alpha_+)}$. In this case the following analogue of Theorem 9.2.1 holds true.

Theorem 9.2.1A *Let* $\lambda_+ \in (0, \infty)$, $\alpha_+ < \infty$ *and* $y = x - n\alpha_+$. *Then, for* $x = n\alpha$ *and any fixed* $\Delta > 0$, *the following representation is valid*:

$$\mathbf{P}\big(S_n \in \Delta[x)\big) = e^{-n\Lambda(\alpha_+)-\lambda_+ y} \int_0^{\Delta} e^{-\lambda_+ z} \mathbf{P}\big(S_n^{(\alpha_+)} - \alpha n \in dz\big). \tag{9.4.2}$$

Proof of Theorem 9.2.1A repeats that of Theorem 9.2.1 the only difference being that, as was already noted, the Cramér transform is now applied at the fixed point λ_+ which does not depend on $\alpha = x/n$. In this case, by (9.2.3),

$$\mathbf{P}(S_n \in dv) = e^{-\lambda_+ v + n \ln \psi(\lambda_+)} \mathbf{P}\big(S_n^{(\alpha_+)} \in dv\big) = e^{-n\Lambda(\alpha_+)+\lambda_+(\alpha_+ n - v)} \mathbf{P}\big(S_n^{(\alpha_+)} \in dv\big).$$

Integrating this equality in v from x to $x + \Delta$, changing the variable $v = x + z$ ($x = n\alpha$), and noting that $\alpha_+ n - v = -y - z$, we obtain (9.4.2).

The theorem is proved. □

Let us return to the proof of Theorem 9.4.1. Assuming that $\Delta = \Delta_n \to 0$, we obtain, by Theorem 9.2.1A, that

$$\mathbf{P}\big(S_n \in \Delta_n[x)\big) = e^{-n\Lambda(\alpha_+)-\lambda_+ y} \mathbf{P}\big(S_n^{(\alpha_+)} - \alpha_+ n \in \Delta_n[y)\big)\big(1 + o(1)\big). \tag{9.4.3}$$

By virtue of (9.4.1), we can apply Theorem 8.7.1 to evaluate the probability on the right-hand side of (9.4.3). This theorem implies that, as $\Delta_n \to 0$ slowly enough,

$$\begin{aligned}
\mathbf{P}\big(S_n^{(\alpha_+)} - \alpha_+ n \in \Delta_n[y)\big) &= \frac{\Delta_n}{\sigma_{\alpha_+}\sqrt{n}} \phi\left(\frac{y}{\sigma_{\alpha_+}\sqrt{n}}\right) + o\left(\frac{1}{\sqrt{n}}\right) \\
&= \frac{\Delta_n}{\sigma_{\alpha_+}\sqrt{2\pi n}} \exp\left\{-\frac{y^2}{\sigma_{\alpha_+}^2 n}\right\} + o\left(\frac{1}{\sqrt{n}}\right)
\end{aligned}$$

uniformly in y. This, together with (9.4.3), proves Theorem 9.4.1. □

9.4.3 The Class of Distributions $\mathcal{ER}$. The Probability of Large Deviations of S_n in an $o(n)$-Vicinity of the Point $\alpha_+ n$ for Distributions $\mathbf{F}$ from the Class $\mathcal{ER}$ in Case $\psi''(\lambda_+)=\infty$

When studying the asymptotics of $\mathbf{P}(S_n \ge \alpha n)$ (or $\mathbf{P}(S_n \in \Delta[\alpha n))$) in the case where $\psi''(\lambda_+) = \infty$ and α is in the vicinity of the point $\alpha_+ < \infty$, we have to impose additional conditions on the distribution $\mathbf{F}$ similarly to what was done in Sect. 8.8 when studying convergence to stable laws.

To formulate these additional conditions it will be convenient to introduce certain classes of distributions. If $\lambda_+ < \infty$, then it is natural to represent the right tails $F_+(t)$ as

$$F_+(t) = e^{-\lambda_+ t} V(t), \tag{9.4.4}$$

where, by the exponential Chebyshev inequality, $V(t) = e^{o(t)}$ as $t \to \infty$.

Definition 9.4.1 We will say that the distribution $\mathbf{F}$ of a random variable ξ (or the random variable ξ itself) belongs to the class $\mathcal{R}$ if its right tail $F_+(t)$ is a regularly varying function, i.e. can be represented as

$$F_+(t) = t^{-\beta} L(t), \tag{9.4.5}$$

where L is a slowly varying function as $t \to \infty$ (see also Sect. 8.8 and Appendix 6).

We will say that the distribution $\mathbf{F}$ (or the random variable ξ) belongs to the class $\mathcal{ER}$ if, in the representation (9.4.4), the function V is regularly varying (which will also be denoted as $V \in \mathcal{R}$).

Distributions from the class $\mathcal{R}$ have already appeared in Sect. 8.8.

The following assertion explains which distributions from $\mathcal{ER}$ correspond to the cases $\alpha_+ = \infty$, $\alpha_+ < \infty$, $\psi''(\lambda_+) = \infty$ and $\psi''(\lambda_+) < \infty$.

Lemma 9.4.1 *Let $\mathbf{F} \in \mathcal{ER}$. For α_+ to be finite it is necessary and sufficient that*

$$\int_1^\infty t V(t)\, dt < \infty.$$

For $\psi''(\lambda_+)$ to be finite, it is necessary and sufficient that

$$\int_1^\infty t^2 V(t)\, dt < \infty.$$

The assertion of the lemma means that $\alpha_+ < \infty$ if $\beta > 2$ in the representation $V(t) = t^{-\beta} L(t)$, where L is an s.v.f. and $\alpha_+ = \infty$ if $\beta < 2$. For $\beta = 2$, the finiteness of α_+ is equivalent to the finiteness of $\int_1^\infty t^{-1} L(t)\, dt$. The same is true for the finiteness of $\psi''(\lambda_+)$.

Proof of Lemma 9.4.1 We first prove the assertion concerning α_+. Since

$$\alpha_+ = \frac{\psi'(\lambda_+)}{\psi(\lambda_+)},$$

we have to estimate the values of $\psi'(\lambda_+)$ and $\psi(\lambda_+)$. The finiteness of $\psi'(\lambda_+)$ is equivalent to that of

$$-\int_1^\infty te^{\lambda_+ t}\,dF_+(t) = \int_1^\infty t\big(\lambda_+ V(t)\,dt - dV(t)\big), \tag{9.4.6}$$

where, for $V(t) = o(1/t)$,

$$-\int_1^\infty t\,dV(t) = V(1) + \int_1^\infty V(t)\,dt.$$

Hence the finiteness of the integral on the left-hand side of (9.4.6) is equivalent to that of the sum

$$\lambda_+ \int_1^\infty tV(t)\,dt + \int_1^\infty V(t)\,dt$$

or, which is the same, to the finiteness of the integral $\int_1^\infty tV(t)\,dt$. Similarly we see that the finiteness of $\psi(\lambda_+)$ is equivalent to that of $\int_1^\infty V(t)\,dt$. This implies the assertion of the lemma in the case $\int_1^\infty V(t)\,dt < \infty$, where one has $V(t) = o(1/t)$. If $\int_1^\infty V(t)\,dt = \infty$, then $\psi(\lambda_+) = \infty$, $\ln\psi(\lambda) \to \infty$ as $\lambda \uparrow \lambda_+$ and hence $\alpha_+ = \lim_{\lambda\uparrow\lambda_+}(\ln\psi(\lambda))' = \infty$.

The assertion concerning $\psi''(\lambda_+)$ can be proved in exactly the same way. The lemma is proved. □

The lemma implies the following:

(a) If $\beta < 2$ or $\beta = 2$ and $\int_1^\infty t^{-1}L(t) = \infty$, then $\alpha_+ = \infty$ and the theorems of the previous section are applicable to $\mathbf{P}(S_n \ge x)$.
(b) If $\beta > 3$ or $\beta = 3$ and $\int_1^\infty t^{-1}L(t)\,dt < \infty$, then $\alpha_+ < \infty$, $\psi''(\lambda_+) < \infty$ and we can apply Theorem 9.4.1.

It remains to consider the case

(c) $\beta \in [2, 3]$, where the integral $\int_1^\infty t^{-1}L(t)\,dt$ is finite for $\beta = 2$ and is infinite for $\beta = 3$.

It is obvious that in case (c) we have $\alpha_+ < \infty$ and $\psi''(\lambda_+) = \infty$.

Put

$$V_+(t) := \frac{\lambda_+ tV(t)}{\beta\psi(\lambda_+)}, \qquad b(n) := V_+^{(-1)}\left(\frac{1}{n}\right),$$

where $V_+^{(-1)}(1/n)$ is the value of the function inverse to V_+ at the point $1/n$.

Theorem 9.4.2 *Let ξ be a non-lattice random variable,* $\mathbf{F} \in \mathcal{ER}$ *and condition* (c) *hold. If $\Delta_n \to 0$ slowly enough as $n \to \infty$, then, for $y = x - \alpha_+ n = o(n)$,*

$$\mathbf{P}\big(S_n \in \Delta_n[x)\big) = \frac{\Delta_n e^{-n\Lambda(\alpha_+)-\lambda_+ y}}{b(n)} \left(f^{(\beta-1,1)}\left(\frac{y}{b(n)}\right) + o(1)\right),$$

where $f^{(\beta-1,1)}$ is the density of the stable law $\mathbf{F}_{(\beta-1,1)}$ with parameters $\beta - 1, 1$, and the remainder term $o(1)$ is uniform in y.

We will see from the proof of the theorem that studying the probabilities of large deviations in the case where $\alpha_+ < \infty$ and $\psi''(\lambda_+) = \infty$ is basically impossible outside the class $\mathcal{ER}$, since it is impossible to find theorems on the limiting distribution of S_n in the case $\operatorname{Var}(\xi) = \infty$ without the conditions $[\mathbf{R}_{\gamma,\rho}]$ of Sect. 8.8 being satisfied.

Proof of Theorem 9.4.2 Condition (c) implies that $\alpha_+ = \mathbf{E}\xi^{(\alpha_+)} < \infty$ and $\operatorname{Var}(\xi^{(\alpha_+)}) = \infty$. We will use Theorem 9.2.1A. For $\Delta_n \to 0$ slowly enough we will obtain, as in the proof of Theorem 9.4.1, that relation (9.4.3) holds true. But now, in contrast to Theorem 9.4.1, in order to calculate the probability on the right-hand side of (9.4.3), we have to employ the integro-local Theorem 8.8.3 on convergence to a stable law. In our case, by the properties of r.v.f.s, one has

$$\begin{aligned}\mathbf{P}\big(\xi^{(\alpha_+)} \geq t\big) &= -\frac{1}{\psi(\lambda_+)} \int_t^\infty e^{\lambda_+ u} dF_+(u) = \frac{1}{\psi(\lambda_+)} \int_t^\infty \big(\lambda_+ V(u)du - dV(u)\big) \\ &= \frac{\lambda_+}{\beta\psi(\lambda_+)} t^{-\beta+1} L_+(t) \sim V_+(t), \end{aligned} \tag{9.4.7}$$

where $L_+(t) \sim L(t)$ is a slowly varying function. Moreover, the left tail of the distribution $\mathbf{F}^{(\alpha_+)}$ decays at least exponentially fast. By virtue of the results of Sect. 8.8, this means that, for $b(n) = V_+^{(-1)}(1/n)$, we have convergence of the distributions of $\frac{S_n^{(\alpha_+)} - \alpha_+ n}{b(n)}$ to the stable law $\mathbf{F}_{\beta-1,1}$ with parameters $\beta - 1 \in [1, 2]$ and 1. It remains to use representation (9.4.3) and Theorem 8.8.3 which implies that, provided $\Delta_n \to 0$ slowly enough, one has

$$\mathbf{P}\big(S_n^{(\alpha_+)} - \alpha_+ n \in \Delta_n[y)\big) = \frac{\Delta_n}{b(n)} f^{(\beta-1,1)}\left(\frac{y}{b(n)}\right) + o\left(\frac{1}{b(n)}\right)$$

uniformly in y. The theorem is proved. □

Theorem 9.4.2 concludes the study of probabilities of large deviations of S_n/n in the vicinity of the point α_+ for distributions from the class $\mathcal{ER}$.

9.4.4 On the Large Deviation Probabilities in the Range $\alpha > \alpha_+$ for Distributions from the Class $\mathcal{ER}$

Now assume that the deviations x of S_n are such that $\alpha = x/n > \alpha_+$, and $y = x - \alpha_+ n$ grows fast enough (faster than $\sqrt{n}$ under the conditions of Theorem 9.4.1 and faster than $b(n)$ under the conditions of Theorem 9.4.2). Then, for the probability

$$\mathbf{P}\big(S^{(\alpha_+)} - \alpha_+ n \in \Delta_n[y)\big), \tag{9.4.8}$$

the deviations y (see representation (9.4.3)) will belong to the zone of large deviations, so applying Theorems 8.7.1 and 8.8.3 to evaluate such probabilities does not make much sense. Relation (9.4.7) implies that, in the case $\mathbf{F} \in \mathcal{ER}$, we have $\mathbf{F}^{(\alpha_+)} \in \mathcal{R}$. Therefore, we will know the asymptotics of the probability (9.4.8) (and hence also of the probability $\mathbf{P}(S_n \in \Delta_n[x))$, see (9.4.3)) if we obtain integro-local theorems for the probabilities of large deviations of the sums S_n, in the case where the summands belong to the class $\mathcal{R}$. Such theorems are also of independent interest in the present chapter, and the next section will be devoted to them. After that, in Sect. 9.6 we will return to the problem on large deviation probabilities in the class $\mathcal{ER}$ mentioned in the title of this section.

9.5 Integral and Integro-Local Theorems on Large Deviation Probabilities for Sums S_n when the Cramér Condition Is not Met

If $\mathbf{E}\xi = 0$ and the right-side Cramér condition is not met ($\lambda_+ = 0$), then the rate function $\Lambda(\alpha)$ degenerates on the right semiaxis: $\Lambda(\alpha) = \lambda(\alpha) = 0$ for $\alpha \geq 0$, and the results of Sects. 9.1–9.4 on the probabilities of large deviations of S_n are of little substance. In this case, in order to find the asymptotics of $\mathbf{P}(S_n \geq x)$ and $\mathbf{P}(S_n \in \Delta[x))$, we need completely different approaches, while finding these asymptotics is only possible under additional conditions on the behaviour of the tail $F_+(t)$ of the distribution $\mathbf{F}$, similarly to what happened in Sect. 8.8 when studying convergence to stable laws.

The above-mentioned additional conditions consist of the assumption that the tail $F_+(t)$ behaves regularly enough. In this section we will assume that $F_+(t) = V(t) \in \mathcal{R}$, where $\mathcal{R}$ is the class of regularly varying functions introduced in the previous section (see also Appendix 6). To make the exposition more homogeneous, we will confine ourselves to the case $\beta > 2$, $\mathrm{Var}(\xi) < \infty$, where $-\beta$ is the power exponent in the function $V \in \mathcal{R}$ (see (9.4.5)). Studying the case $\beta \in [1,2]$ ($\mathrm{Var}(\xi) = \infty$) does not differ much from the exposition below, but it would significantly increase the volume of the exposition and complicate the text, and therefore is omitted. Results for the case $\beta \in (0,2]$ can be found in [8, Chap. 3].

9.5.1 Integral Theorems

Integral theorems for probabilities of large deviations of S_n and maxima $\overline{S}_n = \max_{k\le n} S_k$ in the case $\mathbf{E}\xi = 0$, $\mathrm{Var}(\xi) < \infty$, $\mathbf{F} \in \mathcal{R}$, $\beta > 2$, follow immediately from the bounds obtained in Appendix 8. In particular, Corollaries A8.2.1 and A8.3.1 of Appendix 8 imply the following result.

Theorem 9.5.1 *Let* $\mathbf{E}\xi = 0$, $\mathrm{Var}(\xi) < \infty$, $\mathbf{F} \in \mathcal{R}$ *and* $\beta > 2$. *Then, for* $x \gg \sqrt{n \ln n}$,

$$\mathbf{P}(\overline{S}_n \ge x) \sim \mathbf{P}(S_n \ge x) \sim nV(x). \tag{9.5.1}$$

Under an additional condition $[\mathbf{D}_0]$ to be introduced below, the assertion of this theorem will also follow from the integro-local Theorem 9.5.2 (see below).

Comparing Theorem 9.5.1 with the results of Sects. 9.2–9.4 shows that the nature of the large deviation probabilities is completely different here. Under the Cramér condition and for $\alpha = x/n \in (0, \alpha_+)$, the large deviations of S_n are, roughly speaking, "equally contributed to by all the summands" ξ_k, $k \le n$. This is confirmed by the fact that, for a fixed α, the limiting conditional distribution of ξ_k, $k \le n$, given that $S_n \in \Delta[x)$ (or $S_n \ge x$) for $x = \alpha n$, $\Delta = 1$, as $n \to \infty$ coincides with the distribution $\mathbf{F}^{(\alpha)}$ of the random variable $\xi^{(\alpha)}$. The reader can verify this himself/herself using Theorem 9.3.2. In other words, the conditions $\{S_n \in \Delta[x)\}$ (or $\{S_n \ge x\}$), $x = \alpha n$, change equally (from $\mathbf{F}$ to $\mathbf{F}^{(\alpha)}$) the distributions of all the summands.

However, if the Cramér condition is not met, then under the conditions of Theorem 9.5.1 the large deviations of S_n are essentially due to one large (comparable with x) jump. This is seen from the fact that the value of $nV(x)$ on the right-hand side of (9.5.1) is nothing else but the main term of the asymptotics for $\mathbf{P}(\overline{\xi}_n \ge x)$, where $\overline{\xi}_n = \max_{k\le n} \xi_k$. Indeed, if $nV(x) \to 0$ then

$$\mathbf{P}(\overline{\xi}_n < x) = \big(1 - V(x)\big)^n = 1 - nV(x) + O\big(\big(nV(x)\big)^2\big),$$

$$\mathbf{P}(\overline{\xi}_n \ge x) = nV(x) + O\big(\big(nV(x)\big)^2\big) \sim nV(x).$$

In other words, the probabilities of large deviations of S_n, $\overline{S}_n$ and $\overline{\xi}_n$ are asymptotically the same. The fact that the probabilities of the events $\{\xi_j \ge y\}$ for $y \sim x$ play the determining role in finding the asymptotics of $\mathbf{P}(S_n \ge x)$ can easily be discovered in the bounds from Appendix 8.

Thus, while the asymptotics of $\mathbf{P}(S_n \ge x)$ for $x = \alpha n \gg \sqrt{n}$ in the Cramér case is determined by "the whole distribution $\mathbf{F}$" (as the rate function $\Lambda(\alpha)$ depends on the "the whole distribution $\mathbf{F}$"), these asymptotics in the case $\mathbf{F} \in \mathcal{R}$ are determined by the right tail $F_+(t) = V(t)$ only and do not depend on the "remaining part" of the distribution $\mathbf{F}$ (for the fixed value of $\mathbf{E}\xi = 0$).

9.5.2 Integro-Local Theorems

In this section we will study the asymptotics of $\mathbf{P}(S_n \in \Delta[x))$ in the case where

$$\mathbf{E}\xi = 0, \qquad \operatorname{Var}\xi^2 < \infty, \qquad \mathbf{F} \in \mathcal{R}, \qquad \beta > 2, \qquad x \gg \sqrt{n \ln n}. \qquad (9.5.2)$$

These asymptotics are of independent interest and are also useful, for example, in finding the asymptotics of integrals of type $\mathbf{E}(g(S_n);\ S_n \ge x)$ for $x \gg \sqrt{n \ln n}$ for a wide class of functions g. As was already noted (see Subsection 4.4), in the next section we will use the results from the present section to obtain integro-local theorems under the Cramér condition (for summands from the class $\mathcal{ER}$) for deviations outside the Cramér zone.

In order to obtain integro-local theorems in this section, we will need additional conditions. Besides condition $\mathbf{F} \in \mathcal{R}$, we will also assume that the following holds:

Condition $[\mathbf{D}_0]$ *For each fixed Δ, as $t \to \infty$,*

$$V(t) - V(t + \Delta) = v(t)\big(\Delta + o(1)\big), \qquad v(t) = \frac{\beta V(t)}{t}.$$

It is clear that if the function $L(t)$ in representation (9.4.5) (or the function $V(t)$) is differentiable for t large enough and $L'(t) = o(L(t)/t)$ as $t \to \infty$ (all sufficiently smooth s.v.f.s possess this property; cf. e.g., polynomials of $\ln t$ etc.), then condition $[\mathbf{D}_0]$ will be satisfied, and the derivative $-V'(t) \sim v(t)$ will play the role of the function $v(t)$.

Theorem 9.5.2 *Let conditions* (9.5.2) *and* $[\mathbf{D}_0]$ *be met. Then*

$$\mathbf{P}\big(S_n \in \Delta[x)\big) = \Delta n v(x)\big(1 + o(1)\big), \qquad v(x) = \frac{\beta V(x)}{x},$$

where the remainder term $o(1)$ is uniform in $x \ge N\sqrt{n \ln n}$ and $\Delta \in [\Delta_1, \Delta_2]$ for any fixed $\Delta_2 > \Delta_1 > 0$ and any fixed sequence $N \to \infty$.

Note that in Theorems 9.5.1 and 9.5.2 we do not assume that $n \to \infty$. The assumption that $x \to \infty$ is contained in (9.5.2).

Proof For $y < x$, introduce the events

$$G_n := \big\{S_n \in \Delta[x)\big\}, \qquad B_j := \{\xi_j < y\}, \qquad B := \bigcap_{j=1}^{n} B_j. \qquad (9.5.3)$$

Then

$$\mathbf{P}(G_n) = \mathbf{P}(G_n B) + \mathbf{P}(G_n \overline{B}), \qquad \overline{B} = \bigcup_{j=1}^{n} \overline{B}_j, \qquad (9.5.4)$$

where

$$\sum_{j=1}^{n} \mathbf{P}(G_n\overline{B}_j) \geq \mathbf{P}(G_n\overline{B}) \geq \sum_{j=1}^{n} \mathbf{P}(G_n\overline{B}_j) - \sum_{i<j\leq n} \mathbf{P}(G_n\overline{B}_i\overline{B}_j) \tag{9.5.5}$$

(see property 8 in Sect. 9.2.2).

The proof is divided into three stages: the bounding of $\mathbf{P}(G_nB)$, that of $\mathbf{P}(G_n\overline{B}_i\overline{B}_j)$, $i \neq j$, and the evaluation of $\mathbf{P}(G_n\overline{B}_j)$.

(1) *A bound on* $\mathbf{P}(G_nB)$. We will make use of the rough inequality

$$\mathbf{P}(G_nB) \leq \mathbf{P}(S_n \geq x;\, B) \tag{9.5.6}$$

and Theorem A8.2.1 of Appendix 8 which implies that, for $x = ry$ with a fixed $r > 2$, any $\delta > 0$, and $x \geq N\sqrt{n \ln n}$, $N \to \infty$, we have

$$\mathbf{P}(S_n \geq x;\, B) \leq \big(nV(y)\big)^{r-\delta}. \tag{9.5.7}$$

Here we can always choose r such that

$$\big(nV(x)\big)^{r-\delta} \ll n\Delta v(x) \tag{9.5.8}$$

for $x \gg \sqrt{n}$. Indeed, putting $n := x^2$ and comparing the powers of x on the right-hand and left-hand sides of (9.5.8), we obtain that for (9.5.8) to hold it suffices to choose r such that

$$(2-\beta)(r-\delta) < 1-\beta,$$

which is equivalent, for $\beta > 2$, to the inequality.

$$r > \frac{\beta-1}{\beta-2}.$$

For such r, we will have that, by (9.5.6)–(9.5.8),

$$\mathbf{P}(G_nB) = o\big(n\Delta v(x)\big). \tag{9.5.9}$$

Since $r - \delta > 1$, we see that, for $n \ll x^2$, relations (9.5.8) and (9.5.9) will hold true all the more.

(2) *A bound for* $\mathbf{P}(G_n\overline{B}_i\overline{B}_j)$. It is sufficient to bound $\mathbf{P}(G_n\overline{B}_{n-1}\overline{B}_n)$. Set

$$\delta := \frac{1}{r} < \frac{1}{2}, \qquad H_k := \big\{v : v < (1-k\delta)x + \Delta\big\}, \quad k = 1, 2.$$

Then

$$\mathbf{P}(G_n\overline{B}_{n-1}\overline{B}_n) = \int_{H_2} \mathbf{P}(S_{n-2} \in dz)$$

$$\times \int_{H_1} \mathbf{P}(z+\xi \in dv, \xi \geq \delta x)\mathbf{P}\big(v+\xi \in \Delta[x), \xi \geq \delta x\big). \tag{9.5.10}$$

Since in the domain H_1 we have $x - v > \delta x - \Delta$, the last factor on the right-hand side of (9.5.10) has, by condition $[\mathbf{D}_0]$, the form $\Delta v(x-v)(1+o(1)) \leq c\Delta v(x)$ as $x \to \infty$, so the integral over H_1 in (9.5.10), for x large enough, does not exceed

$$c\Delta v(x)\mathbf{P}(z+\xi \in H_1; \xi \geq \delta x) \leq c\Delta v(x)V(\delta x).$$

The integral over the domain H_2 in (9.5.10) evidently allows a similar bound. Since $nV(x) \to 0$, we obtain that

$$\sum_{i<j\leq n} \mathbf{P}(G_n\overline{B}_i\overline{B}_j) \leq c_1\Delta n^2 v(x)V(x) = o\big(\Delta n v(x)\big). \tag{9.5.11}$$

(3) *The evaluation of* $\mathbf{P}(G_n\overline{B}_j)$ is based on the relation

$$\begin{aligned}\mathbf{P}(G_n\overline{B}_n) &= \int_{H_1} \mathbf{P}(S_{n-1} \in dz)\,\mathbf{P}\big(\xi \in \Delta[x-z), \xi \geq \delta x\big)\\ &\leq \int_{H_1} \mathbf{P}(S_{n-1} \in dz)\,\mathbf{P}\big(\xi \in \Delta[x-z)\big)\\ &= \Delta \int_{H_1} \mathbf{P}(S_{n-1} \in dz)v(x-z)\big(1+o(1)\big),\end{aligned} \tag{9.5.12}$$

which yields

$$\begin{aligned}\mathbf{P}(G_n\overline{B}_n) &\leq \Delta\mathbf{E}\big[v(x-S_{n-1}); S_{n-1} < (1-\delta)x+\Delta\big]\big(1+o(1)\big)\\ &= \Delta v(x)\big(1+o(1)\big).\end{aligned} \tag{9.5.13}$$

The last relation is valid for $x \gg \sqrt{n}$, since, by Chebyshev's inequality, $\mathbf{E}[v(x - S_{n-1}); |S_{n-1}| \leq M\sqrt{n}] \sim v(x)$ as $M \to \infty$, $M\sqrt{n} = o(x)$ and, moreover, the following evident bounds hold:

$$\mathbf{E}\big[v(x-S_{n-1}); S_{n-1} \in \big(M\sqrt{n}, (1-\delta)x+\Delta\big)\big] = o\big(v(x)\big),$$

$$\mathbf{E}\big[v(x-S_{n-1}); S_{n-1} \in (-\infty, -M\sqrt{n})\big] = o\big(v(x)\big)$$

as $M \to \infty$.

Similarly, by (virtue of (9.5.12)) we get

$$\mathbf{P}(G_n\overline{B}_n) \geq \int_{-\infty}^{(1-\delta)x} \mathbf{P}(S_{n-1} \in dz)\,\mathbf{P}\big(\xi \in \Delta[x-z)\big) \sim \Delta v(x). \tag{9.5.14}$$

From (9.5.13) and (9.5.14) we obtain that

$$\mathbf{P}(G_n\overline{B}_n) = \Delta v(x)\bigl(1+o(1)\bigr).$$

This, together with (9.5.4), (9.5.9) and (9.5.11), yields the representation

$$\mathbf{P}(G_n) = \Delta n v(x)\bigl(1+o(1)\bigr).$$

The required uniformity of the term $o(1)$ clearly follows from the preceding argument. The theorem is proved. □

Theorem 9.5.2 implies the following

Corollary 9.5.1 *Let the conditions of Theorem* 9.5.2 *be satisfied. Then there exists a fixed sequence Δ_N converging to zero slowly enough as $N \to \infty$ such that the assertion of Theorem* 9.5.2 *remains true when the segment $[\Delta_1, \Delta_2]$ is replaced in it with $[\Delta_N, \Delta_2]$.*

9.6 Integro-Local Theorems on the Probabilities of Large Deviations of S_n Outside the Cramér Range (Under the Cramér Condition)

We return to the case where the Cramér condition is met. In Sects. 9.3 and 9.4 we obtained integro-local theorems for deviations inside and on the boundary of the Cramér range. It remains to study the asymptotics of $\mathbf{P}(S_n \in \Delta[x))$ outside the Cramér range, i.e. for $\alpha = x/n > \alpha_+$. Preliminary observations concerning this problem were made in Sect. 9.4.4 where it was reduced to integro-local theorems for the sums S_n when Cramér's condition is not satisfied. Recall that in that case we had to restrict ourselves to considering distributions from the class $\mathcal{ER}$ defined in Sect. 9.4.3 (see (9.4.4)).

Theorem 9.6.1 *Let $\mathbf{F} \in \mathcal{ER}$, $\beta > 3$, $\alpha = x/n > \alpha_+$ and $y = x - \alpha_+ n \gg \sqrt{n}$. Then there exists a fixed sequence Δ_N converging to zero slowly enough as $N \to \infty$, such that*

$$\begin{aligned}\mathbf{P}\bigl(S_n \in \Delta_N[x)\bigr) &= e^{-n\Lambda(\alpha_+)-\lambda_+ y} n\Delta_N v_+(y)\bigl(1+o(1)\bigr)\\ &= e^{-n\Lambda(\alpha)} n\Delta_N v_+(y)\bigl(1+o(1)\bigr),\end{aligned}$$

where $v_+(y) = \lambda_+ V(y)/\psi(\lambda_+)$, the remainder term $o(1)$ is uniform in x and n such that $y \gg N\sqrt{n}\ln n$, N being an arbitrary fixed sequence tending to ∞.

Proof By Theorem 9.2.1A there exists a sequence Δ_N converging to zero slowly enough such that (cf. (9.4.3))

$$\mathbf{P}\bigl(S_n \in \Delta_N[x)\bigr) = e^{-n\Lambda(\alpha_+)-\lambda_+ y}\, \mathbf{P}\bigl(S_n^{(\alpha_+)} - \alpha_+ n \in \Delta_N[y)\bigr). \qquad (9.6.1)$$

Since by properties ($\Lambda1$) and ($\Lambda2$) the function $\Lambda(\alpha)$ is linear for $\alpha > \alpha_+$:

$$\Lambda(\alpha) = \Lambda(\alpha_+) + (\alpha - \alpha_+)\lambda_+,$$

the exponent in (9.6.1) can be rewritten as

$$-n\Lambda(\alpha_+) - \lambda_+ y = -n\Lambda(\alpha).$$

The right tail of the distribution of $\xi^{(\alpha_+)}$ has the form (see (9.4.7))

$$\mathbf{P}\big(\xi^{(\alpha_+)} \geq t\big) = \frac{\lambda_+}{\psi(\lambda_+)} \int_t^\infty V(u)\,du + V(t).$$

By the properties of regularly varying functions (see Appendix 6),

$$V(t) - V(t-u) = o\big(\big(V(t)\big)$$

as $t \to \infty$ for any fixed u. This implies that condition [$\mathbf{D}_0$] of Sect. 9.5 is satisfied for the distribution of $\xi^{(\alpha_+)}$.

This means that, in order to calculate the probability on the right-hand side of (9.6.1), we can use Theorem 9.5.2 and Corollary 9.5.1, by virtue of which, as $\Delta_N \to 0$ slowly enough,

$$\mathbf{P}\big(S_n^{(\alpha_+)} - \alpha_+ n \in \Delta_N[y)\big) = n\Delta_N v_+(y)\big(1 + o(1)\big),$$

where the remainder term $o(1)$ is uniform in all x and n such that $y \gg N\sqrt{n \ln n}$, $N \to \infty$.

The theorem is proved. □

Since $\mathbf{P}(S_n \in \Delta_N[x))$ decreases exponentially fast as x (or y) grows (note the factor $e^{-\lambda_+ y}$ in (9.6.1)), Theorem 9.6.1 immediately implies the following integral theorem.

Corollary 9.6.1 *Under the conditions of Theorem* 9.6.1,

$$\mathbf{P}(S_n \geq x) = e^{-n\Lambda(\alpha)} \frac{nV(y)}{\psi(\lambda_+)}\big(1 + o(1)\big).$$

Proof Represent the probability $\mathbf{P}(S_n \geq x)$ as the sum

$$\mathbf{P}(S_n \geq x) = \sum_{k=0}^{\infty} \mathbf{P}\big(S_n \in \Delta_N[x + k\Delta_N)\big)$$

$$\sim e^{-n\Lambda(\alpha)} \frac{n\lambda_+}{\psi(\lambda_+)} \sum_{k=0}^{\infty} \Delta_N V(y + \Delta_N k) e^{-\lambda_+ \Delta_N k}.$$

Here the series on the right-hand side is asymptotically equivalent, as $N \to \infty$, to the integral

$$V(y) \int_0^\infty e^{-\lambda_+ t} dt = \frac{V(y)}{\lambda_+}.$$

The corollary is proved. □

Note that a similar corollary (i.e. the integral theorem) can be obtained under the conditions of Theorem 9.4.2 as well.

In the range of deviations $\alpha = \frac{x}{n} > \alpha_+$, only the case $\mathbf{F} \in \mathcal{ER}$, $\beta \in [2, 3]$ (recall that $\alpha_+ = \infty$ for $\beta < 2$) has not been considered in this text. As we have already said, it could also be considered, but that would significantly increase the length and complexity of the exposition. Results dealing with this case can be found in [8]; one can also find there a more complete study of large deviation probabilities.

Chapter 10
Renewal Processes

Abstract This is the first chapter in the book to deal with random processes in continuous time, namely, with the so-called renewal processes. Section 10.1 establishes the basic terminology and proves the integral renewal theorem in the case of non-identically distributed random variables. The classical Key Renewal Theorem in the arithmetic case is proved in Sect. 10.2, including its extension to the case where random variables can assume negative values. The limiting behaviour of the excess and defect of a random walk at a growing level is established in Sect. 10.3. Then these results are extended to the non-arithmetic case in Sect. 10.4. Section 10.5 is devoted to the Law of Large Numbers and the Central Limit Theorem for renewal processes. It also contains the proofs of these laws for the maxima of sums of independent non-identically distributed random variables that can take values of both signs, and a local limit theorem for the first hitting time of a growing level. The chapter ends with Sect. 10.6 introducing generalised (compound) renewal processes and establishing for them the Central Limit Theorem, in both integral and integro-local forms.

10.1 Renewal Processes. Renewal Functions

10.1.1 Introduction

The sequence of sums of random variables $\{S_n\}$, considered in previous chapters, is often called a *random walk*. It can be considered as the simplest *random process in discrete time* n. The further study of such processes is contained in Chaps. 11, 12 and 20.

In this chapter we consider the simplest processes in *continuous time* t that are also entirely determined by a sequence of independent random variables and do not require, for their construction, any special structures (in the general case such constructions will be needed; see Chap. 18).

Let $\tau_1, \{\tau_j\}_{j=2}^{\infty}$ be a sequence of independent random variables given on a probability space $\langle \Omega, \mathfrak{F}, \mathbf{P} \rangle$ (here we change our conventional notations ξ_j to τ_j for reasons that will become clear in Sect. 10.6, where ξ_j appear again). For the random variables $\tau_2, \tau_3, \ldots$ we will usually assume some homogeneity property: proximity

A.A. Borovkov, *Probability Theory*, Universitext,
DOI 10.1007/978-1-4471-5201-9_10, © Springer-Verlag London 2013

of the expectations or identical distributions. The random variable τ_1 can be arbitrary.

Definition 10.1.1 A *renewal process* is a collection of random variables $\eta(t)$ depending on a parameter t and defined on $\langle\Omega, \mathfrak{F}, \mathbf{P}\rangle$ by the equality

$$\eta(t) := \min\{k \geq 0 : T_k > t\}, \quad t \geq 0, \tag{10.1.1}$$

where

$$T_k := \sum_{j=1}^{k} \tau_j, \qquad T_0 := 0.$$

The variables $\eta(t)$ are not completely defined yet. We do not know what $\eta(t)$ is for ω such that the level t is never reached by the sequence of sums T_k. In that case it is natural to put

$$\eta(t) := \infty \quad \text{if all } T_k \leq t. \tag{10.1.2}$$

Clearly, $\eta(t)$ is a stopping time (see Sect. 4.4).

Usually the random variables $\tau_2, \tau_3, \ldots$ are assumed to be identically distributed with a finite expectation. The distribution of the random variable τ_1 can be arbitrary.

We assume first that all the random variables τ_j are *positive*. Then definition (10.1.1) allows us to consider $\eta(t)$ as a random function that can be described as follows. If we plot the points $T_0 = 0, T_1, T_2, \ldots$ on the real line, then one has $\eta(t) = 0$ on the semi-axis $(-\infty, 0)$, $\eta(t) = 1$ on the semi-interval $[0, T_1)$, $\eta(t) = 2$ on the semi-interval $[T_1, T_2)$ and so on.

The sequence $\{T_k\}_{k=0}^{\infty}$ is also often called a renewal process. Sometimes we will call the sequence $\{T_k\}$ a *random walk*. The quantity $\eta(t)$ can also be called the first passage time of the level t by the random walk $\{T_k\}_{k=0}^{\infty}$.

If, based on the sequence $\{T_k\}$, we construct a random walk $T(x)$ in continuous time:

$$T(x) := T_k \quad \text{for } x \in [k, k+1),\ k \geq 0,$$

then *the renewal process* $\eta(t)$ *will be the generalised inverse of* $T(x)$:

$$\eta(t) = \inf\bigl\{x \geq 0 : T(x) > t\bigr\}.$$

The term "renewal process" is related to the fact that the function $\eta(t)$ and the sequence $\{T_k\}$ are often used to describe the operation of various physical devices comprising replaceable components. If, say, τ_j is the failure-free operating time of such a component, after which the latter requires either replacement or repair ("renewal", which is supposed to happen immediately), then T_k will denote the time of the k-th "renewal" of the component, while $\eta(t)$ will be equal to the number of "renewals" which have occurred by the time t.

Remark 10.1.1 If the j-th renewal of the component does not happen immediately but requires a time $\tau'_j \ge 0$, then, introducing the random variables

$$\tau^*_j := \tau_j + \tau'_j, \qquad T^*_k := \sum_{j=1}^{k} \tau^*_j, \qquad \eta^*(t) := \min\{k : T^*_k > t\},$$

we get an object of the same nature as before, with nearly the same physical meaning. For such an object, a number of additional results can be obtained, see e.g., Remark 10.3.1.

Renewal processes are also quite often used in probabilistic research per se, and also when studying other processes for which there exist so-called "regeneration times" after which the evolution of the process starts anew. Below we will encounter examples of such use of renewal processes.

Now we return to the general case where τ_j may assume both positive and negative values.

Definition 10.1.2 The function

$$H(t) := \mathbf{E}\eta(t), \quad t \ge 0,$$

is called the *renewal function for the sequence* $\{T_k\}_{k=0}^{\infty}$.

In the existing literature, another definition is used more frequently.

Definition 10.1.2A The *renewal function for the sequence* $\{T_k\}_{k=0}^{\infty}$ is defined by

$$U(t) := \sum_{j=0}^{\infty} \mathbf{P}(T_j \le t).$$

The values of $H(u)$ and $T(u)$ can be infinite.

If $\tau_j \ge 0$ *then the above definitions are equivalent*. Indeed, for $t \ge 0$, consider the random variable

$$\nu(t) := \max\{k : T_k \le t\} = \eta(t) - 1.$$

Then clearly

$$\sum_{j=0}^{\infty} \mathrm{I}(T_j \le t) = 1 + \nu(t),$$

where $\mathrm{I}(A)$ is the indicator of the event A, and

$$U(t) = 1 + \mathbf{E}\nu(t) = \mathbf{E}\eta(t) = H(t).$$

The value $U(t) = \mathbf{E}\,\nu(t) + 1$ is the mean time spent by the trajectory $\{T_j\}_{j=0}^{\infty}$ in the interval $[0, t]$.

If τ_j *can take values of different signs* then clearly $\nu(t) \ge \eta(t)$ and, with a positive probability, $\nu(t) > \eta(t)$ (the trajectory $\{T_j\}$, after crossing the level t, can return to the region $(-\infty, t]$). Therefore in that case $U(t) > H(t)$. Thus *for* τ_j *taking values of different signs we have two versions of the renewal function* given in Definitions 10.1.2 and 10.1.2A. We will call them the first and the second versions, respectively. *In the present chapter we will consider the first version only* (Definition 10.1.2). The second version is discussed in Appendix 9.

Note that, for τ_j assuming values of both signs and $t < 0$, we have $H(t) = 0$, $U(t) > 0$, so the function $H(t)$ has a jump of magnitude 1 at the point $t = 0$.

Note also that the functions $H(t)$ and $U(t)$ we defined above are *right-continuous*. In the existing literature, one often considers *left-continuous* versions of renewal functions defined respectively as

$$H(t-0) = \mathbf{E}\min\{k : S_k \geq t\} \quad \text{and} \quad U(t-0) = \sum_{j=0}^{\infty} \mathbf{P}(S_j < t).$$

If all τ_j are identically distributed and $F^{*k}(t)$ is the k-fold convolution of the distribution function $F(t) = \mathbf{P}(\xi_j < t)$, then the second left-continuous version of the renewal function can also be represented in the form

$$\sum_{k=0}^{\infty} F^{*k}(t),$$

where F^{*0} corresponds to the distribution degenerate at zero.

From the point of view of the exposition below, it makes no difference which version of continuity is chosen. For several reasons, in the present chapter it will be more convenient for us to deal with *right-continuous* renewal functions. Everything below will equally apply to left-continuous renewal functions as well.

10.1.2 The Integral Renewal Theorem for Non-identically Distributed Summands

In the case where τ_j, $j \geq 2$, are not necessarily identically distributed and do not possess other homogeneity properties, singling out the random variable τ_1 makes little sense.

Theorem 10.1.1 *Let τ_j, $j \geq 1$, be uniformly integrable from the right, $\mathbf{E}|T_N| < \infty$ for any fixed N and $a_k = \mathbf{E}\tau_k \to a > 0$ as $k \to \infty$. Then the following limit exists*

$$\lim_{t\to\infty} \frac{H(t)}{t} = \frac{1}{a}. \tag{10.1.3}$$

Proof We will need the following definition.

Definition 10.1.3 The random variable

$$\chi(t) = T_{\eta(t)} - t > 0$$

is said to be the *excess* of the level t (or *overshoot* over the level t) for the random walk $\{T_j\}$.

Lemma 10.1.1 *If $a_k \in [a_*, a^*]$, $a_* > 0$, then*

$$\mathbf{E}\eta(t) > \frac{t}{a^*}, \qquad \limsup_{t\to\infty} \frac{\mathbf{E}\eta(t)}{t} \le \frac{1}{a_*}. \tag{10.1.4}$$

Proof By Theorem 4.4.2 (see also Example 4.4.3)

$$\mathbf{E}T_{\eta(t)} = t + \mathbf{E}\chi(t) \le a^* \mathbf{E}\eta(t).$$

This implies the first inequality in (10.1.4). Now introduce truncated random variables $\tau_j^{(s)} := \min(\tau_j, s)$. By virtue of the uniform integrability, one can choose an s such that, for a given $\varepsilon \in (0, a_*)$, we would have

$$a_{j,s} := \mathbf{E}\tau_j^{(s)} \ge a_* - \varepsilon.$$

Then, by Theorem 4.4.2,

$$t + s \ge \mathbf{E}T^{(s)}_{\eta^{(s)}(t)} \ge (a_* - \varepsilon)\mathbf{E}\eta^{(s)},$$

where

$$T_n^{(s)} := \sum_{j=1}^n \tau_j^{(s)}, \qquad \eta^{(s)}(t) := \min\{k : T_k^{(s)} > t\}.$$

Since $\eta(t) \le \eta^{(s)}(t)$, one has

$$H(t) = \mathbf{E}\eta(t) \le \mathbf{E}\eta^{(s)}(t) \le \frac{t+s}{a_* - \varepsilon}. \tag{10.1.5}$$

As $\varepsilon > 0$ can be chosen arbitrarily, we obtain that

$$\limsup_{t\to\infty} \frac{H(t)}{t} \le \frac{1}{a_*}.$$

The lemma is proved. □

We return to the proof of Theorem 10.1.1. For a given $\varepsilon > 0$, find an N such that $a_k \in [a - \varepsilon, a + \varepsilon]$ for all $k > N$ and denote by $H_N(t)$ the renewal function corresponding to the sequence $\{\tau_{N+k}\}_{k=1}^\infty$. Then

$$\begin{aligned} H(t) &= \mathbf{E}\big(\eta(t);\, T_N > t\big) + \int_{-\infty}^t \mathbf{P}(T_N \in du)\big[N + H_N(t-u)\big] \\ &= \mathbf{E}\big[H_N(t - T_N);\, T_N \le t\big] + r_N, \end{aligned} \tag{10.1.6}$$

where

$$r_N := \mathbf{E}\big(\eta(t);\, T_N > t\big) + N\mathbf{P}(T_N \le t) \le N\mathbf{P}(T_N > t) + N\mathbf{P}(T_N \le t) = N.$$

Relation (10.1.5) implies that there exist constants c_1, c_2, such that, for all t,

$$H_N(t) \le c_1 + c_2 t.$$

Therefore, for fixed N and M,

$$\begin{aligned}R_{N,M} &:= \mathbf{E}\big[H_N(t - T_N);\ |T_N| \ge M, T_N \le t\big]\\ &\le (c_1 + c_2 t)\mathbf{P}\big(|T_N| \ge M, T_N \le t\big) + c_2\mathbf{E}|T_N|.\end{aligned}$$

Choose an M such that $c_2\mathbf{P}(|T_N| \ge M) < \varepsilon$. Then

$$\limsup_{t\to\infty} \frac{r_N + R_{N,M}}{t} \le \varepsilon. \tag{10.1.7}$$

To bound $H(t)$ in (10.1.6) it remains to consider, for the chosen N and M, the function

$$H_{N,M}(t) := \mathbf{E}\big[H_N(t - T_N);\ |T_N| < M\big].$$

By Lemma 10.1.1,

$$\begin{aligned}\limsup_{t\to\infty} \frac{H_{N,M}(t)}{t} &\le \frac{1}{a-\varepsilon},\\ \liminf_{t\to\infty} \frac{H_{N,M}(t)}{t} &\ge \frac{\mathbf{P}(|T_N| < M)}{a+\varepsilon} \ge \frac{1+\varepsilon/c_1}{a+\varepsilon}.\end{aligned}$$

This together with (10.1.6) and (10.1.7) yields

$$\limsup_{t\to\infty} \frac{H(t)}{t} \le \varepsilon + \frac{1}{a-\varepsilon}, \qquad \liminf_{t\to\infty} \frac{H(t)}{t} \ge \frac{(1-\varepsilon/c_2)}{a+\varepsilon}.$$

Since ε is arbitrary, the foregoing implies (10.1.3).

The theorem is proved. □

Remark 10.1.2 One can obtain the following generalisation of Theorem 10.1.1, in which no restrictions on $\tau_1 \ge 0$ are imposed. *Let τ_1 be an arbitrary nonnegative random variable, and $\tau_j^* := \tau_{1+j}$ satisfy the conditions of Theorem* 10.1.1. *Then* (10.1.3) *still holds true.*

This assertion follows from the relations

$$H(t) = \mathbf{P}(\tau_1 > t) + \int_0^t \mathbf{P}(\tau_1 \in dv)H^*(t-v), \tag{10.1.8}$$

where $H^*(t)$ corresponds to the sequence $\{\tau_j^*\}$ and, for each fixed N and $v \le N$,

$$\frac{H^*(t-v)}{t} = \frac{H^*(t-v)}{t-v} \cdot \frac{t-v}{t} \to \frac{1}{a}$$

as $t \to \infty$. Therefore

$$\frac{1}{t}\int_0^N \mathbf{P}(\tau_1 \in dv)H^*(t-v) \to \frac{\mathbf{P}(\tau_1 \le N)}{a}.$$

For the remaining part of the integral in (10.1.8), we have

$$\limsup_{t\to\infty} \frac{1}{t}\int_N^t \mathbf{P}(\tau_1 \in dv)H^*(t-v) \le \limsup_{t\to\infty} \frac{H^*(t)}{t}\mathbf{P}(\tau_1 > N) = \frac{\mathbf{P}(\tau_1 > N)}{a}.$$

Since the probability $\mathbf{P}(\tau_1 > N)$ can be made arbitrarily small by the choice of N, the assertion is proved. □

It is not difficult to verify that the condition $\tau_1 \ge 0$ can be relaxed to the condition $\mathbf{E}\min(0,\tau_1) > -\infty$. However, if $\mathbf{E}\min(0,\tau_1) = -\infty$, then $H(t)=\infty$ and relation (10.1.3) does not hold.

Obtaining an analogue of Theorem 10.1.1 for the second version $U(t)$ of the renewal function in the case of *uniformly integrable* τ_j *taking values of both signs* is accompanied by greater technical difficulties and additional conditions. For a fixed $\varepsilon > 0$, split the series $U(t)=\sum_{n=0}^{\infty}\mathbf{P}(T_n \le t)$ into the three parts

$$\sum\nolimits_1 = \sum_{n<\frac{t(1-\varepsilon)}{a}}, \qquad \sum\nolimits_2 = \sum_{|n-\frac{t}{a}|\le\frac{t\varepsilon}{a}}, \qquad \sum\nolimits_3 = \sum_{n>\frac{t(1+\varepsilon)}{a}}.$$

By the law of large numbers (see Corollary 8.3.2),

$$\frac{T_n}{n} \xrightarrow{p} a.$$

Therefore, for $n < \frac{t(1-\varepsilon)}{a}$,

$$\mathbf{P}(T_n \le t) \ge \mathbf{P}\left(T_n \le \frac{na}{1-\varepsilon}\right) \to 1$$

and hence

$$\frac{1}{t}\sum\nolimits_1 \to \frac{1-\varepsilon}{a}.$$

The second sum allows the trivial bound

$$\frac{1}{t}\sum\nolimits_2 < \frac{2\varepsilon}{a},$$

where the right-hand side can be made arbitrarily small by the choice of ε.

The main difficulties are related to estimating $\sum_3$. To illustrate the problems arising here we confine ourselves to the case of identically distributed $\tau_j \overset{d}{=} \tau$. In this case the required estimate for $\sum_3$ can only be obtained under the condition $\mathbf{E}(\tau^-)^2 < \infty$, $\tau^- := \max(0,-\tau)$. Assume without losing generality that $\mathbf{E}\tau^2<\infty$. (If $\mathbf{E}(\tau^+)^2 = \infty$, $\tau^+ := \max(0,\tau)$, then introducing truncated random variables $\tau_j^{(s)} = \min(s,\tau_j)$, we obtain, using obvious conventions concerning notations, that $\mathbf{P}(T_n \le t) \le \mathbf{P}(T_n^{(s)} \le t)$, $U(t) \le U^{(s)}(t)$ and $\sum_3 \le \sum_3^{(s)}$, where $\mathbf{E}(\tau^{(s)})^2 < \infty$ and the value of $\mathbf{E}\tau^{(s)}$ can be made arbitrarily close to a by the choice of s.) In the case $\mathbf{E}\tau^2 < \infty$ we can use Theorem 9.5.1 by virtue of which, for a regularly varying left tail $W(t) = \mathbf{P}(\tau < -t) = t^{-\beta}L(t)$ ($L(t)$ is a slowly varying function) and $n > \frac{t}{a}(1+\varepsilon)$, we have

$$\mathbf{P}(T_n \le t) = \mathbf{P}\big(T_n - an \le -(an-t)\big) \sim nW(an-t).$$

By the properties of slowly varying functions (see Appendix 6), for the values $u = n/t$ comparable to 1, $n > \frac{t}{a}(1+\varepsilon)$ and $t \to \infty$, we have

$$\frac{W(an-t)}{W(\varepsilon t)} \sim \left(\frac{au-1}{\varepsilon}\right)^{-\beta}.$$

Thus for $\beta > 2$, as $t \to \infty$,

$$\sum\nolimits_3 = \sum_{n > \frac{(1+\varepsilon)t}{a}} \mathbf{P}(T_n \le t) \sim \int_{v > \frac{(1+\varepsilon)t}{a}} v W(av - t)\, dv$$

$$\sim t^2 W(\varepsilon t) \int_{\frac{1+\varepsilon}{a}}^{\infty} u \left[\frac{au - 1}{\varepsilon}\right]^{-\beta} du \sim c(\varepsilon) t^2 W(t) = o(1).$$

Summarising, we have obtained that

$$\lim_{t\to\infty} \frac{U(t)}{t} = \frac{1}{a}.$$

Now if $\mathbf{E}(\tau^-)^2 = \infty$ then $U(t) = \infty$ for all t. In this case, instead of $U(t)$ one studies the "local" renewal function

$$U(t,h) = \sum_n \mathbf{P}\big(T_n \in (t, t+h]\big)$$

which is always finite provided that $a > 0$ and has all the properties of the increment $H(t+h) - H(t)$ to be studied below (see e.g. [12]).

In view of the foregoing and since the function $H(t)$ will be of principal interest to us, in what follows we will restrict ourselves to studying the first version of the renewal function, as was noted above. We will mainly pay attention to the asymptotic behaviour of the increments $H(t+h) - H(t)$ as $t \to \infty$. To this is closely related a more general problem that often appears in applications: the problem on the asymptotic behaviour as $t \to \infty$ of integrals (see e.g. Chap. 13)

$$\int_0^t g(t-y)\, dH(y) \tag{10.1.9}$$

for functions $g(v)$ such that

$$\int_0^\infty g(v)\, dv < \infty.$$

Theorems describing the asymptotic behaviour of (10.1.9) will be called the *key renewal theorems*. The next sections and Appendix 9 will be devoted to these theorems. Due to the technical complexity of the mentioned problems, we will confine ourselves to considering only the case where τ_j, $j \ge 2$, are *identically distributed*.

Note that in some special cases the above problems can be solved in a very simple way, since the renewal function $H(t)$ can be found there explicitly. To do this, as it follows from Wald's identity used above, it suffices to find $\mathbf{E}\chi(t)$ in explicit form. If, for instance, τ_j are integer-valued, $\mathbf{P}(\tau_j = 1) > 0$ and $\mathbf{P}(\tau_j \ge 2) = 0$, for all $j \ge 1$, then $\chi(t) \equiv 1$ and Wald's identity yields $H(t) = (t+1)/a$. Similar equalities will hold if $\mathbf{P}(\tau_j > t) = ce^{-\gamma t}$ for $t > 0$ and $\gamma > 0$ (if τ_j are integer-valued, then t takes only integer values in this formula). In that case the distribution of $\chi(t)$ will be exponential and will not depend on t (for more details, see the exposition below and also Chap. 15).

10.2 The Key Renewal Theorem in the Arithmetic Case

We will distinguish between two distribution types for τ_j: *arithmetic in an extended sense* (when the lattice span is not necessary 1; for the definition of arithmetic distributions see Sect. 7.1) and all other distributions that we will call *non-arithmetic*. It is clear that, say, a random variable taking values 1 and $\sqrt{2}$ with positive probabilities cannot be arithmetic.

In the present section, we will consider the arithmetic case. Without loss of generality, we will assume that the lattice span is 1. Then the functions $\mathbf{P}(\tau_j < t)$ and $H(t)$ will be completely determined by their values at integer points $t = k$, $k = 0, 1, 2 \ldots$.

First we consider the case where the τ_j are *positive*, $\tau_j \stackrel{d}{=} \tau$ for $j \geq 2$. In that case, the difference

$$h(k) := H(k) - H(k-1) = \sum_{j=0}^{\infty} \mathbf{P}(T_j = k), \quad k \geq 1,$$

is equal to the expectation of the number of visits of the point k by the walk $\{T_j\}$. Put

$$q_k := \mathbf{P}(\tau_1 = k), \qquad p_k := \mathbf{P}(\tau = k).$$

Definition 10.2.1 A renewal process $\eta(t)$ will be called *homogeneous* and denoted by $\eta_0(t)$ if

$$q_k = \frac{1}{a} \sum_{k}^{\infty} p_j, \quad k = 1, 2, \ldots, \quad a = \mathbf{E}\tau, \quad \left(\text{so that } \sum_{k=1}^{\infty} q_k = 1 \right). \tag{10.2.1}$$

If we denote by $p(z)$ the generating function

$$p(z) = \mathbf{E}z^{\tau} = \sum_{k=1}^{\infty} p_k z^k,$$

then the generating function $q(z) = \mathbf{E}z^{\xi_1} = \sum_{k=1}^{\infty} q_k z^k$ will be equal to

$$q(z) = \frac{1}{a} \sum_{k=1}^{\infty} z^k \sum_{j=k}^{\infty} p_j = \frac{z}{a} \sum_{j=1}^{\infty} p_j \sum_{k=0}^{j-1} z^k = \frac{z}{a} \sum_{j=1}^{\infty} p_j \frac{1 - z^j}{1 - z} = \frac{z(1 - p(z))}{a(1 - z)}.$$

As we will see below, the term "homogeneous" for the process $\eta_0(t)$ is quite justified. One of the reasons for its use is the following exact (non-asymptotic) equality.

Theorem 10.2.1 *For a homogeneous renewal process $\eta_0(t)$, one has*

$$H_0(k) := \mathbf{E}\eta_0(k) = 1 + \frac{k}{a}.$$

Proof Consider the generating function $r(z)$ for the sequence $h_0(k) = H_0(k) - H_0(k-1)$:

$$r(z) = \sum_1^\infty z^k h_0(k) = \sum_{j=1}^\infty \sum_{k=1}^\infty z^k \mathbf{P}(T_j = k)$$
$$= \sum_{j=1}^\infty \mathbf{E} z^{T_j} = q(z) \sum_{j=0}^\infty p^j(z) = \frac{q(z)}{1-p(z)} = \frac{z}{a(1-z)}.$$

This implies that $h_0(k) = 1/a$. Since $H_0(0) = 1$, one has $H_0(k) = 1 + k/a$. The theorem is proved. □

Sometimes the process $\eta_0(t)$ is also called *stationary*. As we will see below, it would be more appropriate to call it a *process with stationary increments* (see Sect. 22.1).

The asymptotic regular behaviour of the function $h(k)$ as $k \to \infty$ persists in the case of arbitrary τ_1 as well.

Denote by d the greatest common divisor (g.c.d.) of the possible values of τ:

$$d := \text{g.c.d.}\{k : p_k > 0\},$$

and let $g(k)$, $k = 0, 1, \ldots$, be an arbitrary sequence such that

$$\sum_{k=0}^\infty \bigl|g(k)\bigr| < \infty.$$

Theorem 10.2.2 (The key renewal theorem) *If $d = 1$, τ_1 is an arbitrary integer-valued random variable and $\tau_j \stackrel{d}{=} \tau > 0$ for $j \ge 2$, then, as $k \to \infty$,*

$$h(k) := H(k) - H(k-1) \to \frac{1}{a}, \qquad \sum_{l=1}^k h(l) g(k-l) \to \frac{1}{a} \sum_{m=0}^\infty g(m).$$

These two relations are equivalent.

The first assertion of the theorem is also called the *local* renewal theorem.

To prove the theorem we will need two auxiliary assertions.

Lemma 10.2.1 *Let all τ_j be identically distributed and $\nu \ge 1$ be a Markov time with respect to the collection of σ-algebras $\{\mathfrak{F}_n\}$, where $\mathfrak{F}_n$ is independent of $\sigma(\tau_{n+1}, \tau_{n+2}, \ldots)$. Then the σ-algebra generated by the random variables ν, $\tau_1, \ldots, \tau_\nu$, and the σ-algebra $\sigma\{\tau_{\nu+1}, \tau_{\nu+2}, \ldots\}$ are independent. The sequence $\{\tau_{\nu+1}, \tau_{\nu+2}, \ldots\}$ has the same distribution as $\{\tau_1, \tau_2, \ldots\}$.*

Thus, in spite of their random numbers, the elements of the sequence $\tau_{\nu+j}$ are distributed as τ_j.

Proof For given Borel sets $B_1, B_2, \ldots, C_1, C_2, \ldots$ put

$$A := \{\nu \in N, \tau_1 \in B_1, \ldots, \tau_\nu \in B_\nu\}, \qquad D_\nu := \{\tau_{\nu+1} \in C_1, \ldots, \tau_{\nu+k} \in C_k\},$$

where $\mathbf{N}$ is a given set of integers and k is arbitrary. Since $\mathbf{P}(D_j) = \mathbf{P}(D_0)$ and the events D_j and $\{\nu = j\}$ are independent, the total probability formula yields

$$\mathbf{P}(D_\nu) = \sum_{j=1}^{\infty} \mathbf{P}(\nu = j, D_j) = \sum_{j=1}^{\infty} \mathbf{P}(\nu = j)\mathbf{P}(D_j) = \mathbf{P}(D_0).$$

Therefore, by Theorem 3.4.3, in order to prove the required independence of the σ-algebras, it suffices to show that $\mathbf{P}(D_\nu A) = \mathbf{P}(D_0)\mathbf{P}(A)$.

By the total probability formula,

$$\mathbf{P}(D_\nu A) = \sum_{j \in \mathbf{N}} \mathbf{P}\big(D_\nu A\{\nu = j\}\big) = \sum_{j \in \mathbf{N}} \mathbf{P}\big(D_j A\{\nu = j\}\big).$$

But the event $A\{\nu = j\}$ belongs to $\mathfrak{F}_j$, whereas $D_j \in \sigma(\tau_{j+1}, \ldots, \tau_{j+k})$. Therefore D_j and $A\{\nu = j\}$ are independent events and

$$\mathbf{P}\big(D_j\, A\{\nu = j\}\big) = \mathbf{P}(D_j)\mathbf{P}\big(A\{\nu = j\}\big) = \mathbf{P}(D_0)\mathbf{P}\big(A\{\nu = j\}\big), \quad j \ge 1.$$

From here it clearly follows that $\mathbf{P}(D_\nu A) = \mathbf{P}(D_0)\mathbf{P}(A)$. The lemma is proved. □

Lemma 10.2.2 *Let $\zeta_1, \zeta_2, \ldots$ be independent arithmetic identically and symmetrically distributed random variables with zero expectation $\mathbf{E}\zeta_j = 0$. Put $Z_n := \sum_{j=1}^{n} \zeta_j$. Then, for any integer k,*

$$\nu_k := \min\{n : Z_n = k\}$$

is a proper random variable: $\mathbf{P}(\nu_k < \infty) = 1$.

The proof of the lemma is given in Sect. 13.3 (see Corollary 13.3.1).

Proof of Theorem 10.2.2 Consider two independent sequences of random variables (we assume that they are given on a common probability space): a sequence $\tau_1, \tau_2, \ldots$, where τ_1 has an arbitrary distribution, and a sequence $\tau'_1, \tau'_2, \ldots$, where $\mathbf{P}(\tau'_1 = k) = q_k$ (see (10.2.1)), and $\mathbf{P}(\tau'_j = k) = \mathbf{P}(\tau_j = k) = p_k$ for $j \ge 2$ (so that $\tau'_j \stackrel{d}{=} \tau_j$ for $j \ge 2$; the process $\eta'(t)$ constructed from the sums $T'_k = \sum_{j=1}^{k} \tau'_j$ is homogeneous (see Definition 10.2.1)).

Set $\nu := \min\{n \ge 1 : T_n = T'_n\}$. It is clearly a Markov time with respect to the sequence $\{\tau_j, \tau'_j\}$. We show that $\mathbf{P}(\nu < \infty) = 1$. Put

$$Z_n := \sum_{j=2}^{n} (\tau_j - \tau'_j) \quad \text{for } n \ge 2,\ Z_1 := 0,\ Z_0 := \tau_1 - \tau'_1.$$

Then

$$\nu = \min\{n \ge 1 : Z_n = -Z_0\}.$$

By Lemma 10.2.2 ($\zeta_j = \tau_j - \tau'_j$ have a symmetric distribution for $j \ge 2$), for each integer k the variable $\nu_k = \min\{n \ge 1 : Z_n = k\}$ is proper. Since Z_n for $n \ge 1$ and Z'_1 are independent, we have

$$\mathbf{P}(\nu < \infty) = \sum_k \mathbf{P}(Z_0 = -k)\mathbf{P}(\nu_k < \infty) = 1.$$

Now we will "glue together" ("couple") the sequences $\{T_n\}$ and $\{T'_n\}$. Since $T_\nu = T'_\nu$ and ν is a Markov time, by Lemma 10.2.1 one can replace $\tau_{\nu+1}, \tau_{\nu+2}, \ldots$ with $\tau'_{\nu+1}, \tau'_{\nu+2}, \ldots$ (and thereby replace $T_{\nu+1}, T_{\nu+2}$ with $T'_{\nu+1}, T'_{\nu+2}, \ldots$) without changing the distribution of the sequence $\{T_n\}$.

Therefore, on the set $\{T_\nu < k\}$ one has $\eta(t) = \eta'(t)$ for $t \ge k-1$ and hence

$$\begin{aligned}
h(k) &= \mathbf{E}\big(\eta(k) - \eta(k-1)\big) \\
&= \mathbf{E}\big[\eta'(k) - \eta'(k-1);\ T_\nu < k\big] + \mathbf{E}\big[\eta(k) - \eta(k-1);\ T_\nu \ge k\big] \\
&= \frac{1}{a} - \mathbf{E}\big[\eta'(k) - \eta'(k-1);\ T_\nu \ge k\big] + \mathbf{E}\big[\eta(k) - \eta(k-1);\ T_\nu \ge k\big].
\end{aligned}$$

Since $|\eta(k) - \eta(k-1)| \le 1$, we have

$$\left| h(k) - \frac{1}{a} \right| \le \mathbf{P}(T_\nu \ge k) \to 0$$

as $k \to \infty$. The first assertion of Theorem 10.2.2 is proved.

Since $h(k) \le 1$, we can make the value of

$$\left| \sum_{l=1}^{k-N} h(l) g(k-l) \right| \le \sum_{l=N+1}^{k-1} |g(l)| \le \sum_{l=N+1}^{\infty} |g(l)|$$

arbitrarily small by choosing an appropriate N. Moreover, by virtue of the first assertion, for any fixed N,

$$\sum_{l=k-N}^{k} h(l) g(k-l) \to \frac{1}{a} \sum_{l=0}^{N} g(l) \quad \text{as } k \to \infty.$$

This implies the second assertion of the theorem. □

Remark 10.2.1 The coupling of $\{T_n\}$ and $\{T'_n\}$ in the proof of Theorem 10.2.2 could be done earlier, at the time $\gamma := \min\{n \ge 1 : T_n \in T'\}$, where T' is the set of points $T' = \{T'_1, T'_2, \ldots\}$.

Theorem 10.2.3 *The assertion of Theorem* 10.2.2 *remains true for arbitrary* (*assuming values of different signs*) τ_j.

Proof We will reduce the problem to the case $\tau_j > 0$. First let all τ_j be identically distributed. Consider the random variable $\chi_1 = \chi(0)$ that we will call the *first positive sum*. We will show in Chap. 12 (see Corollary 12.2.3) that $\mathbf{E}\chi_1 < \infty$ if $a = \mathbf{E}\tau_j < \infty$. According to Lemma 10.2.1, the sequence $\tau_{\eta(0)+1}, \tau_{\eta(0)+2}, \ldots$ will have the same distribution as $\tau_1, \tau_2, \ldots$. Therefore the "second positive sum" χ_2 or, which is the same, the first positive sum of the variables $\tau_{\eta(0)+1}, \tau_{\eta(0)+2}, \ldots$ will have the same distribution as χ_1 and will be independent of it. The same will be true for the subsequent "overshoots" over the already achieved levels $\chi_1, \chi_1 + \chi_2, \ldots$. Now consider the random walk

$$\left\{ H_k = \sum_{i=1}^{k} \chi_i \right\}_{k=1}^{\infty}$$

and put

$$\eta^*(t) := \min\{k : H_k > t\}, \qquad \chi^*(t) := H_{\eta^*(t)} - t, \qquad H^*(t) := \mathbf{E}\eta^*(t).$$

Since $\chi_k > 0$, Theorem 10.2.2 is applicable, and therefore by Wald's identity

$$H^*(k) - H^*(k-1) = \frac{1}{\mathbf{E}\chi_1}\big(1 + \mathbf{E}\chi^*(k) - \mathbf{E}\chi^*(k-1)\big) \to \frac{1}{\mathbf{E}\chi_1},$$
$$\mathbf{E}\chi^*(k) - \mathbf{E}\chi^*(k-1) \to 0.$$

Note now that the distributions of the random variables $\chi(t)$ (see Definition 10.1.3) and $\chi^*(t)$ coincide. Therefore

$$\begin{aligned} H(k) - H(k-1) &= \frac{1}{a}\big(1 + \mathbf{E}\chi(k) - \mathbf{E}\chi(k-1)\big) \\ &= \frac{1}{a}\big(1 + \mathbf{E}\chi^*(k) - \mathbf{E}\chi^*(k-1)\big) \to \frac{1}{a}. \end{aligned}$$

Now let the distributions of τ_1 and τ_j, $j \ge 2$, be different. Then the renewal function $H_1(t)$ for such a walk will be equal to

$$H_1(k) = 1 + \sum_{i=-\infty}^{k} \mathbf{P}(\tau_1 = i)\big[H(k-i) + 1\big] = 1 + \sum_{i=-\infty}^{k} \mathbf{P}(\tau_1 = i)H(k-i),$$

$$h_1(k) = H_1(k) - H_1(k-1) = \sum_{i=-\infty}^{k} \mathbf{P}(\tau_1 = i)h(k-i), \quad k \ge 0,$$

where $H_1(-1) = 0$, $h(0) = H(0)$ and the function $H(t)$ corresponds to identically distributed τ_j. If we had $h(k) < c < \infty$ for all k, that would imply convergence $h_1(k) \to 1/a$ and thus complete the proof of the theorem.

The required inequality $h(k) < c$ actually follows from the following general proposition which is true for arbitrary (not necessarily lattice) random variables τ_j. □

Lemma 10.2.3 *If all τ_j are identically distributed then, for all t and u,*

$$H(t+u) - H(t) \le H(u) \le c_1 + c_2 u.$$

Proof The difference $\eta(t+u) - \eta(t)$ is the number of jumps of the trajectory $\{\widetilde{T}_k\}$ that started at the point $t + \chi(t) > t$ until the first passage of the level $t + u$, where the sequence $\{\widetilde{T}_k\}$ has the same distribution as $\{T_k\}$ and is independent of it (see Lemma 10.2.1). In other words, $\eta(t+u) - \eta(u)$ has the same distribution as $\widetilde{\eta}(t - \chi(t)) \le \widetilde{\eta}(t)$, where $\widetilde{\eta}$ corresponds to $\{\widetilde{T}_k\}$ if $\chi(t) \le u$ and to $\eta(t+u) - \eta(t) = 0$ if $\chi(t) > u$. Therefore $H(t+u) - H(t) \le H(u)$. The inequality for $H(u)$ follows from Theorem 10.2.1. The lemma is proved. □

Theorem 10.2.3 is proved. □

10.3 The Excess and Defect of a Random Walk. Their Limiting Distribution in the Arithmetic Case

Along with the excess $\chi(t) = T_{\eta(t)} - t$ we introduce one more random variable closely related to $\chi(t)$.

Definition 10.3.1 The random variable

$$\gamma(t) := t - T_{\eta(t)-1} = t - T_{\nu(t)}$$

is called the *defect* (or *undershoot*) of the level t in the walk $\{T_n\}$.

The quantity $\chi(t)$ may be thought of as the time during which the component that was working at time t will continue working after that time, while $\gamma(t)$ is the time for which the component has already been working by that time.

One should not think that the sum $\chi(t) + \gamma(t)$ has the same distribution as τ_j—this sum is actually equal to the value of a τ with the *random* subscript $\eta(t)$. In particular, as we will see below, it may turn out that $\mathbf{E}\chi(t) > \mathbf{E}\tau_j$ for large t. The following apparent paradox is related to this fact. A passenger coming to a bus stop at which buses arrive with inter-arrival times $\tau_1 > 0, \tau_2 > 0, \ldots$ ($\mathbf{E}\tau_j = a$), will wait for the arrival of the next bus for a random time χ of which the mean $\mathbf{E}\chi$ could prove to be greater than a.

One of the principal facts of renewal theory is the assertion that, under broad assumptions, the joint distribution of $\chi(t)$ and $\gamma(t)$ has a limit as $t \to \infty$, so that for large t the distribution of $\chi(t)$ does not depend on t any more and becomes stationary. Denote this limiting distribution of $\chi(t)$ by $\mathbf{G}$ and its distribution function by G:

$$G(x) = \lim_{t\to\infty} \mathbf{P}\big(\chi(t) < x\big). \tag{10.3.1}$$

If we take the distribution of τ_1 to be $\mathbf{G}$ then, for such process, by its very construction the distribution of the variable $\chi(t)$ will be independent of t. Indeed, in that case we can think of the positive elements of $\{T_j\}$ as the renewal times for a process which is constructed from the sequence $\{\tau_j\}$ and of which the start is shifted to a point $-N$, where N is very large. Since by virtue of (10.3.1) we can assume that the distributions of $\chi(N)$ and $\chi(N+t)$ coincide with each other, the distribution of the variable $\chi(t)$ (which can be identified with $\chi(N+t)$) is independent of t and coincides with that of τ_1. A formal proof of this fact is omitted, since it will not be used in what follows. However, the reader could carry it out using the explicit form of $G(x)$ from (10.3.1) to be derived below.

In the arithmetic case, the distribution $\mathbf{G}$ is just the law (10.2.1) used to construct the homogeneous renewal process $\eta_0(t)$. We will prove this in our next theorem.

It follows from the fact that, for the process $\eta_0(t)$, the distribution of $\chi(t)$ does not depend on t and coincides with that of τ_1, that the distribution of $\eta_0(t+u) - \eta_0(t)$ coincides with that of $\eta_0(u)$ and hence is also independent of t. It is this property that establishes the stationarity of the increments of the renewal process; we called this property homogeneity. It means that the distribution of the number of renewals over a time interval of length u does not depend on when we start counting, and therefore depends on u only.

Theorems on the limiting distribution of $\chi(t)$ and $\gamma(t)$ are of interest not only from the point of view of their applications. We will need them for a variety of other problems. Again we consider first the case when the variables $\tau_j > 0$ are arithmetic. In that case the "time" can also be assumed discrete and we will denote it, as before, by the letters n and k. Let, as before, $\tau_j \stackrel{d}{=} \tau$ for $j \ge 2$ and $p_k = \mathbf{P}(\tau = k)$.

Theorem 10.3.1 *Let the random variable $\tau > 0$ be arithmetic, $\mathbf{E}\tau = a$ exist, τ_1 be an arbitrary integer random variable, and the g.c.d. of the possible values of τ be equal to 1. Then the following limit exists*

$$\lim_{k\to\infty} \mathbf{P}\big(\gamma(k) = i,\ \chi(k) = j\big) = \frac{p_{i+j}}{a}, \qquad i \ge 0,\ j > 0. \tag{10.3.2}$$

It follows from Theorem 10.3.1 that

$$\begin{aligned} \lim_{k\to\infty} \mathbf{P}\big(\chi(k) = i\big) &= \frac{1}{a}\sum_{j=i}^{\infty} p_j, \quad i > 0; \\ \lim_{k\to\infty} \mathbf{P}\big(\gamma(k) = i\big) &= \frac{1}{a}\sum_{j=i+1}^{\infty} p_j, \quad j \ge 0. \end{aligned} \tag{10.3.3}$$

Proof of Theorem 10.3.1 By the renewal theorem (see Theorem 10.2.2), for $k > i$,

$$\begin{aligned} \mathbf{P}\big(\gamma(k) = i,\ \chi(k) = j\big) &= \sum_{l=1}^{\infty} \mathbf{P}(T_l = k - i,\ \tau_{l+1} = i + j) \\ &= \sum_{l=1}^{\infty} \mathbf{P}(T_l = k - i)\mathbf{P}(\tau = i + j) = h(k-i)p_{i+j} \to \frac{p_{i+j}}{a} \end{aligned}$$

as $k \to \infty$. The theorem is proved. □

If $\mathbf{E}\tau^2 = m_2 < \infty$, then Theorem 10.3.1 allows a refinement of Theorem 10.2.2 (see Theorem 10.3.2 below).

Corollary 10.3.1 *If $m_2 < \infty$, then the random variables $\chi(k)$ are uniformly integrable and*

$$\mathbf{E}\chi(k) \to \frac{1}{a}\sum_{i=0}^{\infty} i \sum_{j=i}^{\infty} p_j = \frac{m_2 + a}{2a} \quad \textit{as } k \to \infty. \tag{10.3.4}$$

Proof The uniform integrability follows from the inequalities $h(k) \le 1$,

$$\mathbf{P}\big(\chi(k) = j\big) = \sum_{i=0}^{k} h(k-i)p_{i+j} \le \sum_{i=j}^{\infty} p_i.$$

This implies (10.3.4) (see Sect. 6.1). □

Now we can state a refined version of the integral theorem that implies Theorem 10.2.2.

Theorem 10.3.2 *If all τ_j are identically distributed and* $\mathbf{E}\tau^2 = m_2 < \infty$, *then*

$$H(k) = \frac{k}{a} + \frac{m_2 + a}{2a^2} + o(1)$$

as $k \to \infty$.

The Proof immediately follows from the Wald identity

$$H(k) = \mathbf{E}\eta(k) = \frac{k + \mathbf{E}\chi(k)}{a}$$

and Corollary 10.3.1. □

Remark 10.3.1 For the process $\eta^*(t)$ corresponding to nonzero times τ'_j required for components' renewals (mentioned in Remark 10.1.1), the reader can easily find, similarly to Theorem 10.3.1, not only the asymptotic value p_{i+j}/a^* of the probability that at time $k \to \infty$ the current component has already worked for time i and will still work for time j, but also the asymptotics of the probability that the component has been "under repair" for time i and will stay in that state for time j, that is given by p'_{i+j}/a^*, where $p'_i = \mathbf{P}(\tau'_j = i)$, $a^* = \mathbf{E}(\tau_j + \tau'_j) = \mathbf{E}\tau^*_j$.

Now consider the question of under what circumstances the distribution of the random variable τ_1 for the homogeneous process (i.e. the distribution of what one could denote by $\chi(\infty)$) will coincide with that of τ_j for $j \ge 2$. Such a coincidence is equivalent to the equality

$$p_i = \frac{1}{a}\sum_{j=i}^{\infty} p_j$$

for $i = 1, 2, \ldots$, or, which is the same, to

$$a(p_i - p_{i-1}) = -p_{i-1}, \qquad p_i = \frac{a-1}{a} p_{i-1}, \qquad p_i = \frac{1}{a-1}\left(\frac{a-1}{a}\right)^i.$$

This means that the renewal process generated by the sequence of independent identically distributed random variables $\tau_1, \tau_2, \ldots$ is homogeneous if and only if τ_j (or, more precisely, τ_{j-1}) have the geometric distribution.

Denote by γ and χ the random variables having distribution (10.3.2). Using (10.3.1), it is not hard to show that γ and χ are independent also only in the case when τ_j, $j \ge 2$, have the geometric distribution. When all τ_j, $j \ge 1$, have such a distribution, $\gamma(n)$ and $\chi(n)$ are also independent, and $\chi(n) \stackrel{d}{=} \tau_1$. These facts can be proved in exactly the same way as for the exponential distribution (see Sect. 10.4).

We now return to the general case and recall that if $\mathbf{E}\tau^2 < \infty$ then (see Corollary 10.3.1)

$$\mathbf{E}\chi = \frac{\mathbf{E}\tau^2}{2a} + \frac{1}{2}.$$

This means, in particular, that if the distribution of τ is such that $\mathbf{E}\tau^2 > 2a^2 - a$, then, for large n, the excess mean value $\mathbf{E}\chi(n)$ will become greater than $\mathbf{E}\tau = a$.

10.4 The Renewal Theorem and the Limiting Behaviour of the Excess and Defect in the Non-arithmetic Case

Recall that in this chapter by the non-arithmetic case we mean that there exists no $h > 0$ such that $\mathbf{P}(\bigcup_k \{\tau = kh\}) = 1$, where k runs over all integers. To state the key renewal theorem in that case, we will need the notion of a *directly integrable function*.

Definition 10.4.1 A function $g(u)$ defined on $[0, \infty)$ is said to be *directly integrable* if:

(1) the function g is Riemann integrable[1] over any finite interval $[0, N]$; and
(2) $\sum_k g(k) < \infty$, where $g_k = \max_{k \le u \le k+1} |g(u)|$.

It is evident that any monotonically decreasing function $g(t) \downarrow 0$ having a finite Lebesgue integral

$$\int_0^\infty g(t)\,dt < \infty$$

is directly integrable. This also holds for differences of such functions.

The notion of directly integrable functions introduced in [12] differs somewhat from the one just defined, although it essentially coincides with it. It will be more convenient for us to use Definition 10.4.1, since it allows us to simplify to some extent the exposition and to avoid auxiliary arguments (see Appendix 9).

Theorem 10.4.1 (The key renewal theorem) *Let $\tau_j \stackrel{d}{=} \tau \ge 0$ for $j \ge 2$ and g be a directly integrable function. If the random variable τ is non-arithmetic, there exists* $\mathbf{E}\tau = a > 0$, *and the distribution of τ_1 is arbitrary, then, as $t \to \infty$,*

$$\int_0^t g(t-u)\,dH(u) \to \frac{1}{a}\int_0^\infty g(u)\,du. \tag{10.4.1}$$

There is a measure $\mathbf{H}$ on $[0, \infty)$ associated with H that is defined by $\mathbf{H}((x, y]) := H(y) - H(x)$. The integral

$$\int_0^t g(t-u)\,dH(u)$$

[1] That is, the sums $n^{-1}\sum_k \underline{g}_k$ and $n^{-1}\sum_k \overline{g}_k$ have the same limits as $n \to \infty$, where $\underline{g}_k = \min_{u\in\Delta_k} g(u)$, $\overline{g}_k = \max_{u\in\Delta_k} g(u)$, $\Delta_k = [k\Delta, (k+1)\Delta)$, and $\Delta = N/n$. The usual definition of Riemann integrability over $[0, \infty)$ assumes that condition (1) of Definition 10.4.1 is satisfied and the limit of $\int_0^N g(u)\,du$ as $N \to \infty$ exists. This approach covers a wider class of functions than in Definition 10.4.1, allowing, for example, the existence of a sequence $t_k \to \infty$ such that $g(t_k) \to \infty$.

in (10.4.1) can also be written as

$$\int_0^t g(t-u)\,\mathbf{H}(du).$$

It follows from (10.4.1), in particular, that, for any fixed u,

$$H(t) - H(t-u) \to \frac{u}{a}. \qquad (10.4.2)$$

It is not hard to see that this relation, which is called the *local renewal theorem*, is equivalent to (10.4.1).

The proof of Theorem 10.4.1 is technically rather difficult, so we have placed it in Appendix 9. One can also find there refinements of Theorem 10.4.1 and its analogue in the case where τ has a density.

The other assertions of Sects. 10.2 and 10.3 can also be extended to the non-arithmetic case without any difficulties. Let all τ_j be nonnegative.

Definition 10.4.2 In the non-arithmetic case, a renewal process $\eta(t)$ is called *homogeneous* (and is denoted by $\eta_0(t)$) if the distribution of the first jump has the form

$$\mathbf{P}(\tau_1 > x) = \frac{1}{a}\int_x^\infty \mathbf{P}(\tau > t)\,dt.$$

The ch.f. of τ_1 equals

$$\varphi_{\tau_1}(\lambda) := \mathbf{E}e^{i\lambda\tau_1} = \frac{1}{a}\int_0^\infty e^{i\lambda x}\mathbf{P}(\tau > x)\,dx.$$

Since here we are integrating over $x > 0$, the integral exists (as well as the function $\varphi(\lambda) = \varphi_\tau(\lambda) := \mathbf{E}e^{i\lambda\tau}$) for all λ with $\operatorname{Im}\lambda > 0$ (for $\lambda = i\alpha + v$, $-\infty < v < \infty$, $\alpha \ge 0$, the factor $e^{i\lambda x}$ is equal to $e^{-\alpha x}e^{ivx}$; see property 6 of ch.f.s). Therefore, for $\operatorname{Im}\lambda \ge 0$,

$$\varphi_{\tau_1}(\lambda) = -\frac{1}{i\lambda a}\left[1 + \int_0^\infty e^{i\lambda x}\,d\mathbf{P}(\tau > x)\right] = \frac{\varphi(\lambda) - 1}{i\lambda a}.$$

Theorem 10.4.2 *For a homogeneous renewal process,*

$$H_0(t) \equiv \mathbf{E}\eta_0(t) = 1 + \frac{t}{a}, \quad t \ge 0.$$

Proof This theorem can be proved in the same way as Theorem 10.2.1. Consider the Fourier–Stieltjes transform of the function $H_0(t)$:

$$r(\lambda) := \int_0^\infty e^{i\lambda x}\,dH_0(x).$$

Note that this transform exists for $\operatorname{Im}\lambda > 0$ and the uniqueness theorem established for ch.f.s remains true for it, since $\varphi^*(v) := r(i\alpha + v)/r(i\alpha)$, $-\infty < v < \infty$ (we put $\lambda = i\alpha + v$ for a fixed $\alpha > 0$) can be considered as the ch.f. of a certain distribution being the "Cramér transform" (see Chap. 9) of $H_0(t)$.

Since $\tau_j \ge 0$, one has

$$H_0(x) = \sum_{j=0}^{\infty} \mathbf{P}(T_j \le x).$$

As $H_0(t)$ has a unit jump at $t = 0$, we obtain

$$r(\lambda) = \int_0^\infty e^{i\lambda x}\, dH_0(x) = 1 + \sum_{j=0}^{\infty} \varphi_{\tau_1}(\lambda)\varphi^j(\lambda) = 1 + \frac{\varphi(\lambda) - 1}{i\lambda a} \frac{1}{1 - \varphi(\lambda)}$$
$$= 1 - \frac{1}{i\lambda a}.$$

It is evident that this transform corresponds to the function $H_0(t) = 1 + t/a$. The theorem is proved. □

In the non-arithmetic case, one has the same connections between the homogeneous renewal process $\eta_0(t)$ and the limiting distribution of $\chi(t)$ and $\gamma(t)$ as we had in the arithmetic case. In the same way as in Sect. 10.3, we can derive from the renewal theorem the following.

Theorem 10.4.3 *If $\tau \ge 0$ is non-arithmetic, $\mathbf{E}\tau = a$, and the distribution of $\tau_1 \ge 0$ is arbitrary, then the following limit exists*

$$\lim_{t\to\infty} \mathbf{P}\bigl(\gamma(t) > u,\, \chi(t) > v\bigr) = \frac{1}{a}\int_{u+v}^\infty \mathbf{P}(\tau > x)\, dx. \tag{10.4.3}$$

Proof For $t > u$, by the total probability formula,

$$\begin{aligned}
&\mathbf{P}\bigl(\gamma(t) > u, \chi(t) > v\bigr)\\
&= \mathbf{P}(\tau_1 > t + v) + \sum_{j=1}^{\infty} \int_0^{t-u} \mathbf{P}\bigl(\eta(t) = j + 1, T_j \in dx, \gamma(t) > u, \chi(t) > v\bigr)\\
&= \mathbf{P}(\tau_1 > t + v) + \sum_{j=1}^{\infty} \int_0^{t-u} \mathbf{P}(T_j \in dx, \tau_{j+1} > t - x + v)\\
&= \mathbf{P}(\tau_1 > t + v) - \mathbf{P}(\tau > t + v) + \int_0^{t-u} dH(x)\mathbf{P}(\tau > t - x + v).
\end{aligned} \tag{10.4.4}$$

Here the first two summands on the right-hand side converge to 0 as $t \to \infty$. By the renewal theorem for $g(x) = \mathbf{P}(\tau > x + u + v)$ (see (10.4.1)), the last integral converges to

$$\frac{1}{a}\int_0^\infty \mathbf{P}(\tau > x + u + v)\, dx.$$

The theorem is proved. □

As was the case in the previous section (see Theorem 10.3.2), in the case $\mathbf{E}\tau^2 = m_2 < \infty$ Theorem 10.4.3 allows us to refine the key renewal theorem.

Theorem 10.4.4 *If all* $\tau_j \stackrel{d}{=} \tau \ge 0$ *are identically distributed and* $\mathbf{E}\tau^2 = m_2 < \infty$, *then, as* $t \to \infty$,

$$H(t) = \frac{t}{a} + \frac{m_2}{2a^2} + o(1).$$

Proof From (10.4.4) for $u = 0$ and Lemma 10.2.3 it follows that $\chi(t)$ are uniformly integrable, for

$$\mathbf{P}\big(\chi(t) > v\big) = \int_0^t dH(x)\,\mathbf{P}(\tau > t - x + v) < (c_1 + c_2)\sum_{k\ge 0}\mathbf{P}(\tau > k + v), \tag{10.4.5}$$

and therefore by (4.4.3)

$$\mathbf{E}\chi(t) \to \frac{1}{a}\int_0^\infty \int_v^\infty \mathbf{P}(\tau > u)\,du\,dv = \frac{m_2}{2a}. \tag{10.4.6}$$

It remains to make use of Wald's identity. The theorem is proved. □

One can add to relation (10.4.6) that, under the conditions of Theorem 10.4.4, one has

$$\mathbf{E}\chi^2(t) = o(t) \tag{10.4.7}$$

as $t \to \infty$. Indeed, (10.4.5) and Lemma 10.2.3 imply

$$\mathbf{P}\big(\chi(t) > v\big) < (c_1 + c_2)\sum_{k\le t}\mathbf{P}(\tau > k + v) < c\int_0^t \mathbf{P}(\tau > z + v)\,dz.$$

Further, integrating by parts, we obtain

$$\begin{aligned}\mathbf{E}\chi^2(t) &= -\int_0^\infty v^2\,d\mathbf{P}\big(\chi(t) > v\big)\\ &= 2\int_0^\infty v\mathbf{P}\big(\chi(t) > v\big)\,dv < 2c\int_0^t \int_0^\infty v\mathbf{P}(\tau > z + v)\,dv\,dz,\end{aligned} \tag{10.4.8}$$

where the inner integral converges to zero as $z \to \infty$:

$$\int_0^\infty v\mathbf{P}(\tau > z + v)\,dv = \frac{1}{2}\int_0^\infty v^2\,d\mathbf{P}(\tau < z + v) < \frac{1}{2}\mathbf{E}\big(\tau^2; \tau > z\big) \to 0.$$

This and (10.4.8) imply (10.4.7).

Note also that if only $\mathbf{E}\tau$ exists, then, by Theorem 10.1.1, we have $\mathbf{E}\chi(t) = o(t)$ and, by Theorem 10.4.1 (or 10.4.3),

$$\mathbf{P}\big(\chi(t) > v\big) \to \frac{1}{a}\int_0^\infty \mathbf{P}(\tau > u + v)\,du.$$

Now let, as before, γ and χ denote random variables distributed according to the limiting distribution (10.4.3). Similarly to the above, it is not hard to establish that if $\mathbf{E}\tau^k < \infty$, $k \ge 1$, then, as $t \to \infty$,

$$\mathbf{E}\chi^{k-1}(t) \to \mathbf{E}\chi^{k-1} < \infty, \qquad \mathbf{E}\chi^k(t) = o(t).$$

Further, it is seen from Theorem 10.4.3 that each of the random variables γ and χ has density equal to $a^{-1}\mathbf{P}(\tau > x)$. The joint distribution of γ and χ may have no density. If τ has density $f(x)$ then there exists a joint density of γ and χ equal to $a^{-1}f(x+y)$. It also follows from Theorem 10.4.3 that γ and χ are independent if and only if

$$\int_x^\infty \mathbf{P}(\tau > u)\,du = \frac{1}{\alpha}e^{-\alpha x}$$

for some $\alpha > 0$, i.e. independence takes place only for the exponential distribution $\tau \in\!\!\!= \boldsymbol{\Gamma}_\alpha$.

Moreover, for *homogeneous* renewal processes the coincidence of $\mathbf{P}(\tau_1 > x)$ and $\mathbf{P}(\tau > x)$ takes place only when $\tau \in\!\!\!= \boldsymbol{\Gamma}_\alpha$. In other words, the renewal process generated by a sequence of identically distributed random variables $\tau_1, \tau, \dots$ will be homogeneous if and only if $\tau_j \in\!\!\!= \boldsymbol{\Gamma}_\alpha$. In that case $\eta_0(t)$ is called (see also Sect. 19.4) a *Poisson process*. This is because for such a process, for each t, the variable $\eta(t) = \eta_0(t)$ has the Poisson distribution with parameter t/α.

The Poisson process has some other remarkable properties as well (see also Sect. 19.4). Clearly, one has $\chi(t) \in\!\!\!= \boldsymbol{\Gamma}_\alpha$ for such a process, and moreover, the variables $\gamma(t)$ and $\chi(t)$ are independent. Indeed, by (10.4.4), taking into account that $H(x)$ has a jump of magnitude 1 at the point $x = 0$, we obtain for $u < t$ that

$$\begin{aligned}\mathbf{P}\big(\gamma(t) > u, \chi(t) > v\big) &= e^{-\alpha(t+v)} + \alpha\int_0^{t-u} e^{-\alpha(t-x+v)}\,dx\\ &= e^{-\alpha(u+v)} = \mathbf{P}\big(\gamma(t) > u\big)\mathbf{P}\big(\chi(t) > v\big);\\ \mathbf{P}\big(\gamma(t) = t, \chi(t) > v\big) &= \mathbf{P}(\tau_1 > t+v) = e^{-\alpha(t+v)} = \mathbf{P}\big(\gamma(t) = t\big)\mathbf{P}\big(\chi(t) > v\big);\\ \mathbf{P}\big(\gamma(t) > t\big) &= 0.\end{aligned}$$

These relations also imply that the random variable $\tau_{\eta(t)} = \gamma(t) + \chi(t)$ has the same distribution as $\min(t, \tau_1) + \tau_2$, where $\tau_j \in\!\!\!= \boldsymbol{\Gamma}_\alpha$, $j = 1, 2$, are independent so that $\tau_{\eta(t)} \Leftrightarrow \boldsymbol{\Gamma}_{\alpha,2}$ as $t \to \infty$.

The fact that $\gamma(t)$ and $\chi(t)$ are independent of each other deserves attention from the point of view of its interpretation. It means the following. The residual lifetime of the component operating at a given time t has the same distribution as the lifetime of a *new* component (recall that $\tau_j \in\!\!\!= \boldsymbol{\Gamma}_\alpha$) and is independent of how long this component has already been working (which at first glance is a paradox). Since the lifetime distributions of devices consisting of large numbers of reliable elements are close to the exponential law (see Theorem 20.3.2), the above-mentioned fact is of significant practical interest.

If τ_i can assume *negative* values as well, the problems related to the distributions of $\gamma(t)$ and $\chi(t)$ become much more complicated. To some extent such problems can be reduced to the case of nonnegative variables, since the distribution of $\chi(t)$ coincides with that of the variable $\chi^*(t)$ constructed from a sequence $\{\tau_j^* \ge 0\}$, where τ_j^* have the same distribution as $\chi(0)$. The distribution of $\chi(0)$ can be found using the methods of Chap. 12.

In particular, for random variables $\tau_1, \tau_2, \ldots$ *taking values of both signs*, Theorems 10.4.1 and 10.4.3 imply the following assertion.

Corollary 10.4.1 *Let $\tau_1, \tau_2, \ldots$ be non-arithmetic independent and identically distributed and $\mathbf{E}\tau_1 = a$. Then the following limit exists*

$$\lim_{t\to\infty} \mathbf{P}\big(\chi(t) > v\big) = \frac{1}{\mathbf{E}\chi(0)} \int_v^\infty \mathbf{P}\big(\chi(0) > t\big)\, dt, \quad v > 0.$$

For arithmetic τ_j,

$$\lim_{k\to\infty} \mathbf{P}\big(\chi(k) = i\big) = \frac{1}{\mathbf{E}\chi(0)} \mathbf{P}\big(\chi(0) > i\big), \quad i > 0.$$

10.5 The Law of Large Numbers and the Central Limit Theorem for Renewal Processes

In this section we return to the general case where τ_j are not necessarily identically distributed (cf. Sect. 10.1).

10.5.1 The Law of Large Numbers

First assume that $\tau_j \ge 0$ and put

$$a_k := \mathbf{E}\tau_k, \qquad A_n := \sum_{k=1}^n a_k.$$

Theorem 10.5.1 *Let $\tau_k \ge 0$ be independent, $\tau_k - a_k$ uniformly integrable, and $n^{-1}A_n \to a > 0$ as $n \to \infty$. Then, as $t \to \infty$,*

$$\frac{\eta(t)}{t} \xrightarrow{p} \frac{1}{a}.$$

Proof The basic relation we shall use is the equality

$$\big\{\eta(t) > n\big\} = \{T_n \le t\}, \tag{10.5.1}$$

which implies

$$\mathbf{P}\left(\frac{\eta(t)}{t} - \frac{1}{a} > \varepsilon\right) = \mathbf{P}\left(\eta(t) > \frac{t}{a}(+\varepsilon)\right) = \mathbf{P}(T_n \le t),$$

where for simplicity we assume that $n = \frac{t}{a}(1+\varepsilon)$ is an integer. Further,

$$\begin{aligned}\mathbf{P}(T_n \le t) &= \mathbf{P}\left(\frac{T_n}{n} \le \frac{a}{1+\varepsilon}\right)\\ &= \mathbf{P}\left(\frac{T_n - A_n}{n} \le \frac{a}{1+\varepsilon} - \frac{A_n}{n}\right) \le \mathbf{P}\left(\frac{T_n - A_n}{n} \le -\frac{a\varepsilon}{2}\right)\end{aligned}$$

for n large enough and ε small enough. Applying the law of large numbers to the right-hand side of this relation (Theorem 8.3.3), we obtain that, for any $\varepsilon > 0$, as $t \to \infty$,

$$\mathbf{P}\left(\frac{\eta(t)}{t} - \frac{1}{a} > \frac{\varepsilon}{a}\right) \to 0.$$

The probability $\mathbf{P}(\frac{\eta(t)}{t} - \frac{1}{a} < -\frac{\varepsilon}{a})$ can be bounded in the same way. The theorem is proved. □

10.5.2 The Central Limit Theorem

Put

$$\sigma_k^2 := \mathbf{E}(\tau_k - a_k)^2 = \operatorname{Var} \tau_k, \qquad B_n^2 := \sum_{k=1}^{n} \sigma_k^2.$$

Theorem 10.5.2 *Let $\tau_k \ge 0$ and the random variables $\tau_k - a_k$ satisfy the Lindeberg condition: for any $\delta > 0$ and $n \to \infty$,*

$$\sum_{k=1}^{n} \mathbf{E}\big(|\tau_k - a_k|^2;\, |\tau_k - a_k| > \delta B_n\big) = o\big(B_n^2\big).$$

Let, moreover, there exist $a > 0$ and $\sigma > 0$ such that, as $n \to \infty$,

$$A_n := \sum_{k=1}^{n} a_k = an + o(\sqrt{n}), \qquad B_n^2 = \sigma^2 n + o(n). \tag{10.5.2}$$

Then

$$\frac{\eta(t) - t/a}{\sigma\sqrt{t/a^3}} \Subset \mathbf{\Phi}_{0,1}. \tag{10.5.3}$$

Proof From (10.5.1) we have

$$\mathbf{P}\big(\eta(t) > n\big) = \mathbf{P}(T_n \le t) = \mathbf{P}\left(\frac{T_n - A_n}{B_n} \le \frac{t - A_n}{B_n}\right). \tag{10.5.4}$$

Let n vary as $t \to \infty$ so that

$$\frac{t - A_n}{B_n} \to v$$

for a fixed v. To find such an n, solve for n the equation

$$\frac{t-an}{\sigma\sqrt{n}}=v.$$

This is a quadratic equation in n, and its solution has the form

$$n=\frac{t}{a}\pm\frac{v\sigma}{a^2}\sqrt{at}\left(1+O\left(\frac{1}{\sqrt{t}}\right)\right). \tag{10.5.5}$$

For such n, by (10.5.2),

$$\frac{t-A_n}{B_n}=\left[\mp\frac{v\sigma}{a}\sqrt{at}+o(\sqrt{t})\right]\frac{(1+o(1))}{\sigma\sqrt{t/a}}=\mp v+o(1).$$

This equality means that we have to choose the minus sign in (10.5.5). Therefore, by (10.5.4) and the central limit theorem,

$$\mathbf{P}\big(\eta(t)>n\big)=\mathbf{P}\left(\frac{\eta(t)-t/a}{\sigma\sqrt{t/a^3}}>-v+o(1)\right)\to\Phi(v)=1-\Phi(-v).$$

Changing $-v$ to u, by the continuity theorems (see Lemma 6.2.2) we get

$$\mathbf{P}\left(\frac{\eta(t)-t/a}{\sigma\sqrt{ta^{-3}}}<u\right)\to\Phi(u).$$

The theorem is proved. □

Remark 10.5.1 In Theorems 10.5.1 and 10.5.2 we considered the case where A_n grows asymptotically linearly as $n\to\infty$. Then the centring parameter t/a for $\eta(t)$ changes asymptotically linearly as well. However, nothing prevents us from considering a more general case where, say, $A_n\sim cn^\alpha$, $\alpha>0$. Then the centring parameter for $\eta(t)$ will be the solution to the equation $cn^\alpha=t$, i.e. the function $(t/c)^{1/\alpha}$ (under the conditions of Theorem 10.5.2, in this case we have to assume that $B_n=o(A_n)$). The asymptotics of the renewal function will have the same form.

In order to extend the assertions of Theorems 10.5.1 and 10.5.2 to τ_j assuming values of both signs, we need some auxiliary assertions that are also of independent interest.

10.5.3 A Theorem on the Finiteness of the Infimum of the Cumulative Sums

In this subsection we will consider *identically distributed* independent random variables $\tau_1,\tau_2,\dots$. We first state the following simple assertion in the form of a lemma.

Lemma 10.5.1 *One has* $\mathbf{E}|\tau|<\infty$ *if and only if*

$$\sum_{j=1}^{\infty}\mathbf{P}\big(|\tau|>j\big)<\infty.$$

The Proof follows in an obvious way from the equality

$$\mathbf{E}|\tau| = \int_0^\infty \mathbf{P}\big(|\tau| > x\big)\,dx$$

and the inequalities

$$\sum_{j=1}^{\infty} \mathbf{P}\big(|\tau| > j\big) \le \int_0^\infty \mathbf{P}\big(|\tau| > x\big)\,dx \le 1 + \sum_{j=1}^{\infty} \mathbf{P}\big(|\tau| > j\big).$$ □

Let, as before,

$$T_n = \sum_{j=1}^{n} \tau_j.$$

Theorem 10.5.3 *If $\tau_j \stackrel{d}{=} \tau$ are identically distributed and independent and* $\mathbf{E}\tau > 0$, *then the random variable $Z := \inf_{k\ge 0} T_k$ is proper (finite with probability* 1).

Proof Let $\eta_1 = \eta(1)$ be the number of the first sum T_k to exceed level 1. Consider the sequence $\{\tau_k^* = \tau_{\eta_1+k}\}$ that, by Lemma 10.2.1, has the same distribution as $\{\tau_k\}$ and is independent of $\eta_1, \tau_1, \ldots, \tau_{\eta_1}$. For this sequence, denote by η_2 the subscript k for which the sum $T_k^* = \sum_{j=1}^{k} \tau_j^*$ first exceeds level 1. It is clear that the random variables η_1 and η_2 are identically distributed and independent. Next, construct for the sequence $\{\tau_k^{**} = \tau_{\eta_1+\eta_2+k}\}$ the random variable η_3 following the same rule, and so on. As a result we will obtain a sequence of Markov times $\eta_1, \eta_2, \ldots$ that determine the times of "renewals" of the original sequence $\{T_k\}$, associated with attaining level 1.

Now set

$$Z_1 := \min_{k<\eta_1} T_k, \qquad Z_2 := \min_{k<\eta_2} T_k^*, \ldots$$

Clearly, the Z_j are identically distributed and

$$Z = \inf\{Z_1, T_{\eta_1} + Z_2, T_{\eta_1+\eta_2} + Z_3, \ldots\},$$

where by definition $T_{\eta_1} > 1$, $T_{\eta_1+\eta_2} > 2$ and so on. Hence

$$\{Z < -N\} = \bigcup_{k=0}^{\infty} \{Z_{k+1} + T_{\eta_1+\cdots+\eta_k} < -N\} \subset \bigcup_{k=0}^{\infty} \{Z_k + k < -N\},$$

$$\mathbf{P}(Z < -N) \le \sum_{k=1}^{\infty} \mathbf{P}(Z_k + k < -N) = \sum_{j=N+1}^{\infty} \mathbf{P}(Z_1 < -j).$$

This expression tends to 0 as $N \to \infty$ provided that $\mathbf{E}|Z_1| < \infty$ (see Lemma 10.5.1). It remains to verify the finiteness of $\mathbf{E}Z_1$, which follows from the finiteness of $\mathbf{E}\eta_1 = \mathbf{E}\eta(1) = H(1) < c$ (see Example 4.4.5) and the relations

$$\mathbf{E}|Z_1| \le \mathbf{E}\sum_{j=1}^{\eta_1} |\tau_j| = \mathbf{E}\eta_1 \mathbf{E}|\tau_1| < \infty$$

(see Theorem 4.4.2). □

10.5.4 Stochastic Inequalities. The Law of Large Numbers and the Central Limit Theorem for the Maximum of Sums of Non-identically Distributed Random Variables Taking Values of Both Signs

In this subsection we extend the assertions of some theorems of Chap. 8 to maxima of sums of random variables with a positive "mean drift". To do this we will have to introduce some additions restrictions that are always satisfied when the summands are identically distributed. Here we will need the notion of stochastic inequalities (or inequalities in distribution). Let ξ and ζ be given random variables.

Definition 10.5.1 We will say that ζ majorises (minorises) ξ in distributionand denote this by $\xi \overset{d}{\le} \zeta$ ($\xi \overset{d}{\ge} \zeta$) if, for all t,

$$\mathbf{P}(\xi \ge t) \le \mathbf{P}(\zeta \ge t) \qquad \big(\mathbf{P}(\xi \ge t) \ge \mathbf{P}(\zeta \ge t)\big).$$

Clearly, if $\xi \overset{d}{\le} \zeta$ then $-\xi \overset{d}{\ge} -\zeta$. We show that stochastic inequalities possess some other properties of ordinary inequalities.

Lemma 10.5.2 *If $\{\xi_k\}_{k=1}^{\infty}$ and $\{\zeta\}_{k=1}^{\infty}$ are sequences of independent (in each sequence) random variables and $\xi_k \overset{d}{\le} \zeta_k$, then, for all n,*

$$S_n \overset{d}{\le} Z_n, \qquad \overline{S}_n \overset{d}{\le} \overline{Z}_n,$$

where

$$S_n = \sum_{k=1}^{n} \xi_k, \qquad Z_n = \sum_{k=1}^{n} \zeta_k, \qquad \overline{S}_n = \max_{k\le n} S_k, \qquad \overline{Z}_n = \max_{k\le n} Z_k.$$

Similarly, if $\xi_k \overset{d}{\ge} \zeta_k$, then $\min_{k\le n} S_k \overset{d}{\ge} \min_{k\le n} Z_k$.

Proof Let $F_k(t) := \mathbf{P}(\xi_k < t)$ and $G_k(t) := \mathbf{P}(\zeta_k < t)$. Using quantile transformations $F_k^{(-1)}$ and $G_k^{(-1)}$ (see Definition 3.2.4) and a sequence of independent random variables $\{\omega_k\}_{k=1}^{\infty}$, $\omega_k \in \mathbf{U}_{0,1}$, we can construct on a common probability space the sequences $\xi_k^* = F_k^{(-1)}(\omega_k)$ and $\zeta_k^* = G_k^{(-1)}(\omega_k)$ such that $\xi_k^* \overset{d}{=} \xi_k$ and $\zeta_k^* \overset{d}{=} \zeta_k$ (the distributions of ξ_k^* and ξ_k and of ζ_k^* and ζ_k^* coincide). Moreover, $\xi_k^* \le \zeta_k^*$, which is a direct consequence of the inequality $F_k(t) \ge G_k(t)$ for all t. Endowing with the superscript * all the notations for sums and maximum of sums of random variables with asterisks, we obviously obtain that

$$S_n \overset{d}{=} S_n^* \le Z_n^* \overset{d}{=} Z_n, \qquad \overline{S}_n \overset{d}{=} \overline{S}_n^* \le \overline{Z}_n^* \overset{d}{=} \overline{Z}_n.$$

The last assertion of the lemma follows from the previous ones. The lemma is proved. □

Below we will need the following corollary of Theorem 10.5.3.

Lemma 10.5.3 *Let ξ_k be independent, $\xi_k \overset{d}{\geq} \zeta$ for all k and $\mathbf{E}\zeta > 0$. Then, for all n, the random variable*

$$D_n := \overline{S}_n - S_n \geq 0$$

is majorised in distribution by the random variable $-Z$: $D_n \overset{d}{\leq} -Z$, where $Z := \inf Z_k$, $Z_k := \sum_{j=1}^{k} \zeta_j$ and ζ_j are independent copies of ζ.

Proof We have

$$\begin{aligned}\overline{S}_n = \max(0, S_1, \dots, S_n) &= S_n + \max(0, -\xi_n, -\xi_n - \xi_{n-1}, \dots, -S_n)\\ &= S_n - \min(0, \xi_n, \xi_n + \xi_{n-1}, \dots, S_n),\end{aligned}$$

where, by the last assertion of Lemma 10.5.2,

$$-D_n \equiv \min(0, \xi_n, \xi_n + \xi_{n-1}, \dots, S_n) \overset{d}{\geq} \min_{k \leq n} Z_k \geq Z, \qquad D_n \overset{d}{\leq} -Z.$$

The fact that Z is a proper random variable follows from Theorem 10.5.3 on the finiteness of the infimum of partial sums. The lemma is proved. □

If $\xi_k \overset{d}{=} \xi$ are identically distributed and $a = \mathbf{E}\xi > 0$, then we can put $\xi = \zeta$. The above reasoning shows that in this case the limit distribution of $\overline{S}_n - S_n$ as $n \to \infty$ exists and coincides with the distribution of the random variable Z (the random variables $\overline{S}_n - S_n$ themselves do not have a limit, and, by the way, neither do the variables $\frac{S_n - an}{\sqrt{n}}$ in the central limit theorem).

Lemma 10.5.3 shows that, for $\xi_k \overset{d}{\geq} \zeta$ and $\mathbf{E}\zeta > 0$, the random variables $\overline{S}_n$ and S_n differ from each other by a proper random variable only. This makes the limit theorems for $\overline{S}_n$ and S_n essentially the same.

We proceed to the law of large numbers and the central limit theorem for $\overline{S}_n$.

Theorem 10.5.4 *Let $a_k = \mathbf{E}\xi_k > 0$, $A_n = \sum_{k=1}^{n} a_k$ and $A_n \sim an$ as $n \to \infty$, $a > 0$. Let, moreover, $\xi_k - a_k$ be uniformly integrable for all k and $\xi_k \overset{d}{\geq} \zeta$ with $\mathbf{E}\zeta > 0$. Then, as $n \to \infty$,*

$$\frac{\overline{S}_n}{n} \xrightarrow{p} a.$$

Note that the *left* uniform integrability of $\xi_k - a_k$ follows from the inequalities $\xi_k \overset{d}{\geq} \zeta$.

Proof By Lemma 10.5.3,

$$\overline{S}_n = S_n + D_n, \quad \text{where } D_n \geq 0,\ D_n \overset{d}{\leq} -Z. \tag{10.5.6}$$

Therefore,

$$\frac{\overline{S}_n}{n} = \frac{S_n - A_n}{n} + \frac{A_n}{n} + \frac{D_n}{n},$$

where by Theorem 8.3.3, as $n \to \infty$,

$$\frac{S_n - A_n}{n} \xrightarrow{p} 0.$$

It is also clear that

$$\frac{A_n}{n} \to a, \qquad \frac{D_n}{n} \xrightarrow{p} 0.$$

The theorem is proved. □

In addition to the notation from Theorem 10.5.3, put

$$\sigma_k^2 := \mathbf{E}(\xi_k - a_k)^2, \qquad B_n^2 := \sum_{k=1}^n \sigma_k^2.$$

Theorem 10.5.5 *Let, for some $a > 0$ and $\sigma > 0$,*

$$A_n = an + o(\sqrt{n}), \qquad B_n^2 = \sigma^2 n + o(n),$$

and let the random variables $\xi_k - a_k$ satisfy the Lindeberg condition, $\xi_k \overset{d}{\geq} \zeta$ with $\mathbf{E}\zeta > 0$. Then

$$\frac{\overline{S}_n - an}{\sigma\sqrt{n}} \Leftrightarrow \boldsymbol{\Phi}_{0,1}. \tag{10.5.7}$$

Proof By virtue of (10.5.6),

$$\frac{\overline{S}_n - an}{\sigma\sqrt{n}} = \frac{S_n - A_n}{B_n} \cdot \frac{B_n}{\sigma\sqrt{n}} + \frac{A_n - an}{\sigma\sqrt{n}} + \frac{D_n}{\sigma\sqrt{n}}, \tag{10.5.8}$$

where, by the central limit theorem,

$$\frac{S_n - A_n}{B_n} \Leftrightarrow \boldsymbol{\Phi}_{0,1}.$$

Moreover,

$$\frac{B_n}{\sigma\sqrt{n}} \xrightarrow{p} 1, \qquad \frac{A_n - an}{\sigma\sqrt{n}} \to 0, \qquad \frac{D_n}{\sigma\sqrt{n}} \xrightarrow{p} 0.$$

This and (10.5.8) imply (10.5.7). The theorem is proved. □

10.5.5 Extension of Theorems 10.5.1 and 10.5.2 to Random Variables Assuming Values of Both Signs

We return to renewal processes and limit theorems for them. In Theorems 10.5.1 and 10.5.2 we obtained the law of large numbers and the central limit theorem for the renewal process $\eta(t)$ defined in (10.1.1) with jumps $\tau_k \ge 0$. Now we drop the last assumption and assume that τ_j can take values of both signs.

Theorem 10.5.6 *Let the conditions of Theorem* 10.5.1 *be met, the condition* $\tau_k \ge 0$ *being replaced with the condition* $\tau_k \overset{d}{\ge} \zeta$ *with* $\mathbf{E}\zeta > 0$. *Then*

$$\frac{\eta(t)}{t} \xrightarrow{p} \frac{1}{a}. \tag{10.5.9}$$

If $\tau_k \overset{d}{=} \tau$ are identically distributed and $\mathbf{E}\tau > 0$, then we can put $\zeta = \tau$. Therefore Theorem 10.5.6 implies the following result.

Corollary 10.5.1 *If* τ_k *are independent and identically distributed and* $\mathbf{E}\tau = a > 0$, *then* (10.5.9) *holds true.*

Proof of Theorem 10.5.6 Here instead of (10.5.1) we should use the relation

$$\{\eta(t) > n\} = \{\overline{T}_n \le t\}, \qquad \overline{T}_n = \max_{k \le n} T_k, \qquad T_k = \sum_{j=1}^{k} \tau_j. \tag{10.5.10}$$

Then we repeat the argument from the proof of Theorem 10.5.1, changing in it T_n to $\overline{T}_n$ and using Theorem 10.5.4, which implies that $\overline{T}_n$ and T_n satisfy the law of large numbers. The theorem is proved. □

Theorem 10.5.7 *Let the conditions of Theorem* 10.5.2 *be met, the condition* $\tau_k \ge 0$ *being replaced with the condition* $\tau_k \overset{d}{\ge} \zeta$ *with* $\mathbf{E}\zeta > 0$. *Then* (10.5.3) *holds true.*

Proof Here we again have to use (10.5.10), instead of (10.5.1), and then repeat the argument proving Theorem 10.5.2 using Theorem 10.5.5, which implies that the distribution of $\frac{\overline{T}_n - an}{\sigma\sqrt{n}}$, as well as the distribution of $\frac{T_n - an}{\sigma\sqrt{n}}$, converges to the standard normal law $\Phi_{0,1}$. The theorem is proved. □

Remark 10.5.2 (An analogue of Remarks 8.3.3, 8.4.1 and 10.1.1) The assertions of Theorems 10.5.6 and 10.5.7 can be generalised as follows. *Let* τ_1 *be an arbitrary random variable and random variables* $\tau_k^* := \tau_{1+k}$, $k \ge 1$, *satisfy the conditions of Theorem* 10.5.6 (Theorem 10.5.7). Then convergence (10.5.9) (10.5.3) still takes place.

Consider, for example, Theorem 10.5.7. Denote by A_x the event

$$A_x := \left\{ \frac{\eta(t) - a/t}{\sigma\sqrt{t/a^3}} < x \right\}.$$

Then the foregoing assertion follows from the relations

$$\mathbf{P}(A_x) = \mathbf{E}\big[\mathbf{P}(A_x|\tau_1);\ |\tau_1| \le N\big] + r_N,$$

where $r_N \le \mathbf{P}(|\tau_1| > N)$ can be made arbitrarily small by the choice of N, and by Theorem 10.5.7

$$\mathbf{P}(A_x|\tau_1) = \mathbf{P}\left(\frac{\eta^*(t-\tau_1) - \frac{t-\tau_1}{a}}{\sigma\sqrt{(t-\tau_1)/a^3}} + O\left(\frac{1}{\sqrt{t}}\right) < x\right) \to \Phi(x)$$

as $t \to \infty$ for each fixed τ_1, $|\tau_1| \le N$. Here $\eta^*(t)$ is the renewal process that corresponds to the sequence $\{\tau_k^*\}$. □

10.5.6 The Local Limit Theorem

If we again narrow our assumptions and return to identically distributed $\tau_k \stackrel{d}{=} \tau \ge 0$ then we can derive local theorems more precise than Theorem 10.5.2. In this subsection we will find an asymptotic representation for $\mathbf{P}(\eta(t) = n)$ as $t \to \infty$. We know from Theorem 10.5.2 what range of values of n the bulk of the distribution of $\eta(t)$ is concentrated in. Therefore we will from the start consider not arbitrary n, but the values of n that can be represented as

$$n = \left[\frac{t}{a} + v\sigma\sqrt{\frac{t}{a^3}}\right], \qquad \sigma^2 = \mathrm{Var}(\tau), \tag{10.5.11}$$

for "proper" values of v ($[s]$ in (10.5.11) is the integer part of s), so that

$$\frac{(t-an)}{\sigma\sqrt{n}} = v + O\left(\frac{1}{\sqrt{t}}\right) \tag{10.5.12}$$

(see (10.5.5)). For the proof, it will be more convenient to consider the probabilities $\mathbf{P}(\eta(t) = n+1)$. Changing $n+1$ to n amends nothing in the argument below.

Theorem 10.5.8 *If $\tau \ge 0$ is either non-lattice or arithmetic and* $\mathrm{Var}(\tau) = \sigma^2 < \infty$, *then, for the values of n defined in* (10.5.11), *as $t \to \infty$,*

$$\mathbf{P}\big(\eta(t) = n+1\big) \sim \frac{a^{3/2}}{\sigma\sqrt{2\pi t}} e^{-v^2/2}, \tag{10.5.13}$$

where in the arithmetic case t is assumed to be integer.

Proof First let, for simplicity, τ have a density and satisfy the conditions of the local limit Theorem 8.7.2. Then

$$\mathbf{P}\big(\eta(t) = n+1\big) = \int_0^t \mathbf{P}(T_n \in du)\mathbf{P}(\tau > t-u), \tag{10.5.14}$$

where by Theorem 8.7.2, as $n \to \infty$,

$$\mathbf{P}\big(T_n - na \in d(u-na)\big) = \frac{du}{\sigma\sqrt{2\pi n}}\left[\exp\left\{-\frac{(u-na)^2}{2n\sigma^2}\right\} + o(1)\right]$$

uniformly in u. Change the variable $u = t - z$. Since for the values of n we are dealing with one has (10.5.12), the exponential

$$\exp\left\{-\frac{(u-na)^2}{2n\sigma^2}\right\} = \exp\left\{-\frac{1}{2}\left(v - \frac{z}{\sigma\sqrt{n}}\right)^2\right\}$$

remains "almost constant" and asymptotically equivalent to $e^{-v^2/2}$ for $|z| < N$, $N \to \infty$, $N = o(\sqrt{n})$. Hence the integral in (10.5.14) is asymptotically equivalent to

$$\frac{1}{\sigma\sqrt{2\pi n}} e^{-v^2/2} \int_0^N \mathbf{P}(\tau > z)\, dz \sim \frac{a}{\sigma\sqrt{2\pi n}} e^{v^2/2}.$$

Since $n \sim t/a$ as $t \to \infty$, we obtain (10.5.13).

If τ has no density, but is non-lattice, then we should use the integro-local Theorem 8.7.1 for small Δ and, in a quite similar fashion, bound the integral in (10.5.14) (with t, which is a multiple of Δ) from above and from below by the sums

$$\sum_{k=0}^{t/\Delta-1} \mathbf{P}\big(T_n \in \Delta[k\Delta)\big)\mathbf{P}\big(\tau > t - (k+1)\Delta\big)$$

and

$$\sum_{k=0}^{t/\Delta-1} \mathbf{P}\big(T_n \in \Delta[k\Delta)\big)\mathbf{P}(\tau > t - k\Delta),$$

respectively. For small Δ both bounds will be close to the right-hand side of (10.5.13).

If τ has an arithmetic distribution then we have to replace integral (10.5.14) with the corresponding sum and, for integer u and t, make use of Theorem 8.7.3.

The theorem is proved. □

If examine the arguments in the proof concerning the behaviour of the correction term, then, in addition to (10.5.13), we can also obtain the representation

$$\mathbf{P}\big(\eta(t) = n\big) = \frac{a^{3/2}}{\sigma\sqrt{2\pi t}} e^{-v^2/2} + o\left(\frac{1}{\sqrt{t}}\right) \tag{10.5.15}$$

uniformly in v (or in n).

10.6 Generalised Renewal Processes

10.6.1 Definition and Some Properties

Let, instead of the sequence $\{\tau_j\}_{j=1}^{\infty}$, there be given a sequence of two-dimensional independent vectors (τ_j, ξ_j), $\tau_j \ge 0$, having the same distribution as (τ, ξ). Let, as before,

$$S_k = \sum_{j=1}^{k} \xi_j, \qquad T_k = \sum_{j=1}^{k} \tau_j, \qquad S_0 = T_0 = 0,$$

$$\eta(t) = \min\{k : T_k > t\}, \qquad \nu(t) = \max\{k : T_k \le t\} = \eta(t) - 1.$$

Definition 10.6.1 The process

$$S_{(\nu)}(t) = qt + S_{\nu(t)}$$

is called *a generalised renewal process with linear drift* q.

The process $S_{(\nu)}(t)$, as well as $\nu(t)$, is *right-continuous*. Clearly, $S_{(\nu)}(t) = qt$ for $t < \tau_1$. At time $t = \tau_1$ the first jump in the process $S_{(\nu)}(t)$ occurs, which is of size ξ_1:

$$S_{(\nu)}(\tau_1 - 0) = q\tau_1, \qquad S_{(\nu)}(\tau_1) = q\tau_1 + \xi_1.$$

After that, on the interval $[T_1, T_2)$ the value of $S_{(\nu)}(t)$ varies linearly with slope q. At the point T_2, the second jump occurs, which is of size ξ_2, and so on.

Generalised renewal processes are evidently a generalisation of random walks S_k (for $\tau_j \equiv 1, q = 0$) and renewal processes $\eta(t) = \nu(t) + 1$ (for $\xi_j \equiv 1, q = 0$). They are widespread in applications, as mathematical models of various physical systems.

Along with the process $S_{(\nu)}(t)$, we will consider generalised renewal processes of the form

$$S(t) = qt + S_{\eta(t)} = S_{(\nu)}(t) + \xi_{\eta(t)},$$

that are in a certain sense more convenient to analyse since $\eta(t)$ is a Markov time with respect to $\mathfrak{F}_n = \sigma(\tau_1, \dots, \tau_n; \xi_1, \dots, \xi_n)$ and has already been well studied.

The fact that the asymptotic properties of the processes $S(t)$ and $S_{(\nu)}(t)$, as $t \to \infty$, (the law of large numbers, the central limit theorem) are identical follows from the next assertion, which shows that the difference $S(t) - S_{(\nu)}(t)$ has a proper limiting distribution.

Lemma 10.6.1 *If* $\mathbf{E}\tau < \infty$, *then the following limiting distribution exists*

$$\lim_{t\to\infty} \mathbf{P}(\xi_{\eta(t)} < v) = \frac{\mathbf{E}(\tau;\ \xi < v)}{\mathbf{E}\tau}.$$

The lemma implies that $\xi_{\eta(t)}/b(t) \xrightarrow{p} 0$ for any function $b(t) \to \infty$ as $t \to \infty$.

Proof By virtue of the key renewal theorem,

$$\begin{aligned}
\mathbf{P}(\xi_{\eta(t)} < v) &= \sum_{k=0}^{\infty} \int_0^t \mathbf{P}(T_k \in du)\mathbf{P}(\tau > t - u, \xi < v) \\
&= \int_0^t dH(t)\mathbf{P}(\tau > t - u, \xi < v) \to \frac{1}{\mathbf{E}\tau} \int_0^\infty \mathbf{P}(\tau > u, \xi < v)\, du \\
&= \frac{\mathbf{E}(\tau;\ \xi < v)}{\mathbf{E}\tau}.
\end{aligned}$$

The lemma is proved. □

As was already noted, $\eta(t)$ is a stopping time with respect to

$$\mathfrak{F}_n = \sigma(\tau_1, \ldots, \tau_n; \xi_1, \ldots, \xi_n).$$

Therefore, if (τ_j, ξ_j) are identically distributed, then by the Wald identity (see Theorem 4.4.2 and Example 4.4.5)

$$\mathbf{E}S(t) = qt + a_\xi \mathbf{E}\eta(t) \sim qt + \frac{a_\xi t}{a} \tag{10.6.1}$$

as $t \to \infty$, where $a_\xi = \mathbf{E}\xi$ and $a = \mathbf{E}\tau$. The second moments of $S(t)$ will be found in Sect. 15.2. The laws of large numbers for $S(t)$ will be established in Sect. 11.5.

10.6.2 The Central Limit Theorem

In order to simplify the exposition, we first assume that the components τ_j and ξ_j of the vectors $(\tau_j, \xi_j) \stackrel{d}{=} (\tau, \xi)$ are independent. Moreover, without losing generality, we assume that $q = 0$.

Theorem 10.6.1 *Let there exist* $\sigma^2 = \operatorname{Var}\tau < \infty$, $\sigma_\xi^2 = \operatorname{Var}(\xi) < \infty$ *with* $\sigma + \sigma_\xi > 0$. *If the coordinates* τ *and* ξ *are independent then, as* $t \to \infty$,

$$\frac{S(t) - rt}{\sigma_S \sqrt{t}} \Subset\Rightarrow \mathbf{\Phi}_{0,1},$$

where $r = a_\xi / a$ *and* $\sigma_S^2 = a^{-1}(\sigma_\xi^2 + r^2\sigma^2) = a^{-1}\operatorname{Var}(\xi - r\tau)$. *The same assertion holds true for* $S_{(\upsilon)}(t)$ *as well.*

Proof If one of the values of σ and σ_ξ is zero, then the assertion of the theorem follows from Theorems 8.2.1 and 10.5.2. Therefore we can assume that $\sigma > 0$ and $\sigma_\xi > 0$. Denote by $\mathfrak{G} = \sigma(\tau_1, \tau_2, \ldots)$ the σ-algebra generated by the sequence $\{\tau_j\}$ and by $A_t \subset \mathfrak{G}$ the set

$$A_t = \bigl\{ \bigl|\eta(t) - t/a\bigr| < t^{1/2+\varepsilon} \bigr\}, \quad \varepsilon \in (0, 1/2).$$

Since by the central limit theorem $\mathbf{P}(A_t) \to 1$ as $t \to \infty$, for any trajectory $\eta(\cdot)$ in A_t we have $\eta(t) \to \infty$ as $t \to \infty$, and the random variables

$$Z(t) = \frac{S(t) - a_\xi \eta(t)}{\sigma_\xi \sqrt{\eta(t)}}$$

are asymptotically normal with parameters $(0, 1)$ by the independence of $\{\xi_j\}$ and $\{\tau_j\}$. In other words, on the sets A_t,

$$\mathbf{E}\bigl(e^{i\lambda Z(t)} \big| \mathfrak{G}\bigr) \to e^{-\lambda^2/2} \quad \text{as } t \to \infty.$$

Since

$$\eta(t) = \frac{t}{a} + \frac{\sigma\sqrt{t}}{a^{3/2}} \zeta_t, \quad \zeta_t \Subset\Rightarrow \mathbf{\Phi}_{0,1}, \quad \text{and} \quad \eta(t) \sim \frac{t}{a}$$

on the sets $A_t \in \mathfrak{G}$, we also have on the sets A_t the relation

$$\mathbf{E}\left(\exp\left\{\frac{i\lambda(S(t) - rt - \frac{a_\xi \sigma \sqrt{t}}{a^{3/2}}\zeta_t)}{\sigma_\xi \sqrt{t/a}}\right\} \middle| \mathfrak{G}\right) \to e^{-\lambda^2/2}.$$

Since the random variables ζ_t and $\eta(t)$ are measurable with respect to $\mathfrak{G}$, the corresponding factor can be taken outside of the conditional expectation, so that

$$\mathbf{E}\left(\exp\left\{\frac{i\lambda(S(t) - rt)}{\sigma_\xi \sqrt{t/a}}\right\} \middle| \mathfrak{G}\right) \sim \exp\left\{-\frac{\lambda^2}{2} + \frac{i\lambda r\sigma}{\sigma_\xi}\zeta_t\right\}.$$

Hence

$$\mathbf{E}\exp\left\{\frac{i\lambda(S(t) - rt)}{\sigma_\xi \sqrt{t/a}}\right\} = o(1) + \mathbf{E}\left(\exp\left\{-\frac{\lambda^2}{2} + \frac{i\lambda\sigma}{\sigma_\xi}\zeta_t\right\};\ A_t\right)$$
$$= o(1) + \exp\left\{-\frac{\lambda^2}{2}\left[1 + \left(\frac{r\sigma}{\sigma_\xi}\right)^2\right]\right\}.$$

This means that

$$\frac{1}{\sqrt{t}}\left(S(t) - \frac{ta_\xi}{a}\right) \Leftrightarrow \mathbf{\Phi}_{0,\sigma_S^2},$$

where

$$\sigma_S^2 = \frac{\sigma_\xi^2}{a}\left[1 + \left(\frac{r\sigma}{\sigma_\xi}\right)^2\right] = a^{-1}\left[\sigma_\xi^2 + r^2\sigma^2\right].$$

The assertion corresponding to $S_{(\nu)}(t)$ follows from Lemma 10.6.1. The theorem is proved. □

Note that Theorems 8.2.1 and 10.5.2 are special cases of Theorem 10.6.1. If $a_\xi = 0$, then $S(t)$ is distributed identically to $S_{[t/a]}$ and is independent of σ.

Now consider the general case where τ and ξ are, generally speaking, dependent. Since $T_{\eta(t)} = t + \chi(t)$, we have the representation

$$S(t) - rt = Z_{\eta(t)} + r\chi(t), \tag{10.6.2}$$

where

$$Z_n = \sum_{j=1}^{n} \zeta_j, \quad \zeta_j = \xi_j - r\tau_j, \quad \mathbf{E}\zeta_j = 0, \quad \frac{\chi(t)}{\sqrt{t}} \xrightarrow{p} 0$$

as $t \to \infty$ ($\chi(t)$ has a proper limiting distribution as $t \to \infty$). Moreover, we will use yet another Wald identity

$$\mathbf{E}Z_{\eta(t)}^2 = d^2\mathbf{E}\eta(t), \quad d^2 = \mathbf{E}\zeta^2, \quad \zeta = \xi - r\tau, \tag{10.6.3}$$

that is derived below in Sect. 15.2.

Theorem 10.6.2 *Let* $(\tau_j, \xi_j) \stackrel{d}{=} (\tau, \xi)$ *be independent identically distributed and such that* $\sigma^2 = \mathrm{Var}(\tau) < \infty$ *and* $\sigma_\xi^2 = \mathrm{Var}(\xi) < \infty$ *exist. Then*

$$\frac{S(t) - rt}{\sigma_S \sqrt{t}} \Leftrightarrow \Phi_{0,1},$$

where $r = a_\xi / a$ *and* $\sigma_S^2 = a^{-1} d^2$. *The random variables* $\frac{S_{(\nu)}(t) - rt}{\sigma_S \sqrt{t}}$ *and* $\frac{Z_{\eta(t)}}{\sigma_S \sqrt{t}}$ *have the same limiting distribution.*

Proof It is seen from (10.6.2) that it suffices to prove that

$$\frac{Z_{\eta(t)}}{\sigma_S \sqrt{t}} \Leftrightarrow \Phi_{0,1}.$$

The main contribution to $Z_{\eta(t)}$ comes from Z_m with $m = [\frac{t}{a} - 2N\sqrt{t}]$, $N \to \infty$, $N = o(\sqrt{t})$, where

$$\frac{\sqrt{a}\, Z_m}{d\sqrt{t}} = \frac{Z_m}{d\sqrt{m}} \sqrt{\frac{ma}{t}} \Leftrightarrow \Phi_{0,1}.$$

The remainder $Z_{\eta(t)} - Z_m$, for each fixed

$$T_m \in I_N := [t - 3aN\sqrt{t},\, t - aN\sqrt{t}\,], \qquad \mathbf{P}(T_m \in I_N) \to 1,$$

has the same distribution as $Z_{\eta(t-T_m)}$, and its variance (see (10.6.3)) is equal to

$$d^2 \mathbf{E}\eta(t - T_m) \sim d^2 \frac{t - T_m}{a} < 3d^2 N\sqrt{t} = o(t).$$

Since $\mathbf{E} Z_{\eta(t-T_m)} = 0$, we have

$$\frac{Z_{\eta(t-T_m)}}{\sqrt{t}} \xrightarrow{p} 0 \tag{10.6.4}$$

as $t \to \infty$. The theorem is proved. □

Note that, for $N \to \infty$ slowly enough, relation (10.6.4) can be derived using not (10.6.3), but the law of large numbers for generalised renewal processes that was obtained in Sect. 11.5.

Theorem 10.6.1 could be proved in a somewhat different way—with the help of the local Theorem 10.5.3. We will illustrate this approach by the proof of the integro-local theorem for $S(t)$.

10.6.3 The Integro-Local Theorem

In this section we will obtain the integro-local theorem for $S(t)$ in the case of non-lattice ξ. In a quite similar way we can obtain local theorems for densities (if they exist) and for the probability $\mathbf{P}(S(t) = k)$ for $q = 0$ for arithmetic ξ_j.

Theorem 10.6.3 *Let the conditions of Theorem* 10.6.1 *hold and, moreover,* ξ *be non-lattice. Then, for any fixed* $\Delta > 0$, *as* $t \to \infty$,

$$\mathbf{P}\big(S(t) - rt \in \Delta[x)\big) = \frac{\Delta}{\sigma_S\sqrt{t}}\phi\left(\frac{x}{\sigma_S\sqrt{t}}\right) + o\left(\frac{1}{\sqrt{t}}\right), \qquad (10.6.5)$$

where the remainder term $o(1/\sqrt{t})$ *is uniform in* x.

Proof Since ξ is non-lattice, one has $\sigma_\xi > 0$. If $\sigma = 0$ then the assertion of the theorem follows from Theorem 8.7.1. Therefore we will assume that $\sigma > 0$. By the independence of $\{\xi_j\}$ and $\{\tau_j\}$,

$$\mathbf{P}\big(S(t) - rt \in \Delta[x)\big) = \sum_{n=1}^{\infty} \mathbf{P}\big(\eta(t) = n\big)\mathbf{P}\big(S_n - rt \in \Delta[x)\big) = \sum_{n \in M_t} + \sum_{n \notin M_t},$$

where $M_t = \{n : |n - t/a| < t^{1/2}N(t)\}$, $N(t) \to \infty$, $N(t) = o(\sqrt{t})$ as $t \to \infty$. We know the asymptotics of both factors of the terms in the sum from Theorems 8.7.1 and 10.5.8 (see also (10.5.15)). It remains to do the summation, which is unfortunately somewhat cumbersome. At the same time, it presents no substantial difficulties, so we will sketch this part of the proof. If we put $an - t =: u$,

$$P_1(t) := \frac{\Delta}{\sigma_\xi\sqrt{2\pi n}} \exp\left\{-\frac{(x - ru)^2}{2n\sigma_\xi^2}\right\}, \qquad P_2(t) := \frac{a^{3/2}}{\sigma\sqrt{2\pi t}} \exp\left\{-\frac{u^2}{2\sigma^2 n}\right\},$$

then

$$\mathbf{P}\big(S_n - rt \in \Delta[x)\big) = P_1(t) + o\left(\frac{1}{\sqrt{n}}\right).$$

Furthermore,

$$\mathbf{P}\big(\eta(t) = n\big) = P_2(t) + o\left(\frac{1}{\sqrt{t}}\right)$$

for $n \in M_t$ and $N(t) \to \infty$ slowly enough as $t \to \infty$. Clearly,

$$\sum_{n \notin M_t} = o\left(\frac{1}{\sqrt{t}}\right).$$

Since the sums of $P_1(t)$ and $P_2(t)$ are bounded in n by a constant, we have

$$\sum_{n \in M_t} = o\left(\frac{1}{\sqrt{t}}\right) + \sum_{n \in M_t} P_1(t)P_2(t).$$

The exponent in the product $P_1(t)P_2(t)$, taken with the negative sign, is equal to

$$\frac{1}{2n}\left[\frac{(x - ru)^2}{\sigma_\xi^2} + \frac{u^2}{\sigma^2}\right] \sim \frac{a}{2t}\left[\frac{(d^2u - rx\sigma^2)^2}{d^2\sigma^2\sigma_\xi^2} + \frac{x^2}{d^2}\right],$$

where $d^2 = r^2\sigma^2 + \sigma_\xi^2$. Since, for $x = o(\sqrt{t}\,N(t))$,

$$\sum_{n \in A_t} \frac{a^{3/2}d}{\sqrt{2\pi t}\sigma\sigma_\xi} \exp\left\{-\frac{a(d^2u - rx\sigma^2)^2}{2td^2\sigma^2\sigma_\xi^2}\right\} \to 1$$

as $t \to \infty$ and this sum does not exceed $1 + o(1)$ for all x (this is an integral sum that corresponds to the integral of the density of the normal law), it is easy to derive (10.6.5) from the foregoing. □

We will continue the study of generalised renewal processes in Sect. 11.5.

Chapter 11
Properties of the Trajectories of Random Walks. Zero-One Laws

Abstract The chapter begins with Sect. 11.1 establishing the Borel–Cantelli and Kolmogorov zero-one laws, and also the zero-one law for exchangeable sequences. The concepts of lower and upper functions are introduced. Section 11.2 contains the first Kolmogorov inequality and several theorems on convergence of random series. Section 11.3 presents Kolmogorov's Strong Law of Large Numbers and Wald's identity for stopping times. Sections 11.4 and 11.5 are devoted to the Strong Law of Large Numbers for independent non-identically distributed random variables, and to the Strong Law of Large Numbers for generalised renewal processes, respectively.

11.1 Zero-One Laws. Upper and Lower Functions

Let, as before, $S_n = \sum_{j=1}^n \xi_j$ be the sums of independent random variables $\xi_1, \xi_2, \dots$. In this chapter we will consider properties of *the "whole" trajectories of random walks* $\{S_n\}$.

The first limit theorem we proved for the distribution of the sums of independent identically distributed random variables was the law of large numbers: $S_n/n \xrightarrow{p} \mathbf{E}\xi$. One could ask whether the *whole trajectory* $S_n/n, S_{n+1}/(n+1), \dots$, starting from some n, will be close to $\mathbf{E}\xi$ with a high probability. That is, whether, for any $\varepsilon > 0$, we will have

$$\lim_{n\to\infty} \mathbf{P}\left(\sup_{k\ge n} \left| \frac{S_k}{k} - \mathbf{E}\xi \right| < \varepsilon \right) = 1. \tag{11.1.1}$$

This is clearly a problem on almost sure convergence, or convergence with probability 1. A similar question arises concerning *generalised renewal processes* discussed in Sect. 10.6.

Assertion (11.1.1), which is called the *strong law of large numbers* and is to be proved in this chapter, is a special case of the so-called *zero-one laws*. As the first such law, we will now present the Borel–Cantelli zero-one law.

11.1.1 Zero-One Laws

Theorem 11.1.1 *Let $\{A_n\}_{n=1}^\infty$ be a sequence of events on a probability space $\langle \Omega, \mathfrak{F}, \mathbf{P}\rangle$, and let A be the event that infinitely many events A_k occur, i.e.*

A.A. Borovkov, *Probability Theory*, Universitext,
DOI 10.1007/978-1-4471-5201-9_11,

$A = \bigcap_{n=1}^{\infty} \bigcup_{k=n}^{\infty} A_k$ *(the event A consists of those ω that belong to infinitely many A_k).*

If $\sum_{k=1}^{\infty} \mathbf{P}(A_k) < \infty$, then $\mathbf{P}(A) = 0$. If $\sum_{k=1}^{\infty} \mathbf{P}(A_k) = \infty$ and the events A_k are independent, then $\mathbf{P}(A) = 1$.

Proof Assume that $\sum_{k=1}^{\infty} \mathbf{P}(A_k) < \infty$. Denote by $\eta = \sum_{k=1}^{\infty} \mathrm{I}(A_k)$ the number of occurrences of events A_k. Then $\mathbf{E}\eta = \sum_{k=1}^{\infty} \mathbf{P}(A_k) < \infty$ which certainly means that η is a proper random variable: $\mathbf{P}(\eta < \infty) = 1 - \mathbf{P}(A) = 1$.

If A_k are independent and $\sum_{k=1}^{\infty} \mathbf{P}(A_k) = \infty$, then, since $\overline{A}_k = \Omega \setminus A_k$ are also independent, we have

$$\begin{aligned} \mathbf{P}(A) &= \lim_{n\to\infty} \mathbf{P}\Biggl(\bigcup_{k=n}^{\infty} A_k\Biggr) = \lim_{n\to\infty} \mathbf{P}\Biggl(\Omega - \bigcap_{k=n}^{\infty} \overline{A}_k\Biggr) \\ &= 1 - \lim_{n\to\infty} \mathbf{P}\Biggl(\bigcap_{k=n}^{\infty} \overline{A}_k\Biggr) = 1 - \lim_{n\to\infty} \lim_{m\to\infty} \mathbf{P}\Biggl(\bigcap_{k=n}^{m} \overline{A}_k\Biggr) \\ &= 1 - \lim_{n\to\infty} \prod_{k=n}^{\infty} \bigl(1 - \mathbf{P}(A_k)\bigr). \end{aligned}$$

Using the inequality $\ln(1-x) \le -x$ we obtain that

$$\prod_{k=n}^{\infty} \bigl(1 - \mathbf{P}(A_k)\bigr) \le \exp\Biggl\{-\sum_{k=n}^{\infty} \mathbf{P}(A_k)\Biggr\}.$$

Hence

$$\prod_{k=n}^{\infty} \bigl(1 - \mathbf{P}(A_k)\bigr) \le e^{-\infty} = 0, \qquad \mathbf{P}(A) = 1.$$

The theorem is proved. □

Remark 11.1.1 It follows from Theorem 11.1.1 that, for independent events A_k, the assertions that $\mathbf{E}\eta < \infty$ and that $\mathbf{P}(\eta < \infty) = 1$ are equivalent to each other. Although in one direction this relationship is obvious, in the opposite direction it is quite meaningful. It implies, in particular, that if $\eta < \infty$ with probability 1, but $\mathbf{E}\eta = \infty$, then A_k are necessarily dependent.

Note also that the argument proving the first part of the theorem has already been used for the same purpose in the proof of Theorem 6.1.1.

Assume that $\{\xi_n\}_{n=1}^{\infty}$ is a sequence of independent random variables given on $\langle \Omega, \mathfrak{F}, \mathbf{P} \rangle$. Denote, as before, by $\sigma(\xi_1, \ldots, \xi_n)$ the σ-algebra generated by the first n random variables $\xi_1, \ldots, \xi_n$, and by $\sigma(\xi_n, \ldots)$ the σ-algebra generated by the random variables $\xi_n, \xi_{n+1}, \xi_{n+2}, \ldots$.

Definition 11.1.1 An event A is said to be a *tail event* if $A \in \sigma(\xi_n, \ldots)$ for any $n > 0$.

For example, the event

$$A = \bigcap_{n=1}^{\infty} \bigcup_{k=n}^{\infty} \{\xi_k > N\}$$

meaning that there occurred infinitely many events $\{\xi_k > N\}$ is clearly a tail event.

Theorem 11.1.2 (Kolmogorov zero-one law) *If A is a tail event, then either* $\mathbf{P}(A) = 0$ *or* $\mathbf{P}(A) = 1$.

Proof Since A is a tail event, $A \in \sigma(\xi_{n+1}, \dots)$, $n \ge 0$. Therefore the event A is independent of the σ-algebra $\sigma(\xi_1, \dots, \xi_n)$ for any n. Hence (see Theorem 3.4.3) the event A is independent of the σ-algebra $\sigma(\xi_1, \dots)$. Since $A \in \sigma(\xi_1, \dots)$, it is independent of itself:

$$\mathbf{P}(A) = \mathbf{P}(AA) = \mathbf{P}(A)\mathbf{P}(A).$$

But this is only possible if $\mathbf{P}(A) = 0$ or 1. The theorem is proved. □

Put $S = \sup\{0, S_1, S_2, \dots\}$, where $S_n = \sum_{k=1}^{n} \xi_k$. An example of an application of the above theorem is given by the following

Corollary 11.1.1 *If ξ_k, $k = 1, 2, \dots$, are independent, then either* $\mathbf{P}(S = \infty) = 1$ *or* $\mathbf{P}(S < \infty) = 1$.

The Proof follows from the fact that $\{S = \infty\}$ is a tail event. Indeed, for any n

$$\begin{aligned}\{S = \infty\} &= \bigl\{\sup(S_{n-1}, S_n, \dots) = \infty\bigr\} \\ &= \bigl\{\sup(0, S_n - S_{n-1}, \dots) = \infty\bigr\} \in \sigma(\xi_n, \dots).\end{aligned}$$ □

Further examples of tail events can be obtained if we consider, for a sequence of independent variables $\xi_1, \xi_2, \dots$, the event $\{$*the series* $\sum_1^{\infty} \xi_k$ *is convergent*$\}$. Theorem 11.1.2 means that the probability of that event can only be 0 or 1.

If we consider the power series $\sum_{k=0}^{\infty} z^k \xi_k$ where ξ_k are independent, we will see that the convergence radius $\rho = \limsup_{k\to\infty} |\xi_k|^{-1/k}$ of this series is a random variable measurable with respect to the σ-algebra $\sigma(\xi_n, \dots)$ for any n ($\{\rho < x\} \in \sigma(\xi_n, \dots)$, $0 \le x \le \infty$). Such random variables are also called *tail random variables*. Since by the foregoing one has $F_\rho(x) = \mathbf{P}(\rho < x) = 0$ or 1, this implies that ρ, as well as any other tail random variable, must be equal to a constant with probability 1.

Under the assumption that the elements of the sequence $\{\xi_k\}_{k=1}^{\infty}$ are not only independent but also identically distributed, Kolmogorov's zero-one law was extended by Hewitt and Savage to a wider class of events.

Let $\omega = (x_1, x_2, \dots)$ be an element of the sample space $\langle \mathbb{R}^\infty, \mathfrak{B}^\infty, \mathbf{P}\rangle$ for the sequence $\xi = (\xi_1, \xi_2, \dots)$ ($\mathbb{R}^\infty$ is a countable direct product of the real lines $\mathbb{R}_k$, $k = 1, 2, \dots$, $\mathfrak{B}^\infty = \sigma(\xi_1, \dots)$ is generated by the sets $\prod_{k=1}^{N} B_k \in \sigma(\xi_1, \dots, \xi_N)$, where $B_k \in \sigma(\xi_k)$ are Borel sets on the lines $\mathbb{R}_k$).

Definition 11.1.2 An event $A \in \mathfrak{B}^\infty$ is said to be *exchangeable* if

$$(x_1, x_2, \ldots, x_{n-1}, x_n, x_{n+1} \ldots) \in A$$

implies that $(x_n, x_2, \ldots, x_{n-1}, x_1, x_{n+1} \ldots) \in A$ for every $n \ge 1$. It is evident that this condition of membership automatically extends to any permutations of finitely many components. Examples of exchangeable events are given by tail events.

Theorem 11.1.3 (Zero-one law for exchangeable events) *If ξ_k are independent and identically distributed and A is an exchangeable event, then either* $\mathbf{P}(A) = 0$ *or* $\mathbf{P}(A) = 1$.

Proof By the approximation theorem (Sect. 3.5), for any $A \in \mathfrak{B}^\infty$ there exists a sequence of events $A_n \in \sigma(\xi_1, \ldots, \xi_n)$ such that

$$\mathbf{P}(\overline{A}_n A \cup \overline{A} A_n) \to 0$$

as $n \to \infty$.

Introduce the transformation

$$T_n\omega = T_n(x_1, x_2, \ldots) = (x_{n+1}, \ldots, x_{2n}, x_1, \ldots, x_n, x_{2n+1} \ldots)$$

and put $B_n = T_n A_n$. If A is exchangeable, then $T_n A = A$ and, for any $B \in \mathfrak{B}^\infty$, one has $\mathbf{P}(T_n B) = \mathbf{P}(B)$ since ξ_j are independent and identically distributed. Therefore $\mathbf{P}(B_n A) = \mathbf{P}(T_n A_n A) = \mathbf{P}(A_n A)$, and hence B_n will also approximate A, which obviously implies that $C_n = A_n B_n$ will have the same approximation property. By independence of A_n and B_n, this means that

$$\mathbf{P}(A) = \lim_{n\to\infty} \mathbf{P}(A_n B_n) = \lim_{n\to\infty} \mathbf{P}^2(A_n) = \mathbf{P}^2(A).$$

The theorem is proved. □

11.1.2 Lower and Upper Functions

Theorem 11.1.3 implies the following interesting fact, the statement of which requires the next definition.

Definition 11.1.3 For a sequence of random variables $\{\eta_n\}_{n=1}^\infty$, a numerical sequence $\{a_n\}_{n=1}^\infty$ is said to be an *upper sequence (function)* if, with probability 1, there occur only finitely many events $\{\eta_n > a_n\}$. A sequence $\{a_n\}_{n=1}^\infty$ is said to be a *lower sequence (function)* if, with probability 1, there occur infinitely many events $\{\eta_n > a_n\}$.

Corollary 11.1.2 *If ξ_k are independent and identically distributed, then any sequence $\{a_n\}$ is either upper or lower for the sequence of sums $\{S_n\}_{n=1}^\infty$ with* $S_n = \sum_{k=1}^n \xi_k$.

In other words, one cannot find an "intermediate" sequence $\{a_n\}$ such that the probability of the event $A = \{S_n > a_n \text{ infinitely often}\}$ would be equal, say, to $1/2$.

Proof To prove the corollary, it suffices to notice that the event A is exchangeable, because swapping ξ_1 and ξ_n in the realisation $(\xi_1, \xi_2, \dots)$ influences the behaviour of the first n sums $S_1, \dots, S_n$ only. □

A similar fact holds, of course, for the sequence of random variables $\{\xi_n\}_{n=1}^{\infty}$ itself, but, unlike the above corollary, that assertion can be proved more easily, since $B = \{\xi_n > a \text{ infinitely often}\}$ is a tail event.

Remark 11.1.2 In regard to the properties of upper and lower sequences for sums $\{S_n\}$ we also note here the following. If $\mathbf{P}(\xi_k = c) \neq 1$, and $\{a_n\}$ is an upper (lower) sequence for $\{S_n\}$, then, for any fixed $k \ge 0$ and v, the sequence $\{b_n = a_{n+k} + v\}_{n=1}^{\infty}$ is also upper (lower) for $\{S_n\}$. This is a consequence of the following relations. Let $v_1 > v_2$ be such that

$$\mathbf{P}(\xi > v_1) > 0, \qquad \mathbf{P}(\xi < v_2) > 0.$$

Then, for the upper sequence $\{a_n\}$ and the event $A = \{S_n > a_n \textit{infinitely many times}\}$, we have

$$\begin{aligned} 0 = \mathbf{P}(A) &\ge \mathbf{P}(\xi_1 > v_1)\mathbf{P}(A \mid \xi_1 > v_1) \\ &\ge \mathbf{P}(\xi_1 > v_1)\mathbf{P}(S_n > a_{n+1} - v_1 \textit{ infinitely many times}). \end{aligned}$$

This implies that the second factor on the right-hand side equals 0, and hence the sequence $\{a_{n+1} - v_1\}$ is also an upper sequence. On the other hand, if $\xi' \stackrel{d}{=} \xi$ is independent of ξ then

$$\begin{aligned} 0 = \mathbf{P}(A) &\ge \mathbf{P}\big(\xi' + S_n > \xi' + a_n \textit{ infinitely many times};\ \xi' < v_2\big) \\ &\ge \mathbf{P}(\xi < v_2)\mathbf{P}(S_{n+1} > a_n + v_2 \textit{ infinitely many times}) \\ &= \mathbf{P}(\xi < v_2)\mathbf{P}(S_n > a_{n-1} + v_2 \textit{ infinitely many times}). \end{aligned}$$

Here the second factor on the right-hand side equals 0, and hence the sequence $\{a_{n-1} + v_2\}$ is also upper. Combining these assertions as many times as necessary, we find that the sequence $\{a_{n+k} + v\}$ is upper for any given k and v. □

From the above remark it follows, in particular, that the quantities $\limsup_{n\to\infty} S_n$ and $\liminf_{n\to\infty} S_n$ cannot both be finite for a sequence of sums of independent identically distributed random variables that are not zeros with probability 1. Indeed, the event $B = \{\limsup_{n\to\infty} S_n \in (a, b)\}$ is exchangeable and therefore $\mathbf{P}(B) = 0$ or $\mathbf{P}(B) = 1$ by virtue of the zero-one law. If $\mathbf{P}(B)$ were equal to 1, $(b, b, \dots)$ would be an upper sequence for $\{S_n\}$. But, by our remark, $(a, a, \dots)$ would then be an upper sequence as well, which would mean that

$$\mathbf{P}\Big(\limsup_{n\to\infty} S_n \le a\Big) = 1,$$

which contradicts the assumption $\mathbf{P}(B) = 1$. □

The reader can also derive from Theorem 11.1.3 that, for any sequences $\{a_n\}$ and $\{b_n\}$, the random variables

$$\limsup_{n\to\infty} \frac{S_n - a_n}{b_n} \quad \text{and} \quad \liminf_{n\to\infty} \frac{S_n - a_n}{b_n}$$

are constant with probability 1.

11.2 Convergence of Series of Independent Random Variables

In the present section we will discuss in more detail convergence of series of independent random variables. We already know that such series converge with probability 1 or 0. We are interested in conditions ensuring convergence.

First of all we answer the following interesting question. It is well known that the series $\sum_{n=1}^{\infty} n^{-\alpha}$ is divergent for $\alpha \le 1$, while the alternating series $\sum_{n=1}^{\infty} (-1)^n n^{-\alpha}$ converges for any $\alpha > 0$ (the difference between neighbouring elements is of order $\alpha n^{-\alpha-1}$). What can be said about the behaviour of the series $\sum_{n=1}^{\infty} \delta_n n^{-\alpha}$, where δ_n are identically distributed and independent with $\mathbf{E}\delta_n = 0$ (for instance, $\delta_n = \pm 1$ with probabilities $1/2$)?

One of the main approaches to studying such problems is based on elucidating the relationship between a.s. convergence and the simpler notion of convergence in probability. It is known that, generally speaking, convergence in probability $\xi_n \xrightarrow{p} \xi$ does not imply a.s. convergence. However, in our situation when $\zeta_n = S_n := \sum_{k=1}^{n} \xi_k$, ξ_k being independent, this is not the case. The main assertion of the present section is the following.

Theorem 11.2.1 *If ξ_k are independent and $S_n = \sum_{k=1}^{n} \xi_k$, then convergence of S_n in probability implies a.s. convergence of S_n.*

We will prove that S_n is a Cauchy sequence. To do this, we will need the following inequality.

Lemma 11.2.1 (The First Kolmogorov inequality) *If ξ_j are independent and, for some $b > 0$ and all $j \le n$,*

$$\mathbf{P}\big(|S_n - S_j| \ge b\big) \le p < 1,$$

then

$$\mathbf{P}\Big(\max_{j\le n} |S_j| \ge x\Big) \le \frac{1}{1-p} \mathbf{P}\big(|S_n| > x - b\big). \tag{11.2.1}$$

Corollary 11.2.1 *If $\mathbf{E}\xi_j = 0$ then*

$$\mathbf{P}\Big(\max_{j\le n} |S_j| \ge x\Big) \le 2\mathbf{P}\big(|S_n| > x - \sqrt{2\,\mathrm{Var}(S_n)}\big).$$

Kolmogorov actually established this last inequality (Lemma 11.2.1 is an insignificant extension of it). It follows from (11.2.1) with $p = 1/2$, since by the Chebyshev inequality

$$\mathbf{P}\big(|S_n - S_j| \geq \sqrt{2\,\mathrm{Var}(S_n)}\big) \leq \frac{\mathrm{Var}(S_n - S_j)}{2\,\mathrm{Var}(S_n)} \leq \frac{1}{2}.$$

Proof of Lemma 11.2.1 Let

$$\eta := \big\{\min k \geq 1 : |S_k| \geq x\big\}.$$

Put $A_j := \{\eta = j\}$, $j = 1, 2, \ldots$. Clearly, A_j are disjoint events and hence

$$\mathbf{P}\big(|S_n| > x - b\big) \geq \sum_{j=1}^{n} \mathbf{P}\big(|S_n| > x - b;\ A_j\big) \geq \sum_{j=1}^{n} \mathbf{P}\big(|S_n - S_j| < b;\ A_j\big).$$

(The last inequality holds because the event $\{|S_n - S_j| < b\}A_j$ implies $\{|S_n| > x - b\}A_j$.) But $A_j \in \sigma(\xi_1, \ldots, \xi_j)$ and $\{|S_n - S_j| < b\} \in \sigma(\xi_{j+1}, \ldots, \xi_n)$. Therefore these two events are independent and

$$\begin{aligned} \mathbf{P}\big(|S_n| > x - b\big) &\geq \sum_{j=1}^{n} \mathbf{P}(A_j)\mathbf{P}\big(|S_n - S_j| < b\big) \\ &\geq (1 - p) \sum_{1}^{n} \mathbf{P}(A_j) = (1 - p)\mathbf{P}\Big(\max_{j \leq n} |S_j| \geq x\Big). \end{aligned}$$

The lemma is proved. □

Proof of Theorem 11.2.1 It suffices to prove that $\{S_n\}$ is a.s. a Cauchy sequence, i.e. that, for any $\varepsilon > 0$,

$$\mathbf{P}\Big(\sup_{n \geq m} |S_n - S_m| > 2\varepsilon\Big) \to 0 \tag{11.2.2}$$

as $m \to \infty$. Let

$$A^{\varepsilon}_{n,m} := \big\{|S_n - S_m| > \varepsilon\big\}, \qquad A^{\varepsilon}_{m} := \bigcup_{n \geq m} A^{\varepsilon}_{n,m}.$$

Then relation (11.2.2) can be written as

$$\mathbf{P}\big(A^{2\varepsilon}_{m}\big) \to 0 \tag{11.2.3}$$

as $m \to \infty$.

Since $\{S_n\}$ is a Cauchy sequence in probability, one has

$$p_{m,M} := \sup_{m \leq n \leq M} \mathbf{P}\big(A^{\varepsilon}_{n,M}\big) \to 0$$

as $m \to \infty$ and $M \to \infty$, so that $p_{m,M} < 1/2$ for all m and M large enough. For such m and M we have by Lemma 11.2.1, for $a = \varepsilon$ and $x = 2\varepsilon$, that

$$\mathbf{P}\Big(\sup_{m\le n\le M}|S_n - S_m| > 2\varepsilon\Big) = \mathbf{P}\Bigg(\bigcup_{n=m+1}^{M} A_{m,n}^{2\varepsilon}\Bigg)$$
$$\le \frac{1}{1-p_{m,M}}\mathbf{P}\big(A_{M,m}^{\varepsilon}\big) \le 2\mathbf{P}\big(A_{M,m}^{\varepsilon}\big).$$

By the properties of probability,

$$\mathbf{P}\big(A_m^{2\varepsilon}\big) = \lim_{M\to\infty}\mathbf{P}\Bigg(\bigcup_{n=m+1}^{M} A_{m,n}^{2\varepsilon}\Bigg) \le 2\limsup_{M\to\infty}\mathbf{P}\big(A_{M,m}^{\varepsilon}\big). \tag{11.2.4}$$

Denote by S the limit (in probability) of the sequence S_n, and

$$B_n^{\varepsilon} := \big\{|S_n - S| > \varepsilon\big\}.$$

Then $\mathbf{P}(B_n^{\varepsilon}) \to 0$ as $n \to \infty$, $A_{M,m}^{\varepsilon} \subset B_M^{\varepsilon/2} \cup B_m^{\varepsilon/2}$, and by (11.2.4)

$$\mathbf{P}\big(A_m^{2\varepsilon}\big) \le 2\mathbf{P}\big(B_m^{\varepsilon/2}\big) \to 0 \quad \text{as } m \to \infty.$$

Relation (11.2.3), and hence the assertion of the theorem, are proved. □

Corollary 11.2.2 *If* $\mathbf{E}\xi_k = 0$ *and* $\sum_1^{\infty}\operatorname{Var}(\xi_k) < \infty$, *then* S_n *converges a.s.*

Proof The assertion follows immediately from Theorem 11.2.1 and the fact that $\{S_n\}$ is a Cauchy sequence in mean quadratic ($\mathbf{E}(S_n - S_m)^2 = \sum_{k=m+1}^{n}\operatorname{Var}(\xi_k) \to 0$ as $m \to \infty$ and $n \to \infty$) and hence in probability.

It turns out that if $\mathbf{E}\xi_k = 0$ and $|\xi_k| < c$ for all k, then the condition $\sum\operatorname{Var}(\xi_k) < \infty$ is necessary and sufficient for a.s. convergence of S_n.[1]

Corollary 11.2.2 also contains an answer to the question posed at the beginning of the section about convergence of $\sum\delta_n n^{-\alpha}$, where δ_n are independent and identically distributed and $\mathbf{E}\delta_n = 0$.

Corollary 11.2.3 *The series* $\sum\delta_n a_n$ *converges with probability* 1 *if* $\operatorname{Var}(\delta_k) = \sigma^2 < \infty$ *and* $\sum a_n^2 < \infty$.

Thus we obtain that the series $\sum\delta_n n^{-\alpha}$, where $\delta_n = \pm 1$ with probabilities $1/2$, is convergent if and only if $\alpha > 1/2$.

An extension of Corollary 11.2.2 is given by the following.

Corollary 11.2.4 (The two series theorem) *A sufficient condition for a.s. convergence of the series* $\sum\xi_n$ *is that the series* $\sum\mathbf{E}\xi_n$ *and* $\sum\operatorname{Var}(\xi_n)$ *are convergent.*

The Proof is obvious, for the sequences $\sum_{k=1}^{n}\mathbf{E}\xi_k$ and $\sum_{k=1}^{n}(\xi_k - \mathbf{E}\xi_k)$ converge a.s. by Corollary 11.2.2. □

[1] For more detail, see e.g. [31].

11.3 The Strong Law of Large Numbers

It is not hard to see that, using the terminology of Sect. 11.1, the strong law of large numbers (11.1.1) means that, for any $\varepsilon > 0$, the sequence $\{\varepsilon n\}_{n=1}^{\infty}$ is an upper one for both sequences $\{S_n\}$ and $\{-S_n\}$ only if $\mathbf{E}\xi_1 = 0$.

We will derive the strong law of large numbers as a corollary of Theorem 10.5.3 on finiteness of the infimum of sums of random variables.

Let, as before, $\xi_1, \xi_2, \ldots$ be independent and identically distributed, $\xi \stackrel{d}{=} \xi_k$.

Theorem 11.3.1 (Kolmogorov's Strong Law of Large Numbers) *A necessary and sufficient condition for $S_n/n \xrightarrow{a.s.} a$ is that there exists $\mathbf{E}\xi_k = a$.*

Proof Sufficiency. Assume, without loss of generality, that $\mathbf{E}\xi_k = 0$. Then it follows from Theorem 10.5.3 that the random variable $Z^{(\varepsilon)} = \inf_{k>0}(S_k + \varepsilon k)$ is proper for any $\varepsilon > 0$ ($S_k + \varepsilon k$ is a sum of random variables $\xi_k + \varepsilon$ with $\mathbf{E}(\xi_k + \varepsilon) > 0$). Therefore,

$$\mathbf{P}\left(\inf_{k\ge n} \frac{S_k}{k} < -2\varepsilon\right) \le \mathbf{P}\left(\bigcup_{k\ge n}\{S_k + \varepsilon k < -\varepsilon n\}\right) \le \mathbf{P}\big(Z^{(\varepsilon)} < -\varepsilon n\big) \to 0$$

as $n \to \infty$. In a similar way we find that

$$\mathbf{P}\left(\sup_{k\ge n} \frac{S_k}{k} > 2\varepsilon\right) \to 0 \quad \text{as } n \to \infty.$$

Since $\mathbf{P}(\sup_{k\ge n} |S_k/k| > 2\varepsilon)$ does not exceed the sum of the above two probabilities, we obtain that $S_n/n \xrightarrow{a.s.} 0$.

Necessity. Note that

$$\frac{\xi_n}{n} = \frac{S_n}{n} - \frac{n-1}{n}\frac{S_{n-1}}{n-1} \xrightarrow{a.s.} 0,$$

so that the event $\{|\xi_n/n| > 1\}$ occurs finitely often with probability 1. By the Borel–Cantelli zero-one law, this means that $\sum_{n=1}^{\infty} \mathbf{P}(|\xi_n/n| > 1) < \infty$ or, which is the same, $\sum \mathbf{P}(|\xi| > n) < \infty$. Therefore, by Lemma 10.5.1, $\mathbf{E}\xi < \infty$ and with necessity $\mathbf{E}\xi = a$. The theorem is proved. □

Thus the condition $\mathbf{E}\xi = 0$ is necessary and sufficient for $\{\varepsilon n\}_{n=1}^{\infty}$ to be an upper sequence for both sequences $\{S_n\}$ and $\{-S_n\}$. In the next chapter, we will derive necessary and sufficient conditions for $\{\varepsilon n\}$ to be an upper sequence for each of the trajectories $\{S_n\}$ and $\{-S_n\}$ separately. Of course, such a condition, say, for the sequence $\{S_n\}$ will be broader than just $\mathbf{E}\xi_k = 0$.

We saw that the above proof of the strong law of large numbers was based on Theorem 10.5.3 on the finiteness of $\inf S_k$ which is based, in turn, on Wald's identity stated as Theorem 4.4.3. There exist other approaches to the proof that are unrelated to Theorem 4.4.3 (see below, e.g. Theorems 11.4.2 and 12.3.1). Now we will show that, using the strong law of large numbers, one can prove Wald's identity for stopping times without any additional restrictions (see e.g. conditions (a)–(d) in Theorem 4.4.3). Furthermore, in our opinion, the proof below better elucidates the nature of the phenomenon we are dealing with.

Consider stopping times ν with respect to a family of σ-algebras of a special kind. In particular, we assume that a sequence $\{\zeta_j\}_{j=1}^{\infty}$ of independent identically distributed random vectors $\zeta_j = (\xi_j, \tau_j)$ is given (where τ_j can also be vectors) and

$$\mathfrak{F}_n := \sigma(\zeta_1, \ldots, \zeta_n). \tag{11.3.1}$$

Theorem 11.3.2 (Wald's identity for stopping times) *Let ν be a stopping time with respect to the family of σ-algebras $\mathfrak{F}_n$ and assume one of the following conditions hold*: (a) $\mathbf{E}\nu < \infty$; *or* (b) $a := \mathbf{E}\xi_j \neq 0$.

Then

$$\mathbf{E}S_\nu = a\mathbf{E}\nu. \tag{11.3.2}$$

The assertion of the theorem means that Wald's identity is true whenever the right-hand side is defined, i.e. only the indefinite case $0 \cdot \infty$ is excluded. Roughly speaking, identity (11.3.2) is valid whenever it makes sense.

This identity implies that, when $\mathbf{E}\nu < \infty$, the condition $a \neq 0$ is superfluous and that, for $a \neq 0$, the finiteness of $\mathbf{E}S_\nu$ implies that of $\mathbf{E}\nu$. If $a = 0$ then the last assertion is not true. The reader can easily illustrate this fact using the fair game discussed in Sect. 4.2.

Proof of Theorem 11.3.2 By the strong law of large numbers, for all large k, the ratio S_k/k lies in the vicinity of the point a. (Here and in what follows, we leave more precise formulations to the reader.) By Lemma 11.2.1, the sequence $\{\zeta_{\nu+k}\}_{k=1}^{\infty}$ has the same distribution as the original sequence $\{\zeta_k\}_{k=1}^{\infty}$. For this "shifted" sequence, consider the stopping time ν_2 defined the same way as ν for the original sequence. Put $\nu_1 := \nu$ and consider the sequence $\{\zeta_{\nu_1+\nu_2+k}\}_{k=1}^{\infty}$ which is again distributed as $\{\zeta_k\}_{k=1}^{\infty}$ (for $\nu_1 + \nu_2$ is again a stopping time). For the new sequence, define the stopping time ν_3, and so on. Clearly, the ν_k are independent and identically distributed, and so are the differences

$$S_{N_k} - S_{N_{k-1}}, \quad k \geq 1, \quad S_0 = 0, \quad \text{where } N_k := \sum_{j=1}^{n} \nu_j.$$

By virtue of the strong law of large numbers, S_{N_k}/N_k also lie in the vicinity of the point a for all large k (or N_k).

If $\mathbf{E}\nu < \infty$ then N_k/k lie in the vicinity of the point $\mathbf{E}\nu$ as $k \to \infty$. Since

$$\frac{S_{N_k}}{N_k} = \frac{S_{N_k}}{k} \cdot \frac{k}{N_k}, \tag{11.3.3}$$

S_{N_k}/k is necessarily in a neighbourhood of the point $a\mathbf{E}\nu$ for all large k. This means that the expectation $\mathbf{E}S_\nu = a\mathbf{E}\nu$ exists.

If $\mathbf{E}\nu = \infty$ then, for $a > 0$, the assumption that the expectation $\mathbf{E}S_\nu$ exists and is finite, together with equality (11.3.3) and the previous argument, leads to a contradiction, since the limit of the left-hand side of (11.3.3) equals $a > 0$, but that of the right-hand side is zero. The contradiction vanishes only if $\mathbf{E}S_\nu = \infty$. The case $a < 0$ is dealt with in the same way. The theorem is proved. □

We now return to the strong law of large numbers and illustrate it by the following example.

Example 11.3.1 Let $\omega = (\omega_1, \omega_2, \dots)$ be a sequence of independent random variables taking the values 1 and 0 with probabilities p and $q = 1 - p$, respectively. To each such sequence, we put into correspondence the number

$$\xi = \xi(\omega) = \sum_{k=1}^{\infty} \omega_k 2^{-k},$$

so that ω is the binary expansion of ξ. It is evident that the possible values of ξ fill the interval [0, 1].

We show that if $p = q = 1/2$ then the distribution of ξ is *uniform*. But if $p \neq 1/2$, then ξ has a singular distribution. Indeed, if $x = \sum_{k=1}^{\infty} \delta_k 2^{-k}$, where δ_k assume the values 0 or 1, then

$$\{\xi < x\} = \{\varepsilon_1 < \delta_1\} \cup \{\omega_1 = \delta_1, \omega_2 < \delta_2\} \cup \{\omega_1 = \delta_1, \omega_2 = \delta_2, \omega_3 < \delta_3\} \cup \cdots.$$

Since the events in this union are disjoint, for $p = 1/2$ we have

$$\begin{aligned} \mathbf{P}(\xi < x) &= \sum_{k=0}^{\infty} \mathbf{P}(\omega_1 = \delta_1, \dots, \omega_k = \delta_k,\ \omega_{k+1} < \delta_{k+1}) \\ &= \sum_{k=0}^{\infty} 2^{-k} \mathbf{P}(\omega_{k+1} < \delta_{k+1}) = \sum_{k=0}^{\infty} 2^{-k-1} \delta_{k+1} = x. \end{aligned}$$

This means that the distribution of ξ is uniform, i.e. for any Borel set $B \subset [0, 1]$, the probability $\mathbf{P}(\xi \in B) = \operatorname{mes} B$ is equal to the Lebesgue measure of B. Put

$$D_n := \sum_{k=1}^{n} \delta_k, \qquad \Omega_n := \sum_{k=1}^{n} \omega_k.$$

Then the set $\{x : \lim_{n\to\infty} D_n/n = p\}$ is Borel measurable and hence

$$\operatorname{mes}\left\{x : \lim_{n\to\infty} \frac{D_n}{n} = \frac{1}{2}\right\} = \mathbf{P}\left(\lim_{n\to\infty} \frac{\Omega_n}{n} = \frac{1}{2}\right).$$

Since by the strong law of large numbers the right-hand side here is equal to one,

$$\operatorname{mes}\left\{x : \lim_{n\to\infty} \frac{D_n}{n} = \frac{1}{2}\right\} = 1.$$

In other words, for almost all $x \in [0, 1]$, the proportion of ones in the binary expansion of x is equal to $1/2$.

Now let $p \neq 1/2$. Then

$$\mathbf{P}\left(\lim_{n\to\infty}\frac{\Omega_n}{n}=p\right)=1,$$

although, as we saw above,

$$\operatorname{mes}\left\{x:\lim_{n\to\infty}\frac{D_n}{n}=p\right\}=0,$$

so that the probability measure is concentrated on a subset of $[0, 1]$ of Lebesgue measure zero. On the other hand, the distribution of the random variable ξ is continuous. This follows from the fact that

$$\{\xi=x\}=\bigcap_{k=1}^{\infty}\{\omega_k=\delta_k\},$$

if x is binary-irrational.

If ξ is binary-rational, i.e. if, for some $r<\infty$, either $\delta_k=0$ for all $k\ge r$ or $\delta_k=1$ for all $k\ge r$, the continuity follows from the inclusion

$$\{\xi=x\}\subset\bigcap_{k=r}^{\infty}\{\omega_k=0\}+\bigcap_{k=r}^{\infty}\{\omega_k=1\},$$

since the probabilities of the two events on the right-hand side are clearly equal to zero. The singularity of $F_\xi(x)$ for $p\neq 1/2$ is proved. □

We suggest the reader to plot the distribution function of ξ.

11.4 The Strong Law of Large Numbers for Arbitrary Independent Variables

Finding necessary and sufficient conditions for convergence

$$S_n/b_n \xrightarrow{a.s.} a$$

when $b_n \uparrow \infty$ and the summands $\xi_1, \xi_2, \dots$ are not identically distributed is a difficult task. We first prove the following theorem.

Theorem 11.4.1 (Kolmogorov's test for almost everywhere convergence) *Assume that ξ_k, $k=1,2,\dots$, are independent,* $\mathbf{E}\xi_k=0$, $\operatorname{Var}(\xi_k)=\sigma_k^2<\infty$ *and, moreover,*

$$\sum_{k=1}^{\infty}\frac{\sigma_k^2}{b_k^2}<\infty. \tag{11.4.1}$$

Then $S_n/b_n \xrightarrow{a.s.} 0$ as $n\to\infty$.

Proof It follows from the conditions of Theorem 11.4.1 that (see Corollary 11.2.2) the series $\sum_{k=1}^{\infty} \xi_k/b_k$ is convergent with probability 1. Therefore the assertion of Theorem 11.4.1 is a consequence of the following well-known lemma from calculus. □

Lemma 11.4.1 *Let $b_n \uparrow \infty$ and a sequence $x_1, x_2, \dots$ be such that the series $\sum_{k=1}^{\infty} x_k$ is convergent. Then, as $n \to \infty$,*

$$\frac{1}{b_n}\sum_{k=1}^{\infty} b_k x_k \to 0.$$

Proof Put $X_n := \sum_{k=n+1}^{\infty} x_k$ so that $X_n \to 0$ as $n \to \infty$, and $X := \max_{n\ge 0}|X_n| < \infty$. Using the Abel transform, we obtain that

$$\begin{aligned}\sum_{k=1}^{n} b_k x_k = \sum_{k=1}^{n} b_k(X_{k-1} - X_k) &= \sum_{k=0}^{n-1} b_{k+1}X_k - \sum_{k=1}^{n} b_k X_k \\ &= \sum_{k=1}^{n-1}(b_{k+1} - b_k)X_k + b_1 X_0 - b_n X_n,\end{aligned}$$

$$\limsup_{n\to\infty} \frac{1}{b_n}\sum_{k=1}^{n} b_k x_k \le \limsup_{n\to\infty} \frac{1}{b_n}\sum_{k=1}^{n-1}(b_{k+1} - b_k)X_k. \tag{11.4.2}$$

Here, for a given $\varepsilon > 0$, we can choose an N such that $|X_k| < \varepsilon$ for $k \ge N$. Therefore

$$\begin{aligned}\sum_{k=1}^{n-1}(b_{k+1} - b_k)X_k &\le \sum_{k=1}^{N-1}(b_{k+1} - b_k)X + \varepsilon\sum_{k=N}^{n-1}(b_{k+1} - b_k) \\ &= X(b_N - b_1) + \varepsilon(b_n - b_N).\end{aligned}$$

From here and (11.4.2) it follows that

$$\limsup_{n\to\infty} \frac{1}{b_n}\sum_{k=1}^{n} b_k x_k \le \varepsilon.$$

Since a similar inequality holds for lim inf, the lemma is proved. □

We could also prove Theorem 11.4.1 directly, using the Kolmogorov inequality, in a way similar to the argument in Theorem 11.2.1.

Example 11.4.1 Assume that ξ_k, $k = 1, 2, \dots$, are independent random variables taking the values $\xi_k = \pm k^{\alpha}$ with probabilities 1/2. As we saw in Example 8.4.1, for $\alpha > -1/2$, the sums S_n of these variables are asymptotically normal with the appropriate normalising factor $n^{-\alpha-1/2}$. Since $\operatorname{Var}(\xi_k) = \sigma_k^2 = k^{2\alpha}$, we see that, for

$\beta > \alpha + 1/2$, $n^{-\beta} S_n$ satisfies the strong law of large numbers because, for $b_k = k^\beta$, the series

$$\sum_{k=1}^{\infty} k^{2\alpha} b_k^{-2} = \sum_{k=1}^{\infty} k^{2\alpha - 2\beta}$$

converges. The "usual" strong law of large numbers (with the normalising factor n^{-1}) holds if the value $\beta = 1$ is admissible, i.e. when $\alpha < 1/2$.

Now we will derive the "usual" strong law of large numbers (with scaling factor $1/n$) under conditions which do not assume the existence of the variances $\mathrm{Var}(\xi_k)$ and are, in a certain sense, minimal. The following generalisation of the "sufficiency part" of Theorem 11.1.3 is valid.

Theorem 11.4.2 *Let* $\mathbf{E}\xi_k = 0$ *and the tails* $\mathbf{P}(|\xi_k| > t)$ *admit a common integrable majorant*:

$$\mathbf{P}\big(|\xi_k| > t\big) \le g(t), \qquad \int_0^\infty g(t)\,dt < \infty. \tag{11.4.3}$$

Then, as $n \to \infty$,

$$\frac{S_n}{n} \xrightarrow{a.s.} 0. \tag{11.4.4}$$

Note that condition (11.4.3) can also be rewritten as $|\xi_k| \overset{d}{\le} \zeta$, $\mathbf{E}\zeta < \infty$. To see this, it suffices to consider a random variable $\zeta \ge 0$ for which $\mathbf{P}(\zeta > t) = \min(1, g(t))$. Here, without loss of generality, we can assume that $g(t)$ is non-increasing (we can take the minimal majorant $g(t) := \sup_k \mathbf{P}(|\xi_k| > t) \downarrow$).

Condition (11.4.3) clearly implies the uniform integrability of ξ_k. The latter was sufficient for the law of large numbers, but is insufficient for the strong law of large numbers. This is shown by the following example.

Example 11.4.2 Let ξ_k be such that, for $t > 0$ and $k > 1$,

$$\mathbf{P}(\xi_k \ge t) = \begin{cases} g(t) + \frac{1}{k \ln k} & \text{if } t \le k, \\ g(t) & \text{if } t > k, \end{cases}$$
$$\mathbf{P}(\xi_k < -t) \le g(t),$$

where $g(t)$ is integrable so that the ξ_k has a positive atom of size $1/(k \ln k)$ at the point k. Evidently, the ξ_k are uniformly integrable. Now suppose that $S_n/n \xrightarrow{a.s} 0$. Since

$$\sum_{k=2}^{\infty} \mathbf{P}(\xi_k \ge k) \ge \sum_{k=2}^{\infty} \frac{1}{k \ln k} = \infty,$$

it follows by the Borel–Cantelli lemma that infinitely many events $\{\xi_k \ge k\}$ occur with probability 1. Since, for any $\varepsilon < 1/2$ and all k large enough, $|S_k| < \varepsilon k$ with probability 1, the events $S_{k+1} = S_k + \xi_{k+1} > k(1 - \varepsilon)$ occur infinitely often. We have obtained a contradiction. □

Proof of Theorem 11.4.2 Represent the random variables ξ_k in the form

$$\xi_k = \xi_k^* + \xi_k^{**}, \qquad \xi_k^* := \xi_k \mathrm{I}\ \big(|\xi_k| < k\big), \qquad \xi_k^{**} := \xi_k \mathrm{I}\ \big(|\xi_k| \ge k\big),$$

and denote by S_n^* and S_n^{**} the respective sums of random variables ξ_k^* and ξ_k^{**}. Then the sum S_n can be written as

$$S_n = \big(S_n^* - \mathbf{E}S_n^*\big) + S_n^{**} - \mathbf{E}S_n^{**}. \tag{11.4.5}$$

Now we will evaluate the three summands on the right-hand side of (11.4.5).

1. Since ξ_k are uniformly integrable, we have

$$\begin{aligned} &\mathbf{E}\xi_k^{**} = o(1) \quad \text{as } k \to \infty, \\ &\mathbf{E}S_n^{**} = o(n), \qquad \frac{\mathbf{E}S_n^{**}}{n} \to 0 \quad \text{as } n \to \infty. \end{aligned} \tag{11.4.6}$$

2. Since

$$\sum \mathbf{P}\big(|\xi_k| > k\big) \le \sum g(k) < \infty,$$

we obtain from Theorem 11.1.1 that, with probability 1, only a finite number of random variables ξ_k^{**} are nonzero and hence, as $n \to \infty$,

$$\frac{S_n^{**}}{n} \xrightarrow{a.s.} 0. \tag{11.4.7}$$

3. To bound the first summand on the right-hand side of (11.4.5) we make use of Theorem 11.4.1. Since

$$\operatorname{Var}\big(\xi_k^*\big) \le \mathbf{E}\big(\xi_k^*\big)^2 = 2\int_0^k u\mathbf{P}\big(|\xi_k^*| \ge u\big)\,du \le 2\int_o^k ug(u)\,du,$$

we see that the series in (11.4.1) for $\xi_k^* - \mathbf{E}\xi_k^*$ and $b_k = k$ admits the upper bound

$$2\sum_{k=1}^{\infty} \frac{1}{k^2}\int_0^k ug(u)\,du. \tag{11.4.8}$$

The last series converges if the integral

$$\int_1^{\infty} \frac{1}{t^2}\left(\int_0^t ug(u)\,du\right)dt$$

converges. Integrating by parts, we obtain

$$-\frac{1}{t}\int_0^t ug(u)\,du\bigg|_1^{\infty} + \int_1^{\infty} g(t)\,dt. \tag{11.4.9}$$

The last summand here is clearly finite. Since $g(u)$ is integrable and monotone, one has

$$ug(u) = o(1) \quad \text{as } u \to \infty, \qquad \int_0^t ug(u)\,du = o(t) \quad \text{as } t \to \infty,$$

and hence the value of the first summand in (11.4.9) is zero at $t = \infty$. We have established that series (11.4.8) converges, and hence, by Theorem 11.4.1, as $n \to \infty$,

$$\frac{S_n^* - \mathbf{E}\,S_n^*}{n} \xrightarrow{a.s.} 0. \tag{11.4.10}$$

Combining (11.4.5)–(11.4.7) and (11.4.10), we obtain (11.4.4). The theorem is proved. □

11.5 The Strong Law of Large Numbers for Generalised Renewal Processes

11.5.1 The Strong Law of Large Numbers for Renewal Processes

Let $\{\tau_j\}$ be a sequence of independent identically distributed variables, $T_n := \sum_{j=1}^{n} \tau_j$ and $\eta(t) := \min\{k : T_k > t\}$.

Theorem 11.5.1 *If $\tau_j \stackrel{d}{=} \tau$ and $\mathbf{E}\tau = a > 0$ exists then, as $t \to \infty$,*

$$\frac{\eta(t)}{t} \xrightarrow{a.s.} \frac{1}{a}, \qquad (11.5.1)$$

i.e., for any $\varepsilon > 0$,

$$\mathbf{P}\left(\left|\frac{\eta(u)}{u} - \frac{1}{a}\right| < \varepsilon \text{ for all } u \ge t\right) \to 1 \qquad (11.5.2)$$

as $t \to \infty$.

Proof First let $\tau \ge 0$. Set

$$A_n := \left\{\left|\frac{T_k}{k} - a\right| < \varepsilon \text{ for all } k \ge n\right\}.$$

The strong law of large numbers for $\{T_k\}$ means that $\mathbf{P}(A_n) \to 1$ as $n \to \infty$.

Consider the function $T(v) := T_{\lfloor v \rfloor}$, where $\lfloor v \rfloor$ is the integer part of v. As was noted in Sect. 10.1, $\eta(t)$ is the generalised inverse function to $T(v)$. In other words, if we plot the graph of the function $T(v)$ as a continuous line (including "vertical" segments corresponding to jumps) then $\eta(t)$ can be regarded as the abscissa of the point of intersection of the graph of $T(v)$ with level t (see Fig. 11.1); for the values of t coinciding with T_k, the intersection will be a segment of length 1, and $\eta(t)$ is then to be taken equal to the right end point of the segment.

Therefore the event that $T(v)$ lies within the limits $v/(a \pm \varepsilon)$ for all sufficiently large v coincides with the event that $\eta(t)$ lies within the limits $t(a \pm \varepsilon)$ for all sufficiently large t. More precisely,

$$A_n \subset B_n := \left\{\frac{u}{a-\varepsilon} > \eta(u) > \frac{u}{a+\varepsilon} \text{ for all } u \ge n(a+\varepsilon)\right\}.$$

This means that

$$\mathbf{P}(B_n) \to 1 \quad \text{as } n \to \infty.$$

This relation is clearly equivalent to (11.5.2).

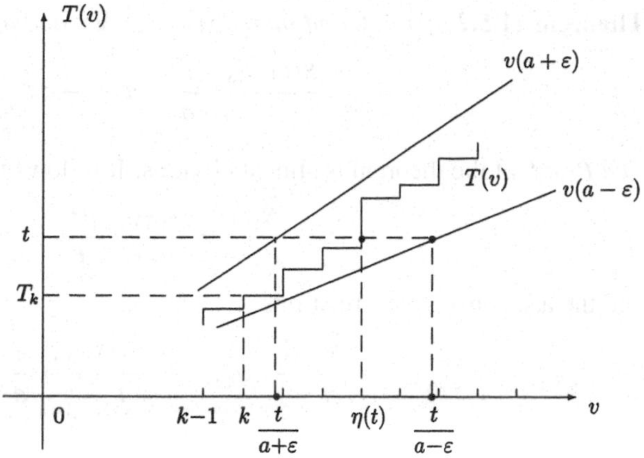

Fig. 11.1 The relative positions of a trajectory of $T(v)$ and the levels $v(a \pm \varepsilon)$ (see the proof of Theorem 11.5.1)

Now suppose τ can also assume negative values. Then $\eta(t) := \min\{k : \overline{T}_k > t\}$, where $\overline{T}_k = \max_{k \le n} T_k$, so that $\eta(t)$ is the generalised inverse function of $\overline{T}(v) := \overline{T}_{[v]}$. Moreover, it is clear that, if $T(v)$ lies within the limits $v(a \pm \varepsilon)$ for all sufficiently large v, then the same is true for the function $\overline{T}(v)$. It remains to repeat the above argument applying it to the processes $\overline{T}(v)$ and $\eta(t)$. The theorem is proved. □

Remark 11.5.1 (An analogue of Remarks 8.3.3, 8.4.1, 10.1.1 and 10.5.1) Convergence (11.5.1) persists if we remove all the restrictions on the random variable τ_1. Namely, the following assertion generalising Theorem 11.5.1 is valid. *Let τ_1 be an arbitrary random variable and the variables $\tau_k^* = \tau_{k+1} \overset{d}{=} \tau$, $k \ge 1$, satisfy the conditions of Theorem* 11.5.1. *Then* (11.5.1) *holds true.*

The Proof of this assertion is quite similar to the proofs of the corresponding assertions in the above mentioned remarks, and we leave it to the reader. □

These assertions show that replacement of one or several terms in the considered sequences of random variables with arbitrary variables changes nothing in the established convergence relations. (The exception is Theorem 11.1.1, in which the condition $\mathbf{E}\min(0, \tau_1) > -\infty$ is essential.) This fact will be used in Chap. 13 devoted to Markov chains.

11.5.2 The Strong Law of Large Numbers for Generalised Renewal Processes

Now let a sequence of independent identically distributed random vectors $(\tau_j, \xi_j) \overset{d}{=} (\tau, \xi)$ be given and $S_n = \sum_{j=1}^n \xi_j$. Our goal is to obtain an analogue of Theorem 11.5.1 for generalised renewal processes $S(t) = S_{\eta(t)}$ (see Sect. 10.6).

Theorem 11.5.2 *If* $\tau > 0$ *and there exist* $a := \mathbf{E}\tau$ *and* $a_\xi := \mathbf{E}\xi$ *then*

$$\frac{S(t)}{t} \xrightarrow{a.s.} \frac{a_\xi}{a} \quad \text{as } t \to \infty.$$

The Proof of the theorem is almost obvious. It follows from the representation

$$\frac{S(t)}{t} = \frac{S_{\eta(t)}}{\eta(t)} \cdot \frac{\eta(t)}{t}$$

and the a.s. convergence relations

$$\frac{S_n}{n} \xrightarrow{a.s.} a_\xi, \qquad \frac{\eta(t)}{t} \xrightarrow{a.s.} \frac{1}{a}.$$ □

Note that the independence of the components τ and ξ is not assumed here.

Chapter 12
Random Walks and Factorisation Identities

Abstract In this chapter, several remarkable and rather useful relations establishing interconnections between different characteristics of random walks (the so-called boundary functionals) are derived, and the arising problems are related to the simplest boundary problems of Complex Analysis. Section 12.1 introduces the concept of factorisation identity and derives two fundamental identities of that kind. Some consequences of these identities, including the trichotomy theorem on the oscillatory behaviour of random walks and a one-sided version of the Strong Law of Large Numbers are presented in Sect. 12.2. Pollaczek–Spitzer's identity and an identity for the global maximum of the random walk are derived in Sect. 12.3, followed by illustrating these results by examples from the ruin theory and the theory of queueing systems in Sect. 12.4. Sections 12.5 and 12.6 are devoted to studying the cases where factorisation components can be obtained in explicit form and so closed form expressions are available for the distributions of a number of important boundary functionals. Sections 12.7 and 12.8 employ factorisation identities to derive the asymptotic properties of the distribution of the excess of a random walk of a high level and that of the global maximum of the walk, and also to analyse the distribution of the first passage time.

In the present chapter we derive several remarkable and rather useful relations establishing interconnections between different characteristics of random walks (the so-called boundary functionals) and also relate the arising problems with the simplest boundary problems of complex analysis.

12.1 Factorisation Identities

12.1.1 Factorisation

On the plane of a complex variable λ, denote by $\varPi$ the real axis $\operatorname{Im}\lambda = 0$ and by $\varPi_+$ ($\varPi_-$) the half-plane $\operatorname{Im}\lambda > 0$ ($\operatorname{Im}\lambda < 0$). Let $\mathfrak{f}(\lambda)$ be a continuous function defined on $\varPi$.

Definition 12.1.1 If there exists a representation

$$\mathfrak{f}(\lambda) = \mathfrak{f}_+(\lambda)\mathfrak{f}_-(\lambda), \quad \lambda \in \varPi, \tag{12.1.1}$$

A.A. Borovkov, *Probability Theory*, Universitext,
DOI 10.1007/978-1-4471-5201-9_12, © Springer-Verlag London 2013

where $\mathfrak{f}_\pm$ are analytic in the domains $\Pi_\pm$ and continuous on $\Pi_\pm \cup \Pi$, then we will say that the function $\mathfrak{f}$ *allows factorisation*. The functions $\mathfrak{f}_\pm$ are called *factorisation components* (*positive* and *negative*, respectively).

Further, denote by $\mathcal{K}$ the class of functions $\mathfrak{f}$ defined on Π that are continuous and such that

$$\sup_{\lambda\in\Pi}\left|\mathfrak{f}(\lambda)\right| < \infty, \qquad \inf_{\lambda\in\Pi}\left|\mathfrak{f}(\lambda)\right| > 0. \tag{12.1.2}$$

Similarly we define the classes $\mathcal{K}_\pm$ of functions analytic in $\Pi_\pm$ and continuous on $\Pi_\pm \cup \Pi$, such that

$$\sup_{\lambda\in\Pi_\pm}\left|\mathfrak{f}_\pm(\lambda)\right| < \infty, \qquad \inf_{\lambda\in\Pi_\pm}\left|\mathfrak{f}_\pm(\lambda)\right| > 0. \tag{12.1.3}$$

Definition 12.1.2 If, for an $\mathfrak{f} \in \mathcal{K}$, there exists a representation (12.1.1), where $\mathfrak{f}_\pm \in \mathcal{K}_\pm$, then we will say that the function $\mathfrak{f}$ allows *canonical factorisation.*

Representations of the form

$$\mathfrak{f}(\lambda) = \mathfrak{f}_+(\lambda)\mathfrak{f}_-(\lambda) f_0, \qquad \mathfrak{f}(\lambda) = \frac{\mathfrak{f}_+(\lambda) f_0}{\mathfrak{f}_-(\lambda)}, \quad \lambda \in \Pi,$$

where $f_0 = \text{const}$ and $\mathfrak{f}_\pm \in \mathcal{K}_\pm$, are also called canonical factorisations.

Lemma 12.1.1 *The components $\mathfrak{f}_\pm$ of a canonical factorisation of a function $\mathfrak{f} \in \mathcal{K}$ are defined uniquely up to a constant factor.*

Proof Together with the canonical factorisation (12.1.1), let there exist another canonical factorisation

$$\mathfrak{f}(\lambda) = \mathfrak{g}_+(\lambda)\mathfrak{g}_-(\lambda), \quad \lambda \in \Pi.$$

Then

$$\mathfrak{f}_+(\lambda)\mathfrak{f}_-(\lambda) = \mathfrak{g}_+(\lambda)\mathfrak{g}_-(\lambda), \quad \lambda \in \Pi,$$

and, by (12.1.2), we can divide both sides of the inequality by $\mathfrak{g}_+(\lambda)\mathfrak{f}_-(\lambda)$. We get

$$\frac{\mathfrak{f}_+(\lambda)}{\mathfrak{g}_+(\lambda)} = \frac{\mathfrak{g}_-(\lambda)}{\mathfrak{f}_-(\lambda)},$$

where, by virtue of (12.1.2), the function $\frac{\mathfrak{f}_+(\lambda)}{\mathfrak{g}_+(\lambda)}$ $(\frac{\mathfrak{g}_-(\lambda)}{\mathfrak{f}_-(\lambda)})$ belongs to the class $\mathcal{K}_+$ $(\mathcal{K}_-)$. We have obtained that the function $\frac{\mathfrak{f}_+(\lambda)}{\mathfrak{g}_+(\lambda)}$, analytical in Π_+, can be analytically continued over the line Π onto the half-plane Π_- (to the function $\frac{\mathfrak{g}_-(\lambda)}{\mathfrak{f}_-(\lambda)}$). After such a continuation, in view of (12.1.3), this function remains bounded on the whole complex plane. By Liouville's theorem, bounded entire functions must be constant, i.e. there exists a constant c, such that, on the whole plane

$$\frac{\mathfrak{f}_+(\lambda)}{\mathfrak{g}_+(\lambda)} = \frac{\mathfrak{g}_-(\lambda)}{\mathfrak{f}_-(\lambda)} = c,$$

holds, so $\mathfrak{f}_+(\lambda) = c\mathfrak{g}_+(\lambda)$, $\mathfrak{f}_-(\lambda) = c^{-1}\mathfrak{g}_-(\lambda)$. The lemma is proved. □

The factorisation problem consists in finding conditions under which a given function $\mathfrak{f}$ admits a factorisation, and in finding the components of the factorisation. This problem has a number of important applications to solving integral equations and is a version of the well-known Cauchy–Riemann boundary-value problem in complex function theory. We will see later that factorisation is also an important tool for studying the so-called boundary problems in probability theory.

12.1.2 The Canonical Factorisation of the Function $\mathfrak{f}_z(\lambda) = 1 - z\varphi(\lambda)$

Let $(\Omega, \mathfrak{F}, \mathbf{P})$ be a probability space on which a sequence $\{\xi_k\}_{k=1}^{\infty}$ of independent identically distributed ($\xi_k \overset{d}{=} \xi$) random variables is given. Put, as before, $S_n := \sum_{k=1}^{n} \xi_k$ and $S_0 = 0$. The sequence $\{S_k\}_{k=0}^{\infty}$ forms a *random walk*.

First of all, note that the function

$$\mathfrak{f}_z(\lambda) := 1 - z\varphi(\lambda), \qquad \varphi(\lambda) := \mathbf{E}e^{i\lambda\xi}, \quad \lambda \in \Pi,$$

belongs to $\mathcal{K}$, for all z with $|z| < 1$ (here z is a complex-valued parameter). This follows from the inequalities $|\varphi(\lambda)| \le 1$ for $\lambda \in \Pi$ and $|z\varphi(\lambda)| < |z| < 1$.

Theorem 12.1.1 (The first factorisation identity) *For* $|z| < 1$, *the function* $\mathfrak{f}_z(\lambda)$ *admits the canonical factorisation*

$$\mathfrak{f}_z(\lambda) = \mathfrak{f}_{z+}(\lambda)C(z)\mathfrak{f}_{z-}(\lambda), \quad \lambda \in \Pi, \tag{12.1.4}$$

where

$$\begin{aligned}
\mathfrak{f}_{+}(\lambda) &= \exp\left\{-\sum_{k=1}^{\infty} \frac{z^k}{k} \mathbf{E}\big(e^{i\lambda S_k};\ S_k > 0\big)\right\} \in \mathcal{K}_+,\\
\mathfrak{f}_{z-}(\lambda) &= \exp\left\{-\sum_{k=1}^{\infty} \frac{z^k}{k} \mathbf{E}\big(e^{i\lambda S_k};\ S_k < 0\big)\right\} \in \mathcal{K}_-,\\
C(z) &= \exp\left\{-\sum_{k=1}^{\infty} \frac{z^k}{k} \mathbf{P}(S_k = 0)\right\}.
\end{aligned} \tag{12.1.5}$$

Proof Since $|z| < 1$, $\ln(1 - z\varphi(\lambda))$ exists, understood in the principal value sense. The following equalities give the desired decomposition:

$$\begin{aligned}
\mathfrak{f}_z(\lambda) &= e^{\ln(1-z\varphi(\lambda))} = \exp\left\{-\sum_{k=1}^{\infty} \frac{z^k\varphi^k(\lambda)}{k}\right\} = \exp\left\{-\sum_{k=1}^{\infty} \frac{z^k}{k} \mathbf{E}e^{i\lambda S_k}\right\}\\
&= \exp\left\{-\sum_{k=1}^{\infty} \frac{z^k}{k} \mathbf{E}\big(e^{i\lambda S_k};\ S_k > 0\big)\right\} \exp\left\{-\sum_{k=1}^{\infty} \frac{z^k}{k} \mathbf{P}(S_k = 0)\right\}\\
&\quad \times \exp\left\{-\sum_{k=1}^{\infty} \frac{z^k}{k} \mathbf{E}\big(e^{i\lambda S_k};\ S_k < 0\big)\right\}.
\end{aligned}$$

Show that $\mathfrak{f}_{z+}(\lambda) \in \mathcal{K}_+$. Indeed, the function $\mathbf{E}(e^{i\lambda S_k}; S_k > 0)$, for every k and $\lambda \in \Pi_+ \cup \Pi$, does not exceed 1 in the absolute value, is analytic in Π_+, and is continuous on $\Pi_+ \cup \Pi$. Analyticity follows from the differentiability of this function at any point $\lambda \in \Pi_+$ (see also Property 6 of ch.f.s in Sect. 7.1). The function $\ln \mathfrak{f}_{z+}(\lambda)$ is a uniformly converging series of functions analytic in Π_+, and hence possesses the same properties together with the function $\mathfrak{f}_{z+}(\lambda)$. The same can be said about the continuity on $\Pi \cup \Pi_+$.

That $\mathfrak{f}_{z-}(\lambda) \in \mathcal{K}_-$ is established in a similar way. The theorem is proved. □

12.1.3 The Second Factorisation Identity

The second factorisation identity is associated with the so-called boundary functionals of the random walk $\{S_k\}$. On the main probability space $(\Omega, \mathfrak{F}, \mathbf{P})$ we define, together with $\{\xi_k\}$, the random variable

$$\eta_+^0 := \min\{k \ge 1;\ S_k \ge 0\}.$$

This is the *first-passage time to zero level*. For the elementary events such that all $S_k < 0$, $k \ge 1$, we put $\eta_+^0 := \infty$. Like the random variable $\eta(0)$ in Sect. 10.1, the variable η_+^0 is a Markov time.

The random variable $\chi_+^0 := S_{\eta_+^0}$ is called the *first nonnegative sum*. It is defined on the set $\{\eta_+^0 < \infty\}$ only.

The *first passing time of zero from the right*

$$\eta_-^0 := \min\{k \ge 1;\ S_k \le 0\}$$

possesses quite similar properties, and so does the *first nonpositive sum* $\chi_-^0 := S_{\eta_-^0}$.

Studying the properties of the introduced random variables, which are called boundary functionals of the random walk $\{S_k\}$, is of significant independent interest. For instance, the variable η_+^0 is a stopping time, and understanding its nature is essential for studying stopping times in many more complex problems (see e.g. the problems of the renewal theory in Chap. 10, the problems of statistical control described in Sect. 4.4 and so on). Moreover, the variables η_+^0 and χ_+^0 will be needed to describe the extrema

$$\zeta := \sup(S_1, S_2, \dots) \quad \text{and} \quad \gamma := \inf(S_1, S_2, \dots),$$

which are also termed boundary functionals and play an important role in the problems of mathematical statistics, queueing theory (see Sect. 12.4), etc.

Put, as before, $\varphi(\lambda) := \varphi_\xi(\lambda) = \mathbf{E}e^{i\lambda\xi}$.

Theorem 12.1.2 (The second factorisation identity) *For the ch.f. of the joint distributions of the introduced random variables, for* $|z| < 1$ *and* $\operatorname{Im}\lambda = 0$, *the canonical factorisation*

$$\begin{aligned}\mathfrak{f}_z(\lambda) &:= 1 - z\varphi(\lambda) \\ &= \left[1 - \mathbf{E}\left(e^{i\lambda\chi_+^0} z^{\eta_+^0};\ \eta_+^0 < \infty\right)\right] D^{-1}(z) \left[1 - \mathbf{E}\left(e^{i\lambda\chi_-^0} z^{\eta_-^0};\ \eta_-^0 < \infty\right)\right],\end{aligned}$$

of $\mathfrak{f}_z(\lambda)$ *holds true, where*

$$D(z) := 1 - \mathbf{E}\left(z^{\eta_+^0};\ \chi_+^0 = 0,\ \eta_+^0 < \infty\right) = 1 - \mathbf{E}\left(z^{\eta_-^0};\ \chi_-^0 = 0,\ \eta_-^0 < \infty\right).$$

Proof Set $\zeta_n := \max\{S_1, \dots, S_n\}$. We have

$$\begin{aligned}\varphi^n(\lambda) = \mathbf{E}e^{i\lambda S_n} &= \sum_{k=1}^n \mathbf{E}\left(e^{i\lambda S_n};\ \eta_+^0 = k\right) + \mathbf{E}\left(e^{i\lambda S_n};\ \zeta_n < 0\right) \\ &= \sum_{k=1}^n \mathbf{E}\left(e^{i\lambda(S_n - S_k)} e^{i\lambda S_k} \mathrm{I}\left(\eta_+^0 = k\right)\right) + M_n, \end{aligned} \qquad (12.1.6)$$

where $M_n = \mathbf{E}(e^{i\lambda S_n};\ \zeta_n < 0)$ and $\mathrm{I}(A)$ is the indicator of the event A. For each fixed k, the random variables $S_n - S_k$ and $S_k \mathrm{I}(\eta_+^0 = k) = \chi_+^0 \mathrm{I}(\eta_+^0 = k)$ are independent. Hence,

$$\varphi^n(\lambda) = \sum_{k=1}^n \varphi^{n-k}(\lambda) \mathbf{E}\left(e^{i\lambda\chi_+^0};\ \eta_+^0 = k\right) + M_n.$$

Now multiply both sides by z^n, $n = 0, 1, \dots$, and then sum up over n. We will use the convention that, for $n = 0$,

$$\sum_{k=1}^n = 0, \qquad M_n = 1.$$

For the convolution of two sequences $c_n = \sum_{k=1}^n a_k b_{n-k}$, we have

$$\sum_{n=0}^\infty c_n z^n = \sum_{n=1}^\infty a_n z^n \sum_{n=0}^\infty b_n z^n,$$

provided that the series in this equality converges absolutely. Since $|z| < 1$ and $|\varphi(\lambda)| \le 1$ for $\operatorname{Im}\lambda = 0$, one has

$$\begin{aligned}\sum_{n=0}^\infty z^n \varphi^n(\lambda) = \frac{1}{1 - z\varphi(\lambda)} &= \sum_{n=0}^\infty z^n \sum_{k=1}^n \varphi^{n-k}(\lambda) \mathbf{E}\left(e^{i\lambda\chi_+^0};\ \eta_+^0 = k\right) + \sum_{n=0}^\infty z^n M_n \\ &= \sum_{k=1}^\infty z^k \mathbf{E}\left(e^{i\lambda\chi_+^0};\ \eta_+^0 = k\right) \sum_{n=0}^\infty z^n \varphi^n(\lambda) + \sum_{n=0}^\infty z^n M_n \\ &= \frac{1}{1 - z\varphi(\lambda)} \mathbf{E}\left(e^{i\lambda\chi_+^0} z^{\eta_+^0};\ \eta_+^0 < \infty\right) + \sum_{n=0}^\infty z^n M_n,\end{aligned}$$

or, which is the same,

$$\mathfrak{f}_z(\lambda) = 1 - z\varphi(\lambda) = \frac{1 - \mathbf{E}(e^{i\lambda\chi_+^0} z^{\eta_+^0};\ \eta_+^0 < \infty)}{\sum_{n=0}^\infty z^n \mathbf{E}(e^{i\lambda S_n};\ \zeta_n < 0)} = \frac{\mathfrak{a}_{z+}(\lambda)}{\mathfrak{a}_{z-}(\lambda)}, \qquad (12.1.7)$$

where $\mathfrak{a}_{z\pm}(\lambda)$ denote the numerator and denominator of the ratio obtained for $\mathfrak{f}_z(\lambda)$.

It is easy to see that, if we put

$$\gamma_n := \min(S_1, \dots, S_n)$$

then, repeating the above arguments, we will arrive at the equality

$$\mathfrak{f}_z(\lambda) = \frac{1 - \mathbf{E}(e^{i\lambda\chi_-^0} z^{\eta_-^0};\ \eta_-^0 < \infty)}{\sum_{n=0}^{\infty} z^n \mathbf{E}(e^{i\lambda S_n};\ \gamma_n > 0)} = \frac{\mathfrak{b}_{z-}(\lambda)}{\mathfrak{b}_{z+}(\lambda)}, \tag{12.1.8}$$

where, similarly to the above, $\mathfrak{b}_{z\mp}(\lambda)$, respectively, denote the numerator and denominator in relation (12.1.8).

Now we show that $\mathfrak{a}_{z\pm}(\lambda) \in \mathcal{K}$ and $\mathfrak{b}_{z\pm}(\lambda) \in \mathcal{K}$ for $|z| < 1$. Indeed, for $|z| < 1$ and $\operatorname{Im} \lambda = 0$,

$$\left|\mathbf{E}\big(e^{i\lambda\chi_+^0} z^{\eta_+^0};\ \eta_+^0 < \infty\big)\right| \le \mathbf{E}\big(|z|^{\eta_+^0};\ \eta_+^0 < \infty\big) < 1$$

and therefore

$$\sup_{\lambda\in\Pi} \left|\mathfrak{a}_{z+}(\lambda)\right| < \infty, \qquad \inf_{\lambda\in\Pi} \left|\mathfrak{a}_{z+}(\lambda)\right| > 0.$$

Since $\mathfrak{f}_z(\lambda) \in \mathcal{K}$, this also implies that $\mathfrak{a}_{z-}(\lambda) \in \mathcal{K}$. In the same way we obtain that $\mathfrak{b}_{z\mp}(\lambda) \in \mathcal{K}$. By equating the right-hand sides of (12.1.7) and (12.1.8) and multiplying them by $\mathfrak{a}_{z-}(\lambda)\mathfrak{b}_{z+}(\lambda)$, we get

$$\mathfrak{a}_{z+}(\lambda)\mathfrak{b}_{z+}(\lambda) = \mathfrak{a}_{z-}(\lambda)\mathfrak{b}_{z-}(\lambda), \qquad \lambda \in \Pi. \tag{12.1.9}$$

Further, the functions $\mathfrak{a}_{z+}(\lambda)$ and $\mathfrak{b}_{z+}(\lambda)$ are bounded and analytic in Π_+ for the same reasons as the function $\mathfrak{f}_{z+}(\lambda)$ (see the proof of Theorem 12.1.1). Similarly, $\mathfrak{a}_{z-}(\lambda)$ and $\mathfrak{b}_{z-}(\lambda)$ are bounded and analytic in Π_-. We obtain that the function $\mathfrak{a}_{z+}(\lambda)\mathfrak{b}_{z+}(\lambda)$ is bounded and analytic in Π_+ and, by (12.1.9), has an entire bounded analytic continuation over the boundary Π to the whole complex plane. This means that this function necessarily equals a constant c, and $\mathfrak{b}_{z+}(\lambda) = c\mathfrak{a}_{z+}^{-1}(\lambda) \in \mathcal{K}_+$, $\mathfrak{a}_{z-}(\lambda) = c\mathfrak{b}_{z-}^{-1}(\lambda) \in \mathcal{K}_-$, so relations (12.1.7) and (12.1.8) deliver a canonical factorisation of $\mathfrak{f}_z(\lambda)$.

Further, $e^{i\lambda x} \to 0$ as $\operatorname{Im} \lambda \to -\infty$, $x < 0$, and therefore

$$\mathfrak{b}_{z-}(-i\infty) = 1 - \mathbf{E}\big(z^{\eta_-^0};\ \chi_-^0 = 0,\ \eta_-^0 < \infty\big), \qquad \mathfrak{a}_{z-}(-i\infty) = 1,$$

$$\mathfrak{a}_{z-}(\lambda)\mathfrak{b}_{z-}(\lambda) = \mathfrak{a}_{z-}(-i\infty)\mathfrak{b}_{z-}(-i\infty) = 1 - \mathbf{E}\big(z^{\eta_-^0};\ \chi_-^0 = 0,\ \eta_-^0 < \infty\big) = D(z).$$

Substituting into (12.1.7) the value $\mathfrak{a}_{z-}(\lambda) = D(z)/\mathfrak{b}_{z-}(\lambda)$ derived from this equality, we obtain the assertion of the theorem. The second relation for $D(z)$ follows from the equality $D(z) = \mathfrak{a}_{z+}(i\infty)\mathfrak{b}_{z+}(i\infty)$. The theorem is proved. □

In the proof of Theorem 12.1.2 we used, in formula (12.1.6), a decomposition of $\mathbf{E}e^{i\lambda S_n}$ into summands corresponding to the disjoint events

$$\left\{\bigcup_{k=1}^{n} \{\eta_+^0 = k\}\right\} = \{\zeta_n \ge 0\} \quad \text{and} \quad \{\zeta_n < 0\}.$$

But the scheme of the proof will still work if we consider the partition of Ω into the events $\{\zeta_n > 0\}$ and $\{\zeta_n \le 0\}$. In order to do this, we introduce the random variables

$$\eta_+ := \min\{k : S_k > 0\}$$

($\eta_+ = \infty$ if $\zeta \le 0$; note that $\eta_+ = \eta(0)$ in the notation of Sect. 10.1),

$$\begin{aligned} \chi_+ &:= S_{\eta_+}, \\ \eta_- &:= \min\{k : S_k < 0\} \quad (\eta_- = \infty \text{ if } \gamma \ge 0), \\ \chi_- &:= S_{\eta_-}. \end{aligned}$$

The variable η_+ (η_-) is called the *time of the first positive (negative) sum* χ_+ (χ_-). Now we can write, together with equalities (12.1.7) and (12.1.8), the relations

$$\begin{aligned} \mathfrak{f}_z(\lambda) = 1 - z\varphi(\lambda) &= \frac{1 - \mathbf{E}(e^{i\lambda\chi_+} z^{\eta_+};\ \eta_+ < \infty)}{\sum_{n=0}^{\infty} z^n \mathbf{E}(e^{i\lambda S_n};\ \zeta_n \le 0)} \\ &= \frac{1 - \mathbf{E}(e^{i\lambda\chi_-} z^{\eta_-};\ \eta_- < \infty)}{\sum_{n=0}^{\infty} z^n \mathbf{E}(e^{i\lambda S_n};\ \gamma_n \ge 0)}. \end{aligned} \tag{12.1.10}$$

Combining these relations with (12.1.7) and (12.1.8), we will use below the same argument as above to prove the following assertion.

Theorem 12.1.3

$$\begin{aligned} 1 - \mathbf{E}\left(e^{i\lambda\chi_+^0} z^{\eta_+^0};\ \eta_+^0 < \infty\right) &= D(z)\left[1 - \mathbf{E}\left(e^{i\lambda\chi_+} z^{\eta_+};\ \eta_+ < \infty\right)\right], \\ 1 - \mathbf{E}\left(e^{i\lambda\chi_-^0} z^{\eta_-^0};\ \eta_-^0 < \infty\right) &= D(z)\left[1 - \mathbf{E}\left(e^{i\lambda\chi_-} z^{\eta_-};\ \eta_- < \infty\right)\right]. \end{aligned} \tag{12.1.11}$$

Here the function $D(z)$ defined in Theorem 12.1.2 *also satisfies the relations*

$$D^{-1}(z) = \sum_{n=0}^{\infty} z^n \mathbf{P}(S_n = 0,\ \zeta_n \le 0) = \sum_{n=0}^{\infty} z^n \mathbf{P}(S_n = 0,\ \gamma_n \ge 0). \tag{12.1.12}$$

Clearly, from Theorem 12.1.3 one can obtain some other versions of the factorisation identity. For instance, one has

$$\mathfrak{f}_z(\lambda) = \left[1 - \mathbf{E}(e^{i\lambda\chi_+} z^{\eta_+};\ \eta_+ < \infty)\right]\left[1 - \mathbf{E}\left(e^{i\lambda\chi_-^0} z^{\eta_-^0};\ \eta_-^0 < \infty\right)\right]. \tag{12.1.13}$$

Representations (12.1.12) for $D(z)$ imply, in particular, that

$$\mathbf{P}(S_n = 0,\ \zeta_n \le 0) = \mathbf{P}(S_n = 0,\ \gamma_n \ge 0)$$

and that $D(z) \equiv 1$ if $\mathbf{P}(S_n = 0) = 0$ for all $n \ge 1$.

Proof of Theorem 12.1.3 Let us derive the first relation in (12.1.11). Comparing (12.1.8) with (12.1.10) we find, as above, that

$$\left[1 - \mathbf{E}\left(e^{i\lambda\chi_+} z^{\eta_+};\ \eta_+ < \infty\right)\right]\mathfrak{b}_{z+}(\lambda) = \text{const} = 1, \tag{12.1.14}$$

since the product equals 1 for $\lambda = i\infty$. Therefore we obtain (12.1.13) by virtue of (12.1.8). It remains to compare (12.1.13) with the identity of Theorem 12.1.2.

Expressions (12.1.12) for $D(z)$ follow if we recall (see (12.1.8) and (12.1.10)) that the left-hand side of (12.1.14) equals

$$\left[\sum_{n=0}^{\infty} z^n \mathbf{E}\left(e^{i\lambda S_n};\ \zeta_n \le 0\right)\right]\left[1 - \mathbf{E}\left(e^{i\lambda\chi_-^0} z^{\eta_-^0};\ \eta_-^0 < \infty\right)\right].$$

Since this product also equals 1, letting $\lambda = -i\infty$ here and in the second identity of (12.1.11) we get the first equality in (12.1.12). The second equality is proved in a similar way. □

Remark 12.1.1 It is important to note that Theorems 12.1.2 and 12.1.3, as well as proving the existence of the identities, also provide a means of finding the characteristic function of the joint distribution of χ and η. That is, if we manage somehow to get a representation for $\mathfrak{f}_z(\lambda) = 1 - z\varphi(\lambda)$ of the form $\mathfrak{h}_{z+}(\lambda)\mathfrak{h}_{z-}(\lambda)$, where $\mathfrak{h}_{z\pm}(\lambda) \in \mathcal{K}_\pm$, then by uniqueness of the canonical factorisation we can, for instance, claim that, up to a constant factor, the function $1 - \mathbf{E}(e^{i\lambda\chi_+} z^{\eta_+};\ \eta_+)$ coincides with $\mathfrak{h}_{z+}(\lambda)$. For examples of how such arguments can be used, see Sects. 12.5 and 12.6.

12.2 Some Consequences of Theorems 12.1.1–12.1.3

12.2.1 Direct Consequences

Theorems 12.1.1–12.1.3 (and also their modifications of the form (12.1.13)) and the uniqueness of the canonical factorisation (see Lemma 12.1.1) directly imply the next result.

Corollary 12.2.1 *In the notation of Theorems* 12.1.1 *and* 12.1.2 *one has the following equalities.*

$$1 - \mathbf{E}\left(e^{i\lambda\chi_+} z^{\eta_+};\ \eta_+ < \infty\right) = \mathfrak{f}_{z+}(\lambda);$$
$$D(z) = C(z);$$
$$1 - \mathbf{E}\left(e^{i\lambda\chi_-} z^{\eta_-};\ \eta_- < \infty\right) = \mathfrak{f}_{z-}(\lambda).$$

Now we will obtain, as corollaries of Theorems 12.1.1–12.1.3, some further identities in which the parameter z is fixed and equal to 1.

Corollary 12.2.2 *Letting* $z \to 1$ *in* (12.1.13) *we obtain*

$$\mathfrak{f}_1(\lambda) := 1 - \varphi(\lambda) = \left[1 - \mathbf{E}\left(e^{i\lambda\chi_+};\ \eta_+ < \infty\right)\right]\left[1 - \mathbf{E}\left(e^{i\lambda\chi_-^0};\ \eta_-^0 < \infty\right)\right]. \quad (12.2.1)$$

It is obvious that one can similarly write other identities of such type corresponding to the identities that can be derived from Theorems 12.1.1–12.1.3.

Clearly, identity (12.2.1) delivers a factorisation of the function $\mathfrak{f}_1(\lambda) = 1 - \varphi(\lambda)$, but this factorisation is not canonical since $\mathfrak{f}_1(0) = 0$ and $\mathfrak{f}_1(\lambda) \notin \mathcal{K}$.

Corollary 12.2.3 *If there exists* $\mathbf{E}\xi = a < 0$ *then* $\mathbf{P}(\eta_-^0 < \infty) = 1$, $\mathbf{E}\chi_-^0$ *exists, and* $\mathbf{P}(\zeta \le 0) = a/\mathbf{E}\chi_-^0 > 0$.

Proof The first relation follows from the law of large numbers, because

$$\mathbf{P}(\eta_-^0 > n) < \mathbf{P}(S_n > 0) \to 0$$

as $n \to \infty$. Therefore, in the case under consideration, one has

$$\mathbf{E}(e^{i\lambda\chi_-^0}; \eta_-^0 < \infty) = \mathbf{E}e^{i\lambda\chi_-^0}.$$

The existence of $\mathbf{E}\chi_-^0$ follows from Wald's identity $\mathbf{E}\chi_-^0 = a\mathbf{E}\eta_-^0$ and the theorems of Chap. 10, which imply that $\mathbf{E}\eta_-^0 \le \mathbf{E}\eta_- < \infty$, since $\mathbf{E}\eta_-$ is the value of the corresponding renewal function at 0.

Finally, dividing both sides of the identity in Corollary 12.2.2 by λ and taking the limit as $\lambda \to 0$, we obtain

$$a = \big(1 - \mathbf{P}(\eta_+ < \infty)\big)\mathbf{E}\chi_-^0 = \mathbf{P}(\zeta \le 0)\mathbf{E}\chi_-^0. \qquad \square$$

It is interesting to note that, as a consequence of this assertion, we can obtain the *strong law of large numbers*. Indeed, since $\{\zeta < \infty\}$ is a tail event and $\mathbf{P}(\zeta < \infty) \ge \mathbf{P}(\zeta \le 0)$, Corollary 12.2.3 implies that $\mathbf{P}(\zeta < \infty) = 1$ for $a < 0$. This means that the assertion of Theorem 10.5.3 holds, and it was this assertion that the strong law of large numbers was derived from.

Based on factorisation identities, we will obtain below a generalisation of this law.

In the remaining part of this chapter, to avoid trivial complications, we will be assuming that ξ takes, with positive probability, both positive and negative values.

Corollary 12.2.4 *If* $a = \mathbf{E}\xi = 0$ *then* $\mathbf{P}(\eta_+ < \infty) = \mathbf{P}(\eta_-^0 < \infty) = 1$, *so that*

$$1 - \varphi(\lambda) = \big(1 - \mathbf{E}e^{i\lambda\chi_+}\big)\big(1 - \mathbf{E}\,e^{i\lambda\chi_-^0}\big). \tag{12.2.2}$$

If, moreover, $\mathbf{E}\xi^2 = \sigma^2 < \infty$ *then there exist* $\mathbf{E}\chi_+$ *and* $\mathbf{E}\chi_-^0$, *and*

$$\mathbf{E}\chi_+\mathbf{E}\chi_-^0 = -\frac{\sigma^2}{2}.$$

Proof Consider the sequence $\widetilde{\xi}_k = \xi_k - \varepsilon$, $\varepsilon > 0$. Denoting by $\widetilde{\zeta}$, $\widetilde{\chi}_-^0$ and $\widetilde{a}$ the corresponding characteristics for the newly introduced sequence, we obtain by Corollary 12.2.3 that

$$\mathbf{P}(\zeta \le 0) < \mathbf{P}(\widetilde{\zeta} \le 0) = \frac{\widetilde{a}}{\mathbf{E}\widetilde{\chi}_-^0} = -\frac{\varepsilon}{\widetilde{\chi}_-^0},$$

where

$$\mathbf{E}\widetilde{\chi}_-^0 \le \mathbf{E}(\widetilde{\xi}_1;\ \widetilde{\xi}_1 \le 0) = \mathbf{E}(\xi - \varepsilon;\ \xi \le \varepsilon) \le \mathbf{E}(\xi;\ \xi \le 0) < 0.$$

So we can make the probability $\mathbf{P}(\zeta \le 0)$ arbitrarily small by choosing an appropriate ε, and thus $\mathbf{P}(\zeta \le 0) = \mathbf{P}(\eta_+ = \infty) = 0$. Similarly, we find that $\mathbf{P}(\gamma \ge 0) = 0$ and hence

$$\mathbf{P}(\eta_-^0 = \infty) \le \mathbf{P}(\eta_- = \infty) = \mathbf{P}(\gamma \ge 0) = 0.$$

The obtained relations and Corollary 12.2.2 yield identity (12.2.2).

In order to prove the second assertion of the corollary, divide both sides of identity (12.2.2) by $\lambda^2 = -(i\lambda)^2$ and let $\lambda \in \Pi$ tend to zero. Then the limit of the left-hand side will be equal to $\sigma^2/2$ (see (7.1.1)), whereas that of the right-hand side will be equal to $-\mathbf{E}\chi_+\mathbf{E}\chi_-^0$, where $\mathbf{E}\chi_+ > 0$, $|\mathbf{E}\chi_-^0| > 0$. The corollary is proved. □

Corollary 12.2.5

1. *We always have* $\sum \frac{\mathbf{P}(S_k=0)}{k} < \infty$.
2. *The following three conditions are equivalent*:

   (a) $\mathbf{P}(\zeta < \infty) = 1$;
   (b) $\mathbf{P}(\zeta \le 0) = \mathbf{P}(\eta_+ = \infty) > 0$;
   (c) $\sum_{k=1}^{\infty} \frac{\mathbf{P}(S_k>0)}{k} < \infty$ *or* $\sum_{k=1}^{\infty} \frac{\mathbf{P}(S_k\ge 0)}{k} < \infty$.

Proof To obtain the first assertion, one should let $z \to 1$ in the second equality in Corollary 12.2.1 and recall that

$$D(1) = 1 - \mathbf{P}\big(\chi_+^0 = 0,\ \eta_+^0 < \infty\big) > \mathbf{P}(\xi > 0) > 0.$$

The equivalence of (b) and (c) follows from the equality

$$1 - \mathbf{P}(\eta_+ < \infty) = \mathbf{P}(\zeta \le 0) = \exp\left\{-\sum_{k=1}^{\infty} \frac{\mathbf{P}(S_k > 0}{k}\right\},$$

which is derived by putting $\lambda = 0$ and letting $z \to 1$ in the first identity of Corollary 12.2.1.

Now we will establish the equivalence of (b) and (c). If $\mathbf{P}(\zeta \le 0) > 0$ then $\mathbf{P}(\zeta < \infty) > 0$ and hence $\mathbf{P}(\zeta < \infty) = 1$, since $\{\zeta < \infty\}$ is a tail event. Conversely, let ζ be a proper random variable. Choose an N such that $\mathbf{P}(\zeta \le N) > 0$, and $b > 0$ such that $k = N/b$ is an integer and $\mathbf{P}(\xi < -b) > 0$. Then

$$\{\zeta \le 0\} \supset \Big\{\xi_1 < -b, \ldots, \xi_k < -b,\ \sup_{j\ge 1}(-bk + \xi_{k+1} + \cdots + \xi_{k+j}) \le 0\Big\}.$$

Since the sequence $\xi_{k+1}, \xi_{k+2}, \ldots$ is distributed identically to $\xi_1, \xi_2, \ldots$, one has

$$\mathbf{P}(\zeta \le 0) \ge \big[\mathbf{P}(\xi < -b)\big]^k \mathbf{P}(\zeta \le bk) > 0.$$ □

Corollary 12.2.6

1. $\mathbf{P}(\zeta < \infty, \gamma > -\infty) = 0$.

2. *If there exists* $\mathbf{E}\xi = a < 0$ *then*

$$\mathbf{P}(\eta_+ < \infty) < 1, \qquad \mathbf{P}(\zeta < \infty, \gamma = -\infty) = 1,$$
$$\left(\sum_{k=1}^{\infty} \frac{\mathbf{P}(S_k > 0)}{k} < \infty, \ \sum_{k=1}^{\infty} \frac{\mathbf{P}(S_k < 0)}{k} = \infty\right).$$

3. *If there exists* $\mathbf{E}\xi = a = 0$ *then*

$$\mathbf{P}(\zeta = \infty, \gamma = -\infty) = 1,$$
$$\left(\sum_{k=1}^{\infty} \frac{\mathbf{P}(S_k > 0)}{k} = \infty, \ \sum_{k=1}^{\infty} \frac{\mathbf{P}(S_k < 0)}{k} = \infty\right).$$

Here we do not consider the case $a > 0$ since it is "symmetric" to the case $a < 0$.

Proof The first assertion follows from the fact that at least one of the two series $\sum_{k=1}^{\infty} \frac{\mathbf{P}(S_k<0)}{k}$ and $\sum_{k=1}^{\infty} \frac{\mathbf{P}(S_k\ge 0)}{k}$ diverges. Therefore, by Corollary 12.2.5 either $\mathbf{P}(\gamma = -\infty) = 1$ or $\mathbf{P}(\zeta = \infty) = 1$.

The second and third assertions follow from Corollaries 12.2.3–12.2.5 in an obvious way. □

12.2.2 A Generalisation of the Strong Law of Large Numbers

The above mentioned generalisation of the strong law of large numbers consists of the following.

Theorem 12.2.1 (The one-sided law of large numbers) *Convergence of the series*

$$\sum_{k=1}^{\infty} \frac{\mathbf{P}(S_k > \varepsilon k)}{k}$$

for every $\varepsilon > 0$ *is a necessary and sufficient condition for*

$$\mathbf{P}\left(\limsup_{n\to\infty} \frac{S_n}{n} \le 0\right) = 1. \tag{12.2.3}$$

Proof Sufficiency. If the series converges then by Corollary 12.2.5 we have

$$\mathbf{P}\Big(\sup_k \{S_k - \varepsilon k\} < \infty\Big) = 1.$$

Hence $\{\varepsilon n\}$ is an upper sequence for $\{S_n\}$ and

$$\mathbf{P}\left(\limsup_{k\to\infty} \frac{S_k}{k} \le \varepsilon\right) = 1.$$

But since ε is arbitrary, we see that

$$\mathbf{P}\Big(\limsup_{k\to\infty}\frac{S_k}{k}\le 0\Big)=1.$$

Necessity. Conversely, if equality (12.2.3) holds then, for any $\varepsilon>0$, with probability 1 we have $S_n/n<\varepsilon$ for all n large enough. This means that $\sup_k(S_k-\varepsilon k)<\infty$ with probability 1, and hence by Corollary 12.2.5 the series $\sum_{k=1}^{\infty}\frac{\mathbf{P}(S_k>\varepsilon k)}{k}$ converges. The theorem is proved. □

Corollary 12.2.7 *With probability* 1 *we have*

$$\limsup_{n\to\infty}\frac{S_n}{n}=\alpha,$$

where

$$\alpha=\inf\Big\{b:\sum\frac{\mathbf{P}(S_k>bk)}{k}<\infty\Big\}.$$

Proof For any $b>\alpha$, the series in the definition of the number α converges. Since $\{\limsup_{n\to\infty}S_n/n\le b\}$ is a tail event and $S_n'=S_n-bn$ again form a sequence of sums of independent identically distributed random variables, Theorem 12.2.1 immediately implies that

$$\mathbf{P}\Big(\limsup\frac{S_n}{n}\le b\Big)=1,$$

$$\mathbf{P}\Big(\limsup\frac{S_n}{n}\le\alpha\Big)=\mathbf{P}\Big(\bigcap_{k=1}^{\infty}\Big\{\limsup\frac{S_n}{n}\le\alpha+\frac{1}{k}\Big\}\Big)=1.$$

If we assume that $\mathbf{P}(\limsup S_n/n\le\alpha^*)=1$ for $\alpha^*<\alpha$ then, for $\xi_k^*=\xi_k-\alpha^*$ and $S_k^*=\sum_{j=1}^{k}\xi_j^*$, we will have $\limsup\frac{S_n^*}{n}\le 0$, and

$$\sum_{k=1}^{\infty}\frac{\mathbf{P}(S_k>(\alpha^*+\varepsilon)k)}{k}<\infty$$

for any $\varepsilon>0$, which contradicts the definition of α. The corollary is proved. □

In order to derive the conventional law of large numbers from Theorem 12.2.1 it suffices to use Corollary 12.2.7 and assertion 2 of Corollary 12.2.6. We obtain that in the case $\mathbf{E}\xi=0$ the value of α in Corollary 12.2.7 is 0 and hence $\limsup S_n/n=0$ with probability 1. One can establish in the same way that $\liminf S_n/n=0$. □

12.3 Pollaczek–Spitzer's Identity. An Identity for $S=\sup_{k\ge 0}S_k$

It is important to note that, besides Theorems 12.1.1 and 12.1.2, there exist a number of factorisation identities that give explicit representations (in terms of factorisation components) for ch.f.s of the so-called *boundary functionals* of the trajectory of the random walk $\{S_k\}$, i.e. functionals associated with the crossing by the trajectory of $\{S_k\}$ of certain levels (not just the zero level, as in Theorems 12.1.1–12.1.3). The functionals

$$\overline{S}_n = \max_{k\le n} S_k, \qquad \theta_n = \min\{k : S_k = \overline{S}_n\}$$

and some others are also among the boundary functionals. For instance, for the triple transform of the joint distribution of $(\overline{S}_n, \theta_n)$, the following representation is valid.

For $|z| < 1$, $|\rho| < 1/|z|$ and $\operatorname{Im}\lambda \ge 0$, one has

$$(1-z)\sum_{n=0}^{\infty} z^n \mathbf{E}\big(\rho^{\theta_n} e^{i\lambda \overline{S}_n}\big) = \frac{\mathfrak{f}_{z+}(0)}{\mathfrak{f}_{z\rho+}(\lambda)}.$$

(For more detail on factorisation identities, see [3].)

Among many consequences of this identity we will highlight two results that can also be established using the already available Theorems 12.1.1–12.1.3.

12.3.1 Pollaczek–Spitzer's Identity

So far we have obtained several factorisation identities as relations for numerators in representations (12.1.7), (12.1.8) and (12.1.9). Now we turn to the denominators. We will obtain one more identity playing an important role in studying the distributions of

$$\overline{S}_n = \max(0, \zeta_n) = \max(0, S_1, \dots, S_n).$$

This is the so-called *Pollaczek–Spitzer identity* relating the ch.f.s of $\overline{S}_n$, $n = 1, 2, \dots$, with those of $\max(0, S_n)$, $n = 1, 2, \dots$.

Theorem 12.3.1 *For* $|z| < 1$ *and* $\operatorname{Im}\lambda \ge 0$,

$$\sum_{n=0}^{\infty} z^n \mathbf{E} e^{i\lambda \overline{S}_n} = \exp\left\{\sum_{n=0}^{\infty} \frac{z^k}{k} \mathbf{E} e^{i\lambda \max(0, S_k)}\right\}.$$

Using the notation of Theorem 12.1.1, one could write the right-hand side of this identity as

$$\frac{\mathfrak{f}_{z+}(0)}{(1-z)\mathfrak{f}_{z+}(\lambda)}$$

(see the last relation in the proof of the theorem).

Proof Theorems 12.1.1–12.1.3 (as well as their modifications of the form (12.1.13)) and the uniqueness of the canonical factorisation imply that

$$\sum_{k=0}^{\infty} z^n \mathbf{E}\big(e^{i\lambda S_n};\ \zeta_n < 0\big) = \big[1 - \mathbf{E}\big(e^{i\lambda \chi_-} z^{\eta_-};\ \eta_- < \infty\big)\big]^{-1} = \mathfrak{f}_{z-}^{-1}(\lambda),$$

where we assume that $\mathbf{E}(e^{i\lambda S_0};\ \zeta_0 < 0) = 1$, so all the functions in the above relation turn into 1 at $\lambda = -i\infty$. Set

$$S_k^* := S_{n-k} - S_n, \qquad \theta_n^* := \min\{k : S_k^* = \overline{S}_n^* := \max(0, S_1^*, \ldots, S_n^*)\}$$

(θ_n^* is time of the first maximum in the sequence $0, S_1^*, \ldots, S_n^*$). Then the event $\{S_n \in dx, \zeta_n < 0\}$ can be rewritten as $\{S_n^* \in -dx, \theta_n^* = n\}$. This implies that

$$\begin{aligned} &\mathbf{E}\big(e^{i\lambda S_n};\ \zeta_n < 0\big) = \mathbf{E}\big(e^{-i\lambda S_n^*};\ \theta_n^* = n\big), \\ &\sum_{n=0}^{\infty} z^n \mathbf{E}\big(e^{i\lambda S_n^*};\ \theta_n^* = n\big) = \mathfrak{f}_{z-}^{-1}(-\lambda). \end{aligned} \tag{12.3.1}$$

But the sequence $S_1^*, \ldots, S_n^*$ is distributed identically to the sequence of sums $\xi_1^*, \xi_1^* + \xi_2^*, \ldots, \xi_1^* + \cdots + \xi_n^*$, where $\xi_k^* = -\xi_k$. If we put $\theta_n := \min\{k : S_k = \overline{S}_n\}$ then identity (12.3.1) can be equivalently rewritten as

$$\sum_{n=0}^{\infty} z^n \mathbf{E}\big(e^{i\lambda S_n};\ \theta_n = n\big) = \big(\mathfrak{f}_{z-}^*(-\lambda)\big)^{-1},$$

where $\mathfrak{f}_{z-}^*(\lambda)$ is the negative factorisation component of the function $1 - z\varphi^*(\lambda) = 1 - z\varphi(-\lambda)$ corresponding to the random variable $-\xi$. Since

$$1 - z\varphi(-\lambda) = \mathfrak{f}_{z+}(-\lambda)C(z)\mathfrak{f}_{z-}(-\lambda)$$

and the function $\mathfrak{f}_{z+}(-\lambda)$ possesses all the properties of the negative component $\mathfrak{f}_{z-}^*(\lambda)$ of the factorisation of $1 - z\varphi^*(\lambda)$, while the function $\mathfrak{f}_{z-}(-\lambda)$ has all the properties of a positive component, we see that $\mathfrak{f}_{z-}^*(\lambda) = \mathfrak{f}_{z+}(-\lambda)$ and

$$\sum_{n=0}^{\infty} z^n \mathbf{E}\big(e^{i\lambda S_n};\ \theta_n = n\big) = \frac{1}{\mathfrak{f}_{z+}(\lambda)}.$$

Now we note that

$$\begin{aligned} \mathbf{E}e^{i\lambda \overline{S}_n} &= \sum_{k=0}^{n} \mathbf{E}\big(e^{i\lambda \overline{S}_n};\ \theta_n = k\big) \\ &= \sum_{k=0}^{n} \mathbf{E}\big(e^{i\lambda S_k};\ \theta_k = k,\ S_{k+1} - S_k \le 0, \ldots, S_n - S_k \le 0\big) \\ &= \sum_{k=0}^{n} \mathbf{E}\big(e^{i\lambda S_k};\ \theta_k = k\big)\mathbf{P}(\overline{S}_{n-k} = 0). \end{aligned}$$

Since the right-hand side is the convolution of two sequences, we obtain that

$$\sum_{n=0}^{\infty} z^n \mathbf{E}\, e^{i\lambda \overline{S}_n} = \sum_{n=0}^{\infty} z^n \mathbf{P}(\overline{S}_n = 0) \frac{1}{\mathfrak{f}_{z+}(\lambda)}.$$

Putting $\lambda = 0$ we get

$$\frac{1}{1-z} = \sum_{n=0}^{\infty} z^n \mathbf{P}(\overline{S}_n = 0) \frac{1}{\mathfrak{f}_{z+}(0)}.$$

Therefore,

$$\begin{aligned}\sum_{n=0}^{\infty} z^n \mathbf{E}\, e^{i\lambda \overline{S}_n} &= \frac{\mathfrak{f}_{z+}(0)}{(1-z)\mathfrak{f}_{z+}(\lambda)}\\ &= \exp\left\{-\ln(1-z)+\sum_{k=1}^{\infty}\frac{z^k}{k}\mathbf{E}\big(e^{i\lambda S_k};\, S_k>0\big)-\sum_{k=1}^{\infty}\frac{z^k}{k}\mathbf{P}(S_k>0)\right\}\\ &= \exp\left\{\sum_{k=1}^{\infty}\frac{z^k}{k}\mathbf{E}\big(e^{i\lambda S_k};\, S_k>0\big)+\sum_{k=1}^{\infty}\frac{z^k}{k}\mathbf{P}(S_k\le 0)\right\}\\ &= \exp\left\{\sum_{k=1}^{\infty}\frac{z^k}{k}\mathbf{E}\big(e^{i\lambda \max(0,S_k)}\big)\right\}.\end{aligned}$$

The theorem is proved. □

12.3.2 An Identity for $S=\sup_{k\ge 0} S_k$

The second useful identity to be discussed in this subsection is associated with the distribution of the random variable $S=\sup_{k\ge 0} S_k=\max(0,\zeta)$ (of course, we deal here with the cases when $\mathbf{P}(S<\infty)=1$). This distribution is of interest in many applications. Two such illustrative applications will be discussed in the next subsection.

We will establish the relationship of the distribution of S with that of the vector (χ_+,η_+) and with the factorisation components of the function $1-z\varphi(\lambda)$.

First of all, note that the random variable η_+ is a Markov time. For such variables, one can easily see (cf. Lemma 10.2.1) that the sequence $\xi_1^*=\xi_{\eta_++1}, \xi_2^*=\xi_{\eta_++2},\dots$ on the set $\{\omega:\eta_+<\infty\}$ (or given that $\eta_+<\infty$) is distributed identically to $\xi_1,\xi_2,\dots$ and does not depend on $(\eta_+,\xi_1,\dots,\xi_{\eta_+})$. Indeed,

$$\begin{aligned}&\mathbf{P}\big(\xi_1^*\in B_1,\dots,\xi_k^*\in B_k \,\big|\, \eta_+=j, \xi_1\in A_1,\dots,\xi_{\eta_+}\in A_{\eta_+}\big)\\ &\quad=\mathbf{P}(\xi_{j+1}\in B_1,\dots,\xi_{j+k}\in B_k \mid \xi_1\in A_1,\dots,\xi_j\in A_j;\ \eta_+=j)\\ &\quad=\mathbf{P}(\xi_1\in B_1,\dots,\xi_k\in B_k).\end{aligned}$$

Considering the new sequence $\{\xi_k^*\}_{k=1}^{\infty}$ we note that it will exceed level 0 (the level χ_+ for the original sequence) with probability $p=\mathbf{P}(\eta_+<\infty)$, and that the distribution of $\zeta^*=\sup_{k\ge 1}(\xi_1^*+\cdots+\xi_k^*)$ coincides with the distribution of $\zeta=\sup_{k\ge 1} S_k$.

Thus, with $S^*:=\max(0,\zeta^*)$, we have

$$S=S(\omega)=\begin{cases}0 & \text{on } \{\omega:\eta_+=\infty\},\\ S_{\eta_+}+S^*=\chi_++S^* & \text{on } \{\omega:\eta_+<\infty\}.\end{cases}$$

Since, as has already been noted, S^* does not depend on χ_+ and η_+, and the distribution of S^* coincides with that of S, we have

$$\mathbf{E}e^{i\lambda S} = \mathbf{P}(\eta_+ = \infty) + \mathbf{E}\big(e^{i\lambda(\chi_+ + S^*)};\ \eta_+ < \infty\big)$$
$$= (1-p) + \mathbf{E}e^{i\lambda S}\mathbf{E}\big(e^{i\lambda\chi_+};\ \eta_+ < \infty\big).$$

This implies the following result.

Theorem 12.3.2 *If* $\sum \frac{\mathbf{P}(S_k>0)}{k} < \infty$ *or, which is the same,* $p = \mathbf{P}(\eta_+ < \infty) < 1$, *then*

$$\mathbf{E}e^{i\lambda S} = \frac{1-p}{1-\mathbf{E}(e^{i\lambda\chi_+}, \eta_+ < \infty)} = \frac{1-p}{\mathfrak{f}_{1+}(\lambda)}.$$

In exactly the same way we can obtain the relation

$$\mathbf{E}e^{i\lambda S} = \frac{1-p_0}{1-\mathbf{E}(e^{i\lambda\chi_+^0}, \eta_+^0 < \infty)}, \tag{12.3.2}$$

where $p_0 = \mathbf{P}(\eta_+^0 < \infty) < 1$.

In this case, one can write a factorisation identity in following form:

$$1-\varphi(\lambda) = \frac{(1-p_0)(1-\mathbf{E}e^{i\lambda\chi_-})}{\mathbf{E}e^{i\lambda S}} = \frac{(1-p)(1-\mathbf{E}e^{i\lambda\chi_-^0})}{\mathbf{E}e^{i\lambda S}}. \tag{12.3.3}$$

In Sects. 12.5–12.7 we will discuss the possibility of finding the explicit form and the asymptotic properties of the distribution of S.

12.4 The Distribution of S in Insurance Problems and Queueing Theory

In this section we show that the need to analyse the distribution of the variable S considered in Sect. 12.3 arises in insurance problems and also when studying queueing systems.

12.4.1 Random Walks in Risk Theory

Consider the following simplified model of an insurance business operation. Denote by x the initial surplus of the company and consider the daily dynamics of the surplus. During the k-th day the company receives insurance premiums at the rate $\xi_k^+ \ge 0$ and pays out claims made by insured persons at the rate $\xi_k^- \ge 0$ (in case of a fire, a traffic accident, and so on). The amounts $\xi_k = \xi_k^- - \xi_k^+$ are random since they depend on the number of newly insured persons, the size of premiums, claim amounts and so on. For a foreseeable "homogeneous" time period, the amount ξ_k can be assumed to be independent and identically distributed. If we put $S_n := \sum_{k=1}^n \xi_k$ then the company's surplus after n days will be $Z_n = x - S_n$, provided that we allow it to be negative. But if we assume that the company ruins at the time when Z_n first becomes negative, then the probability of no ruin during the first n days equals

$$\mathbf{P}\Big(\min_{k\le n} Z_k \ge 0\Big) = \mathbf{P}(\overline{S}_n \le x),$$

where, as above, $\overline{S}_n = \max_{k\le n} S_k$. Accordingly, the probability of ruin within n days is equal to $\mathbf{P}(\overline{S}_n > x)$, and the probability of ruin in the long run can be identified with $\mathbf{P}(S > x)$. It follows that, for the probability of ruin to be less than 1, it is necessary that $\mathbf{E}\xi_k < 0$ or, which is the same, that $\mathbf{E}\xi_k^- < \mathbf{E}\xi_k^+$. When this condition is satisfied, in order to make the probability of ruin small enough, one has to make the initial surplus x large enough. In this connection it is of interest to find the explicit form of the distribution of S, or at least the asymptotic behaviour of $\mathbf{P}(S > x)$ as $x \to \infty$. Sections 12.5–12.7 will be focused on this.

12.4.2 Queueing Systems

Imagine that "customers" who are to be served by a certain system arrive with time intervals $\tau_1, \tau_2, \dots$ between successive arrivals. These could be phone calls, planes landing at an airport, clients in a shop, messages to be processed by a computer, etc. Assume that serving the k-th customer (the first customer arrived at time 0, the second at time τ_1, and so on) requires time s_k, $k = 1, 2, \dots$ If, at the time of the k-th customer's arrival, the system was busy serving one of the preceding customers, the newly arrived customer joins the "queue" and waits for service which starts immediately after the system has finished serving all the preceding customers. The problem is to find the distribution of the waiting time w_n of the n-th customer—the time spent waiting for the service.

Let us find out how the quantities w_{n+1} and w_n are related to each other. The $(n+1)$-th customer arrived τ_n time units after the n-th customer, but will have to wait for an extra s_n time units during the service of the n-th customer. Therefore,

$$w_{n+1} = w_n - \tau_n + s_n,$$

only if $w_n - \tau_n + s_n \ge 0$. If $w_n - \tau_n + s_n < 0$ then clearly $w_{n+1} = 0$. Thus, if we put $\xi_{n+1} := s_n - \tau_n$, then

$$w_{n+1} = \max(0, w_n + \xi_{n+1}), \quad n \ge 1, \tag{12.4.1}$$

with the initial value of $w_1 \ge 0$. Let us find the solution to this recurrence equation. Let, as above, $S_n = \sum_{k=1}^n \xi_k$. Denote by $\theta(n)$ the time when the trajectory of $0, S_1, \dots, S_n$ first attains its minimum:

$$\theta(n) := \min\{k : S_k = \underline{S}_n\}, \qquad \underline{S}_n := \min_{0\le j\le n} S_j.$$

Then clearly (for $w_0 := w_1$)

$$w_{n+1} = w_1 + S_n \quad \text{if } w_{\theta(n)} = w_1 + \underline{S}_n > 0 \tag{12.4.2}$$

(since in this case the right-hand side of (12.4.1) does not vanish and $w_{k+1} = w_k + \xi_k$ for all $k \le n$), and

$$w_{n+1} = S_n - S_{\theta(n)} \quad \text{if } w_1 + \underline{S}_n \le 0 \tag{12.4.3}$$

($w_{\theta(n)} = 0$ and $w_{k+1} = w_k + \xi_k$ for all $k \ge \theta(n)$). Put

$$S_{n,j} := \sum_{k=n-j+1}^{n} \xi_k, \qquad \overline{S}_{n,n} := \max_{0 \le j \le n} S_{n,j},$$

so that

$$S_{n,0} = 0, \qquad S_{n,n} = S_n.$$

Then

$$S_n - S_{\theta(n)} = S_n - \underline{S}_n = \max_{0 \le j \le n} (S_n - S_j) = \overline{S}_{n,n},$$

so that $w_1 + \underline{S}_n = w_1 + S_n - \overline{S}_{n,n}$ and the inequality $w_1 + \overline{S}_n \le 0$ in (12.4.3) is equivalent to the inequality $\overline{S}_{n,n} = S_n - S_{\theta(n)} \ge w_1 + S_n$. Therefore (12.4.2) and (12.4.3) can be rewritten as

$$w_{n+1} = \max(\overline{S}_{n,n}, w_1 + S_n). \tag{12.4.4}$$

This implies that, for each fixed $x > 0$,

$$\mathbf{P}(w_{n+1} > x) = \mathbf{P}(\overline{S}_{n,n} > x) + \mathbf{P}(\overline{S}_{n,n} \le x,\ w_1 + S_n > x).$$

Now assume that $\xi_k \overset{d}{=} \xi$ are independent and identically distributed with $\mathbf{E}\xi < 0$. Then $\overline{S}_{n,n} \overset{d}{=} \overline{S}_n$ and, as $n \to \infty$, we have $S_n \xrightarrow{a.s.} -\infty$, $\mathbf{P}(w_1 + S_n > x) \to 0$ and $\mathbf{P}(\overline{S}_n > x) \uparrow \mathbf{P}(S > x)$. We conclude that, for any initial value w_1, the following limit exists

$$\lim_{n \to \infty} \mathbf{P}(w_n > x) = \mathbf{P}(S > x).$$

This distribution is called the *stationary waiting time distribution*. We already know that it will be proper if $\mathbf{E}\xi = \mathbf{E}s_1 - \mathbf{E}\tau_1 < 0$. As in the previous section, here arises the problem of finding the distribution of S. If, on the other hand, $\mathbf{E}s_1 > \mathbf{E}\tau_1$ or $\mathbf{E}s_1 = \mathbf{E}\tau_1$ and $s_1 \not\equiv \tau_1$ then the "stationary" waiting time will be infinite.

12.4.3 Stochastic Models in Continuous Time

In the theory of queueing systems and risk theory one can equally well employ stochastic models in *continuous time*, when, instead of random walks $\{S_n\}$, one uses generalised renewal processes $Z(t)$ as described in Sect. 10.6. For a given sequence of independent identically distributed random vectors (τ_j, ζ_j), the process $Z(t)$ is defined by the equality

$$Z(t) := Z_{\nu(t)},$$

where

$$Z_n := \sum_{j=1}^{n} \zeta_j, \qquad \nu(t) := \max\{k : T_k \le t\}, \quad T_k := \sum_{j=1}^{k} \tau_j.$$

For instance, in risk theory, the capital inflow during time t that comes from regular premium payments can be described by the function qt, $q > 0$. The insurer covers claims of sizes $\zeta_1, \zeta_2, \dots$ with time intervals $\tau_1, \tau_2, \dots$ between them (the first claim is covered at time τ_1). Thus, if the initial surplus is x, then the surplus at time t will be

$$x + qt - Z_{\nu(t)} = x + qt - Z(t).$$

The insurer ruins if $\inf_t (x + qt - Z(t)) < 0$ or, which is the same,

$$\sup_t \bigl(Z(t) - qt\bigr) > x.$$

It is not hard to see that

$$\sup_t (Z_{\nu(t)} - qt) = \sup_{k \ge 0} S_k =: S,$$

where $S_k = \sum_{j=1}^{k} \xi_j$, $\xi_j = \zeta_j - q\tau_j$. Thus the continuous-time version of the ruin problem for an insurance company also reduces to finding the distribution of the maximums of the cumulative sums.

12.5 Cases Where Factorisation Components Can Be Found in an Explicit Form. The Non-lattice Case

As was already noted, the boundary functionals of random walks that were considered in Sects. 12.1–12.3 appear in many applied problems (see e.g., Sect. 12.4). This raises the question: in what cases can one find, in an explicit form, the factorisation components and hence the explicit form of the boundary functionals distributions we need? Here we will deal with factorisation of the function $1 - \varphi(\lambda)$ and will be interested in the boundary functionals $\chi_\pm$ and S.

12.5.1 Preliminary Notes on the Uniqueness of Factorisation

As was already mentioned, the factorisation of the function $1 - \varphi(\lambda)$ obtained in Corollaries 12.2.2 and 12.2.4 is not canonical since that function vanishes at $\lambda = 0$. In this connection arises the question of whether a factorisation is unique. In other words, if, say, in the case $\mathbf{E}\xi < 0$, we obtained a factorisation

$$1 - \varphi(\lambda) = \mathfrak{f}_+(\lambda)\mathfrak{f}_-(\lambda),$$

where $\mathfrak{f}_\pm$ are analytic on $\Pi_\pm$ and continuous on $\Pi_\pm \cup \Pi$, then under what conditions can we state that

$$\mathbf{E}e^{i\lambda S} = \frac{\mathfrak{f}_+(0)}{\mathfrak{f}_+(\lambda)}$$

(cf. Theorem 12.3.2)? In order to answer this question, in contrast to the above, we will have to introduce here restrictions on the distribution of ξ.

1. We will assume that $\mathbf{E}\xi$ exists, and in the case $\mathbf{E}\xi = 0$ that $\mathbf{E}\xi^2$ also exists.
2. Regarding the structure of the distribution of ξ we will assume that either

   (a) the distribution $\mathbf{F}$ is non-lattice and the Cramér condition on ch.f. holds:

$$\limsup_{\substack{|\lambda|\to\infty \\ \operatorname{Im}\lambda=0}} \bigl|\varphi(\lambda)\bigr| < 1, \qquad (12.5.1)$$

   or

   (b) the distribution $\mathbf{F}$ is arithmetic.

Condition (12.5.1) always holds once the distribution $\mathbf{F}$ has a nonzero absolutely continuous component. Indeed, if $\mathbf{F} = \mathbf{F}_a + \mathbf{F}_s + \mathbf{F}_d$ is the decomposition of $\mathbf{F}$ into the absolutely continuous, singular and discrete components then, by the Lebesgue theorem, $\int e^{i\lambda x}\mathbf{F}_a(dx) \to 0$ as $|\lambda| \to \infty$ on $\operatorname{Im} \lambda = 0$, and so

$$\limsup_{|\lambda|\to\infty} \bigl|\varphi(\lambda)\bigr| \le \mathbf{F}_s\bigl((-\infty,\infty)\bigr) + \mathbf{F}_d\bigl((-\infty,\infty)\bigr) < 1.$$

For lattice distributions concentrated at the points $a + hk$, k being an integer, condition (12.5.1) is evidently not satisfied since, for $\lambda = 2\pi j/h$, we have $|\varphi(\lambda)| = |e^{i2\pi a/h}| = 1$ for all integers j. The condition is also not met for any discrete distribution, since any "part" of such a distribution, concentrated on a finite number of points, can be approximated arbitrarily well by a lattice distribution. For singular distributions, condition (12.5.1) can yet be satisfied.

Since, for non-lattice distributions, $|\varphi(\lambda)| < 1$ for $\lambda \ne 0$, under condition (12.5.1) one has

$$\sup_{|\lambda|>\varepsilon} \bigl|\varphi(\lambda)\bigr| < 1 \qquad (12.5.2)$$

for any $\varepsilon > 0$. This means that the function $\mathfrak{f}(\lambda) = 1 - \varphi(\lambda)$ has no zeros on the real line Π (completed by the points $\pm\infty$) except at the point $\lambda = 0$.

In case (b), when the distribution of $\mathbf{F}$ is arithmetic, one can consider the ch.f. $\varphi(\lambda)$ on the segment $[0, 2\pi]$ only or, which is the same, consider the generating function $p(z) = \mathbf{E}z^\xi$, in which case we will be interested in the factorisation of the function $1 - p(z)$ on the unit circle $|z| = 1$.

Under the aforementioned conditions, we can "tweak" the function $1 - \varphi(\lambda)$ so that it allows canonical factorisation.

In this section we will confine ourselves to the non-lattice case. The arithmetic case will be considered in Sect. 12.6.

Lemma 12.5.1 *Let the distribution* $\mathbf{F}$ *be non-lattice and condition* (12.5.1) *hold. Then*:

1. *If* $\mathbf{E}\xi < 0$ *then the function*

$$\mathfrak{v}(\lambda) := \frac{1-\varphi(\lambda)}{i\lambda}(i\lambda + 1) \tag{12.5.3}$$

belongs to $\mathcal{K}$ *and allows a unique canonical factorisation*

$$\mathfrak{v}(\lambda) = \mathfrak{v}_+(\lambda)\mathfrak{v}_-(\lambda),$$

where

$$\mathfrak{v}_+(\lambda) := 1 - \mathbf{E}\big(e^{i\lambda\chi_+};\, \eta_+ < \infty\big) = \frac{1-p}{\mathbf{E}e^{i\lambda S}}, \tag{12.5.4}$$

$$\mathfrak{v}_-(\lambda) := \frac{1-\mathbf{E}e^{i\lambda\chi_-^0}}{i\lambda}(i\lambda + 1). \tag{12.5.5}$$

2. *If* $\mathbf{E}\xi = 0$ *and* $\mathbf{E}\xi^2 < \infty$ *then the function*

$$\mathfrak{v}^0(\lambda) := \frac{1-\varphi(\lambda)}{\lambda^2}\big(\lambda^2 + 1\big) \tag{12.5.6}$$

belongs to $\mathcal{K}$ *and allows a unique canonical factorisation*

$$\mathfrak{v}^0(\lambda) = \mathfrak{v}_+^0(\lambda)\mathfrak{v}_-^0(\lambda),$$

where

$$\begin{aligned} \mathfrak{v}_+^0(\lambda) &:= \frac{1-\mathbf{E}e^{i\lambda\chi_+}}{i\lambda}(i\lambda - 1), \\ \mathfrak{v}_-^0(\lambda) &:= \frac{1-\mathbf{E}e^{i\lambda\chi_-^0}}{i\lambda}(i\lambda + 1) \end{aligned} \tag{12.5.7}$$

(*cf. Corollaries* 12.2.2 *and* 12.2.4).

Here we do not consider the case $\mathbf{E}\xi > 0$ since it is "symmetric" to the case $\mathbf{E}\xi < 0$ and the corresponding assertion can be derived from the assertion 1 of the lemma by applying it to the random variables $-\xi_k$ (or by changing λ to $-\lambda$ in the identities), so that in the case $\mathbf{E}\xi > 0$, the function $\frac{1-\varphi(\lambda)}{i\lambda}(i\lambda - 1)$ will allow a unique canonical factorisation.

The uniqueness of the canonical factorisation immediately implies the following result.

Corollary 12.5.1 *If, for* $\mathbf{E}\xi < 0$, *we have a canonical factorisation*

$$\mathfrak{v}(\lambda) = \mathfrak{w}_+(\lambda)\mathfrak{w}_-(\lambda),$$

then

$$\mathbf{E}e^{i\lambda S} = \frac{\mathfrak{w}_+(0)}{\mathfrak{w}_+(\lambda)}. \tag{12.5.8}$$

Proof of Lemma 12.5.1 Let $\mathbf{E}\xi < 0$. Since

$$\frac{1-\varphi(\lambda)}{i\lambda} \to -\mathbf{E}\xi > 0$$

as $\lambda \to 0$ and (12.5.1) is satisfied, we see that $\mathfrak{v}(\lambda)$ is bounded and continuous on Π and is bounded away from zero. This means that $\mathfrak{v}(\lambda) \in \mathcal{K}$.

Further, by Corollary 12.2.2 (see (12.2.1))

$$\mathfrak{v}(\lambda) = \frac{(1 - \mathbf{E}e^{i\lambda\chi_-^0})(i\lambda + 1)}{i\lambda}\left[1 - \mathbf{E}\left(e^{i\lambda\chi_+};\ \eta_+ < \infty\right)\right],$$

where $\mathbf{E}\chi_-^0 \in (-\infty, 0)$. Therefore, similarly to the above, we find that

$$\mathfrak{v}_-(\lambda) := \frac{(1 - \mathbf{E}e^{i\lambda\chi_-^0})(i\lambda + 1)}{i\lambda} \in \mathcal{K}.$$

Furthermore, $\mathfrak{v}_-(\lambda) \in \mathcal{K}_-$ (the factor $i\lambda + 1$ has a zero at the point $\lambda = i \in \Pi_+$). Evidently, we also have

$$\mathfrak{v}_+(\lambda) = 1 - \mathbf{E}\left(e^{i\lambda\chi_+};\ \eta_+ < \infty\right) \in \mathcal{K} \cap \mathcal{K}_+.$$

This proves the first assertion of the lemma. The last equality in (12.5.4) follows from Theorem 12.3.2. The uniqueness follows from Lemma 12.1.1.

The second assertion is proved in a similar way using Corollary 12.2.5, which implies that $\mathbf{E}\chi_+ \in (0, \infty)$, $\mathbf{E}\chi_-^0 \in (-\infty, 0)$, and

$$\mathfrak{v}^0(\lambda) = \left[\frac{(1 - \mathbf{E}e^{i\lambda\chi_+})(i\lambda - 1)}{i\lambda}\right]\left[\frac{(1 - \mathbf{E}e^{i\lambda\chi_-^0})(i\lambda + 1)}{i\lambda}\right], \qquad (12.5.9)$$

where, as before, we can show that $\mathfrak{v}^0(\lambda) \in \mathcal{K}$ and the factors on the right-hand side of (12.5.9) belong to $\mathcal{K} \cap \mathcal{K}_\pm$, correspondingly. The lemma is proved. □

12.5.2 Classes of Distributions on the Positive Half-Line with Rational Ch.F.s

As we saw in Example 7.1.5, the ch.f. of the exponential distribution with density $\beta e^{-\beta x}$ on $(0, \infty)$ is $\beta/(\beta - i\lambda)$. The j-th power of this ch.f. ocrresponds to the gamma-distribution $\mathbf{\Gamma}_{\beta,j}$ (the j-th convolution of the exponential distribution) with density (see Sect. 7.7)

$$\frac{\beta^k x^{j-1} e^{-\beta x}}{(j-1)!}, \quad x \ge 0.$$

This means that a density of the form

$$\sum_{k=1}^{K}\sum_{j=1}^{l_k} a_{kj} x^{j-1} e^{-\beta_k x} \qquad (12.5.10)$$

on $(0, \infty)$ (where all $\beta_k > 0$ are different) can then be considered as a mixture of gamma-distributions and its ch.f. will be a rational function $P_m(\lambda)/Q_n(\lambda)$, where

P_m and Q_n are polynomials of degrees m and n, respectively (for definiteness, we can put)

$$Q_n(\lambda) = \prod_{k=1}^{K} (\beta_k - i\lambda)^{l_k}, \tag{12.5.11}$$

and necessarily $m < n$ (see Property 7.1.8) with $n = \sum_{k=1}^{K} l_k$. Here all the zeros of the polynomial Q_n are real. But not only densities of the form (12.5.10) can have rational ch.f.s. Clearly, the Fourier transform of the function $e^{-\beta x} \cos \gamma x$, which can be rewritten as

$$\frac{1}{2} e^{-\beta x} \left(e^{i\gamma x} + e^{-i\gamma x}\right), \tag{12.5.12}$$

will also be a rational function. Complex-valued functions of this kind will have poles that are *symmetric* with respect to the imaginary line (in our case, at the points $\lambda = -i\beta \pm \gamma$). Convolutions of functions of the form (12.5.12) will have a more complex form but will not go beyond representation (12.5.10), where β_k are "symmetric" complex numbers. Clearly, densities of the form (12.5.10), where β_k are either real and positive or complex and symmetric, $\operatorname{Re} \beta_k > 0$, exhaust all the distributions with rational ch.f.s (the coefficients of the "conjugate" complex-valued exponentials must coincide to avoid the presence of irremovable complex terms).

It is obvious that the converse is also true: rational ch.f.s $P_m(\lambda)/Q_n(\lambda)$ correspond to densities of the form (12.5.10). In order to show this it suffices to decompose $P_m(\lambda)/Q_n(\lambda)$ into partial fractions, for which the inverse Fourier transforms are known.

We will call densities of the form (12.5.10) on $(0, \infty)$ *exponential polynomials* with exponents β_k. We will call the number l_k the *multiplicity* of the exponent β_k — it corresponds to the multiplicity of the pole of the Fourier transform at the point $\lambda = -i\beta_k$ (recall that $Q_n(\lambda) = \prod_{k=1}^{K} (\beta_k - i\lambda)^{l_k}$). One can approximate an arbitrary distribution on $(0, \infty)$ by exponential polynomials (for more details, see [3]).

12.5.3 *Explicit Canonical Factorisation of the Function* $\mathfrak{v}(\lambda)$ *in the Case when the Right Tail of the Distribution* **F** *Is an Exponential Polynomial*

Consider a distribution **F** on the whole real line $(-\infty, \infty)$ with $\mathbf{E}\xi < 0$ and such that, for $x > 0$, the distribution has a density that is an exponential polynomial (12.5.10). Denote by $\mathcal{EP}$ the class of all such distributions. The ch.f. of a distribution $\mathbf{F} \in \mathcal{EP}$ can be represented as

$$\varphi(\lambda) = \varphi^+(\lambda) + \varphi^-(\lambda),$$

where the function

$$\varphi^-(\lambda) = \mathbf{E}\left(e^{i\lambda\xi};\ \xi \le 0\right), \quad \xi \in \mathbf{F},$$

is analytic on Π_- and continuous on $\Pi_- \cup \Pi$, and $\varphi^+(\lambda)$ is a rational function

$$\varphi^+(\lambda) = \frac{P_m(\lambda)}{Q_n(\lambda)}, \qquad m < n, \tag{12.5.13}$$

analytic on Π_+. Here $\varphi^+(\lambda)$ is a ch.f. up to the factor $\mathbf{P}(\xi > 0) > 0$.

It is important to note that, for real μ, the equality

$$\psi^+(\mu) := \varphi^+(-i\mu) = \mathbf{E}\big(e^{\mu\xi};\ \xi > 0\big) = \frac{P_m(-i\mu)}{Q_n(-i\mu)} \tag{12.5.14}$$

only makes sense for $\mu < \beta_1$, where β_1 is the minimal zero of the polynomial $Q_n(-i\mu)$ (i.e. the pole of $\psi^+(\lambda)$). It is necessarily a simple and real root since the function $\psi^+(\mu)$ is real and monotonically increasing. Further, $\psi^+(\mu) = \infty$ for $\mu \ge \beta_1$. Therefore the function $\mathbf{E}(e^{i\lambda\xi};\ \xi > 0)$ is undefined for $\operatorname{Re} i\lambda \ge \beta_1$ ($\operatorname{Im}\lambda \le -\beta_1$). However, the right-hand side of (12.5.14) (and hence $\varphi^+(\lambda)$) can be analytically continued onto the lower half-plane $\operatorname{Im}\lambda \le -\beta_1$ to a function defined on the whole complex plane. In what follows, when we will be discussing zeros of the function $1 - \varphi(\lambda)$ on Π_-, we will mean *zeros of this analytical continuation*, i.e. of the function $\varphi^-(\lambda) + P_m(\lambda)/Q_n(\lambda)$.

Further, note that, for distributions from the class $\mathcal{EP}$, the Cramér condition (12.5.1) on ch.f.s always holds, since $\varphi^+(\lambda) \to 0$ as $|\lambda| \to \infty$, and

$$\limsup_{|\lambda|\to\infty,\ \lambda\in\Pi} \big|\varphi(\lambda)\big| = \limsup_{|\lambda|\to\infty,\ \lambda\in\Pi} \big|\varphi_-(\lambda)\big| \le \mathbf{P}(\xi \le 0) < 1.$$

For a distribution $\mathbf{F} \in \mathcal{EP}$, the canonical factorisation of the functions $\mathfrak{v}(\lambda)$ and $\mathfrak{v}^0(\lambda)$ (see (12.5.3) and (12.5.6)) can be obtained in explicit form expressed in terms of the zeros of the function $1 - \varphi(\lambda)$.

Theorem 12.5.1 *Let there exist* $\mathbf{E}\xi < 0$. *In order for the positive component* $\mathfrak{w}_+(\lambda)$ *of a canonical factorisation*

$$\mathfrak{v}(\lambda) = \mathfrak{w}_+(\lambda)\mathfrak{w}_-(\lambda), \qquad \mathfrak{w}_\pm \in \mathcal{K}_\pm \cap \mathcal{K},$$

to be a rational function, it is necessary and sufficient that the function

$$\varphi^+(\lambda) = \mathbf{E}\big(e^{i\lambda\xi};\ \xi > 0\big)$$

is rational.

If $\varphi^+ = P_m/Q_n$ *is an uncancellable ratio of polynomials* P_m *and* Q_n *of degrees* m *and* n, *respectively,* $m < n$, *then the function* $1 - \varphi(\lambda)$ *has precisely* n *zeros on* Π_- *(we denote them by* $-i\mu_1, \dots, -i\mu_n$*), and*

$$\mathfrak{w}_+(\lambda) = \frac{\prod_{k=1}^n (\mu_k - i\lambda)}{Q_n(\lambda)}, \tag{12.5.15}$$

where $Q_n(-i\mu_k) \ne 0$ *(i.e. ratio (12.5.15) is uncancellable).*

If all zeros $-i\mu_k$ *are arranged in descending order of their imaginary parts*:

$$\operatorname{Re}\mu_1 \le \operatorname{Re}\mu_2 \le \cdots \le \operatorname{Re}\mu_n,$$

then the zero $-i\mu_1$ *will be simple and purely imaginary,* $\mu_1 < \min(\operatorname{Re}\mu_2, \beta_1)$, *where* β_1 *is the minimal zero of* $Q_n(-i\mu)$.

The theorem implies that the component $\mathfrak{w}_-(\lambda)$ can also be found in an explicit form:

$$\mathfrak{w}_-(\lambda) = \frac{(1-\varphi(\lambda))(i\lambda+1)Q_n(\lambda)}{i\lambda\prod_{k=1}^n(\mu_k - i\lambda)}.$$

From Corollary 12.5.1 we obtain the following assertion.

Corollary 12.5.2 *If* $\mathbf{E}\xi < 0$ *and* $\varphi = P_m/Q_m$ *then*

$$\mathbf{E}e^{i\lambda S} = \frac{\mathfrak{w}_+(0)}{\mathfrak{w}_+(\lambda)} = \frac{Q_n(\lambda)}{\prod_{k=1}^n(\mu_k - i\lambda)}\,\frac{\prod_{k=1}^n \mu_k}{Q_n(0)}.$$

By Theorem 12.5.1 and (12.3.3) we also have

$$\begin{aligned} \mathbf{E}e^{i\lambda\chi_-^0} &= 1 - \frac{(1-\varphi(\lambda))}{1-p}\,\frac{Q_n(\lambda)}{\prod_{k=1}^\infty(\mu_k - i\lambda)}\,\frac{\prod_{k=1}^n \mu_k}{Q_n(0)}, \\ \mathbf{E}\big(e^{i\lambda\chi_+};\, \eta_+ < \infty\big) &= 1 - \frac{(1-p)\prod_{k=1}^n(\mu_k - i\lambda)Q_n(0)}{Q_n(\lambda)\prod_{k=1}^n \mu_k}. \end{aligned} \qquad (12.5.16)$$

Proof of Theorem 12.5.1 The proof of *sufficiency* will be divided into several stages.

1. In the vicinity of the point $\lambda = 0$ on the line Π, the value of

$$\mathfrak{v}(\lambda) = \frac{(1-\varphi(\lambda))(i\lambda+1)}{i\lambda}$$

lies in the vicinity of the point $-\mathbf{E}\xi > 0$. By virtue of (12.5.2), outside a neighbourhood of zero one has

$$\arg\big(1-\varphi(\lambda)\big) \in \left(-\frac{\pi}{2}, \frac{\pi}{2}\right), \qquad \arg\frac{i\lambda+1}{i\lambda} \in \left(-\frac{\pi}{2}, \frac{\pi}{2}\right), \qquad (12.5.17)$$

where, for a complex number $z = |z|e^{i\gamma}$, $\arg z$ denotes the exponent γ. In (12.5.17) $\arg z$ means the principal value of the argument from $(-\pi, \pi]$. Clearly, $\arg z_1 z_2 = \arg z_1 + \arg z_2$. This implies that, when λ changes from $-T$ to T for large T, the values of $\arg \mathfrak{v}(\lambda)$ do not leave the interval $(-\pi, \pi)$ and do not come close to its boundaries. Moreover, the initial and final values of $\mathfrak{v}(\lambda)$ lie in the sector $\arg z \in (-\frac{\pi}{2}, \frac{\pi}{2})$. This means that, for any T, the following relation is valid for the *index* of the function $\mathfrak{v}$ on $[-T, T]$:

$$\operatorname{ind}_T \mathfrak{v} := \frac{1}{2\pi}\int_{-T}^{T} d\big(\arg \mathfrak{v}(\lambda)\big) \in \left(-\frac{b}{2}, \frac{b}{2}\right), \quad b < 1. \qquad (12.5.18)$$

(If the distribution $\mathbf{F}$ has a density on $(-\infty, 0]$ as well then $\varphi(\pm T) \to 0$ and $\operatorname{ind}_T \mathfrak{v} \to 0$ as $T \to \infty$.)

2. Represent the function $\mathfrak{v}$ as the product $\mathfrak{v}(\lambda) = \mathfrak{v}_1(\lambda)\mathfrak{v}_2(\lambda)$, where

$$\begin{aligned} \mathfrak{v}_1(\lambda) &= \frac{(i\lambda+1)^n}{Q_n(\lambda)}, \\ \mathfrak{v}_2(\lambda) &= \frac{Q_n(\lambda)(1-\varphi(\lambda))}{i\lambda(i\lambda+1)^{n-1}} = \frac{Q_n(\lambda) - P_m(\lambda) - Q_n(\lambda)\varphi^-(\lambda)}{i\lambda(i\lambda+1)^{n-1}}. \end{aligned} \qquad (12.5.19)$$

We show that

$$|n + \mathrm{ind}_T\, \mathfrak{v}_2| < \frac{1}{2}. \tag{12.5.20}$$

In order to do this, we first note that the function $\mathfrak{v}_1$ is analytic on Π_+ and has there a zero of multiplicity n at the point $\lambda = i$. Consider a closed contour $\mathcal{T}_T^+$ consisting of the segment $[-T, T]$ and the semicircle $|\lambda| = T$ lying in Π_+. According to the argument principle in complex function theory, the number of zeros of the function $\mathfrak{v}_1$ inside $\mathcal{T}_T$ equals the increment of the argument of $\mathfrak{v}_1(\lambda)$ divided by 2π when moving along the contour $\mathcal{T}_T^+$ in the positive direction, i.e.

$$\frac{1}{2\pi}\int_{\mathcal{T}_T^+} d \arg \mathfrak{v}_1(\lambda) = n.$$

As, moreover, $\mathfrak{v}_1(\lambda) \to (-1)^n = \mathrm{const}$ as $|\lambda| \to \infty$ (see (12.5.11) and (12.5.19)), we see that the increment of $\arg \mathfrak{v}_1$ on the semicircle tends to 0 as $T \to \infty$, and hence

$$\mathrm{ind}_T\, \mathfrak{v}_1 \to \mathrm{ind}\, \mathfrak{v}_1 := \frac{1}{2\pi}\int_{-\infty}^{\infty} d \arg \mathfrak{v}_1(\lambda) = n.$$

It remains to note that $\mathrm{ind}_T\, \mathfrak{v} = \mathrm{ind}_T\, \mathfrak{v}_1 + \mathrm{ind}_T\, \mathfrak{v}_2$ and make use of (12.5.18).

3. We show that $1 - \varphi(\lambda)$ has precisely n zeros in Π_-. To this end, we first show that the function $\mathfrak{v}_2(\lambda)$, which is analytic in Π_- and continuous on $\Pi_- \cup \Pi$, has n zeros in Π_-. Consider the positively oriented closed contour $\mathcal{T}_T^-$ consisting of the segment $[-T, T]$ (traversed in the negative direction) and the lower half of the circle $|\lambda| = T$, and compute

$$\frac{1}{2}\int_{\mathcal{T}_T^-} d \arg \mathfrak{v}_2(\lambda). \tag{12.5.21}$$

Since $\mathfrak{v}_2(\lambda) \sim (-1)^n(1 - \varphi^-(\lambda))$ (see (12.5.11) and (12.5.19)), $|\varphi^-(\lambda)| < 1$ as $|\lambda| \to \infty$, $\mathrm{Im}\, \lambda \le 0$, for large T the part of integral (12.5.21) over the semicircle will be less than $1/2$ in absolute value. Comparing this with (12.5.20) we obtain that integral (12.5.21), being an integer, is necessarily equal to n. This means that $\mathfrak{v}_2(\lambda)$ has exactly n zeros in Π_-, which we will denote by $-i\mu_1, \ldots, -i\mu_n$. Since $Q_n(-i\mu_k) \ne 0$ (otherwise we would have, by (12.5.19), $P_m(-i\mu_k) = 0$, which would mean cancellability of the fraction P_m/Q_n), the function $1 - \varphi(\lambda)$ has in Π_- the same zeros as $\mathfrak{v}_2(\lambda)$ (see (12.5.19)).

4. It remains to put

$$\mathfrak{w}_+(\lambda) = \frac{\prod_{k=1}^n(\mu_k - i\lambda)}{Q_n(\lambda)},$$

$$\mathfrak{w}_-(\lambda) = \frac{(Q_n(\lambda) - P_m(\lambda) - Q_n(\lambda)\varphi^-(\lambda))(i\lambda + 1)}{i\lambda \prod_{k=1}^n(\mu_k - i\lambda)}$$

and note that $\mathfrak{w}_\pm \in \mathcal{K}_\pm \cap \mathcal{K}$.

The last assertion of the theorem follows from the fact that the real function $\psi(\mu) = \varphi(-i\mu)$ for $\mathrm{Im}\, \mu = 0$ is convex on $[0, \beta_1)$, $\psi'(0) = \mathbf{E}\xi < 0$ and

$\psi(\mu) \to \infty$ as $\mu \to \beta_1$. Therefore on $[0, \beta_1)$ there exists a unique real solution to the equation $\psi(\mu) = 1$. There are no complex zeros in the half-plane $\operatorname{Re}\mu \le \mu_1$ since in this region, for $\operatorname{Im}\mu \ne 0$, one has

$$\left|\psi(\mu)\right| < \psi(\operatorname{Re}\mu) \le \psi(\mu_1) = 1$$

because of the presence of an absolutely continuous component.

Necessity. Now let $\mathfrak{w}_+(\lambda)$ be rational. This means that

$$\mathfrak{w}_+(\lambda) = c_1 + \int_0^\infty e^{i\lambda x} g(x)\,dx,$$

where $c_1 = \mathfrak{w}_+(i\infty)$ and $g(x)$ is an exponential polynomial. It follows from the equality (see (12.5.5))

$$\begin{aligned} 1 - \varphi(\lambda) = \mathfrak{w}_+(\lambda)\frac{\mathfrak{w}_-(\lambda) i\lambda}{i\lambda + 1} &= c_2 \mathfrak{w}_+(\lambda)\bigl(1 - \mathbf{E}e^{i\lambda\chi_-^0}\bigr) \\ &= c_2\left(c_1 + \int_0^\infty e^{i\lambda x} g(x)\,dx\right) \int_{-\infty}^0 e^{i\lambda x}\,dW(x), \end{aligned}$$

where $W(x) = -\mathbf{P}(\chi_-^0 < x)$ for $x < 0$, $c_2 = \text{const}$, that ξ has a density for $x > 0$ that is equal to

$$\int_{-\infty}^0 dW(t)\, g(x-t).$$

Since the integral

$$\int_{-\infty}^0 dW(t)\,(x-t)^k e^{-\beta(x-t)} = e^{-\beta x} \sum_{j=0}^k (-1)^j \binom{k}{j} x^j c_{kj},$$

$$c_{kj} = \int_{-\infty}^0 dW(t)\, t^{k-j} e^{\beta t},$$

is an exponential polynomial, the integral $\int_{-\infty}^0 dW(t)\,g(x-t)$ is also an exponential polynomial, which implies the rationality of $\mathbf{E}(e^{i\lambda\xi};\, \xi > 0)$. The theorem is proved. □

Example 12.5.1 Let the distribution $\mathbf{F}$ be exponential on the positive half-line:

$$\mathbf{P}(\xi > x) = q e^{-\beta x}, \quad \beta > 0,\ q < 1.$$

Then $\varphi^+(\lambda) = q\beta/(\beta - i\lambda)$ and we can put $m = 0$, $n = 1$, $P_0(\lambda) = q\beta$, $Q_1(\lambda) = \beta - i\lambda$. The equation $\psi(\mu) := \mathbf{E}e^{\mu\xi_1} = 1$ has, in the half-plane $\operatorname{Re}\mu > 0$, the unique solution μ_1,

$$\mathfrak{w}_+(\lambda) = \frac{\mu_1 - i\lambda}{Q_1(\lambda)}$$

(see (12.5.15)). By Corollary 12.5.2,

$$\mathbf{E}e^{i\lambda S} = \frac{\mu_1}{Q_1(0)} \frac{Q_1(\lambda)}{(\mu_1 - i\lambda)} = \frac{\mu_1(\beta - i\lambda)}{\beta(\mu_1 - i\lambda)} = \frac{\mu_1}{\beta} + \frac{\beta - \mu_1}{\beta} \frac{\mu_1}{\mu_1 - i\lambda}.$$

This yields $\mathbf{P}(S=0)=\mu_1/\beta$,

$$\frac{\mathbf{P}(S\in dx)}{dx}=\left(1-\frac{\mu_1}{\beta}\right)\mu_1 e^{-\mu_1 x}\quad \text{for } x>0,$$

i.e. the distribution of S is exponential on $(0,\infty)$ with parameter μ_1 and has a positive atom at zero.

Example 12.5.2 (A generalisation of Example 12.5.1) Let $\mathbf{F}$ have, on the positive half-line, the density $\sum_{k=1}^n a_k e^{-\beta_k x}$ (a sum of exponentials), where $0<\beta_1<\beta_2<\dots<\beta_n$, $a_k>0$. Then

$$Q_n(\lambda)=\prod_{k=1}^n(\beta_k-i\lambda).$$

As was already noted in Theorem 12.5.1, the equation $\psi(\mu):=\varphi(-i\lambda)=1$ has, on the interval $(0,\beta_1)$, a unique zero μ_1. The function $\psi^-(\mu):=\varphi^-(-i\mu)$ is continuous, positive, and bounded for $\mu>0$. On each interval (β_k,β_{k+1}), $k=1,\dots,n-1$, the function

$$\psi^+(\mu):=\varphi^+(-i\mu)=\sum_{k=1}^n\frac{a_k\beta_k}{(\beta_k-\mu)}$$

is continuous and changes from $-\infty$ to ∞. Therefore, on each of these intervals, there exists at least one root μ_{k+1} of the equation $\psi(\mu)=1$. Since by Theorem 12.5.1 there are only n roots of this equation in $\operatorname{Re}\mu>0$, we obtain that μ_{k+1} is the unique root in (β_k,β_{k+1}) and

$$\mathfrak{w}_+(\lambda)=\frac{\prod_{k=1}^n(\mu_k-i\lambda)}{Q_n(\lambda)},\qquad \mathbf{E}e^{i\lambda S}=\prod_{k=1}^n\frac{(\beta_k-i\lambda)\mu_k}{(\mu_k-i\lambda)\beta_k}. \tag{12.5.22}$$

This means that $1-p:=\mathbf{P}(S=0)=\prod_{k=1}^n\frac{\mu_k}{\beta_k}$, and

$$\frac{\mathbf{P}(S\in dx)}{dx}=\sum_{k=1}^n b_k e^{-\mu_k x}\quad \text{for } x>0,$$

where $\mu_k\in(\beta_{k-1},\beta_k)$, $k=1,\dots,n$, $\beta_0=0$, and the coefficients b_k are defined by the decomposition of (12.5.22) into partial fractions.

By (12.5.16),

$$\mathbf{E}\big(e^{i\lambda\chi_+};\,\eta_+<\infty\big)=1-\frac{1-p}{\mathbf{E}e^{i\lambda S}}=1-\prod_{k=1}^n\frac{(\mu_k-i\lambda)}{(\beta_k-i\lambda)}, \tag{12.5.23}$$

so the conditional distribution of χ_+ given $\chi_+<\infty$ has a density which is equal to

$$\sum_{k=1}^n c_k e^{-\beta_k x}, \tag{12.5.24}$$

where the coefficients c_k, similarly to the above, are defined by the expansion of the right-hand side of (12.5.23) into partial fractions. Relation (12.5.24) means that the density of χ_+ has the same "structure" as the density of ξ does for $x > 0$, but differs in coefficients of the exponentials only. By (12.5.16) this property of the density of χ_+ holds in the general case as well.

12.5.4 Explicit Factorisation of the Function $\mathfrak{v}(\lambda)$ when the Left Tail of the Distribution F Is an Exponential Polynomial

Now consider the case where the *left* tail of the distribution **F** has a density which is an exponential polynomial (belongs to the class $\mathcal{EP}$). In this case,

$$\varphi^-(\lambda) = \mathbf{E}\big(e^{i\lambda\xi};\ \xi < 0\big) = \frac{P_m(\lambda)}{Q_n(\lambda)},$$

where

$$Q_n(\lambda) = \prod_{k=1}^{K} (\beta_k - i\lambda)^{l_k}, \quad n = \sum_{k=1}^{K} l_k, \quad \operatorname{Re}\beta_k < 0,\ m < n.$$

Theorem 12.5.2 *Let there exist* $\mathbf{E}\xi < 0$. *For the positive component of the canonical factorisation* $\mathfrak{v}(\lambda) = \mathfrak{w}_+(\lambda)\mathfrak{w}_-(\lambda)$ *of the function*

$$\mathfrak{v}(\lambda) = \frac{(1-\varphi(\lambda))(i\lambda+1)}{i\lambda}$$

to be representable as

$$\mathfrak{w}_+(\lambda) = \big(1-\varphi(\lambda)\big)R(\lambda),$$

where $R(\lambda)$ *is a rational function, it is necessary and sufficient that the function* $\varphi^-(\lambda)$ *is rational. If* $\varphi^-(\lambda) = P_m(\lambda)/Q_n(\lambda)$ *then the function* $1-\varphi(\lambda)$ *has precisely* $n-1$ *zeros in the half-plane on* $\operatorname{Im}\lambda > 0$ *which we denote by* $-i\mu_1, \dots, -i\mu_{n-1}$, *and*

$$R(\lambda) = \frac{Q_n(\lambda)}{i\lambda \prod_{k=1}^{n-1}(\mu_k - i\lambda)}.$$

Theorem 12.5.2, Corollary 12.5.1 and (12.3.3) imply the following assertion.

Corollary 12.5.3 *If* $\mathbf{E}\xi < 0$ *and* $\varphi^-(\lambda) = P_m(\lambda)/Q_n(\lambda)$ *then*

$$\mathbf{E}e^{i\lambda S} = \frac{\mathfrak{w}_+(0)}{\mathfrak{w}_+(\lambda)} = -\frac{\mathbf{E}\xi\, Q_n(0) i\lambda \prod_{k=1}^{n-1}(\mu_k - i\lambda)}{(1-\varphi(\lambda))Q_n(\lambda)\prod_{k=1}^{n-1}\mu_k},$$

$$\mathbf{E}e^{i\lambda\chi_-} = 1 + \frac{(1-p_0)\mathbf{E}\xi\, Q_n(0) i\lambda \prod_{k=1}^{n-1}(\mu_k - i\lambda)}{\prod_{k=1}^{n-1}\mu_k Q_n(\lambda)}.$$

Here the density of χ_- has the same "structure" as the density of ξ does for $x<\infty$.

Proof of Theorem 12.5.2 The proof is close to that of Theorem 12.5.1, but unfortunately is not its direct consequence. We present here a brief proof of Theorem 12.5.2 under the simplifying assumption that the distribution $\mathbf{F}$ is absolutely continuous. Using the scheme of the proof of Theorem 12.5.1, the reader can easily reconstruct the argument in the general case.

Sufficiency. As in Theorem 12.5.1, we verify that the trajectory of $\mathfrak{v}(\lambda)$, $-\infty<\lambda<\infty$, does not intersect the ray $\arg\mathfrak{v}=-\pi$, so in our case there exists

$$\operatorname{ind}\mathfrak{v}:=\lim_{T\to\infty}\operatorname{ind}_T\mathfrak{v}=0.$$

Put $\mathfrak{v}:=\mathfrak{v}_1\mathfrak{v}_2$, where

$$\mathfrak{v}_1:=\frac{Q_n-P_m-Q_n\varphi^+}{i\lambda(i\lambda-1)^{n-1}},\qquad \mathfrak{v}_2:=\frac{(i\lambda+1)(i\lambda-1)^{n-1}}{Q_n}.$$

Clearly, $\mathfrak{v}_2\in\mathcal{K}_-\cap\mathcal{K}$ and has exactly $n-1$ zeros in Π_-. Hence, by the argument principle, $\operatorname{ind}\mathfrak{v}_2=-(n-1)$, and

$$\operatorname{ind}\mathfrak{v}_1=-\operatorname{ind}\mathfrak{v}_2=n-1.$$

Since $\mathfrak{v}_1\in\mathcal{K}_+\cap\mathcal{K}$, again using the argument principle we obtain that $\mathfrak{v}_1$, as well as $1-\varphi$, has exactly $n-1$ zeros $-i\mu_1,\dots,-i\mu_{n-1}$ in Π_+. Putting

$$\mathfrak{w}_+:=\frac{(1-\varphi)Q_n}{i\lambda\prod_{k=1}^{n-1}(\mu_k-i\lambda)},\qquad \mathfrak{w}_-:=\frac{(i\lambda+1)\prod_{k=1}^{n-1}(\mu_k-i\lambda)}{Q_n},$$

we obtain a canonical factorisation.

Necessity. Similarly to the preceding arguments, the necessity follows from the factorisation identity

$$\begin{aligned}1-\varphi(\lambda)&=c_1\bigl(1-\mathbf{E}\bigl(e^{i\lambda\chi_+};\ \eta_+<\infty\bigr)\bigr)\mathfrak{w}_-(\lambda)\\&=c_1\int_0^\infty e^{i\lambda x}\,dV(x)\biggl(c_2+\int_{-\infty}^0 e^{i\lambda x}g(x)\,dx\biggr),\end{aligned}$$

where $V(x)=\mathbf{P}(\chi_+>x;\ \eta_+<\infty)$ for $x>0$, $c_i=\text{const}$ and $g(x)$ is an exponential polynomial. The theorem is proved. □

As in Sect. 12.5.1, we do not consider the case $\mathbf{E}\xi>0$ since it reduces to applying the aforementioned argument to the random variable $-\xi$.

12.5.5 Explicit Canonical Factorisation for the Function $\mathfrak{v}^0(\lambda)$

The goal of this subsection, as it was in Sects. 12.5.3 and 12.5.4, is to find an explicit form of the components $\mathfrak{w}_\pm^0(\lambda)$ in the canonical factorisation of the function

$\mathfrak{v}^0(\lambda) = \frac{1-\varphi(\lambda)}{\lambda^2}(\lambda^2+1)$ in (12.5.6) in terms of the zeros of the function $1-\varphi(\lambda)$ in the case where $\mathbf{E}\xi = 0$ and either $\varphi^+(\lambda)$ or $\varphi^-(\lambda)$ is a rational function. When $\mathbf{E}\xi = 0$, it is sufficient to consider the case where $\varphi^+(\lambda)$ is rational, i.e. the distribution $\mathbf{F}$ has on the positive half-line a density which is an exponential polynomial, so that

$$\varphi^+(\lambda) = \frac{P_m(\lambda)}{Q_n(\lambda)}, \qquad Q_n(\lambda) = \prod_{k=1}^{K}(\beta_k - i\lambda)^{l_k}, \quad n = \sum_{k=1}^{K} l_k.$$

The case where it is the function $\varphi^-(\lambda)$ that is rational is treated by switching to random variable $-\xi$.

Theorem 12.5.3 *Let $\mathbf{E}\xi = 0$ and $\mathbf{E}\xi^2 = \sigma^2 < \infty$. For the positive component $\mathfrak{w}_+^0(\lambda)$ of the canonical factorisation*

$$\mathfrak{v}^0(\lambda) = \mathfrak{w}_+^0(\lambda)\mathfrak{w}_-^0(\lambda), \qquad \mathfrak{w}_\pm \in \mathcal{K}_\pm \cap \mathcal{K},$$

to be a rational function it is necessary and sufficient that the function $\varphi^+(\lambda) = \mathbf{E}(e^{i\lambda\xi};\ \xi > 0)$ is rational. If $\varphi^+(\lambda) = P_m(\lambda)/Q_n(\lambda)$ is an uncancellable ratio of polynomials of degrees m and n, respectively, $m < n$, then the function $1-\varphi(\lambda)$ has exactly $n-1$ zeros in Π_- which we denote by $-i\mu_1, \ldots, -i\mu_{n-1}$, and

$$\mathfrak{w}_+^0(\lambda) = \frac{\prod_{k=1}^{n-1}(\mu_k - i\lambda)(i\lambda - 1)}{Q_n(\lambda)}, \qquad \mathfrak{w}_-^0(\lambda) = \frac{(1-\varphi(\lambda))(i\lambda+1)Q_n(\lambda)}{\lambda^2\prod_{k=1}^{n-1}(\mu_k - i\lambda)}. \tag{12.5.25}$$

Relation (12.5.3) and the uniqueness of canonical factorisation imply the following representation.

Corollary 12.5.4 *Under the conditions of Theorem* 12.5.3,

$$\mathbf{E}e^{i\lambda\chi_+} = 1 - \frac{i\lambda\mathbf{E}\chi_+ Q_n(0)\prod_{k=1}^{n-1}(1 - \frac{i\lambda}{\mu_k})}{Q_n(\lambda)}.$$

Proof The corollary follows from (12.5.7), (12.5.25), the uniqueness of canonical factorisation and the equalities

$$\mathfrak{v}_+^0(0) = \mathbf{E}\chi_+, \qquad \mathfrak{v}_+^0(\lambda) = \frac{\mathfrak{w}_+^0(\lambda)\mathbf{E}\chi_+}{\mathfrak{w}_+^0(0)},$$

$$1 - \mathbf{E}e^{i\lambda\chi_+} = \frac{\mathfrak{v}_+^0(\lambda)i\lambda}{i\lambda - 1} = \frac{i\lambda\mathbf{E}\chi_+ Q_n(0)\prod_{k=1}^{n-1}(\mu_k - i\lambda)}{Q_n(\lambda)\prod_{k=1}^{n-1}\mu_k}.$$

Thus, here the "structure" of the density of χ_+ again repeats the structure of the density of ξ for $x > 0$. □

Proof of Theorem 12.5.3 The proof is similar to that of Theorem 12.5.1.

Sufficiency.

1. In the vicinity of the point $\lambda = 0$, $\lambda \in \Pi$, the value of $\mathfrak{v}^0(\lambda)$ lies in the vicinity of the point $\sigma^2/2 > 0$ by Property 7.1.5 of ch.f.s. Outside of a neighbourhood of

zero, similarly to (12.5.17), we have

$$\arg\big(1-\varphi(\lambda)\big)\in\left(-\frac{\pi}{2},\frac{\pi}{2}\right),\qquad \arg\frac{\lambda^2+1}{\lambda^2}=0.$$

This, analogously to (12.5.18), implies

$$\operatorname{ind}_T \mathfrak{v}^0 := \frac{1}{2\pi}\int_{-T}^{T} d\big(\arg\mathfrak{v}^0(\lambda)\big)\in(-b/2,b/2),\quad b<1.$$

2. Represent $\mathfrak{v}^0$ as $\mathfrak{v}^0=\mathfrak{v}_1\mathfrak{v}_2$, where

$$\begin{aligned}\mathfrak{v}_1 &:= \frac{(i\lambda+1)^n}{Q_n},\\ \mathfrak{v}_2 &:= \frac{(1-\varphi)(1-i\lambda)\,Q_n}{\lambda^2(i\lambda+1)^{n-1}}=\frac{(Q_n-P_m-Q_n\varphi^-)(1-i\lambda)}{\lambda^2(i\lambda+1)^{n-1}}.\end{aligned}\tag{12.5.26}$$

Then, similarly to (12.5.20), we find that

$$\operatorname{ind}_T \mathfrak{v}_1\to n\quad\text{as } T\to\infty,\qquad |n+\operatorname{ind}_T \mathfrak{v}_2|<\frac{1}{2}.$$

3. We show that $1-\varphi(\lambda)$ has exactly $n-1$ zeros in Π_-. To this end, note that the function $\mathfrak{v}_2$, which is analytic in Π_- and continuous on $\Pi_-\cup\Pi$ has exactly n zeros in Π_-. As in the proof of Theorem 12.5.1, consider the contour $\mathfrak{T}_T^-$. In the same way as in the argument in this proof, we obtain that

$$\frac{1}{2\pi}\int_{\mathfrak{T}_T^-} d\big(\arg\mathfrak{v}_2(\lambda)\big)=n,$$

so that $\mathfrak{v}_2$ has exactly n zeros in Π_-. Further, by (12.5.26) we have $\mathfrak{v}_2=\mathfrak{v}_3\mathfrak{v}_4$, where the function $\mathfrak{v}_3=(1-i\lambda)/(i\lambda+1)$ has one zero in Π_- at the point $\lambda=-i$. Therefore the function

$$\mathfrak{v}_4=\frac{(Q_n-P_m-Q_n\varphi^-)}{\lambda^2(i\lambda+1)^{n-2}},$$

which is analytic in Π_-, has $n-1$ zeros there. Since the zeros of $1-\varphi(\lambda)$ and those of $\mathfrak{v}_4(\lambda)$ in Π_- coincide, the assertion concerning the zeros of $1-\varphi(\lambda)$ is proved.

4. It remains to put

$$\mathfrak{w}_+^0(\lambda):=\frac{\prod_{k=1}^{n-1}(\mu_k-i\lambda)(1-i\lambda)}{Q_n(\lambda)},\qquad \mathfrak{w}_-^0(\lambda):=\frac{(1-\varphi(\lambda))(i\lambda+1)Q_n(\lambda)}{\lambda^2\prod_{k=1}^{n-1}(\mu_k-i\lambda)}$$

and note that $\mathfrak{w}_\pm^0\in\mathcal{K}_\pm\cap\mathcal{K}$.

Necessity is proved in exactly the same way as in Theorems 12.5.1 and 12.5.2. The theorem is proved. □

12.6 Explicit Form of Factorisation in the Arithmetic Case

The content of this section is similar to that of Sect. 12.5 and has the same structure, but there are also some significant differences.

12.6.1 Preliminary Remarks on the Uniqueness of Factorisation

As was already noted in Sect. 12.5, for arithmetic distributions defined by collections of probabilities $p_k = \mathbf{P}(\xi = k)$, we should use, instead of the ch.f.s $\varphi(\lambda)$, the generating functions

$$p(z) = \mathbf{E}z^{\xi} = \sum_{k=-\infty}^{\infty} z^k p_k$$

defined on the unit circle $|z| = 1$, which will be denoted by Π, as the axis $\operatorname{Im}\lambda = 0$ was in Sect. 12.5. The symbols Π_+ (Π_-) will denote the interior (exterior) of Π. For arithmetic distributions we will discuss the factorisation

$$1 - p(z) = \mathfrak{f}_+(z)\mathfrak{f}_-(z)$$

on the unit circle, where $\mathfrak{f}_\pm$ are analytic on $\Pi_\pm$ and continuous including the boundary Π. Similarly to the non-lattice case, the classes of such functions, that, moreover, are bounded and bounded away from zero on $\Pi_\pm$, we will denote by $\mathcal{K}_\pm$. Continuous bounded functions on Π, which are also bounded away from zero, form the class $\mathcal{K}$. The notion of *canonical factorisation* on Π is introduced in exactly the same way as above. Factorisation components must belong to the classes $\mathcal{K}_\pm$. The uniqueness of factorisation components (up to a constant factor) is proved in the same way as in Lemma 12.1.1.

We now show that if, similarly to the above, we "tweak" the function $1 - p(z)$ then it will admit a canonical factorisation. We will denote the tweaked function and its factorisation components by the same symbols as in Sect. 12.5. This will not lead to any confusion.

Lemma 12.6.1 1. *If* $\mathbf{E}\xi < 0$ *then the function*

$$\mathfrak{v}(z) := \frac{(1 - p(z))z}{1 - z}$$

belongs to $\mathcal{K}$ *and admits a unique canonical factorisation*

$$\mathfrak{v}(z) = \mathfrak{v}_+(z)\mathfrak{v}_-(z),$$

where

$$\mathfrak{v}_+(z) := 1 - \mathbf{E}\big(z^{\chi_+};\ \eta_+ < \infty\big) = \frac{1 - p}{\mathbf{E}z^S}, \qquad p := \mathbf{P}(\eta_+ < \infty),$$

$$\mathfrak{v}_-(z) := \frac{(1 - \mathbf{E}z^{\chi_-^0})z}{1 - z}.$$

2. *If* $\mathbf{E}\xi = 0$ *and* $\mathbf{E}\xi^2 < \infty$ *then the function*

$$\mathfrak{v}^0(z) := \frac{(1 - p(z))z}{(1 - z)^2}$$

belongs to $\mathcal{K}$ *and admits a unique canonical factorisation*

$$\mathfrak{v}^0(z) = \mathfrak{v}_+^0(z)\mathfrak{v}_-^0(z),$$

where

$$\mathfrak{v}_+^0 := \frac{1-\mathbf{E}z^{\chi_+}}{1-z}, \qquad \mathfrak{v}_-^0 := \frac{(1-\mathbf{E}z^{\chi_-^0})z}{1-z}.$$

Here we do not discuss the case $\mathbf{E}\xi > 0$ since it reduces to the case $\mathbf{E}\xi < 0$. We will also not present an analogue of Corollary 12.5.1 in view of its obviousness.

Proof of Lemma 12.6.1 Let $\mathbf{E}\xi < 0$. Since

$$\frac{(1-p(z))z}{1-z} \to -\mathbf{E}\xi > 0$$

as $z \to 1$, $p(z)$ is continuous on the compact Π and, furthermore, $|p(z)| < 1$ for $z \neq 1$, we see that $\mathfrak{v}(z)$ is bounded away from zero on Π and bounded, and hence belongs to $\mathcal{K}$. Further, by Corollary 12.2.2 (see (12.2.1) for $i\lambda = z$),

$$\mathfrak{v}(z) = \frac{(1-\mathbf{E}z^{\chi_-^0})z}{1-z}\left[1-\mathbf{E}\left(z^{\chi_+};\ \eta_+ < \infty\right)\right],$$

where $\mathbf{E}\chi_-^0 \in (-\infty, 0)$. Therefore, similarly to the above, we get

$$\mathfrak{v}_-(z) = \frac{(1-\mathbf{E}z^{\chi_-^0})z}{1-z} \in \mathcal{K}.$$

Moreover, it is obvious that $\mathfrak{v}_-(z) \in \mathcal{K}_-$. In the same way as above, we obtain that

$$\mathfrak{v}_+(z) = 1-\mathbf{E}\left(z^{\chi_+};\ \eta_+ < \infty\right) \in \mathcal{K}_+ \cap \mathcal{K}.$$

This proves the first assertion of the lemma.

The second assertion is proved similarly by using Corollary 12.2.4, by which

$$\mathfrak{v}^0(z) = \frac{1-\mathbf{E}z^{\chi_+}}{1-z} \cdot \frac{(1-\mathbf{E}z^{\chi_-^0})z}{1-z}.$$

Next, as before, we establish that $\mathfrak{v}^0 \in \mathcal{K}$ and that the factors on the right-hand side, denoted by $\mathfrak{v}_\pm^0(z)$, belong to $\mathcal{K}_\pm \cap \mathcal{K}$. The lemma is proved. □

12.6.2 The Classes of Distributions on the Positive Half-Line with Rational Generating Functions

The content of Sect. 12.5.2 is mostly preserved here. Now by exponential polynomials we mean the sequences

$$p_x = \sum_{k=1}^{K}\sum_{j=1}^{l_k} a_{kj} x^{j-1} q_k^x, \quad x = 1, 2, \ldots, \tag{12.6.1}$$

where $q_k < 1$ are different (cf. (12.5.10)). To probabilities p_x of such type will correspond rational functions

$$p^+(z) = \mathbf{E}(z^\xi;\ \xi > 0) = \sum_{x=1}^{\infty} z^x p_x = \frac{P_m(z)}{Q_n(z)},$$

where $1 \le m < n$, $n = \sum_{k=1}^{K} l_k$, and, for definiteness, we put

$$Q_n(z) = \prod_{k=1}^{K} (1 - q_k z)^{l_k}. \tag{12.6.2}$$

Here a significant difference from the non-lattice case is that, for $p^+(z)$ to be rational, we do not need (12.6.1) to be valid *for all* $x > 0$. It is sufficient that (12.6.1) holds for all x, *starting from some* $r + 1 \ge 1$. The first r probabilities $p_1, \dots, p_r$ can be *arbitrary*. In this case $p^+(z)$ will have the form

$$p^+(z) = \frac{P_m(z)}{Q_n(z)} + T_r(z) = \frac{P_M(z)}{Q_n(z)}, \tag{12.6.3}$$

where T_r is a polynomial of degree r (for $r = 0$ we put $T_0 = 0$), so that p^+ is again a rational function, but now the degree of the polynomial P_M

$$M = \begin{cases} m, & \text{if } r = 0, \\ n + r, & \text{if } r \ge 1 \end{cases} \tag{12.6.4}$$

in the numerator can be greater than the degree n of the polynomial in the denominator. In what follows, we only assume that $n + r > 0$, so that *the value* $n = 0$ *is allowed* (in this case there will be no exponential part in (12.6.1)). In that case we will assume that $Q_0 = 1$ and $P_m = 0$. The distributions corresponding to (12.6.3) will also be called exponential polynomials.

12.6.3 Explicit Canonical Factorisation of the Function $\mathfrak{v}(z)$ in the Case when the Right Tail of the Distribution F Is an Exponential Polynomial

Consider an arithmetic distribution $\mathbf{F}$ on the whole real line $(-\infty, \infty)$, $\mathbf{E}\xi < 0$, which is an exponential polynomial on the half-line $x > 0$. As before, denote the class of all such distributions by $\mathcal{EP}$. The generating function $p(z)$ of the distribution $\mathbf{F} \in \mathcal{EP}$ can be represented as

$$p(z) = p^+(z) + p^-(z),$$

where the function

$$p^-(z) = \mathbf{E}(z^\xi;\ \xi \le 0)$$

is analytic in Π_- and continuous including the boundary Π, and $p^+(z)$ is a rational function

$$p^+(z) = \mathbf{E}(z^\xi;\ \xi > 0) = \frac{P_M(z)}{Q_n(z)}$$

analytic in Π_+.

As above, in this case the canonical factorisation of the function

$$\mathfrak{v}(z) = \frac{(1 - p(z))z}{1 - z}$$

can be found in explicit form in terms of the zeros of the function $1 - p(z)$.

Theorem 12.6.1 *Let there exist* $\mathbf{E}\xi < 0$. *For the positive component* $\mathfrak{w}_+(z)$ *of the canonical factorisation*

$$\mathfrak{v}(z) = \mathfrak{w}_+(z)\mathfrak{w}_-(z), \qquad w_\pm \in \mathcal{K}_\pm \cap \mathcal{K},$$

to be a rational function it is necessary and sufficient that $p^+(z) = \mathbf{E}(z^\xi;\ \xi > 0)$ *is a rational function.*

If $p^+ = P_M/Q_n$, *where* M *is defined in* (12.6.4), *is an uncancellable ratio of polynomials then the function* $1 - p(z)$ *has in* Π_- *exactly* $n + r$ *zeros, which will be denoted by* $z_1, \dots, z_{n+r}$, *and*

$$\mathfrak{w}_+(z) = \frac{\prod_{k=1}^{n+r}(z_k - z)}{Q_n(z)},$$

where $Q_n(z_k) \neq 0$.

If we arrange the zeros $\{z_k\}$ *according to the values of* $|z_k|$ *in ascending order, then the point* $z_1 > 1$ *is a simple real zero.*

The theorem implies that

$$\mathfrak{w}_-(z) = \frac{(1 - p(z))z\, Q_n(z)}{(1 - z)\prod_{k=1}^{n+r}(z_k - z)}.$$

By Lemma 12.6.1, from Theorem 12.6.1 we obtain the following representation.

Corollary 12.6.1 *If* $\mathbf{E}\xi < 0$ *and* $p^+ = P_M/Q_n$ *then*

$$\mathbf{E}z^S = \frac{\mathfrak{w}_+(1)}{\mathfrak{w}_+(z)} = \frac{Q_n(z)\prod_{k=1}^{n+r}(z_k - 1)}{Q_n(1)\prod_{k=1}^{n+r}(z_k - z)}.$$

Similarly to (12.5.16), we can also write down the explicit form of $\mathbf{E}z^{\chi_-^0}$ and $\mathbf{E}(z^{\chi_+};\ \eta_+ < \infty)$ as well.

Proof of Theorem 12.6.1 The proof is similar to that of Theorem 12.5.1.

Sufficiency.

1. In the vicinity of the point $z = 1$ in Π the value of $-\mathfrak{v}(z)$ lies in the vicinity of the point $-\mathbf{E}\xi > 0$. Outside a neighbourhood of the point $z = 1$ we have for $z \in \Pi$,

$$\arg(1-p(z)) \in \left(-\frac{\pi}{2}, \frac{\pi}{2}\right), \quad \arg\left(\frac{z}{z-1}\right) = -\arg\left(1-\frac{1}{z}\right) \in \left(-\frac{\pi}{2}, \frac{\pi}{2}\right).$$

This implies that, for $z \in \Pi$,

$$\arg(-\mathfrak{v}(z)) \in (-\pi, \pi),$$

and hence the trajectory of $-\mathfrak{v}(z)$, $z \in \Pi$, never intersects the ray $\arg \mathfrak{v} = -\pi$,

$$\operatorname{ind} \mathfrak{v} := \frac{1}{2\pi} \int_0^{2\pi} d\left(\arg \mathfrak{v}\left(e^{i\lambda}\right)\right) = 0.$$

2. Represent the function $\mathfrak{v}$ as $\mathfrak{v} = \mathfrak{v}_1 \mathfrak{v}_2$, where

$$\mathfrak{v}_1(z) := \frac{z^{n+r}}{Q_n(z)}, \qquad \mathfrak{v}_2(z) := \frac{Q_n(z) - P_M(z) - p^-(z) Q_n(z)}{(1-z) z^{n+r-1}}.$$

We show that

$$\operatorname{ind} \mathfrak{v}_2 = -n - r. \tag{12.6.5}$$

In order to do this, we first note that the function $\mathfrak{v}_1$ is analytic in Π_+ and has there a zero of multiplicity $n+r$. Hence by the argument principle $\operatorname{ind} \mathfrak{v}_1 = n + r$. Since $0 = \operatorname{ind} \mathfrak{v} = \operatorname{ind} \mathfrak{v}_1 + \operatorname{ind} \mathfrak{v}_2$, we obtain the desired relation.

3. We show that $1 - p(z)$ has exactly $n + r$ zeros in Π_-. The function $\mathfrak{v}_2(z)$ is analytic on Π_- and continuous including the boundary Π. The positively oriented contour Π, which contains Π_+, corresponds to the negatively oriented contour with respect to Π_-. By (12.6.5) this means that $\mathfrak{v}_2(z)$ has precisely $n + r$ zeros on Π_- while the point $z = \infty$ is not a zero since the numerator and the denominator of $\mathfrak{v}(z)$ grow as $|z|^{n+r}$ as $|z| \to \infty$.

4. Denote the zeros of $\mathfrak{v}_2$ by $z_1, \ldots, z_{n+r}$ and put

$$\mathfrak{w}_+(z) := \frac{\prod_{k=1}^{n+r}(z_k - z)}{Q_n(z)}, \qquad \mathfrak{w}_-(z) := \frac{Q_n(z)(1-p(z))z}{(1-z)\prod_{k=1}^{n+r}(z_k - z)}.$$

It is easy to see that $\mathfrak{w}_\pm \in \mathcal{K}_\pm \cap \mathcal{K}$. The fact that $Q_n(z_k) \neq 0$ and z_1 is a simple real zero of $1 - p(z)$ is proved in the same way as in Theorem 12.5.1.

Necessity is also established in the same fashion as in Theorem 12.5.1. The theorem is proved. □

Clearly, in the arithmetic case we have complete analogues of Examples 12.5.1 and 12.5.2. In particular, if

$$\mathbf{P}(\xi = k) = c q^{k-1}, \quad c < (1-q),\ k = 1, 2, \ldots,$$

then

$$\mathfrak{w}_+(z) = \frac{z_1 - z}{1 - qz}, \qquad \mathbf{E} z^S = \frac{(1-qz)(z_1 - 1)}{(z_1 - z)(1-q)},$$

$$\mathbf{P}(S = 0) = \frac{1 - z_1^{-1}}{1-q}, \qquad \mathbf{P}(S = k) = \frac{(z_1^{-1} - q)(z_1 - 1) z_1^k}{1-q}, \quad k \geq 1.$$

In contrast to Sect. 12.5, here one can give another example where the distribution of S is geometric.

Example 12.6.1 Let $\mathbf{P}(\xi=1)=p_1>0$ and $\mathbf{P}(\xi\ge 2)=0$. In this case $\chi_+\equiv 1$ on the set $\{\eta_+<\infty\}$, and to find the distribution of S there is no need to use Theorem 12.6.1. Indeed, $\mathbf{P}(S=0)=1-p=\mathbf{P}(\eta_+=\infty)$. If $\eta_+<\infty$ then the trajectory $\xi_{\eta_++1},\xi_{\eta_+=2},\dots$ is distributed identically to $\xi_1,\xi_2,\dots$ and hence

$$S=\begin{cases}0 & \text{with probability } 1-p,\\ \chi_+ + S_{(1)} & \text{with probability } p,\end{cases}$$

where the variable $S_{(1)}$ is distributed identically to S, $\chi_+\equiv 1$. This yields

$$\mathbf{E}z^S=(1-p)+pz\mathbf{E}z^S,\qquad \mathbf{E}z^S=\frac{1-p}{1-pz},$$

$$\mathbf{P}(S=k)=(1-p)p^k,\quad k=0,1,\dots$$

By virtue of identity (12.3.3) (for $e^{i\lambda}=z$) the point $z_1=p^{-1}$ is necessarily a zero of the function $1-p(z)$.

12.6.4 Explicit Canonical Factorisation of the Function $\mathfrak{v}(z)$ when the Left Tail of the Distribution F Is an Exponential Polynomial

We now consider the case where the distribution $\mathbf{F}$ on the *negative* half-line can be represented as an exponential polynomial, up to the values of $\mathbf{P}(\xi=-k)$ at finitely many points $0,-1,-2,\dots,-r$. In this case, the value of $p^-(z)$ is derived similarly to that of $p^+(z)$ in (12.6.3) by replacing z with z^{-1}:

$$p^-(z)=\mathbf{E}\big(z^\xi;\ \xi<0\big)=\frac{z^{n-M}P_M(z)}{Q_n(z)},$$

where Q_n and P_M are polynomials (which differ from (12.6.3)),

$$M=\begin{cases}m, & \text{if } r=0,\\ n+r, & \text{if } r\ge 1,\end{cases}\qquad Q_n(z)=\prod_{k=1}^{K}(z-q_k)^{l_k},$$

and all $q_k<1$ are distinct.

Theorem 12.6.2 *Let there exist* $\mathbf{E}\xi<0$. *For the positive component of the canonical factorisation*

$$\mathfrak{v}(z)=\mathfrak{w}_+(z)\mathfrak{w}_-(z)$$

to be representable as

$$\mathfrak{w}_+(z)=\big(1-p(z)\big)R(z),$$

where $R(z)$ *is a rational function, it is necessary and sufficient that* $p^-(z)$ *is a rational function. If*

$$p^-(z)=\frac{z^{n-M}P_M(z)}{Q_n(z)},$$

where P_M *and* Q_n *are defined in* (12.6.2) *and* (12.6.3), *then the function* $1 - p(z)$ *has in* Π_+ *exactly* $n + r - 1$ *zeros that we denote by* $z_1, \dots, z_{n+r-1}$, *and*

$$R(z) := \frac{Q_n(z)}{(1-z)\prod_{k=1}^{n+r-1}(z - z_k)}.$$

Proof The proof is very close to that of Theorems 12.5.2 and 12.6.1. Therefore we will only present a brief proof of *sufficiency*.

1. As in Theorem 12.6.1, one can verify that

$$\text{ind}\, \mathfrak{v} = 0.$$

2. Represent $\mathfrak{v}(z)$ as $\mathfrak{v} = \mathfrak{v}_1 \mathfrak{v}_2$, where

$$\mathfrak{v}_1 := \frac{(Q_n(z) - z^{n-M} P_M(z) - p^+(z) Q_n(z)) z^r}{(1-z)}, \qquad \mathfrak{v}_2(z) := \frac{z^{1-r}}{Q_n(z)}.$$

The function $\mathfrak{v}_2$ is analytic in Π_-, continuous including the boundary Π, and has a zero at $z = \infty$ of multiplicity $n + r - 1$, so that

$$\text{ind}\, \mathfrak{v}_2 = n + r - 1.$$

The function $\mathfrak{v}_1$ is analytic in Π_+ and, by the argument principle, has there $n + r - 1$ zeros $z_1, \dots, z_{n+r-1}$. The function $1 - p(z)$ has the same zeros.

3. By putting

$$\mathfrak{w}_+(z) := \frac{(1 - p(z)) Q_n(z)}{(1-z)\prod_{k=1}^{n+r-1}(z - z_k)}, \qquad \mathfrak{w}_-(z) := \frac{z\prod_{k=1}^{n+r-1}(z - z_k)}{Q_n(z)}$$

we obtain $\mathfrak{w}_\pm \in \mathcal{K}_\pm \cap \mathcal{K}$. The theorem is proved. □

12.6.5 Explicit Factorisation of the Function $\mathfrak{v}^0(z)$

By virtue of the remarks at the beginning of Sect. 12.5.5 it is sufficient to consider factorisation of the function

$$\mathfrak{v}^0(z) := \frac{(1 - p(z)) z}{(1-z)^2}$$

for $\mathbf{E}\xi = 0$ and $\mathbf{E}\xi^2 < \infty$ just in the case when the function

$$p^+(z) = \mathbf{E}\big(z^\xi;\ \xi > 0\big) = \frac{P_M(z)}{Q_n(z)}$$

is rational, where $Q_n(z) = \prod_{k=1}^K (1 - q_k z)^{l_k}$, $n = \sum_{k=1}^K l_k$ (see (12.6.2), (12.6.3)).

Theorem 12.6.3 *Let* $\mathbf{E}\xi = 0$ *and* $\mathbf{E}\xi^2 = \sigma^2 < \infty$. *For the positive component* $\mathfrak{w}_+^0(z)$ *of the canonical factorisation*

$$\mathfrak{v}^0(z) = \mathfrak{w}_+^0(z) \mathfrak{w}_-^0(z), \qquad w_\pm \in \mathcal{K}_\pm \cap \mathcal{K},$$

to be rational, it is necessary and sufficient that the function $\mathrm{p}^+(z)$ *is rational. If* $\mathrm{p}^+(z) = P_M(z)/Q_n(z)$, *where M is defined in* (12.6.4), *is an uncancellable ratio of polynomials then the function* $1 - p(z)$ *has in* Π_- *exactly* $n + r - 1$ *zeros that we denote by* $z_1, \ldots, z_{n+r-1}$, *and*

$$\mathfrak{w}_+^0(z) = \frac{\prod_{k=1}^{n+r-1}(z_k - z)}{Q_n(z)}, \qquad \mathfrak{w}_-^0(z) = \frac{(1 - p(z)) z Q_n(z)}{(1 - z)^2 \prod_{k=1}^{n+r-1}(z_k - z)}.$$

Corollaries similar to Corollary 12.5.4 hold true here as well.

Proof of Theorem 12.6.3 The proof is similar to those of Theorems 12.5.3, 12.6.1 and 12.6.2. Therefore, as in the previous theorem, we restrict ourselves to the key elements of the proof of sufficiency.

1. In the vicinity of the point $z = 1$ in Π, the value of $-\mathfrak{v}^0(z)$ lies in the vicinity of the point $\sigma^2/2 > 0$. Outside of a neighbourhood of the point $z = 1$, for $z \in \Pi$ we have

$$\arg\bigl(1 - p(z)\bigr) \in \left(-\frac{\pi}{2}, \frac{\pi}{2}\right),$$

$$\arg \frac{-z}{(1 - z)^2} = -\arg\left((1 - z)\left(1 - \frac{1}{z}\right)\right) = -\arg\left(2 - z - \frac{1}{z}\right) = 0.$$

Hence

$$\operatorname{ind} \mathfrak{v}^0 := \frac{1}{2\pi} \int_0^{2\pi} d\bigl(\arg \mathfrak{v}^0\bigl(e^{i\lambda}\bigr)\bigr) = 0.$$

2. Represent the function $\mathfrak{v}^0(z)$ as

$$\mathfrak{v}^0(z) = \mathfrak{v}_1(z)\mathfrak{v}_2(z),$$

where

$$\mathfrak{v}_1(z) := \frac{z^{n+r-1}}{Q_n(z)}, \qquad \mathfrak{v}_2(z) := \frac{Q_n - P_M - \mathrm{p}^-(z) Q_n}{(1 - z)^2 z^{n+r-2}}.$$

As before, we show that $\operatorname{ind} \mathfrak{v}_1 = n + r - 1$ and that $1 - p(z)$ has, on Π_-, exactly $n + r - 1$ zeros, which are denoted by $z_1, \ldots, z_{n+r-1}$. It remains to put

$$\mathfrak{w}_+^0(z) = \frac{\prod_{k=1}^{n+r-1}(z_k - z)}{Q_n(z)}, \qquad \mathfrak{w}_-^0(z) = \frac{Q_n(z)(1 - p(z)) z}{(1 - z)^2 \prod_{k=1}^{n+r-1}(z_k - z)}.$$

The theorem is proved. □

12.7 Asymptotic Properties of the Distributions of $\chi_\pm$ and S

We saw in the previous sections that one can find the distributions of the variables S and $\chi_\pm$ in explicit form only in some special cases. Meanwhile, in applied problems of, say, risk theory (see Sect. 12.4) one is interested in the values of $\mathbf{P}(S > x)$ for large x (corresponding to small ruin probabilities). In this connection there arises the problem on the asymptotic behaviour of $\mathbf{P}(S > x)$ as $x \to \infty$, as well as related problems on the asymptotics of $\mathbf{P}(|\chi_\pm| > x)$. It turns out that these problems can be solved under rather broad conditions.

12.7.1 *The Asymptotics of* $\mathbf{P}(\chi_+ > x \mid \eta_+ < \infty)$ *and* $\mathbf{P}(\chi_-^0 < -x)$ *in the Case* $\mathbf{E}\xi \le 0$

We introduce some classes of functions that will be used below.

Definition 12.7.1 A function $G(t)$ is called (asymptotically) *locally constant* (l.c.) if, for any fixed v,

$$\frac{G(t+v)}{G(t)} \to 1 \quad \text{as } t \to \infty. \tag{12.7.1}$$

It is not hard to see that, say, the functions $G(t) = t^\alpha[\ln(1+t)]^\gamma$, $t > 0$, are l.c.

We denote the class of all l.c. functions by $\mathcal{L}$. The properties of functions from $\mathcal{L}$ are studied in Appendix 6. In particular, it is established that (12.7.1) holds uniformly in v on any fixed segment, and that $G(t) = e^{o(t)}$ and $G(t) = o(G^I(t))$ as $t \to \infty$, where

$$G^I(t) := \int_t^\infty G(u)\,du. \tag{12.7.2}$$

Denote by $\mathcal{E}$ the class of distributions satisfying the right-hand side Cramér condition (the exponential class). The class $\mathcal{E}^* \subset \mathcal{E}$ of distributions $\mathbf{G}$ whose "tails" $G(t) = \mathbf{G}((t,\infty))$ satisfy, for any fixed $v > 0$, the relation

$$\frac{G(t+v)}{G(t)} \to 0 \quad \text{as } t \to \infty, \tag{12.7.3}$$

could be called the "superexponential" class. For example, the normal distribution belongs to $\mathcal{E}^*$. In the arithmetic case, one has to put $v = 1$ in (12.7.3) and consider integer-valued t.

In the case $\mathbf{E}\xi \le 0$ it is convenient to introduce a random variable χ with the distribution

$$\mathbf{P}(\chi \in dv) = \mathbf{P}(\chi_+ \in dv \mid \eta_+ < \infty) = \frac{\mathbf{P}(\chi_+ \in dv;\ \eta_+ < \infty)}{p}, \qquad p = \mathbf{P}(\eta_+ < \infty).$$

If $\mathbf{E}\xi = 0$ then the distributions of χ and χ_+ coincide. In the sequel we will confine ourselves to non-lattice ξ (then $\chi_\pm$ will also be non-lattice). In the arithmetic case everything will look quite similar.

Denote by $F_+(t)$ the right "tail" of the distribution $\mathbf{F}$: $F_+(t) := \mathbf{F}((t,\infty))$ and put

$$F_+^I(t) := \int_t^\infty F_+(u)\,du.$$

Theorem 12.7.1 *Let there exist* $\mathbf{E}\xi \le 0$ *and, in the case* $\mathbf{E}\xi = 0$, *assume* $\mathbf{E}\xi^2 < \infty$ *holds.*

1. *If* $F_+(t) = o(F_+^I(t))$ *as* $t \to \infty$ *then, as* $x \to \infty$,

$$\mathbf{P}(\chi > x) \sim -\frac{F_+^I(x)}{p\mathbf{E}\chi_-^0}. \tag{12.7.4}$$

2. *If* $F_+(t) = V(t)e^{-\beta t}$, $\beta > 0$, $V \in \mathcal{L}$ *then*

$$\mathbf{P}(\chi > x) \sim \frac{F_+(x)}{p(1 - \mathbf{E}e^{\beta \chi_-^0})}. \tag{12.7.5}$$

3. *If* $F_+ \in \mathcal{E}^*$ *then*

$$\mathbf{P}(\chi > x) \sim \frac{F_+(x)}{p\mathbf{P}(\chi_-^0 < 0)}. \tag{12.7.6}$$

Proof The proof is based on identity (12.2.1) of Corollary 12.2.2, which can be rewritten as

$$1 - p\mathbf{E}e^{i\lambda\chi} = \frac{1 - \varphi(\lambda)}{1 - \varphi_-^0(\lambda)}, \qquad \varphi_-^0(\lambda) := \mathbf{E}e^{i\lambda\chi_-^0}. \tag{12.7.7}$$

Introduce the renewal function $H_-(t)$ corresponding to the random variable $\chi_-^0 \le 0$:

$$H_-(t) = \sum_{k=0}^{\infty} \mathbf{P}(H_k \ge t), \qquad H_k = \chi_-^{(1)} + \cdots + \chi_-^{(k)},$$

where $\chi_-^{(k)}$ are independent copies of χ_-^0, $a_- := \mathbf{E}\chi_-^0 > -\infty$. As was noted in Sect. 10.1, the function $1/(1 - \varphi_-^0(\lambda))$ can be represented as

$$\frac{1}{1 - \varphi_-^0(\lambda)} = -\int_{-\infty}^{0} e^{i\lambda t}\, dH_-(t)$$

(the function $H_-(t)$ decreases). Therefore, for $x > 0$ and any $N > 0$, we obtain from (12.7.7) that

$$p\mathbf{P}(\chi > x) = -\int_{-\infty}^{0} dH_-(t)\, F_+(x - t) = -\int_{-N}^{0} - \int_{-\infty}^{-N}. \tag{12.7.8}$$

Here, by the condition of assertion 1,

$$-\int_{-N}^{0} \le F_+(x)\bigl[H_-(-N) - H_-(0)\bigr] = o\bigl(F_+^I(x)\bigr) \quad \text{as } x \to \infty.$$

Evidently, this relation will still be true when $N \to \infty$ slowly enough as $x \to \infty$. Furthermore, by the local renewal theorem, as $N \to \infty$,

$$-\int_{-\infty}^{-N} dH_-(t)\, F_+(x - t) \sim \int_{-\infty}^{-N} F_+(x - t)\frac{dt}{|a_-|} = \frac{F_+^I(x + N)}{|a_-|}. \tag{12.7.9}$$

For a formal justification of this relation, the interval $(-\infty, -N]$ should be divided into small intervals $(-N_{k+1}, -N_k]$, $k = 0, 1, \ldots$, $N_0 = N$, $N_{k+1} > N_k$, on each of which we use the local renewal theorem, so that

$$\frac{F_+(x-N_k)(N_{k+1}-N_k)}{|a_-|}\big(1+o(1)\big) \le -\int_{-N_{k+1}}^{-N_k} dH_-(t)\,F_+(x-t)$$
$$\le \frac{F_+(x-N_{k+1})(N_{k+1}-N_k)}{|a_-|}\big(1+o(1)\big).$$

From here it is not difficult to obtain the required bounds for the left-hand side of (12.7.9) that are asymptotically equivalent to the right-hand side. Since, for N growing slowly enough,

$$F_+^I(x) - F_+^I(x+N) = \int_x^{x+N} F_+(u)\,du < F_+(x)N = o\big(F_+^I(x)\big)$$

one has $F_+^I(x+N) \sim F_+^I(x)$, and we finally obtain the relation

$$p\mathbf{P}(\chi > x) \sim \frac{F_+^I(x+N)}{|a_-|}.$$

This proves (12.7.4).

If $F_+(t) = V(t)e^{-\beta t}$, $V \in \mathcal{L}$, then we find from (12.7.8) that

$$p\mathbf{P}(\chi > x) \sim -V(x)e^{-\beta x}\int_{-\infty}^0 dH_-(t)\,e^{t\beta} = \frac{F_+(x)}{1-\mathbf{E}e^{\beta\chi_-^0}}.$$

This proves (12.7.5).

Now let $F_+ \in \mathcal{E}^*$. If we denote by $h_0 > 0$ the jump of the function $H_-(t)$ at the point 0 then, clearly,

$$-\int_{-\infty}^0 dH_-(t)\,\frac{F_+(x-t)}{F_+(x)} \to h_0 \quad \text{as } x \to \infty,$$

and hence

$$p\mathbf{P}(\chi > x) \sim F_+(x)h_0.$$

If we put $q := \mathbf{P}(\chi_-^0 = 0)$ then h_0, being the average time spent by the random walk $\{H_k\}$ at the point 0, equals

$$h_0 = \sum_{k=0}^{\infty} q^k = \frac{1}{1-q}.$$

The theorem is proved. □

Now consider the asymptotics of $\mathbf{P}(\chi_-^0 < -x)$ as $x \to \infty$.

Put $F_-(t) := \mathbf{F}((-\infty,-t)) = \mathbf{P}(\xi < -t)$.

Theorem 12.7.2 *Let* $\mathbf{E}\xi < 0$.

1. *If* $F_- \in \mathcal{L}$ *then, as* $x \to \infty$,

$$\mathbf{P}\big(\chi_-^0 < -x\big) \sim \frac{F_-(x)}{1-p}.$$

2. *If* $F_-(t) = e^{-\gamma t}V(t)$, $V(t) \in \mathcal{L}$, *then*

$$\mathbf{P}(\chi_-^0 < -x) \sim \frac{\mathbf{E}e^{-\gamma S} F_-(x)}{1-p}.$$

3. *If* $F_- \in \mathcal{E}^*$ *then*

$$\mathbf{P}(\chi_-^0 < -x) \sim \frac{F_-(x)\mathbf{P}(S=0)}{1-p}.$$

Proof Making use of identity (12.3.3):

$$1 - \varphi_-^0(\lambda) = \frac{(1-\varphi(\lambda))\mathbf{E}\, e^{i\lambda S}}{1-p}, \qquad \varphi_-^0(\lambda) = \mathbf{E}e^{i\lambda \chi_-^0}.$$

This implies that $\mathbf{P}(\chi_-^0 < -x)$ is the weighted mean of the value $F_-(x+t)$ with the weight function $\mathbf{P}(S \in dt)/(1-p)$:

$$\mathbf{P}(\chi_-^0 < -x) = \frac{1}{1-p}\int_0^\infty \mathbf{P}(S \in dt)\, F_-(x+t).$$

From here the assertions of the theorem follow in an obvious way. □

If $\mathbf{E}\xi = 0$ then the asymptotics of $\mathbf{P}(\chi_-^0 < -x)$ will be different.

12.7.2 The Asymptotics of $\mathbf{P}(S > x)$

We will study the asymptotics of $\mathbf{P}(S > x)$ in the two non-overlapping and mutually complementary cases where $F_+ \in \mathcal{E}$ (the Cramér condition holds) and where F_+^I belongs to the class $\mathcal{S}$ of subexponential functions.

Definition 12.7.2 A distribution $\mathbf{G}$ on $[0,\infty)$ with the tail $G(t) := \mathbf{G}([t,\infty))$ belongs to the class $\mathcal{S}_+$ of *subexponential distributions on the positive half-line* if

$$G^{2*}(t) \sim 2G(t) \quad \text{as } t \to \infty. \tag{12.7.10}$$

A distribution $\mathbf{G}$ on the whole real line belongs to the class $\mathcal{S}$ of *subexponential distributions* if the distribution $\mathbf{G}^+$ of the positive part $\zeta^+ = \max\{0,\zeta\}$ of a random variable $\zeta \Subset \mathbf{G}$ belongs to $\mathcal{S}_+$. A random variable is called *subexponential* if its distribution is subexponential.

As we will see later (see Theorem A6.4.3 in Appendix 6), the subexponentiality distribution $\mathbf{G}$ is in essence a property of the asymptotics of the *tail* of $G(t)$ as $t \to \infty$. Therefore we can also talk about *subexponential functions*. A nonincreasing function $G_1(t)$ on $(0,\infty)$ is called *subexponential* if the distribution $\mathbf{G}$ with a tail $G(t)$ such that $G(t) \sim cG_1(t)$ as $t \to \infty$ for some $c > 0$ is subexponential. (For example, distributions with tails $G_1(t)/G_1(0)$ or $\min(1, G_1(t))$ if $G_1(0) > 1$.)

The properties of subexponential distributions are studied in Appendix 6. In particular, it is established that $\mathcal{S} \subset \mathcal{L}$, $\mathcal{R} \subset \mathcal{S}$ ($\mathcal{R}$ is the class of regularly varying functions) and that $G(t) = o(G^I(t))$ if $G^I \in \mathcal{S}$.

Theorem 12.7.3 *If $F_+^I(t) \in \mathcal{S}$ and $a = \mathbf{E}\xi < 0$, then, as $x \to \infty$,*

$$\mathbf{P}(S > x) \sim \frac{1}{|a|} F_+^I(x). \tag{12.7.11}$$

Proof Making use of the identity from Theorem 12.3.2:

$$\mathbf{E}e^{i\lambda S} = \frac{1-p}{1-p\varphi_\chi(\lambda)}, \qquad \varphi_\chi(\lambda) := \mathbf{E}e^{i\lambda\chi}, \tag{12.7.12}$$

it follows that

$$\mathbf{E}e^{i\lambda S} = (1-p)\sum_{k=0}^{\infty} p^k \varphi_\chi^k(\lambda),$$

and hence, for $x > 0$,

$$\mathbf{P}(S > x) = (1-p)\sum_{k=1}^{\infty} p^k \mathbf{P}(H_k > x), \qquad H_k := \sum_{j=1}^{k} \chi_j, \tag{12.7.13}$$

where χ_j are independent copies of χ. By assertion 1 of Theorem 12.7.1 the distribution of χ is subexponential, while by Theorem A6.4.3 of Appendix 6, as $x \to \infty$, for each fixed k one has

$$\mathbf{P}(H_k > x) \sim k\mathbf{P}(\chi > x). \tag{12.7.14}$$

Moreover, again by Theorem A6.4.3 of Appendix 6, for any $\varepsilon > 0$, there exists a $b = b(\varepsilon)$ such that, for all x and $k \ge 2$,

$$\frac{\mathbf{P}(H_k > x)}{\mathbf{P}(\chi > x)} < b(1+\varepsilon)^k.$$

Therefore, for $(1+\varepsilon)p < 1$, the series

$$\sum_{k=1}^{\infty} p^k \frac{\mathbf{P}(H_k > x)}{\mathbf{P}(\chi > x)}$$

converges uniformly in x. Passing to the limit as $x \to \infty$, by virtue of (12.7.14) we obtain that

$$\lim_{x\to\infty} \frac{\mathbf{P}(S > x)}{\mathbf{P}(\chi > x)} = (1-p)\sum_{k=1}^{\infty} kp^k = \frac{p}{1-p}$$

or, which is the same, that

$$\mathbf{P}(S > x) \sim \frac{p\mathbf{P}(\chi > x)}{1-p} \quad \text{as } x \to \infty,$$

where, by Theorem 12.7.1,

$$\mathbf{P}(\chi > x) \sim -\frac{F_+^I(x)}{p\mathbf{E}\chi_-^0}.$$

Since, by Corollary 12.2.3,

$$(1-p)\mathbf{E}\chi_-^0 = \mathbf{E}\xi,$$

we obtain (12.7.11). The theorem is proved. □

Now consider the case when $\mathbf{F}$ satisfies the right-hand side Cramér condition ($\mathbf{F}_+ \in \mathcal{E}$). For definiteness, we will again assume that the distribution $\mathbf{F}$ is non-lattice. Furthermore, we will assume that there exists an $\mu_1 > 0$ such that

$$\psi(\mu_1) := \mathbf{E}e^{\mu_1\xi} = 1, \qquad b := \mathbf{E}\xi e^{\mu_1\xi} = \psi'(\mu_1) < \infty. \tag{12.7.15}$$

In this case the Cramér transform of the distribution of $\mathbf{F}$ at the point μ_1 will be of the form

$$\mathbf{F}_{(\mu_1)}(dt) = \frac{e^{\mu_1 t}\mathbf{F}(dt)}{\psi(\mu_1)} = e^{\mu_1 t}\mathbf{F}(dt). \tag{12.7.16}$$

A random variable $\xi_{(\mu_1)}$ with the distribution $\mathbf{F}_{(\mu_1)}$ has, by (12.7.15), a finite expectation equal to b. Denote the size of the first overshoot of the level x by a random walk with jumps $\xi_{(\mu_1)}$ by $\chi_{(\mu_1)}(x)$. By Corollary 10.4.1, the distribution of $\chi_{(\mu_1)}(x)$ converges, as $x \to \infty$, to the limiting distribution: $\chi_{(\mu_1)}(x) \Rightarrow \chi_{(\mu_1)}$, so that

$$\mathbf{E}e^{-\mu_1\chi_{(\mu_1)}(x)} \to \mathbf{E}e^{-\mu_1\chi_{(\mu_1)}}. \tag{12.7.17}$$

Theorem 12.7.4 *Let* $\mathbf{F}_+ \in \mathcal{E}$ *and* (12.7.15) *be satisfied. Then, as* $x \to \infty$,

$$\mathbf{P}(S > x) \sim ce^{-\mu_1 x}, \tag{12.7.18}$$

where $c = \mathbf{E}e^{-\mu_1\chi_{(\mu_1)}} < 1$.

There is a somewhat different interpretation of the constant c in Remark 15.2.3. Exact upper and lower bounds for $e^{\mu_1 x}\mathbf{P}(S > x)$ are contained in Theorem 15.3.5.

Note that the finiteness of $\mathbf{E}\xi < 0$ is not assumed in Theorem 12.7.4. In the arithmetic case, we have to consider only integer x.

Proof Put $\eta(x) := \min\{n \ge 1 : S_n > x\}$, $X_n := x_1 + \cdots + x_n$ and $\overline{X}_n := \max_{k \le n} X_k$. Then

$$\mathbf{P}(S > x) = \mathbf{P}\big(\eta(x) < \infty\big) = \sum_{n=1}^{\infty} \mathbf{P}\big(\eta(x) = n\big), \tag{12.7.19}$$

where

$$\begin{aligned}
\mathbf{P}\big(\eta(x) = n\big) &= \underbrace{\int \cdots \int}_{n} \mathbf{F}(dx_1) \ldots \mathbf{F}(dx_n) \mathrm{I}(\overline{X}_{n-1} \le x, X_n > x) \\
&= \underbrace{\int \cdots \int}_{n} \mathbf{F}_{(\mu_1)}(dx_1) \ldots \mathbf{F}_{(\mu_1)}(dx_n) e^{-\mu_1 X_n} \mathrm{I}(\overline{X}_{n-1} \le x, X_n > x) \\
&= \mathbf{E}_{(\mu_1)} e^{-\mu_1 S_n} \mathrm{I}\big(\eta(x) = n\big).
\end{aligned} \tag{12.7.20}$$

Here $\mathbf{E}_{(\mu_1)}$ denotes the expectation when taken assuming that the distribution of the summands ξ_i is $\mathbf{F}_{(\mu_1)}$. By the convexity of the function $\psi(\mu)=\mathbf{E}e^{\mu\xi}$,

$$\mathbf{E}_{(\mu_1)}\xi=\int xe^{\mu_1 x}\mathbf{F}(dx)=\psi'(\mu_1)=b>0,$$

and hence

$$\mathbf{P}_{(\mu_1)}\bigl(\eta(x)<\infty\bigr)=1.$$

Therefore, returning to (12.7.19), we obtain

$$\mathbf{P}(S>x)=\mathbf{E}_{(\mu_1)}\sum_{k=1}^{\infty}e^{-\mu_1 S_n}\,\mathrm{I}\bigl(\eta(x)=n\bigr)=\mathbf{E}_{(\mu_1)}e^{-\mu_1 S_{\eta(x)}}, \qquad (12.7.21)$$

where $S_{\eta(x)}=x+\chi_{(\mu_1)}(x)$ and, by (12.7.17),

$$e^{\mu_1 x}\mathbf{P}(S>x)\to c=\mathbf{E}e^{-\mu_1\chi_{(\mu_1)}}<1.$$

This proves (12.7.18). For arithmetic ξ the proof is the same. We only have to replace $\mathbf{F}(dt)$ in (12.7.15) and (12.7.16) by $p_k=\mathbf{P}(\xi=k)$, as well as integration by summation. The theorem is proved. □

Corollary 12.7.1 *If, in the arithmetic case,* $\mathbf{E}\xi<0$, $p_1=\mathbf{P}(\xi=1)>0$, $\mathbf{P}(\xi\ge 2)=0$ *then the conditions of Theorem* 12.7.4 *are satisfied and one has*

$$\mathbf{P}(S>x)=e^{-\mu_1(k+1)},\quad k\ge 0.$$

Proof The proof follows immediately from (12.7.21) if we note that, in the case under consideration, $\chi_{(\mu_1)}(x)\equiv 1$ and $S_{\eta(x)}=x+1$. This assertion repeats the result of Example 12.6.1. □

Remark 12.7.1 The asymptotics (12.7.18), obtained by a probabilistic argument, admits a simple analytic interpretation. From (12.7.18) it follows that, as $\mu\uparrow\mu_1$, we have

$$\mathbf{E}e^{\mu S}\sim\frac{c\mu_1}{\mu_1-\mu}.$$

But that $\mathbf{E}e^{\mu S}$ has precisely this form follows from identity (12.3.3):

$$\mathbf{E}e^{\mu S}=\frac{(1-p)(1-\mathbf{E}e^{\mu\chi_-^0})}{1-\psi(\mu)}.$$

Indeed, since, by assumption, $\psi(\mu)=\mathbf{E}e^{\mu\xi}$ is left-differentiable at the point μ_1 and

$$\psi(\mu)=1-b(\mu_1-\mu)+o\bigl((\mu_1-\mu)\bigr), \qquad (12.7.22)$$

one has

$$\mathbf{E}e^{\mu S}\sim\frac{(1-p)(1-\mathbf{E}e^{\mu_1\chi_-^0})}{b(\mu_1-\mu)} \qquad (12.7.23)$$

as $\mu \uparrow \mu_1$. This implies, in particular, yet another representation for the constant c in (12.7.18):

$$c = \frac{(1-p)(1-\mathbf{E}e^{\mu_1 \chi_-^0})}{b}.$$

Since

$$\mathbf{E}e^{\mu S} = \frac{\mathfrak{w}_+(0)}{\mathfrak{w}_+(\lambda)}$$

and $\mathfrak{w}_+(\lambda)$ has a zero at the point μ_1, we can obtain representations similar to (12.7.22) and (12.7.23) in terms of the values of $\mathfrak{w}_+(0)$ and $\mathfrak{w}'(\mu_1)$.

We should also note that the proof of asymptotics (12.7.18) with the help of relations of the form (12.7.23) is based on certain facts from mathematical analysis and is relatively simple only under the additional condition (12.5.1).

There are other ways to prove (12.7.18), but they also involve additional restrictions. For instance, (12.3.3) implies

$$\mathbf{E}e^{i\lambda S} = (1-p)\sum_{k=0}^{\infty} \bigl[\varphi^k(\lambda) - \varphi^k(\lambda)\mathbf{E}\,e^{i\lambda\chi_-^0}\bigr],$$

$$\begin{aligned}
\mathbf{P}(S > x) &= (1-p)\sum_{k=0}^{\infty} \bigl[\mathbf{P}(S_k > x) - \mathbf{P}\bigl(S_k + \chi_-^0 > x\bigr)\bigr] \\
&= (1-p)\int_0^{\infty} \mathbf{P}\bigl(\chi_-^0 \in dt\bigr)\sum_{k=0}^{\infty} \mathbf{P}\bigl(S_k \in (x, x+t]\bigr),
\end{aligned}$$

and the problem now reduces to integro-local theorems for large deviations of S_k (see Chap. 9) or to local theorems for the renewal function in the region where the function converges to zero.

12.7.3 The Distribution of the Maximal Values of Generalised Renewal Processes

Let $\{(\tau_i, \zeta_i)\}_{j=1}^{\infty}$ be a sequence of independent identically distributed random vectors,

$$Z(t) = Z_{\nu(t)},$$

where

$$Z_n := \sum_{j=1}^{n} \zeta_j, \qquad \nu(t) := \max\{k : T_k \le t\}, \quad T_k := \sum_{j=1}^{k} \tau_j.$$

In Sect. 12.4.3 we reduced the problem of finding the distribution of $\sup_t (Z(t) - qt)$ to that of the distribution of $S := \sup_{k \ge 0} S_k$, $S_k := \sum_{j=1}^{k} \xi_j$, $\xi_j := \zeta_j - q\tau_j$ in the case $q > 0$, $\zeta_k \ge 0$. We show that such a reduction takes place in the general case as well. If $q \ge 0$ and the ζ_k can take values of both signs, then the reduction is the same as in Sect. 12.4.3. Now if $q < 0$ then

$$\sup_t (Z_{\nu(t)} - qt) = \sup(-qT_1, Z_1 - qT_2, Z_2 - qT_3, \ldots)$$
$$= -q\tau_1 + \sup_{k \ge 1} \big[Z_{k-1} - q(T_k - \tau_1)\big] \stackrel{d}{=} S - q\tau,$$

where the random variables τ_1 and S are independent.

12.8 On the Distribution of the First Passage Time

12.8.1 The Properties of the Distributions of the Times $\eta_\pm$

In this section we will establish a number of relations between the random variables $\eta_\pm$ and the time θ when the global maximum $S = \sup S_k$ is attained for the first time:

$$\theta := \min\{k : S_k = S\} \quad (\text{if } S < \infty \text{ a.s.}).$$

Put

$$P(z) := \sum_{k=0}^{\infty} z^k \mathbf{P}(\eta_-^0 > k), \qquad q(z) := \mathbf{E}\big(z^{\eta_+} \big| \eta_+ < \infty\big),$$
$$D_+ := \sum_{k=1}^{\infty} \frac{\mathbf{P}(S_k > 0)}{k}.$$

Further, let η be a random variable with the distribution

$$\mathbf{P}(\eta = k) = \mathbf{P}(\eta_+ = k \mid \eta_+ < \infty)$$

(and the generating function $q(z)$), $\eta_1, \eta_2, \ldots$ be independent copies of η,

$$H_k := \eta_1 + \cdots + \eta_k, \qquad H_0 = 0,$$

and ν be a random variable independent of $\{\eta_k\}$ with the geometric distribution $\mathbf{P}(\nu = k) = (1-p)p^k$, $k \ge 0$.

Theorem 12.8.1 *If $p = \mathbf{P}(\eta_+ < \infty) < 1$ then*

1.
$$1 - p = \frac{1}{\mathbf{E}\eta_-^0} = e^{-D_+}. \tag{12.8.1}$$

2.
$$P(z) = \frac{1}{1 - pq(z)} = \frac{\mathbf{E}z^\theta}{1-p}. \tag{12.8.2}$$

3.
$$\mathbf{P}(\eta_-^0 > n) = (1-p)\mathbf{P}(H_\nu = n) > \mathbf{P}(\eta_+ = n) \tag{12.8.3}$$

for all $n \ge 0$.

Recall that, for the condition $p < 1$ to hold, it is sufficient that $\mathbf{E}\xi < 0$ (see Corollary 12.2.6).

The second assertion of the theorem implies that the distributions of η_-^0, η_+ and θ uniquely determine each other, so that if at least one of them is known then, to find the other two, it is not necessary to know the original distribution $\mathbf{F}$. In particular, $\mathbf{P}(\theta = n) = (1-p)\mathbf{P}(\chi_-^0 > n)$.

Proof of Theorem 12.8.1 The arguments in this subsection are based on the following identities which follow from Theorems 12.1.1–12.1.3 if we put there $\lambda = 0$ and $|z| < 1$:

$$1 - z = \bigl[1 - \mathbf{E}z^{\eta_-^0}\bigr]\bigl[1 - \mathbf{E}\bigl(z^{\eta_+};\ \eta_+ < \infty\bigr)\bigr], \tag{12.8.4}$$

$$1 - \mathbf{E}z^{\eta_-^0} = \exp\Biggl\{-\sum_{k=1}^{\infty} \frac{z^k}{k}\mathbf{P}(S_k \le 0)\Biggr\}, \tag{12.8.5}$$

$$1 - \mathbf{E}\bigl(z^{\eta_+};\ \eta_+ < \infty\bigr) = \exp\Biggl\{-\sum_{k=1}^{\infty} \frac{z^k}{k}\mathbf{P}(S_k > 0)\Biggr\}. \tag{12.8.6}$$

Since

$$\frac{1 - \mathbf{E}z^{\eta_-^0}}{1 - z} = P(z), \qquad P(1) = \mathbf{E}\eta_-^0$$

we obtain from (12.8.4) the first equalities in (12.8.1) and (12.8.2). The second equality in (12.8.1) follows from (12.8.6).

To prove the second equality in (12.8.2), we make use of the relation

$$\theta = \begin{cases} 0 & \text{on } \{\omega : \eta_+ = \infty\}, \\ \eta_+ + \theta^* & \text{on } \{\omega : \eta_+ < \infty\}, \end{cases}$$

where θ^* is distributed on $\{\eta_+ < \infty\}$ identically to θ and does not depend on η_+. It follows that

$$\mathbf{E}z^{\theta} = (1-p) + \mathbf{E}z^{\theta}\mathbf{E}\bigl(z^{\eta_+};\ \eta_+ < \infty\bigr).$$

This implies the second equality in (12.8.2). The last assertion of the theorem follows from the first equality in (12.8.2), which implies

$$\begin{aligned} P(z) &= \sum_{k=0}^{\infty} p^k q^k(z) = (1-p)\sum_{k=0}^{\infty} \mathbf{P}(\nu = k) \sum_{n=0}^{\infty} \mathbf{P}(H_k = n) z^n \\ &= (1-p)\sum_{n=0}^{\infty} z^n \mathbf{P}(H_\nu = n). \end{aligned}$$

The theorem is proved. □

The second equality in (12.8.2) and identity (12.7.12) mean that the representations

$$\theta = \eta_1 + \cdots + \eta_\nu \quad \text{and} \quad S = \chi_1 + \cdots + \chi_\nu,$$

respectively, hold true, where ν has the geometric distribution $\mathbf{P}(\nu = k) = (1-p)p^k$, $k \ge 0$, and does not depend on $\{\eta_j\}$, $\{\chi_j\}$.

Note that the probabilities $\mathbf{P}(S_k > 0) = \mathbf{P}(S_k - ak > -ak)$ on the right-hand sides of (12.8.5) and (12.8.6) are, for large k and $a = \mathbf{E}\xi < 0$, the probabilities of large deviations that were studied in Chap. 9. The results of that chapter on the asymptotics of these probabilities together with relations (12.8.5) and (12.8.6) give us an opportunity to find the asymptotics of $\mathbf{P}(\eta_+ = n)$ and $\mathbf{P}(\eta_-^0 = n)$ as $n \to \infty$ (see [8]).

Now consider the case where the both random variables η_-^0 and η_+ are proper. That is always the case if $\mathbf{E}\xi = 0$ (see Corollary 12.2.6). Here identities (12.8.4)–(12.8.6) hold true (with $\mathbf{P}(\eta_+ < \infty) = 1$). As before, (12.8.4) implies that the distributions of η_-^0 and η_+ uniquely determine each other.

Let $\eta_1, \eta_2, \ldots$ be independent copies of η_+, $H_k = \eta_1 + \cdots + \eta_k$ and $H_0 = 0$. For the sums H_k, define the local renewal function

$$h_n := \sum_{n=0}^{\infty} \mathbf{P}(H_k = n).$$

Theorem 12.8.2 *If* $\mathbf{P}(\eta_-^0 < \infty) = \mathbf{P}(\eta_+ < \infty) = 1$ *then*:

1. $\mathbf{E}\eta_-^0 = \mathbf{E}\eta_+ = \infty$.
2. $\mathbf{P}(\eta_-^0 > n) = h_n$.

Proof From (12.8.4) it follows that

$$P(z) = \frac{1 - \mathbf{E}z^{\eta_-^0}}{1 - z} = \frac{1}{1 - \mathbf{E}z^{\eta_+}} \to \infty \tag{12.8.7}$$

as $z \to 1$. Since $P(z) \to \mathbf{E}\eta_-^0$ as $z \to 1$, we have proved that $\mathbf{E}\eta_-^0$ is infinite. That $\mathbf{E}\eta_+$ is also infinite is shown in the same way. The second assertion also follows from (12.8.7) since the right-hand side of (12.8.7) is $\sum_{n=0}^{\infty} z^n h_n$. The theorem is proved. □

Now we turn to the important class of symmetric distributions. We will say that the *distribution of a random variable* ξ *is symmetric* if it coincides with the distribution of $-\xi$, and will call the *distribution of* ξ *continuous* if the distribution function of ξ is continuous. For such random variables, $\mathbf{E}\xi = 0$ (if $\mathbf{E}\xi$ exists), the distributions of S_n are also symmetric continuous for all n, and

$$\mathbf{P}(S_n > 0) = \mathbf{P}(S_n < 0) = \frac{1}{2}, \qquad \mathbf{P}(S_n = 0) = 0,$$

and hence $D(z) \equiv 1$, $\mathbf{P}(\chi_+^0 = 0) = 0$, and $\eta_+ = \eta_+^0$, $\chi_+ = \chi_+^0$ with probability 1.

Theorem 12.8.3 *If the distribution of* ξ *is symmetric and continuous then*

$$\begin{aligned} \mathbf{P}(\eta_+ = n) = \mathbf{P}\big(\eta_-^0 = n\big) &= \frac{(2n)!}{(2n-1)(n!)^2 2^{2n}} \sim \frac{1}{2\sqrt{\pi}\, n^{3/2}}, \\ \mathbf{P}(\gamma_n > 0) = \mathbf{P}(\zeta_n < 0) &\sim \frac{1}{\sqrt{\pi n}} \end{aligned} \tag{12.8.8}$$

as $n \to \infty$ *(*γ_n *and* ζ_n *are defined in Section* 12.1.3*).*

Proof Since $\mathbf{E}z^{\eta_-^0} = \mathbf{E}z^{\eta_+}$, by virtue of (12.8.4) one has

$$1 - \mathbf{E}z^{\eta_+} = \sqrt{1-z}.$$

Expanding $\sqrt{1-z}$ into a series, we obtain the second equality in (12.8.8). The asymptotic equivalence follows from Stirling's formula.

The second assertion of the theorem follows from the first one and the equality

$$\mathbf{P}(\zeta_n < 0) = \sum_{k=n+1}^{\infty} \mathbf{P}(\eta_+ = k).$$

The assertions concerning η_-^0 and γ_n follow by symmetry.

The theorem is proved. □

Note that, under the conditions of Theorem 12.8.3, the distributions of the variables η_+, η_-, γ_n, ζ_n do not depend on the distribution of ξ. Also note that the asymptotics

$$\mathbf{P}(\eta_+ = n) \sim \frac{1}{2\sqrt{\pi}\, n^{3/2}}$$

persists in the case of non-symmetric distributions as well provided that $\mathbf{E}\xi = 0$ and $\mathbf{E}\xi^2 < \infty$ (see [8]).

12.8.2 The Distribution of the First Passage Time of an Arbitrary Level x by Arithmetic Skip-Free Walks

The main object in this section is the time

$$\eta(x) = \min\{k : S_k \ge x\}$$

of the first passage of the level x by the random walk $\{S_k\}$. Below we will consider the class of arithmetic random walks for which $\chi_+ \equiv 1$.

By an *arithmetic skip-free walk* we will call a sequence $\{S_k\}_{k=0}^{\infty}$, where the distribution of ξ is arithmetic and $\max_\omega \xi(\omega) = 1$ (i.e. $p_1 > 0$ and $p_k = 0$ for $k \ge 2$, where $p_k = \mathbf{P}(\xi = k)$). The term "skip-free walk" appears due to the fact that the walk $\{S_k\}$, $k = 0, 1, \ldots$, cannot skip any integer level $x > 0$: if $S_n > x$ then necessarily there is a $k < n$ such that $S_k = x$.

As we already know from Example 12.6.1, for skip-free walks with $\mathbf{E}\xi < 0$ the distribution of S is geometric:

$$\mathbf{P}(S = k) = (1-p)p^k, \quad k = 0, 1, \ldots,$$

where $p = \mathbf{P}(\eta_+ < \infty)$ and $z_1 = p^{-1}$ is the zero of the function $1 - p(z)$ with $p(z) = \sum_k p_k z^k$.

It turns out that one can find many other explicit formulas for skip-free walks. In this section we will be interested in the distribution of the maximum

$\overline{S}_n = \max(0, S_1, \ldots, S_n)$; as we already noted, knowing the distribution is important for many problems of mathematical statistics, queueing theory, etc. Note that finding the distribution of $\overline{S}_n$ is the same as finding the distribution of $\eta(x)$, since

$$\{\overline{S}_n < x\} = \{\eta(x) > n\}. \tag{12.8.9}$$

Here we put $\eta(x) := \infty$ if $S < x$.

The Pollaczek–Spitzer identity (see Theorem 12.3.1) provides the double transform of the distribution of $\overline{S}_n$. Analysing this identity shows that the distribution of $\overline{S}_n$ (or $\eta(x)$) itself typically cannot be expressed in terms of the distribution of ξ_k in explicit form. However, for discrete skip-free walks one has remarkable "duality" relations which we will now prove with the help of Pollaczek–Spitzer's identity.

Theorem 12.8.4 *If ξ is integer-valued then $\mathbf{P}(\xi_k \geq 2) = 0$ is a necessary and sufficient condition for*

$$n\mathbf{P}\big(\eta(x) = n\big) = x\mathbf{P}(S_n = x), \quad x \geq 1. \tag{12.8.10}$$

Using the Wald identity, it is also not hard to verify that if the expectation $\mathbf{E}\xi_1 = a > 0$ exists then the *walk $\{S_n\}$ will be skip-free if and only if* $\mathbf{E}\eta(x) = x/a$. (Note that the definition of $\eta(x)$ in this section somewhat differs from that in Chap. 10. One obtains it by changing x to $x+1$ on the right-hand side of the definition of $\eta(x)$ from Chap. 10.)

The asymptotics of the local probabilities $\mathbf{P}(S_n = x)$ was studied in Chap. 9 (see e.g., Theorem 9.3.4). This together with (12.8.10) enables us to find the asymptotics of $\mathbf{P}(\eta(x) = n)$.

Proof of Theorem 12.8.4 Set

$$r_x := \mathbf{P}\big(\eta(x) = \infty\big) = \mathbf{P}(S < x), \qquad q_{x,n} := \mathbf{P}\big(\eta(x) = n\big),$$

$$Q_{x,n} := \mathbf{P}\big(\eta(x) > n\big) = \sum_{k=n+1}^{\infty} q_{x,k} + r_x.$$

Since for each y, $0 \leq y \leq x$,

$$\big\{\eta(x) = n\big\} \subset \bigcup_{k=0}^{n} \big\{\eta(y) = k\big\},$$

using the fact that the walk is skip-free, by the total probability formula one has

$$q_{x,n} = \sum_{k=0}^{n} q_{y,k} q_{x-y,n-k},$$

where $q_{0,0} = 1$, and $q_{y,0} = 0$ for $y > 0$. Hence for $|z| \leq 1$ using convolution we have

$$q_x(z) := \sum_{k=0}^{\infty} q_{x,n} z^n = \mathbf{E}\big(z^{\eta(x)};\ \eta(x) < \infty\big) = q_y(z) q_{x-y}(z).$$

Putting $y=1$ and $q_0(z)=1$, we obtain

$$q_x(z)=q(z)q_{x-1}(z)=q^x(z),\quad x\ge 0.$$

From here one can find the generating function $Q_x(z)$ of the sequence $Q_{x,n}$:

$$\begin{aligned}Q_x(z):=\sum_{n=0}^{\infty}z^n\left(r_x+\sum_{k=n+1}^{\infty}q_{x,k}\right)&=\frac{r_x}{1-z}+\sum_{n=1}^{\infty}q_{x,n}\sum_{k=0}^{n-1}z^k\\ =\frac{r_x}{1-z}+\sum_{n=1}^{\infty}q_{x,n}\frac{1-z^n}{1-z}&=\frac{r_x}{1-z}+\frac{q_x(1)-q_x(z)}{1-z}=\frac{1-q_x(z)}{1-z}.\end{aligned}$$

Note that here the quantity $q_x(1)=\mathbf{P}(\eta(x)<\infty)=\mathbf{P}(S\ge x)$ can be less than 1. Using (12.8.9) we obtain that

$$\mathbf{P}(\overline{S}_n=x)=\mathbf{P}\big(\eta(x+1)>n\big)-\mathbf{P}\big(\eta(x)>n\big),$$

$$\sum_{n=0}^{\infty}z^n\mathbf{P}(\overline{S}_n=x)=\frac{(1-q^{x+1}(z))-(1-q^x(z))}{1-z}=\frac{q^x(z)(1-q(z))}{1-z}.$$

Finally, making use of the absolute summability of the series below, we find that, for $|v|<1$ and $|z|<1$,

$$\sum_{n=0}^{\infty}z^n\mathbf{E}v^{\overline{S}_n}=\sum_{x=0}^{\infty}v^x\sum_{n=0}^{\infty}z^n\mathbf{P}(\overline{S}_n=x)=\frac{1-q(z)}{(1-z)(1-vq(z))}.$$

Turning now to the Pollaczek–Spitzer formula, we can write that

$$\sum_{n=1}^{\infty}\frac{z^n}{n}\mathbf{E}v^{\max(0,S_n)}=\ln\frac{1-q(z)}{1-z}-\ln\big(1-vq(z)\big)=\ln\frac{1-q(z)}{1-z}+\sum_{x=1}^{\infty}\frac{(vq(z))^x}{x}.$$

Comparing the coefficients of v^x, $x\ge 1$, we obtain

$$\sum_{n=1}^{\infty}\frac{z^n}{n}\mathbf{P}(S_n=x)=\frac{q^x(z)}{x},\quad x\ge 1. \tag{12.8.11}$$

Taking into account that $q^x(z)=q_x(z)$ and comparing the coefficients of z^n, $n\ge 1$, in (12.8.11) we get

$$\frac{1}{n}\mathbf{P}(S_n=x)=\frac{1}{x}\mathbf{P}(\eta_x=n),\quad x\ge 1,\ n\ge 1.$$

Sufficiency is proved.

The necessity of the condition $\mathbf{P}(\xi\ge 2)=0$ follows from equality (12.8.10) for $x=n=1$:

$$p_1=q_{1,1}=\sum_{k=1}^{\infty}p_k,\qquad \sum_{k=2}^{\infty}p_k=\mathbf{P}(\xi\ge 2)=0.$$

The theorem is proved. □

Using the obtained formulas one can, for instance, find in Example 4.2.3 the distribution of the time to ruin in a game with an infinitely rich adversary (the total capital being infinite). If the initial capital of the first player is x then, for the time $\eta(x)$ of his ruin, we obtain

$$\mathbf{P}\big(\eta(x)=n\big)=\frac{x}{n}\mathbf{P}(S_n=x),$$

where

$$S_n=\sum_{j=1}^{n}\xi_j; \qquad \mathbf{P}(\xi_j=1)=q, \qquad \mathbf{P}(\xi_j=-1)=p$$

(p is the probability for the first player to win in a single play). Therefore, if n and x are both either odd or even then

$$\mathbf{P}\big(\eta(x)=n\big)=\frac{x}{n}\binom{n}{(n-x)/2}q^{(n+x)/2}p^{(n-x)/2}, \tag{12.8.12}$$

and $\mathbf{P}(\eta(x)=n)=0$ otherwise.

It is interesting to ask how fast $\mathbf{P}(\eta(x)>n)$ decreases as n grows in the case when the player will be ruined with probability 1, i.e. when $\mathbf{P}(\eta(x)<\infty)=1$. As we already know, this happens if and only if $p\le q$. (The assertion also follows from the results of Sect. 13.3.)

Applying Stirling's formula, as was done when proving the local limit theorem for the Bernoulli scheme, it is not difficult to obtain from (12.8.12) that, for each fixed x, as $n\to\infty$ (n and x having the same parity), for $p\le q$,

$$\mathbf{P}\big(\eta(x)=n\big)\sim\frac{x}{n^{3/2}}\sqrt{\frac{2}{\pi}}(4pq)^{n/2}\left(\frac{q}{p}\right)^{x/2};$$

$$\mathbf{P}\big(\eta(x)\ge n\big)\sim\frac{x}{n^{3/2}(p-q)^2}\sqrt{\frac{2}{\pi}}(4pq)^{n/2}\left(\frac{q}{p}\right)^{x/2} \quad \text{for } p<q$$

and

$$\mathbf{P}\big(\eta(x)\ge n\big)\sim x\sqrt{\frac{2}{\pi n}} \quad \text{for } p=q.$$

The last relation allowed us, under the conditions of Sect. 8.8, to obtain the limiting distribution for the number of intersections of the trajectory $S_1,\dots,S_n$ with the strip $[u,v]$ (see (8.8.24)). Up to the normalising constants, this assertion also remains true for arbitrary random walks such that $\mathbf{E}\xi_k=0$ and $\mathbf{E}\xi_k^2<\infty$. However, even in the case of a skip-free walk, the proof of this assertion requires additional efforts, despite the fact that, for such walks, an upward intersection of the line $x=0$ by the trajectory $\{S_n\}$ divides the trajectory, as in Sect. 8.8, into independent identically distributed cycles.

Chapter 13
Sequences of Dependent Trials. Markov Chains

Abstract The chapter opens with in Sect. 13.1 presenting the key definitions and first examples of countable Markov chains. The section also contains the classification of states of the chain. Section 13.2 contains necessary and sufficient conditions for recurrence of states, the Solidarity Theorem for irreducible Markov chains and a theorem on the structure of a periodic Markov chain. Key theorems on random walks on lattices are presented in Sect. 13.3, along with those for a general symmetric random walk on the real line. The ergodic theorem for general countable homogeneous chains is established in Sect. 13.4, along with its special case for finite Markov chains and the Law of Large Numbers and the Central Limit Theorem for the number of visits to a given state. This is followed by a short Sect. 13.5 detailing the behaviour of transition probabilities for reducible chains. The last three sections are devoted to Markov chains with arbitrary state spaces. First the ergodicity of such chains possessing a positive atom is proved in Sect. 13.6, then the concept of Harris Markov chains is introduced and conditions of ergodicity of such chains are established in Sect. 13.7. Finally, the Laws of Large Numbers and the Central Limit Theorem for sums of random variables defined on a Markov chain are obtained in Sect. 13.8.

13.1 Countable Markov Chains. Definitions and Examples. Classification of States

13.1.1 Definition and Examples

So far we have studied sequences of independent trials. Now we will consider the simplest variant of a sequence of *dependent* trials.

Let G be an experiment having a finite or countable set of outcomes $\{E_1, E_2, \dots\}$. Suppose we keep repeating the experiment G. Denote by X_n the number of the outcome of the n-th experiment.

In general, the probabilities of different values of E_{X_n} can depend on what events occurred in the previous $n-1$ trials. If this probability, *given a fixed outcome* $E_{X_{n-1}}$ *of the* $(n-1)$*-st trial*, does not depend on the outcomes of the preceding $n-2$ trials, then one says that this sequence of trials forms a Markov chain.

A.A. Borovkov, *Probability Theory*, Universitext,
DOI 10.1007/978-1-4471-5201-9_13,

To give a precise definition of a Markov chain, consider a sequence of integer-valued random variables $\{X_n\}_{n=0}^{\infty}$. If the n-th trial resulted in outcome E_j, we set $X_n := j$.

Definition 13.1.1 A sequence $\{X_n\}_0^{\infty}$ forms a *Markov chain* if

$$\mathbf{P}(X_n = j | X_0 = k_0, X_1 = k_1, \ldots, X_{n-2} = k_{n-2}, X_{n-1} = i)$$
$$\mathbf{P}(X_n = j | X_{n-1} = i) =: p_{ij}^{(n)}. \qquad (13.1.1)$$

These are the so-called countable (or discrete) Markov chains, i.e. Markov chains *with countable state spaces.*

Thus, a Markov chain may be thought of as a system with possible states $\{E_1, E_2, \ldots\}$. Some "initial" distribution of the variable X_0 is given:

$$\mathbf{P}(X_0 = j) = p_j^0, \quad \sum p_j^0 = 1.$$

Next, at integer time epochs the system changes its state, the conditional probability of being at state E_j at time n given the previous history of the system only being dependent on the state of the system at time $n-1$. One can briefly characterise this property as follows: *given the present, the future and the past of the sequence X_n are independent.*

For example, the branching process $\{\zeta_n\}$ described in Sect. 7.7, where ζ_n was the number of particles in the n-th generation, is a Markov chain with possible states $\{0, 1, 2, \ldots\}$.

In terms of conditional expectations or conditional probabilities (see Sect. 4.8), the *Markov property* (as we shall call property (13.1.1)) can also be written as

$$\mathbf{P}\big(X_n = j \,\big|\, \sigma(X_0, \ldots, X_{n-1})\big) = \mathbf{P}\big(X_n \,\big|\, \sigma(X_{n-1})\big),$$

where $\sigma(\cdot)$ is the σ-algebra generated by random variables appearing in the argument, or, which is the same,

$$\mathbf{P}(X_n = j \,|\, X_0, \ldots, X_{n-1}) = \mathbf{P}(X_n \mid X_{n-1}).$$

This definition allows immediate extension to the case of a Markov chain with a more general state space (see Sects. 13.6 and 13.7).

The problem of the existence of a sequence $\{X_n\}_0^{\infty}$ which is a Markov chain with given transition probabilities $p_{ij}^{(n)}$ ($p_{ij}^{(n)} \ge 0$, $\sum_j p_{ij}^{(n)} = 1$) and a given "initial" distribution $\{p_k^0\}$ of the variable X_0 can be solved in the same way as for independent random variables. It suffices to apply the Kolmogorov theorem (see Appendix 2) and specify consistent joint distributions by

$$\mathbf{P}(X_0 = k_0, X_1 = k_1, \ldots, X_n = k_n) := p_{k_0}^0 p_{k_0 k_1}^{(1)} p_{k_1 k_2}^{(2)} \cdots p_{k_{n-1} k_n}^{(n)},$$

which are easily seen to satisfy the Markov property (13.1.1).

Definition 13.1.2 A Markov chain $\{X_n\}_0^\infty$ is said to be *homogeneous* if the probabilities $p_{ij}^{(n)}$ do not depend on n.

We consider several examples.

Example 13.1.1 (Walks with absorption and reflection) Let $a > 1$ be an integer. Consider a walk of a particle over integers between 0 and a. If $0 < k < a$, then from the point k with probabilities $1/2$ the particle goes to $k-1$ or $k+1$. If k is equal to 0 or a, then the particle remains at the point k with probability 1. This is the so-called walk with *absorption*. If X_n is a random variable which is equal to the coordinate of the particle at time n, then the sequence $\{X_n\}$ forms a Markov chain, since the conditional expectation of the random variable X_n given $X_0, X_1, \dots, X_{n-1}$ depends only on the value of X_{n-1}. It is easy to see that this chain is homogeneous.

This walk can be used to describe a fair game (see Example 4.2.3) in the case when the total capital of both gamblers equals a. Reaching the point a means the ruin of the second gambler.

On the other hand, if the particle goes from the point 0 to the point 1 with probability 1, and from the point a to the point $a-1$ with probability 1, then we have a walk with *reflection*. It is clear that in this case the positions X_n of the particle also form a homogeneous Markov chain.

Example 13.1.2 Let $\{\xi_k\}_{k=0}^\infty$ be a sequence of independent integer-valued random variables and $d > 0$ be an integer. The random variables $X_n := \sum_{k=0}^n \xi_k \pmod d$ obtained by adding ξ_k modulo d ($X_n = \sum_{k=0}^n \xi_k - jd$, where j is such that $0 \le X_n < d$) form a Markov chain. Indeed, we have $X_n = X_{n-1} + \xi_n \pmod d$, and therefore the conditional distribution of X_n given $X_1, X_2, \dots, X_{n-1}$ depends only on X_{n-1}.

If, in addition, $\{\xi_k\}$ are identically distributed, then this chain is homogeneous.

Of course, all the aforesaid also holds when $d = \infty$, i.e. for the conventional summation. The only difference is that the set of possible states of the system is in this case infinite.

From the definition of a homogeneous Markov chain it follows that the probabilities $p_{ij}^{(n)}$ of transition from state E_i to state E_j on the n-th step do not depend on n. Denote these probabilities by p_{ij}. They form the *transition* matrix $P = \|p_{ij}\|$ with the properties

$$p_{ij} \ge 0, \quad \sum_j p_{ij} = 1.$$

The second property is a consequence of the fact that the system, upon leaving the state E_i, enters with probability 1 one of the states $E_1, E_2, \dots$.

Matrices with the above properties are said to be *stochastic*.

The matrix P *completely describes the law of change of the state of the system after one step*. Now consider the change of the state of the system after k steps. We introduce the notation $p_{ij}(k) := \mathbf{P}(X_k = j|X_0 = i)$. For $k > 1$, the total probability formula yields

$$p_{ij}(k) = \sum_s \mathbf{P}(X_{k-1} = s|X_0 = i)p_{sj} = \sum_s p_{is}(k-1)p_{sj}.$$

Summation here is carried out over all states. If we denote by $P(k) := \|p_{ij}(k)\|$ the matrix of transition probabilities $p_{ij}(k)$, then the above equality means that $P(k) = P(k-1)P$ or, which is the same, that $P(k) = P^k$. Thus the matrix P *uniquely determines transition probabilities for any number of steps*. It should be added here that, for a homogeneous chain,

$$\mathbf{P}(X_{n+k} = j|X_n = i) = \mathbf{P}(X_k = j|X_0 = i) = p_{ij}(k).$$

We see from the aforesaid that the "distribution" of a chain will be completely determined by the matrix P and the initial distribution $p_k^0 = \mathbf{P}(X_0 = k)$.

We leave it to the reader as an exercise to verify that, for an arbitrary $k \geq 1$ and sets $B_1, \ldots, B_{n-k}$,

$$\mathbf{P}(X_n = j|X_{n-k} = i;\, X_{n-k-1} \in B_1, \ldots, X_0 \in B_{n-k}) = p_{ij}(k).$$

To prove this relation one can first verify it for $k = 1$ and then make use of induction.

It is obvious that a sequence of independent integer-valued identically distributed random variables X_n forms a Markov chain with $p_{ij} = p_j = \mathbf{P}(X_n = j)$. Here one has $P(k) = P^k = P$.

13.1.2 Classification of States[1]

Definition 13.1.3

K1. A state E_i is called *inessential* if there exist a state E_j and an integer $t_0 > 0$ such that $p_{ij}(t_0) > 0$ and $p_{ji}(t) = 0$ for every integer t.

Otherwise the state E_i is called *essential*.

K2. Essential states E_i and E_j are called *communicating* if there exist such integers $t > 0$ and $s > 0$ that $p_{ij}(t) > 0$ and $p_{ji}(s) > 0$.

Example 13.1.3 Assume a system can be in one of the four states $\{E_1, E_2, E_2, E_4\}$ and has the transition matrix

$$P = \begin{pmatrix} 0 & 1/2 & 1/2 & 0 \\ 1/2 & 0 & 0 & 1/2 \\ 0 & 0 & 1/2 & 1/2 \\ 0 & 0 & 1/2 & 1/2 \end{pmatrix}.$$

[1] Here and in Sect. 12.2 we shall essentially follow the paper by A.N. Kolmogorov [23].

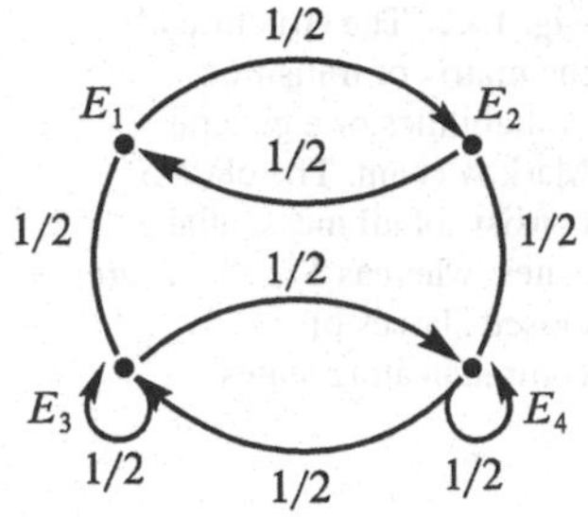

Fig. 13.1 Possible transitions and their probabilities in Example 13.1.3

In Fig. 13.1 the states are depicted by dots, transitions from state to state by arrows, numbers being the corresponding probabilities. In this chain, the states E_1 and E_2 are inessential while E_3 and E_4 are essential and communicating.

In the walk with absorption described in Example 13.1.1, the states $1, 2, \ldots, a-1$ are inessential. The states 0 and a are essential but non-communicating, and it is natural to call them *absorbing*. In the walk with reflection, all states are essential and communicating.

Let $\{X_n\}_{n=0}^{\infty}$ be a homogeneous Markov chain. We distinguish the class S^0 of all inessential states. Let E_i be an essential state. Denote by S_{E_i} the class of states comprising E_i and all states communicating with it. If $E_j \in S_{E_i}$, then E_j is essential and communicating with E_i, and $E_i \in S_{E_j}$. Hence $S_{E_i} = S_{E_j}$. Thus, the whole set of essential states can be decomposed into disjoint classes of communicating states which will be denoted by $S^1, S^2, \ldots$

Definition 13.1.4 If the class S_{E_i} consists of the single state E_i, then this state is called *absorbing*.

It is clear that after a system has hit an essential state E_i, it can never leave the class S_{E_i}.

Definition 13.1.5 A Markov chain consisting of a single class of essential communicating states is said to be *irreducible*. A Markov chain is called *reducible* if it contains more than one such class.

If we enumerate states so that the states from S^0 come first, next come states from S^1 and so on, then the matrix of transition probabilities will have the form shown in Fig. 13.2. Here the submatrices marked by zeros have only zero entries. The cross-hatched submatrices are stochastic.

Each such submatrix corresponds to some irreducible chain. If, at some time, the system is at a state of such an irreducible chain, then the system will never leave this chain in the future. Hence, to study the dynamics of an arbitrary Markov chain, it is sufficient to study the dynamics of irreducible chains. Therefore one of the basic objects of study in the theory of Markov chains is *irreducible Markov chains*. We will consider them now.

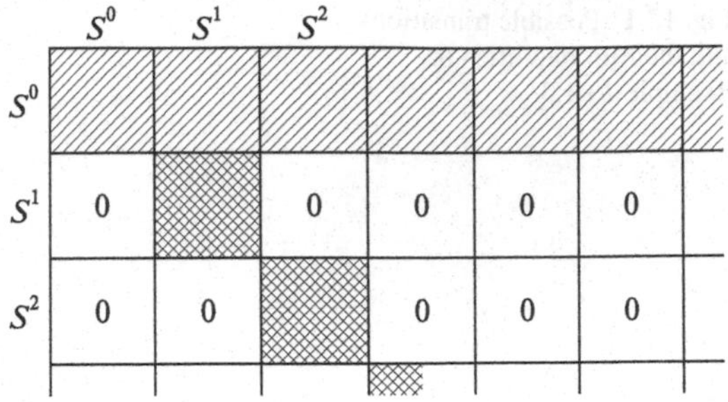

Fig. 13.2 The structure of the matrix of transition probabilities of a general Markov chain. The class S^0 consists of all inessential states, whereas $S^1, S^2, \dots$ are closed classes of communicating states

We introduce the following notation:

$$f_j(n) := \mathbf{P}(X_n = j, X_{n-1} \neq j, \dots, X_1 \neq j \mid X_0 = j), \qquad F_j := \sum_{n=1}^{\infty} f_j(n);$$

$f_j(n)$ is the probability that the system leaving the j-th state will return to it for the first time after n steps. The probability that the system leaving the j-th state will eventually return to it is equal to F_j.

Definition 13.1.6

K3. A state E_j is said to be *recurrent* (or *persistent*) if $F_j = 1$, and *transient* if $F_j < 1$.
K4. A state E_j is called *null* if $p_{jj}(n) \to 0$ as $n \to \infty$, and *positive* otherwise.
K5. A state E_j is called *periodic* with period d_j if the recurrence with this state has a positive probability only when the number of steps is a multiple of $d_j > 1$, and d_j is the maximum number having such property.

In other words, $d_j > 1$ is the greatest common divisor (g.c.d.) of the set of numbers $\{n : f_j(n) > 0\}$. Note that one can always choose from this set a finite subset $\{n_1, \dots, n_k\}$ such that d_j is the greatest common divisor of these numbers. It is also clear that $p_{jj}(n) = f_j(n) = 0$ if $n \neq 0 \pmod{d_j}$.

Example 13.1.4 Consider a walk of a particle over integer points on the real line defined as follows. The particle either takes one step to the right or remains on the spot with probabilities 1/2. Here $f_j(1) = 1/2$, and if $n > 1$ then $f_j(n) = 0$ for any point j. Therefore $F_j < 1$ and all the states are transient. It is easily seen that $p_{jj}(n) = 1/2^n \to 0$ as $n \to \infty$ and hence every state is null.

On the other hand, if the particle jumps to the right with probability 1/2 and with the same probability jumps to the left, then we have a chain with period 2, since recurrence to any particular state is only possible in an even number of steps.

13.2 Necessary and Sufficient Conditions for Recurrence of States. Types of States in an Irreducible Chain. The Structure of a Periodic Chain

Recall that the function

$$a(z) = \sum_{n=0}^{\infty} a_n z^n$$

is called the generating function of the sequence $\{a_n\}_{n=0}^{\infty}$. Here z is a complex variable. If the sequence $\{a_n\}$ is bounded, then this series converges for $|z| < 1$.

Theorem 13.2.1 *A state E_j is recurrent if and only if $P_j = \sum_{n=1}^{\infty} p_{jj}(n) = \infty$. For a transient E_j,*

$$F_j = \frac{P_j}{1 + P_j}. \tag{13.2.1}$$

The assertion of this theorem is a kind of expansion of the Borel–Cantelli lemma to the case of dependent events $A_n = \{X_n = j\}$. With probability 1 there occur infinitely many events A_n if and only if

$$\sum_{n=1}^{\infty} \mathbf{P}(A_n) = P_j = \infty.$$

Proof By the total probability formula we have

$$p_{jj}(n) = f_j(1)p_{jj}(n-1) + f_j(2)p_{jj}(n-2) + \cdots + f_j(n-1)p_{jj}(1) + f_j(n) \cdot 1.$$

Introduce the generating functions of the sequences $\{p_{jj}(n)\}_{n=0}^{\infty}$ and $\{f_j(n)\}_{n=0}^{\infty}$:

$$P_j(z) := \sum_{n=1}^{\infty} p_{jj}(n) z^n, \qquad F_j(z) := \sum_{n=1}^{\infty} f_j(n) z^n.$$

Both series converge inside the unit circle and represent analytic functions. The above formula for $p_{jj}(n)$, after multiplying both sides by z^n and summing up over n, leads (by the rule of convolution) to the equality

$$P_j(z) = z f_1(1)\bigl(1 + P_j(z)\bigr) + z^2 f_1(2)\bigl(1 + P_j(z)\bigr) + \cdots = \bigl(1 + P_j(z)\bigr) F_j(z).$$

Thus

$$F_j(z) = \frac{P_j(z)}{1 + P_j(z)}, \qquad P_j(z) = \frac{F_j(z)}{1 + F_j(z)}.$$

Assume that $P_j = \infty$. Then $P_j(z) \to \infty$ as $z \uparrow 1$ and therefore $F_j(z) \to 1$. Since $F_j(z) < F_j$ for real $z < 1$, we have $F_j = 1$ and hence E_j is recurrent.

Now suppose that $F_j = 1$. Then $F_j(z) \to 1$ as $z \uparrow 1$, and so $P_j(z) \to \infty$. Therefore $P_j(z) = \infty$.

If E_j is transient, it follows from the above that $P_j(z) < \infty$, and setting $z := 1$ we obtain equality (13.2.1). □

The quantity $P_j = \sum_{n=1}^{\infty} p_{jj}(n)$ can be interpreted as the mean number of visits to the state E_j, provided that the initial state is also E_j. It follows from the fact that the number of visits to the state E_j can be represented as $\sum_{n=1}^{\infty} I(X_n = j)$, where, as before, $I(A)$ is the indicator of the event A. Therefore the expectation of this number is equal to

$$\mathbf{E}\sum_{n=1}^{\infty} I(X_n = j) = \sum_{n=1}^{\infty} \mathbf{E}I(X_n = j) = \sum_{n=1}^{\infty} p_{jj}(n) = P_j.$$

Theorem 13.2.1 implies the following result.

Corollary 13.2.1 *A transient state is always null.*

This is obvious, since it immediately follows from the convergence of the series $\sum p_{jj}(n) < \infty$ that $p_{jj}(n) \to 0$.

Thus, based on definitions K3–K5, we could distinguish, in an irreducible chain, 8 possible types of states (each of the three properties can either be present or not). But in reality there are only 6 possible types since transient states are automatically null, and positive states are recurrent. These six types are generated by:

1) Classification by the asymptotic properties of the probabilities $p_{jj}(n)$ (transient, recurrent null and positive states).

2) Classification by the arithmetic properties of the probabilities $p_{jj}(n)$ or $f_j(n)$ (periodic or aperiodic).

Theorem 13.2.2 (Solidarity Theorem) *In an irreducible homogeneous Markov chain all states are of the same type: if one is recurrent then all are recurrent, if one is null then all are null, if one state is periodic with period d then all states are periodic with the same period d.*

Proof Let E_k and E_j be two different states. There exist numbers N and M such that

$$p_{kj}(N) > 0, \quad p_{jk}(M) > 0.$$

The total probability formula

$$p_{kk}(N + M + n) = \sum_{l,s} p_{kl}(N) p_{ls}(n) p_{sk}(M)$$

implies the inequality

$$p_{kk}(N + M + n) \ge p_{kj}(N) p_{jj}(n) p_{jk}(M) = \alpha\beta p_{jj}(n).$$

Here $n > 0$ is an arbitrary integer, $\alpha = p_{jj}(N) > 0$, and $\beta = p_{jj}(M) > 0$. In the same way one can obtain the inequality

$$p_{jj}(N+M+n) \geq \alpha\beta p_{kk}(n).$$

Hence

$$\frac{1}{\alpha\beta} p_{kk}(N+M+n) \geq p_{kk}(n) \geq \alpha\beta p_{kk}(n-M-N). \qquad (13.2.2)$$

We see from these inequalities that the asymptotic properties of $p_{kk}(n)$ and $p_{jj}(n)$ are the same. If E_k is null, then $p_{kk}(n) \to 0$, therefore $p_{jj}(n) \to 0$ and E_j is also null. If E_k is recurrent or, which is equivalent, $P_k = \sum_{n=1}^{\infty} p_{kk}(n) = \infty$, then

$$\sum_{n=M+N+1}^{\infty} p_{jj}(n) \geq \alpha\beta \sum_{n=M+N+1}^{\infty} p_{kk}(n-M-N) = \infty,$$

and E_j is also recurrent.

Suppose now that E_k is a periodic state with period d_k. If $p_{kk}(n) > 0$, then d_k divides n. We will write this as $d_k \mid n$. Since $p_{kk}(M+N) \geq \alpha\beta > 0$, then $d_k \mid (M+N)$.

We now show that the state E_j is also periodic and its period d_j is equal to d_k. Indeed, if $p_{jj}(n) > 0$ for some n, then by virtue of (13.2.2), $p_{kk}(n+M+N) > 0$. Therefore $d_k \mid (n+M+N)$, and since $d_k \mid (M+N)$, $d_k \mid n$ and hence $d_k \leq d_j$. In a similar way one can prove that $d_j \leq d_k$. Thus $d_j = d_k$. □

If the states of an irreducible Markov chain are periodic with period $d > 1$, then the chain is called *periodic*.

We will now show that the study of periodic chains can essentially be reduced to the study of aperiodic chains.

Theorem 13.2.3 *If a Markov chain is periodic with period d, then the set of states can be split into d subclasses $\Psi_0, \Psi_1, \dots, \Psi_{d-1}$ such that, with probability* 1, *in one step the system passes from Ψ_k to Ψ_{k+1}, and from Ψ_{d-1} the system passes to Ψ_0.*

Proof Choose some state, say, E_1. Based on this we will construct the subclasses $\Psi_0, \Psi_1, \dots, \Psi_{d-1}$ in the following way: $E_i \in \Psi_\alpha$, $0 \leq \alpha \leq d-1$, if there exists an integer $k > 0$ such that $p_{1i}(kd+\alpha) > 0$.

We show that no state can belong to two subclasses simultaneously. To this end it suffices to prove that if $E_i \in \Psi_\alpha$ and $p_{1i}(s) > 0$ for some s, then $s = \alpha \pmod{d}$.

Indeed, there exists a number $t_1 > 0$ such that $p_{i1}(t_1) > 0$. So, by the definition of Ψ_α, we have $p_{11}(kd+\alpha+t_1) > 0$. Moreover, $p_{11}(s+t_1) > 0$. Hence $d \mid (kd + \alpha + t_1)$ and $d \mid (s+t_1)$. This implies $\alpha = s \pmod{d}$.

Since starting from the state E_1 it is possible with positive probability to enter any state E_i, the union $\bigcup_\alpha \Psi_\alpha$ contains all the states.

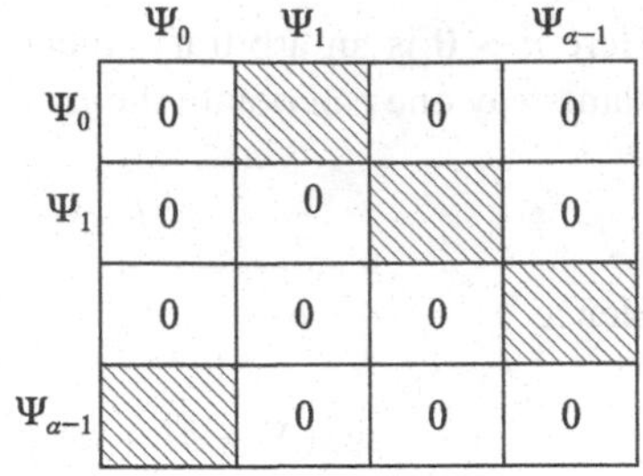

Fig. 13.3 The structure of the matrix of transition probabilities of a periodic Markov chain: an illustration to the proof of Theorem 13.2.3

We now prove that in one step the system goes from Ψ_α with probability 1 to $\Psi_{\alpha+1}$ (here the sum $\alpha + 1$ is modulo d). We have to show that, for $E_i \in \Psi_\alpha$,

$$\sum_{E_j \in \Psi_{\alpha+1}} p_{ij} = 1.$$

To do this, it suffices to prove that $p_{ij} = 0$ when $E_i \in \Psi_\alpha$, $E_j \notin \Psi_{\alpha+1}$. If we assume the opposite ($p_{ij} > 0$) then, taking into account the inequality $p_{1i}(kd + \alpha) > 0$, we have $p_{1j}(kd + \alpha + 1) > 0$ and consequently $E_j \in \Psi_{\alpha+1}$. This contradiction completes the proof of the theorem. □

We see from the theorem that the matrix of a periodic chain has the form shown in Fig. 13.3 where non-zero entries can only be in the shaded cells.

From a periodic Markov chain with period d one can construct d new Markov chains. The states from the subset Ψ_α will be the states of the α-th chain. Transition probabilities are given by

$$p_{ij}^\alpha := p_{ij}(d).$$

By virtue of Theorem 13.2.3, $\sum_{E_j \in \Psi_\alpha} p_{ij}^\alpha = 1$. The new chains, to which one can reduce in a certain sense the original one, will have no subclasses.

13.3 Theorems on Random Walks on a Lattice

1. A random walk on integer points on the line. Imagine a particle moving on integer points of the real line. Transitions from one point to another occur in equal time intervals. In one step, from point k the particle goes with a positive probability p to the point $k + 1$, and with positive probability $q = 1 - p$ it moves to the point $k - 1$. As was already mentioned, to this physical system there corresponds the following Markov chain:

$$X_n = X_{n-1} + \xi_n = X_0 + S_n,$$

where ξ_n takes values 1 and -1 with probabilities p and q, respectively, and $S_n = \sum_{k=1}^n \xi_k$. The states of the chain are integer points on the line.

It is easy to see that returning to a given point with a positive probability is only possible after an even number of steps, and $f_0(2) = 2pq > 0$. Therefore this chain is periodic with period 2.

We now establish conditions under which the random walk forms a recurrent chain.

Theorem 13.3.1 *The random walk $\{X_n\}$ forms a recurrent Markov chain if and only if $p = q = 1/2$.*

Proof Since $0 < p < 1$, the random walk is an irreducible Markov chain. Therefore by Theorem 13.2.2 it suffices to examine the type of any given point, for example, zero.

We will make use of Theorem 13.2.1. In order to do this, we have to investigate the convergence of the series $\sum_{n=1}^{\infty} p_{00}(n)$. Since our chain is periodic with period 2, one has $p_{00}(2k+1) = 0$. So it remains to compute $\sum_1^{\infty} p_{00}(2k)$. The sum S_n is the coordinate of the walking particle after n steps ($X_0 = 0$). Therefore $p_{00}(2k) = \mathbf{P}(S_{2k} = 0)$. The equality $S_{2k} = 0$ holds if k of the random variables ξ_j are equal to 1 and the other k are equal to -1 (k steps to the right and k steps to the left). Therefore, by Theorem 5.2.1,

$$\mathbf{P}(S_{2k} = 0) \sim \frac{1}{\sqrt{\pi k}} e^{-2kH(1/2)} = \frac{1}{\sqrt{\pi k}} (4pq)^k.$$

We now elucidate the behaviour of the function $\beta(p) = 4pq = 4p(1-p)$ on the interval $[0, 1]$. At the point $p = 1/2$ the function $\beta(p)$ attains its only extremum, $\beta(1/2) = 1$. At all the other points of $[0, 1]$, $\beta(p) < 1$. Therefore $4pq < 1$ for $p \neq 1/2$, which implies convergence of the series $\sum_{k=1}^{\infty} p_{00}(2k)$ and hence the transience of the Markov chain. But if $p = 1/2$ then $p_{00}(2k) \sim 1/\sqrt{\pi k}$ and the series $\sum_{k=1}^{\infty} p_{00}(2k)$ diverges, which implies, in turn, that all the states of the chain are recurrent. The theorem is proved. □

Theorem 13.3.1 allows us to make the following remark. If $p \neq 1/2$, then the mean number of recurrences to 0 is finite, as it is equal to $\sum_{k=1}^{\infty} p_{00}(2k)$. This means that, after a certain time, the particle will never return to zero. The particle will "drift" to the right or to the left depending on whether p is greater than $1/2$ or less. This can easily be obtained from the law of large numbers.

If $p = 1/2$, then the mean number of recurrences to 0 is infinite; the particle has no "drift". It is interesting to note that the increase in the mean number of recurrences is not proportional to the number of steps. Indeed, the mean number of recurrences over the first $2n$ steps is equal to $\sum_{k=1}^{n} p_{00}(2k)$. From the proof of Theorem 13.3.1 we know that $p_{00}(2k) \sim 1/\sqrt{\pi k}$. Therefore, as $n \to \infty$,

$$\sum_{k=1}^{n} p_{00}(2k) \sim \sum_{k=1}^{n} \frac{1}{\sqrt{\pi k}} \sim \frac{2\sqrt{n}}{\sqrt{\pi}}.$$

Thus, in the fair game considered in Example 4.2.2, the proportion of ties rapidly decreases as the number of steps increases, and deviations are growing both in magnitude and duration.

13.3.1 Symmetric Random Walks in $\mathbb{R}^k$, $k \geq 2$

Consider the following random walk model in the k-dimensional Euclidean space $\mathbb{R}^k$. If the walking particle is at point $(m_1, \ldots, m_k)$, then it can move with probabilities $1/2^k$ to any of the 2^k vertices of the cube $|x_j - m_j| = 1$, i.e. the points with coordinates $(m_1 \pm 1, \ldots, m_k \pm 1)$. It is natural to call this walk symmetric. Denoting by X_n the position of the particle after the n-th jump, we have, as before, a sequence of k-dimensional random variables forming a homogeneous irreducible Markov chain. We shall show that all states of the walk on the plane are, as in the one-dimensional case, recurrent. In the three-dimensional space, the states will turn out to be transient. Thus we shall prove the following assertion.

Theorem 13.3.2 *The symmetric random walk is recurrent in spaces of one and two dimensions and transient in spaces of three or more dimensions.*

In this context, W. Feller made the sharp comment that the proverb "all roads lead to Rome" is true only for two-dimensional surfaces. The assertion of Theorem 13.3.2 is adjacent to the famous theorem of Pólya on the transience of symmetric walks in $\mathbb{R}^k$ for $k > 2$ when the particle jumps to neighbouring points along the coordinate axes (so that ξ_j assumes $2k$ values with probabilities $1/2k$ each). We now turn to the proof of Theorem 13.3.2.

Proof of Theorem 13.3.2 Let $k = 2$. It is not difficult to see that our walk X_n can be represented as a sum of two independent components

$$X_n = \left(X_n^{-1}, 0\right) + \left(0, X_n^2\right), \quad \left(X_0^1, X_0^2\right) = X_0,$$

where X_n^i, $i = 1, 2, \ldots$, are scalar (one-dimensional) sequences describing symmetric independent random walks on the respective lines (axes). This is obvious, for the two-dimensional sequence admits the representation

$$X_{n+1} = X_n + \xi_n, \tag{13.3.1}$$

where ξ_n assumes 4 values $(\pm 1, 0) + (0, \pm 1) = (\pm 1, \pm 1)$ with probabilities $1/4$ each.

With the help of representation (13.3.1) we can investigate the asymptotic behaviour of the transition probabilities $p_{ij}(n)$. Let X_0 coincide with the origin $(0, 0)$. Then

$$p_{00}(2n) = \mathbf{P}\left(X_{2n} = (0, 0) | X_0 = (0, 0)\right)$$

$$= \mathbf{P}(X_{2n}^1 = 0 | X_0^1 = 0)\mathbf{P}(X_{2n}^2 = 0 | X_0^2 = 0) \sim (1/\sqrt{\pi n})^2 = 1/(\pi n).$$

From this it follows that the series $\sum_{n=0}^{\infty} p_{00}(n)$ diverges and so all the states of our chain are recurrent.

The case $k = 3$ should be treated in a similar way. Represent the sequence X_n as a sum of three independent components

$$X_n = (X_n^1, 0, 0) + (0, X_n^2, 0) + (0, 0, X_n^3),$$

where the X_n^i are, as before, symmetric random walks on the real line. If we set $X_0 = (0, 0, 0)$, then

$$p_{00}(2n) = (\mathbf{P}(X_{2n}^1 = 0 | X_0^1 = 0))^3 \sim 1/(\pi n)^{3/2}.$$

The series $\sum_{n=1}^{\infty} p_{00}(n)$ is convergent here, and hence the states of the chain are transient. In contrast to the straight line and plane cases, a particle leaving the origin will, with a positive probability, never come back.

It is evident that a similar situation takes place for walks in k-dimensional space with $k \ge 3$, since $\sum_{n=1}^{\infty} (\pi n)^{-k/2} < \infty$ for $k \ge 3$. The theorem is proved. □

13.3.2 Arbitrary Symmetric Random Walks on the Line

Let, as before,

$$X_n = X_0 + \sum_1^n \xi_j, \tag{13.3.2}$$

but now ξ_j are arbitrary independent identically distributed integer-valued random variables. Theorem 13.3.1 may be generalised in the following way:

Theorem 13.3.3 *If the ξ_j are symmetric and the expectation $\mathbf{E}\xi_j$ exists (and hence $\mathbf{E}\xi_j = 0$) then the random walk X_n forms a recurrent Markov chain with null states.*

Proof It suffices to verify that

$$\sum_{n=1}^{\infty} \mathbf{P}(S_n = 0) = \infty,$$

where $S_n = \sum_1^n \xi_j$, and that $\mathbf{P}(S_n = 0) \to 0$ as $n \to \infty$. Put

$$p(z) := \mathbf{E}z^{\xi_1} = \sum_{k=-\infty}^{\infty} z^k \mathbf{P}(\xi_1 = k).$$

Then the generating function of S_n will be equal to $\mathbf{E}z^{S_n} = p^n(z)$, and by the inversion formula (see Sect. 7.7)

$$\mathbf{P}(S_n = 0) = \frac{1}{2\pi i} \int_{|z|=1} p^n z^{-1} dz, \tag{13.3.3}$$

$$\sum_{n=0}^{\infty} \mathbf{P}(S_n = 0) = \frac{1}{2\pi i} \int_{|z|=1} \frac{dz}{z(1 - p(z))} = \frac{1}{\pi} \int_0^{\pi} \frac{dt}{1 - p(e^{it})}.$$

The last equality holds since the real function $p(r)$ is even and is obtained by substituting $z = e^{it}$.

Since $\mathbf{E}\xi_1 = 0$, one has $1 - p(e^{it}) = o(t)$ as $t \to 0$ and, for sufficiently small δ and $0 \le t < \delta$,

$$0 \le 1 - p(e^{it}) < t$$

(the function $p(e^{it})$ is real by virtue of the symmetry of ξ_1). This implies

$$\int_0^{\pi} \frac{dt}{1 - p(e^{it})} \ge \int_0^{\delta} \frac{dt}{t} = \infty.$$

Convergence $\mathbf{P}(S_n = 0) \to 0$ is a consequence of (13.3.3) since, for all z on the circle $|z| = 1$, with the possible exclusion of finitely many points, one has $p(z) < 1$ and hence $p^n(z) \to 0$ as $n \to \infty$. The theorem is proved. □

Theorem 13.3.3 can be supplemented by the following assertion.

Theorem 13.3.4 *Under the conditions of Theorem* 13.3.3, *if the g.c.d. of the possible values of ξ_j equals* 1 *then the set of values of $\{X_n\}$ constitutes a single class of essential communicating states. This class coincides with the set of all integers.*

The assertion of the theorem follows from the next lemma.

Lemma 13.3.1 *If the g.c.d. of integers $a_1 > 0, \dots, a_r > 0$ is equal to* 1, *then there exists a number K such that every natural $k \ge K$ can be represented as*

$$k = n_1 a_1 + \cdots + n_r a_r,$$

where $n_i \ge 0$ are some integers.

Proof Consider the function $L(\mathbf{n}) = n_1 a_1 + \cdots + n_r a_r$, where $\mathbf{n} = (n_1, \dots, n_r)$ is a vector with integer (possibly negative) components. Let $d > 0$ be the minimal natural number for which there exists a vector $\mathbf{n}^0$ such that

$$d = L(\mathbf{n}^0).$$

We show that every natural number that can be represented as $L(\mathbf{n})$ is divisible by d. Suppose that this is not true. Then there exist $\mathbf{n}$, k and $0 < \alpha < d$ such that

$$L(\mathbf{n}) = kd + \alpha.$$

But since the function $L(\mathbf{n})$ is linear,

$$L\big(\mathbf{n} - kx^0\big) = kd + \alpha - kd = \alpha < d,$$

which contradicts the minimality of d in the set of positive integer values of $L(\mathbf{n})$.

The numbers $a_1, \dots, a_r$ are also the values of the function $L(\mathbf{n})$, so they are divisible by d. The greatest common divisor of these numbers is by assumption equal to one, so that $d = 1$.

Let k be an arbitrary natural number. Denoting by $\theta < A$ the remainder after dividing k by $A := a_1 + \cdots + a_r$, we can write

$$\begin{aligned} k &= m(a_1 + \cdots + a_r) + \theta = m(a_1 + \cdots + a_r) + \theta L\big(\mathbf{n}^0\big) \\ &= a_1\big(m + \theta n_1^0\big) + a_2\big(m + \theta n_2^0\big) + \cdots + a_r\big(m + \theta n_r^0\big), \end{aligned}$$

where $n_i := m + \theta n_i^0 > 0$, $i = 1, \dots, r$, for sufficiently large k (or m).

The lemma is proved. □

Proof of Theorem 13.3.4 Put $q_j := \mathbf{P}(\xi = a_j) > 0$. Then, for each $k \ge K$, there exists an $\mathbf{n}$ such that $n_j \ge 0$, $\sum_{j=1}^r a_j n_j = k$, and hence, for $n = \sum_{j=1}^r n_j$, we have

$$p_{0k}(n) \ge q_1^{n_1} \cdots q_r^{n_r} > 0.$$

In other words, all the states $k \ge K$ are reachable from 0. Similarly, all the states $k \le -K$ are reachable from 0. The states $k \in [-K, K]$ are reachable from the point $-2K$ (which is reachable from 0). The theorem is proved. □

Corollary 13.3.1 *If the conditions of Theorems* 13.3.3 *and* 13.3.4 *are satisfied, then the chain* (13.3.2) *with an arbitrary initial state X_0 visits every state k infinitely many times with probability* 1. *In particular, for any X_0 and k, the random variable $\nu = \min\{n : X_n = k\}$ will be proper.*

If we are interested in investigating the periodicity of the chain (13.3.2), then more detailed information on the set of possible values of ξ_j is needed. We leave it to the reader to verify that, for example, if this set is of the form $\{a + a_k d\}$, $k = 1, 2, \dots$, $d \ge 1$, g.c.d. $(a_1, a_2, \dots) = 1$, g.c.d. $(a, d) = 1$, then the chain will be periodic with period d.

13.4 Limit Theorems for Countable Homogeneous Chains

13.4.1 Ergodic Theorems

Now we return to arbitrary countable homogeneous Markov chains. We will need the following conditions:

(I) There exists a state E_0 such that the recurrence time $\tau^{(s)}$ to E_s ($\mathbf{P}(\tau^{(s)}=n)=f_s(n)$) has finite expectation $\mathbf{E}\tau^{(s)}<\infty$.
(II) The chain is irreducible.
(III) The chain is aperiodic.

We introduce the so-called "taboo probabilities" $P_i(n,j)$ of transition from E_i to E_j in n steps without visiting the "forbidden" state E_i:

$$P_i(n,j):=\mathbf{P}(X_n=j;X_1\neq i,\dots,X_{n-1}\neq i\mid X_0=i).$$

Theorem 13.4.1 (The ergodic theorem) *Conditions* (I)–(III) *are necessary and sufficient for the existence, for all i and j, of the positive limits*

$$\lim_{n\to\infty}p_{ij}(n)=\pi_j>0,\quad i,j=0,1,2,\dots. \tag{13.4.1}$$

The sequence of values $\{\pi_j\}$ is the unique solution of the system

$$\begin{cases}\sum_{j=0}^{\infty}\pi_j=1,\\ \pi_j=\sum_{k=0}^{\infty}\pi_k p_{kj},\quad j=0,1,2,\dots,\end{cases} \tag{13.4.2}$$

in the class of absolutely convergent series.

Moreover, $\mathbf{E}\tau^{(j)}<\infty$ for all j, and the quantities $\pi_j=(\mathbf{E}\tau^{(j)})^{-1}$ admit the representation

$$\pi_j=\big(\mathbf{E}\tau^{(j)}\big)^{-1}=\big(\mathbf{E}\tau^{(s)}\big)^{-1}\sum_{k=1}^{\infty}P_s(k,j) \tag{13.4.3}$$

for any s.

Definition 13.4.1 A chain possessing property (13.4.1) is called *ergodic*.

The numbers π_j are essentially the probabilities that the system will be in the respective states E_j after a long period of time has passed. It turns out that these probabilities lose dependence on the initial state of the system. The system "forgets" where it began its motion. The distribution $\{\pi_j\}$ is called *stationary* or *invariant*. Property (13.4.2) expresses the invariance of the distribution with respect to the transition probabilities p_{ij}. In other words, if $\mathbf{P}(X_n=k)=\pi_k$, then $\mathbf{P}(X_{n+1}=k)=\sum\pi_j p_{jk}$ is also equal to π_k.

Proof of Theorem 13.4.1 *Sufficiency* in the first assertion of the theorem. Consider the "trajectory" of the Markov chain starting at a fixed state E_s. Let $\tau_1 \ge 1$, $\tau_2 \ge 1$, ... be the time intervals between successive returns of the system to E_s. Since after each return the evolution of the system begins anew from the same state, by the Markov property the durations τ_k of the cycles (as well as the cycles themselves) are independent and identically distributed, $\tau_k \stackrel{d}{=} \tau^{(s)}$. Moreover, it is obvious that

$$\mathbf{P}(\tau_k = n) = \mathbf{P}\big(\tau^{(s)} = n\big) = f_s(n).$$

Recurrence of E_s means that the τ_k are proper random variables. Aperiodicity of E_s means that the g.c.d. of all possible values of τ_k is equal to 1. Since

$$p_{ss}(n) = \mathbf{P}\big(\gamma(n) = 0\big),$$

where $\gamma(n)$ is the defect of level n for the renewal process $\{T_k\}$,

$$T_k = \sum_{i=1}^{k} \tau_i,$$

by Theorem 10.3.1 the following limit exists

$$\lim_{n\to\infty} p_{ss}(n) = \lim_{n\to\infty} \mathbf{P}\big(\gamma(n) = 0\big) = \frac{1}{\mathbf{E}\tau_1} > 0. \tag{13.4.4}$$

Now prove the existence of $\lim_{n\to\infty} p_{sj}(n)$ for $j \ne s$. If $\gamma(n)$ is the defect of level n for the walk $\{T_k\}$ then, by the total probability formula,

$$p_{sj}(n) = \sum_{k=1}^{n} \mathbf{P}\big(\gamma(n) = k\big)\mathbf{P}\big(X_n = j \mid X_0 = s, \gamma(n) = k\big). \tag{13.4.5}$$

Note that the second factors in the terms on the right-hand side of this formula do not depend on n by the Markov property:

$$\begin{aligned}
&\mathbf{P}\big(X_n = j \mid X_0 = s, \gamma(n) = k\big) \\
&\quad = \mathbf{P}(X_n = j \mid X_0 = s, X_{n-1} \ne s, \dots, X_{n-k+1} \ne s, X_{n-k} = s) \\
&\quad = \mathbf{P}(X_k = j \mid X_0 = s, X_1 \ne s, \dots, X_{k-1} \ne s) = \frac{P_s(k, j)}{\mathbf{P}(\tau_1 \ge k)},
\end{aligned} \tag{13.4.6}$$

since, for a fixed $X_0 = s$,

$$\begin{aligned}
\mathbf{P}(X_k = j \mid X_1 \ne s, \dots, X_{k-1} \ne s) &= \frac{\mathbf{P}(X_k = j, X_1 \ne s, \dots, X_{k-1} \ne s)}{\mathbf{P}(X_1 \ne s, \dots, X_{k-1} \ne s)} \\
&= \frac{P_s(k, j)}{\mathbf{P}(\tau^{(s)} \ge k)}.
\end{aligned}$$

For the sake of brevity, put $\mathbf{P}(\tau_1 > k) = P_k$. The first factors in (13.4.5) converge, as $n \to \infty$, to $P_{k-1}/\mathbf{E}\tau_1$ and, by virtue of the equality

$$\mathbf{P}\big(\gamma(n) = k\big) = \mathbf{P}\big(\gamma(n-k) = 0\big) P_{k-1} \le P_{k-1}, \tag{13.4.7}$$

are dominated by the convergent sequence P_{k-1}. Therefore, by the dominated convergence theorem, the following limit exists

$$\lim_{n\to\infty} p_{sj}(n) = \sum_{k=1}^{\infty} \frac{P_{k-1}}{\mathbf{E}\tau_1} \frac{P_s(k,j)}{\mathbf{P}(\tau_1 \ge k)} = \frac{1}{\mathbf{E}\tau_1} \sum_{k=1}^{\infty} P_s(k,j) =: \pi_j, \tag{13.4.8}$$

and we have, by (13.4.5)–(13.4.7),

$$p_{sj}(n) \le \sum_{k=1}^{n} P_s(k,j) \le \sum_{k=1}^{\infty} P_s(k,j) = \pi_j \mathbf{E}\tau_1. \tag{13.4.9}$$

To establish that, for any i,

$$\lim_{n\to\infty} p_{ij}(n) = \pi_j > 0,$$

we first show that the system departing from E_i will, with probability 1, eventually reach E_s.

In other words, if $f_{is}(n)$ is the probability that the system, upon leaving E_i, hits E_s for the first time on the n-th step then

$$\sum_{n=1}^{\infty} f_{is}(n) = 1.$$

Indeed, both states E_i and E_s are recurrent. Consider the cycles formed by subsequent visits of the system to the state E_i. Denote by A_k the event that the system is in the state E_s at least once during the k-th cycle. By the Markov property the events A_k are independent and $\mathbf{P}(A_k) > 0$ does not depend on k. Therefore, by the Borel–Cantelli zero–one law (see Sect. 11.1), with probability 1 there will occur infinitely many events A_k and hence $\mathbf{P}(\bigcup A_k) = 1$.

By the total probability formula,

$$p_{ij}(n) = \sum_{k=1}^{n} f_{is}(k) p_{sj}(n-k),$$

and the dominated convergence theorem yields

$$\lim_{n\to\infty} p_{ij}(n) = \sum_{n=1}^{\infty} f_{is}(k) \pi_j = \pi_j.$$

Representation (13.4.3) follows from (13.4.8).

Now we will prove the *necessity* in the first assertion of the theorem. That conditions (II)–(III) are necessary is obvious, since $p_{ij}(n) > 0$ for every i and j if n is large enough. The necessity of condition (I) follows from the fact that equalities (13.4.4) are valid for E_s. The first part of the theorem is proved.

It remains to prove the second part of the theorem. Since

$$\sum p_{sj}(n) = 1,$$

one has $\sum_j \pi_j \le 1$. By virtue of the inequalities $p_{sj}(n) \le \pi_j \mathbf{E}\tau_1$ (see (13.4.9)), we can use the dominated convergence theorem both in the last equality and in the equality $p_{sj}(n+1) = \sum_{k=0}^{\infty} p_{sk}(n) p_{kj}$ which yields

$$\sum \pi_j = 1, \quad \pi_j = \sum_{k=0}^{\infty} \pi_k p_{kj}.$$

It remains to show that the system has a unique solution. Let the numbers $\{q_j\}$ also satisfy (13.4.2) and assume the series $\sum |q_j|$ converges. Then, changing the order of summation, we obtain that

$$q_j = \sum_k q_k p_{kj} = \sum_k p_{kj} \Big(\sum_l p_{lk} q_l \Big) = \sum_l q_l \sum_k p_{lk} p_{kj} = \sum_l q_l p_{lj}(2)$$
$$= \sum_l p_{lj}(2) \Big(\sum_m p_{ml} q_m \Big) = \sum_m q_m p_{mj}(3) = \cdots = \sum_k q_k p_{kj}(n)$$

for any n. Since $\sum q_k = 1$, passing to the limit as $n \to \infty$ gives

$$q_j = \sum_k q_k \pi_j = \pi_j.$$

The theorem is proved. □

If a Markov chain is periodic with period d, then $p_{ij}(t) = 0$ for $t \ne kd$ and every pair of states E_i and E_j belonging to the same subclass (see Theorem 13.2.3). But if $t = kd$, then from the theorem just proved and Theorem 13.2.3 it follows that the limit $\lim_{k\to\infty} p_{ij}(kd) = \pi_j > 0$ exists and does not depend on i.

Verifying conditions (II)–(III) of Theorem 13.4.1 usually presents no serious difficulties. The main difficulties would be related to verifying condition (I). For finite Markov chains, this condition is always met.

Theorem 13.4.2 *Let a Markov chain have finitely many states and satisfy conditions* (II)–(III). *Then there exist $c > 0$ and $q < 1$ such that, for the recurrence time τ to an arbitrary fixed state, one has*

$$\mathbf{P}(\tau > n) < cq^n, \quad n \ge 1. \tag{13.4.10}$$

These equalities clearly mean that *condition* (I) *is always met for finite chains and hence the ergodic theorem for them holds if and only if conditions* (II)–(III) *are satisfied.*

Proof Consider a state E_s and put

$$r_j(n) := \mathbf{P}(X_k \neq s, k = 1, 2, \dots, n | X_0 = j).$$

Then, if the chain has m states one has $r_j(m) < 1$ for any j. Indeed, $r_j(n)$ does not grow as n increases. Let N be the smallest number satisfying $r_j(N) < 1$. This means that there exists a sequence of states $E_j, E_{j_1}, \dots, E_{j_N}$ such that $E_{j_N} = E_s$ and the probability of this sequence $p_{jj_1} \cdots p_{j_{N-1}j_N}$ is positive. But it is easy to see that $N \le m$, since otherwise this sequence would contain at least two identical states. Therefore the cycle contained between these states could be removed from the sequence which could only increase its probability. Thus

$$r_j(m) < 1, \quad r(m) = \max_j r_j(m) < 1.$$

Moreover, $r_j(n_1 + n_2) \le r_j(n_1) r(n_2) \le r(n_1) r(n_2)$.

It remains to note that if τ is the recurrence time to E_s, then $\mathbf{P}(\tau > nm) = r_s(nm) \le r(m)^n$. The statement of the theorem follows. □

Remark 13.4.1 Condition (13.4.10) implies the *exponential* rate of convergence of the differences $|p_{ij}(n) - \pi_j|$ to zero. One can verify this by making use of the analyticity of the function

$$F_s(z) = \sum_{n=1}^{\infty} f_s(n) z^n$$

in the domain $|z| < q^{-1}$, $q^{-1} > 1$, and of the equality

$$P_s(z) = \sum p_{ss}(n) z^n = \frac{1}{1 - F_s(z)} - 1 \tag{13.4.11}$$

(see Theorem 13.2.1; we assume that the τ in condition (13.4.10) refers to the state E_s, so that $f_s(n) = \mathbf{P}(\tau = n)$). Since $F_s'(1) = \mathbf{E}\tau = 1/\pi_s$, one has

$$F_s(z) = 1 + \frac{(z-1)}{\pi_s} + \cdots,$$

and from (13.4.11) it follows that the function

$$P_s(z) - \frac{z\pi_s}{1-z} = \sum_{n=1}^{\infty} \big(p_{ss}(n) - \pi_s\big) z^n$$

is analytic in the disk $|z| \le 1 + \varepsilon$, $\varepsilon > 0$. It evidently follows from this that

$$|p_{ss}(n) - \pi_s| < c(1+\varepsilon)^{-n}, \quad c = \text{const}.$$

Now we will give two examples of finite Markov chains.

Example 13.4.1 Suppose that the behaviour of two chess players A and B playing in a multi-player tournament can be described as follows. Independently of the outcomes of the previous games, player A wins every new game with probability p, loses with probability q, and makes a tie with probability $r = 1 - p - q$. Player B is less balanced. He wins a game with probabilities $p + \varepsilon$, p and $p - \varepsilon$, respectively, if he won, made a tie, or lost in the previous one. The probability that he loses behaves in a similar way: in the above three cases, it equals $q - \varepsilon$, q and $q + \varepsilon$, respectively. Which of the players A and B will score more points in a long tournament?

To answer this question, we will need to compute the stationary probabilities π_1, π_2, π_3 of the states E_1, E_2, E_3 which represent a win, tie, and loss in a game, respectively (cf. the law of large numbers at the end of this section).

For player A, the Markov chain with states E_1, E_2, E_3 describing his performance in the tournament will have the matrix of transition probabilities

$$P_A = \begin{pmatrix} p & r & q \\ p & r & q \\ p & r & q \end{pmatrix}.$$

It is obvious that $\pi_1 = p, \pi_2 = r, \pi_3 = q$ here.

For player B, the matrix of transition probabilities is equal to

$$P_B = \begin{pmatrix} p+\varepsilon & r & q-\varepsilon \\ p & r & q \\ p-\varepsilon & r & q+\varepsilon \end{pmatrix}.$$

Equations for stationary probabilities in this case have the form

$$\pi_1(p+\varepsilon) + \pi_2 p + \pi_3(p-\varepsilon) = \pi_1,$$
$$\pi_1 r + \pi_2 r + \pi_3 r = \pi_2,$$
$$\pi_1 + \pi_2 + \pi_3 = 1.$$

Solving this system we find that

$$\pi_2 - r = 0, \qquad \pi_1 - p = \varepsilon \frac{p-q}{1-2\varepsilon}.$$

Thus, the long run proportions of ties will be the same for both players, and B will have a greater proportion of wins if $\varepsilon > 0$, $p > q$ or $\varepsilon < 0$, $p < q$. If $p = q$, then the stationary distributions will be the same for both A and B.

Example 13.4.2 Consider the summation of independent integer-valued random variables $\xi_1, \xi_2, \dots$ modulo some $d > 1$ (see Example 13.1.2). Set $X_0 := 0$, $X_1 := \xi_1 - \lfloor \xi_1/d \rfloor d$, $X_2 := X_1 + \xi_2 - \lfloor (X_1 + \xi_2)/d \rfloor d$ etc. (here $\lfloor x \rfloor$ denotes the integral part of x), so that X_n is the remainder of the division of $X_{n-1}+\xi_n$ by d. Such summation is sometimes also called summation on a circle (points 0 and d are glued together in a single point). Without loss of generality, we can evidently suppose that ξ_k takes the values $0, 1, \dots, d-1$ only. If $\mathbf{P}(\xi_k = j) = p_j$ then

$$p_{ij} = \mathbf{P}(X_n = j | X_{n-1} = i) = \begin{cases} p_{j-i} & \text{if } j \ge i, \\ p_{d+j-i} & \text{if } j < i. \end{cases}$$

Assume that the set of all indices k with $p_k > 0$ has a g.c.d. equal to 1. Then it is clear that the chain $\{X_n\}$ has a single class of essential states without subclasses, and there will exist the limits

$$\lim_{n\to\infty} p_{ij}(n) = \pi_j$$

satisfying the system $\sum_i \pi_i p_{ij} = \pi_j$, $\sum \pi_j = 1$, $j = 0, \dots, d-1$. Now note that the stochastic matrix of transition probabilities $\|p_{ij}\|$ has in this case the following property:

$$\sum_i p_{ij} = \sum_j p_{ij} = 1.$$

Such matrices are called *doubly stochastic*. Stationary distributions for them are always uniform, since $\pi_j = 1/d$ satisfy the system for final probabilities.

Thus summation of arbitrary random variables on a circle leads to the *uniform limit distribution*. The rate of convergence of $p_{ij}(k)$ to the stationary distribution is exponential.

It is not difficult to see that the convolution of two uniform distributions under addition modulo d is also uniform. The uniform distribution is in this sense stable. Moreover, the convolution of an arbitrary distribution with the uniform distribution will also be uniform. Indeed, if η is uniformly distributed and independent of ξ_1 then (addition and subtraction are modulo d, $p_j = \mathbf{P}(\xi_1 = j)$)

$$\mathbf{P}(\xi_1 + \eta = k) = \sum_{j=0}^{d-1} p_j \mathbf{P}(\eta = k - j) = \sum_{j=0}^{d-1} p_j \frac{1}{d} = \frac{1}{d}.$$

Thus, if one transmits a certain signal taking d possible values (for example, letters) and (uniform) "random" noise is superimposed on it, then the received signal will also have the uniform distribution and therefore will contain no information about the transmitted signal. This fact is widely used in cryptography.

This example also deserves attention as a simple illustration of laws that appear when summing random variables taking values not in the real line but in some group (the set of numbers $0, 1, \dots, d-1$ with addition modulo d forms a finite Abelian group). It turns out that the phenomenon discovered in the example—the uniformity of the limit distribution—holds for a much broader class of groups.

We return to arbitrary countable chains. We have already mentioned that the main difficulties when verifying the conditions of Theorem 13.4.1 are usually related to condition (I). We consider this problem in Sect. 13.7 in more detail for a wider class of chains (see Theorems 13.7.2–13.7.3 and corollaries thereafter). Sometimes condition (I) can easily be verified using the results of Chaps. 10 and 12.

Example 13.4.3 We saw in Sect. 12.5 that waiting times in the queueing system satisfy the relationships

$$X_{n+1} = \max(X_n + \xi_{n+1}, 0), \qquad w_1 = 0,$$

where the ξ_n are independent and identically distributed. Clearly, X_n form a homogeneous Markov chain with the state space $\{0, 1, \dots\}$, provided that the ξ_k are integer-valued. The sequence X_n may be interpreted as a *walk with a delaying screen* at the point 0. If $\mathbf{E}\xi_k < 0$ then it is not hard to derive from the theorems of Chap. 10 (see also Sect. 13.7) that the recurrence time to 0 has finite expectation. Thus, applying the ergodic theorem we can, independently of Sect. 11.4, come to the conclusion that there exists a limiting (stationary) distribution for X_n as $n \to \infty$ (or, taking into account what we said in Sect. 11.4, conclude that $\sup_{k\ge 0} S_k$ is finite, where $S_k = \sum_{j=1}^{k} \xi_j$, which is essentially the assertion of Theorem 10.2.1).

Now we will make several remarks allowing us to state one more criterion for ergodicity which is related to the existence of a solution to Eq. (13.4.2).

First of all, note that Theorem 13.2.2 (the solidarity theorem) can now be complemented as follows. A state E_j is said to be *ergodic* if, for any i, $p_{ij}(n) \to \pi_j > 0$ as $n \to \infty$. A state E_j is said to be *positive recurrent* if it is recurrent and non-null (in that case, the recurrence time $\tau^{(j)}$ to E_j has finite expectation $E\tau^{(j)} < \infty$). It follows from Theorem 13.4.1 that, for an irreducible aperiodic chain, a *state E_j is ergodic if and only if it is positive recurrent. If at least one state is ergodic, all states are.*

Theorem 13.4.3 *Suppose a chain is irreducible and aperiodic* (*satisfies conditions* (II)–(III)). *Then only one of the following two alternatives can take place*: *either all the states are null or they are all ergodic. The existence of an absolutely convergent solution to system* (13.4.2) *is necessary and sufficient for the chain to be ergodic.*

Proof The first assertion of the theorem follows from the fact that, by the local renewal Theorem 10.2.2 for the random walk generated by the times of the chain's hitting the state E_j, the limit $\lim_{n\to\infty} p_{jj}(n)$ always exists and equals $(E\tau^{(j)})^{-1}$.

Therefore, to prove sufficiency in the second assertion (the necessity follows from Theorem 13.4.1) we have, in the case of the existence of an absolutely convergent solution $\{\pi_j\}$, to exclude the existence of null states. Assume the contrary, $p_{ij}(n) \to 0$. Choose j such that $\pi_j > 0$. Then

$$0 < \pi_j = \sum \pi_i p_{ij}(n) \to 0$$

as $n \to \infty$ by dominated convergence. This contradiction completes the proof of the theorem. □

13.4.2 *The Law of Large Numbers and the Central Limit Theorem for the Number of Visits to a Given State*

In conclusion of this section we will give two assertions about the limiting behaviour, as $n \to \infty$, of the number $m_j(n)$ of visits of the system to a fixed state E_j by the time n. Let $\tau^{(j)}$ be the recurrence time to the state E_j.

Theorem 13.4.4 *Let the chain be ergodic and, at the initial time epoch, be at an arbitrary state E_s. Then, as $n \to \infty$,*

$$\frac{\mathbf{E}m_j(n)}{n} \to \pi_j, \qquad \frac{m_j(n)}{n} \xrightarrow{a.s.} \pi_j.$$

If additionally $\operatorname{Var}(\tau^{(j)}) = \sigma_j^2 < \infty$ *then*

$$\mathbf{P}\left(\frac{m_j(n) - n\pi_j}{\sigma_j\sqrt{n\pi_j^3}} < x | X_0 = s\right) \to \Phi(x)$$

as $n \to \infty$, where $\Phi(x)$ is, as before, the distribution function of the normal law with parameters $(0, 1)$.

Proof Note that the sequence $m_j(n) + 1$ coincides with the renewal process formed by the random variables $\tau_1, \tau_2, \tau_3, \dots$, where τ_1 is the time of the first visit to the state E_j by the system which starts at E_s and $\tau_k \stackrel{d}{=} \tau^{(j)}$ for $k \ge 2$. Clearly, by the Markov property all τ_j are independent. Since $\tau_1 \ge 0$ is a proper random variable, Theorem 13.4.4 is a simple consequence of the generalisations of Theorems 10.1.1, 11.5.1, and 10.5.2 that were stated in Remarks 10.1.1, 11.5.1 and 10.5.1, respectively.

The theorem is proved. □

Summarising the contents of this section, one can note that studying the sequences of dependent trials forming homogeneous Markov chains with discrete sets of states can essentially be carried out with the help of results obtained for sequences of independent random variables. Studying other types of dependent trials requires, as a rule, other approaches.

13.5* The Behaviour of Transition Probabilities for Reducible Chains

Now consider a *finite* Markov chain *of the general type*. As we saw, its state space consists of the class of inessential states S^0 and several classes $S^1, \dots, S^l$ of essential states. To clarify the nature of the asymptotic behaviour of $p_{ij}(n)$ for such

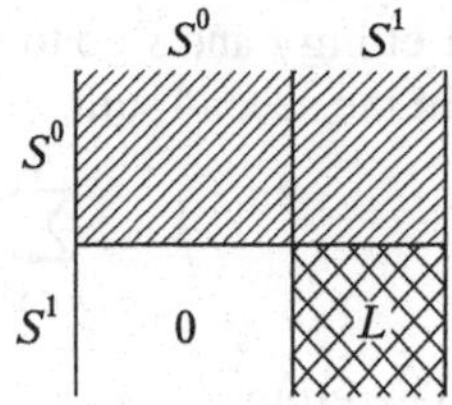

Fig. 13.4 The structure of the matrix of transition probabilities of a periodic Markov chain with the class S^0 of inessential states: an illustration to the proof of Theorem 13.2.3

chains, it suffices to consider the case where essential states constitute a single class without subclasses ($l = 1$). Here, the matrix of transition probabilities $p_{ij}(n)$ has the form depicted in Fig. 13.4.

By virtue of the ergodic theorem, the entries of the submatrix L have positive limits π_j. Thus it remains to analyse the behaviour of the entries in the upper part of the matrix.

Theorem 13.5.1 *Let $E_i \in S^0$. Then*

$$\lim_{t\to\infty} p_{ij}(t) = \begin{cases} 0, & \text{if } E_j \in S^0, \\ \pi_j > 0, & \text{if } E_j \in S^1. \end{cases}$$

Proof Let $E_j \in S^0$. Set

$$A_j(t) := \max_{E_i \in S^0} p_{ij}(t).$$

For any essential state E_r there exists an integer t_r such that $p_{ir}(t_r) > 0$. Since transition probabilities in L are all positive starting from some step, there exists an s such that $p_{il}(s) > 0$ for $E_i \in S^0$ and all $E_l \in S^1$. Therefore, for sufficiently large t,

$$p_{ij}(t) = \sum_{E_k \in S^0} p_{ik}(s)p_{kj}(t-s) \le A_j(t-s) \sum_{E_k \in S^0} p_{ik}(s),$$

where

$$q(i) := \sum_{E_k \in S^0} p_{ik}(s) = 1 - \sum_{E_k \in S^1} p_{ik}(s) < 1.$$

If we put $q := \max_{E_i \in S^0} q(i)$, then the displayed inequality implies that

$$A_j(t) \le q A_j(t-s) \le \cdots \le q^{[t/s]}.$$

Thus $\lim_{t\to\infty} p_{ij}(t) \le \lim_{t\to\infty} A_j(t) = 0$.

Now let $E_i \in S^0$ and $E_j \in S^1$. One has

$$p_{ij}(t+s) = \sum_k p_{ik}(t)p_{kj}(s) = \sum_{E_k \in S^0} p_{ik}(t)p_{kj}(s) + \sum_{E_k \in S^1} p_{ik}(t)p_{kj}(s).$$

Letting t and s go to infinity, we see that the first sum in the last expression is $o(1)$. In the second sum,

$$\sum_{E\in S^1} p_{ik}(t) = 1 + o(1); \qquad p_{kj}(t) = \pi_j + o(1).$$

Therefore

$$p_{ij}(t+s) = \pi_j \sum_{E_k\in S} p_{ik}(t) + o(t) = \pi_j + o(1)$$

as $t, s \to \infty$. The theorem is proved. □

Using Theorem 13.5.1, it is not difficult to see that the existence of the limit

$$\lim_{t\to\infty} p_{ij}(n) = \pi_j \ge 0$$

is a necessary and sufficient condition for the chain to have two classes S^0 and S^1, of which S^1 contains no subclasses.

13.6 Markov Chains with Arbitrary State Spaces. Ergodicity of Chains with Positive Atoms

13.6.1 Markov Chains with Arbitrary State Spaces

The Markov chains $X = \{X_n\}$ considered so far have taken values in the countable sets $\{1, 2, \ldots\}$ or $\{0, 1, \ldots\}$; such chains are called *countable* (*denumerable*) or *discrete*. Now we will consider Markov chains with values in an arbitrary set of states $\mathcal{X}$ endowed with a σ-algebra $\mathfrak{B}_{\mathcal{X}}$ of subsets of $\mathcal{X}$. The pair $(\mathcal{X}, \mathfrak{B}_{\mathcal{X}})$ forms a (measurable) *state space* of the chain $\{X_n\}$. Further let $(\Omega, \mathfrak{F}, \mathbf{P})$ be the underlying probability space. A measurable mapping Y of the space $(\Omega, \mathfrak{F})$ into $(\mathcal{X}, \mathfrak{B}_{\mathcal{X}})$ is called an *$\mathcal{X}$-valued random element.* If $\mathcal{X} = \mathbb{R}$ and $\mathfrak{B}_{\mathcal{X}}$ is the σ-algebra of Borel sets on the line, then Y will be a conventional random variable. The mapping Y could be the identity, in which case $(\Omega, \mathfrak{F}) = (\mathcal{X}, \mathfrak{B}_{\mathcal{X}})$ is also called a *sample space*.

Consider a sequence $\{X_n\}$ of $\mathcal{X}$-valued random elements and denote by $\mathfrak{F}_{k,m}$, $m \ge k$, the σ-algebra generated by the elements $X_k, \ldots, X_m$ (i.e. by events of the form $\{X_k \in B_k\}, \ldots, \{X_m \in B_m\}$, $B_i \in \mathfrak{B}_{\mathcal{X}}$, $i = k, \ldots, m$). It is evident that $\mathfrak{F}_n := \mathfrak{F}_{0,n}$ form a non-decreasing sequence $\mathfrak{F}_0 \subset \mathfrak{F}_1 \ldots \subset \mathfrak{F}_n \ldots$. The conditional expectation $\mathbf{E}(\xi\,|\,\mathfrak{F}_{k,m})$ will sometimes also be denoted by $\mathbf{E}(\xi\,|\,X_k, \ldots, X_m)$.

Definition 13.6.1 An *$\mathcal{X}$-valued Markov chain* is a sequence of $\mathcal{X}$-valued elements X_n such that, for any $B \in \mathfrak{B}_{\mathcal{X}}$,

$$\mathbf{P}(X_{n+1} \in B \mid \mathfrak{F}_n) = \mathbf{P}(X_{n+1} \in B \mid X_n) \quad \text{a.s.} \tag{13.6.1}$$

In the sequel, the words "almost surely" will, as a rule, be omitted.

By the properties of conditional expectations, relation (13.6.1) is clearly equivalent to the condition: for any measurable function $f : \mathcal{X} \to \mathbb{R}$, one has

$$\mathbf{E}\big(f(X_{n+1}) \mid \mathfrak{F}_n\big) = \mathbf{E}\big(f(X_{n+1}) \mid X_n\big). \tag{13.6.2}$$

Definition 13.6.1 is equivalent to the following.

Definition 13.6.2 A sequence $X = \{X_n\}$ forms a Markov chain if, for any $A \in \mathfrak{F}_{n+1,\infty}$,

$$\mathbf{P}(A|\mathfrak{F}_n) = \mathbf{P}(A|X_n) \tag{13.6.3}$$

or, which is the same, for any $\mathfrak{F}_{n+1,\infty}$-measurable function $f(\omega)$,

$$\mathbf{E}\big(f(\omega)|\mathfrak{F}_n\big) = \mathbf{E}\big(f(\omega)|X_n\big). \tag{13.6.4}$$

Proof of equivalence We have to show that (13.6.2) implies (13.6.3). First take any $B_1, B_2 \in \mathfrak{B}_{\mathcal{X}}$ and let $A := \{X_{n+1} \in B_1, X_{n+2} \in B_2\}$. Then, by virtue of (13.6.2),

$$\begin{aligned}\mathbf{P}(A|\mathfrak{F}_n) &= \mathbf{E}\big[\mathrm{I}(X_{n+1} \in B_1)\mathbf{P}(X_{n+2} \in B_2|\mathfrak{F}_{n+1})|\mathfrak{F}_n\big] \\ &= \mathbf{E}\big[\mathrm{I}(X_{n+1} \in B_1)\mathbf{P}(X_{n+2} \in B_2|X_{n+1})|\mathfrak{F}_n\big] \\ &= \mathbf{E}(A|X_n).\end{aligned}$$

This implies inequality (13.6.3) for any $A \in \mathcal{A}_{n+1,n+2}$, where $\mathcal{A}_{k,m}$ is the algebra generated by sets $\{X_k \in B_k, \dots, X_m \in B_m\}$. It is clear that $\mathcal{A}_{n+1,n+2}$ generates $\mathfrak{F}_{n+1,n+2}$. Now let $A \in \mathfrak{F}_{n+1,n+2}$. Then, by the approximation theorem, there exist $A_k \in \mathcal{A}_{n+1,n+2}$ such that $d(A, A_k) \to 0$ (see Sect. 3.4). From this it follows that $\mathrm{I}(A_k) \xrightarrow{p} \mathrm{I}(A)$ and, by the properties of conditional expectations (see Sect. 4.8.2),

$$\mathbf{P}\big(A_k|\mathfrak{F}^*\big) \xrightarrow{p} \mathbf{P}\big(A|\mathfrak{F}^*\big),$$

where $\mathfrak{F}^* \subset \mathfrak{F}$ is some σ-algebra. Put $P_A = P_A(\omega) := \mathbf{P}(A|X_n)$. We know that, for $A_k \in \mathcal{A}_{n+1,n+2}$,

$$\mathbf{E}(P_{A_k}; B) = \mathbf{P}(A_k B) \tag{13.6.5}$$

for any $B \in \mathfrak{F}_n$ (this just means that $P_{A_k}(\omega) = \mathbf{P}(A_k|\mathfrak{F}_n)$). Again making use of the properties of conditional expectations (the dominated convergence theorem, see Sect. 4.8.2) and passing to the limit in (13.6.5), we obtain that $\mathbf{E}(P_A; B) = \mathbf{P}(AB)$. This proves (13.6.3) for $A \in \mathfrak{F}_{n+1,n+2}$.

Repeating the above argument m times, we prove (13.6.3) for $A \in \mathfrak{F}_{n+1,m}$. Using a similar scheme, we can proceed to the case of $A \in \mathfrak{F}_{n+1,\infty}$. □

Note that (13.6.3) can easily be extended to events $A \in \mathfrak{F}_{n,\infty}$. In the above proof of equivalence, one could work from the very beginning with $A \in \mathfrak{F}_{n,\infty}$ (first with $A \in \mathcal{A}_{n,n+2}$, and so on).

We will give one more equivalent definition of the Markov property.

Definition 13.6.3 A sequence $\{X_n\}$ forms a Markov chain if, for any events $A \in \mathfrak{F}_n$ and $B \in \mathfrak{F}_{n,\infty}$,

$$\mathbf{P}(AB|X_n) = \mathbf{P}(A|X_n)\mathbf{P}(B|X_n). \tag{13.6.6}$$

This property means that the future is conditionally independent of the past given the present (conditional independence of $\mathfrak{F}_n$ and $\mathfrak{F}_{n,\infty}$ given X_n).

Proof of the equivalence Assume that (13.6.4) holds. Then, for $A \in \mathfrak{F}_n$ and $B \in \mathfrak{F}_{n,\infty}$,

$$\begin{aligned}\mathbf{P}(AB|X_n) &= \mathbf{E}\big[\mathbf{E}(\mathrm{I}_A\mathrm{I}_B|\mathfrak{F}_n)|X_n\big] = \mathbf{E}\big[\mathrm{I}_A\mathbf{E}(\mathrm{I}_B|\mathfrak{F}_n)|X_n\big] \\ &= \mathbf{E}\big[\mathrm{I}_A\mathbf{E}(\mathrm{I}_B|X_n)|X_n\big] = \mathbf{E}(\mathrm{I}_B|X_n)\mathbf{E}(\mathrm{I}_A|X_n),\end{aligned}$$

where I_A is the indicator of the event A.

Conversely, let (13.6.6) hold. Then

$$\begin{aligned}\mathbf{P}(AB) &= \mathbf{E}\mathbf{P}(AB|X_n) = \mathbf{E}\mathbf{P}(A|X_n)\mathbf{P}(B|X_n) \\ &= \mathbf{E}\mathbf{E}[\mathrm{I}_A\mathbf{P}(B|X_n)|X_n] = \mathbf{E}\mathrm{I}_A\mathbf{P}(B|X_n).\end{aligned} \tag{13.6.7}$$

On the other hand,

$$\mathbf{P}(AB) = \mathbf{E}\mathrm{I}_A\mathrm{I}_B = \mathbf{E}\mathrm{I}_A\,\mathbf{P}(B|\mathfrak{F}_n). \tag{13.6.8}$$

Since (13.6.7) and (13.6.8) hold for any $A \in \mathfrak{F}_n$, this means that

$$\mathbf{P}(B|X_n) = \mathbf{P}(B|\mathfrak{F}_n). \qquad \square$$

Thus, let $\{X_n\}$ be an $\mathcal{X}$-valued Markov chain. Then, by the properties of conditional expectations,

$$\mathbf{P}(X_{n+1} \in B|X_n) = P_{(n)}(X_n, B),$$

where the function $P_n(x, B)$ is, for each $B \in \mathfrak{B}_{\mathcal{X}}$, measurable in x with respect to the σ-algebra $\mathfrak{B}_{\mathcal{X}}$. In what follows, we will assume that the functions $P_{(n)}(x, B)$ are conditional distributions (see Definition 4.9.1), i.e., for each $x \in \mathcal{X}$, $P_{(n)}(x, B)$ is a probability distribution in B. Conditional distributions $P_{(n)}(x, B)$ always exist if the σ-algebra $\mathfrak{B}_{\mathcal{X}}$ is countably-generated, i.e. generated by a countable collection of subsets of $\mathcal{X}$ (see [27]). This condition is always met if $\mathcal{X} = \mathbb{R}^k$ and $\mathfrak{B}_{\mathcal{X}}$ is the σ-algebra of Borel sets. In our case, there is an additional problem that the "null probability" sets $\mathcal{N} \subset \mathcal{X}$, on which one can arbitrarily vary $P_{(n)}(x, B)$, can depend on the distribution of X_n, since the "null probability" is with respect to the distribution of X_n.

Definition 13.6.4 A Markov chain $X = \{X_n\}$ is called *homogeneous* if there exist conditional distributions $P_{(n)}(x, B) = P(x, B)$ independent of n and the initial value X_0 (or the distributions of X_n). The function $P(x, B)$ is called the *transition* *probability* (or *transition function*) of the homogeneous Markov chain. It can be graphically written as

$$P(x, B) = \mathbf{P}(X_1 \in B \mid X_0 = x). \tag{13.6.9}$$

If the Markov chain is countable, $\mathcal{X} = \{1, 2, \ldots\}$, then, in the notation of Sect. 13.1, one has $P(i, \{j\}) = p_{ij} = p_{ij}(1)$.

The transition probability and initial distribution (of X_0) completely determine the joint distribution of $X_0, \ldots, X_n$ for any n. Indeed, by the total probability formula and the Markov property

$$\begin{aligned} &\mathbf{P}(X_0 \in B_0, \ldots, X_n \in B_n) \\ &\quad = \int_{y_0 \in B_0} \cdots \int_{y_n \in B_n} \mathbf{P}(X_0 \in dy_0)\, P(y_0, dy_1) \cdots P(y_{n-1}, dy_n). \end{aligned} \tag{13.6.10}$$

A Markov chain with the initial value $X_0 = x$ will be denoted by $\{X_n(x)\}$.

In applications, Markov chains are usually given by their conditional distributions $P(x, B)$ or—in a "stronger form"—by explicit formulas expressing X_{n+1} in terms X_n and certain "control" elements (see Examples 13.4.2, 13.4.3, 13.6.1, 13.6.2, 13.7.1–13.7.3) which enable one to immediately write down transition probabilities. In such cases, as we already mentioned, the joint distribution of $(X_0, \ldots, X_n)$ can be defined in terms of the initial distribution of X_0 and the transition function $P(x, B)$ by formula (13.6.10). It is easily seen that the sequence $\{X_n\}$ with so defined joint distributions satisfy all the definitions of a Markov chain and has transition function $P(x, B)$. In what follows, wherever it is needed, we will assume condition (13.6.10) is satisfied. It can be considered as one more definition of a Markov chain, but a stronger one than Definitions 13.6.2–13.6.4, for it explicitly gives (or uses) the transition function $P(x, B)$.

One of the main objects of study will be the asymptotic behaviour of the n step transition probability:

$$P(x, n, B) := \mathbf{P}\big(X_n(x) \in B\big) = \mathbf{P}(X_n \in B \mid X_0 = x).$$

The following recursive relation, which follows from the total probability formula (or from (13.6.10)), holds for this function:

$$\mathbf{P}(X_{n+1} \in B) = \mathbf{E}\mathbf{E}\big(\mathrm{I}(X_{n+1} \in B) \mid \mathfrak{F}_n\big) = \int \mathbf{P}(X_n \in dy) P(y, B),$$

$$P(x, n+1, B) = \int P(x, n, dy) P(y, B). \tag{13.6.11}$$

Now note that the Markov property (13.6.3) of homogeneous chains can also be written in the form

$$\mathbf{P}(X_{n+k} \in B_k \mid \mathfrak{F}_n) = P(X_n, k, B_k),$$

or, more generally,

$$\mathbf{P}(X_{n+1} \in B_1, \ldots, X_{n+k} \in B_k | \mathfrak{F}_n) = \mathbf{P}\big(X_1^{\text{new}}(X_n) \in B_1, \ldots, X_k^{\text{new}}(X_n) \in B_k\big), \tag{13.6.12}$$

where $\{X_k^{\text{new}}(x)\}$ is a Markov chain independent of $\{X_n\}$ and having the same transition function as $\{X_n\}$ and the initial value x. Property (13.6.12) can be extended to a random time n. Recall the definition of a stopping time.

Definition 13.6.5 A random variable $\nu \ge 0$ is called a *Markov* or *stopping time* with respect to $\{\mathfrak{F}_n\}$ if $\{\nu \le n\} \in \mathfrak{F}_n$. In other words, that the event $\{\nu \le n\}$ occurred or not is completely determined by the trajectory segment $X_0, X_1, \ldots, X_n$.

Note that, in Definition 13.6.5, by $\mathfrak{F}_n$ one often understands wider σ-algebras, the essential requirements being the relations $\{\nu \le n\} \in \mathcal{F}_n$ and measurability of $X_0, \ldots, X_n$ with respect to $\mathfrak{F}_n$.

Denote by $\mathfrak{F}_\nu$ the σ-algebra of events B such that $B \cap \{\nu = k\} \in \mathfrak{F}_k$. In other words, $\mathcal{F}_\nu$ can be thought of as the σ-algebra generated by the sets $\{\nu = k\}B_k$, $B_k \in \mathfrak{F}_k$, i.e. by the trajectory of $\{X_n\}$ until time ν.

Lemma 13.6.1 (The Strong Markov Property) *For any $k \ge 1$ and $B_1, \ldots, B_k \in \mathfrak{B}_\mathcal{X}$,*

$$\mathbf{P}(X_{\nu+1} \in B_1, \ldots, X_{\nu+k} \in B_k | \mathfrak{F}_\nu) = \mathbf{P}\big(X_1^{\text{new}}(X_\nu) \in B_1, \ldots, X_k^{\text{new}}(X_\nu) \in B_k\big),$$

where the process $\{X_k^{\text{new}}\}$ is defined in (13.6.12).

Thus, after a random stopping time ν, the trajectory $X_{\nu+1}, X_{\nu+2}, \ldots$ will evolve according to the same laws as $X_1, X_2, \ldots$, but with the initial condition X_ν. This property is called the *strong Markov property*. It will be used below for the first hitting times $\nu = \tau_V$ of certain sets $V \subset \mathcal{X}$ by $\{X_n\}$. We have already used this property tacitly in Sect. 13.4, when the set V coincided with a point, which allowed us to cut the trajectory of $\{X_n\}$ into independent cycles.

Proof of Lemma 13.6.1 For the sake of simplicity, consider one-dimensional distributions. We have to prove that

$$\mathbf{P}(X_{\nu+1} \in B_1 | \mathfrak{F}_\nu) = P(X_\nu, B_1).$$

For any $A \in \mathfrak{F}_\nu$,

$$\begin{aligned}
\mathbf{E}\big(P(X_\nu, B_1); A\big) &= \sum_n \mathbf{E}\big(P(X_n, B_1); A\{\nu = n\}\big) \\
&= \sum_n \mathbf{E}\mathbf{E}\big(I\big(A\{\nu = n\}\{X_{n+1} \in B_1\}\big) | \mathfrak{F}_n\big) \\
&= \sum_n \mathbf{P}\big(A\{\nu = n\}\{X_{n+1} \in B_1\}\big) = \mathbf{P}\big(A\{X_{\nu+1} \in B_1\}\big).
\end{aligned}$$

But this just means that $P(X_\nu, B_1)$ is the required conditional expectation. The case of multi-dimensional distributions is dealt with in the same way, and we leave it to the reader. □

Now we turn to consider the asymptotic properties of distributions $P(x, n, B)$ as $n \to \infty$.

Definition 13.6.6 A distribution $\pi(\cdot)$ on $(\mathcal{X}, \mathfrak{B}_\mathcal{X})$ is called *invariant* if it satisfies the equation

$$\pi(B) = \int \pi(dy)P(y, B), \quad B \in \mathfrak{B}_\mathcal{X}. \qquad (13.6.13)$$

It follows from (13.6.11) that if $X_n \in \pi$, then $X_{n+1} \in \pi$. The distribution π is also called *stationary*.

For Markov chains in arbitrary state spaces $\mathcal{X}$, a simple and complete classification similar to the one carried out for countable chains in Sect. 13.1 is not possible, although some notions can be extended to the general case.

Such natural and important notions for countable chains as, say, irreducibility of a chain, take in the general case another form.

Example 13.6.1 Let $X_{n+1} = X_n + \xi_n \pmod 1$ (X_{n+1} is the fractional part of $X_n + \xi_n$), ξ_n be independent and identically distributed and take with positive probabilities the two values 0 and $\sqrt{2}$. In this example, the chain "splits", according to the initial state x, into a continual set of "subchains" with state spaces of the form $M_x = \{x + k\sqrt{2} \pmod 1),\ k = 0, 1, 2 \ldots\}$. It is evident that if $x_1 - x_2$ is not a multiple of $\sqrt{2} \pmod 1$, then M_{x_1} and M_{x_2} are disjoint, $\mathbf{P}(X_n(x_1) \in M_{x_2}) = 0$ and $\mathbf{P}(X_n(x_2) \in M_{x_1}) = 0$ for all n. Thus the chain is clearly reducible. Nevertheless, it turns out that the chain is ergodic in the following sense: for any x, $X_n(x) \Leftrightarrow U_{0,1}$ ($P(x, n, [0, t]) \to t$) as $n \to \infty$ (see, e.g., [6], [18]). For the most commonly used irreducibility conditions, see Sect. 13.7.

Definition 13.6.7 A chain is called *periodic* if there exist an integer $d \geq 2$ and a set $\mathcal{X}_1 \subset \mathcal{X}$ such that, for $x \in \mathcal{X}_1$, one has $P(x, n, \mathcal{X}_1) = \mathbf{P}(X_n(x) \in \mathcal{X}_1) = 1$ for $n = kd$, $k = 1, 2, \ldots$, and $P(x, n, \mathcal{X}_1) = 0$ for $n \neq kd$.

Periodicity means that the whole set of states $\mathcal{X}$ is decomposed into subclasses $\mathcal{X}_1, \ldots, \mathcal{X}_d$, such that $\mathbf{P}(X_1(x) \in \mathcal{X}_{k+1}) = 1$ for $x \in \mathcal{X}_k$, $k = 1, \ldots, d$, $\mathcal{X}_{d+1} = \mathcal{X}_1$. In the absence of such a property, the chain will be called *aperiodic*.

A state $x_0 \in \mathcal{X}$ is called an *atom* of the chain X if, for any $x \in \mathcal{X}$,

$$\mathbf{P}\left(\bigcup_{n=1}^{\infty} \{X_n(x) = x_0\}\right) = 1.$$

Example 13.6.2 Let $X_0 \geq 0$ and, for $n \geq 0$,

$$X_{n+1} = \begin{cases} (X_n + \xi_{n+1})^+ & \text{if } X_n > 0, \\ \eta_{n+1} & \text{if } X_n = 0, \end{cases}$$

where ξ_n and $\eta_n \geq 0$, $n = 1, 2, \ldots$, are two sequences of independent random variables, identically distributed in each sequence. It is clear that $\{X_n\}$ is a Markov chain and, for $\mathbf{E}\xi_k < 0$, by the strong law of large numbers, this chain has an atom at the point $x_0 = 0$:

$$\mathbf{P}\left(\bigcup_{n=1}^{\infty} \{X_n(x) = 0\}\right) = \mathbf{P}\Big(\inf_k S_k \leq -x\Big) = 1,$$

where $S_k = \sum_{j=1}^k \xi_j$. This chain is a generalisation of the Markov chain from Example 13.4.3.

Markov chains in an arbitrary state space $\mathcal{X}$ are rather difficult to study. However, if a chain has an atom, the situation may become much simpler, and the ergodic theorem on the asymptotic behaviour of $P(x, n, B)$ as $n \to \infty$ can be proved using the approaches considered in the previous sections.

13.6.2 Markov Chains Having a Positive Atom

Let x_0 be an atom of a chain $\{X_n\}$. Set

$$\tau := \min\bigl\{k > 0 : X_k(x_0) = x_0\bigr\}.$$

This is a proper random variable ($\mathbf{P}(\tau < \infty) = 1$).

Definition 13.6.8 The atom x_0 is said to be *positive* if $\mathbf{E}\tau < \infty$.

In the terminology of Sect. 13.4, x_0 is a recurrent non-null (positive) state.

To characterise convergence of distributions in arbitrary spaces, we will need the notions of the total variation distance and convergence in total variation. If $\mathbf{P}$ and $\mathbf{Q}$ are two distributions on $(\mathcal{X}, \mathfrak{B}_{\mathcal{X}})$, then the *total variation distance* between them is defined by

$$\|\mathbf{P} - \mathbf{Q}\| = 2 \sup_{B \in \mathfrak{B}_{\mathcal{X}}} \bigl|\mathbf{P}(B) - Q(B)\bigr|.$$

One says that a sequence of distributions $\mathbf{P}_n$ on $(\mathcal{X}, \mathfrak{B}_{\mathcal{X}})$ converges in total variation to $\mathbf{P}$ ($\mathbf{P}_n \xrightarrow{TV} \mathbf{P}$) if $\|\mathbf{P}_n - \mathbf{P}\| \to 0$ as $n \to \infty$. For more details, see Sect. 3.6.2 of Appendix 3.

As in Sect. 13.4, denote by $P_{x_0}(k, B)$ the "taboo probability"

$$P_{x_0}(k, B) := \mathbf{P}\bigl(X_k(x_0) \in B, X_1(x_0) \neq x_0, \ldots, X_{k-1}(x_0) \neq x_0\bigr)$$

of transition from x_0 into B in k steps without visiting the "forbidden" state x_0.

Theorem 13.6.1 *If the chain $\{X_n\}$ has a positive atom and the g.c.d. of the possible values of τ is 1, then the chain is ergodic in the convergence in total variation sense:*

there exists a unique invariant distribution π *such that, for any* $x \in \mathfrak{X}$, *as* $n \to \infty$,

$$\left\| P(x, n, \cdot) - \pi(\cdot) \right\| \to 0. \tag{13.6.14}$$

Moreover, for any $B \in \mathfrak{B}_{\mathfrak{X}}$,

$$\pi(B) = \frac{1}{\mathbf{E}\tau} \sum_{k=1}^{\infty} P_{x_0}(k, B). \tag{13.6.15}$$

If we denote by $X_n(\boldsymbol{\mu}_0)$ *a Markov chain with the initial distribution* $\boldsymbol{\mu}_0$ ($X_0 \Subset \boldsymbol{\mu}_0$) *and put*

$$P(\boldsymbol{\mu}_0, n, B) := \mathbf{P}\big(X_n(\boldsymbol{\mu}_0) \in B\big) = \int \boldsymbol{\mu}_0(dx)\, P(x, n, B),$$

then, as well as (13.6.14), *we will also have that, as* $n \to \infty$,

$$\left\| P(\boldsymbol{\mu}_0, n, \cdot) - \pi(\cdot) \right\| \to 0 \tag{13.6.16}$$

for any initial distribution $\boldsymbol{\mu}_0$.

The condition that there exists a positive atom is an analogue of conditions (I) and (II) of Theorem 13.4.1. A number of conditions sufficient for the finiteness of $\mathbf{E}\tau$ can be found in Sect. 13.7. The condition on the g.c.d. of possible values of τ is the aperiodicity condition.

Proof We will effectively repeat the proof of Theorem 13.4.1. First let $X_0 = x_0$. As in Theorem 13.4.1 (we keep the notation of that theorem), we find that

$$\begin{aligned}
&P(x_0, n, B)\\
&\quad = \sum_{k=1}^{n} \mathbf{P}\big(\gamma(n) = k\big)\, \mathbf{P}(X_n \in B \,|\, X_{n-k} = x_0, X_{n-k+1} \neq x_0, \dots, X_{n-1} \neq x_0)\\
&\quad = \sum_{k=1}^{n} \frac{\mathbf{P}(\gamma(n) = k)}{\mathbf{P}(\tau \geq k)} \mathbf{P}(\tau \geq k)\, \mathbf{P}(X_k \in B \,|\, X_0 = x_0, X_1 \neq x_0, \dots, X_{k-1} \neq x_0)\\
&\quad = \sum_{k=1}^{n} \frac{\mathbf{P}(\gamma(n) = k)}{\mathbf{P}(\tau \geq k)} P_{x_0}(k, B).
\end{aligned}$$

For the measure π defined in (13.6.15) one has

$$\begin{aligned}
&P(x_0, n, B) - \pi(B)\\
&\quad = \sum_{k=1}^{n} \left(\frac{\mathbf{P}(\gamma(n) = k)}{\mathbf{P}(\tau \geq k)} - \frac{1}{\mathbf{E}\tau} \right) P_{x_0}(k, B) - \frac{1}{\mathbf{E}\tau} \sum_{k>n} P_{x_0}(k, B).
\end{aligned}$$

Since $\mathbf{P}(\gamma(n)=k) \le \mathbf{P}(\tau \ge k)$ and $P_{x_0}(k,B) \le \mathbf{P}(\tau \ge k)$ (see the proof of Theorem 13.4.1), one has, for any N,

$$\sup_B \bigl|P(x_0,n,B)-\pi(B)\bigr| \le \sum_{k=1}^{N}\biggl(\frac{\mathbf{P}(\gamma(n)=k)}{\mathbf{P}(\tau\ge k)}-\frac{1}{\mathbf{E}\tau}\biggr)+2\sum_{k>N}\mathbf{P}(\tau \ge k). \tag{13.6.17}$$

Further, since

$$\mathbf{P}\bigl(\gamma(n)=k\bigr)\to \mathbf{P}(\tau\ge k)/\mathbf{E}\tau, \qquad \sum_{k=1}^{\infty}\mathbf{P}(\tau\ge k)=\mathbf{E}\tau<\infty,$$

the right-hand side of (13.6.17) can be made arbitrarily small by choosing N and then n. Therefore,

$$\lim_{n\to\infty}\sup_B\bigl|P(x_0,n,B)-\pi(B)\bigr|=0.$$

Now consider an arbitrary initial state $x\in\mathcal{X}$, $x\ne x_0$. Since x_0 is an atom, for the probabilities

$$F(x,k,x_0):=\mathbf{P}\bigl(X_k(x)=x_0,\,X_1\ne x_0,\ldots,X_{k-1}\ne x_0\bigr)$$

of hitting x_0 for the first time on the k-th step, one has

$$\sum_k F(x,k,x_0)=1, \qquad P(x,n,B)=\sum_{k=1}^{n}F(x,k,x_0)P(x_0,n-k,B),$$

$$\begin{aligned}&\bigl\|P(x,n,\cdot)-\pi(\cdot)\bigr\|\\ &\quad\le \sum_{k\le n/2}F(x,k,x_0)\bigl\|P(x_0,n-k,\cdot)-\pi(\cdot)\bigr\|+2\sum_{k>n/2}F(x,k,x_0)\to 0\end{aligned}$$

as $n\to\infty$.

Relation (13.6.16) follows from the fact that

$$\bigl\|P(\mu_0,n,\cdot)-\pi(\cdot)\bigr\|\le\int\mu_0(dx)\bigl\|P(x,n,\cdot)-\pi(\cdot)\bigr\|\to 0$$

by the dominated convergence theorem.

Further, from the convergence of $P(x,n,\cdot)$ in total variation it follows that

$$\int P(x,n,dy)P(y,B)\to\int\pi(dy)P(y,B).$$

Since the left hand-side of this relation is equal to $P(x,n+1,B)$ by virtue of (13.6.11) and converges to $\pi(B)$, one has (13.6.13), and hence π is an invariant measure.

Now assume that π_1 is another invariant distribution. Then

$$\pi_1(\cdot) = \mathbf{P}(\pi_1, n, \cdot) \xrightarrow{TV} \pi(\cdot), \quad \pi_1 = \pi.$$

The theorem is proved. □

Returning to Example 13.6.2, we show that the conditions of Theorem 13.6.1 are met provided that $\mathbf{E}\xi_k < 0$ and $\mathbf{E}\eta_k < \infty$. Indeed, put

$$\eta(-x) := \min\left\{k \ge 1 : S_k = \sum_{j=1}^{k} \xi_j \le -x\right\}.$$

By the renewal Theorem 10.1.1,

$$H(x) = \mathbf{E}\eta(-x) \sim \frac{x}{|\mathbf{E}\xi_1|} \quad \text{as} \quad x \to \infty$$

for $\mathbf{E}\xi_1 < 0$, and therefore there exist constants c_1 and c_2 such that $H(x) < c_1 + c_2 x$ for all $x \ge 0$. Hence, for the atom $x_0 = 0$, we obtain that

$$\mathbf{E}\tau = \int_0^\infty \mathbf{P}(\eta_1 \in dx)\, H(x) \le c_1 + c_2 \int_0^\infty x\mathbf{P}(\eta_1 \in dx) = c_1 + c_2 \mathbf{E}\eta_1 < \infty.$$

13.7* Ergodicity of Harris Markov Chains

13.7.1 The Ergodic Theorem

In this section we will consider the problem of establishing ergodicity of Markov chains in arbitrary state spaces $(\mathfrak{X}, \mathfrak{B}_\mathfrak{X})$. A lot of research has been done on this problem, the most important advancements being associated with the names of W. Döblin, J.L. Doob, T.E. Harris and E. Omey. Until recently, this research area had been considered as a rather difficult one, and not without reason. However, the construction of an artificial atom suggested by K.B. Athreya, P.E. Ney and E. Nummelin (see, e.g. [6, 27, 29]) greatly simplified considerations and allowed the proof of ergodicity by reducing the general case to the special case discussed in the last section.

In what follows, the notion of a "Harris chain" will play an important role. For a fixed set $V \in \mathfrak{B}_\mathfrak{X}$, define the random variable

$$\tau_V(x) = \min\{k \ge 1 : X_k(x) \in V\},$$

the time of the first hitting of V by the chain starting from the state x (we put $\tau_V(x) = \infty$ if all $X_k(x) \notin V$).

Definition 13.7.1 A Markov chain $X = \{X_n\}$ in $(\mathcal{X}, \mathfrak{B}_{\mathcal{X}})$ is said to be a *Harris chain* (or *Harris irreducible*) if there exists a set $V \in \mathfrak{B}_{\mathcal{X}}$, a probability measure $\boldsymbol{\mu}$ on $(\mathcal{X}, \mathfrak{B}_{\mathcal{X}})$, and numbers $n_0 \ge 1$, $p \in (0, 1)$ such that

(I_0) $\mathbf{P}(\tau_V(x) < \infty) = 1$ for all $x \in \mathcal{X}$; and
(II) $P(x, n_0, B) \ge p\boldsymbol{\mu}(B)$ for all $x \in V$, $B \in \mathfrak{B}_{\mathcal{X}}$.

Condition (I_0) plays the role of an irreducibility condition: starting from any point $x \in \mathcal{X}$, the trajectory of X_n will sooner or later visit the set V. Condition (II) guarantees that, after n_0 steps since hitting V, the distribution of the walking particle will be minorised by a common "distribution" $p\boldsymbol{\mu}(\cdot)$. This condition is sometimes called a "mixing condition"; it ensures a "partial loss of memory" about the trajectory's past. This is not the case for the chain from Example 13.6.1 for which condition (II) does not hold for any V, $\boldsymbol{\mu}$ or n_0 ($P(x, \cdot)$ form a collection of mutually singular distributions which are singular with respect to Lebesgue measure).

If a chain has an atom x_0, then conditions (I_0) and (II) are always satisfied for $V = \{x_0\}$, $n_0 = 1$, $p = 1$, and $\boldsymbol{\mu}(\cdot) = P(x_0, \cdot)$, so that such a chain is a Harris chain.

The set V is usually chosen to be a "compact" set (if $\mathcal{X} = \mathbb{R}^k$, it will be a bounded set), for otherwise one cannot, as a rule, obtain inequalities in (II). If the space $\mathcal{X}$ is "compact" itself (a finite or bounded subset of $\mathbb{R}^k$), condition (II) can be met for $V = \mathcal{X}$ (condition (I_0) then always holds). For example, if $\{X_n\}$ is a finite, irreducible and aperiodic chain, then by Theorem 13.4.2 there exists an n_0 such that $P(i, n_0, j) \ge p > 0$ for all i and j. Therefore condition (II) holds for $V = \mathcal{X}$ if one takes $\boldsymbol{\mu}$ to be a uniform distribution on $\mathcal{X}$.

One could interpret condition (II) as that of the presence, in all distributions $P(x, n_0, \cdot)$ for $x \in V$, of a component which is absolutely continuous with respect to the measure $\boldsymbol{\mu}$:

$$\inf_{x \in V} \frac{P(x, n_0, dy)}{\boldsymbol{\mu}(dy)} \ge p > 0.$$

We will also need a condition of "positivity" (positive recurrence) of the set V (or that of "positivity" of the chain):

(I) $\sup_{x \in V} \mathbf{E}\tau_V(x) < \infty$,

and the aperiodicity condition which will be written in the following form. Let $X_k(\boldsymbol{\mu})$ be a Markov chain with an initial value $X_0 \Subset \boldsymbol{\mu}$, where $\boldsymbol{\mu}$ is from condition (II). Put

$$\tau_V(\boldsymbol{\mu}) := \min\bigl\{k \ge : X_k(\boldsymbol{\mu}) \in V\bigr\}.$$

It is evident that $\tau_V(\boldsymbol{\mu})$ is, by virtue of (I_0), a proper random variable. Denote by $n_1, n_2, \ldots$ the possible values of $\tau_V(\boldsymbol{\mu})$, i.e. the values for which

$$\mathbf{P}\bigl(\tau_V(\boldsymbol{\mu}) = n_k\bigr) > 0, \quad k = 1, 2, \ldots.$$

Then the aperiodicity condition will have the following form.

(III) *There exists a* $k \ge 1$ *such that*

$$\text{g.c.d.}\{n_0 + n_1, n_0 + n_2, \ldots, n_0 + n_k\} = 1,$$

where n_0 *is from condition* (II).

Condition (III) is always satisfied if (II) holds for $n_0 = 1$ and $\mu(V) > 0$ (then $n_1 = 0$, $n_0 + n_1 = 1$).

Verifying condition (I) usually requires deriving bounds for $\mathbf{E}\tau_V(x)$ for $x \notin V$ which would automatically imply (I_0) (see the examples below).

Theorem 13.7.1 *Suppose conditions* (I_0), (I), (II) *and* (III) *are satisfied for a Markov chain* X, *i.e. the chain is an aperiodic positive Harris chain. Then there exists a unique invariant distribution* π *such that, for any initial distribution* μ_0, *as* $n \to \infty$,

$$\bigl\| P(\mu_0, n, \cdot) - \pi(\cdot) \bigr\| \to 0. \qquad (13.7.1)$$

The proof is based on the use of the above-mentioned construction of an "artificial atom" and reduction of the problem to Theorem 13.6.1. This allows one to obtain, in the course of the proof, a representation for the invariant measure π similar to (13.6.15) (see (13.7.5)).

A remarkable fact is that the conditions of Theorem 13.7.1 are necessary for convergence (13.7.1) (for more details, see [6]).

Proof of Theorem 13.7.1 For simplicity's sake, assume that $n_0 = 1$. First we will construct an "extended" Markov chain $X^* = \{X_n^*\} = \{\widetilde{X}_n, \omega(n)\}$, $\omega(n)$ being a sequence of independent identically distributed random variables with

$$\mathbf{P}\bigl(\omega(n) = 1\bigr) = p, \qquad \mathbf{P}\bigl(\omega(n) = 0\bigr) = 1 - p.$$

The joint distribution of $(\widetilde{X}(n), \omega(n))$ in the state space

$$\mathcal{X}^* := \mathcal{X} \times \{0, 1\} = \bigl\{x^* = (x, \delta) : x \in \mathcal{X}; \delta = 0, 1\bigr\}$$

and the transition function P^* of the chain X^* are defined as follows (the notation $X_n^*(x^*)$ has the same meaning as $X_n(x)$):

$$\mathbf{P}\bigl(X_1^*(x^*) \in (B, \delta)\bigr) =: P^*\bigl(x^*, (B, \delta)\bigr) = P(x, B)\, \mathbf{P}\bigl(\omega(1) = \delta\bigr) \quad \text{for } x \notin V$$

(i.e., for $\widetilde{X}_n \notin V$, the components of X_{n+1}^* are "chosen at random" independently with the respective marginal distributions). But if $x \in V$, the distribution of $X^*(x^*, 1)$ is given by

$$\mathbf{P}(X_1^*\bigl((x, 1) \in (B, \delta)\bigr) = P^*\bigl((x, 1), (B, \delta)\bigr) = \mu(B)\, \mathbf{P}\bigl(\omega(1) = \delta\bigr),$$
$$\mathbf{P}(X_1^*\bigl((x, 0) \in (B, \delta)\bigr) = P^*\bigl((x, 0), (B, \delta)\bigr) = Q(x, B)\, \mathbf{P}\bigl(\omega(1) = \delta\bigr),$$

where

$$Q(x, B) := \big(P(x, B) - p\mu(B)\big)/(1 - p),$$

so that, for any $B \in \mathfrak{B}_{\mathcal{X}}$,

$$p\mu(B) + (1 - p)Q(x, B) = P(x, B). \tag{13.7.2}$$

Thus $\mathbf{P}(\omega(n + 1) = 1| X_n^*) = p$ for any values of X_n^*. However, when "choosing" the value $\widetilde{X}_{n+1}$ there occurs (only when $\widetilde{X}_n \in V$) a partial randomisation (or splitting): for $\widetilde{X}_n \in V$, we let $\mathbf{P}(\widetilde{X}_{n+1} \in B| X_n^*)$ be equal to the value $\mu(B)$ (not depending on $\widetilde{X}_n \in V$!) provided that $\omega(n) = 1$. If $\omega(n) = 0$, then the value of the probability is taken to be $Q(\widetilde{X}_n, B)$. It is evident that, by virtue of condition (II) (for $n_0 = 1$), $\mu(B)$ and $Q(x, B)$ are probability distributions, and by equality (13.7.2) the first component $\widetilde{X}_n$ of the process X_n^* has the property $\mathbf{P}(\widetilde{X}_{n+1} \in B| \widetilde{X}_n) = P(\widetilde{X}_n, B)$, and therefore the distributions of the sequences X and $\widetilde{X}$ coincide.

As we have already noted, the "extended" process $X^*(n)$ possesses the following property: the conditional distribution $\mathbf{P}(X_{n+1}^* \in (B, \delta)| X_n^*)$ does not depend on $X^*(n)$ on the set $X_n^* \in V^* := (V, 1)$ and is there the known distribution $\mu(B)\mathbf{P}(\omega(1) = \delta)$. This just means that visits of the chain X^* to the set V^* divide the trajectory of X^* into independent cycles, in the same way as it happens in the presence of a positive atom.

We described above how one constructs the distribution of X^* from that of X. Now we will give obvious relations reconstructing the distribution of X from that of the chain X^*:

$$\mathbf{P}\big(X_n(x) \in B\big) = p\,\mathbf{P}(X_n^*\big((x, 1) \in B^*\big) + (1 - p)\,\mathbf{P}\big(X_n^*(x, 0) \in B^*\big), \tag{13.7.3}$$

where $B^* := (B, 0) \cup (B, 1)$. Note also that, if we consider $X_n = \widetilde{X}_n$ as a component of X_n^*, we need to write it as a function $X_n(x^*)$ of the initial value $x^* \in \mathcal{X}^*$.

Put

$$\tau^* := \min\big\{k \ge 1 : X_k^*\,\big(x^*\big) \in V^*\big\}, \quad x^* \in V^* = (V, 1).$$

It is obvious that τ^* does not depend on the value $x^* = (x, 1)$, since $X_1(x^*)$ has the distribution μ for any $x \in V$. This property allows one to identify the set V^* with a single point. In other words, one needs to consider one more state space $\mathcal{X}^{**}$ which is obtained from $\mathcal{X}^*$ if we replace the set $V^* = (V, 1)$ by a point to be denoted by x_0. In the new state space, we construct a chain X^{**} equivalent to X^* using the obvious relations for the transition probability P^{**}:

$$P^{**}\big(x^*, (B, \delta)\big) := P^*\big(x^*, (B, \delta)\big) \quad \text{for } x^* \ne (V, 1) = V^*,\ (B, d) \ne V^*,$$

$$P^{**}\big(x_0, (B, \delta)\big) := p\mu(B), \qquad P^{**}\big(x^*, x_0\big) := P^*\big(x^*, V^*\big).$$

Thus we have constructed a chain X^{**} with the transition function P^{**}, and this chain has atom x_0. Clearly, $\tau^* = \min\{k \ge 1 : X_k^{**}(x_0) = x_0\}$. We now prove that this atom is positive. Put

$$E := \sup_{x\in V} \mathbf{E}\tau_V(x).$$

Lemma 13.7.1 $\mathbf{E}\tau^* \le \frac{2}{p}E$.

Proof Consider the evolution of the first component $X_k(x^*)$ of the process $X_k^*(x^*)$, $x^* \in V^*$. Partition the time axis $k \ge 0$ into intervals by hitting the set V by $X_k(x^*)$. Let $\tau_1 \ge 1$ be the first such hitting time (recall that $X_1(x^*) \stackrel{d}{=} X_0(\boldsymbol{\mu})$ has the distribution $\boldsymbol{\mu}$, so that $\tau_1 = 1$ if $\boldsymbol{\mu}(V) = 1$). Prior to time τ_1 (in the case $\tau_1 > 1$) transitions of $X_k(x^*)$, $k \ge 2$, were governed by the transition function $P(y, B)$, $y \in V^c = \mathcal{X} \setminus V$. At time τ_1, according to the definition of X^*, one carries out a Bernoulli trial independent of the past history of the process with success (which is the event $\omega(\tau_1) = 1$) probability p. If $\omega(\tau_1) = 1$ then $\tau^* = \tau_1$. If $\omega(\tau_1) = 0$ then the transition from $X_{\tau_1}(x^*)$ to $X_{\tau_1+1}(x^*)$ is governed by the transition function $Q(y, B) = (P(y, B) - p\boldsymbol{\mu}(B))/(1 - p)$, $y \in V$. The further evolution of the chain is similar: if $\tau_1 + \tau_2$ is the time of the second visit of $X(x^*, k)$ to V (in the case $\omega(\tau_1) = 0$) then in the time interval $[\tau_1 + 1, \tau_2]$ transitions of $X(x^*, k)$ occur according to the transition function $P(y, B)$, $y \in V^c$. At time $\tau_1 + \tau_2$ one carries out a new Bernoulli trial with the outcome $\omega(\tau_1 + \tau_2)$. If $\omega(\tau_1 + \tau_2) = 1$, then $\tau^* = \tau_1 + \tau_2$. If $\omega(\tau_1 + \tau_2) = 0$, then the transition from $X(x^*, \tau_1 + \tau_2)$ to $X(x^*, \tau_1 + \tau_2 + 1)$ is governed by $Q(y, B)$, and so on.

In other words, the evolution of the component $X_k(x^*)$ of the process $X_k^*(x^*)$ is as follows. Let $\widetilde{X} = \{\widetilde{X}_k\}$, $k = 1, 2, \ldots$, be a Markov chain with the distribution $\boldsymbol{\mu}$ at time $k = 1$ and transition probability $Q(x, B)$ at times $k \ge 2$,

$$Q(x, B) = \begin{cases} (P(x, B) - p\boldsymbol{\mu}(B))/(1 - p) & \text{if } x \in V, \\ P(x, B) & if x \in V^c. \end{cases}$$

Define T_i as follows:

$$T_0 := 0, \qquad T_1 = \tau_1 = \min\{k \ge 1 : \widetilde{X}_k \in V\},$$
$$T_i := \tau_1 + \cdots + \tau_i = \min\{k > T_{i-1} : \widetilde{X}_k \in V\}, \quad i \ge 2.$$

Let, further, ν be a random variable independent of $\widetilde{X}$ and having the geometric distribution

$$\mathbf{P}(\nu = k) = (1 - p)^{k-1} p, \quad k \ge 1, \quad \nu = \min\bigl\{k \ge 1 : \omega(T_k) = 1\bigr\}. \tag{13.7.4}$$

Then it follows from the aforesaid that the distribution of $X_1(x^*), \ldots, X_{\tau^*}(x^*)$ coincides with that of $\widetilde{X}_1, \ldots, \widetilde{X}_\nu$; in particular, $\tau^* = T_\nu$, and

$$\mathbf{E}\tau^* = \sum_{k=1}^{\infty} p(1 - p)^{k-1} \mathbf{E}T_k.$$

Further, since $\boldsymbol{\mu}(B) \le P(x, B)/p$ for $x \in V$, then, for any $x \in V$,

$$\mathbf{E}\tau_1 = \boldsymbol{\mu}(V) + \int_{V^c} \boldsymbol{\mu}(du)\bigl(1 + \mathbf{E}\tau_V(u)\bigr)$$

$$\le \frac{1}{p}\left[P(x,V)+\int_{V^c}P(x,du)\bigl(1+\mathbf{E}\tau_V(u)\bigr)\right]=\frac{\mathbf{E}\tau_V(x)}{p}\le\frac{E}{p}.$$

To bound $\mathbf{E}\tau_i$ for $i\ge 2$, we note that $Q(x,B)\le(1-p)^{-1}P(x,B)$ for $x\in V$. Therefore, if we denote by $\mathfrak{F}_{(i)}$ the σ-algebra generated by $\{\widetilde{X}_k,\,\omega(\tau_k)\}$ for $k\le T_i$, then

$$\begin{aligned}\mathbf{E}(\tau_i|\mathfrak{F}_{(i-1)}) &\le \sup_{x\in V}\left[Q(x,V)+\int_{V^c}Q(x,du)\bigl(1+\mathbf{E}\tau_V(u)\bigr)\right]\\ &\le \frac{1}{1-p}\sup_{x\in V}\left[P(x,V)+\int_{V^c}P(x,du)\bigl(1+\mathbf{E}\tau_V(u)\bigr)\right]\\ &=(1-p)^{-1}\sup_{x\in V}\mathbf{E}\tau_V(x)=E(1-p)^{-1}.\end{aligned}$$

This implies the inequality $\mathbf{E}T_k\le E(1/p+(k-1)/(1-p))$, from which we obtain that

$$\mathbf{E}\tau^*\le E\left(1/p+p\sum_{k=1}^{\infty}(k-1)(1-p)^{k-2}\right)=2E/p.$$

The lemma is proved. □

We return to the proof of the theorem. To make use of Theorem 13.6.1, we now have to show that $\mathbf{P}(\tau^*(x^*)<\infty)=1$ for any $x^*\in\mathcal{X}^*$, where

$$\tau^*(x^*):=\min\bigl\{k\ge 1: X_k^*(x^*)\in V^*\bigr\}.$$

But the chain X visits V with probability 1. After ν visits to V (ν was defined in (13.7.4)), the process $X^*=(X(n),\omega(n))$ will be in the set V^*.

The aperiodicity condition for $n_0=1$ will be met if $\boldsymbol{\mu}(V)>0$. In that case we obtain by virtue of Theorem 13.6.1 that there exists a unique invariant measure π^* such that, for any $x^*\in\mathcal{X}^*$,

$$\begin{gathered}\bigl\|P^*(x^*,n,\cdot)-\boldsymbol{\pi}^*(\cdot)\bigr\|\to 0,\qquad \boldsymbol{\pi}^*\bigl((B,\delta)\bigr)=\frac{1}{\mathbf{E}\tau^*}\sum_{k=1}^{\infty}P_{V^*}^*\bigl(k,(B,d)\bigr),\\ P_{V^*}^*\bigl(k,(B,\delta)\bigr)=\mathbf{P}\bigl(X_k^*(x^*)\in(B,\delta),X_1^*(x^*)\notin V^*,\dots,X_{k-1}^*(x^*)\notin V^*\bigr).\end{gathered}\tag{13.7.5}$$

In the last equality, we can take any point $x^*\in V^*$; the probability does not depend on the choice of $x^*\in V^*$.

From this and the "inversion formula" (13.7.3) we obtain assertion (13.7.1) and a representation for the invariant measure π of the process X.

The proof of the convergence $\|P(\boldsymbol{\mu}_0,n,\cdot)-\pi(\cdot)\|\to 0$ and uniqueness of the invariant measure is exactly the same as in Theorem 13.6.1 (these facts also follow from the respective assertions for X^*).

Verifying the conditions of Theorem 13.6.1 in the case where $n_0 > 1$ or $\mu(V) = 0$ causes no additional difficulties and we leave it to the reader.

The theorem is proved. □

Note that in a way similar to that in the proof of Theorem 13.4.1, one could also establish the uniqueness of the solution to the integral equation for the invariant measure (see Definition 13.6.6) in a wider class of signed finite measures.

The main and most difficult to verify conditions of Theorem 13.7.1 are undoubtedly conditions (I) and (II). Condition (I_0) is usually obtained "automatically", in the course of verifying condition (I), for the latter requires bounding $\mathbf{E}\tau_V(x)$ for all x. Verifying the aperiodicity condition (III) usually causes no difficulties. If, say, recurrence to the set V is possible in m_1 and m_2 steps and g.c.d. $(m_1, m_2) = 1$, then the chain is aperiodic.

13.7.2 On Conditions (I) and (II)

Now we consider in more detail the main conditions (I) and (II). Condition (II) is expressed directly in terms of local characteristics of the chain (transition probabilities in one or a fixed number of steps $n_0 > 1$), and in this sense it could be treated as a "final" one. One only needs to "guess" the most appropriate set V and measure μ (of course, if there are any). For example, for multi-dimensional Markov chains in $\mathcal{X} = \mathbb{R}^d$, condition (II) will be satisfied if at least one of the following two conditions is met.

(II_a) *The distribution of $X_{n_0}(x)$ has, for some n_0 and $N > 0$ and all $x \in V_N := \{y : |y| \le N\}$, a component which is absolutely continuous with respect to Lebesgue measure (or to the sum of the Lebesgue measures on $\mathbb{R}^d$ and its "coordinate" subspaces) and is "uniformly" positive on the set V_M for some $M > 0$.* In this case, one can take μ to be the uniform distribution on V_M.

(II_l) *$\mathcal{X} = \mathbb{Z}^d$ is the integer lattice in $\mathbb{R}^d$.* In this case the chain is countable and everything simplifies (see Sect. 13.4).

We have already noted that, in the cases when a chain has a positive atom, which is the case in Example 13.6.2, no assumptions about the structure (smoothness) of the distribution of $X_{n_0}(x)$ are needed.

The "positivity" condition (I) is different. It is given in terms of rather complicated characteristics $\mathbf{E}\tau_V(x)$ requiring additional analysis and a search for conditions in terms of local characteristics which would ensure (I). The rest of the section will mostly be devoted to this task.

First of all, we will give an "intermediate" assertion which will be useful for the sequel. We have already made use of such an assertion in Example 13.6.2.

Theorem 13.7.2 *Suppose there exists a nonnegative measurable function $g : \mathcal{X} \to \mathbb{R}$ such that the following conditions* (I^g) *are met*:

(I^g)$_1$ *$\mathbf{E}\tau_V(x) \le c_1 + c_2 g(x)$ for $x \in V^c = \mathcal{X} \setminus V$, $c_1, c_2 = const.$*

$(\mathrm{I}^g)_2$ $\sup_{x\in V}$ $\mathbf{E}g(X_1(x)) < \infty$.
Then conditions (I_0) *and* (I) *are satisfied.*

The function g from Theorem 13.7.2 is often called the *test*, or *Lyapunov, function*. For brevity's sake, put $\tau_V(x) := \tau(x)$.

Proof If (I^g) holds then, for $x \in V$,

$$\begin{aligned}\mathbf{E}\tau(x) &\le 1+\mathbf{E}\big[\tau\big(X_1(x)\big); X_1(x)\in V^c\big]\\ &\le 1+\mathbf{E}\big(\mathbf{E}\big[\tau\big(X_1(x)\big)|X_1(x)\big]; X_1(x)\in V^c\big)\\ &\le 1+\mathbf{E}\big(c_1+c_2 g\big(X_1(x)\big); X_1(x)\in V^c\big)\\ &\le 1+c_1+c_2\sup_{x\in V}\mathbf{E}g\big(X_1(x)\big)<\infty.\end{aligned}$$

The theorem is proved. □

Note that condition $(\mathrm{I}^g)_2$, like condition (II), refers to "local" characteristics of the system, and in that sense it can also be treated as a "final" condition (up to the choice of function g).

We now consider conditions ensuring $(\mathrm{I}^g)_1$. The processes

$$\{X_n\}=\big\{X_n(x)\big\},\quad X_0(x)=x,$$

to be considered below (for instance, in Theorem 13.7.3) do not need to be Markovian. We will only use those properties of the processes which will be stated in conditions of assertions.

We will again make use of nonnegative trial functions $g:\mathcal{X}\to R$ and consider a set V "induced" by the function g and a set U which in most cases will be a bounded interval of the real line:

$$V:=g^{-1}(U)=\big\{x\in\mathcal{X}:g(x)\in U\big\}.$$

The notation $\tau(x)=\tau_U(x)$ will retain its meaning:

$$\tau(x):=\min\big\{k\ge 1:g\big(X_k(x)\big)\in U\big\}=\min\big\{k\ge 1:X_k(x)\in V\big\}.$$

The next assertion is an essential element of Lyapunov's (or the test functions) approach to the proof of positive recurrence of a Markov chain.

Theorem 13.7.3 *If* $\{X_n\}$ *is a Markov chain and, for* $x\in V^c$,

$$\mathbf{E}g\big(X_1(x)\big)-g(x)\le -\varepsilon, \tag{13.7.6}$$

then $\mathbf{E}\tau(x)\le g(x)/\varepsilon$ *and therefore* $(\mathrm{I}^g)_1$ *holds.*

To prove the theorem we need

Lemma 13.7.2 *If, for some* $\varepsilon > 0$, *all* $n = 0, 1, 2, \ldots$, *and any* $x \in V^c$,

$$\mathbf{E}\big(g(X_{n+1}) - g(X_n)|\tau(x) > n\big) \le -\varepsilon, \tag{13.7.7}$$

then

$$\mathbf{E}\tau(x) \le \frac{g(x)}{\varepsilon}, \quad x \in V^c,$$

and therefore $(\mathrm{I}^g)_1$ *holds.*

Proof Put $\tau(x) := \tau$ for brevity and set

$$\tau_{(N)} := \min(\tau, N), \qquad \Delta(n) := g(X_{n+1}) - g(X_n).$$

We have

$$\begin{aligned}
-g(x) &= -\mathbf{E}g(X_0) \le \mathbf{E}\big(g(X_{\tau_{(N)}}) - g(X_0)\big) \\
&= \mathbf{E}\sum_{n=0}^{\tau_{(N)}-1} \Delta(n) = \sum_{n=0}^{N} \mathbf{E}\Delta(n) I(\tau > n) \\
&= \sum_{n=0}^{N} \mathbf{P}(\tau > n)\mathbf{E}\big(\Delta(n)|\tau > n\big) \le -\varepsilon \sum_{n=0}^{N} \mathbf{P}(\tau > n).
\end{aligned}$$

This implies that, for any N,

$$\sum_{n=0}^{N} \mathbf{P}(\tau > n) \le \frac{g(x)}{\varepsilon}.$$

Therefore this inequality will also hold for $N = \infty$, so that $\mathbf{E}\tau \le g(x)/\varepsilon$. The lemma is proved. □

Proof of Theorem 13.7.3 The proof follows in an obvious way from the fact that, by (13.7.6) and the homogeneity of the chain, $\mathbf{E}(g(X_{n+1}) - g(X_n)|\, X_n) \le -\varepsilon$ holds on $\{X_n \in V^c\}$, and from inclusion $\{\tau > n\} \subset \{X_n \in V^c\}$, so that

$$\mathbf{E}\big(g(X_{n+1}) - g(X_n); \tau > n\big) = \mathbf{E}\big[\mathbf{E}\big(g(X_{n+1}) - g(X_n)|X_n\big); \tau > n\big] \le -\varepsilon\mathbf{P}(\tau > n).$$

The theorem is proved. □

Theorem 13.7.3 is a modification of the positive recurrence criterion known as the Foster–Moustafa–Tweedy criterion (see, e.g., [6, 27]).

Consider some applications of the obtained results. Let X be a Markov chain on the real half-axis $\mathbb{R}_+ = [0, \infty)$. For brevity's sake, put $\xi(x) := X_1(x) - x$. This is the one-step increment of the chain starting at the point x; we could also define $\xi(x)$ as a random variable with the distribution

$$\mathbf{P}\big(\xi(x) \in B\big) = P(x, B - x) \quad \big(B - x = \{y \in \mathcal{X} : y + x \in B\}\big).$$

Corollary 13.7.1 *If, for some $N \ge 0$ and $\varepsilon > 0$,*

$$\sup_{x \le N} \mathbf{E}\xi(x) < \infty, \qquad \sup_{x > N} \mathbf{E}\xi(x) \le -\varepsilon, \tag{13.7.8}$$

then conditions (I_0) *and* (I) *hold for* $V = [0, N]$.

Proof Make use of Theorems 13.7.2, 13.7.3 and Corollary 13.3.1 with $g(x) \equiv x$, $V = [0, N]$. Conditions $(\mathrm{I}^g)_2$ and (13.7.6) are clearly satisfied. □

Thus the presence of a "negative drift" in the region $x > N$ guarantees positivity of the chain. However, that condition (I) is met could also be ensured when the "drift" $\mathbf{E}\xi(x)$ vanishes as $x \to \infty$.

Corollary 13.7.2 *Let* $\sup_x \mathbf{E}\xi^2(x) < \infty$ *and*

$$\mathbf{E}\xi^2(x) \le \beta, \quad \mathbf{E}\xi(x) \le -\frac{c}{x} \quad \text{for } x > N.$$

If $2c > \beta$ *then conditions* (I_0) *and* (I) *hold for* $V = [0, N]$.

Proof We again make use of Theorems 13.7.2 and 13.7.3, but with $g(x) = x^2$. We have for $x > N$:

$$\mathbf{E}g\big(X_1(x)\big) - g(x) = \mathbf{E}\big(2x\xi(x) + \xi^2(x)\big) \le -2c + \beta < 0.$$ □

Before proceeding to examples related to ergodicity we note the following. The "larger" the set V the easier it is to verify condition (I), and the "smaller" that set, the easier it is to verify condition (II). In this connection there arises the question of when one can consider two sets: a "small" set W and a "large" set $V \supset W$ such that if (I) holds for V and (II) holds for W then both (I) and (II) would hold for W. Under conditions of Sect. 13.6 one can take W to be a "one-point" atom x_0.

Lemma 13.7.3 *Let sets V and W be such that the condition*

$$(\mathrm{I}_V) \quad E := \sup_{x \in V} \mathbf{E}\tau_V(x) < \infty$$

holds and there exists an m such that

$$\inf_{x \in V} \mathbf{P}\left(\bigcup_{j=1}^{m} \big\{X_j(x) \in W\big\} \right) \ge q > 0.$$

Then the following condition is also met:

$$(\mathrm{I}_W) \quad \sup_{x \in W} \mathbf{E}\tau_W(x) \le \sup_{x \in V} \mathbf{E}\tau_W(x) \le \frac{mE}{q}.$$

Thus, under the assumptions of Lemma 13.7.3, if condition (I) holds for V and condition (II) holds for W, then conditions (I) and (II) hold for W.

To prove Lemma 13.7.3, we will need the following assertion extending (in the form of an inequality) the well-known Wald identity.

Assume we are given a sequence of nonnegative random variables $\tau_1, \tau_2, \ldots$ which are measurable with respect to σ-algebras $\mathfrak{U}_1 \subset \mathfrak{U}_2 \subset \cdots$, respectively, and let $T_n := \tau_1 + \cdots + \tau_n$. Furthermore, let ν be a given stopping time with respect to $\{\mathfrak{U}_n\}$: $\{\nu \le n\} \in \mathfrak{U}_n$.

Lemma 13.7.4 *If* $\mathbf{E}(\tau_n | \mathfrak{U}_{n-1}) \le a$ *then* $\mathbf{E}T_\nu \le a\mathbf{E}\nu$.

Proof We can assume without loss of generality that $\mathbf{E}\nu < \infty$ (otherwise the inequality is trivial). The proof essentially repeats that of Theorem 4.4.1. One has

$$\mathbf{E}\tau_\nu = \sum_{k=1}^{\infty} \mathbf{E}(T_k;\, \nu = k) = \sum_{k=1}^{\infty} \mathbf{E}(\tau_k, \nu \ge k). \tag{13.7.9}$$

Changing the summation order here is well-justified, for the summands are nonnegative. Further, $\{\nu \le k-1\} \in \mathfrak{U}_{k-1}$ and hence $\{\nu \ge k\} \in \mathfrak{U}_{k-1}$. Therefore

$$\mathbf{E}(\tau_k;\, \nu \ge k) = \mathbf{E}\mathrm{I}(\nu \ge k)\mathbf{E}(\tau_k | \mathfrak{U}_{k-1}) \le a\mathbf{P}(\nu \ge k).$$

Comparing this with (13.7.9) we get

$$\mathbf{E}T_\nu \le a \sum_{k=1}^{\infty} \mathbf{P}(\nu \ge k) = a\mathbf{E}nu.$$

The lemma is proved. □

Proof of Lemma 13.7.3 Suppose the chain starts at a point $x \in V$. Consider the times $T_1, T_2, \ldots$ of successive visits of X to V, $T_0 = 0$. Put $Y_0 := x$, $Y_k := X_{T_k}(x)$, $k = 1, 2, \ldots$. Then, by virtue of the strong Markov property, the sequence (Y_k, T_k) will form a Markov chain. Set $\mathfrak{U}_k := \sigma(T_1, \ldots, T_k; Y_1, \ldots, Y_k)$, $\tau_k := T_k - T_{k-1}$, $k = 1, 2 \ldots$. Then $\nu := \min\{k : Y_k \in W\}$ is a stopping time with respect to $\{\mathfrak{U}_k\}$. It is evident that $\mathbf{E}(\tau_k | \mathfrak{U}_{k-1}) \le E$. Bound $\mathbf{E}\nu$. We have

$$p_k := \mathbf{P}(\nu \ge km) \le \mathbf{P}\left(\bigcap_{j=1}^{T_{km}} \{X_j \notin W\} \right)$$

$$= \mathbf{E}\mathrm{I}\left(\bigcap_{j=1}^{T_{(k-1)m}} \{X_j \notin W\} \right) \mathbf{E}\left(\mathrm{I}\left(\bigcap_{j=T_{(k-1)m}+1}^{T_{km}} \{X_j \notin W\} \right) \Big| \mathfrak{U}_{(k-1)m} \right).$$

Since $\tau_j \geq 1$, the last factor, by the assumptions of the lemma and the strong Markov property, does not exceed

$$\mathbf{P}\left(\bigcap_{j=1}^{m}\{X_j^{\text{new}}(X_{T_{(k-1)m}}) \notin W\}\right) \leq (1-q),$$

where, as before, $X_k^{\text{new}}(x)$ is a chain with the same distribution as $X_k(x)$ but independent of the latter chain. Thus $p_k \leq (1-q)p_{k-1} \leq (1-q)^k$, $\mathbf{E}\nu \leq m/q$, and by Lemma 13.7.4 we have $\mathbf{E}T_\nu \leq E_m/q$. It remains to notice that $\tau_W(x) = T_\nu$. The lemma is proved. □

Example 13.7.1 A random walk with reflection. Let $\xi_1, \xi_2, \ldots$ be independent identically distributed random variables,

$$X_{n+1} := |X_n + \xi_{n+1}|, \quad n = 0, 1, \ldots. \tag{13.7.10}$$

If the ξ_k and hence the X_k are non-arithmetic, then the chain X has, generally speaking, no atoms. If, for instance, ξ_k have a density $f(t)$ with respect to Lebesgue measure then $\mathbf{P}(X_k(x) = y) = 0$ for any $x, y, k \geq 1$. We will assume that a broader condition (A) holds:

(A). *In the decomposition*

$$\mathbf{P}(\xi_k < t) = p_a\, F_a(t) + p_c\, F_c(t)$$

of the distribution of ξ_k *into the absolutely continuous* (F_a) *and singular* (F_c) *(including discrete) components, one has* $p_a > 0$.

Corollary 13.7.3 *If condition* (A) *holds,* $a = \mathbf{E}\xi_k < 0$, *and* $\mathbf{E}|\xi_k| < \infty$, *then the Markov chain defined in* (13.7.10) *satisfies the conditions of Theorem* 13.7.2 *and therefore is ergodic in the sense of convergence in total variation.*

Proof We first verify that the chain satisfies the conditions of Corollary 13.7.1. Since in our case $|X_1(x) - x| \leq |\xi_1|$, the first of conditions (13.7.8) is satisfied. Further,

$$\mathbf{E}\xi(x) = \mathbf{E}|x + \xi_1| - x = \mathbf{E}(\xi_1; \xi_1 \geq -x) - \mathbf{E}(2x + \xi_1; \xi_1 < -x) \to \mathbf{E}\xi_1$$

as $x \to \infty$, since

$$x\mathbf{P}(\xi_1 < -x) \leq \mathbf{E}\big(|\xi_1|, |\xi_1| > x\big) \to 0.$$

Hence there exists an N such that $\mathbf{E}\xi(x) \leq a/2 < 0$ for $x \geq N$. This proves that conditions (I_0) and (I) hold for $V = [0, N]$.

Now verify that condition (II) holds for the set $W = [0, h]$ with some h. Let $f(t)$ be the density of the distribution F_a from condition (A). There exist an $f_0 > 0$ and a segment $[t_1, t_2]$, $t_2 > t_1$, such that $f(t) > f_0$ for $t \in [t_1, t_2]$. The density of $x + \xi_1$

will clearly be greater than f_0 on $[x+t_1, x+t_2]$. Put $h := (t_2 - t_1)/2$. Then, for $0 \le x \le h$, one will have $[t_2 - h, t_2] \subset [x+t_1, x+t_2]$.

Suppose first that $t_2 > 0$. The aforesaid will then mean that the density of $x + \xi_1$ will be greater than f_0 on $[(t_2 - h)^+, t_2]$ for all $x \le h$ and, therefore,

$$\inf_{x \le h} \mathbf{P}\big(X_1(x) \in B\big) \ge p_1 \int_B f_0(t)\, dt,$$

where

$$f_0(t) = \begin{cases} f_0 & \text{if} t \in [(t_2 - h)^+, t_2], \\ 0 & \text{otherwise.} \end{cases}$$

This means that condition (II) is satisfied on the set $W = [0, h]$. The case $t_2 \le 0$ can be considered in a similar way.

It remains to make use of Lemma 13.7.3 which implies that condition (I) will hold for the set W. The condition of Lemma 13.7.3 is clearly satisfied (for sufficiently large m, the distribution of $X_m(x)$, $x \le N$, will have an absolutely continuous component which is positive on W). For the same reason, the chain X cannot be periodic. Thus all conditions of Theorem 13.7.2 are met. The corollary is proved. □

Example 13.7.2 An oscillating random walk. Suppose we are given two independent sequences $\xi_1, \xi_2, \dots$ and $\eta_1, \eta_2, \dots$ of independent random variables, identically distributed in each of the sequences. Put

$$X_{n+1} := \begin{cases} X_n + \xi_{n+1} & \text{if } X_n \ge 0, \\ X_n + \eta_{n+1} & \text{if } X_n < 0. \end{cases} \tag{13.7.11}$$

Such a random walk is called *oscillating*. It clearly forms a Markov chain in the state space $\mathcal{X} = (-\infty, \infty)$.

Corollary 13.7.4 *If at least one of the distributions of ξ_k or η_k satisfies condition* (A) *and $-\infty < \mathbf{E}\xi_k < 0$, $\infty > \mathbf{E}\eta_k > 0$, then the chain* (13.7.11) *will satisfy the conditions of Theorem* 13.7.2 *and therefore will be ergodic.*

Proof The argument is quite similar to the proof of Corollary 13.7.3. One just needs to take, in order to verify condition (I), $g(x) = |x|$ and $V = [-N, N]$. After that it remains to make use of Lemma 13.7.3 with $W = [0, h]$ if condition (A) is satisfied for ξ_k (and with $W = [-h, 0)$ if it is met for η_k). □

Note that condition (A) in Examples 13.7.1 and 13.7.2 can be relaxed to that of the existence of an absolutely continuous component for the distribution of the sum $\sum_{j=1}^m \xi_j$ (or $\sum_{j=1}^m \eta_j$) for some m. On the other hand, if the distributions of these sums are singular for all m, then convergence of distributions $P(x, n, \cdot)$ in total variation cannot take place. If, for instance, one has $\mathbf{P}(\xi_k = -\sqrt{2}) = \mathbf{P}(\xi_k = 1) = 1/2$ in Example 13.7.1, then $\mathbf{E}\xi_k < 0$ and condition (I) will be met, while condition (II) will not. Convergence of $P(x,n,\cdot)$ in total variation to the limiting distribution π is also impossible. Indeed, it follows from the equation for the invariant distribution π that this distribution is necessarily continuous. On the other hand, say, the distributions $P(0,n,\cdot)$ are concentrated on the countable set $\mathcal{N}$ of the numbers $|-k\sqrt{2}+l|$; $k,l=1,2,\ldots$. Therefore $\mathbf{P}(0,n,\mathcal{N})=1$ for all n, $\pi(\mathcal{N})=0$. Hence only weak convergence of the distributions $\mathbf{P}(x,n,\cdot)$ to $\pi(\cdot)$ may take place. And although this convergence does not raise any doubts, we know no reasonably simple proof of this fact.

Example 13.7.3 (continuation of Examples 13.4.2 and 13.6.1) Let $\mathcal{X}=[0,1]$, $\xi_1,\xi_2,\ldots$ be independent and identically distributed, and $X_{n+1}:=X_n+\xi_{n+1}$ (mod 1) or, which is the same, $X_{n+1}:=\{X_n+\xi_{n+1}\}$, where $\{x\}$ denotes the fractional part of x. Here, condition (I) is clearly met for $V=\mathcal{X}=[0,1]$. If the ξ_k satisfy condition (A) then, as was the case in Example 13.7.1, condition (II) will be met for the set $W=[0,h]$ with some $h>0$, which, together with Lemma 13.7.3, will mean, as before, that the conditions of Theorem 13.7.2 are satisfied. The invariant distribution π will in this example be uniform on [0, 1]. For simplicity's sake, we can assume that the distribution of ξ_k has a density $f(t)$, and without loss of generality we can suppose that $\xi_k\in[0,1]$ ($f(t)=0$ for $t\notin[0,1]$). Then the density $p(x)\equiv 1$ of the invariant measure π will satisfy the equation for the invariant measure:

$$p(x)=1=\int_0^x dy\, f(x-y)+\int_x^1 dy\, f(x-y+1)=\int_0^1 f(y)\,dy.$$

Since the stationary distribution is unique, one has $\pi=U_{0,1}$. Moreover, by Theorem A3.4.1 of Appendix 3, along with convergence of $P(x,n,\cdot)$ to $U_{0,1}$ in total variation, convergence of the densities $P(x,n,dt)/dt$ to 1 in (Lebesgue) measure will take place.

The fact that the invariant distribution is uniform remains true for arbitrary non-lattice distributions of ξ_k. However, as we have already mentioned in Example 13.6.1, in the general case (without condition (A)) only weak convergence of the distributions $P(x,n,\cdot)$ to the uniform distribution is possible (see [6, 18]).

13.8 Laws of Large Numbers and the Central Limit Theorem for Sums of Random Variables Defined on a Markov Chain

13.8.1 Random Variables Defined on a Markov Chain

Let, as before, $X=\{X_n\}$ be a Markov Chain in an arbitrary measurable state space $\langle\mathcal{X},\mathfrak{B}_{\mathcal{X}}\rangle$ defined in Sect. 13.6, and let a measurable function $f\colon\mathcal{X}\to\mathbb{R}$ be given on $\langle\mathcal{X},\mathfrak{B}_{\mathcal{X}}\rangle$. The sequence of sums

$$S_n:=\sum_{k=1}^n f(X_k) \tag{13.8.1}$$

is a generalisation of the random walks that were studied in Chaps. 8 and 11. One can consider an even more general problem on the behaviour of *sums of random variables defined on a Markov chain*. Namely, we will assume that a collection of distributions $\{\mathbf{F}_x\}$ is given which depend on the parameter $x \in \mathcal{X}$. If $F_x^{(-1)}(t)$ is the quantile transform of $\mathbf{F}_x$ and $\omega \in\!\!\!\in \mathbf{U}_{0,1}$, then $\xi_x := F_x^{(-1)}(\omega)$ will have the distribution $\mathbf{F}_x$ (see Sect. 3.2.4).

The mapping $\mathbf{F}_x$ of the space $\mathcal{X}$ into the set of distributions is assumed to be such that the function $\xi_x(t) = F_x^{(-1)}(t)$ is measurable on $\mathcal{X} \times \mathbb{R}$ with respect to $\mathfrak{B}_{\mathcal{X}} \times \mathfrak{B}$, where $\mathfrak{B}$ is the σ-algebra of Borel sets on the real line. In this case, $\xi_x(\omega)$ will be a random variable such that the moments

$$\mathbf{E}\xi_x^s = \int_{-\infty}^{\infty} v^s d\mathbf{F}_x(v) = \int_0^1 \left[F_x^{(-1)}(u)\right]^s du$$

are measurable with respect to $\mathfrak{B}_{\mathcal{X}}$ (and hence will be random variables themselves if we set a distribution on $\langle \mathcal{X}, \mathfrak{B}_{\mathcal{X}} \rangle$).

Definition 13.8.1 If $\omega_i \in\!\!\!\in \mathbf{U}_{0,1}$ are independent then the sequence

$$\xi_{X_n} := F_{X_n}^{(-1)}(\omega_n), \quad n = 0, 1, \ldots,$$

is called a *sequence of random variables defined on the Markov chain* $\{X_n\}$.

The basic objects of study in this section are the asymptotic properties of the distributions of the sums

$$S_n := \sum_{k=0}^{n} \xi_{X_k}. \tag{13.8.2}$$

If the distribution $\mathbf{F}_x$ is degenerate and concentrated at the point $f(x)$ then (13.8.2) turns into the sum (13.8.1). If the chain X is countable with states $E_0, E_1, \ldots$ and $f(x) = \mathrm{I}(E_j)$ then $S_n = m_j(n)$ is the number of visits to the state E_j by the time n considered in Theorem 13.4.4.

13.8.2 *Laws of Large Numbers*

In this and the next subsection we will confine ourselves to Markov chains satisfying the ergodicity conditions from Sects. 13.6 and 13.7. As was already noticed, ergodicity conditions for Harris chains mean, in essence, the existence of a positive atom (possibly in the extended state space). Therefore, for the sake of simplicity, we will assume from the outset that the chain X has a positive atom at a point x_0 and put, as before,

$$\tau(x) := \min\{k \ge 0 : X_k(x) = x_0\}, \quad \tau(x_0) = \tau.$$

Summing up the conditions sufficient for (I_0) and (I) to hold (the finiteness of $\tau(x)$ and $\mathbf{E}\tau$) studied in Sect. 13.7, we obtain the following assertion in our case.

Corollary 13.8.1 *Let there exist a set* $V \in \mathfrak{B}_{\mathcal{X}}$ *such that, for the stopping time* $\tau_V(x) := \min\{k : X_k(x) \in V\}$, *we have*

$$E := \sup_{x \in V} \mathbf{E}\tau_V(x) < \infty. \tag{13.8.3}$$

Furthermore, let there exist an $m \ge 1$ *such that*

$$\inf_{x \in V} \mathbf{P}\left(\bigcup_{j=1}^{m} \{X_j(x) = x_0\}\right) \ge q > 0.$$

Then

$$\mathbf{E}\tau \le \frac{mE}{q}.$$

This assertion follows from Lemma 13.7.2. One can justify conditions (I_0) and (13.8.3) by the following assertion.

Corollary 13.8.2 *Let there exist an* $\varepsilon > 0$ *and a nonnegative measurable function* $g : \mathcal{X} \to \mathbb{R}$ *such that*

$$\sup_{x \in V} \mathbf{E}g\big(X_1(x)\big) < \infty$$

and, for $x \in V^c$,

$$\mathbf{E}g\big(X_1(x)\big) - g(x) \le -\varepsilon.$$

Then conditions (I_0) *and* (13.8.3) *are met.*

In order to formulate and prove the law of large numbers for the sums (13.8.2), we will use the notion of the increment of the sums (13.8.2) on a cycle between consequent visits of the chain to the atom x_0. Divide the trajectory $X_0, X_1, X_2, \dots, X_n$ of the chain X on the time interval $[0, n]$ into segments of lengths $\tau_1 := \tau(x), \tau_2, \tau_3, \dots$ ($\tau_j \stackrel{d}{=} \tau$ for $j \ge 2$) corresponding to the visits of the chain to the atom x_0. Denote the increment of the sum S_n on the k-th cycle (on $(T_{k-1}, T_k]$) by ζ_k:

$$\zeta_1 := \sum_{j=0}^{\tau_1} \xi_{X_j},$$

$$\zeta_k := \sum_{j=T_{k-1}+1}^{T_k} \xi_{X_j}, \quad k \ge 2, \quad \text{where } T_k := \sum_{j=1}^{k} \tau_j, \; k \ge 1, \; T_0 = 0. \tag{13.8.4}$$

The vectors (τ_k, ζ_k), $k \ge 2$, are clearly independent and identically distributed. For brevity, the index k will sometimes be omitted: $(\tau_k, \zeta_k) \stackrel{d}{=} (\tau, \zeta)$ for $k \ge 2$.

Now we can state the law of large numbers for the sums (13.8.2).

Theorem 13.8.1 *Let* $\mathbf{P}(\tau(x) < \infty) = 1$ *for all* x, $\mathbf{E}\tau < \infty$, $\mathbf{E}|\zeta| < \infty$, *and the g.c.d. of all possible values of* τ *equal* 1. *Then*

$$\frac{S_n}{n} = \frac{1}{n}\sum_{k=1}^{n} \xi_{X_k} \xrightarrow{p} \frac{\mathbf{E}\zeta}{\mathbf{E}\tau} \quad \textit{as} \quad n \to \infty.$$

Proof Put

$$\nu(n) := \max\{k : T_k \le n\}.$$

Then the sum S_n can be represented as

$$S_n = \zeta_1 + Z_{\nu(n)} + z_n, \tag{13.8.5}$$

where

$$Z_k := \sum_{j=2}^{k} \zeta_j, \qquad z_n := \sum_{j=T_{\nu(n)}+1}^{n} \xi_{X_j}.$$

Since τ_1 and ζ_1 are proper random variables, we have, as $n \to \infty$,

$$\frac{\zeta_1}{n} \xrightarrow{a.s.} 0. \tag{13.8.6}$$

The sum z_n consists of $\gamma(n) := n - T_{\nu(n)}$ summands. Theorem 10.3.1 implies that the distribution of $\gamma(n)$ converges to a proper limiting distribution, and the same is true for z_n. Hence, as $n \to \infty$,

$$\frac{z_n}{n} \xrightarrow{p} 0. \tag{13.8.7}$$

The sums $Z_{\nu(n)}$, being the main part of (13.8.5), are nothing else but a generalised renewal process corresponding to the vectors (τ, ζ) (see Sect. 10.6).

Since $\mathbf{E}\tau < \infty$, by Theorem 11.5.2, as $n \to \infty$,

$$\frac{Z_{\nu(n)}}{n} \xrightarrow{p} \frac{\mathbf{E}\zeta}{\mathbf{E}\tau}. \tag{13.8.8}$$

Together with (13.8.6) and (13.8.7) this means that

$$\frac{S_n}{n} \xrightarrow{p} \frac{\mathbf{E}\zeta}{\mathbf{E}\tau}. \tag{13.8.9}$$

The theorem is proved. □

As was already noted, sufficient conditions for $\mathbf{P}(\tau(x) < \infty) = 1$ and $\mathbf{E}\tau < \infty$ to hold are contained in Corollaries 13.8.1 and 13.8.2. It is more difficult to find conditions sufficient for $\mathbf{E}\zeta < \infty$ that would be adequate for the nature of the problem.

Below we will obtain certain relations which clarify, to some extent, the connection between the distributions of ζ and τ and the stationary distribution of the chain X.

Theorem 13.8.2 (A generalisation of the Wald identity) *Assume* $\mathbf{E}\tau < \infty$, *the g.c.d. of all possible values of* τ *be* 1, $\boldsymbol{\pi}$ *be the stationary distribution of the chain* X, *and*

$$\mathbf{E}_\pi \mathbf{E}|\xi_x| := \int \mathbf{E}|\xi_x|\boldsymbol{\pi}(dx) < \infty. \tag{13.8.10}$$

Then

$$\mathbf{E}\zeta = \mathbf{E}\tau \mathbf{E}_\pi \mathbf{E}\xi_x. \tag{13.8.11}$$

The value of $\mathbf{E}_\pi \mathbf{E}\xi_x$ is the "doubly averaged" value of the random variable ξ_x: over the distribution $\mathbf{F}_x$ and over the stationary distribution $\boldsymbol{\pi}$.

Theorem 13.8.2 implies that the condition $\sup_x \mathbf{E}|\xi_x| < \infty$ is sufficient for the finiteness of $\mathbf{E}|\zeta|$.

Proof [of Theorem 13.8.2] First of all, we show that condition (13.8.10) implies the finiteness of $\mathbf{E}|\zeta|$. If $\xi_x \ge 0$ then $\mathbf{E}\zeta$ is always well-defined. If we assume that $\mathbf{E}\zeta = \infty$ then, repeating the proof of Theorem 13.8.1, we would easily obtain that, in this case, $S_n/n \xrightarrow{p} \infty$, and hence necessarily $\mathbf{E}S_n/n \to \infty$ as $n \to \infty$. But

$$\mathbf{E}S_n = \sum_{j=0}^{n} \mathbf{E}\xi_{X_j} = \sum_{j=0}^{n} \int (\mathbf{E}\xi_x)\mathbf{P}(X_j \in dx),$$

where the distribution $\mathbf{P}(X_j \in \cdot)$ converges in total variation to $\boldsymbol{\pi}(\cdot)$ as $j \to \infty$,

$$\int (\mathbf{E}\xi_x)\mathbf{P}(X_j \in dx) \to \int (\mathbf{E}\xi_x)\boldsymbol{\pi}(dx),$$

and hence

$$\frac{1}{n}\mathbf{E}S_n \to \mathbf{E}_\pi \mathbf{E}\xi_x < \infty. \tag{13.8.12}$$

This contradicts the above assumption, and therefore $\mathbf{E}\zeta < \infty$. Applying the above argument to the random variables $|\xi_x|$, we conclude that condition (13.8.10) implies $\mathbf{E}|\zeta| < \infty$.

Let, as above, $\eta(n) := \nu(n) + 1 = \min\{k : T_k > n\}$. We will need the following.

Lemma 13.8.5 *If* $\mathbf{E}|\zeta| < \infty$ *then*

$$\mathbf{E}\zeta_{\eta(n)} = o(n). \tag{13.8.13}$$

If $\mathbf{E}\zeta^2 < \infty$ *then*

$$\mathbf{E}\zeta^2_{\eta(n)} = o(n) \tag{13.8.14}$$

as $n \to \infty$.

Proof Without losing generality, assume that $\xi_x \ge 0$ and $\zeta \ge 0$. Since $\tau_j \ge 1$, we have

$$h(k) := \sum_{j=0}^{k} \mathbf{P}(T_j = k) \le 1 \quad \text{for all } k.$$

Therefore,

$$\mathbf{P}(\zeta_{\eta(n)} > v) = \sum_{k=0}^{n} h(k)\mathbf{P}(\zeta > v, \tau > n-k) \le \sum_{k=0}^{n} \mathbf{P}(\zeta > v, \tau > k).$$

If $\mathbf{E}\zeta < \infty$ then

$$\mathbf{E}\zeta_{\eta(n)} \le \sum_{k=0}^{n} \int_0^\infty \mathbf{P}(\zeta > v; \tau > k)\, dv = \sum_{k=0}^{n} \mathbf{E}(\zeta; \tau > k), \tag{13.8.15}$$

where $\mathbf{E}(\zeta; \tau > k) \to 0$ as $k \to \infty$. This follows from Lemma A3.2.3 of Appendix 3. Together with (13.8.15) this proves (13.8.13).

Similarly, for $\mathbf{E}\zeta^2 < \infty$,

$$\mathbf{E}\zeta^2_{\eta(n)} \le 2\sum_{k=0}^{n} \int_0^\infty v\mathbf{P}(\zeta > v, \tau > k)\, dv = \sum_{k=0}^{n} \mathbf{E}(\zeta^2, \tau > k) = o(n).$$

The lemma is proved. □

Now we continue the proof of Theorem 13.8.2. Consider representation (13.8.5) for $X_0 = x_0$ and assume again that $\xi_x \ge 0$. Then $\zeta_1 = \xi_{x_0}$,

$$S_n = \zeta_1 + Z_{\eta(n)} + z_n - \zeta_{\eta(n)},$$

where by the Wald identity

$$\mathbf{E}Z_{\eta(n)} = \mathbf{E}\eta(n)\mathbf{E}\zeta \sim n\frac{\mathbf{E}\zeta}{\mathbf{E}\tau}.$$

Since $\pi(\{x_0\}) = 1/\mathbf{E}\tau > 0$, we have, by (13.8.10), $\mathbf{E}|\xi_{x_0}| < \infty$. Moreover, for $\xi_x \ge 0$,

$$|\zeta_{\eta(n)} - z_n| < \zeta_{\eta(n)}.$$

Hence, by Lemma 13.8.5,

$$\mathbf{E}S_n = n\frac{\mathbf{E}\zeta}{\mathbf{E}\tau} + o(n). \tag{13.8.16}$$

Combining this with (13.8.12), we obtain the assertion of the theorem.

It remains to consider the case where ξ_x can take values of both signs. Introduce new random variables ξ_x^* on the chain X, defined by the equalities $\xi_x^* := |\xi_x|$, and endow with the superscript $*$ all already used notations that will correspond to the new random variables. Since all $\xi^*_x \ge 0$, by condition (13.8.10) we can apply to them all the above assertions and, in particular, obtain that

$$\mathbf{E}\zeta^* < \infty, \qquad \mathbf{E}\zeta^*_{\eta(n)} = o(n). \tag{13.8.17}$$

Since

$$|\zeta| \le \zeta^*, \qquad |\zeta_{\eta(n)}| \le \zeta^*_{\eta(n)}, \qquad |\zeta_{\eta(n)} - z_n| < \zeta^*_{\eta(n)},$$

it follows from (13.8.17) that

$$\mathbf{E}|\zeta| < \infty, \qquad \mathbf{E}|\zeta_{\eta(n)} - z_n| = o(n)$$

and relation (13.8.16) is valid along with identity (13.8.11).

The theorem is proved. □

Now we will prove the strong law of large numbers.

Theorem 13.8.3 *Let the conditions of Theorem* 13.8.1 *be satisfied. Then*

$$\frac{S_n}{n} \xrightarrow{a.s.} \mathbf{E}_\pi \mathbf{E}\xi_x \quad \text{as } n \to \infty.$$

Proof Since in representation (13.8.5) one has $\zeta_1/n \xrightarrow{a.s.} 0$ as $n \to \infty$, we can neglect this term in (13.8.5).

The strong laws of large numbers for $\{Z_k\}$ and $\{T_k\}$ mean that, for a given $\varepsilon > 0$, the trajectory of $\{S_{T_k}\}$ will lie within the boundaries $k\mathbf{E}\zeta(1 \pm \varepsilon)$ and $\frac{\mathbf{E}\zeta}{\mathbf{E}\tau} T_k(1 \pm 2\varepsilon)$ for all $k \ge n$ and n large enough. (We leave a more formal formulation of this to the reader.)

We will prove the theorem if we verify that the probability of the event that, between the times T_k, $k \ge n$, the trajectory of S_j will cross at least once the boundaries $rj(1 \pm 3\varepsilon)$, where $r = \dfrac{\mathbf{E}\zeta}{\mathbf{E}\tau}$, tends to zero as $n \to \infty$. Since

$$\max_{T_{k-1} < j \le T_k} |S_j - S_{T_k}| \le \zeta^*_k \tag{13.8.18}$$

(in the notation of the proof of Theorem 13.8.1), it is sufficient to verify that $\mathbf{P}(A_n) \to 0$ as $n \to \infty$, where $A_n := \bigcup_{k=n}^\infty \{\zeta^*_k > \varepsilon r T_k\}$. But

$$\mathbf{P}(A_n) = \mathbf{P}(A_n B_n) + \mathbf{P}(A_n \overline{B}_n), \tag{13.8.19}$$

where

$$B_n = \bigcap_{k=n}^\infty \bigl\{T_k > k\mathbf{E}\tau(1-\varepsilon)\bigr\}, \qquad \mathbf{P}(\overline{B}_n) \to 0 \quad \text{as } n \to \infty,$$

so the second summand in (13.8.19) tends to zero. The first summand on the right-hand side of (13.8.19) does not exceed (for $c = \varepsilon(1-\varepsilon)\mathbf{E}\zeta$)

$$\mathbf{P}\left(\bigcup_{k=n}^{\infty}\{\zeta_k^* > \varepsilon\mathbf{E}\zeta k(1-\varepsilon)\}\right) \le \sum_{k=n}^{\infty}\mathbf{P}(\zeta_k^* > ck) \to 0$$

as $n \to \infty$, since $\mathbf{E}\zeta^* < \infty$ (see (13.8.17)). The theorem is proved. □

13.8.3 The Central Limit Theorem

As in Theorem 13.8.1, first we will prove the main assertion under certain conditions on the moments of ζ and τ, and then we will establish a connection of these conditions to the stationary distribution of the chain X. Below we retain the notation of the previous section.

Theorem 13.8.4 *Let $\mathbf{P}(\tau(x) < \infty) = 1$ for any x, $\mathbf{E}\tau^2 < \infty$, the g.c.d. of all possible values of τ is 1, and $\mathbf{E}\zeta^2 < \infty$. Then, as $n \to \infty$,*

$$\frac{S_n - rn}{d\sqrt{n/a}} \Leftrightarrow \Phi_{0,1},$$

where $r := a_\zeta/a$, $a_\zeta := \mathbf{E}\zeta$, $a := \mathbf{E}\tau$ and $d^2 := \mathbf{D}(\zeta - r\tau)$.

Proof We again make use of representation (13.8.5), where clearly

$$\frac{\zeta_1}{\sqrt{n}} \xrightarrow{p} 0, \qquad \frac{z_n}{\sqrt{n}} \xrightarrow{p} 0$$

(see the proof of Theorem 13.8.1). This means that the problem reduces to that of finding the limiting distribution of $Z_{\nu(n)} = Z_{\eta(n)} - \zeta_{\eta(n)}$, where by Lemma 10.6.1 $\zeta_{\eta(n)}$ has a proper limiting distribution, and so $\zeta_{\eta(n)}/\sqrt{n} \xrightarrow{p} 0$ as $n \to \infty$. Furthermore, by Theorem 10.6.3,

$$\frac{Z_{\eta(n)}}{\sigma_S\sqrt{n}} \Leftrightarrow \Phi_{0,1},$$

where $\sigma_S^2 := a^{-1}\mathbf{D}(\zeta - r\tau)$, $r = \frac{\mathbf{E}\zeta}{\mathbf{E}\tau}$. The theorem is proved. □

Now we will establish relations between the moment characteristics used for normalising S_n and the stationary distribution $\boldsymbol{\pi}$. The answer for the number r was given in Theorem 13.8.2: $r = \mathbf{E}_\pi\mathbf{E}\xi_x$. For the number σ_S^2 we have the following result.

Theorem 13.8.5 *Let*

$$\sigma^2 := \int \mathbf{D}\xi_x \pi(dx) + 2\sum_{j=1}^{\infty} \mathbf{E}(\xi_{X_0} - r)(\xi_{X_j} - r)$$

be well-defined and finite, where $X_0 \Subset \pi$. *Then*

$$\sigma_S^2 := a^{-1} d^2 = \sigma^2.$$

Note that here the expectation under the sum sign is a "triple averaging": over the distribution $\pi(dy)\mathbf{P}(y, j, dz)$ and the distributions of ξ_y and ξ_z.

Proof We have

$$\mathbf{E}(S_n - rn)^2 = \mathbf{E}\left[\sum_{k=0}^{n}(\xi_{X_k} - r)\right]^2$$

$$= \sum_{k=0}^{n} \mathbf{E}(\xi_{X_k} - r)^2 + 2\sum_{k<j} \mathbf{E}(\xi_{X_k} - r)(\xi_{X_j} - r), \quad (13.8.20)$$

where

$$\sum_{k=0}^{n} \mathbf{E}(\xi_{X_k} - r)^2 = \sum_{k=0}^{n} \mathbf{E}(\xi_{X_k} - \mathbf{E}\xi_{X_k})^2 + \sum_{k=0}^{n} (\mathbf{E}\xi_{X_k} - r)^2. \quad (13.8.21)$$

The summands in the first sum on the right-hand side of (13.8.21) converge to $\sigma_\xi^2 := \int \mathbf{D}\xi_x \pi(dx)$, the summands in the second sum converging to zero. Therefore, the left-hand side of (13.8.21) is asymptotically equivalent to $n\sigma_\xi^2$.

Further,

$$\sum_{k<j} \mathbf{E}(\xi_{X_k} - r)(\xi_{X_j} - r) = \sum_{k=0}^{n} \sum_{j \ge k+1} \mathbf{E}(\xi_{X_k} - r)(\xi_{X_j} - r), \quad (13.8.22)$$

where the distribution of X_k converges in total variation to the stationary distribution π of the chain. Hence the inner sums on the right-hand side of (13.8.22), for large k and $n - k$ (say, for $\sqrt{n} < k < n - \sqrt{n}$ when $n \to \infty$), will be close to

$$E := \sum_{j=1}^{\infty} \mathbf{E}(\xi_{X_0} - r)(\xi_{X_j} - r),$$

where $X_0 = \pi$ and the whole sum on the right-hand side of (13.8.22) is asymptotically equivalent, as $n \to \infty$, to nE (or will be $o(n)$ if $E = 0$).

Thus

$$\frac{1}{n}\mathbf{E}(S_n - rn)^2 \sim \sigma_\xi^2 + 2E. \quad (13.8.23)$$

We now show that the existence of σ_ξ^2 and E implies the finiteness of $d^2 = \mathbf{E}(\zeta - r\tau)^2$.

Consider the truncated random variables

$$\xi_x^{(N)} := \begin{cases} \xi_x & \text{if } \xi_x \in [-N, N], \\ N & \text{if } \xi_x > N, \\ -N & \text{if } \xi_x < -N. \end{cases}$$

Since $\sigma_\xi^2 < \infty$, we have $\mathbf{E}\xi_x^2 < \infty$ (a.e. with respect to the measure π) and

$$r^{(N)} \to r, \quad \bigl(\sigma_\xi^{(N)}\bigr)^2 \to \sigma_\xi^2, \quad E^{(N)} \to E \quad \text{as } N \to \infty,$$

where the superscript (N) means that the notation corresponds to the truncated random variables. By virtue of Theorem 13.8.4,

$$\liminf_{n\to\infty} \frac{1}{n} \mathbf{E}\bigl(S_n^{(N)} - r^{(N)}\bigr)^2 \ge a^{-1}\bigl(d^{(N)}\bigr)^2.$$

If we assume that $d = \infty$ then we will get that the lim inf on the left-hand side of this relation is infinite. But this contradicts relation (13.8.23), by which the above lim inf equals $(\sigma_\xi^{(N)})^2 + 2\mathbf{E}^{(N)}$ and remains bounded. We have obtained a contradiction, which shows that $d < \infty$.

On the other hand, for $d < \infty$, $\mathbf{E}\zeta^2 < \infty$ and, for the initial value x_0, by (13.8.5) we have

$$\begin{aligned} \mathbf{E}(S_n - rn)^2 &= \mathbf{E}(Z_{\nu(n)} + z_n - rn)^2 \\ &= \mathbf{E}(Z_{\eta(n)} - rn)^2 + 2\mathbf{E}(Z_{\eta(n)} - rn)(z_n - \zeta_{\eta(n)}) + \mathbf{E}(z_n - \zeta_{\eta(n)})^2, \end{aligned} \tag{13.8.24}$$

where $n = T_{\eta(n)} - \chi(n)$. Therefore, putting $Y_n := Z_n - rT_n = \sum_{k=1}^n (\zeta_k - r\tau_k)$, we obtain

$$\mathbf{E}(Z_{\eta(n)} - rn)^2 = \mathbf{E}Y_{\eta(n)}^2 - 2\mathbf{E}Y_{\eta(n)}\chi(n) + \mathbf{E}\chi^2(n).$$

By virtue of (10.4.7), $\mathbf{E}\chi^2(n) = o(n)$. By (10.6.4) (with a somewhat different notation),

$$\mathbf{E}Y_{\eta(n)}^2 = d^2\mathbf{E}\eta(n),$$

where $d^2 := \mathbf{D}(\zeta - r\tau)$, $\mathbf{E}\eta(n) \sim n/a$ and $a = \mathbf{E}\tau$. Hence, applying the Cauchy–Bunjakovsky inequality, we get

$$\bigl|\mathbf{E}Y_{\eta(n)}\chi(n)\bigr| = o(n), \qquad \mathbf{E}(Z_{\eta(n)} - rn)^2 \sim nd^2a^{-1}. \tag{13.8.25}$$

It remains to estimate the last two terms on the right-hand side of (13.8.24). But

$$\bigl|\zeta_{\eta(n)} - z_n\bigr| \le \zeta_{\eta(n)}^*,$$

where ζ^* corresponds to the summands $\xi^*_{X_k} = |\xi_{X_k}|$ and where, by Lemma 13.8.5 applied to $\xi^*_x = |\xi_x|$, we have

$$\mathbf{E}\big(\zeta^*_{\eta(n)}\big)^2 = o(n).$$

Therefore $\mathbf{E}(\zeta_{\eta(n)} - z_n) = o(n)$ and, by the Cauchy–Bunjakovsky inequality and relation (13.8.25), the same relation is valid for the shifted moment in (13.8.24). Thus,

$$\mathbf{E}(S_n - rn)^2 \sim a^{-1}d^2 n.$$

Combining this relation with (13.8.23), we obtain the assertion of the theorem. □

Chapter 14
Information and Entropy

Abstract Section 14.1 presents the definitions and key properties of information and entropy. Section 14.2 discusses the entropy of a (stationary) finite Markov chain. The Law of Large Numbers is proved for the amount of information contained in a message that is a long sequence of successive states of a Markov chain, and the asymptotic behaviour of the number of the most common states in a sequence of successive values of the chain is established. Applications of this result to coding are discussed.

14.1 The Definitions and Properties of Information and Entropy

Suppose one conducts an experiment whose outcome is not predetermined. The term "experiment" will have a broad meaning. It may be a test of a new device, a satellite launch, a football match, a referendum and so on. If, in a football match, the first team is stronger than the second, then the occurrence of the event A that the first team won carries little significant information. On the contrary, the occurrence of the complementary event $\overline{A}$ contains a lot of information. The event B that a leading player of the first team was injured does contain information concerning the event A. But if it was the first team's doctor who was injured then that would hardly affect the match outcome, so such an event B carries no significant information about the event A.

The following quantitative measure of information is conventionally adopted. Let A and B be events from some probability space $\langle \Omega, \mathfrak{F}, \mathbf{P} \rangle$.

Definition 14.1.1 The *amount of information about the event A contained in the event (message) B* is the quantity

$$I(A|B) := \log \frac{\mathbf{P}(A|B)}{\mathbf{P}(A)}.$$

The notions of the "amount of information" and "entropy" were introduced by C.E. Shannon in 1948. For some special situations the notion of amount of information had also been considered in earlier papers (e.g., by R.V.L. Hartley, 1928). The exposition in Sect. 14.2 of this chapter is substantially based on the paper of A.Ya. Khinchin [21].

A.A. Borovkov, *Probability Theory*, Universitext,
DOI 10.1007/978-1-4471-5201-9_14,

The occurrence of the event $B = A$ may be interpreted as the message that A took place.

Definition 14.1.2 The number $I(A) := I(A|A)$ is called the *amount of information contained in the message* A:

$$I(A) := I(A|A) = -\log \mathbf{P}(A).$$

We see from this definition that the larger the probability of the event A, the smaller $I(A)$. As a rule, the logarithm to the base 2 is used in the definition of information. Thus, say, the message that a boy (or girl) was born in a family carries a unit of information (it is supposed that these events are equiprobable, and $-\log_2 p = 1$ for $p = 1/2$). Throughout this chapter, we will write just $\log x$ for $\log_2 x$.

If the events A and B are independent, then $I(A|B) = 0$. This means that the event B does not carry any information about A, and vice versa. It is worth noting that we always have

$$I(A|B) = I(B|A).$$

It is easy to see that if the events A and B are independent, then

$$I(AB) = I(A) + I(B). \tag{14.1.1}$$

Consider an example. Let a chessman be placed at random on one of the squares of a chessboard. The information that the chessman is on square number k (the event A) is equal to $I(A) = \log 64 = 6$. Let B_1 be the event that the chessman is in the i-th row, and B_2 that the chessman is in the j-th column. The message A can be transmitted by transmitting B_1 first and then B_2. We have

$$I(B_1) = \log 8 = 3 = I(B_2).$$

Therefore

$$I(B_1) + I(B_2) = 6 = I(A),$$

so that transmitting the message A "by parts" requires communicating the same amount of information (which is equal to 6) as transmitting A itself. One could give other examples showing that the introduced numerical characteristics are quite natural.

Let G be an experiment with outcomes $E_1, \dots, E_N$ occurring with probabilities $p_1, \dots, p_N$.

The information resulting from the experiment G is a random variable $J_G = J_G(\omega)$ assuming the value $-\log p_j$ on the set E_j, $j = 1, \dots, N$.

Thus, if in the probability space $\langle \Omega, \mathfrak{F}, \mathbf{P} \rangle$ corresponding to the experiment G, Ω coincides with the set $(E_1, \dots, E_N)$, then $J_G(\omega) = I(\omega)$.

Definition 14.1.3 The expectation of the information obtained in the experiment G, $\mathbf{E} J_G = -\sum p_j \log p_j$, is called the *entropy* of the experiment. We shall denote it by

$$H_{\mathbf{p}} = H(G) := -\sum_{j=1}^{N} p_j \log p_j,$$

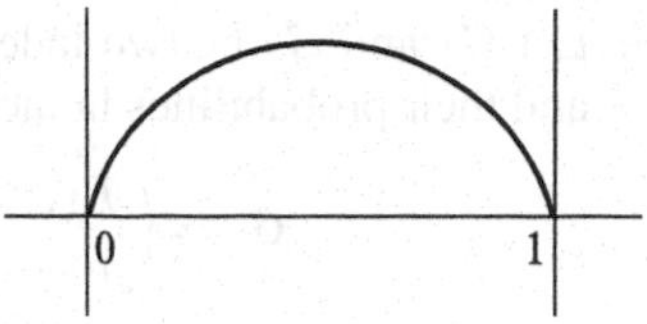

Fig. 14.1 The plot of the entropy $f(p)$ of a random experiment with two outcomes

where $\mathbf{p} = (p_1, \dots, p_N)$. For $p_j = 0$, by continuity we set $p_j \log p_j$ to be equal to zero.

The entropy of an experiment is, in a sense, a measure of its uncertainty. Let, for example, our experiment have two outcomes A and B with probabilities p and $q = 1 - p$, respectively. The entropy of the experiment is equal to

$$H_{\mathbf{p}} = -p \log p - (1 - p) \log(1 - p) = f(p).$$

The graph of this function is depicted in Fig. 14.1.

The only maximum of $f(p)$ equals $\log 2 = 1$ and is attained at the point $p = 1/2$. This is the case of maximum uncertainty. If p decreases, then the uncertainty also decreases together with $H_{\mathbf{p}}$, and $H_{\mathbf{p}} = 0$ for $\mathbf{p} = (0, 1)$ or $(0, 1)$.

The same properties can easily be seen in the general case as well.

The properties of entropy.

1. $H(G) = 0$ *if and only if there exists a* j, $1 \le j \le N$, *such that* $p_j = \mathbf{P}(E_j) = 1$.
2. $H(G)$ *attains its maximum when* $p_j = 1/N$ *for all* j.

Proof The second derivative of the function $\beta(x) = x \log x$ is positive on $[0, 1]$, so that $\beta(x)$ is convex. Therefore, for any $q_i \ge 0$ such that $\sum_{i=1}^N q_i = 1$, and any $x_i \ge 0$, one has the inequality

$$\beta\left(\sum_{i=1}^N q_i x_i\right) \le \sum_{i=1}^N q_i \beta(x_i).$$

If we take $q_i = 1/N$, $x_i = p_i$, then

$$\left(\frac{1}{N}\sum_{i=1}^N p_i\right) \log\left(\frac{1}{N}\sum_{i=1}^N p_i\right) \le \sum_{i=1}^N \frac{1}{N} p_i \log p_i.$$

Setting $\mathbf{u} := (\frac{1}{N}, \dots, \frac{1}{N})$ we obtain from this that

$$-\log \frac{1}{N} = \log N = H_{\mathbf{u}} \ge -\sum_{i=1}^N p_i \log p_i = H_{\mathbf{p}}. \qquad \square$$

Note that if the entropy $H(G)$ equals its maximum value $H(G) = \log N$, then $J_G(\omega) = \log N$ with probability 1, i.e. the information $J_G(\omega)$ becomes constant.

3. Let G_1 and G_2 be two independent experiments. We write down the outcomes and their probabilities in these experiments in the following way:

$$G_1 = \begin{pmatrix} E_1, \dots, E_N \\ p_1, \dots, p_N \end{pmatrix}, \qquad G_2 = \begin{pmatrix} A_1, \dots, A_M \\ q_1, \dots, q_M \end{pmatrix}.$$

Combining the outcomes of these two experiments we obtain a new experiment

$$G = G_1 \times G_2 = \begin{pmatrix} E_1A_1, E_1A_2, \dots, E_NA_M \\ p_1q_1, p_1q_2, \dots, p_Nq_M \end{pmatrix}.$$

The information J_G obtained as a result of this experiment is a random variable taking values $-\log p_iq_j$ with probabilities p_iq_j, $i = 1, \dots, N$; $j = 1, \dots, M$. But the sum $J_{G_1} + J_{G_2}$ of two independent random variables equal to the amounts of information obtained in the experiments G_1 and G_2, respectively, clearly has the same distribution. Thus the *information obtained in a sequence of independent experiments is equal to the sum of the information from these experiments*. Since in that case clearly

$$\mathbf{E}J_G = \mathbf{E}J_{G_1} + \mathbf{E}J_{G_2},$$

we have that *for independent G_1 and G_2 the entropy of the experiment G is equal to the sum of the entropies of the experiments G_1 and G_2*:

$$H(G) = H(G_1) + H(G_2).$$

4. If the experiments G_1 and G_2 are *dependent*, then the experiment G can be represented as

$$G = \begin{pmatrix} E_1A_1, E_1A_2, \dots, E_NA_M \\ q_{11}, q_{12}, \dots, q_{NM} \end{pmatrix}$$

with $q_{ij} = p_i p_{ij}$, where p_{ij} is the conditional probability of the event A_j given E_i, so that

$$\sum_{j=1}^{M} q_{ij} = p_i = \mathbf{P}(E_i), \quad i = 1, \dots, N;$$

$$\sum_{j=1}^{N} q_{ij} = q_j = \mathbf{P}(A_i), \quad j = 1, \dots, M.$$

In this case the equality $J_G = J_{G_1} + J_{G_2}$, generally speaking, does not hold. Introduce a random variable J_2^* which is equal to $-\log p_{ij}$ on the set E_iA_j. Then evidently $J_G = J_{G_1} + J_2^*$. Since

$$\mathbf{P}(A|E_i) = p_{ij},$$

the quantity J_2^* for a fixed i can be considered as the information from the experiment G_2 given the event E_i occurred. We will call the quantity

$$\mathbf{E}(J_2^*|E_i) = -\sum_{j=1}^{M} p_{ij} \log p_{ij}$$

the *conditional entropy $H(G_2|E_i)$ of the experiment G_2 given E_i*, and the quantity

$$\mathbf{E}J_2^* = -\sum_{i,j} q_{ij} \log p_{ij} = \sum_i p_i H(G_2|E_i)$$

the *conditional entropy $H(G_2|G_1)$ of the experiment G_2 given G_1*. In this notation, we obviously have

$$H(G) = H(G_1) + H(G_2|G_1).$$

We will prove that in this equality we always have

$$H(G_2|G_1) \le H(G_2),$$

i.e. *for two experiments G_1 and G_2 the entropy $H(G)$ never exceeds the sum of the entropies $H(G_1)$ and $H(G_2)$*:

$$H(G) = H(G_1 \times G_2) \le H(G_1) + H(G_2).$$

Equality takes place here only when $q_{ij} = p_i q_j$, i.e. when G_1 and G_2 are independent.

Proof First note that, for any two distributions $(u_1, \dots, u_n)$ and $(v_1, \dots, v_n)$, one has the inequality

$$-\sum_i u_i \log u_i \le -\sum_i u_i \log v_i, \tag{14.1.2}$$

equality being possible here only if $v_i = u_i$, $i = 1, \dots, n$. This follows from the concavity of the function $\log x$, since it implies that, for any $a_i > 0$,

$$\sum_i u_i \log a_i \le \log\left(\sum_i u_i a_i\right),$$

equality being possible only if $a_1 = a_2 = \dots = a_n$. Putting $a_i = v_i/u_i$, we obtain relation (14.1.2).

Next we have

$$H(G_1) + H(G_2) = -\sum_{i,j} q_{ij}(\log p_i + \log q_j) = -\sum_{i,j} q_{ij} \log p_i q_j,$$

and because $\{p_i q_j\}$ is obviously a distribution, by virtue of (14.1.2)

$$-\sum q_{ij} \log p_i q_j \ge -\sum q_{ij} \log q_{ij} = H(G)$$

holds, and equality is possible here only if $q_{ij} = p_i q_j$. □

5. As we saw when considering property 3, the information obtained as a result of the experiment G_1^n consisting of n independent repetitions of the experiment G_1 is equal to

$$J_{G_1^n} = -\sum_{j=1}^{N} \nu_j \log p_j,$$

where ν_j is the number of occurrences of the outcome E_j. By the law of large numbers, $\nu_j/n \xrightarrow{p} p_j$ as $n \to \infty$, and hence

$$\frac{1}{n} J_{G_1^n} \xrightarrow{p} H(G_1) = H_p.$$

To conclude this section, we note that the measure of the amount of information resulting from an experiment we considered here can be *derived* as the only possible one (up to a constant multiplier) if one starts with a few simple requirements that are natural to impose on such a quantity.[1]

It is also interesting to note the connections between the above-introduced notions and large deviation probabilities. As one can see from Theorems 5.1.2 and 5.2.4, the difference between the "biased" entropy $-\sum p_j^* \ln p_j$ and the entropy $-\sum p_j^* \ln p_j^*$ ($p_j^* = \nu_j/n$ are the relative frequencies of the outcomes E_j) is an analogue of the deviation function (see Sect. 8.8) in the multi-dimensional case.

14.2 The Entropy of a Finite Markov Chain. A Theorem on the Asymptotic Behaviour of the Information Contained in a Long Message; Its Applications

14.2.1 The Entropy of a Sequence of Trials Forming a Stationary Markov Chain

Let $\{X_k\}_{k=1}^{\infty}$ be a stationary finite Markov chain with one class of essential states without subclasses, $E_1, \dots, E_N$ being its states. Stationarity of the chain means that $\mathbf{P}(X_1 = j) = \pi_j$ coincide with the stationary probabilities. It is clear that

$$\mathbf{P}(X_2 = j) = \sum_k \pi_k p_{kj} = \pi_j, \qquad \mathbf{P}(X_3 = j) = \pi_j, \qquad \text{and so on.}$$

Let G_k be an experiment determining the value of X_k (i.e. the state the system entered on the k-th step). If $X_{k-1} = i$, then the entropy of the k-th step equals

$$H(G_k | X_{k-1} = i) = -\sum_j p_{ij} \log p_{ij}.$$

By definition, the entropy of a stationary Markov chain is equal to

$$H = \mathbf{E} H(G_k | X_{k-1}) = H(G_k | G_{k-1}) = -\sum_i \pi_i \sum_j p_{ij} \log p_{ij}.$$

Consider the first n steps $X_1, \dots, X_n$ of the Markov chain. By the Markov property, the entropy of this composite experiment $G^{(n)} = G_1 \times \cdots \times G_n$ is equal to

[1] See, e.g., [11].

$$H\left(G^{(n)}\right) = H(G_1) + H(G_2|G_1) + \cdots + H(G_n|G_{n-1})$$
$$= -\sum \pi_j \log \pi_j + (n-1)H \sim nH$$

as $n \to \infty$. If X_k were independent then, as we saw, we would have exact equality here.

14.2.2 The Law of Large Numbers for the Amount of Information Contained in a Message

Now consider a finite sequence $(X_1, \ldots, X_n)$ as a message (event) C_n and denote, as before, by $I(C_n) = -\log \mathbf{P}(C_n)$ the amount of information contained in C_n. The value of $I(C_n)$ is a function on the space of elementary outcomes equal to the information $J_{G^{(n)}}$ contained in the experiment $G^{(n)}$. We now show that, with probability close to 1, this information behaves asymptotically as nH, as was the case for independent X_k. Therefore H is essentially the average information per trial in the sequence $\{X_k\}_{k=1}^{\infty}$.

Theorem 14.2.1 *As* $n \to \infty$,

$$\frac{I(C_n)}{n} = \frac{-\log \mathbf{P}(C_n)}{n} \xrightarrow{a.s.} H.$$

This means that, for any $\delta > 0$, the set of all messages C_n can be decomposed into two classes. For the first class, $|I(C_n)/n - H| < \delta$, and the sum of the probabilities of the elements of the second class tends to 0 as $n \to \infty$.

Proof Construct from the given Markov chain a new one $\{Y_k\}_{k=1}^{\infty}$ by setting $Y_k := (X_k, X_{k+1})$. The states of the new chain are pairs of states (E_i, E_j) of the chain $\{X_k\}$ with $p_{ij} > 0$. The transition probabilities are obviously given by

$$p_{(i,j)(k,l)} = \begin{cases} 0, & j \neq k, \\ p_{kl}, & j = k. \end{cases}$$

Note that one can easily prove by induction that

$$p_{(i,j)(k,l)}(n) = p_{jk}(n-1)p_{kl}. \tag{14.2.1}$$

From the definition of $\{Y_k\}$ it follows that the ergodic theorem holds for this chain. This can also be seen directly from (14.2.1), the stationary probabilities being

$$\lim_{n\to\infty} p_{(i,j)(k,l)}(n) = \pi_k p_{kl}.$$

Now we will need the law of large numbers for the number of visits $m_{(k,l)}(n)$ of the chain $\{Y_k\}_{k=1}^{\infty}$ to state (k,l) over time n. By virtue of this law (see Theorem 13.4.4),

$$\frac{m_{(k,l)}(n)}{n} \xrightarrow{a.s.} \pi_k p_{kl} \quad \text{as } n \to \infty.$$

Consider the random variable $\mathbf{P}(C_n)$:

$$\mathbf{P}(C_n) = \mathbf{P}(E_{X_1} E_{X_2} \cdots E_{X_n}) = \mathbf{P}(E_{X_1})\mathbf{P}(E_{X_2}|E_{X_1}) \cdots \mathbf{P}(E_{X_n}|E_{X_{n-1}})$$
$$= \pi_{X_1} p_{X_1 X_2} \cdots p_{X_{n-1} X_n} = \pi_{X_1} \prod_{(k,l)} p_{kl}^{m_{(k,l)}(n-1)}.$$

The product here is taken over all pairs (k,l). Therefore ($\pi_i = \mathbf{P}(X_1 = i)$)

$$\log \mathbf{P}(C_n) = \log \pi_{X_1} + \sum_{k,l} m_{(k,l)}(n-1) \log p_{kl},$$

$$\frac{1}{n} \log \mathbf{P}(C_n) \xrightarrow{p} \sum_{k,l} \pi_k p_{kl} \log p_{kl} = -H.$$

□

14.2.3 The Asymptotic Behaviour of the Number of the Most Common Outcomes in a Sequence of Trials

Theorem 14.2.1 has an important corollary. Rank all the messages (words) C_n of length n according to the values of their probabilities in descending order. Next pick the most probable words one by one until the sum of their probabilities exceeds a prescribed level α, $0 < \alpha < 1$. Denote the number (and also the set) of the selected words by $M_\alpha(n)$.

Theorem 14.2.2 *For each* $0 < \alpha < 1$, *there exists one and the same limit*

$$\lim_{n\to\infty} \frac{\log M_\alpha(n)}{n} = H.$$

Proof Let $\delta > 0$ be a number, which can be arbitrarily small. We will say that C_n falls into category K_1 if its probability $\mathbf{P}(C_n) > 2^{-n(H-\delta)}$, and into category K_2 if

$$2^{-n(H+\delta)} < \mathbf{P}(C_n) \le 2^{-n(H-\delta)}.$$

Finally, C_n belongs to the third category K_3 if

$$\mathbf{P}(C_n) \le 2^{-n(H+\delta)}.$$

Since, by Theorem 14.2.1, $\mathbf{P}(C_n \in K_1 \cup K_3) \to 0$ as $n \to \infty$, the set $M_\alpha(n)$ contains only the words from K_1 and K_2, and the last word from $M_\alpha(n)$ (i.e. having the smallest probability)—we denote it by $C_{\alpha,n}$—belongs to K_2. This means that

$$M_\alpha(n) 2^{-n(H+\delta)} < \sum_{C_n \in M_\alpha(n)} \mathbf{P}(C_n) < \alpha + \mathbf{P}(C_{\alpha,n}) < \alpha + 2^{-n(H-\delta)}.$$

This implies

$$\frac{\log M_\alpha(n)}{n} < \frac{(\alpha + 2^{-n(H-\delta)})}{n} + H + \delta.$$

Since δ is arbitrary, we have

$$\limsup_{n\to\infty} \frac{\log M_\alpha(n)}{n} \le H.$$

On the other hand, the words from K_2 belonging to $M_\alpha(n)$ have total probability $\ge \alpha - \mathbf{P}(K_1)$. If $M_\alpha^{(2)}(n)$ is the number of these messages then

$$M_\alpha^{(2)}(n)2^{-n(H-\delta)} \ge \alpha - \mathbf{P}(K_1),$$

and, consequently,

$$M_\alpha(n)2^{-n(H-\delta)} \ge \alpha - \mathbf{P}(K_1).$$

Since $\mathbf{P}(K_1) \to 0$ as $n \to \infty$, for sufficiently large n one has

$$\frac{\log M_\alpha(n)}{n} \ge H - \delta + \frac{1}{n}\log\frac{\alpha}{2}.$$

It follows that

$$\limsup_{n\to\infty} \frac{\log M_\alpha(n)}{n} \ge H.$$

The theorem is proved. □

Now one can obtain a useful interpretation of this theorem. Let N be the number of the chain states. Suppose for simplicity's sake that $N = 2^m$. Then the number of different words of length n (chains C_n) will be equal to $N^n = 2^{nm}$. Suppose, further, that these words are transmitted using a binary code, so that m binary symbols are used to code every state. Thus, with such transmission method—we will call it *direct coding*—the length of the messages will be equal to nm. (For example, one can use Markov chains to model the Russian language and take $N = 32$, $m = 5$.) The assertion of Theorem 14.2.2 means that, for large n, with probability $1 - \varepsilon$, $\varepsilon > 0$, only 2^{nH} of the totality of 2^{nm} words will be transmitted. The probability of transmitting all the remaining words will be small if ε is small. From this it is easy to establish the existence of another more economical code requiring, with a large probability, a smaller number of digits to transmit a word. Indeed, one can enumerate the selected 2^{nH} most likely words using, say, a binary code again, and then transmit only the number of the word. This clearly requires only nH digits. Since we always have $H \le \log N = m$, the length of the message will be $m/H \ge 1$ times smaller.

This is a special case of the so-called *basic coding theorem* for Markov chains: for large n, there exists a code for which, with a high probability, the original message C_n can be transmitted by a sequence of signals which is m/H times shorter than in the case of the direct coding.

The above coding method is rather an oversimplified example than a recipe for efficiently compressing the messages. It should be noted that finding a really efficient coding method is a rather difficult task. For example, in Morse code it is reasonable to encode more frequent letters by shorter sequences of dots and dashes.

However, the text reduction by m/H times would not be achieved. Certain compression techniques have been used in this book as well. For example, we replaced the frequently encountered words "characteristic function" by "ch.f." We could achieve better results if, say, shorthand was used. The structure of a code with a high compression coefficient will certainly be very complicated. The theorems of the present chapter give an upper bound for the results we can achieve.

Since $H = \sum \frac{1}{n} \log N = m$, for a sequence of *independent* equiprobable symbols, such a text is *incontractible*. This is why the proximity of "new" messages (encoded using a new alphabet) to a sequence of equiprobable symbols could serve as a criterion for constructing new codes.

It should be taken into account, however, that the text "redundancy" we are "fighting" with is in many cases a useful and helpful phenomenon. Without such redundancy, it would be impossible to detect misprints or reconstruct omissions as easily as we, say, restore the letter "r" in the word "info · mation".

The reader might know how difficult it is to read a highly abridged and formalised mathematical text. While working with an ideal code no errors would be admissible (even if we could find any), since it is impossible to reconstruct an omitted or distorted symbol in a sequence of equiprobable digits. In this connection, there arises one of the basic problems of information theory: to find a code with the smallest "redundancy" which still allows one to eliminate the transmission noise.

Chapter 15
Martingales

Abstract The definitions, simplest properties and first examples of martingales and sub/super-martingales are given in Sect. 15.1. Stopping (Markov) times are introduced in Sect. 15.2, which also contains Doob's theorem on random change of time and Wald's identity together with a number of its applications to boundary crossing problems and elsewhere. This is followed by Sect. 15.3 presenting fundamental martingale inequalities, including Doob's inequality with a number of its consequences, and an inequality for the number of strip crossings. Section 15.4 begins with Doob's martingale convergence theorem and also presents Lévy's theorem and an application to branching processes. Section 15.5 derives several important inequalities for the moments of stochastic sequences.

15.1 Definitions, Simplest Properties, and Examples

In Chap. 13 we considered sequences of dependent random variables $X_0, X_1, \dots$ forming Markov chains. Dependence was described there in terms of transition probabilities determining the distribution of X_{n+1} given X_n. That enabled us to investigate rather completely the properties of Markov chains.

In this chapter we consider another type of sequence of dependent random variables. Now dependence will be characterised only by the mean value of X_{n+1} given the whole "history" $X_0, \dots, X_n$. It turns out that one can also obtain rather general results for such sequences.

Let a probability space $\langle \Omega, \mathfrak{F}, \mathbf{P} \rangle$ be given together with a sequence of random variables $X_0, X_1, \dots$ defined on it and an increasing family (or flow) of σ-algebras $\{\mathfrak{F}_n\}_{n\ge 0}$: $\mathfrak{F}_0 \subseteq \mathfrak{F}_1 \subseteq \cdots \subseteq \mathfrak{F}_n \subseteq \cdots \subseteq \mathfrak{F}$.

Definition 15.1.1 A sequence of pairs $\{X_n, \mathfrak{F}_n;\ n \ge 0\}$ is called a *stochastic sequence* if, for each $n \ge 0$, X_n is $\mathfrak{F}_n$-measurable. A stochastic sequence is said to be a *martingale* (one also says that $\{X_n\}$ is a *martingale with respect to the flow of σ-algebras* $\{\mathfrak{F}_n\}$) if, for every $n \ge 0$,

(1)

$$\mathbf{E}|X_n| < \infty, \tag{15.1.1}$$

A.A. Borovkov, *Probability Theory*, Universitext,
DOI 10.1007/978-1-4471-5201-9_15, © Springer-Verlag London 2013

(2) X_n is measurable with respect to $\mathfrak{F}_n$,

(3)

$$\mathbf{E}(X_{n+1} \mid \mathfrak{F}_n) = X_n. \tag{15.1.2}$$

A stochastic sequence $\{X_n, \mathfrak{F}_n;\ n \geq 0\}$ is called a *submartingale* (*supermartingale*) if conditions (1)–(3) hold with the sign "=" replaced in (15.1.2) with "$\geq$" ("$\leq$", respectively).

We will say that a sequence $\{X_n\}$ forms a *martingale* (*submartingale, supermartingale*) if, for $\mathfrak{F}_n = \sigma(X_0, \dots, X_n)$, the pairs $\{X_n, \mathfrak{F}_n\}$ form a sequence with the same name. Submartingales and supermartingales are often called *semimartingales*.

It is evident that relation (15.1.2) persists if we replace X_{n+1} on its left-hand side with X_m for any $m > n$. Indeed, by virtue of the properties of conditional expectations,

$$\mathbf{E}(X_m|\mathfrak{F}_n) = \mathbf{E}\big[\mathbf{E}(X_m|\mathfrak{F}_{m-1})\big|\mathfrak{F}_n\big] = \mathbf{E}(X_{m-1}|\mathfrak{F}_n) = \cdots = X_n.$$

A similar assertion holds for semimartingales.

If $\{X_n\}$ is a martingale, then $\mathbf{E}(X_{n+1}|\sigma(X_0, \dots, X_n)) = X_n$, and, by a property of conditional expectations,

$$\mathbf{E}\big(X_{n+1}\big|\sigma(X_n)\big) = \mathbf{E}\big[\mathbf{E}\big(X_{n+1}\big|\sigma(X_0, \dots, X_n)\big)\big|\sigma(X_n)\big] = \mathbf{E}\big(X_n\big|\sigma(X_n)\big) = X_n.$$

So, for martingales, as for Markov chains, we have

$$\mathbf{E}\big(X_{n+1}\big|\sigma(X_0, \dots, X_n)\big) = \mathbf{E}\big(X_{n+1}\big|\sigma(X_n)\big).$$

The similarity, however, is limited to this relation, because for a martingale, the equality does not hold for distributions, but the additional condition

$$\mathbf{E}\big(X_{n+1}\big|\sigma(X_n)\big) = X_n$$

is imposed.

Example 15.1.1 Let $\xi_n, n \geq 0$ be independent. Then $X_n = \xi_1 + \cdots + \xi_n$ form a martingale (submartingale, supermartingale) if $\mathbf{E}\xi_n = 0$ ($\mathbf{E}\xi_n \geq 0$, $\mathbf{E}\xi_n \leq 0$). It is obvious that X_n also form a Markov chain. The same is true of $X_n = \prod_{k=0}^{n} \xi_k$ if $\mathbf{E}\xi_n = 1$.

Example 15.1.2 Let $\xi_n, n \geq 0$, be independent. Then

$$X_n = \sum_{k=1}^{n} \xi_{k-1}\xi_k, \quad n \geq 1, \qquad X_0 = \xi_0,$$

form a martingale if $\mathbf{E}\xi_n = 0$, because

$$\mathbf{E}\big(X_{n+1}\big|\sigma(X_0, \dots, X_n)\big) = X_n + \mathbf{E}\big(\xi_n\xi_{n+1}\big|\sigma(\xi_n)\big) = X_n.$$

Clearly, $\{X_n\}$ is not a Markov chain here. An example of a sequence which is a Markov chain but not a martingale can be obtained, say, if we consider a random walk on a segment with reflection at the endpoints (see Example 13.1.1).

As well as $\{0, 1, \dots\}$ we will use other sets of indices for X_n, for example, $\{-\infty < n < \infty\}$ or $\{n \le -1\}$, and also sets of integers including infinite values $\pm\infty$, say, $\{0 \le n \le \infty\}$. We will denote these sets by a common symbol $\mathcal{N}$ and write martingales (semimartingales) as $\{X_n, \mathfrak{F}_n;\ n \in \mathcal{N}\}$. By $\mathfrak{F}_{-\infty}$ we will understand the σ-algebra $\bigcap_{n\in\mathcal{N}} \mathfrak{F}_n$, and by $\mathfrak{F}_\infty$ the σ-algebra $\sigma(\bigcup_{n\in\mathcal{N}} \mathfrak{F}_n)$ generated by $\bigcup_{n\in\mathcal{N}} \mathfrak{F}_n$, so that $\mathfrak{F}_{-\infty} \subseteq \mathfrak{F}_n \subseteq \mathfrak{F}_\infty \subseteq \mathfrak{F}$ for any $n \in \mathcal{N}$.

Definition 15.1.2 A stochastic sequence $\{X_n, \mathfrak{F}_n;\ n \in \mathcal{N}\}$ is called a *martingale* (*submartingale, supermartingale*), if the conditions of Definition 15.1.1 hold for any $n \in \mathcal{N}$.

If $\{X_n, \mathfrak{F};\ n \in \mathcal{N}\}$ is a martingale and the left boundary n_0 of $\mathcal{N}$ is finite (for example, $\mathcal{N} = \{0, 1, \dots\}$), then the martingale $\{X_n, \mathfrak{F}_n\}$ can be always extended "to the whole axis" by setting $\mathfrak{F}_n := \mathfrak{F}_{n_0}$ and $X_n := X_{n_0}$ for $n < n_0$. The same holds for the right boundary as well. Therefore if a martingale (semimartingale) $\{X_n, \mathfrak{F}_n;\ n \in \mathcal{N}\}$ is given, then without loss of generality we can always assume that one is actually given a martingale (semimartingale) $\{X_n, \mathfrak{F}_n;\ -\infty \le n \le \infty\}$.

Example 15.1.3 Let $\{\mathfrak{F}_n,\ -\infty \le n \le \infty\}$ be a given sequence of increasing σ-algebras, and ξ a random variable on $\langle \Omega, \mathfrak{F}, \mathbf{P}\rangle$, $\mathbf{E}|\xi| < \infty$. Then $\{X_n, \mathfrak{F}_n;\ -\infty \le n \le \infty\}$ with $X_n = \mathbf{E}(\xi\,|\mathfrak{F}_n)$ forms a martingale.

Indeed, by the property of conditional expectations, for any $m \le \infty$ and $m > n$,

$$\mathbf{E}(X_m|\mathfrak{F}_n) = \mathbf{E}\big[\mathbf{E}(\xi\,|\mathfrak{F}_m)\big|\mathfrak{F}_n\big] = \mathbf{E}(\xi\,|\mathfrak{F}_n) = X_n.$$

Definition 15.1.3 The martingale of Example 15.1.3 is called a *martingale generated by the random variable* ξ *(and the family* $\{\mathfrak{F}_n\}$*)*.

Definition 15.1.4 A set $\mathcal{N}_+$ is called the *right closure of* $\mathcal{N}$ if:

(1) $\mathcal{N}_+ = \mathcal{N}$ when the maximal element of $\mathcal{N}$ is finite;
(2) $\mathcal{N}_+ = \mathcal{N} \cup \{\infty\}$ if $\mathcal{N}$ is not bounded from the right.

If $\mathcal{N} = \mathcal{N}_+$ then we say that $\mathcal{N}$ is *right closed*. A *martingale* (*semimartingale*) $\{X_n, \mathfrak{F};\ n \in \mathcal{N}\}$ is said to be *right closed* if $\mathcal{N}$ is right closed.

Lemma 15.1.1 *A martingale* $\{X_n, \mathfrak{F};\ n \in \mathcal{N}\}$ *is generated by a random variable if and only if it is right closed.*

The Proof of the lemma is trivial. In one direction it follows from Example 15.1.3, and in the other from the equality

$$\mathbf{E}(X_N|\mathfrak{F}_n) = X_n, \qquad N = \sup\{k;\ k \in \mathcal{N}\},$$

which implies that $\{X_n, \mathfrak{F}\}$ is generated by X_N. The lemma is proved. □

Now we consider an interesting and more concrete example of a martingale generated by a random variable.

Example 15.1.4 Let $\xi_1, \xi_2, \ldots$ be independent and identically distributed and assume $\mathbf{E}|\xi_1| < \infty$. Set

$$S_n = \xi_1 + \cdots + \xi_n, \quad X_{-n} = S_n/n, \quad \mathfrak{F}_{-n} = \sigma(S_n, S_{n+1}, \ldots) = \sigma(S_n, \xi_{n+1}, \ldots).$$

Then $\mathfrak{F}_{-n} \subset \mathfrak{F}_{-n+1}$ and, for any $1 \le k \le n$, by symmetry

$$\mathbf{E}(\xi_k|\mathfrak{F}_{-n}) = \mathbf{E}(\xi_1|\mathfrak{F}_{-n}).$$

From this it follows that

$$S_n = \mathbf{E}(S_n|\mathfrak{F}_{-n}) = \sum_{k=1}^{n} \mathbf{E}(\xi_k|\mathfrak{F}_{-n}) = n\mathbf{E}(\xi_1|\mathfrak{F}_{-n}), \qquad \frac{S_n}{n} = \mathbf{E}(\xi_1|\mathfrak{F}_{-n}).$$

This means that $\{X_n, \mathfrak{F}_n;\ n \le 1\}$ forms a martingale generated by ξ_1.

We will now obtain a series of auxiliary assertions giving the simplest properties of martingales and semimartingales. When considering semimartingales, we will confine ourselves to submartingales only, since the corresponding properties of supermartingales will follow immediately if one considers the sequence $Y_n = -X_n$, where $\{X_n\}$ is a submartingale.

Lemma 15.1.2

(1) *The property that* $\{X_n, \mathfrak{F}_n;\ n \in \mathcal{N}\}$ *is a martingale is equivalent to invariability in* $m \ge n$ *of the set functions (integrals)*

$$\mathbf{E}(X_m;\ A) = \mathbf{E}(X_n;\ A) \tag{15.1.3}$$

for any $A \in \mathfrak{F}_n$. *In particular,* $\mathbf{E}X_m = \text{const}$.

(2) *The property that* $\{X_n, \mathfrak{F}_n;\ n \in \mathcal{N}\}$ *is a submartingale is equivalent to the monotone increase in* $m \ge n$ *of the set functions*

$$\mathbf{E}(X_m;\ A) \ge \mathbf{E}(X_n;\ A) \tag{15.1.4}$$

for every $A \in \mathfrak{F}_n$. *In particular,* $\mathbf{E}X_m \uparrow$.

The Proof follows immediately from the definitions. If (15.1.3) holds then, by the definition of conditional expectation, $X_n = \mathbf{E}(X_m|\mathfrak{F}_n)$, and vice versa. Now let (15.1.4) hold. Put $Y_n = \mathbf{E}(X_m|\mathfrak{F}_n)$. Then (15.1.4) implies that $\mathbf{E}(Y_n;\ A) \ge \mathbf{E}(X_n;\ A)$ and $\mathbf{E}(Y_n - X_n;\ A) \ge 0$ for any $A \in \mathfrak{F}_n$. From this it follows that $Y_n = \mathbf{E}(X_m|\mathfrak{F}_n) \ge X_n$ with probability 1. The converse assertion can be obtained as easily as the direct one. The lemma is proved. □

Lemma 15.1.3 *Let* $\{X_n, \mathfrak{F}_n;\ n \in \mathcal{N}\}$ *be a martingale,* $g(x)$ *be a convex function, and* $\mathbf{E}|g(X_n)| < \infty$. *Then* $\{g(X_n), \mathfrak{F}_n;\ n \in \mathcal{N}\}$ *is a submartingale.*

If, in addition, $g(x)$ *is nondecreasing, then the assertion of the theorem remains true when* $\{X_n, \mathfrak{F}_n;\ n \in \mathcal{N}\}$ *is a submartingale.*

The Proof of both assertions follows immediately from Jensen's inequality

$$\mathbf{E}\big(g(X_{n+1})\big|\mathfrak{F}_n\big) \ge g\big(\mathbf{E}(X_{n+1}|\mathfrak{F}_n)\big) \ge g\big(\mathbf{E}(X_n|\mathfrak{F}_n)\big). \qquad \square$$

Clearly, the function $g(x) = |x|^p$ for $p \ge 1$ satisfies the conditions of the first part of the lemma, and the function $g(x) = e^{\lambda x}$ for $\lambda > 0$ meets the conditions of the second part of the lemma.

Lemma 15.1.4 *Let $\{X_n, \mathfrak{F}_n;\ n \in \mathcal{N}\}$ be a right closed submartingale. Then, for $X_n(a) = \max\{X_n, a\}$ and any a, $\{X_n(a), \mathfrak{F}_n;\ n \in \mathcal{N}\}$ is a uniformly integrable submartingale.*

If $\{X_n, \mathfrak{F}_n;\ n \in \mathcal{N}\}$ is a right closed martingale, then it is uniformly integrable.

Proof Let $N := \sup\{k : k \in \mathcal{N}\}$. Then, by Lemma 15.1.3, $\{X_n(a), \mathfrak{F}_n;\ n \in \mathcal{N}\}$ is a submartingale. Hence, for any $c > 0$,

$$c\mathbf{P}\big(X_n(a) > c\big) \le \mathbf{E}\big(X_n(a);\ X_n(a) > c\big) \le \mathbf{E}\big(X_N(a);\ X_n(a) > c\big) \le \mathbf{E}X_N^+(a)$$

(here $X^+ = \max(0, X)$) and so

$$\mathbf{P}\big(X_n(a) > c\big) \le \frac{1}{c}\mathbf{E}\big(X_N^+(a)\big) \to 0,$$

uniformly in n as $c \to \infty$. Therefore we get the required uniform integrability:

$$\sup_n \mathbf{E}\big(X_n(a);\ X_n(a) > c\big) \le \sup_n \mathbf{E}\big(X_N(a);\ X_n(a) > c\big) \to 0,$$

since $\sup_n \mathbf{P}(X_n(a) > c) \to 0$ as $c \to \infty$ (see Lemma A3.2.3 in Appendix 3; by truncating at the level a we avoided estimating the "negative tails").

If $\{X_n, \mathfrak{F}_n;\ n \in \mathcal{N}\}$ is a martingale, then its uniform integrability will follow from the first assertion of the lemma applied to the submartingale $\{|X_n|, \mathfrak{F}_n;\ n \in \mathcal{N}\}$. The lemma is proved. $\square$

The nature of martingales can be clarified to some extent by the following example.

Example 15.1.5 Let $\xi_1, \xi_2, \ldots$ be an arbitrary sequence of random variables, $\mathbf{E}|\xi_k| < \infty$, $\mathfrak{F}_n = \sigma(\xi_1, \ldots, \xi_n)$ for $n \ge 1$, $\mathfrak{F}_0 = (\varnothing, \Omega)$ (the trivial σ-algebra),

$$S_n = \sum_{k=1}^n \xi_k, \qquad Z_n = \sum_{k=1}^n \mathbf{E}(\xi_k|\mathfrak{F}_{k-1}), \qquad X_n = S_n - Z_n.$$

Then $\{X_n, \mathfrak{F}_n;\ n \ge 1\}$ is a martingale. This is a consequence of the fact that

$$\mathbf{E}(S_{n+1} - Z_{n+1}|\mathfrak{F}_n) = \mathbf{E}\big(X_n + \xi_{n+1} - \mathbf{E}(\xi_{n+1}|\mathfrak{F}_n)\big|\mathfrak{F}_n\big) = X_n.$$

In other words, for an arbitrary sequence $\{\xi_n\}$, the sequence S_n can be "compensated" by a so-called "predictable" (in the sense that its value is determined by $S_1, \ldots, S_{n-1}$) sequence Z_n so that $S_n - Z_n$ will be a martingale.

15.2 The Martingale Property and Random Change of Time. Wald's Identity

Throughout this section we assume that $\mathcal{N} = \{n \ge 0\}$. Recall the definition of a stopping time.

Definition 15.2.1 A random variable ν will be called a *stopping time* or a *Markov time* (with respect to an increasing family of σ-algebras $\{\mathfrak{F}_n;\ n \ge 0\}$) if, for any $n \ge 0$, $\{\nu \le n\} \in \mathfrak{F}_n$.

It is obvious that a constant $\nu \equiv m$ is a stopping time. If ν is a stopping time, then, for any fixed m, $\nu(m) = \min(\nu, m)$, is also a stopping time, since for $n \ge m$ we have

$$\nu(m) \le m \le n, \quad \bigl\{\nu(m) \le n\bigr\} = \Omega \in \mathfrak{F}_n,$$

and if $n < m$ then

$$\bigl\{\nu(m) \le n\bigr\} = \{\nu \le n\} \in \mathfrak{F}_n.$$

If ν is a stopping time, then

$$\{\nu = n\} = \{\nu \le n\} - \{\nu \le n-1\} \in \mathfrak{F}_n, \quad \{\nu \ge n\} = \Omega - \{\nu \le n-1\} \in \mathfrak{F}_{n-1}.$$

Conversely, if $\{\nu = n\} \in \mathfrak{F}_n$, then $\{\nu \le n\} \in \mathfrak{F}_n$ and therefore ν is a stopping time.

Let a martingale $\{X_n, \mathfrak{F}_n;\ n \ge 0\}$ be given. A typical example of a stopping time is the time ν at which X_n first hits a given measurable set B:

$$\nu = \inf\{n \ge 0 : X_n \in B\}$$

($\nu = \infty$ if all $X_n \notin B$). Indeed,

$$\{\nu = n\} = \{X_0 \notin B, \dots, X_{n-1} \notin B,\ X_n \in B\} \in \mathfrak{F}_n.$$

If ν is a proper stopping time ($\mathbf{P}(\nu < \infty) = 1$), then X_ν is a random variable, since

$$X_\nu = \sum_{n=0}^{\infty} X_n \mathrm{I}_{\{\nu = n\}}.$$

By $\mathfrak{F}_\nu$ we will denote the σ-algebra of sets $A \in \mathfrak{F}$ such that $A \cap \{\nu = n\} \in \mathfrak{F}_n$, $n = 0, 1, \dots$ This σ-algebra can be thought of as being generated by the events $\{\nu \le n\} \cap B_n$, $n = 0, 1, \dots$, where $B_n \in \mathfrak{F}_n$. Clearly, ν and X_ν are $\mathfrak{F}_\nu$-measurable. If ν_1 and ν_2 are two stopping times, then $\{\nu_2 \ge \nu_1\} \in \mathfrak{F}_{\nu_1}$ and $\{\nu_2 \ge \nu_1\} \in \mathfrak{F}_{\nu_2}$, since $\{\nu_2 \ge \nu_1\} = \bigcup_n [\{\nu_2 = n\} \cap \{\nu_1 \le n\}]$.

We already know that if $\{X_n, \mathfrak{F}_n\}$ is a martingale then $\mathbf{E}X_n$ is constant for all n. Will this property remain valid for $\mathbf{E}X_\nu$ if ν is a stopping time? From Wald's identity we know that this is the case for the martingale from Example 15.1.1. In the general case one has the following.

Theorem 15.2.1 (Doob) *Let $\{X_n, \mathfrak{F}_n;\ n \ge 0\}$ be a martingale (submartingale) and ν_1, ν_2 be stopping times such that*

$$\mathbf{E}|X_{\nu_i}| < \infty, \quad i = 1, 2, \tag{15.2.1}$$

$$\liminf_{n\to\infty} \mathbf{E}\big(|X_n|;\ \nu_2 \ge n\big) = 0. \tag{15.2.2}$$

Then, on the set $\{\nu_2 \ge \nu_1\}$,

$$\mathbf{E}(X_{\nu_2}|\mathfrak{F}_{\nu_1}) = X_{\nu_1} \quad (\ge X_{\nu_1}). \tag{15.2.3}$$

This theorem extends the martingale (submartingale) property to random time.

Corollary 15.2.1 *If $\nu_2 = \nu \ge 0$ is an arbitrary stopping time, then putting $\nu_1 = n$ (also a stopping time) we have that, on the set $\nu \ge n$,*

$$\mathbf{E}(X_\nu|\mathfrak{F}_n) = X_n, \qquad \mathbf{E}X_\nu = \mathbf{E}X_0,$$

or, which is the same, for any $A \in \mathfrak{F}_n \cap \{\nu \ge n\}$,

$$\mathbf{E}(X_\nu;\ A) = \mathbf{E}(X_n;\ A).$$

For submartingales substitute "=" by "≥".

Proof of Theorem 15.2.1 To prove (15.2.3) it suffices to show that, for any $A \in \mathfrak{F}_{\nu_1}$,

$$\mathbf{E}\big(X_{\nu_2};\ A \cap \{\nu_2 \ge \nu_1\}\big) = \mathbf{E}\big(X_{\nu_1};\ A \cap \{\nu_2 \ge \nu_1\}\big). \tag{15.2.4}$$

Since the random variables ν_i are discrete, we just have to establish (15.2.4) for sets $A_n = A \cap \{\nu_1 = n\} \in \mathfrak{F}_n$, $n = 0, 1, \dots$, i.e. to establish the equality

$$\mathbf{E}\big(X_{\nu_2};\ A_n \cap \{\nu_2 \ge n\}\big) = \mathbf{E}\big(X_n;\ A_n \cap \{\nu_2 \ge n\}\big). \tag{15.2.5}$$

Thus the proof is reduced to the case $\nu_1 = n$. We have

$$\begin{aligned}\mathbf{E}\big(X_n;\ A_n \cap \{\nu_2 \ge n\}\big) &= \mathbf{E}\big(X_n;\ A_n \cap \{\nu_2 = n\}\big) + \mathbf{E}\big(X_n;\ A_n \cap \{\nu_2 \ge n+1\}\big)\\ &= \mathbf{E}\big(X_{\nu_2};\ A_n \cap \{\nu_2 = n\}\big) + \mathbf{E}\big(X_{n+1};\ A_n \cap \{\nu_2 \ge n+1\}\big).\end{aligned}$$

Here we used the fact that $\{\nu_2 \ge n_1\} \in \mathfrak{F}_n$ and the martingale property (15.1.3).

Applying this equality $m - n$ times we obtain that

$$\begin{aligned}&\mathbf{E}\big(X_{\nu_2};\ A_n \cap \{n \le \nu_2 < m\}\big)\\ &\quad = \mathbf{E}\big(X_n;\ A_n \cap \{\nu_2 \ge n\}\big) - \mathbf{E}\big(X_m;\ A_n \cap \{\nu_2 \ge m\}\big).\end{aligned} \tag{15.2.6}$$

By (15.2.2) the last expression converges to zero for some sequence $m \to \infty$. Since

$$A_{n,m} := A_n \cap \{n \le \nu_2 < m\} \uparrow B_n = A_n \cap \{n \le \nu_2\},$$

by the property of integrals and by virtue of (15.2.6),

$$\mathbf{E}\big(X_{\nu_2};\ A_n \cap \{n \le \nu_2\}\big) = \lim_{m\to\infty} \mathbf{E}(X_{\nu_2};\ A_{n,m}) = \mathbf{E}\big(X_n;\ A_n \cap \{\nu_2 \ge n\}\big).$$

Thus we proved equality (15.2.5) and hence Theorem 15.2.1 for martingales. The proof for submartingales can be obtained by simply changing the equality signs in certain places to inequalities. The theorem is proved. □

The conditions of Theorem 15.2.1 are far from always being met, even in rather simple cases. Consider, for instance, a fair game (see Examples 4.2.3 and 4.4.5) versus an infinitely rich adversary, in which $z + S_n$ is the fortune of the first gambler after n plays (given he has not been ruined yet). Here $z > 0$, $S_n = \sum_{k=1}^n \xi_k$, $\mathbf{P}(\xi_k = \pm 1) = 1/2$, $\eta(z) = \min\{k : S_k = -z\}$ is obviously a Markov (stopping) time, and the sequence $\{S_n;\ n \ge 0\}$, $S_0 = 0$, is a martingale, but $S_{\eta(z)} = -z$. Hence $\mathbf{E}S_{\eta(z)} = -z \ne \mathbf{E}S_n = 0$, and equality (15.2.5) does not hold for $\nu_1 = 0$, $\nu_2 = \eta(z)$, $z > 0$, $n > 0$. In this example, this means that condition (15.2.2) is not satisfied (this is related to the fact that $\mathbf{E}\eta(z) = \infty$).

Conditions (15.2.1) and (15.2.2) of Theorem 15.2.1 can, generally speaking, be rather hard to verify. Therefore the following statements are useful in applications.

Put for brevity

$$\xi_n := X_n - X_{n-1}, \qquad \xi_0 := X_0, \qquad Y_n := \sum_{k=0}^{n} |\xi_k|, \quad n = 0, 1, \ldots$$

Lemma 15.2.1 *The condition*

$$\mathbf{E}Y_\nu < \infty \tag{15.2.7}$$

is sufficient for (15.2.1) *and* (15.2.2) (*with* $\nu_i = \nu$).

The Proof is almost evident since $|X_\nu| \le Y_\nu$ and

$$\mathbf{E}(|X_n|;\ \nu > n) \le \mathbf{E}(Y_\nu;\ \nu > n).$$

Because $\mathbf{P}(\nu > n) \to 0$ and $\mathbf{E}Y_\nu < \infty$, it remains to use the property of integrals by which $\mathbf{E}(\eta;\ A_n) \to 0$ if $\mathbf{E}|\eta| < \infty$ and $\mathbf{P}(A_n) \to 0$. □

We introduce the following notation:

$$a_n := \mathbf{E}\big(|\xi_n|\,\big|\mathfrak{F}_{n-1}\big), \quad \sigma_n^2 := \mathbf{E}\big(\xi_n^2\big|\mathfrak{F}_{n-1}\big), \quad n = 0, 1, 2, \ldots,$$

where $\mathfrak{F}_{-1}$ can be taken to be the trivial σ-algebra.

Theorem 15.2.2 *Let* $\{X_n;\ n \ge 0\}$ *be a martingale (submartingale) and* ν *be a stopping time (with respect to* $\{\mathfrak{F}_n = \sigma(X_0, \ldots, X_n)\}$).

(1) *If*

$$\mathbf{E}\nu < \infty \tag{15.2.8}$$

and, for all $n \ge 0$, *on the set* $\{\nu \ge n\} \in \mathfrak{F}_{n-1}$ *one has*

$$a_n \le c = \text{const}, \tag{15.2.9}$$

then

$$\mathbf{E}|X_\nu| < \infty, \quad \mathbf{E}X_\nu = \mathbf{E}X_0 \quad (\ge \mathbf{E}X_0). \tag{15.2.10}$$

(2) *If, in addition,* $\mathbf{E}\sigma_n^2 = \mathbf{E}\xi_n^2 < \infty$ *then*

$$\mathbf{E}X_\nu^2 = \mathbf{E}\sum_{k=1}^{\nu} \sigma_k^2. \tag{15.2.11}$$

Proof By virtue of Theorem 15.2.1, Corollary 15.2.1 and Lemma 15.2.1, to prove (15.2.10) it suffices to verify that conditions (15.2.8) and (15.2.9) imply (15.2.7). Quite similarly to the proof of Theorem 4.4.1, we have

$$\mathbf{E}|Y_\nu| = \sum_{n=0}^{\infty}\left(\sum_{k=0}^{n} \mathbf{E}\big(|\xi_k|;\ \nu = n\big)\right) = \sum_{k=0}^{\infty}\sum_{n=k}^{\infty} \mathbf{E}\big(|\xi_k|;\ \nu = n\big) = \sum_{k=0}^{\infty} \mathbf{E}\big(|\xi_k|;\ \nu \ge k\big).$$

Here $\{\nu \ge k\} = \Omega \setminus \{\nu \le k-1\} \in \mathfrak{F}_{k-1}$. Therefore, by condition (15.2.9),

$$\mathbf{E}\big(|\xi_k|;\ \nu \ge k\big) = \mathbf{E}\big(\mathbf{E}\big(|\xi_k|\,\big|\mathfrak{F}_{k-1}\big);\ \nu \ge k\big) \le c\,\mathbf{P}(\nu \ge k).$$

This means that

$$\mathbf{E}Y_\nu \le c\sum_{k=0}^{\infty} \mathbf{P}(\nu \ge k) = c\,\mathbf{E}\nu < \infty.$$

Now we will prove (15.2.11). Set $Z_n := X_n^2 - \sum_0^n \sigma_k^2$. One can easily see that Z_n is a martingale, since

$$\mathbf{E}\big(X_{n+1}^2 - X_n^2 - \sigma_{n+1}^2\big|\mathfrak{F}_n\big) = \mathbf{E}\big(2X_n\xi_{n+1} + \xi_{n+1}^2 - \sigma_{n+1}^2\big|\mathfrak{F}_n\big) = 0.$$

It is also clear that $\mathbf{E}|Z_n| < \infty$ and $\nu(n) = \min(\nu, n)$ is a stopping time. By virtue of Lemma 15.2.1, conditions (15.2.1) and (15.2.2) always hold for the pair $\{Z_k\}$, $\nu(n)$. Therefore, by the first part of the theorem,

$$\mathbf{E}Z_{\nu(n)} = 0, \qquad \mathbf{E}X_{\nu(n)}^2 = \mathbf{E}\sum_{k=1}^{\nu(n)} \sigma_k^2. \tag{15.2.12}$$

It remains to verify that

$$\lim_{n\to\infty} \mathbf{E}X_{\nu(n)}^2 = \mathbf{E}X_\nu^2, \qquad \lim_{n\to\infty} \mathbf{E}\sum_{k=1}^{\nu(n)} \sigma_k^2 = \mathbf{E}\sum_{k=1}^{\nu} \sigma_k^2. \tag{15.2.13}$$

The second equality follows from the monotone convergence theorem ($\nu(n) \uparrow \nu$, $\sigma_k^2 \ge 0$). That theorem implies the former equality as well, for $X_{\nu(n)}^2 \xrightarrow{a.s.} X_\nu^2$ and $X_{\nu(n)}^2 \uparrow$. To verify the latter claim, note that $\{X_n^2, \mathfrak{F}_n;\ n \ge 0\}$ is a martingale, and therefore, for any $A \in \mathfrak{F}_n$,

$$\begin{aligned}
\mathbf{E}\big(X_{\nu(n)}^2;\ A\big) &= \mathbf{E}\big(X_\nu^2;\ A \cap \{\nu \le n\}\big) + \mathbf{E}\big(X_n^2;\ A \cap \{\nu > n\}\big) \\
&\le \mathbf{E}\big(X_\nu^2;\ A \cap \{\nu \le n\}\big) + \mathbf{E}\big(\mathbf{E}\big(X_{n+1}^2\big|\mathfrak{F}_n\big);\ A \cap \{\nu > n\}\big) \\
&= \mathbf{E}\big(X_\nu^2;\ A \cap \{\nu < n+1\}\big) + \mathbf{E}\big(X_{n+1}^2;\ A \cap \{\nu \ge n+1\}\big) \\
&= \mathbf{E}\big(X_{\nu(n+1)}^2;\ A\big).
\end{aligned}$$

Thus (15.2.12) and (15.2.13) imply (15.2.11), and the theorem is completely proved. □

The main assertion of Theorem 15.2.2 for martingales (submartingales):

$$\mathbf{E}X_\nu = \mathbf{E}X_0 \quad (\geq \mathbf{E}X_0) \tag{15.2.14}$$

was obtained as a consequence of Theorem 15.2.1. However, we could get it directly from some rather transparent relations which, moreover, enable one to extend it to improper stopping times ν.

A stopping time ν is called *improper* if $0 < \mathbf{P}(\nu < \infty) = 1 - \mathbf{P}(\nu = \infty) < 1$. To give an example of an improper stopping time, consider independent identically distributed random variables ξ_k, $a = \mathbf{E}\xi_k < 0$, $X_n = \sum_{k=1}^n \xi_k$, and put

$$\nu = \eta(x) := \min\{k \geq 1 : X_k > x\}, \quad x \geq 0.$$

Here ν is finite only for such trajectories $\{X_k\}$ that $\sup_k X_k > x$. If the last inequality does not hold, we put $\nu = \infty$. Clearly,

$$\mathbf{P}(\nu = \infty) = \mathbf{P}\Big(\sup_k X_k \leq x\Big) > 0.$$

Thus, for an arbitrary (possibly improper) stopping time, we have

$$\mathbf{E}(X_\nu;\ \nu < \infty) = \sum_{k=0}^{\infty} \mathbf{E}(X_k;\ \nu = k) = \sum_{k=0}^{\infty} \big[\mathbf{E}(X_k;\ \nu \geq k) - \mathbf{E}(X_k;\ \nu \geq k+1)\big]. \tag{15.2.15}$$

Assume now that changing the order of summation is justified here. Then, by virtue of the relation $\{\nu \geq k+1\} \in \mathfrak{F}_k$, we get

$$\begin{aligned}\mathbf{E}(X_\nu;\ \nu < \infty) &= \mathbf{E}X_0 + \sum_{k=0}^{\infty} \mathbf{E}(X_{k+1} - X_k;\ \nu \geq k+1)\\ &= \mathbf{E}X_0 + \sum_{k=0}^{\infty} \mathbf{E}\mathrm{I}(\nu \geq k+1)\mathbf{E}(X_{k+1} - X_k | \mathfrak{F}_k).\end{aligned} \tag{15.2.16}$$

Since for martingales (submartingales) the factors $\mathbf{E}(X_{k+1} - X_k | \mathfrak{F}_k) = 0\ (\geq 0)$, we obtain the following.

Theorem 15.2.3 *If the change of the order of summation in* (15.2.15) *and* (15.2.16) *is legitimate then, for martingales* (*submartingales*),

$$\mathbf{E}(X_\nu;\ \nu < \infty) = \mathbf{E}X_0 \quad (\geq \mathbf{E}X_0). \tag{15.2.17}$$

Assumptions (15.2.8) and (15.2.9) of Theorem 15.2.2 are nothing else but conditions ensuring the absolute convergence of the series in (15.2.15) (see the proof of Theorem 15.2.2) and (15.2.16), because the sum of the absolute values of the terms in (15.2.16) is dominated by

$$\sum_{k=1}^{\infty} a_k \mathbf{P}(\nu \geq k+1) \leq a\mathbf{E}\nu < \infty,$$

where, as before, $a_k = \mathbf{E}(|\xi_k| \mid \mathfrak{F}_{k-1})$ with $\xi_k = X_k - X_{k-1}$. This justifies the change of the order of summation.

There is still another way of proving (15.2.17) based on (15.2.15) specifying a simple condition ensuring the required justification. First note that identity (15.2.17) assumes that the expectation $\mathbf{E}(X_\nu;\ \nu < \infty)$ exists, i.e. both values $\mathbf{E}(X_\nu^\pm;\ \nu < \infty)$ are finite, where $x^\pm = \max(\pm x, 0)$.

Theorem 15.2.4 1. *Let $\{X_n, \mathfrak{F}_n\}$ be a martingale. Then the condition*

$$\lim_{n\to\infty} \mathbf{E}(X_n;\ \nu > n) = 0 \tag{15.2.18}$$

is necessary and sufficient for the relation

$$\lim_{n\to\infty} \mathbf{E}(X_n;\ \nu \le n) = \mathbf{E}X_0. \tag{15.2.19}$$

A necessary and sufficient condition for (15.2.17) *is that* (15.2.18) *holds and at least one of the values $\mathbf{E}(X_\nu^\pm;\ \nu < \infty)$ is finite.*

2. *If $\{X_n, \mathfrak{F}_n\}$ is a supermartingale and*

$$\liminf_{n\to\infty} \mathbf{E}(X_n;\ \nu > n) \ge 0, \tag{15.2.20}$$

then

$$\limsup_{n\to\infty} \mathbf{E}(X_n;\ \nu \le n) \le \mathbf{E}X_0.$$

If, in addition, at least one of the values $\mathbf{E}(X_\nu^\pm;\ \nu < \infty)$ is finite then

$$\mathbf{E}(X_\nu;\ \nu < \infty) \le \mathbf{E}X_0.$$

3. *If, in conditions* (15.2.18) *and* (15.2.20), *we replace the quantity $\mathbf{E}(X_n;\ \nu > n)$ with $\mathbf{E}(X_n;\ \nu \ge n)$, the first two assertions of the theorem will remain true.*

The corresponding symmetric assertions hold for submartingales.

Proof As we have already mentioned, for martingales, $\mathbf{E}(\xi_k;\ \nu \ge k) = 0$. Therefore, by virtue of (15.2.18)

$$\mathbf{E}X_0 = \lim_{n\to\infty} \left[\mathbf{E}X_0 + \sum_{k=1}^{n} \mathbf{E}(\xi_k;\ \nu \ge k) - \mathbf{E}(X_n,\ \nu \ge n+1) \right].$$

Here

$$\begin{aligned}
\sum_{k=1}^{n} \mathbf{E}(\xi_k;\ \nu \ge k) &= \sum_{k=1}^{n} \mathbf{E}(X_k;\ \nu \ge k) - \sum_{k=1}^{n} \mathbf{E}(X_{k-1};\ \nu \ge k) \\
&= \sum_{k=1}^{n} \mathbf{E}(X_k;\ \nu \ge k) - \sum_{k=1}^{n-1} \mathbf{E}(X_k;\ \nu \ge k+1).
\end{aligned}$$

Hence

$$\mathbf{E}X_0 = \lim_{n\to\infty}\sum_{k=0}^{n}\bigl[\mathbf{E}(X_k;\ \nu\ge k) - \mathbf{E}(X_k;\ \nu\ge k+1)\bigr]$$
$$= \lim_{n\to\infty}\sum_{k=0}^{n}\mathbf{E}(X_k;\ \nu=k) = \lim_{n\to\infty}\mathbf{E}(X_\nu;\ \nu\le n).$$

These equalities also imply the necessity of condition (15.2.18).

If at least one of the values $\mathbf{E}(X_\nu^{\pm};\ \nu<\infty)$ is finite, then by the monotone convergence theorem

$$\lim_{n\to\infty}\mathbf{E}(X_n;\ \nu\le n) = \lim_{n\to\infty}\mathbf{E}\bigl(X_n^+;\ \nu\le n\bigr) - \lim_{n\to\infty}\mathbf{E}\bigl(X_n^-;\ \nu\le n\bigr)$$
$$= \mathbf{E}\bigl(X_\nu^+;\ \nu<\infty\bigr) - \mathbf{E}\bigl(X_\nu^-;\ \nu<\infty\bigr) = \mathbf{E}(X_\nu;\ \nu<\infty).$$

The third assertion of the theorem follows from the fact that the stopping time $\nu(n)=\min(\nu,n)$ satisfies the conditions of the first part of the theorem (or those of Theorems 15.2.1 and 15.2.3), and therefore, for the martingale $\{X_n\}$,

$$\mathbf{E}X_0 = \mathbf{E}X_{\nu(n)} = \mathbf{E}(X_\nu;\ \nu<n) + \mathbf{E}(X_\nu;\ \nu\ge n),$$

so that (15.2.19) implies the convergence $\mathbf{E}(X_n;\ \nu\ge n)\to 0$ and vice versa.

The proof for semimartingales is similar. The theorem is proved. □

That assertions (15.2.17) and (15.2.19) are, generally speaking, not equivalent even when (15.2.18) holds (i.e., $\lim_{n\to\infty}\mathbf{E}(X_\nu;\ \nu\le n)=\mathbf{E}(X_\nu;\ \nu<\infty)$ is not always the case), can be illustrated by the following example. Let ξ_k be independent random variables with

$$\mathbf{P}\bigl(\xi_k=3^k\bigr) = \mathbf{P}\bigl(\xi_k=-3^k\bigr) = 1/2,$$

ν be independent of $\{\xi_k\}$, and $\mathbf{P}(\nu=k)=2^{-k}$, $k=1,2,\dots$. Then $X_0=0$, $X_k=X_{k-1}+\xi_k$ for $k\ge 1$ is a martingale,

$$\mathbf{E}X_n=0,\quad \mathbf{P}(\nu<\infty)=1,\quad \mathbf{E}(X_n;\ \nu>n)=\mathbf{E}X_n\mathbf{P}(\nu>n)=0$$

by independence, and condition (15.2.18) is satisfied. By virtue of (15.2.19), this means that $\lim_{n\to\infty}\mathbf{P}(X_\nu;\ \nu\le n)=0$ (one can also verify this directly). On the other hand, the expectation $\mathbf{E}(X_\nu;\ \nu<\infty)=\mathbf{E}X_\nu$ is not defined, since $\mathbf{E}X_\nu^+ = \mathbf{E}X_\nu^- = \infty$. Indeed, clearly

$$X_{k-1}\ge -\frac{3^k-3}{2},\qquad \bigl\{\xi_k=3^k\bigr\}\subset\left\{X_k\ge\frac{3^k+3}{2}\right\},\qquad \mathbf{P}\left(X_k\ge\frac{3^k+3}{2}\right)\ge\frac{1}{2},$$
$$\mathbf{E}X_k^+\ge\frac{3^k+3}{4},\qquad \mathbf{E}X_\nu^+=\sum_{k=1}^{\infty}2^{-k}\mathbf{E}X_k^+\ge\sum_{k=1}^{\infty}2^{-k-2}3^k=\infty.$$

By symmetry, we also have $\mathbf{E}X_\nu^-=\infty$.

Corollary 15.2.2 1. *If* $\{X_n,\mathfrak{F}_n\}$ *is a nonnegative martingale, then condition* (15.2.18) *is necessary and sufficient for* (15.2.17).

2. *If* $\{X_n, \mathfrak{F}_n\}$ *is a nonnegative supermartingale and* ν *is an arbitrary stopping time, then*

$$\mathbf{E}(X_\nu;\ \nu < \infty) \le \mathbf{E}X_0. \tag{15.2.21}$$

Proof The assertion follows in an obvious way from Theorem 15.2.4 since one has $\mathbf{E}(X_\nu^-;\ \nu < \infty) = 0$. □

Theorem 15.2.2 implies the already known Wald's identity (see Theorem 4.4.3) supplemented with another useful statement.

Theorem 15.2.5 (Wald's identity) *Let* $\zeta_1, \zeta_2, \ldots$ *be independent identically distributed random variables,* $S_n = \zeta_1 + \cdots + \zeta_n$, $S_0 = 0$, *and assume* $\mathbf{E}\zeta_1 = a$. *Let, further,* ν *be a stopping time with* $\mathbf{E}\nu < \infty$. *Then*

$$\mathbf{E}S_\nu = a\mathbf{E}\nu. \tag{15.2.22}$$

If, moreover, $\sigma^2 = \operatorname{Var}\zeta_k < \infty$, *then*

$$\mathbf{E}[S_\nu - \nu a]^2 = \sigma^2 \mathbf{E}\nu. \tag{15.2.23}$$

Proof It is clear that $X_n = S_n - na$ forms a martingale and conditions (15.2.8) and (15.2.9) are met. Therefore $\mathbf{E}X_\nu = \mathbf{E}X_0 = 0$, which is equivalent to (15.2.22), and $\mathbf{E}X_\nu^2 = \mathbf{E}\nu\sigma^2$, which is equivalent to (15.2.23). □

Example 15.2.1 Consider a generalised renewal process (see Sect. 10.6) $S(t) = S_{\eta(t)}$, where $S_n = \sum_{j=1}^n \xi_j$ (in this example we follow the notation of Chap. 10 and change the meaning of the notation S_n from the above), $\eta(t) = \min\{k : T_k > t\}$, $T_n = \sum_{j=1}^n \tau_j$ and (τ_j, ξ_j) are independent vectors distributed as (τ, ξ), $\tau > 0$. Set $a_\xi = \mathbf{E}\xi$, $a = \mathbf{E}\tau$, $\sigma_\xi^2 = \operatorname{Var}\xi$ and $\sigma^2 = \operatorname{Var}\tau$. As we know from Wald's identity in Sect. 4.4,

$$\mathbf{E}\eta(t) = \frac{t + \mathbf{E}\chi(t)}{a}, \qquad \mathbf{E}S(t) = a_\xi \mathbf{E}\eta(t),$$

where $\mathbf{E}\chi(t) = o(t)$ as $t \to \infty$ (see Theorem 10.1.1) and, in the non-lattice case, $\mathbf{E}\chi(t) \to \frac{\sigma^2 + a^2}{2a^2}$ if $\sigma^2 < \infty$ (see Theorem 10.4.3).

We now find $\operatorname{Var}\eta(t)$ and $\operatorname{Var}S(t)$. Omitting for brevity's sake the argument t, we can write

$$\begin{aligned} a^2 \operatorname{Var}\eta(t) &= a^2 \operatorname{Var}\eta = \mathbf{E}(a\eta - a\mathbf{E}\eta)^2 = \mathbf{E}(a\eta - T_\eta + T_\eta - a\mathbf{E}\eta)^2 \\ &= \mathbf{E}(T_\eta - a\eta)^2 + \mathbf{E}(T_\eta - a\mathbf{E}\eta)^2 - 2\mathbf{E}(T_\eta - a\eta)(T_\eta - a\mathbf{E}\eta). \end{aligned}$$

The first summand on the right-hand side is equal to

$$\sigma^2 \mathbf{E}\eta = \frac{\sigma^2 t}{a} + O(1)$$

by Theorem 15.2.3. The second summand equals, by (10.4.8) ($\chi(t) = T_{\eta(t)} - t$),

$$\mathbf{E}\big(t + \chi(t) - a\mathbf{E}\eta\big)^2 = \mathbf{E}\big(\chi(t) - \mathbf{E}\chi(t)\big)^2 \le \mathbf{E}\chi^2(t) = o(t).$$

The last summand, by the Cauchy–Bunjakovsky inequality, is also $o(t)$. Finally, we get

$$\operatorname{Var}\eta(t) = \frac{\sigma^2 t}{a^3} + o(t).$$

Consider now (with $r = a_\xi/a$; $\zeta_j = \xi_j - r\tau_j$, $\mathbf{E}\zeta_j = 0$)

$$\begin{aligned}\operatorname{Var} S(t) &= \mathbf{E}(S_\eta - a_\xi \mathbf{E}\eta)^2 = \mathbf{E}\big[S_\eta - rT_\eta + r(T_\eta - a\mathbf{E}\eta)\big]^2 \\ &= \mathbf{E}\left(\sum_{j=1}^{\eta}\zeta_j\right)^2 + r^2\mathbf{E}(T_\eta - a\mathbf{E}\eta)^2 + 2r\mathbf{E}\left(\sum_{j=1}^{\eta}\zeta_j\right)(T_\eta - a\mathbf{E}\eta).\end{aligned}$$

The first term on the right-hand side is equal to

$$\mathbf{E}\eta \operatorname{Var}\zeta = \frac{t \operatorname{Var}\zeta}{a} + O(1)$$

by Theorem 15.2.3. The second term has already been estimated above. Therefore, as before, the sum of the last two terms is $o(t)$. Thus

$$\operatorname{Var} S(t) = \frac{t}{a}\mathbf{E}(\xi - r\tau)^2 + o(t).$$

This corresponds to the scaling used in Theorem 10.6.2.

Example 15.2.2 Examples 4.4.4 and 4.5.5 referring to the fair game situation with $\mathbf{P}(\zeta_k = \pm 1) = 1/2$ and $\nu = \min\{k : S_k = z_2 \text{ or } S_k = -z_1\}$ (z_1 and z_2 being the capitals of the gamblers) can also illustrate the use of Theorem 15.2.5.

Now consider the case $p = \mathbf{P}(\zeta_k = 1) \neq 1/2$. The sequence $X_n = (q/p)^{S_n}$, $n \geq 0$, $q = 1 - p$ is a martingale, since

$$\mathbf{E}(q/p)^{\zeta_k} = p(q/p) + q(p/q) = 1.$$

By Theorem 15.2.5 (the probabilities P_1 and P_2 were defined in Example 4.4.5),

$$\mathbf{E}X_\nu = \mathbf{E}X_0 = 1, \qquad P_1(q/p)^{z_2} + P_2(q/p)^{z_1} = 1.$$

From this relation and equality $P_1 + P_2 = 1$ we have

$$P_1 = \frac{(q/p)^{z_1} - 1}{(q/p)^{z_1} - (q/p)^{z_2}}, \qquad P_2 = 1 - P_1.$$

Using Wald's identity again, we also obtain that

$$\mathbf{E}\nu = \frac{\mathbf{E}S_\nu}{\mathbf{E}\zeta_1} = \frac{P_1 z_2 - P_2 z_1}{p - q}.$$

Note that these equalities could have been obtained by elementary methods[1] but this would require lengthy calculations.

[1] See, e.g., [12].

In the cases when the nature of S_ν is simple enough, the assertions of the type of Theorems 15.2.1–15.2.2 enable one to obtain (or estimate) the distribution of the random variable ν itself. In such situations, the following assertion is rather helpful.

Suppose that the conditions of Theorem 15.2.5 are met, but, instead of conditions on the moments of ζ_n, the Cramér condition (cf. Chap. 9) is assumed to be satisfied:

$$\psi(\lambda) := \mathbf{E}e^{\lambda\zeta} < \infty$$

for some $\lambda \neq 0$.

In other words, if

$$\lambda_+ := \sup\bigl(\lambda : \psi(\lambda) < \infty\bigr) \geq 0, \qquad \lambda_- := \inf\bigl(\lambda : \psi(\lambda) < \infty\bigr) \leq 0,$$

then $\lambda_+ - \lambda_- > 0$. Everywhere in what follows we will only consider the values

$$\lambda \in B := \bigl\{\psi(\lambda) < \infty\bigr\} \subseteq [\lambda_-, \lambda_+]$$

for which $\psi'(\lambda) < \infty$. For such λ, the positive martingale

$$X_n = \frac{e^{\lambda S_n}}{\psi^n(\lambda)}, \qquad X_0 = 1,$$

is well-defined so that $\mathbf{E}X_n = 1$.

Theorem 15.2.6 *Let ν be an arbitrary stopping time and $\lambda \in B$. Then*

$$\mathbf{E}\biggl(\frac{e^{\lambda S_\nu}}{\psi(\lambda)^\nu};\ \nu < \infty\biggr) \leq 1 \tag{15.2.24}$$

and, for any $s > 1$ and $r > 1$ such that $1/r + 1/s = 1$,

$$\mathbf{E}\bigl(e^{\lambda S_\nu};\ \nu < \infty\bigr) \leq \bigl\{\mathbf{E}\bigl[\psi^{r\nu/s}(\lambda s);\ \nu < \infty\bigr]\bigr\}^{1/r}. \tag{15.2.25}$$

A necessary and sufficient condition for

$$\mathbf{E}\biggl(\frac{e^{\lambda S_\nu}}{\psi(\lambda)^\nu};\ \nu < \infty\biggr) = 1 \tag{15.2.26}$$

is that

$$\lim_{n\to\infty} \mathbf{E}\biggl(\frac{e^{\lambda S_n}}{\psi(\lambda)^n};\ \nu > n\biggr) = 0. \tag{15.2.27}$$

Remark 15.2.1 Relation (15.2.26) is known as the *fundamental Wald identity*. In the literature it is usually considered for a.s. finite ν (when $\mathbf{P}(\nu < \infty) = 1$) being in that case an extension of the obvious equality $\mathbf{E}e^{\lambda S_n} = \psi^n(\lambda)$ to the case of random ν. Originally, identity (15.2.26) was established by A. Wald in the special case where ν is the exit time of the sequence $\{S_n\}$ from a finite interval (see Corollary 15.2.3), and was accompanied by rather restrictive conditions. Later, these conditions were removed (see e.g. [13]). Below we will obtain a more general assertion for the problem on the first exit of the trajectory $\{S_n\}$ from a strip with curvilinear boundaries.

Remark 15.2.2 The fundamental Wald identity shows that, although the nature of a stopping time could be quite general, there exists a stiff functional constraint (15.2.26) on the joint distribution of ν and S_ν (the distribution of ζ_k is assumed to be known). In the cases where one of these variables can somehow be "computed" or "eliminated" (see Examples 15.2.2–15.2.4) Wald's identity turns into an explicit formula for the Laplace transform of the distribution of the other variable. If ν and S_ν prove to be independent (which rarely happens), then (15.2.26) gives the relationship

$$\mathbf{E}e^{\lambda S_\nu} = \left[\mathbf{E}\psi(\lambda)^{-\nu}\right]^{-1}$$

between the Laplace transforms of the distributions of ν and S_ν.

Proof of Theorem 15.2.6 As we have already noted, for

$$X_n = e^{\lambda S_n}\psi^{-n}(\lambda), \qquad \mathfrak{F}_n = \sigma(\zeta_1, \ldots, \zeta_n),$$

$\{X_n, \mathfrak{F}_n; n \ge 0\}$ is a positive martingale with $X_0 = 1$ and $\mathbf{E}X_n = 1$. Corollary 15.2.2 immediately implies (15.2.24).

Inequality (15.2.25) is a consequence of Hölder's inequality and (15.2.24):

$$\begin{aligned}\mathbf{E}\big(e^{(\lambda/s)S_\nu}; \nu < \infty\big) &= \mathbf{E}\left[\left(\frac{e^{\lambda S_\nu}}{\psi^\nu(\lambda)}\right)^{1/s}\psi^{\nu/s}(\lambda); \nu < \infty\right] \\ &\le \left[\mathbf{E}\big(\psi^{\nu r/s}(\lambda); \nu < \infty\big)\right]^{1/r}.\end{aligned}$$

The last assertion of the theorem (concerning the identity (15.2.26)) follows from Theorem 15.2.4. □

We now consider several important special cases. Note that $\psi(\lambda)$ is a convex function ($\psi''(\lambda) > 0$), $\psi(0) = 1$, and therefore there exists a unique point λ_0 at which $\psi(\lambda)$ attains its minimum value $\psi(\lambda_0) \le 1$ (see also Sect. 9.1).

Corollary 15.2.3 *Assume that we are given a sequence $g(n)$ such that*

$$g^+(n) := \max\big(0, g(n)\big) = o(n) \quad \textit{as } n \to \infty.$$

If $S_n \le g(n)$ holds on the set $\{\nu > n\}$, then (15.2.26) *holds for $\lambda \in (\lambda_0, \lambda_+] \cap B$, $B = \{\lambda : \psi(\lambda) < \infty\}$.*

The random variable $\nu = \nu_g = \inf\{k \ge 1 : S_k > g(k)\}$ for $g(k) = o(k)$ obviously satisfies the conditions of Corollary 15.2.3. For stopping times ν_g one could also consider the case $g(n)/n \to c \ge 0$ as $n \to \infty$, which can be reduced to the case $g(n) = o(n)$ by introducing the random variables

$$\zeta_k^* := \zeta_k - c, \qquad S_k^* := \sum_{j=1}^k \zeta_j^*,$$

for which $\nu_g = \inf\{k \ge 1 : S_k^* > g(k) - ck\}$.

Proof of Corollary 15.2.3 For $\lambda > \lambda_0$, $\lambda \in B$, we have

$$\begin{aligned}\mathbf{E}\left(\frac{e^{\lambda S_n}}{\psi^n(\lambda)}; \nu > n\right) &\le \psi^{-n}(\lambda)\mathbf{E}\big(e^{\lambda S_n}; S_n \le g(n)\big)\\ &= \psi^{-n}(\lambda)\mathbf{E}\big(e^{(\lambda-\lambda_0)S_n}\cdot e^{\lambda_0 S_n}; S_n \le g(n)\big)\\ &\le \psi^{-n}(\lambda)e^{(\lambda-\lambda_0)g(n)}\mathbf{E}\big(e^{\lambda_0 S_n}; S_n \le g(n)\big)\\ &\le \psi^{-n}(\lambda)e^{(\lambda-\lambda_0)g^+(n)}\mathbf{E}e^{\lambda_0 S_n} = \left(\frac{\psi(\lambda_0)}{\psi(\lambda)}\right)^n e^{(\lambda-\lambda_0)g^+(n)} \to 0\end{aligned}$$

as $n \to \infty$, because $(\lambda - \lambda_0)g^+(n) = o(n)$. It remains to use Theorem 15.2.6. The corollary is proved. □

We now return to Theorem 15.2.6 for arbitrary stopping times. It turns out that, based on the Cramér transform introduced in Sect. 9.1, one can complement its assertions without using any martingale techniques.

Together with the original distribution $\mathbf{P}$ of the sequence $\{\zeta_k\}_{k=1}^{\infty}$ we introduce the family of distributions $\mathbf{P}_\lambda$ of this sequence in $\langle\mathbb{R}^\infty, \mathfrak{B}^\infty\rangle$ (see Sect. 5.5) generated by the finite-dimensional distributions

$$\mathbf{P}_\lambda(\zeta_k \in dx_k) = \frac{e^{\lambda x_k}}{\psi(\lambda)}\mathbf{P}(\zeta_k \in dx_k),$$

$$\mathbf{P}_\lambda(\zeta_k \in dx_1, \dots, \zeta_n \in dx_n) = \prod_{k=1}^{n}\mathbf{P}_\lambda(\zeta_k \in dx_k).$$

This is the *Cramér transform* of the distribution $\mathbf{P}$.

Theorem 15.2.7 *Let ν be an arbitrary stopping time. Then, for any $\lambda \in B$,*

$$\mathbf{E}\left(\frac{e^{\lambda S_\nu}}{\psi^\nu(\lambda)}; \nu < \infty\right) = \mathbf{P}_\lambda(\nu < \infty). \tag{15.2.28}$$

Proof Since $\{\nu = n\} \in \sigma(\zeta_1, \dots, \zeta_n)$, there exists a Borel set $D_n \subset \mathbb{R}^n$, such that

$$\{\nu = n\} = \big\{(\zeta_1, \dots, \zeta_n) \in D_n\big\}.$$

Further,

$$\mathbf{E}\left(\frac{e^{\lambda S_\nu}}{\psi^\nu(\lambda)}; \nu < \infty\right) = \sum_{n=0}^{\infty}\mathbf{E}\left(\frac{e^{\lambda S_n}}{\psi^n(\lambda)}; \nu = n\right),$$

where

$$\begin{aligned}\mathbf{E}\left(\frac{e^{\lambda S_n}}{\psi^n(\lambda)}; \nu = n\right) &= \int_{(x_1,\dots,x_n)\in D_n}\frac{e^{\lambda(x_1+\dots+x_n)}}{\psi^n(\lambda)}\mathbf{P}(\zeta_1 \in dx_1, \dots, \zeta_n \in dx)\\ &= \int_{(x_1,\dots,x_n)\in D_n}\mathbf{P}_\lambda(\zeta_1 \in dx_1, \dots, \zeta_n \in dx_n) = \mathbf{P}_\lambda(\nu = n).\end{aligned}$$

This proves the theorem. □

For a given function $g(n)$, consider now the stopping time

$$\nu = \nu_g = \inf\{k : S_k \ge g(k)\}$$

(cf. Corollary 15.2.3). The assertion of Theorem 15.2.7 can be obtained in that case in the following way. Denote by $\mathbf{E}_\lambda$ the expectation with respect to the distribution $\mathbf{P}_\lambda$.

Corollary 15.2.4 1. *If $g^+(n) = \max(0, g(n)) = o(n)$ as $n \to \infty$ and $\lambda \in (\lambda_0, \lambda_+] \cap B$, then one has $\mathbf{P}_\lambda(\nu_g < \infty) = 1$ in relation* (15.2.28).

2. *If $g(n) \ge 0$ and $\lambda < \lambda_0$, then $\mathbf{P}_\lambda(\nu_g < \infty) < 1$.*

3. *For $\lambda = \lambda_0$, the distribution $\mathbf{P}_{\lambda_0}$ of the variable ν can either be proper* (*when one has $\mathbf{P}_{\lambda_0}(\nu_g < \infty) = 1$*) *or improper* ($\mathbf{P}_{\lambda_0}(\nu_g < \infty) < 1$). *If $\lambda_0 \in (\lambda_-, \lambda_+)$, $g(n) < (1-\varepsilon)\sigma(2\log\log n)^{1/2}$ for all $n \ge n_0$, starting from some n_0, and $\sigma^2 = \mathbf{E}_{\lambda_0}\zeta_1^2$, then $\mathbf{P}_\lambda(\nu_g < \infty) = 1$.*

But if $\lambda \in (\lambda_-, \lambda_+)$, $g(n) \ge 0$, and $g(n) \ge (1+\varepsilon)\sigma(2\log\log n)^{1/2}$ for $n \ge n_0$, then $\mathbf{P}_\lambda(\nu_g < \infty) < 1$ (*we exclude the trivial case $\zeta_k \equiv 0$*).

Proof Since $\mathbf{E}_\lambda\zeta_k = \frac{\psi'(\lambda)}{\psi(\lambda)}$, the expectation $\mathbf{E}_\lambda\zeta_k$ is of the same sign as the difference $\lambda - \lambda_0$, and $\mathbf{E}_{\lambda_0}\zeta_k = 0$ ($\psi'(\lambda_0) = 0$ if $\lambda_0 \in (\lambda_-, \lambda_+)$). Hence the first assertion follows from the relations

$$\mathbf{P}_\lambda(\nu = \infty) = \mathbf{P}_\lambda\big(X_n < g(n) \text{ for all } n\big) < \mathbf{P}\big(X_n < g^+(n)\big) \to 0$$

as $n \to \infty$ by the law of large numbers for the sums $X_n = \sum_{k=1}^n \zeta_k$, since $\mathbf{E}_\lambda\zeta_k > 0$.

The second assertion is a consequence of the strong law of large numbers since $\mathbf{E}_\lambda\zeta_k < 0$ and hence $\mathbf{P}_\lambda(\nu = \infty) = \mathbf{P}(\sup_n X_n \le 0) > 0$.

The last assertion of the corollary follows from the law of the iterated logarithm which we prove in Sect. 20.2. The corollary is proved. □

The condition $g(n) \ge 0$ of part 2 of the corollary can clearly be weakened to the condition $g(n) = o(n)$, $\mathbf{P}(\nu > n) > 0$ for any $n > 0$. The same is true for part 3.

An assertion similar to Corollary 15.2.4 is also true for the (stopping) time ν_{g_-,g_+} of the first passage of one of the two boundaries $g_\pm(n) = o(n)$:

$$\nu_{g_-,g_+} := \inf\{k \ge 1 : S_k \ge g_+(k) \text{ or } S_k \le g_-(k)\}.$$

Corollary 15.2.5 *For $\lambda \in B \backslash \{\lambda_0\}$, we have $\mathbf{P}_\lambda(\nu_{g_-,g_+} < \infty) = 1$.*

If $\lambda = \lambda_0 \in (\lambda_-, \lambda_+)$, then the $\mathbf{P}_\lambda$-distribution of ν may be either proper or improper.

If, for some $n_0 > 2$,

$$g_\pm(n) \lesseqgtr \pm(1-\varepsilon)\sigma\sqrt{2\ln\ln n}$$

for $n \ge n_0$ then $\mathbf{P}_{\lambda_0}(\nu_{g_-,g_+} < \infty) = 1$.

If $g_\pm(n) \gtrless 0$ and, additionally,

$$g_\pm(n) \gtrless \pm(1+\varepsilon)\sigma\sqrt{2\ln\ln n}$$

for $n \ge n_0$ then $\mathbf{P}_{\lambda_0}(\nu_{g_-,g_+} < \infty) < 1$.

Proof The first assertion follows from Corollary 15.2.4 applied to the sequences $\{\pm X_n\}$. The second is a consequence of the law of the iterated logarithm from Sect. 20.2. □

We now consider several relations following from Corollaries 15.2.3, 15.2.4 and 15.2.5 (from identity (15.2.26)) for the random variables $\nu = \nu_g$ and $\nu = \nu_{g_-,g_+}$.

Let $a < 0$ and $\psi(\lambda_+) \ge 1$. Since $\psi'(0) = a < 0$ and the function $\psi(\lambda)$ is convex, the equation $\psi(\lambda) = 1$ will have a unique root $\mu > 0$ in the domain $\lambda > 0$. Setting $\lambda = \mu$ in (15.2.26) we obtain the following.

Corollary 15.2.6 *If $a < 0$ and $\psi(\lambda_+) \ge 1$ then, for the stopping times $\nu = \nu_g$ and $\nu = \nu_{g_-,g_+}$, we have the equality*

$$\mathbf{E}\big(e^{\mu S_\nu};\ \nu < \infty\big) = 1.$$

Remark 15.2.3 For an $x > 0$, put (as in Chap. 10) $\eta(x) := \inf\{k : S_k > 0\}$. Since $S_{\eta(x)} = x + \chi(x)$, where $\chi(x) := S_{\eta(x)} - x$ is the value of overshoot over the level x, Corollary 15.2.6 implies

$$\mathbf{E}\big(e^{\mu(x+\chi(x))};\ \eta(x) < \infty\big) = 1. \tag{15.2.29}$$

Note that $\mathbf{P}(\eta(x) < \infty) = \mathbf{P}(S > x)$, where $S = \sup_{k\ge 0} S_k$. Therefore, Theorem 12.7.4 and (15.2.29) imply that, as $x \to \infty$,

$$e^{\mu x}\mathbf{P}\big(\eta(x) < \infty\big) = \big[\mathbf{E}\big(e^{\mu\chi(x)}\,\big|\,\eta(x) < \infty\big)\big]^{-1} \to c. \tag{15.2.30}$$

The last convergence relation corresponds to the fact that the limiting conditional distribution (as $x \to \infty$) $\mathbf{G}$ of $\chi(x)$ exists given $\eta(x) < \infty$. If we denote by χ a random variable with the distribution $\mathbf{G}$ then (15.2.30) will mean that $c = [\mathbf{E}\,e^{\mu\chi}]^{-1} < 1$. This provides an interpretation of the constant c that is different from the one in Theorem 12.7.4.

In Corollary 15.2.6 we "eliminated" the "component" $\psi^\nu(\lambda)$ in identity (15.2.26). "Elimination" of the other component $e^{\lambda S_\nu}$ is possible only in some special cases of random walks, such as the so-called skip-free walks (see Sect. 12.8) or walks with exponentially (or geometrically) distributed $\zeta_k^+ = \max(0, \zeta_k)$ or $\zeta_k^- = -\min(0, \zeta_k)$. We will illustrate this with two examples.

Example 15.2.3 We return to the ruin problem discussed in Example 15.2.2. In that case, Corollary 15.2.4 gives, for $g_-(n) := -z_1$ and $g_+(n) = z_2$, that

$$e^{\lambda z_2}\mathbf{E}\big(\psi(\lambda)^{-\nu};\ S_\nu = z_2\big) + e^{-\lambda z_1}\mathbf{E}\big(\psi(\lambda)^{-\nu};\ S_\nu = -z_1\big) = 1.$$

In particular, for $z_1 = z_2 = z$ and $p = 1/2$, we have by symmetry that

$$\mathbf{E}\big(\psi(\lambda)^{-\nu};\ S_\nu = z\big) = \frac{1}{e^{\lambda z} + e^{-\lambda z}}, \qquad \mathbf{E}\big(\psi(\lambda)^{-\nu}\big) = \frac{2}{e^{\lambda z} + e^{-\lambda z}}. \tag{15.2.31}$$

Let $\lambda(s)$ be the unique positive solution of the equation $s\psi(\lambda)=1$, $s\in(0,1)$. Since here $\psi(\lambda)=\frac{1}{2}(e^{\lambda}+e^{-\lambda})$, solving the quadratic equation yields

$$e^{\lambda(s)}=\frac{1+\sqrt{1-s^2}}{s}.$$

Identity (15.2.31) now gives

$$\mathbf{E}s^{\nu}=2\big(e^{\lambda(s)z}+e^{-\lambda(s)z}\big).$$

We obtain an explicit form of the generating function of the random variable ν, which enables us to find the probabilities $\mathbf{P}(\nu=n)$, $n=1,2,\dots$ by expanding elementary functions into series.

Example 15.2.4 Simple explicit formulas can also be obtained from Wald's identity in the problem with one boundary, where $\nu=\nu_g$, $g(n)=z$. In that case, the class of distributions of ζ_k could be wider than in Example 15.2.3. Suppose that one of the two following conditions holds (cf. Sect. 12.8).

1. *The transform walk is arithmetic and skip-free, i.e.* ζ_k *are integers,* $\mathbf{P}(\xi_k=1)>0$ *and* $\mathbf{P}(\zeta_k\ge 2)=0$.
2. *The walk is right exponential, i.e.*

$$\mathbf{P}(\zeta_k>t)=ce^{-\alpha t} \tag{15.2.32}$$

either for all $t>0$ *or for* $t=0,1,2,\dots$ *if the walk is integer-valued (the geometric distribution).*

The random variable ν_g will be proper if and only if $\mathbf{E}\xi_k=\psi'(0)\ge 0$ (see Chaps. 9 and 12). For skip-free random walks, Wald's identity (15.2.26) yields $(g(n)=z>0,\ S_\nu=z)$

$$e^{\lambda z}\mathbf{E}\psi^{-\nu}(\lambda)=1,\quad \lambda>\lambda_0. \tag{15.2.33}$$

For $s\le 1$, the equation $\psi(\lambda)=s^{-1}$ (cf. Example 15.2.3) has in the domain $\lambda>\lambda_0$ a unique solution $\lambda(s)$. Therefore identity (15.2.33) can be written as

$$\mathbf{E}s^{\nu}=e^{-z\lambda(s)}. \tag{15.2.34}$$

This statement implies a series of results from Chaps. 9 and 12. Many properties of the distribution of $\nu:=\nu_z$ can be derived from this identity, in particular, the asymptotics of $\mathbf{P}(\nu_z=n)$ as $z\to\infty$, $n\to\infty$. We already know one of the ways to find this asymptotics. It consists of using Theorem 12.8.4, which implies

$$\mathbf{P}(\nu_z=n)=\frac{x}{n}\mathbf{P}(S_n=z), \tag{15.2.35}$$

and the local Theorem 9.3.4 providing the asymptotics of $\mathbf{P}(S_n=z)$. Using relation (15.2.34) and the inversion formula is an alternative approach to studying the asymptotics of $\mathbf{P}(\nu_z=n)$. If we use the inversion formula, there will arise an integral of the form

$$\int_{|s|=1}s^{-n}e^{-z\mu(s)}\,ds, \tag{15.2.36}$$

where the integrand $s^{-n}e^{-z\mu(s)}$, after the change of variable $\mu(s)=\lambda$ (or $s=\psi(\lambda)^{-1}$), takes the form

$$\exp -\{z\lambda - n\ln\psi(\lambda)\}.$$

The integrand in the inversion formula for the probability $\mathbf{P}(S_n=z)$ has the same form. This probability has already been studied quite well (see Theorem 9.3.4); its exponential part has the form $e^{-n\Lambda(\alpha)}$, where $\alpha=z/n$, $\Lambda(\alpha)=\sup_\lambda(\alpha\lambda-\ln\psi(\lambda))$ is the large deviation rate function (see Sect. 9.1 and the footnote for Definition 9.1.1). A more detailed study of the inversion formula (15.2.36) allows us to obtain (15.2.35).

Similar relations can be obtained for random walks with exponential right distribution tails. Let, for example, (15.2.32) hold for all $t>0$. Then the conditional distribution $\mathbf{P}(S_\nu>t\,|\,\nu=n,\,S_{n-1}=x)$ coincides with the distribution

$$\mathbf{P}(\zeta_n>z-x+t\,|\,\zeta_n>z-x)=e^{-\alpha t}$$

and clearly depends neither on n nor on x. This means that ν and S_ν are independent, $S_\nu=z+\gamma$, $\gamma\in\boldsymbol{\Gamma}_\alpha$,

$$\mathbf{E}\psi(\lambda)^{-\nu}=\frac{1}{\mathbf{E}e^{(z+\gamma)\lambda}}=e^{-\lambda z}\frac{\alpha-\lambda}{\alpha},\quad \lambda_0<\lambda<\alpha;\qquad \mathbf{E}s^\nu=e^{-z\lambda(s)}\frac{\alpha-\lambda(s)}{\alpha},$$

where $\lambda(s)$ is, as before, the only solution to the equation $\psi(\lambda)=s^{-1}$ in the domain $\lambda>\lambda_0$. This implies the same results as (15.2.34).

If $\mathbf{P}(\zeta_k>t)=c_1e^{-\alpha t}$ and $\mathbf{P}(\zeta_k<-t)=c_2e^{-\beta t}$, $t>0$, then, in the problem with two boundaries, we obtain for $\nu=\nu_{g_-,g_+}$, $g_+(n)=z_2$ and $g_-(n)=-z_1$ in exactly the same way from (15.2.26) that

$$\frac{\alpha e^{\lambda z_2}}{\alpha-\lambda}\mathbf{E}\bigl(\psi^{-\nu}(\lambda);\,S_\nu\ge z_2\bigr)+\frac{\beta e^{-\lambda z_1}}{\beta+\lambda}\mathbf{E}\bigl(\psi^{-\nu}(\lambda);\,S_\nu\le -z_1\bigr)=1,\quad \lambda\in(-\beta,\alpha).$$

15.3 Inequalities

15.3.1 Inequalities for Martingales

First of all we note that the property $\mathbf{E}X_n\le 1$ of the sequence $X_n=e^{\lambda S_n}\psi_0(\lambda)^{-n}$ forming a supermartingale for an appropriate function $\psi_0(\lambda)$ remains true when we replace n with a stopping time ν (an analogue of inequality (15.2.24)) in a much more general case than that of Theorem 15.2.6. Namely, ζ_k may be dependent.

Let, as before, $\{\mathfrak{F}_n\}$ be an increasing sequence of σ-algebras, and ζ_n be $\mathfrak{F}_n$-measurable random variables. Suppose that a.s.

$$\mathbf{E}\bigl(e^{\lambda\zeta_n}\,|\,\mathfrak{F}_{n-1}\bigr)\le\psi_0(\lambda). \tag{15.3.1}$$

This condition is always met if a.s.

$$\mathbf{P}(\zeta_n\ge x\,|\,\mathfrak{F}_{n-1})\le G(x),\qquad \psi_0(\lambda)=-\int e^{\lambda x}\,dG(x)<\infty.$$

In that case the sequence $X_n = e^{\lambda S_n}\psi_0^{-n}(\lambda)$ forms a supermartingale:

$$\mathbf{E}(X_n|\mathfrak{F}_{n-1}) \le X_{n-1}, \quad \mathbf{E}X_n \le 1.$$

Theorem 15.3.1 *Let* (15.3.1) *hold and* ν *be a stopping time. Then inequalities* (15.2.24) *and* (15.2.25) *will hold true with* ψ *replaced by* ψ_0.

The Proof of the theorem repeats almost verbatim that of Theorem 15.2.6. □

Now we will obtain inequalities for the distribution of

$$\overline{X}_n = \max_{k\le n} X_k \quad \text{and} \quad X_n^* = \max_{k\le n} |X_k|,$$

X_n being an arbitrary submartingale.

Theorem 15.3.2 (Doob) *Let* $\{X_n, \mathfrak{F}_n;\ n \ge 0\}$ *be a nonnegative submartingale. Then, for all* $x \ge 0$ *and* $n \ge 0$,

$$\mathbf{P}(\overline{X}_n > x) \le \frac{1}{x}\mathbf{E}X_n.$$

Proof Let

$$\nu = \eta(x) := \inf\{k \ge 0 :\ X_k > x\}, \qquad \nu(n) := \min(\nu, n).$$

It is obvious that n and $\nu(n)$ are stopping times, $\nu(n) \le n$, and therefore, by Theorem 15.2.1 (see (15.2.3) for $\nu_2 = n$, $\nu_1 = \nu(n)$),

$$\mathbf{E}X_n \ge \mathbf{E}X_{\nu(n)}.$$

Observing that $\{\overline{X}_n > x\} = \{X_{\nu(n)} > x\}$, we have from Chebyshev's inequality that

$$\mathbf{P}(\overline{X}_n > x) = \mathbf{P}(X_{\nu(n)} > x) \le \frac{1}{x}\mathbf{E}X_{\nu(n)} \le \frac{1}{x}\mathbf{E}X_n.$$

The theorem is proved. □

Theorem 15.3.2 implies the following.

Theorem 15.3.3 (The second Kolmogorov inequality) *Let* $\{X_n, \mathfrak{F}_n;\ n \ge 0\}$ *be a martingale with a finite second moment* $\mathbf{E}X_n^2$. *Then* $\{X_n^2, \mathfrak{F}_n;\ n \ge 0\}$ *is a submartingale and by Theorem* 15.3.2

$$\mathbf{P}\big(X_n^* > x\big) \le \frac{1}{x^2}\mathbf{E}X_n^2.$$

Originally A.N. Kolmogorov established this inequality for sums $X_n = \xi_1 + \cdots + \xi_n$ of independent random variables ξ_n. Theorem 15.3.3 extends Kolmogorov's proof to the case of submartingales and refines Chebyshev's inequality.

The following generalisation of Theorem 15.3.3 is also valid.

Theorem 15.3.4 *If $\{X_n, \mathfrak{F}_n;\ n \ge 0\}$ is a martingale and $\mathbf{E}|X_n|^p < \infty$, $p \ge 1$, then $\{|X_n|^p, \mathfrak{F}_n;\ n \ge 0\}$ forms a nonnegative submartingale and, for all $x > 0$,*

$$\mathbf{P}\big(X_n^* \ge x\big) \le \frac{1}{x^p}\mathbf{E}|X_n|^p.$$

If $\{X_n, \mathfrak{F}_n;\ n \ge 0\}$ is a submartingale, $\mathbf{E}e^{\lambda X_n} < \infty$, $\lambda > 0$, then $\{e^{\lambda X_n}, \mathfrak{F}_n;\ n \ge 0\}$ also forms a nonnegative submartingale,

$$\mathbf{P}(\overline{X}_n \ge x) \le e^{-\lambda x}\mathbf{E}e^{\lambda X_n}.$$

Both Theorem 15.3.4 and Theorem 15.3.3 immediately follow from Lemma 15.1.3 and Theorem 15.3.2.

If $X_n = S_n = \sum_{k=1}^n \zeta_k$, where ζ_k are independent, identically distributed and satisfy the Cramér condition: $\lambda_+ = \sup\{\lambda : \psi(\lambda) < \infty\} > 0$, then, with the help of the fundamental Wald identity, one can obtain sharper inequalities for $\mathbf{P}(\overline{X}_n > x)$ in the case $a = \mathbf{E}\xi_k < 0$.

Recall that, in the case $a = \psi'(0) < 0$, the function $\psi(\lambda) = \mathbf{E}e^{\lambda\zeta_k}$ decreases in a neighbourhood of $\lambda = 0$, and, provided that $\psi(\lambda_+) \ge 1$, the equation $\psi(\lambda) = 1$ has a unique solution μ in the domain $\lambda > 0$.

Let ζ be a random variable having the same distribution as ζ_k. Put

$$\psi_+ := \sup_{t>0} \mathbf{E}\big(e^{\mu(\zeta-t)}\big|\zeta > t\big), \qquad \psi_- := \inf_{t>0} \mathbf{E}\big(e^{\mu(\zeta-t)}\big|\zeta > t\big).$$

If, for instance, $\mathbf{P}(\zeta > t) = ce^{-\alpha t}$ for $t > 0$ (in this case necessarily $\alpha > \mu$ in (15.2.32)), then

$$\mathbf{P}(\zeta - t > v|\zeta > t) = \frac{\mathbf{P}(\zeta > t + v)}{\mathbf{P}(\zeta > t)} = e^{-\alpha v}, \qquad \psi_+ = \psi_- = \frac{\alpha}{\alpha - \mu}.$$

A similar equality holds for integer-valued ξ with a geometric distribution.

For other distributions, one has $\psi_+ > \psi_-$.

Under the above conditions, one has the following assertion which supplements Theorem 12.7.4 for the distribution of the random variable $S = \sup_k S_k$.

Theorem 15.3.5 *If $a = \mathbf{E}\zeta < 0$ then*

$$\psi_+^{-1}e^{-\mu x} \le \mathbf{P}(S > x) \le \psi_-^{-1}e^{-\mu x}, \quad x > 0. \tag{15.3.2}$$

This theorem implies that, in the case of exponential right tails of the distribution of ζ (see (15.2.32)), inequalities (15.3.2) become the exact equality

$$\mathbf{P}(S > x) = \frac{\alpha - \mu}{\alpha}e^{-\mu x}.$$

(The same result was obtained in Example 12.5.1.) This means that inequalities (15.3.2) are unimprovable. Since $\overline{S}_n = \max_{k\le n} S_k \le S$, relation (15.3.2) implies that, for any n,

$$\mathbf{P}(\overline{S}_n > x) \le \psi_-^{-1}e^{-\mu x}.$$

Proof of Theorem 15.3.5 Set $\nu := \infty$ if $S = \sup_{k\ge 0} S_k \le x$, and put $\nu := \eta(x) = \min\{k : S_k > x\}$ otherwise. Further, let $\chi(x) := S_{\eta(x)} - x$ be the excess of the level x. We have

$$\begin{aligned}
\mathbf{P}\big(\chi(x) > v;\ \nu < \infty\big) &= \sum_{k=1}^{\infty} \int_{-\infty}^{x} \mathbf{P}(\overline{S}_{k-1} \le x,\ S_{k-1} \in du,\ \zeta_k > x - u + v) \\
&= \sum_{k=1}^{\infty} \int_{-\infty}^{x} \mathbf{P}(\overline{S}_{k-1} \le x,\ S_{k-1} \in du,\ \zeta_k > x - u) \\
&\quad \times \mathbf{P}(\zeta_k > x - u + v | \zeta_k > x - u), \\
\mathbf{E}\big(e^{\mu\chi(x)};\ \nu < \infty\big) &\le \sum_{k=1}^{\infty} \int_{-\infty}^{x} \mathbf{P}(\overline{S}_{k-1} \le x,\ S_{k-1} \in du,\ \zeta_k > x - u)\, \psi_+ \\
&= \psi_+ \sum_{k=1}^{\infty} \mathbf{P}(\nu = k) = \psi_+ \mathbf{P}(\nu < \infty).
\end{aligned}$$

Similarly,

$$\mathbf{E}\big(e^{\mu\chi(x)};\ \nu < \infty\big) \ge \psi_- \mathbf{P}(\nu < \infty).$$

Next, by Corollary 15.2.6,

$$1 = \mathbf{E}\big(e^{\mu S_\nu};\ \nu < \infty\big) = e^{\mu x}\mathbf{E}\big(e^{\mu\chi(x)};\ \nu < \infty\big) \le e^{\mu x}\psi_+ \mathbf{P}(\nu < \infty).$$

Because $\mathbf{P}(\nu < \infty) = \mathbf{P}(S > x)$, we get from this the right inequality of Theorem 15.3.5. The left inequality is obtained in the same way. The theorem is proved. □

Remark 15.3.1 We proved Theorem 15.3.5 with the help of the fundamental Wald identity. But there is a direct proof based on the following relations:

$$\begin{aligned}
\psi^n(\lambda) = \mathbf{E}e^{\lambda S_n} &\ge \sum_{k=1}^{n} \mathbf{E}\big(e^{(S_k + S_n - S_k)\lambda};\ \nu = k\big) \\
&= \sum_{k=1}^{n} \mathbf{E}\big(e^{(x+\chi(x))\lambda} e^{(S_n - S_k)\lambda};\ \nu = k\big).
\end{aligned} \tag{15.3.3}$$

Here the random variables $e^{\lambda\chi(x)}\mathrm{I}(\nu = k)$ and $S_n - S_k$ are independent and, as before,

$$\mathbf{E}\big(e^{\lambda\chi(x)};\ \nu = k\big) \ge \psi_-\,\mathbf{P}(\nu = k).$$

Therefore, for all λ such that $\psi(\lambda) \le 1$,

$$\psi^n(\lambda) \ge e^{\lambda x}\psi_- \sum_{k=1}^{n} \psi^{n-k}(\lambda)\,\mathbf{P}(\nu = k) \ge \psi_- e^{\lambda x}\psi^n(\lambda)\,\mathbf{P}(\nu \le n).$$

Hence we obtain

$$\mathbf{P}(\overline{S}_n > x) = \mathbf{P}(\nu \le n) \le \psi_-^{-1} e^{-\lambda x}.$$

Since the right-hand side does not depend on n, the same inequality also holds for $\mathbf{P}(S > x)$. The lower bound is obtained in a similar way. One just has to show that, in the original equality (cf. (15.3.3))

$$\psi^n(\lambda) = \sum_{k=1}^{n} \mathbf{E}\big(e^{\lambda S_n};\ \nu = k\big) + \mathbf{E}\big(e^{\lambda S_n};\ \nu > n\big),$$

one has $\mathbf{E}(e^{\lambda S_n};\ \nu > n) = o(1)$ as $n \to \infty$ for $\lambda = \mu$, which we did in Sect. 15.2.

15.3.2 Inequalities for the Number of Crossings of a Strip

We now return to arbitrary submartingales X_n and prove an inequality that will be necessary for the convergence theorems of the next section. It concerns the number of crossings of a strip by the sequence X_n. Let $a < b$ be given numbers. Set $\nu_0 = 0$,

$$\nu_1 := \min\{n > 0 : X_n \le a\}, \qquad \nu_2 := \min\{n > \nu_1 : X_n \ge b\},$$
$$\cdots\cdots\cdots \qquad\qquad \cdots\cdots\cdots$$
$$\nu_{2k-1} := \min\{n > \nu_{2k-2} : X_n \le a\}, \quad \nu_{2k} := \min\{n > \nu_{2k-1} : X_n \ge b\}.$$

We put $\nu_m := \infty$ if the path $\{X_n\}$ for $n \ge \nu_{m-1}$ never crosses the corresponding level. Using this notation, one can define the number of upcrossings of the strip (interval) $[a, b]$ by the trajectory $X_0, \ldots, X_n$ as the random variable

$$\nu(a, b; n) := \begin{cases} \max\{k : \nu_{2k} \le n\} & \text{if } \nu_2 \le n, \\ 0 & \text{if } \nu_2 > n. \end{cases}$$

Set $(a)^+ = \max(0, a)$.

Theorem 15.3.6 (Doob) *Let $\{X_n, \mathfrak{F}_n;\ n \ge 0\}$ be a submartingale. Then, for all n,*

$$\mathbf{E}\nu(a, b; n) \le \frac{\mathbf{E}(X_n - a)^+}{b - a}. \tag{15.3.4}$$

It is clear that inequality (15.3.4) assumes by itself that only the submartingale $\{X_n, \mathfrak{F}_n; 0 \le k \le n\}$ is given.

Proof The random variable $\nu(a, b;\ n)$ coincides with the number of upcrossings of the interval $[0, b - a]$ by the sequence $(X_n - a)^+$. Now $\{(X_n - a)^+, \mathfrak{F}_n;\ n \ge 0\}$ is a nonnegative submartingale (see Example 15.1.4) and therefore, without loss of generality, one can assume that $a = 0$ and $X_n \ge 0$, and aim to prove that

$$\mathbf{E}\nu(0, b;\ n) \le \frac{\mathbf{E}X_n}{b}.$$

Let

$$\eta_j := \begin{cases} 1 & \text{if } \nu_k < j \le \nu_{k+1} \text{ for some odd } k, \\ 0 & \text{if } \nu_k < j \le \nu_{k+1} \text{ for some even } k. \end{cases}$$

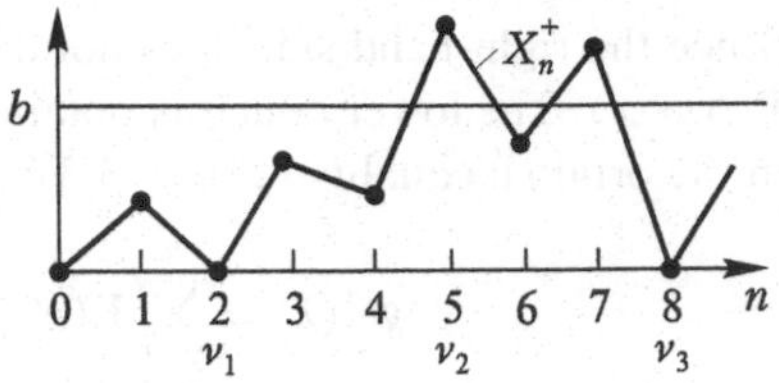

Fig. 15.1 Illustration to the proof of Theorem 15.3.6 showing the locations of the random times ν_1, ν_2, and ν_3 (here $a = 0$)

In Fig. 15.1, $\nu_1 = 2$, $\nu_2 = 5$, $\nu_3 = 8$; $\eta_j = 0$ for $j \le 2$, $\eta_j = 1$ for $3 \le j \le 5$ etc. It is not hard to see (using the Abel transform) that (with $X_0 = 0$, $\eta_0 = 1$)

$$\eta_0 X_0 + \sum_1^n \eta_j (X_j - X_{j-1}) = \sum_0^{n-1} X_j (\eta_j - \eta_{j+1}) + \eta_n X_n \ge b\nu(0, b; n).$$

Moreover (here $\mathcal{N}_1$ denotes the set of odd numbers),

$$\{\eta_j = 1\} = \bigcup_{k \in \mathcal{N}_1} \{\nu_k < j \le \nu_{k+1}\} = \bigcup_{k \in \mathcal{N}_1} \big[\{\nu_k \le j-1\} - \{\nu_{k+1} \le j-1\}\big] \in \mathfrak{F}_{j-1}.$$

Therefore, by virtue of the relation $\mathbf{E}(X_j | \mathfrak{F}_{j-1}) - X_{j-1} \ge 0$, we obtain

$$\begin{aligned}
b\mathbf{E}\nu(0, b; n) &\le \mathbf{E}\sum_1^n \eta_j (X_j - X_{j-1}) = \sum_1^n \mathbf{E}(X_j - X_{j-1};\ \eta_j = 1) \\
&= \sum_1^n \mathbf{E}\big[\mathbf{E}(X_j - X_{j-1} | \mathfrak{F}_{j-1});\ \eta_j = 1\big] = \sum_1^n \mathbf{E}\big[\mathbf{E}(X_j | \mathfrak{F}_{j-1}) - X_{j-1};\ \eta_j = 1\big] \\
&\le \sum_1^n \mathbf{E}\big[\mathbf{E}(X_j | \mathfrak{F}_{j-1}) - X_{j-1}\big] = \sum_1^n \mathbf{E}(X_j - X_{j-1}) = \mathbf{E}X_n.
\end{aligned}$$

The theorem is proved. □

15.4 Convergence Theorems

Theorem 15.4.1 (Doob's martingale convergence theorem) *Let*

$$\{X_n, \mathfrak{F}_n;\ -\infty < n < \infty\}$$

be a submartingale. Then

(1) *The limit* $X_{-\infty} := \lim_{n \to -\infty} X_n$ *exists a.s.,* $\mathbf{E}X_{-\infty}^+ < \infty$*, and the process* $\{X_n, \mathfrak{F}_n;\ -\infty \le n < \infty\}$ *is a submartingale.*
(2) *If* $\sup_n \mathbf{E}X_n^+ < \infty$ *then* $X_\infty := \lim_{n \to \infty} X_n$ *exists a.s. and* $\mathbf{E}X_\infty^+ < \infty$*. If, moreover,* $\sup_n \mathbf{E}|X_n| < \infty$ *then* $\mathbf{E}|X_\infty| < \infty$*.*
(3) *The random sequence* $\{X_n, \mathfrak{F}_n;\ -\infty \le n \le \infty\}$ *forms a submartingale if and only if the sequence* $\{X_n^+\}$ *is uniformly integrable.*

Proof (1) Since

$$\{\limsup X_n > \liminf X_n\} = \bigcup_{\substack{\text{rational}\\ a,b}} \{\limsup X_n > b > a > \liminf X_n\}$$

(here the limits are taken as $n \to -\infty$), the assumption on divergence with positive probability

$$\mathbf{P}(\limsup X_n > \liminf X_n) > 0$$

means that there exist rational numbers $a < b$ such that

$$\mathbf{P}(\limsup X_n > b > a > \liminf X_n) > 0. \tag{15.4.1}$$

Let $\nu(a, b;\, m)$ be the number of upcrossings of the interval $[a, b]$ by the sequence $Y_1 = X_{-m}, \ldots, Y_m = X_{-1}$ and $\nu(a, b) = \lim_{m\to\infty} \nu(a, b;\, m)$. Then (15.4.1) means that

$$\mathbf{P}\big(\nu(a, b) = \infty\big) > 0. \tag{15.4.2}$$

By Theorem 15.3.6 (applied to the sequence $Y_1, \ldots, Y_m$),

$$\mathbf{E}\nu(a, b;\, m) \le \frac{\mathbf{E}(X_{-1} - a)^+}{b - a} \le \frac{\mathbf{E}X_{-1}^+ + |a|}{b - a}, \tag{15.4.3}$$

$$\mathbf{E}\nu(a, b) \le \frac{\mathbf{E}X_{-1}^+ + |a|}{b - a}. \tag{15.4.4}$$

Inequality (15.4.4) contradicts (15.4.2) and hence proves that

$$\mathbf{P}(\limsup X_n = \liminf X_n) = 1.$$

Moreover, by the Fatou–Lebesgue theorem ($X_{-\infty}^+ := \liminf X_n^+$),

$$\mathbf{E}X_{-\infty}^+ \le \liminf X_n^+ \le \mathbf{E}X_{-1}^+ < \infty. \tag{15.4.5}$$

Here the second inequality follows from the fact that $\{X_n^+, \mathfrak{F}_n\}$ is also a submartingale (see Lemma 15.1.3) and therefore $\mathbf{E}X_n^+ \uparrow$.

By Lemma 15.1.2, to prove that $\{X_n, \mathfrak{F}_n;\, -\infty \le n < \infty\}$ is a submartingale, it suffices to verify that, for any $A \in \mathfrak{F}_{-\infty} \subset \mathfrak{F}$,

$$\mathbf{E}(X_{-\infty};\, A) \le \mathbf{E}(X_n;\, A). \tag{15.4.6}$$

Set $X_n(a) := \max(X_n, a)$. By Lemma 15.1.4, $\{X_n(a), \mathfrak{F}_n;\, n \le 0\}$ is a uniformly integrable submartingale. Therefore, for any $-\infty < k < n$,

$$\begin{aligned} \mathbf{E}\big(X_k(a);\, A\big) &\le \mathbf{E}\big(X_n(a);\, A\big), \\ \mathbf{E}\big(X_{-\infty}(a);\, A\big) &= \lim_{k\to-\infty} \mathbf{E}\big(X_k(a);\, A\big) \le \mathbf{E}\big(X_n(a);\, A\big). \end{aligned} \tag{15.4.7}$$

Letting $a \to -\infty$ we obtain (15.4.6) from the monotone convergence theorem.

(2) The second assertion of the theorem is proved in the same way. One just has to replace the right-hand sides of (15.4.3) and (15.4.4) with $\mathbf{E}X_n^+$ and $\sup_n \mathbf{E}X_n^+$, respectively. Instead of (15.4.5) we get (the limits here are as $n \to \infty$)

$$\mathbf{E}X_\infty^+ \le \liminf \mathbf{E}X_n^+ < \infty,$$

and if $\sup_n \mathbf{E}|X_n| < \infty$ then

$$\mathbf{E}|X_\infty| \le \liminf \mathbf{E}|X_n| < \infty.$$

(3) The last assertion of the theorem is proved in exactly the same way as the first one—the uniform integrability enables us to deduce along with (15.4.7) that, for any $A \in \mathfrak{F}_n$,

$$\mathbf{E}\big(X_\infty(a);\ A\big) = \lim_{k\to\infty} \mathbf{E}\big(X_k(a);\ A\big) \ge \mathbf{E}\big(X_n(a);\ A\big).$$

The converse part of the third assertion of the theorem follows from Lemma 15.1.4. The theorem is proved. □

Now we will obtain some consequences of Theorem 15.4.1.

So far (see Sect. 4.8), while studying convergence of conditional expectations, we dealt with expectations of the form $\mathbf{E}(X_n|\mathfrak{F})$. Now we can obtain from Theorem 15.4.1 a useful theorem on convergence of conditional expectations of another type.

Theorem 15.4.2 (Lévy) *Let a nondecreasing family $\mathfrak{F}_1 \subseteq \mathfrak{F}_2 \subseteq \cdots \subseteq \mathfrak{F}$ of σ-algebras and a random variable ξ, with $\mathbf{E}|\xi| < \infty$, be given on a probability space $\langle \Omega, \mathfrak{F}, \mathbf{P}\rangle$. Let, as before, $\mathfrak{F}_\infty := \sigma(\bigcup_n \mathfrak{F}_n)$ be the σ-algebra generated by events from $\mathfrak{F}_1, \mathfrak{F}_2, \ldots$. Then, as $n \to \infty$,*

$$\mathbf{E}(\xi|\mathfrak{F}_n) \xrightarrow{a.s.} \mathbf{E}(\xi|\mathfrak{F}_\infty). \tag{15.4.8}$$

Proof Set $X_n := \mathbf{E}(\xi|\mathfrak{F}_n)$. We already know (see Example 15.1.3) that the sequence $\{X_n, \mathfrak{F}_n;\ 1 < n \le \infty\}$ is a martingale and therefore, by Theorem 15.4.1, the limit $\lim_{n\to\infty} X_n = X_{(\infty)}$ exists a.s. It remains to prove that $X_{(\infty)} = \mathbf{E}(\xi|\mathfrak{F}_\infty)$ (i.e., that $X_{(\infty)} = X_\infty$). Since $\{X_n, \mathfrak{F}_n;\ 1 \le n \le \infty\}$ is by Lemma 15.1.4 a uniformly integrable martingale,

$$\mathbf{E}(X_{(\infty)};\ A) = \lim_{n\to\infty} \mathbf{E}(X_n;\ A) = \lim_{n\to\infty} \mathbf{E}\big(\mathbf{E}(\xi|\mathfrak{F}_n);\ A\big) = \mathbf{E}(\xi;\ A)$$

for $A \in \mathfrak{F}_k$ and any $k = 1, 2, \ldots$ This means that the left- and right-hand sides of the last relation, being finite measures, coincide on the algebra $\bigcup_{n=1}^\infty \mathfrak{F}_n$. By the theorem on extension of a measure (see Appendix 1), they will coincide for all $A \in \sigma(\bigcup_{n=1}^\infty \mathfrak{F}_n) = \mathfrak{F}_\infty$. Therefore, by the definition of conditional expectation,

$$X_{(\infty)} = \mathbf{E}(\xi|\mathfrak{F}_\infty) = X_\infty.$$

The theorem is proved. □

We could also note that the uniform integrability of $\{X_n, \mathfrak{F}_n;\ 1 \le n \le \infty\}$ implies that $\xrightarrow{a.s.}$ in (47) can be replaced by $\xrightarrow{(1)}$.

Theorem 15.4.1 *implies the strong law of large numbers*. Indeed, turn to our Example 15.1.4. By Theorem 15.4.1, the limit $X_{-\infty} = \lim_{n\to-\infty} X_n = \lim_{n\to\infty} n^{-1}S_n$ exists a.s. and is measurable with respect to the tail (trivial) σ-algebra, and therefore it is constant with probability 1. Since $\mathbf{E}X_{-\infty} = \mathbf{E}\xi_1$, we have $n^{-1}S_n \xrightarrow{a.s.} \mathbf{E}\xi_1$.

One can also obtain some extensions of the theorems on series convergence of Chap. 11 to the case of dependent variables. Let

$$X_n = S_n = \sum_{k=1}^{n} \xi_k$$

and X_n form a submartingale ($\mathbf{E}(\xi_{n+1}|\mathfrak{F}_n) \ge 0$). Let, moreover, $\mathbf{E}|X_n| < c$ for all n and for some $c < \infty$. Then the limit $S_\infty = \lim_{n\to\infty} S_n$ exists a.s. (As well as Theorem 15.4.1, this assertion is a generalisation of the monotone convergence theorem. The crucial role is played here by the condition that $\mathbf{E}|X_n|$ is bounded.) In particular, if ξ_k are independent, $\mathbf{E}\xi_k = 0$, and the variances σ_k^2 of ξ_k are such that $\sum_{k=1}^{\infty} \sigma_k^2 < \sigma^2 < \infty$, then

$$\mathbf{E}|X_n| \le \big(\mathbf{E}X_n^2\big)^{1/2} \le \left(\sum_{k=1}^{n} \sigma_k^2\right)^{1/2} \le \sigma < \infty,$$

and therefore $S_n \xrightarrow{a.s.} S_\infty$. Thus we obtain, as a consequence, the Kolmogorov theorem on series convergence.

Example 15.4.1 Consider a *branching process* $\{Z_n\}$ (see Sect. 7.7). We know that Z_n admits a representation

$$Z_n = \zeta_1 + \cdots + \zeta_{Z_{n-1}},$$

where the ζ_k are identically distributed integer-valued random variables independent of each other and of Z_{n-1}, ζ_k being the number of descendants of the k-th particle from the $(n-1)$-th generation. Assuming that $Z_0 = 1$ and setting $\mu := \mathbf{E}\zeta_k$, we obtain

$$\mathbf{E}(Z_n|Z_{n-1}) = \mu Z_{n-1}, \qquad \mathbf{E}Z_n = \mu \mathbf{E}Z_{n-1} = \mu^n.$$

This implies that $X_n = Z_n/\mu^n$ is a martingale, because

$$\mathbf{E}(X_n|X_{n-1}) = \mu^{1-n} Z_{n-1} = X_{n-1}.$$

For branching processes we have the following.

Theorem 15.4.3 *The sequence $X_n = \mu^{-n} Z_n$ converges almost surely to a proper random variable X with $\mathbf{E}X < \infty$. The ch.f. $\varphi(\lambda)$ of the random variable X satisfies the equation*

$$\varphi(\mu\lambda) = p\big(\varphi(\lambda)\big),$$

where $p(v) = \mathbf{E}v^{\zeta_k}$.

Theorem 15.4.3 means that $\mu^{-n} Z_n$ has a proper limiting distribution as $n \to \infty$.

Proof Since $X_n \ge 0$ and $\mathbf{E}X_n = 1$, the first assertion follows immediately from Theorem 15.4.1.

Since $\mathbf{E}z^{Z_n}$ is equal to the n-th iteration of the function $f(z)$, for the ch.f. of Z_n we have ($\varphi_\eta(\lambda) := \mathbf{E}e^{i\lambda\eta}$)

$$\varphi_{Z_n}(\lambda) = p\big(\varphi_{Z_{n-1}}(\lambda)\big),$$
$$\varphi_{X_n}(\lambda) = \varphi_{Z_n}\big(\mu^{-n}\lambda\big) = p\big(\varphi_{Z_{n-1}}(\mu^{-n}\lambda)\big) = p\left(\varphi_{X_{n-1}}\left(\frac{\lambda}{\mu}\right)\right).$$

Because $X_n \Rightarrow X$ and the function p is continuous, from this we obtain the equation for the ch.f. of the limiting distribution X:

$$\varphi(\lambda) = p\left(\varphi\left(\frac{\lambda}{\mu}\right)\right).$$

The theorem is proved. □

In Sect. 7.7 we established that in the case $\mu \le 1$ the process Z_n becomes extinct with probability 1 and therefore $\mathbf{P}(X = 0) = 1$. We verify now that, for $\mu > 1$, the distribution of X is nondegenerate (not concentrated at zero). It suffices to prove that $\{X_n, 0 \le n \le \infty\}$ forms a martingale and consequently

$$\mathbf{E}X = \mathbf{E}X_n \neq 0.$$

By Theorem 15.4.1, it suffices to verify that the sequence X_n is uniformly integrable. To simplify the reasoning, we suppose that $\operatorname{Var}(\zeta_k) = \sigma^2 < \infty$ and show that then $\mathbf{E}X_n^2 < c < \infty$ (this certainly implies the required uniform integrability of X_n, see Sect. 6.1). One can directly verify the identity

$$Z_n^2 - \mu^{2n} = \sum_{k=1}^{n} \big[Z_k^2 - (\mu Z_{k-1})^2\big]\mu^{2n-2k}.$$

Since $\mathbf{E}[Z_k^2 - (\mu Z_{k-1})^2 | Z_{k-1}] = \sigma^2 Z_{k-1}$ (recall that $\operatorname{Var}(\eta) = \mathbf{E}(\eta^2 - (\mathbf{E}\eta)^2)$), we have

$$\operatorname{Var}(Z_n) = \mathbf{E}\big(Z_n^2 - \mu^{2n}\big) = \sum_{k=1}^{n} \mu^{2n-2k}\sigma^2 \mathbf{E}Z_{k-1}$$
$$= \mu^{2n}\sigma^2 \sum_{k=1}^{n} \mu^{-k-1} = \frac{\sigma^2\mu^n(\mu^n - 1)}{\mu(\mu-1)},$$
$$\mathbf{E}X_n^2 = \mu^{-2n}\mathbf{E}Z_n^2 = 1 + \frac{\sigma^2(1-\mu^{-n})}{\mu(\mu-1)} \le 1 + \frac{\sigma^2}{\mu(\mu-1)}.$$

Thus we have proved that X is a nondegenerate random variable,

$$\mathbf{E}X = 1, \quad \operatorname{Var}(X_n) \to \frac{\sigma^2}{\mu(\mu-1)}.$$

From the last relation one can easily obtain that $\operatorname{Var}(X) = \frac{\sigma^2}{\mu(\mu-1)}$. To this end one can, say, prove that X_n is a Cauchy sequence in mean quadratic and hence (see Theorem 6.1.3) $X_n \xrightarrow{(2)} X$.

15.5 Boundedness of the Moments of Stochastic Sequences

When one uses convergence theorems for martingales, conditions ensuring boundedness of the moments of stochastic sequences $\{X_n, \mathfrak{F}_n\}$ are of significant interest (recall that the boundedness of $\mathbf{E}X_n$ is one of the crucial conditions for convergence of submartingales). The boundedness of the moments, in turn, ensures that X_n is stochastically bounded, i.e., that $\sup_n \mathbf{P}(X_n > N) \to 0$ as $N \to \infty$. The last boundedness is also of independent interest in the cases where one is not able to prove, for the sequence $\{X_n\}$, convergence or any other ergodic properties.

For simplicity's sake, we confine ourselves to considering nonnegative sequences $X_n \ge 0$. Of course, if we could prove convergence of the distributions of X_n to a limiting distribution, as was the case for Markov chains or submartingales in Theorem 15.4.1, then we would have a more detailed description of the asymptotic behaviour of X_n as $n \to \infty$. This convergence, however, requires that the sequence X_n satisfies stronger constraints than will be used below.

The basic and rather natural elements of the boundedness conditions to be considered below are: the boundedness of the moments of $\xi_n = X_n - X_{n-1}$ of the respective orders and the presence of a negative "drift" $\mathbf{E}(\xi_n|\mathfrak{F}_{n-1})$ in the domain $X_{n-1} > N$ for sufficiently large N. Such a property has already been utilised for Markov chains; see Corollary 13.7.1 (otherwise the trajectory of X_n may go to ∞).

Let us begin with exponential moments. The simplest conditions ensuring the boundedness of $\sup_n \mathbf{E}e^{\lambda X_n}$ for some $\lambda > 0$ are as follows: *for all $n \ge 1$ and some $\lambda > 0$ and $N < \infty$,*

$$\mathbf{E}\big(e^{\lambda\xi_n}\big|\mathfrak{F}_{n-1}\big)\,\mathrm{I}(X_{n-1} > N) \le \beta(\lambda) < 1, \tag{15.5.1}$$

$$\mathbf{E}\big(e^{\lambda\xi_n}\big|\mathfrak{F}_{n-1}\big)\,\mathrm{I}(X_{n-1} \ge N) \le \psi(\lambda) < \infty. \tag{15.5.2}$$

Theorem 15.5.1 *If conditions* (15.5.1) *and* (15.5.2) *hold then*

$$\mathbf{E}\big(e^{\lambda X_n}\big|\mathfrak{F}_0\big) \le \beta(\lambda)e^{\lambda X_0} + \frac{\psi(\lambda)\,e^{\lambda N}}{1-\beta(\lambda)}. \tag{15.5.3}$$

Proof Denote by A_n the left-hand side of (15.5.3). Then, by virtue of (15.5.1) and (15.5.2), we obtain

$$\begin{aligned}
A_n &= \mathbf{E}\big\{\mathbf{E}\big[e^{\lambda X_n}\big(\mathrm{I}(X_{n-1} > N) + \mathrm{I}(X_{n-1} \le N)\big)\big|\mathfrak{F}_{n-1}\big]\big|\mathfrak{F}_0\big\} \\
&\le \mathbf{E}\big[e^{\lambda X_{n-1}}\big(\beta(\lambda)\,\mathrm{I}(X_{n-1} > N) + \psi(\lambda)\,\mathrm{I}(X_{n-1} \le N)\big)\big|\mathfrak{F}_0\big] \\
&\le \beta(\lambda)A_{n-1} + e^{\lambda N}\psi(\lambda).
\end{aligned}$$

This immediately implies that

$$A_n \le A_0\beta^n(\lambda) + e^{\lambda N}\psi(\lambda)\sum_{k=0}^{n-1}\beta^k(\lambda) \le A_0\beta^n(\lambda) + \frac{e^{\lambda N}\psi(\lambda)}{1-\beta(\lambda)}.$$

The theorem is proved. □

The conditions

$$\mathbf{E}(\xi_n|\mathfrak{F}_{n-1}) \le -\varepsilon < 0 \quad \textit{on the } \omega\textit{-set} \quad \{X_{n-1} > N\}, \tag{15.5.4}$$

$$\mathbf{E}\big(e^{\lambda|\xi_n|}\big|\mathfrak{F}_{n-1}\big) \le \psi_1(\lambda) < \infty \quad \textit{for some } \lambda > 0 \tag{15.5.5}$$

are sufficient for (15.5.1) *and* (15.5.2).

The first condition means that $Y_n := (X_n + \varepsilon n)\,\mathrm{I}(X_{n-1} > N)$ is a supermartingale.

We now prove sufficiency of (15.5.4) and (15.5.5). That (15.5.2) holds is clear. Further, make use of the inequality

$$e^x \le 1 + x + \frac{x^2}{2} e^{|x|},$$

which follows from the Taylor formula for e^x with the remainder in the Cauchy form:

$$e^x = 1 + x + \frac{x^2}{2} e^{\theta x}, \quad \theta \in [0, 1].$$

Then, on the set $\{X_{n-1} > N\}$, one has

$$\mathbf{E}\big(e^{\lambda\xi_n}\big|\mathfrak{F}_{n-1}\big) \le 1 - \lambda\varepsilon + \frac{\lambda^2}{2}\mathbf{E}\big(\xi_n^2 e^{\lambda|\xi_n|}\big|\mathfrak{F}_{n-1}\big).$$

Since $x^2 < e^{\lambda x/2}$ for sufficiently large x, by the Hölder inequality it follows that, together with (15.5.5), we will have

$$\mathbf{E}\big(\xi_n^2 e^{\lambda|\xi_n|/2}\big|\mathfrak{F}_{n-1}\big) \le \psi_2(\lambda) < \infty.$$

This implies that, for sufficiently small λ, one has on the set $\{X_{n-1} > N\}$ the inequality

$$\mathbf{E}\big(e^{\lambda\xi_n}\big|\mathfrak{F}_{n-1}\big) \le 1 - \lambda\varepsilon + \frac{\lambda^2}{2}\psi_2(\lambda) =: \beta(\lambda) \le 1 - \frac{\lambda\varepsilon}{2} < 1.$$

This proves (15.5.1). □

Corollary 15.5.1 *If, in addition to the conditions of Theorem* 15.5.1, *the distribution of X_n converges to a limiting distribution*: $\mathbf{P}(X_n < t) \Rightarrow \mathbf{P}(X < t)$, *then*

$$\mathbf{E}e^{\lambda X} \le \frac{e^{\lambda N}\psi(\lambda)}{1 - \beta(\lambda)}.$$

The corollary follows from the Fatou–Lebesgue theorem (see also Lemma 6.1.1):

$$\mathbf{E}e^{\lambda X} \le \liminf_{n\to\infty} \mathbf{E}e^{\lambda X_n}.$$ □

We now obtain bounds for "conventional" moments. Set

$$M^l(n) := \mathbf{E}X_n^l,$$

$$m(0) := 1, \quad m(1) := \sup_{n\ge 1}\ \sup_{\omega\in\{X_{n-1}>N\}} \mathbf{E}(\xi_n|\mathfrak{F}_{n-1}),$$

$$m(l) := \sup_{n\ge 1}\sup_{\omega} \mathbf{E}\big(|\xi_n|^l\big|\mathfrak{F}_{n-1}\big), \quad l > 1.$$

Theorem 15.5.2 *Assume that* $\mathbf{E}X_0^s < \infty$ *for some* $s > 1$ *and there exist* $N \ge 0$ *and* $\varepsilon > 0$ *such that*

$$m(1) \le -\varepsilon, \tag{15.5.6}$$

$$m(s) < c < \infty. \tag{15.5.7}$$

Then

$$\liminf_{n\to\infty} M^{s-1}(n) < \infty. \tag{15.5.8}$$

If, moreover,

$$M^s(n+1) > M^s(n) - c_1 \tag{15.5.9}$$

for some $c_1 > 0$, *then*

$$\sup_n M^{s-1}(n) < \infty. \tag{15.5.10}$$

Corollary 15.5.2 *If conditions* (15.5.6) *and* (15.5.7) *are met and the distribution of* X_n *converges weakly to a limiting distribution*: $\mathbf{P}(X_n < t) \Rightarrow \mathbf{P}(X < t)$, *then* $\mathbf{E}X^{s-1} < \infty$.

This assertion follows from the Fatou–Lebesgue theorem (see also Lemma 6.1.1), which implies

$$\mathbf{E}X^{s-1} \le \liminf_{n\to\infty} \mathbf{E}X_n^{s-1}. \qquad \square$$

The assertion of Corollary 15.5.2 is unimprovable. One can see this from the example of the sequence $X_n = (X_{n-1} + \zeta_n)^+$, where $\zeta_k \overset{d}{=} \zeta$ are independent and identically distributed. If $\mathbf{E}\zeta_k < 0$ then the limiting distribution of X_n coincides with the distribution of $S = \sup_k S_k$ (see Sect. 12.4). From factorisation identities one can derive that $\mathbf{E}S^{s-1}$ is finite if and only if $\mathbf{E}(\zeta^+)^s < \infty$. An outline of the proof is as follows. Theorem 12.3.2 implies that $\mathbf{E}S^k = c\,\mathbf{E}(\chi_+^k;\ \eta_+ < \infty)$, $c = \text{const} < \infty$. It follows from Corollary 12.2.2 that

$$1 - \mathbf{E}\big(e^{i\lambda\chi_+};\ \eta_+ < \infty\big) = \big(1 - \mathbf{E}e^{i\lambda\zeta}\big) \int_0^\infty e^{-i\lambda x}\, dH(x),$$

where $H(x)$ is the renewal function for the random variable $-\chi_-^0 \ge 0$. Since

$$a_1 + b_1 x \le H(x) \le a_2 + b_2 x$$

(see Theorem 10.1.1 and Lemma 10.1.1; a_i, b_i are constants), integrating the convolution

$$\mathbf{P}(\chi_+ > x,\ \eta_+ < \infty) = \int_0^\infty \mathbf{P}(\zeta > v + x)\, dH(v)$$

by parts we verify that, as $x \to \infty$, the left-hand side has the same order of magnitude as $\int_0^\infty \mathbf{P}(\zeta > v + x)\, dv$. Hence the required statement follows.

We now return to Theorem 15.5.2. Note that in all of the most popular problems the sequence $M^{s-1}(n)$ behaves "regularly": either it is bounded or $M^{s-1}(n) \to \infty$. Assertion (15.5.8) means that, under the conditions of Theorem 15.5.2, the second possibility is excluded. Condition (15.5.9) ensuring (15.5.10) is also rather broad.

Proof of Theorem 15.5.2 Let for simplicity's sake $s > 1$ be an integer. We have

$$\begin{aligned}\mathbf{E}(X_n^s;\, X_{n-1} > N) &= \int_N^\infty \mathbf{E}\big((x+\xi_n)^s;\, X_{n-1} \in dx\big)\\ &= \sum_{l=0}^{s} \binom{s}{l} \int_N^\infty x^l \mathbf{E}\big(\xi_n^{s-l};\, X_{n-1} \in dx\big).\end{aligned}$$

If we replace ξ_n^{s-l} for $s-l \ge 2$ with $|\xi_n|^{s-l}$ then the right-hand side can only increase. Therefore,

$$\mathbf{E}(X_n^s;\, X_{n-1} > N) \le \sum_{l=0}^{s} \binom{s}{l} m(s-l) M_N^l(n-1),$$

where

$$M_N^l(n) = \mathbf{E}(X_n^l;\, X_n > N).$$

The moments $M^s(n) = \mathbf{E}X_n^s$ satisfy the inequalities

$$\begin{aligned}M^s(n) &\le \mathbf{E}\big[(N+|\xi_n|)^s;\, X_{n-1} \le N\big] + \sum_{l=0}^{s} \binom{s}{l} m(s-l) M_N^l(n-1)\\ &\le 2^s\big[N^s + c\big] + \sum_{l=0}^{s} \binom{s}{l} m(s-l) M_N^l(n-1).\end{aligned} \tag{15.5.11}$$

Suppose now that (15.5.8) does not hold: $M^{s-1}(n) \to \infty$. Then all the more $M^s(n) \to \infty$ and there exists a subsequence n' such that $M^s(n') > M^s(n'-1)$. Since $M^l(n) \le [M^{l+1}(n)]^{l/l+1}$, we obtain from (15.5.6) and (15.5.11) that

$$\begin{aligned}M^s(n') &\le \mathrm{const} + M^s(n'-1) + sM^{s-1}(n'-1)m(1) + o\big(M^{s-1}(n'-1)\big)\\ &\le M^s(n'-1) - \frac{1}{2} s\varepsilon M^{s-1}(n'-1)\end{aligned}$$

for sufficiently large n'. This contradicts the assumption that $M^s(n) \to \infty$ and hence proves (15.5.8).

We now prove (15.5.10). If this relation is not true then there exists a sequence n' such that $M^{s-1}(n') \to \infty$ and $M^s(n') > M^s(n'-1) - c_1$. It remains to make use of the above argument.

We leave the proof for a non-integer $s > 1$ to the reader (the changes are elementary). The theorem is proved. □

Remark 15.5.1 (1) The assertions of Theorems 15.5.1 and 15.5.2 will remain valid if one requires inequalities (15.5.4) or $\mathbf{E}(\xi_n+\varepsilon|\mathfrak{F}_{n-1})\mathrm{I}(X_{n-1}>N)\le 0$ to hold not for all n, but only for $n\ge n_0$ for some $n_0>1$.

(2) As in Theorem 15.5.1, condition (15.5.6) means that the sequence of random variables $(X_n+\varepsilon n)\mathrm{I}(X_{n-1}>N)$ forms a supermartingale.

(3) The conditions of Theorems 15.5.1 and 15.5.2 may be weakened by replacing them with "averaged" conditions. Consider, for instance, condition (15.5.1). By integrating it over the set $\{X_{n-1}>x>N\}$ we obtain

$$\mathbf{E}\big(e^{\lambda\xi_n};\,X_{n-1}>x\big)\le\beta(\lambda)\mathbf{P}(X_{n-1}>x)$$

or, which is the same,

$$\mathbf{E}\big(e^{\lambda\xi_n}\big|X_{n-1}>x\big)\le\beta(\lambda). \tag{15.5.12}$$

The converse assertion that (15.5.12) for all $x>N$ implies relation (15.5.1) is obviously false, so that condition (15.5.12) is weaker than (15.5.1). A similar remark is true for condition (15.5.4).

One has the following generalisations of Theorems 15.5.1 and 15.5.2 to the case of "averaged conditions".

Theorem 15.5.1A *Let, for some $\lambda>0$, $N>0$ and all $x\ge N$,*

$$\mathbf{E}\big(e^{\lambda\xi_n}\big|X_{n-1}>x\big)\le\beta(\lambda)<1,\qquad \mathbf{E}\big(e^{\lambda\xi_n};\,X_{n-1}\le N\big)\le\psi(\lambda)<\infty.$$

Then

$$\mathbf{E}e^{\lambda X_n}\le\beta^n(\lambda)\,\mathbf{E}e^{\lambda X(0)}+\frac{e^{\lambda N}\psi(\lambda)}{1-\beta(\lambda)}.$$

Put

$$\overline{m}(1):=\sup_{n\ge1}\sup_{x\ge N}\mathbf{E}(\xi_n|X_{n-1}>x),$$

$$\overline{m}(l):=\sup_{n\ge1}\sup_{x\ge N}\mathbf{E}\big(|\xi_n|^l\big|X(n)>x\big),\quad l>1.$$

Theorem 15.5.2A *Let $\mathbf{E}X_0^s<\infty$ and there exist $N\ge0$ and $\varepsilon>0$ such that*

$$\overline{m}(1)\le-\varepsilon,\qquad \overline{m}(s)<\infty,\qquad \mathbf{E}\big(|\xi_n|^s;\,X_{n-1}\le N\big)<c<\infty.$$

Then (15.5.8) *holds true. If, in addition,* (15.5.9) *is valid, then* (15.5.10) *is true.*

The proofs of Theorems 15.5.1A and 15.5.2A are quite similar to those of Theorems 15.5.1 and 15.5.2. The only additional element in both cases is integration by parts. We will illustrate this with the proof of Theorem 15.5.1A. Consider

$$
\begin{aligned}
\mathbf{E}\big(e^{\lambda X_n};\,X_{n-1}>N\big) &= \int_N^\infty e^{\lambda x}\mathbf{E}\big(e^{\lambda\xi_n};\,X_{n-1}\in dx\big)\\
&= \mathbf{E}\big(e^{\lambda(N+\xi_n)};\,X_{n-1}>N\big)+\int_N^\infty \lambda e^{\lambda x}\mathbf{E}\big(e^{\lambda\xi_n};\,X_{n-1}>x\big)\,dx\\
&\le \mathbf{E}\big(e^{\lambda(N+\xi_n)};\,X_{n-1}>N+\beta(\lambda)\big)\int_N^\infty \lambda e^{\lambda x}\mathbf{P}(X_{n-1}>x)\,dx\\
&= e^{\lambda N}\mathbf{E}\big(e^{\lambda\xi_n}-\beta(\lambda);\,X_{n-1}>N\big)+\beta(\lambda)\int_N^\infty e^{\lambda x}\mathbf{P}(X_{n-1}\in dx)\\
&\le \beta(\lambda)\mathbf{E}\big(e^{\lambda X_{n-1}};\,X_{n-1}>N\big).
\end{aligned}
$$

From this we find that

$$
\begin{aligned}
\beta_n(\lambda) := \mathbf{E}e^{\lambda X_n} &\le \mathbf{E}\big(e^{\lambda(X_{n-1}+\xi_n)};\,X_{n-1}\le N\big)+\mathbf{E}\big(e^{\lambda X_n};\,X_{n-1}>N\big)\\
&\le e^{\lambda N}\psi(\lambda)+\beta(\lambda)\,\mathbf{E}\big(e^{\lambda X_{n-1}};\,X_{n-1}>N\big)\\
&\le e^{\lambda N}\psi(\lambda)-\mathbf{P}(X_{n-1}\le N)\beta(\lambda)+\beta(\lambda)\beta_n(\lambda);\\
\beta_n(\lambda) &\le \beta^n(\lambda)\beta_0(\lambda)+\frac{e^{\lambda N}\psi(\lambda)}{1-\beta(\lambda)}.
\end{aligned}
$$

□

Note that Theorem 13.7.2 and Corollary 13.7.1 on "positive recurrence" can also be referred to as theorems on boundedness of stochastic sequences.

Chapter 16
Stationary Sequences

Abstract Section 16.1 contains the definitions and a discussion of the concepts of strictly stationary sequences and measure preserving transformations. It also presents Poincaré's theorem on the number of visits to a given set by a stationary sequence. Section 16.2 discusses invariance, ergodicity, mixing and weak dependence. The Birkhoff–Khintchin ergodic theorem is stated and proved in Sect. 16.3.

16.1 Basic Notions

Let $\langle \Omega, \mathfrak{F}, \mathbf{P}\rangle$ be a probability space and $\boldsymbol{\xi} = (\xi_0, \xi_1, \ldots)$ an infinite sequence of random variables given on it.

Definition 16.1.1 A sequence $\boldsymbol{\xi}$ is said to be strictly *stationary* if, for any k, the distribution of the vector $(\xi_n, \ldots, \xi_{n+k})$ does not depend on n, $n \geq 0$.

Along with the sequence $\boldsymbol{\xi}$, consider the sequence $(\xi_n, \xi_{n+1}, \ldots)$. Since the finite-dimensional distributions of these sequences (i.e. the distributions of the vectors $(\xi_m, \ldots, \xi_{m+k})$) coincide, the distributions of the sequences will also coincide (one has to make use of the measure extension theorem (see Appendix 1) or the Kolmogorov theorem (see Sect. 3.5). In other words, for a stationary sequence $\boldsymbol{\xi}$, for any n and $\mathbf{B} \in \mathfrak{B}^\infty$ (for notation see Sect. 3.5), one has

$$\mathbf{P}(\boldsymbol{\xi} \in \mathbf{B}) = \mathbf{P}\big((\xi_n, \xi_{n+1}, \ldots) \in \mathbf{B}\big).$$

The simplest example of a stationary sequence is given by a sequence of independent identically distributed random variables $\boldsymbol{\zeta} = (\zeta_0, \zeta_1, \ldots)$. It is evident that the sequence $\xi_k = \alpha_0\zeta_k + \cdots + \alpha_s\zeta_{k+s}$, $k = 0, 1, 2, \ldots$, will also be stationary, but the variables ξ_k will no longer be independent. The same holds for sequences of the form

$$\xi_k = \sum_{j=0}^{\infty} \alpha_j \zeta_{k+j},$$

provided that $\mathbf{E}|\zeta_j| < \infty$, $\sum |\alpha_j| < \infty$, or if $\mathbf{E}\zeta_k = 0$, $\operatorname{Var}(\zeta_k) < \infty$, $\sum \alpha_j^2 < \infty$ (the latter ensures a.s. convergence of the series of random variables, see Sect. 10.2). In a

A.A. Borovkov, *Probability Theory*, Universitext,
DOI 10.1007/978-1-4471-5201-9_16,

similar way one can consider stationary sequences $\xi_k = g(\zeta_k, \zeta_{k+1}, \dots)$ "generated" by $\boldsymbol{\zeta}$, where $g(x)$ is an arbitrary measurable functional $\mathbb{R}^\infty \mapsto \mathbb{R}$.

Another example is given by stationary Markov chains. If $\{X_n\}$ is a real-valued Markov chain with invariant measure $\boldsymbol{\pi}$ and transition probability $P(\cdot,\cdot)$ then the chain $\{X_n\}$ with $X \Subset \boldsymbol{\pi}$ will form a stationary sequence, because the distribution

$$\mathbf{P}(X_n \in B_0, \dots, X_{n+k} \in B_k) = \int_{B_0} \pi(dx_0) \int_{B_1} P(x_0, dx_1) \cdots \int_{B_k} P(x_{k-1}, dx_k)$$

will not depend on n.

Any stationary sequence $\boldsymbol{\xi} = (\xi_0, \xi_1, \dots)$ can always be extended to a stationary sequence $\overline{\boldsymbol{\xi}} = (\dots \xi_{-1}, \xi_0, \xi_1, \dots)$ given on the "whole axis".

Indeed, for any n, $-\infty < n < \infty$, and $k \ge 0$ define the joint distributions of $(\xi_n, \dots, \xi_{n+k})$ as those of $(\xi_0, \dots, \xi_k)$. These distributions will clearly be consistent (see Sect. 3.5) and by the Kolmogorov theorem there will exist a unique probability distribution on $\mathbb{R}^\infty_{-\infty} = \prod_{k=-\infty}^{\infty} \mathbb{R}_k$ with the respective σ-algebra such that any finite-dimensional distribution is a projection of that distribution on the corresponding subspace. It remains to take the random element $\overline{\boldsymbol{\xi}}$ to be the identity mapping of $\mathbb{R}^\infty_{-\infty}$ onto itself.

In some of the subsequent sections it will be convenient for us to use stationary sequences given on the whole axis.

Let $\boldsymbol{\xi}$ be such a sequence. Define a transformation θ of the space $\mathbb{R}^\infty_{-\infty}$ onto itself with the help of the relations

$$(\theta \boldsymbol{x})_k = (\boldsymbol{x})_{k+1} = x_{k+1}, \tag{16.1.1}$$

where $(\boldsymbol{x})_k$ is the k-th component of the vector $\boldsymbol{x} \in \mathbb{R}^\infty_{-\infty}$, $-\infty < k < \infty$. The transformation θ clearly has the following properties:

1. It is a one-to-one mapping, θ^{-1} is defined by

$$\left(\theta^{-1} \boldsymbol{x}\right)_k = x_{k-1}.$$

2. The sequence $\theta \overline{\boldsymbol{\xi}}$ is also stationary, its distribution coinciding with that of $\overline{\boldsymbol{\xi}}$:

$$\mathbf{P}(\theta \overline{\boldsymbol{\xi}} \in \mathbf{B}) = \mathbf{P}(\overline{\boldsymbol{\xi}} \in \mathbf{B}).$$

It is natural to call the last property of the transformation θ the "measure preserving" property.

The above remarks explain to some extent why historically exploring the properties of stationary sequences followed the route of studying measure preserving transforms. Studies in that area constitute a substantial part of the modern analysis. In what follows, we will relate the construction of stationary sequences to measure preserving transformations, and it will be more convenient to regard the latter as "primary" objects.

Definition 16.1.2 Let $\langle \Omega, \mathfrak{F}, \mathbf{P} \rangle$ be the basic probability space. A transformation T of Ω into itself is said to be *measure preserving* if:

(1) T is measurable, i.e. $T^{-1}A = \{\omega : T\omega \in A\} \in \mathfrak{F}$ for any $A \in \mathfrak{F}$; and
(2) T preserves the measure: $\mathbf{P}(T^{-1}A) = \mathbf{P}(A)$ for any $A \in \mathfrak{F}$.

Let T be a measure preserving transformation, T^n its n-th iteration and $\xi = \xi(\omega)$ be a random variable. Put $U\xi(\omega) = \xi(T\omega)$, so that U is a *transformation of random variables*, and $U^k\xi(\omega) = \xi(T^k\omega)$. *Then*

$$\boldsymbol{\xi} = \{U^n\xi(\omega)\}_0^\infty = \{\xi(T^n\omega)\}_0^\infty \tag{16.1.2}$$

is a stationary sequence of random variables.

Proof Indeed, let $A = \{\omega : \boldsymbol{\xi} \in \mathbf{B}\}$, $\mathbf{B} \in \mathfrak{B}^\infty$ and $A_1 = \{\omega : \theta\boldsymbol{\xi} \in \mathbf{B}\}$. We have

$$\boldsymbol{\xi} = (\xi(\omega), \xi(T\omega), \ldots), \qquad \theta\boldsymbol{\xi} = (\xi(T\omega), \xi(T^2\omega), \ldots).$$

Therefore $\omega \in A_1$ if and only if $T\omega \in A$, i.e. when $A_1 = T^{-1}A$. But $\mathbf{P}(T^{-1}A) = \mathbf{P}(A)$ and hence $\mathbf{P}(A_1) = \mathbf{P}(A)$, so that $\mathbf{P}(A_n) = \mathbf{P}(A)$ for any $n \ge 1$ as well, where $A_n = \{\omega : \theta^n\boldsymbol{\xi} \in B\}$. □

Stationary sequences defined by (16.1.2) will be referred to as *sequences generated by the transformation T*.

To be able to construct stationary sequences on the whole axis, we will need measure preserving transformations acting both in "positive" and "negative" directions.

Definition 16.1.3 A transformation T is said to be *bidirectional measure preserving* if:

(1) T is a one-to-one transformation, the domain and range of T coincide with the whole Ω;
(2) the transformations T and T^{-1} are measurable, i.e.

$$T^{-1}A = \{\omega : T\omega \in A\} \in \mathfrak{F}, \qquad TA = \{T\omega : \omega \in A\} \in \mathfrak{F}$$

for any $A \in \mathfrak{F}$;
(3) the transformation T preserves the measure: $\mathbf{P}(T^{-1}A) = \mathbf{P}(A)$, and therefore $\mathbf{P}(A) = \mathbf{P}(TA)$ for any $A \in \mathfrak{F}$.

For such transformations we can, as before, construct stationary sequences $\boldsymbol{\xi}$ defined on the whole axis:

$$\boldsymbol{\xi} = \{U^n\xi(\omega)\}_{-\infty}^\infty = \{\xi(T^n\omega)\}_{-\infty}^\infty.$$

The argument before Definition 16.1.2 shows that this approach "exhausts" all stationary sequences given on $\langle \Omega, \mathfrak{F}, \mathbf{P} \rangle$, i.e. to any stationary sequence $\boldsymbol{\xi}$ we can relate a measure preserving transformation T and a random variable $\xi = \xi_0$ such that $\xi_k(\omega) = \xi_0(T^k\omega)$. In this construction, we consider the "sample probability space" $\langle \mathbb{R}^\infty, \mathfrak{B}^\infty, \mathbf{P} \rangle$ for which $\boldsymbol{\xi}(\omega) = \omega$, $\theta = T$. The transformation $\theta = T$ (that is, transformation (16.1.1)) will be called the *pathwise shift transformation*. It always exists and "generates" any stationary sequence.

Now we will give some simpler examples of (bidirectional) measure preserving transformations.

Example 16.1.1 Let $\Omega = \{\omega_1, \dots, \omega_d\}$, $d \ge 2$, be a finite set, $\mathfrak{F}$ be the σ-algebra of all its subsets, $T\omega_i = \omega_{i+1}$, $1 \le i \le d-1$ and $T\omega_d = \omega_1$. If $\mathbf{P}(\omega_i) = 1/d$ then T and T^{-1} are measure preserving transformations.

Example 16.1.2 Let $\Omega = [0, 1)$, $\mathfrak{F}$ be the σ-algebra of Borel sets, $\mathbf{P}$ the Lebesgue measure and s a fixed number. Then $T\omega = \omega + s \pmod 1$ is a bidirectional measure preserving transformation.

In these examples, the spaces Ω are rather small, which allows one to construct on them only stationary sequences with deterministic or almost deterministic dependence between their elements. If we choose in Example 16.1.1 the variable ξ so that all $\xi(\omega_i)$ are different, then the value $\xi_k(\omega) = \xi(T^k\omega)$ will uniquely determine $T^k\omega$ and thereby $T^{k+1}\omega$ and $\xi_{k+1}(\omega)$. The same can be said of Example 16.1.2 in the case when $\xi(\omega)$, $\omega \in [0, 1)$, is a monotone function of ω.

As our argument at the beginning of the section shows, the space $\Omega = \mathbb{R}^\infty$ is large enough to construct on it any stationary sequence.

Thus, we see that the concept of a measure preserving transformation arises in a natural way when studying stationary processes. But not only in that case. It also arises, for instance, while studying the dynamics of some physical systems. Indeed, the whole above argument remains valid if we consider on $\langle \Omega, \mathfrak{F} \rangle$ an arbitrary measure $\boldsymbol{\mu}$ instead of the probability $\mathbf{P}$. For example, for $\Omega = \mathbb{R}^\infty$, the value $\boldsymbol{\mu}(A)$, $A \in \mathfrak{F}$, could be the Lebesgue measure (volume) of the set A. The measure preserving property of the transformation T will mean that any set A, after the transform T has acted on it (which, say, corresponds to the change of the physical system's state in one unit of time), will retain its volume. This property is rather natural for incompressible liquids. Many laws to be established below will be equally applicable to such physical systems.

Returning to probabilistic models, i.e. to the case when the measure is a probability distribution, it turns out that, in that case, for any set A with $\mathbf{P}(A) > 0$, the "trajectory" $T^n\omega$ will visit A infinitely often for almost all (with respect to the measure $\mathbf{P}$) $\omega \in A$.

Theorem 16.1.1 (Poincaré) *Let T be a measure preserving transformation and $A \in \mathfrak{F}$. Then, for almost all $\omega \in A$, the relation $T^n\omega \in A$ holds for infinitely many $n \ge 1$.*

Proof Put $N := \{\omega \in A : T^n\omega \notin A \text{ for all } n \ge 1\}$. Because $\{\omega : T^n\omega \in A\} \in \mathfrak{F}$, it is not hard to see that $N \in \mathfrak{F}$. Clearly, $N \cap T^{-n}N = \varnothing$ for any $n \ge 1$, and $T^{-m}N \cap T^{-(m+n)}N = T^{-m}(N \cap T^{-n}N) = \varnothing$. This means that we have infinitely many sets $T^{-n}N$, $n = 0, 1, 2, \dots$, which are disjoint and have one and the same probability. This evidently implies that $\mathbf{P}(N) = 0$.

Thus, for each $\omega \in A \setminus N$, there exists an $n_1 = n_1(\omega)$ such that $T^{n_1}\omega \in A$. Now we apply this assertion to the measure preserving mapping $T_k = T^k$, $k \ge 1$. Then, for each $\omega \in A \setminus N_k$, $\mathbf{P}(N_k) = 0$, there exists an $n_k = n_k(\omega) \ge 1$ such that $(T^k)^{n_k}\omega \in A$. Since $kn_k \ge k$, the theorem is proved. □

Corollary 16.1.1 *Let $\xi(\omega) \ge 0$ and $A = \{\omega : \xi(\omega) > 0\}$. Then, for almost all $\omega \in A$,*

$$\sum_{n=0}^{\infty} \xi(T^n\omega) = \infty.$$

Proof Put $A_k = \{\omega : \xi(\omega) \ge 1/k\} \subset A$. Then by Theorem 16.1.1 the above series diverges for almost all $\omega \in A_k$. It remains to notice that $A = \bigcup_k A_k$. □

Remark 16.1.1 Formally, one does not need condition $\mathbf{P}(A) > 0$ in Theorem 16.1.1 and Corollary 16.1.1. However, in the absence of that condition, the assertions may become meaningless, since the set $A \setminus N$ in the proof of Theorem 16.1.1 can turn out to be empty. Suppose, for example, that in the conditions of Example 16.1.2, A is a one-point set: $A = \{\omega\}$, $\omega \in [0, 1)$. If s is irrational, then $T^k\omega$ will never be in A for $k \ge 1$. Indeed, if we assume the contrary, then we will infer that there exist integers k and m such that $\omega + sk - m = \omega$, $s = m/k$, which contradicts the irrationality of s.

16.2 Ergodicity (Metric Transitivity), Mixing and Weak Dependence

Definition 16.2.1 A set $A \in \mathfrak{F}$ is said to be *invariant* (with respect to a measure preserving transformation T) if $T^{-1}A = A$. A set $A \in \mathfrak{F}$ is said to be *almost invariant* if the sets $T^{-1}A$ and A differ from each other by a set of probability zero: $\mathbf{P}(A \oplus T^{-1}A) = 0$, where $A \oplus B = A\overline{B} \cup \overline{A}B$ is the symmetric difference.

It is evident that the class of all invariant (almost invariant) sets forms a σ-algebra which will be denoted by $\mathfrak{I}$ ($\mathfrak{I}^*$).

Lemma 16.2.1 *If A is an almost invariant set then there exists an invariant set B such that* $\mathbf{P}(A \oplus B) = 0$.

Proof Put $B = \limsup_{n\to\infty} T^{-n}A$ (recall that $\limsup_{n\to\infty} A_n = \bigcap_{n=1}^{\infty} \bigcup_{k=n}^{\infty} A_k$ is the set of all points which belong to infinitely many sets A_k). Then

$$T^{-1}B = \limsup_{n\to\infty} T^{-(n+1)}A = B,$$

i.e. $B \in \mathfrak{J}$. It is not hard to see that

$$A \oplus B \subset \bigcup_{k=0}^{\infty} \bigl(T^{-k} A \oplus T^{-(k+1)} A\bigr).$$

Since

$$\mathbf{P}\bigl(T^{-k} A \oplus T^{-(k+1)} A\bigr) = \mathbf{P}\bigl(A \oplus T^{-1} A\bigr) = 0,$$

we have $\mathbf{P}(A \oplus B) = 0$. The lemma is proved. □

Definition 16.2.2 A measure preserving transformation T is said to be *ergodic* (or *metric transitive*) if each invariant set has probability zero or one.

A stationary sequence $\{\xi_k\}$ associated with such T (i.e. the sequence which generated T or was generated by T) is also said to be ergodic (metric transitive).

Lemma 16.2.2 *A transformation T is ergodic if and only if each almost invariant set has probability* 0 *or* 1.

Proof Let T be ergodic and $A \in \mathfrak{J}^*$. Then by Lemma 16.2.1 there exists an invariant set B such that $\mathbf{P}(A \oplus B) = 0$. Because $\mathbf{P}(B) = 0$ or 1, the probability $\mathbf{P}(A) = 0$ or 1. The converse assertion is obvious. □

Definition 16.2.3 A random variable $\zeta = \zeta(\omega)$ is said to be *invariant* (*almost invariant*) if $\zeta(\omega) = \zeta(T\omega)$ for all $\omega \in \Omega$ (for almost all $\omega \in \Omega$).

Theorem 16.2.1 *Let T be a measure preserving transformation. The following three conditions are equivalent*:

(1) *T is ergodic*;
(2) *each almost invariant random variable is a.s. constant*;
(3) *each invariant random variable is a.s. constant.*

Proof (1) $\Rightarrow$ (2). Assume that T is ergodic and ξ is almost invariant, i.e. $\xi(\omega) = \xi(T\omega)$ a.s. Then, for any $v \in \mathbb{R}$, we have $A_v := \{\omega : \xi(\omega) \le v\} \in \mathfrak{J}^*$ and, by Lemma 16.2.2, $\mathbf{P}(A_v)$ equals 0 or 1. Put $V := \sup\{v : \mathbf{P}(A_v) = 0\}$. Since $A_v \uparrow \Omega$ as $v \uparrow \infty$ and $A_v \downarrow \varnothing$ as $v \downarrow -\infty$, one has $|V| < \infty$ and

$$\mathbf{P}\bigl(\xi(\omega) < V\bigr) = \mathbf{P}\left(\bigcup_{n=1}^{\infty} \left\{\xi(\omega) < V - \frac{1}{n}\right\}\right) = 0.$$

Similarly, $\mathbf{P}(\xi(\omega) > V) = 0$. Therefore $\mathbf{P}(\xi(\omega) = V) = 1$.

(2) $\Rightarrow$ (3). Obvious.

(3) $\Rightarrow$ (1). Let $A \in \mathfrak{J}$. Then the indicator function I_A is an invariant random variable, and since it is constant, one has either $\mathrm{I}_A = 0$ or $\mathrm{I}_A = 1$ a.s. This implies that $\mathbf{P}(A)$ equals 0 or 1. The theorem is proved. □

The assertion of the theorem clearly remains valid if one considers in (3) only bounded random variables. Moreover, if ξ is invariant, then the truncated variable $\xi_{(N)} = \min(\xi, N)$ is also invariant.

Returning to Examples 16.1.1 and 16.1.2, in Example 16.1.1,

$$\Omega = (\omega_1, \dots, \omega_d), \qquad T\omega_i = \omega_{i+1 \pmod d}, \qquad \mathbf{P}(\omega_i) = 1/d.$$

The transformation T is obviously metric transitive.

In Example 16.1.2, $\Omega = [0, 1)$, $T\omega = \omega + s \pmod 1$, and $\mathbf{P}$ is the Lebesgue measure. We will now show that T is *ergodic if and only if s is irrational.*

Consider a square integrable random variable $\xi = \xi(\omega) : \mathbf{E}\xi^2(\omega) < \infty$. Then by the Parseval equality, the Fourier series

$$\xi(\omega) = \sum_{n=0}^{\infty} a_n e^{2\pi i n\omega}$$

for this function has the property $\sum_{n=0}^{\infty} |c_n^2| < \infty$. Assume that s is irrational, while ξ is invariant. Then

$$\begin{aligned} a_n &= \mathbf{E}\xi(\omega)e^{-2\pi i n\omega} = \mathbf{E}\xi(T\omega)e^{-2\pi i n T\omega} \\ &= e^{-2\pi i n s}\mathbf{E}\xi(T\omega)e^{-2\pi i n\omega} = e^{-2\pi i n s}\mathbf{E}\xi(\omega)e^{-2\pi i n\omega} = e^{-2\pi i n s} a_n. \end{aligned}$$

For irrational s, this equality is only possible when $a_n = 0$, $n \ge 1$, and $\xi(\omega) = a_0 =$ const. By Theorem 16.2.1 this means that T is ergodic.

Now let $s = m/n$ be rational (m and n are integers). Then the set

$$A = \bigcup_{k=0}^{n-1} \left\{ \omega : \frac{2k}{2n} \le \omega < \frac{2k+1}{2n} \right\}$$

will be invariant and $\mathbf{P}(A) = 1/2$. This means that T is not ergodic. □

Definition 16.2.4 A measure preserving transformation T is called *mixing* if, for any $A_1, A_2 \in \mathfrak{F}$, as $n \to \infty$,

$$\mathbf{P}\big(A_1 \cap T^{-n}A_2\big) \to \mathbf{P}(A_1)\mathbf{P}(A_2). \qquad (16.2.1)$$

Now consider the stationary sequence $\boldsymbol{\xi} = (\xi_0, \xi_1, \dots)$ generated by the transformation $T : \xi_k(\omega) = \xi_0(T^k\omega)$.

Definition 16.2.5 A stationary sequence $\boldsymbol{\xi}$ is said to be *weakly dependent* if ξ_k and ξ_{k+n} are asymptotically independent as $n \to \infty$, i.e. for any $B_1, B_2 \in \mathfrak{B}$

$$\mathbf{P}(\xi_k \in B_1, \xi_{k+n} \in B_2) \to \mathbf{P}(\xi_0 \in B_1)\mathbf{P}(\xi_0 \in B_2). \qquad (16.2.2)$$

Theorem 16.2.2 *A measure preserving transformation T is mixing if and only if any stationary sequence $\boldsymbol{\xi}$ generated by T is weakly dependent.*

Proof Let T be mixing. Put $A_i := \xi_0^{-1}(B_i)$, $i = 1, 2$, and set $k = 0$ in (16.2.2). Then

$$\mathbf{P}(\xi_0 \in B_1, \xi_n \in B_2) = \mathbf{P}\big(A_1 \cap T^{-n} A_2\big) \to \mathbf{P}(A_1)\mathbf{P}(A_2).$$

Now assume any sequence generated by T is weakly dependent. For any given $A_1, A_2 \in \mathfrak{F}$, define the random variable

$$\xi(\omega) = \begin{cases} 0 & \text{if } \omega \notin A_1 \cup A_2; \\ 1 & \text{if } \omega \in A_1 \overline{A}_2; \\ 2 & \text{if } \omega \in A_1 A_2; \\ 3 & \text{if } \omega \in \overline{A}_1 A_2; \end{cases}$$

and put $\xi_k(\omega) := \xi(T^k \omega)$. Then, as $n \to \infty$,

$$\begin{aligned} \mathbf{P}\big(A_1 \cap T^{-n} A_2\big) &= \mathbf{P}(0 < \xi_0 < 3, \xi_n > 2) \to \mathbf{P}(0 < \xi_0 < 3)\mathbf{P}(\xi_0 > 2) \\ &= \mathbf{P}(A_1)\mathbf{P}(A_2). \end{aligned}$$

The theorem is proved. □

Let $\{X_n\}$ be a stationary real-valued Markov chain with an invariant distribution π that satisfies the conditions of the ergodic theorem, i.e. such that, for any $B \in \mathfrak{B}$ and $x \in \mathbb{R}$, as $n \to \infty$,

$$\mathbf{P}(X_n \in B \mid X_0 = x) \to \pi(B).$$

Then $\{X_n\}$ is weakly dependent, and therefore, by Theorem 16.2.2, the respective transformation T is mixing. Indeed,

$$\mathbf{P}(X_0 \in B_1, X_n \in B_2) = \mathbf{E}\mathrm{I}(X_0 \in B_1)\mathbf{P}(X_n \in B_2 \mid X_0),$$

where the last factor converges to $\pi(B_2)$ for each X_0. Therefore the above probability tends to $\pi(B_2)\pi(B_1)$.

Further characterisations of the mixing property will be given in Theorems 16.2.4 and 16.2.5.

Now we will introduce some notions that are somewhat broader than those from Definitions 16.2.4 and 16.2.5.

Definition 16.2.6 A transformation T is called *mixing on the average* if, as $n \to \infty$,

$$\frac{1}{n}\sum_{k=1}^{n} \mathbf{P}\big(A_1 \cap T^{-k} A_2\big) \to \mathbf{P}(A_1)\mathbf{P}(A_2). \tag{16.2.3}$$

A stationary sequence $\boldsymbol{\xi}$ is said to be *weakly dependent on the average* if

$$\frac{1}{n}\sum_{k=1}^{n} \mathbf{P}(\xi_0 \in B_1, \xi_k \in B_2) \to \mathbf{P}(\xi_0 \in B_1)\mathbf{P}(\xi_0 \in B_2). \tag{16.2.4}$$

Theorem 16.2.3 *A measure preserving transformation T is mixing on the average if and only if any stationary sequence ξ generated by T is weakly dependent on the average.*

The Proof is the same as for Theorem 16.2.2, and is left to the reader. □

If $\{X_n\}$ is a periodic real-valued Markov chain with period d such that each of the embedded sub-chains $\{X_{i+nd}\}_{n=0}^{\infty}$, $i = 0, \dots, d-1$, satisfies the ergodicity conditions with invariant distributions $\pi^{(i)}$ on disjoint sets $\mathcal{X}_0, \dots, \mathcal{X}_{d-1}$, then the "common" invariant distribution π will be equal to $d^{-1}\sum_{i=0}^{d-1}\pi^{(i)}$, and the chain $\{X_n\}$ will be weakly dependent on the average. At the same time, it will clearly not be weakly dependent for $d > 1$.

Theorem 16.2.4 *A measure preserving transformation T is ergodic if and only if it is mixing on the average.*

Proof Let T be mixing on the average, and $A_1 \in \mathfrak{F}$, $A_2 \in \mathfrak{I}$. Then $A_2 = T^{-k}A_2$ and hence $\mathbf{P}(A_1 \cap T^{-k}A_2) = \mathbf{P}(A_1A_2)$ for all $k \geq 1$. Therefore, (16.2.3) means that $\mathbf{P}(A_1A_2) = \mathbf{P}(A_1)\mathbf{P}(A_2)$. For $A_1 = A_2$ we get $\mathbf{P}(A_2) = \mathbf{P}^2(A_2)$, and consequently $\mathbf{P}(A_2)$ equals 0 or 1.

We postpone the proof of the converse assertion until the next section. □

Now we will give one more important property of ergodic transforms.

Theorem 16.2.5 *A measure preserving transformation T is ergodic if and only if, for any $A \in \mathfrak{F}$ with $\mathbf{P}(A) > 0$, one has*

$$\mathbf{P}\left(\bigcup_{n=0}^{\infty} T^{-n}A\right) = 1. \tag{16.2.5}$$

Note that property (16.2.5) means that the sets $T^{-n}A$, $n = 0, 1, \dots$, "exhaust" the whole space Ω, which associates well with the term "mixing".

Proof Let T be ergodic. Put $B := \bigcup_{n=0}^{\infty} T^{-n}A$. Then $T^{-1}B \subset B$. Because T is measure preserving, one also has that $\mathbf{P}(T^{-1}B) = \mathbf{P}(B)$. From this it follows that $T^{-1}B = B$ up to a set of measure 0 and therefore B is almost invariant. Since T is ergodic, $\mathbf{P}(B)$ equals 0 or 1. But $\mathbf{P}(B) \geq \mathbf{P}(A) > 0$, and hence $\mathbf{P}(B) = 1$.

Conversely, if T is not ergodic, then there exists an invariant set A such that $0 < \mathbf{P}(A) < 1$ and, therefore, for this set $T^{-n}A = A$ holds and

$$\mathbf{P}(B) = \mathbf{P}(A) < 1.$$

The theorem is proved. □

Remark 16.2.1 In Sects. 16.1 and 16.2 we tacitly or explicitly assumed (mainly for the sake of simplicity of the exposition) that the components ξ_k of the stationary sequence $\boldsymbol{\xi}$ are real. However, we never actually used this, and so we could, as we did while studying Markov chains, assume that the state space $\mathcal{X}$, in which ξ_k take their values, is an arbitrary measurable space. In the next section we will substantially use the fact that ξ_k are real- or vector-valued.

16.3 The Ergodic Theorem

For a sequence $\xi_1, \xi_2, \ldots$ of independent identically distributed random variables we proved in Chap. 11 the strong law of large numbers:

$$\frac{S_n}{n} \xrightarrow{a.s.} \mathbf{E}\xi_1, \quad \text{where } S_n = \sum_{k=0}^{n-1} \xi_k.$$

Now we will prove the same assertion under much broader assumptions—for stationary ergodic sequences, i.e. for sequences that are weakly dependent on the average.

Let $\{\xi_k\}$ be an arbitrary strictly stationary sequence, T be the associated measure preserving transformation, and $\mathfrak{I}$ be the σ-algebra of invariant sets.

Theorem 16.3.1 (Birkhoff–Khintchin) *If* $\mathbf{E}|\xi_0| < \infty$ *then*

$$\frac{1}{n}\sum_{k=0}^{n-1} \xi_k \xrightarrow{a.s.} \mathbf{E}(\xi_0 \mid \mathfrak{I}). \tag{16.3.1}$$

If the sequence $\{\xi_k\}$ *(or transformation* T*) is ergodic, then*

$$\frac{1}{n}\sum_{k=0}^{n-1} \xi_k \xrightarrow{a.s.} \mathbf{E}\xi_0. \tag{16.3.2}$$

Below we will be using the representation $\xi_k = \xi(T^k\omega)$ for $\xi = \xi_0$. We will need the following auxiliary result.

Lemma 16.3.1 *Set*

$$S_n(\omega) := \sum_{k=0}^{n-1} \xi\bigl(T^k\omega\bigr), \qquad M_k(\omega) := \max\bigl\{0, S_1(\omega), \ldots, S_k(\omega)\bigr\}.$$

Then, under the conditions of Theorem 16.3.1,

$$\mathbf{E}\bigl[\xi(\omega)\mathrm{I}_{\{M_n>0\}}(\omega)\bigr] \ge 0$$

for any $n \ge 1$.

Proof For all $k \le n$, one has $S_k(T\omega) \le M_n(T\omega)$, and hence

$$\xi(\omega) + M_n(T\omega) \ge \xi(\omega) + S_k(T\omega) = S_{k+1}(\omega).$$

Because $\xi(\omega) \ge S_1(\omega) - M_n(T\omega)$, we have

$$\xi(\omega) \ge \max\bigl\{\max(S_1(\omega), \dots, S_n(\omega)\bigr\} - M_n(T\omega).$$

Further, since

$$\bigl\{M_n(\omega) > 0\bigr\} = \bigl\{\max\bigl(S_1(\omega), \dots, S_n(\omega)\bigr) > 0\bigr\},$$

we obtain that

$$\begin{aligned}\mathbf{E}\bigl[\xi(\omega)\mathrm{I}_{\{M_n>0\}}(\omega)\bigr] &\ge \mathbf{E}\bigl(\max\bigl(S_1(\omega), \dots, S_n(\omega)\bigr) - M_n(T\omega)\bigr)\,\mathrm{I}_{\{M_n>0\}}(\omega)\\ &\ge \mathbf{E}\bigl(M_n(\omega) - M_n(T\omega)\bigr)\mathrm{I}_{\{M_n>0\}}(\omega)\\ &\ge \mathbf{E}\bigl(M_n(\omega) - M_n(T\omega)\bigr) = 0.\end{aligned}$$

The lemma is proved. □

Proof of Theorem 16.3.1 Assertion (16.3.2) is an evident consequence of (16.3.1), because, for ergodic T, the σ-algebra $\mathfrak{I}$ is trivial and $\mathbf{E}(\xi|\mathfrak{I}) = \mathbf{E}\xi$ a.s. Hence, it suffices to prove (16.3.1).

Without loss of generality, we can assume that $\mathbf{E}(\xi|\mathfrak{I}) = 0$, for one can always consider $\xi - \mathbf{E}(\xi|\mathfrak{I})$ instead of ξ.

Let $\overline{S} := \limsup_{n\to\infty} n^{-1}S_n$ and $\underline{S} := \liminf_{n\to\infty} n^{-1}S_n$. To prove the theorem, it suffices to establish that

$$0 \le \underline{S} \le \overline{S} \le 0 \quad \text{a.s.} \tag{16.3.3}$$

Since $\overline{S}(\omega) = \overline{S}(T\omega)$, the random variable $\overline{S}$ is invariant and hence the set $A - \varepsilon = \{\overline{S}(\omega) > \varepsilon\}$ is also invariant for any $\varepsilon > 0$. Introduce the variables

$$\begin{aligned}\xi^*(\omega) &:= \bigl(\xi(\omega) - \varepsilon\bigr)\mathrm{I}_{A_\varepsilon}(\omega),\\ S_k^*(\omega) &:= \xi^*(\omega) + \cdots + \xi^*\bigl(T^{k-1}\omega\bigr),\\ M_k^*(\omega) &:= \max\bigl(0, S_1^*, \dots, S_k^*\bigr).\end{aligned}$$

Then, by Lemma 16.3.1, for any $n \ge 1$, one has

$$\mathbf{E}\xi^*\mathrm{I}_{\{M_n^*>0\}} \ge 0.$$

But, as $n \to \infty$,

$$\bigl\{M_n^* > 0\bigr\} = \Bigl\{\max_{1\le k\le n} S_k^* > 0\Bigr\} \uparrow \Bigl\{\sup_{k\ge 1} S_k^* > 0\Bigr\}$$

$$= \left\{ \sup_{k\ge 1} \frac{S_k^*}{k} > 0 \right\} = \left\{ \sup_{k\ge 1} \frac{S_k}{k} > \varepsilon \right\} \cap A_\varepsilon = A_\varepsilon.$$

The last equality follows from the observation that

$$A_\varepsilon = \{\overline{S} > \varepsilon\} \subset \left\{ \sup_{k\ge 1} \frac{S_k}{k} > \varepsilon \right\}.$$

Further, $\mathbf{E}|\xi^*| \le \mathbf{E}|\xi| + \varepsilon$. Hence, by the dominated convergence theorem,

$$0 \le \mathbf{E}\xi^* \mathrm{I}_{\{M_n^*>0\}} \to \mathbf{E}\xi^* \mathrm{I}_{A_\varepsilon}.$$

Consequently,

$$\begin{aligned} 0 \le \mathbf{E}\xi^* \mathrm{I}_{A_\varepsilon} &= \mathbf{E}(\xi - \varepsilon)\mathrm{I}_{A_\varepsilon} = \mathbf{E}\xi \mathrm{I}_{A_\varepsilon} - \varepsilon \mathbf{P}(A_\varepsilon) \\ &= \mathbf{E}\mathrm{I}_{A_\varepsilon} \mathbf{E}(\xi \mid \mathfrak{I}) - \varepsilon \mathbf{P}(A_\varepsilon) = -\varepsilon \mathbf{P}(A_\varepsilon). \end{aligned}$$

This implies that $\mathbf{P}(A_\varepsilon) = 0$ for any $\varepsilon > 0$, and therefore $\mathbf{P}(\overline{S} \le 0) = 1$.

In a similar way, considering the variables $-\xi$ instead of ξ, we obtain that

$$\limsup_{n\to\infty} \left(-\frac{S_n}{n} \right) = -\liminf_{n\to\infty} \frac{S_n}{n} = -\underline{S},$$

and $\mathbf{P}(-\underline{S} \le 0) = 1$, $\mathbf{P}(\underline{S} \ge 0) = 1$. The required inequalities (16.3.3), and therefore the theorem itself, are proved. □

Now we can complete the

Proof of Theorem 16.2.4 It remains to show that the ergodicity of T implies mixing on the average. Indeed, let T be ergodic and $A_1, A_2 \in \mathfrak{F}$. Then, by Theorem 16.3.1, we have

$$\zeta_n = \frac{1}{n} \sum_{k=1}^{n} \mathrm{I}\bigl(T^{-k} A_2\bigr) \xrightarrow{a.s.} \mathbf{P}(A_2), \qquad \mathrm{I}(A_1)\zeta_n \xrightarrow{a.s.} \mathrm{I}(A_1)\mathbf{P}(A_2).$$

Since $\zeta_n \mathrm{I}(A_1)$ are bounded, one also has the convergence

$$\mathbf{E}\zeta_n \mathrm{I}(A_1) \to \mathbf{P}(A_2) \cdot \mathbf{P}(A_1).$$

Therefore

$$\frac{1}{n} \sum_{k=1}^{n} \mathbf{P}\bigl(A_1 \cap T^{-k} A_2\bigr) = \mathbf{E}\mathrm{I}(A_1)\zeta_n \to \mathbf{P}(A_1)\mathbf{P}(A_2).$$

The theorem is proved. □

Now we will show that convergence in mean also holds in (16.3.1) and (16.3.2).

Theorem 16.3.2 *Under the assumptions of Theorem* 16.3.1, *one has along with* (16.3.1) *and* (16.3.2) *that, respectively,*

$$\mathbf{E}\left|\frac{1}{n}\sum_{k=0}^{n-1}\xi_k - \mathbf{E}(\xi_0|\mathfrak{I})\right| \to 0 \tag{16.3.4}$$

and

$$\mathbf{E}\left|\frac{1}{n}\sum_{k=0}^{n-1}\xi_k - \mathbf{E}\xi_0\right| \to 0 \tag{16.3.5}$$

as $n \to \infty$.

Proof The assertion of the theorem follows in an obvious way from Theorems 16.3.1, 6.1.7 and the uniform integrability of the sums

$$\frac{1}{n}\sum_{k=0}^{n-1}\xi_k,$$

which follows from Theorem 6.1.6. □

Corollary 16.3.1 *If* $\{\xi_k\}$ *is a stationary metric transitive sequence and* $a = \mathbf{E}\xi_k < 0$, *then* $S(\omega) = \sup_{k\ge 0} S_k(\omega)$ *is a proper random variable.*

The proof is obvious since, for $0 < \varepsilon < -a$, one has $S_k < (a+\varepsilon)k < 0$ for all $k \ge n(\omega) < \infty$. □

An unusual feature of Theorem 16.3.1 when compared with the strong law of large numbers from Chap. 11 is that the limit of

$$\frac{1}{n}\sum_{k=0}^{n-1}\xi_k$$

can be a random variable. For instance, let $T\omega_k := \omega_{k+2}$ and $d = 2l$ be even in the situation of Example 16.1.1. Then the transformation T will not be ergodic, since the set $A = \{\omega_1, \omega_3, \dots, \omega_{d-1}\}$ will be invariant, while $\mathbf{P}(A) = 1/2$.

On the other hand, it is evident that, for any function $\xi(\omega)$, the sum

$$\frac{1}{n}\sum_{k=0}^{n-1}\xi\left(T^k\omega\right)$$

will converge with probability $1/2$ to

$$\frac{2}{d}\sum_{j=0}^{l-1}\xi(\omega_{2j+1})$$

(if $\omega = \omega_i$ and i is odd) and with probability $1/2$ to

$$\frac{2}{d}\sum_{j=1}^{l}\xi(\omega_{2j})$$

(if $\omega = \omega_i$ and i is even). This limiting distribution is just the distribution of $\mathbf{E}(\xi \,|\, \mathfrak{I})$.

Chapter 17
Stochastic Recursive Sequences

Abstract The chapter begins with introducing the concept of stochastic random sequences in Sect. 17.1. The idea of renovating events together with the key results on ergodicity of stochastic random sequences and the boundedness thereof is presented in Sect. 17.2, whereas the Loynes ergodic theorem for the case of monotone functions specifying the recursion is proved in Sect. 17.3. Section 17.4 establishes ergodicity conditions for contracting in mean Lipschitz transformations.

17.1 Basic Concepts

Consider two measurable state spaces $\langle\mathcal{X},\mathfrak{B}_{\mathcal{X}}\rangle$ and $\langle\mathcal{Y},\mathfrak{B}_{\mathcal{Y}}\rangle$, and let $\{\xi_n\}$ be a sequence of random elements taking values in $\mathcal{Y}$. If $\langle\Omega,\mathfrak{F},\mathbf{P}\rangle$ is the underlying probability space, then $\{\omega:\xi_k\in B\}\in\mathfrak{F}$ for any $B\in\mathfrak{B}_{\mathcal{Y}}$. Assume, moreover, that a measurable function $f:\mathcal{X}\times\mathcal{Y}\to\mathcal{X}$ is given on the measurable space $\langle\mathcal{X}\times\mathcal{Y},\mathfrak{B}_{\mathcal{X}}\times\mathfrak{B}_{\mathcal{Y}}\rangle$, where $\mathfrak{B}_{\mathcal{X}}\times\mathfrak{B}_{\mathcal{Y}}$ denotes the σ-algebra generated by sets $A\times B$ with $A\in\mathfrak{B}_{\mathcal{X}}$ and $B\in\mathfrak{B}_{\mathcal{Y}}$.

For simplicity's sake, by $\mathcal{X}$ and $\mathcal{Y}$ we can understand the real line $\mathbb{R}$, and by $\mathfrak{B}_{\mathcal{X}}$, $\mathfrak{B}_{\mathcal{Y}}$ the σ-algebras of Borel sets.

Definition 17.1.1 A sequence $\{X_n\}$, $n=0,1,\ldots$, taking values in $\langle\mathcal{X},\mathfrak{B}_{\mathcal{X}}\rangle$ is said to be a *stochastic recursive sequence* (s.r.s.) driven by the sequence $\{\xi_n\}$ if X_n satisfies the relation

$$X_{n+1}=f(X_n,\xi_n) \tag{17.1.1}$$

for all $n\geq 0$. For simplicity's sake we will assume that the initial state X_0 is independent of $\{\xi_n\}$.

The distribution of the sequence $\{X_n,\xi_n\}$ on $\langle(\mathcal{X}\times\mathcal{Y})^\infty,(B_{\mathcal{X}}\times B_{\mathcal{Y}})^\infty\rangle$ can be constructed in an obvious way from finite-dimensional distributions similarly to the manner in which we constructed on $\langle\mathcal{X}^\infty,\mathfrak{B}^\infty_{\mathcal{X}}\rangle$ the distribution of a Markov chain X with values in $\langle\mathcal{X},\mathfrak{B}_{\mathcal{X}}\rangle$ from its transition function $P(x,B)=\mathbf{P}(X_1(x)\in B)$. The finite-dimensional distributions of $\{(X_0,\xi_0),\ldots,(X_k,\xi_k)\}$ for the s.r.s. are given by the relations

A.A. Borovkov, *Probability Theory*, Universitext,
DOI 10.1007/978-1-4471-5201-9_17, © Springer-Verlag London 2013

$$\mathbf{P}(X_l \in A_l,\, \xi_l \in B_l;\, l=0,\dots,k)$$
$$= \int_{B_0} \cdots \int_{B_k} \mathbf{P}(\xi_l \in dy_l,\, l=0,\dots,k) \prod_{l=1}^{k} \mathrm{I}\big(f_l(X_0, y_0, \dots, y_l) \in A_l\big),$$

where $f_1(x, y_0) := f(x, y_0)$, $f_l(x, y_0, \dots, y_l) := f(f_{l-1}(x, y_0, \dots, y_{l-1}), y_l)$.

Without loss of generality, the sequence $\{\xi_n\}$ can be assumed to be given for all $-\infty < n < \infty$ (as we noted in Sect. 16.1, for a stationary sequence, the required extension to $n < 0$ can always be achieved with the help of Kolmogorov's theorem).

A *stochastic recursive sequence is a more general object than a Markov chain*. It is evident that if ξ_k are independent, then the X_n form a Markov chain. A stronger assertion is true as well: under broad assumptions about the space $\langle \mathcal{X}, \mathfrak{B}_{\mathcal{X}} \rangle$, for any Markov chain $\{X_n\}$ in $\langle \mathcal{X}, \mathfrak{B}_{\mathcal{X}} \rangle$ one can construct a function f and a sequence of independent identically distributed random variables $\{\xi_n\}$ such that (17.1.1) holds. We will elucidate this statement in the simplest case when both $\mathcal{X}$ and $\mathcal{Y}$ coincide with the real line $\mathbb{R}$. Let $P(x, B)$, $B \in \mathfrak{B}$, be the transition function of the chain $\{X_n\}$, and $F_x(t) = P(x, (-\infty, t))$ the distribution function of $X_1(x)$ ($X_0 = x$). Then if $F_x^{-1}(t)$ is the function inverse (in t) to $F_x(t)$ and $\alpha \in \mathbf{U}_{0,1}$ is a random variable, then, as we saw before (see e.g. Sect. 6.2), the random variable $F_x^{-1}(\alpha)$ will have the distribution function $F_x(t)$. Therefore, if $\{\alpha_n\}$ is a sequence of independent random variables uniformly distributed over [0, 1], then the sequence $X_{n+1} = F_{X_n}^{-1}(\alpha_n)$ will have the same distribution as the original chain $\{X_n\}$. Thus the Markov chain is an s.r.s. with the function $f(x, y) = F_x^{-1}(y)$ and driving sequence $\{\alpha_n\}$, $\alpha_n \in \mathbf{U}_{0,1}$.

For more general state spaces $\mathcal{X}$, a similar construction is possible if the σ-algebra $\mathfrak{B}_{\mathcal{X}}$ is countably-generated (i.e. is generated by a countable collection of sets from $\mathcal{X}$). This is always the case for Borel σ-algebras in $\mathcal{X} = \mathbb{R}^d$, $d \ge 1$ (see [22]).

One can always consider $f(\cdot, \xi_n)$ as a sequence of *random mappings* of the space $\mathcal{X}$ into itself. The principal problem we will be interested in is again (as in Chap. 13) that of the existence of the limiting distribution of X_n as $n \to \infty$.

In the following sections we will consider three basic approaches to this problem.

17.2 Ergodicity and Renovating Events. Boundedness Conditions

17.2.1 Ergodicity of Stochastic Recursive Sequences

We introduce the σ-algebras

$$\mathfrak{F}_{l,n}^{\xi} := \sigma\{\xi_k;\, l \le k \le n\},$$
$$\mathfrak{F}_{n}^{\xi} := \sigma\{\xi_k;\, k \le n\} = \mathfrak{F}_{-\infty,n}^{\xi},$$

$$\mathfrak{F}^{\xi} := \sigma\{\xi_k;\ -\infty < k < \infty\} = \mathfrak{F}^{\xi}_{-\infty,\infty}.$$

In the sequel, for the sake of definiteness and simplicity, we will assume the initial value X_0 to be constant unless otherwise stated.

Definition 17.2.1 An event $A \in \mathfrak{F}^{\xi}_{n+m}$, $m \ge 0$, is said to be *renovating* for the s.r.s. $\{X_n\}$ on the segment $[n, n+m]$ if there exists a measurable function $g : \mathcal{Y}^{m+1} \to \mathcal{X}$ such that, on the set A (i.e. for $\omega \in A$),

$$X_{n+m+1} = g(\xi_n, \ldots, \xi_{n+m}). \qquad (17.2.1)$$

It is evident that, for $\omega \in A$, relations of the form $X_{n+m+k+1} = g_k(\xi_n, \ldots, \xi_{n+m+k})$ will hold for all $k \ge 0$, where g_k is a function depending on its arguments only and determined by the event A.

The *sequence of events* $\{A_n\}$, $A_n \in \mathfrak{F}^{\xi}_{n+m}$, where the integer m is fixed, is said to be *renovating* for the s.r.s. $\{X_n\}$ if there exists an integer $n_0 \ge 0$ such that, for $n \ge n_0$, one has relation (17.2.1) for $\omega \in A_n$, the function g being common for all n.

We will be mainly interested in "positive" renovating events, i.e. renovating events having positive probabilities $\mathbf{P}(A_n) > 0$.

The simplest example of a renovating event is the hitting by the sequence X_n of a fixed point x_0 : $A_n = \{X_n = x_0\}$ (here $m = 0$), although such an event could be of zero probability. Below we will consider a more interesting example.

The motivation behind the introduction of renovating events is as follows. After the trajectory $\{X_k, \xi_k\}$, $k \le n+m$, has entered a renovating set $A \in \mathfrak{F}^{\xi}_{n+m}$, the future evolution of the process will not depend on the values $\{X_k\}$, $k \le n+m$, but will be determined by the values of $\xi_k, \xi_{k+1}, \ldots$ only. It is not a complete "regeneration" of the process which we dealt with in Chap. 13 while studying Markov chains (first of all, because the ξ_k are now, generally speaking, dependent), but it still enables us to establish ergodicity of the sequence X_n (in approximately the same sense as in Chap. 13).

Note that, generally speaking, the event A and hence the function g may depend on the initial value X_0. If X_0 is random then a renovating event is to be taken from the σ-algebra $\mathfrak{F}^{\xi}_{n+m} \times \sigma(X_0)$.

In what follows it will be assumed that the sequence $\{\xi_n\}$ *is stationary*. The symbol U will denote the measure preserving shift transformation of $\mathfrak{F}^{\xi}$-measurable random variables generated by $\{\xi_n\}$, so that $U\xi_n = \xi_{n+1}$, and the symbol T will denote the shift transformation of sets (events) from the σ-algebra $\mathfrak{F}^{\xi}$: $\xi_{n+1}(\omega) = \xi_n(T\omega)$. The symbols U^n and T^n, $n \ge 0$, will denote the powers (iterations) of these transformations respectively (so that $U^1 = U$, $T^1 = T$; U^0 and T^0 are identity transformations), while U^{-n} and T^{-n} are transformations inverse to U^n and T^n, respectively.

A sequence of events $\{A_k\}$ is said to be *stationary* if $A_k = T^k A_0$ for all k.

Example 17.2.1 Consider a real-valued sequence

$$X_{n+1} = (X_n + \xi_n)^+, \quad X_0 = \text{const} \ge 0, \quad n \ge 0, \qquad (17.2.2)$$

where $x^+ = \max(0, x)$ and $\{\xi_n\}$ is a stationary metric transitive sequence. As we already know from Sect. 12.4, the sequence $\{X_n\}$ describes the dynamics of waiting times for customers in a single-channel service system. The difference is that in Sect. 12.4 the initial value has subscript 1 rather than 0, and that now the sequence $\{\xi_n\}$ has a more general nature. Furthermore, it was established in Sect. 12.4 that Eq. (17.2.2) has the solution

$$X_{n+1} = \max(\overline{S}_{n,n}, X_0 + S_n), \tag{17.2.3}$$

where

$$S_n := \sum_{k=0}^{n} \xi_k, \qquad \overline{S}_{n,k} := \max_{-1 \le j \le k} S_{n,j}, \qquad S_{n,j} := \sum_{k=n-j}^{n} \xi_k, \qquad S_{n,-1} := 0 \tag{17.2.4}$$

(certain changes in the subscripts in comparison to (17.2.4) are caused by different indexing of the initial values). From representation (17.2.3) one can see that the event

$$B_n := \{X_0 + S_n \le 0, \overline{S}_{n,n} = 0\} \in \mathfrak{F}_n^{\xi}$$

implies the event $\{X_{n+1} = 0\}$ and so is renovating for $m = 0$, $g(y) \equiv 0$. If $X_{n+1} = 0$ then

$$X_{n+2} = g_1(\xi_n, \xi_{n+1}) := \xi_{n+1}^+, \quad X_{n+3} = g_2(\xi_n, \xi_{n+1}, \xi_{n+2}) := \left(\xi_{n+1}^+ + \xi_{n+2}\right)^+,$$

and so on do not depend on X_0.

Now consider, for some $n_0 > 1$ and any $n \ge n_0$, the narrower event

$$A_n := \Big\{X_0 + \sup_{j \ge n_0} S_{n,j} \le 0,\ \overline{S}_{n,\infty} := \sup_{j \ge -1} S_{n,j} = 0\Big\}$$

(we assume that the sequence $\{\xi_n\}$ is defined on the whole axis). Clearly, $A_n \subset B_n \subset \{X_{n+1} = 0\}$, so A_n is a renovating event as well. But, unlike B_n, the renovating event A_n is *stationary*: $A_n = T^n A_0$.

We assume now that $\mathbf{E}\xi_0 < 0$ and show that in this case $\mathbf{P}(A_0) > 0$ for sufficiently large n_0. In order to do this, we first establish that $\mathbf{P}(\overline{S}_{0,\infty} = 0) > 0$. Since, by the ergodic theorem, $S_{0,j} \xrightarrow{a.s.} -\infty$ as $j \to \infty$, we see that $\overline{S}_{0,\infty}$ is a proper random variable and there exists a v such that $\mathbf{P}(\overline{S}_{0,\infty} < v) > 0$. By the total probability formula,

$$0 < \mathbf{P}(\overline{S}_{0,\infty} < v) = \sum_{j=0}^{\infty} \mathbf{P}\Big(\overline{S}_{0,j-1} < S_{0,j} < v,\ \sup_{k \ge j}(S_{0,k} - S_{0,j}) = 0\Big).$$

Therefore there exists a j such that

$$\mathbf{P}\Big(\sup_{k \ge j}(S_{0,k} - S_{0,j}) = 0\Big) > 0.$$

But the supremum in the last expression has the same distribution as $\overline{S}_{0,\infty}$. This proves that $p := \mathbf{P}(\overline{S}_{0,\infty} = 0) > 0$. Next, since $S_{0,j} \xrightarrow{a.s.} -\infty$, one also has $\sup_{j\ge k} S_{0,j} \xrightarrow{a.s.} -\infty$ as $k \to \infty$. Therefore, $\mathbf{P}(\sup_{j\ge k} S_{0,j} < -X_0) \to 1$ as $k \to \infty$, and hence there exists an n_0 such that

$$\mathbf{P}\Bigl(\sup_{j\ge n_0} S_{0,j} < -X_0\Bigr) > 1 - \frac{p}{2}.$$

Since $\mathbf{P}(AB) \ge \mathbf{P}(A) + \mathbf{P}(B) - 1$ for any events A and B, the aforesaid implies that $\mathbf{P}(A_0) \ge p/2 > 0$.

In the assertions below, we will use the existence of stationary renovating events A_n with $\mathbf{P}(A_0) > 0$ as a condition insuring convergence of the s.r.s. X_n to a stationary sequence. However, in the last example such convergence can be established directly. Let $\mathbf{E}\xi_0 < 0$. Then by (17.2.3), for any fixed v,

$$\mathbf{P}(X_{n+1} > v) = \mathbf{P}(\overline{S}_{n,n} > v) + \mathbf{P}(\overline{S}_{n,n} \le v,\, X_0 + S_n > v),$$

where evidently

$$\mathbf{P}(X_0 + S_n > v) \to 0, \quad \mathbf{P}(\overline{S}_{n,n} > v) \uparrow \mathbf{P}(\overline{S}_{0,\infty} > v)$$

as $n \to \infty$. Hence the following limit exists

$$\lim_{n\to\infty} \mathbf{P}(X_n > v) = \mathbf{P}(\overline{S}_{0,\infty} > v). \tag{17.2.5}$$

Recall that in the above example the sequence of events A_n becomes renovating for $n \ge n_0$. But we can define other renovating events C_n along with a number m and function $g : \mathbb{R}^{m+1} \to \mathbb{R}$ as follows:

$$m := n_0, \qquad C_n := T^m A_n, \qquad g(y_0, \dots, y_m) := 0.$$

The events $C_n \in \mathfrak{F}^{\xi}_{n+m}$ are renovating for $\{X_n\}$ on the segment $[n, n+m]$ *for all* $n \ge 0$, so in this case the n_0 in the definition of a renovating sequence will be equal to 0.

A similar argument can also be used in the general case for arbitrary renovating events. Therefore we will assume in the sequel that the number n_0 from the definition of renovating events is equal to zero.

In the general case, the following assertion is valid.

Theorem 17.2.1 *Let $\{\xi_n\}$ be an arbitrary stationary sequence and for the s.r.s. $\{X_n\}$ there exists a sequence of renovating events $\{A_n\}$ such that*

$$\mathbf{P}\Biggl(\bigcup_{j=1}^{n} A_j T^{-s} A_{j+s}\Biggr) \to 1 \quad \text{as } n \to \infty \tag{17.2.6}$$

uniformly in $s \ge 1$. Then one can define, on a common probability space with $\{X_n\}$, a stationary sequence $\{X^n := U^n X^0\}$ satisfying the equations $X^{n+1} = f(X^n, \xi_n)$ and such that

$$\mathbf{P}\{X_k = X^k \text{ for all } k \ge n\} \to 1 \quad \text{as } n \to \infty. \tag{17.2.7}$$

If the sequence $\{\xi_n\}$ is metric transitive and the events A_n are stationary, then the relations $\mathbf{P}(A_0) > 0$ and $\mathbf{P}(\bigcup_{n=0}^{\infty} A_n) = 1$ are equivalent and imply (17.2.6) *and* (17.2.7).

Note also that if we introduce the measure $\pi(B) = \mathbf{P}(X^0 \in B)$ (as we did in Chap. 13), then (17.2.7) will imply convergence in total variation:

$$\sup_{B \in \mathfrak{B}_{\mathcal{X}}} \left|\mathbf{P}(X_n \in B) - \pi(B)\right| \to 0 \quad \text{as } n \to \infty.$$

Proof of Theorem 17.2.1 First we show that (17.2.6) implies that

$$\mathbf{P}\left(\bigcap_{k=0}^{\infty} \{X_{n+k} \ne U^{-s} X_{n+k+s}\}\right) \to 0 \quad \text{as } n \to \infty \tag{17.2.8}$$

uniformly in $s \ge 0$. For a fixed $s \ge 1$, consider the sequence $X_j^s = U^{-s} X_{s+j}$. It is defined for $j \ge -s$, and

$$X_{-s}^s = X_0, \qquad X_{-s+1}^s = f\left(X_{-s}^s, \xi_{-s}\right) = f(X_0, \xi_{-s})$$

and so on. It is clear that the event

$$\left\{X_j = X_j^s \text{ for some } j \in [0, n]\right\}$$

implies the event

$$\left\{X + n + k = X_{n+k}^s \text{ for all } k \ge 0\right\}.$$

We show that

$$\mathbf{P}\left(\bigcup_{j=1}^{n} \{X_j = X_j^s\}\right) \to 1 \quad \text{as } n \to \infty.$$

For simplicity's sake put $m = 0$. Then, for the event $X_{j+1} = X_{j+1}^s$ to occur, it suffices that the events A_j and $T^{-s} A_{j+s}$ occur simultaneously. In other words,

$$\bigcup_{j=0}^{n-1} A_j T^{-s} A_{j+s} \subset \bigcup_{j=1}^{n} \{X_j = X_j^s\} \subset \bigcap_{k=0}^{\infty} \{X_{n+k} = X_{n+k}^s\}.$$

Therefore (17.2.6) implies (17.2.8) and convergence

$$\mathbf{P}\left(X_k^n \ne X_k^{n+s}\right) \to 0 \quad \text{as } n \to \infty$$

uniformly in $k \ge 0$ and $s \ge 0$. If we introduce the metric ρ putting $\rho(x, y) := 1$ for $x \ne y$, $\rho(x, x) = 0$, then the aforesaid means that, for any $\delta > 0$, there exists an N such that

$$\mathbf{P}\big(\rho\big(X_k^n, X_k^{n+s}\big) > \delta\big) = \mathbf{P}\big(\rho\big(X_k^n, X_k^{n+s}\big) \ne 0\big) < \delta$$

for $n \ge N$ and any $k \ge 0$, $s \ge 0$, i.e. X_k^n is a Cauchy sequence with respect to convergence in probability for each k. Because any space $\mathcal{X}$ is complete with such a metric, there exists a random variable X^k such that $X_k^n \xrightarrow{p} X^k$ as $n \to \infty$ (see Lemma 4.2). Due to the specific nature of the metric ρ this means that

$$\mathbf{P}\big(X_k^n \ne X^k\big) \to 0 \quad \text{as } n \to \infty. \tag{17.2.9}$$

The sequence X^k is stationary. Indeed, as $n \to \infty$,

$$\mathbf{P}\big(X^{k+1} \ne U X^k\big) = \mathbf{P}\big(X_{k+1}^n \ne U X_k^n\big) + o(1) = \mathbf{P}\big(X_{k+1}^n \ne X_{k+1}^{n-1}\big) + o(1) = o(1).$$

Since the probability $\mathbf{P}(X^{k+1} \ne U X^k)$ does not depend on n, $X^{k+1} = U X^k$ a.s.

Further, $X_{n+k+1} = f(X_{n+k}, \xi_{n+k})$, and therefore

$$X_{k+1}^n = U^{-n} f(X_{n+k}, \xi_{n+k}) = f\big(X_k^n, \xi_k\big). \tag{17.2.10}$$

The left and right-hand sides here converge in probability to X^{k+1} and $f(X^k, \xi_k)$, respectively. This means that $X^{k+1} = f(X^k, \xi_k)$.

To prove convergence (17.2.7) it suffices to note that, by virtue of (17.2.10), the values X_k^n and X^k, after having become equal for some k, will never be different for greater values of k. Therefore, as well as (17.2.9) one has the relation

$$\mathbf{P}\Bigg(\bigcup_{k\ge 0}\big\{X_k^n \ne X^k\big\}\Bigg) = \mathbf{P}\Bigg(\bigcup_{k\ge 0}\big\{X_{k+n} \ne X^{k+n}\big\}\Bigg) \to 0 \quad \text{as } n \to \infty,$$

which is equivalent to (17.2.7).

The last assertion of the theorem follows from Theorem 16.2.5. The theorem is proved. □

Remark 17.2.1 It turns out that condition (17.2.6) is also a necessary one for convergence (17.2.7) (see [6]). For more details on convergence of stochastic recursive sequences and their generalisations, and also on the relationship between (17.2.6) and conditions (I) and (II) from Chap. 13, see [6].

In Example 17.2.1 the sequence X^k was actually found in an explicit form (see (17.2.3) and (17.2.5)):

$$X^k = \overline{S}_{k,-\infty} = \sup_{j\ge 0} S_{k-1}^j. \tag{17.2.11}$$

These random variables are proper by Corollary 16.3.1. It is not hard to also see that, for $X_0 = 0$, one has (see (17.2.3))

$$U^{-1} X_{n+k} \uparrow X^k. \tag{17.2.12}$$

17.2.2 Boundedness of Random Sequences

Consider now conditions of boundedness of an s.r.s. in spaces $\mathcal{X} = [0, \infty)$ and $\mathcal{X} = (-\infty, \infty)$. Assertions about boundedness will be stated in terms of existence of stationary majorants, i.e. stationary sequences M_n such that

$$X_n \le M_n \quad \text{for all } n.$$

Results of this kind will be useful for constructing stationary renovating sequences.

Majorants will be constructed for a class of random sequences more general than stochastic recursive sequences. Namely, we will consider the class of random sequences satisfying the inequalities

$$X_{n+1} \le \bigl(X_n + h(X_n, \xi_n)\bigr)^+, \tag{17.2.13}$$

where the measurable function h will in turn be bounded by rather simple functions of X_n and ξ_n. The sequence $\{\xi_n\}$ will be assumed given on the whole axis.

Theorem 17.2.2 *Assume that there exist a number $N > 0$ and a measurable function g_1 with $\mathbf{E} g_1(\xi_n) < 0$ such that* (17.2.13) *holds with*

$$h(x, y) \le \begin{cases} g_1(y) & \text{for } x > N, \\ g_1(y) + N - x & \text{for } x \le N. \end{cases} \tag{17.2.14}$$

If $X_0 \le M < \infty$, then the stationary sequence

$$M_n = \max(M, N) + \sup_{j \ge -1} S_{n-1,j}, \tag{17.2.15}$$

where $S_{n,-1} = 0$ and $S_{k,j} = g_1(\xi_k) + \cdots + g_1(\xi_{k-j})$ for $j \ge 0$, is a majorant for X_n.

Proof For brevity's sake, put $\zeta_i := g_1(\xi_i)$, $Z := \max(M, N)$, and $Z_n := X_n - Z$. Then Z_n will satisfy the following inequalities:

$$Z_{n+1} \le \begin{cases} (Z_n + Z + \zeta_n)^+ - Z \le (Z_n + \zeta_n)^+ & \text{for } Z_n > N - Z, \\ (N + \zeta_n)^+ - Z \le \zeta_n^+ & \text{for } Z_n \le N - Z. \end{cases}$$

Consider now a sequence $\{Y_n\}$ defined by the relations $Y_0 = 0$ and

$$Y_{n+1} = (Y_n + \zeta_n)^+.$$

Assume that $Z_n \le Y_n$. If $Z_n > N - Z$ then

$$Z_{n+1} \le (Z_n + \zeta_n)^+ \le (Y_n + \zeta_n)^+ = Y_{n+1}.$$

If $Z_n \le N - Z$ then

$$Z_{n+1} \le \zeta_n^+ \le (Y_n + \zeta_n)^+ = Y_{n+1}.$$

Because $Z_0 \le 0 = Y_0$, it is evident that $Z_n \le Y_n$ for all n. But we know the solution of the equation for Y_n and, by virtue of (17.2.11) and (17.2.13),

$$X_n - Z \le \sup_{j \ge -1} S_{n-1,j}.$$

The theorem is proved. □

Theorem 17.2.2A *Assume that there exist a number $N > 0$ and measurable functions g_1 and g_2 such that*

$$\mathbf{E} g_1(\xi_n) < 0, \qquad \mathbf{E} g_2(\xi_n) < 0 \tag{17.2.16}$$

and

$$h(x,y) \le \begin{cases} g_1(y) & \text{for } x > N, \\ g_1(y) + g_2(y) & \text{for } x \le N. \end{cases} \tag{17.2.17}$$

If $Z_0 \le M < \infty$, then the conditions of Theorem 17.2.2 *are satisfied (possibly for other N and g_1) and for X_n there exists a stationary majorant of the form* (17.2.15).

Proof We set $g := -\mathbf{E} g_1(\xi_n) > 0$ and find $L > 0$ such that $\mathbf{E}(g_2(\xi_n);\ g_2(\xi_n) > L) \le g/2$. Introduce the function

$$g_1^*(y) := g_1(y) + g_2(y) \mathrm{I}\big(g_2(y) > L\big).$$

Then $\mathbf{E} g_1^*(\xi_n) \le -g/2 < 0$ and

$$\begin{aligned} h(x,y) &\le g_1(y) + g_2(y)\mathrm{I}(x \le N) \\ &\le g_1^*(y) + g_2(y)\mathrm{I}(x \le N) - g_2(y)\mathrm{I}\big(g_2(y) > L\big) \\ &\le g_1^*(y) + L\mathrm{I}(x \le N) \le g_1^*(y) + (L + N - x)\mathrm{I}(x \le N) \\ &\le g_1^*(y) + (L + N - x)\mathrm{I}(x \le L + N). \end{aligned}$$

This means that inequalities (17.2.14) hold with N replaced with $N^* = N + L$. The theorem is proved. □

Note again that in Theorems 17.2.2 and 17.2.2A we did not assume that $\{X_n\}$ is an s.r.s.

The reader will notice the similarity of the conditions of Theorems 17.2.2 and 17.2.2A to the boundedness condition in Sect. 15.5, Theorem 13.7.3 and Corollary 13.7.1.

The form of the assertions of Theorems 17.2.2 and 17.2.2A enables one to construct stationary renovating events for a rather wide class of nonnegative stochastic recursive sequences (so that $\mathcal{X} = [0, \infty)$) having, say, a "positive atom" at 0. It is convenient to write such sequences in the form

$$X_{n+1} = \big(X_n + h(X_n, \xi_n)\big)^+. \tag{17.2.18}$$

Example 17.2.2 Let an s.r.s. (see (17.1.1)) be described by Eq. (17.2.18) and satisfy conditions (17.2.14) or (17.2.17), where the function h is sufficiently "regular" to ensure that

$$B_{n,T} = \bigcap_{t \le T} \big\{h(t, \xi_n) \le -t\big\}$$

is an event for any T. (For instance, it is enough to require $h(t, v)$ to have at most a countable set of discontinuity points t. Then the set $B_{n,T}$ can be expressed as the intersection of countably many events $\bigcap_k \{h(t_k, \xi_n) \le -t_k\}$, where $\{t_k\}$ form a countable set dense on $[0, T]$.) Furthermore, let there exist an $L > 0$ such that

$$\mathbf{P}(M_n < L, B_{n,L}) > 0 \tag{17.2.19}$$

(M_n was defined in (17.2.15)). Then the event $A_n = \{M_n < L\} B_{n,L}$ is clearly a positive stationary renovating event with the function $g(y) = (h(0, y))^+$, $m = 0$. (On the set $A_n \in \mathfrak{F}_n^{\xi}$ we have $X_{n+1} = 0$, $X_{n+2} = h(0, \xi_{n+1})^+$ and so on.) Therefore, an s.r.s. satisfying (17.2.18) satisfies the conditions of Theorem 17.2.1 and is ergodic in the sense of assertion (17.2.7).

It can happen that, from a point $t \le L$, it would be impossible to reach the point 0 in one step, but it could be done in $m > 1$ steps. If B is the set of sequences $(\xi_n, \dots, \xi_{n+m})$ that effect such a transition, and $\mathbf{P}(M_n < L)$, then $A_n = \{M_n < L\} B$ will also be stationary renovating events.

17.3 Ergodicity Conditions Related to the Monotonicity of f

Now we consider ergodicity conditions for stochastic recursive sequences that are related to the analytic properties of the function f from (17.1.1). As we already noted, the sequence $f(x, \xi_k)$, $k = 1, 2, \dots$, may be considered as a sequence of random transformations of the space $\mathcal{X}$. Relation (17.1.1) shows that X_{n+1} is the result of the application of $n + 1$ random transformations $f(\cdot, \xi_k)$, $k = 0, 1, \dots, n$, to the initial value $X_0 = x \in \mathcal{X}$. Denoting by ξ_n^{n+k} the vector $\xi_n^{n+k} = (\xi_n, \dots, \xi_{n+k})$

and by f_k the k-th iteration of the function f: $f_1(x, y_1) = f(x, y_1)$, $f_2(x, y_1, y_2) = f(f(x, y_1), y_2)$ and so on, we can re-write (17.1.1) for $X_0 = x$ in the form

$$X_{n+1} = X_{n+1}(x) = f_{n+1}\big(x, \xi_0^n\big),$$

so that the "forward" and "backward" equations hold true:

$$f_{n+1}\big(x, \xi_0^n\big) = f\big(f_n\big(x, \xi_0^{n-1}\big), \xi_n\big) = f_n\big(f(x, \xi_0), \xi_1^n\big). \qquad (17.3.1)$$

In the present section we will be studying stochastic recursive sequences for which the function f from representation (17.1.1) is monotone in the first argument. To this end, we need to assume that a partial order relation "$\geq$" is defined in the space $\mathcal{X}$. In the space $\mathcal{X} = \mathbb{R}^d$ of vectors $x = (x(1), \dots, x(d))$ (or its subspaces) the order relation can be introduced in a natural way by putting $x_1 \geq x_2$ if $x_1(k) \geq x_2(k)$ for all k.

Furthermore, we will assume that, for each non-decreasing sequence $x_1 \leq x_2 \leq \dots \leq x_n \leq \dots$, there exists a limit $x \in \mathcal{X}$, i.e. the smallest element $x \in \mathcal{X}$ for which $x_k \leq x$ for all k. In that case we will write $x_k \uparrow x$ or $\lim_{k\to\infty} x_k = x$. In $\mathcal{X} = \mathbb{R}^d$ such convergence will mean conventional convergence. To facilitate this, we will need to complete the space $\mathbb{R}^d$ by adding points with infinite components.

Theorem 17.3.1 (Loynes) *Suppose that the transformation $f = f(x, y)$ and space $\mathcal{X}$ satisfy the following conditions*:

(1) *there exists an $x_0 \in \mathcal{X}$ such that $f(x_0, y) \geq x_0$ for all $y \in \mathcal{Y}$*;
(2) *the function f is monotone in the first argument*: $f(x_1, y) \geq f(x_2, y)$ *if* $x_1 \geq x_2$;
(3) *the function f is continuous in the first argument with respect to the above convergence*: $f(x_n, y) \uparrow f(x, y)$ *if* $x_n \uparrow x$.

Then there exists a stationary random sequence $\{X_n\}$ satisfying Eq. (17.1.1): $X^{n+1} = UX^n = f(X^n, \xi_n)$, *such that*

$$U^{-n} X_{n+s}(x)) \uparrow X^s \quad \textit{as } n \to \infty, \qquad (17.3.2)$$

where convergence takes place for all elementary outcomes.

Since the distributions of X_n and $U^{-n}X_n$ coincide, in the case where convergence of random variables $\eta_n \uparrow \eta$ means convergence (in a certain sense) of their distributions (as is the case when $\mathcal{X} = \mathbb{R}^d$), Theorem 17.2.1 also implies convergence of the distributions of X_n to that of X^0 as $n \to \infty$.

Remark 17.3.1 A substantial drawback of this theorem is that it holds only for a single initial value $X_0 = x_0$. This drawback disappears if the point x_0 is accessible with probability 1 from any $x \in \mathcal{X}$, and ξ_k are independent. In that case x_0 is likely to be a positive atom, and Theorem 13.6.1 for Markov chains is also applicable.

The limiting sequence X^s in (17.3.2) can be "improper" (in spaces $\mathcal{X}=\mathbb{R}^d$ it may assume infinite values). The sequence X^s will be proper if the s.r.s. X_n satisfies, say, the conditions of the theorems of Sect. 15.5 or the conditions of Theorem 17.2.2.

Proof of Theorem 17.3.1 Put

$$v_s^{-k} := f_{k+s}\big(x_0, \xi_{-k}^{s-1}\big) = U^{-k} f_{k+s}\big(x_0, \xi_0^{s+k-1}\big) = U^{-k} X_{k+s}(x_0).$$

Here the superscript $-k$ indicates the number of the element of the driving sequence $\{\xi_n\}_{n=-\infty}^{\infty}$ such that the elements of this sequence starting from that number are used for constructing the s.r.s. The subscript s is the "time epoch" at which we observe the value of the s.r.s. From the "backward" equation in (17.3.1) we get that

$$v_s^{-k-1} = f_{k+s}\big(f(x_0, \xi_{-k-1}), \xi_{-k}^{s-1}\big) \ge f_{k+s}\big(x_0, \xi_{-k}^{s-1}\big) = v_s^{-k}.$$

This means that the sequence v_s^{-k} increases as k grows, and therefore there exists a random variable $X^s \in \mathcal{X}$ such that

$$v_s^{-k} = U^{-k} X_{k+s}(x_0) \uparrow X^s \quad \text{as } k \to \infty.$$

Further, v_s^{-k} is a function of ξ_{-k}^{s-1}. Therefore, X^s is a function of $\xi_{-\infty}^{s-1}$:

$$X^s = G\big(\xi_{-\infty}^{s-1}\big).$$

Hence

$$U X^s = U G\big(\xi_{-\infty}^{s-1}\big) = G\big(\xi_{-\infty}^{s}\big) = X^{s+1},$$

which means that $\{X^s\}$ is stationary. Using the "forward" equation from (17.3.1), we obtain that

$$v_s^{-k-1} = f\big(f_{k+s}\big(x_0, \xi_{-k-1}^{s-2}\big), \xi_{s-1}\big) = f\big(v_{s-1}^{-k-1}, \xi_{s-1}\big).$$

Passing to the limit as $k \to \infty$ gives, since f is continuous, that

$$X^s = f\big(X^{s-1}, \xi_{s-1}\big).$$

The theorem is proved. □

Example 17.2.1 clearly satisfies all the conditions of Theorem 17.3.1 with $\mathcal{X} = [0, \infty)$, $x_0 = 0$, and $f(x, y) = (x + y)^+$.

17.4 Ergodicity Conditions for Contracting in Mean Lipschitz Transformations

In this section we will assume that $\mathcal{X}$ is a complete separable metric space with metric ρ. Consider the following conditions on the iterations $X_k(x) = f_k(x, \xi_0^{k-1})$.

Condition (B) (boundedness). *For some $x_0 \in \mathcal{X}$ and any $\delta > 0$, there exists an $N = N_\delta$ such that, for all $n \ge 1$,*

$$\mathbf{P}\big(\rho\big(x_0, X_n(x_0) > N\big)\big) = \mathbf{P}\big(\rho\big(x_0, f_n\big(x_0, \xi_0^{n-1}\big)\big) > N\big) < \delta.$$

It is not hard to see that condition **(B)** holds (possibly with a different N) as soon as we can establish that, for some $m \ge 1$, the above inequality holds for all $n \ge m$.

Condition **(B)** is clearly met for stochastic random sequences satisfying the conditions of Theorems 17.2.2 and 17.2.2A or the theorems of Sect. 15.5.

Condition (C) (contraction in mean). *The function f is continuous in the first argument and there exist $m \ge 1$, $\beta > 0$ and a measurable function $q : \mathbb{R}^m \to \mathbb{R}$ such that, for any x_1 and x_2,*

$$\rho\big(f_m\big(x_1, \xi_0^{m-1}\big), f_m\big(x_2, \xi_0^{m-1}\big)\big) \le q\big(\xi_0^{m-1}\big)\rho(x_1, x_2),$$
$$m^{-1}\mathbf{E}\ln q\big(\xi_0^{m-1}\big) \le -\beta < 0.$$

Observe that conditions **(B)** and **(C)** are, generally speaking, not related to each other. Let, for instance, $\mathcal{X} = \mathbb{R}$, $X_0 \ge 0$, $\xi_n \ge 0$, $\rho(x, y) = |x - y|$, and $f(x, y) = bx + y$, so that

$$X_{n+1} = bX_n + \xi_n.$$

Then condition **(C)** is clearly satisfied for $0 < b < 1$, since

$$\big|f(x_1, y) - f(x_2, y)\big| = b|x_1 - x_2|.$$

At the same time, condition **(B)** will be satisfied if and only if $\mathbf{E}\ln\xi_0 < \infty$. Indeed, if $\mathbf{E}\ln\xi_0 = \infty$, then the event $\{\ln\xi_k > -2k\ln b\}$ occurs infinitely often a.s. But X_{n+1} has the same distribution as

$$b^{n+1}X_0 + \sum_{k=0}^{n} b^k\xi_k = b^{n+1}X_0 + \sum_{k=0}^{n} \exp\{k\ln b + \ln\xi_k\},$$

where, in the sum on the right-hand side, the number of terms exceeding $\exp\{-k\ln b\}$ increases unboundedly as n grows. This means that $X(n+1) \xrightarrow{p} \infty$ as $n \to \infty$. The case $\mathbf{E}\ln\xi_0 < \infty$ is treated in a similar way. The fact that **(B)**, generally speaking, does not imply **(C)** is obvious.

As before, we will assume that the "driving" stationary sequence $\{\xi_n\}_{n=-\infty}^{\infty}$ is given on the whole axis. Denote by U the respective distribution preserving shift operator.

Convergence in probability and a.s. of a sequence of $\mathcal{X}$-valued random variables $\eta_n \in \mathcal{X}$ ($\eta_n \xrightarrow{p} \eta$, $\eta_n \xrightarrow{a.s.} \eta$) is defined in the natural way by the relations $\mathbf{P}(\rho(\eta_n, \eta) > \delta) \to 0$ as $n \to \infty$ and $\mathbf{P}(\rho(\eta_k, \eta) > \delta \textit{ for some } k \ge n) \to 0$ as $n \to \infty$ for any $\delta > 0$, respectively.

Theorem 17.4.1 *Assume that conditions* **(B)** *and* **(C)** *are met. Then there exists a stationary sequence* $\{X_n\}$ *satisfying* (17.1.1):

$$X^{n+1} = UX^n = f(X^n, \xi_n)$$

such that, for any fixed x,

$$U^{-n}X_{n+s}(x) \xrightarrow{a.s.} X^s \quad as \quad n \to \infty. \tag{17.4.1}$$

This convergence is uniform in x *over any bounded subset of* $\mathcal{X}$.

Theorem 17.2.2 implies the weak convergence, as $n \to \infty$, of the distributions of $X_n(x)$ to that of X^0. Condition **(B)** is clearly necessary for ergodicity. As the example of a generalised autoregressive process below shows, condition **(C)** is also necessary in some cases.

Set $Y_n := U^n X_n(x_0)$, where x_0 is from condition **(B)**. We will need the following auxiliary result.

Lemma 17.4.1 *Assume that conditions* **(B)** *and* **(C)** *are met and the stationary sequence* $\{q(\xi_{km}^{km+m-1})\}_{k=-\infty}^{\infty}$ *is ergodic. Then, for any* $\delta > 0$, *there exists an* n_δ *such that, for all* $k \geq 0$,

$$\sup_{k\geq 0} \mathbf{P}\big(\rho(Y_{n+k}, Y_n) < \delta \textit{ for all } n \geq n_\delta\big) \geq 1 - \delta. \tag{17.4.2}$$

For ergodicity of $\{q(\xi^{km+m-1})\}_{k=-\infty}^{\infty}$ *it suffices that the transformation* T^m *is metric transitive.*

The lemma means that, with probability 1, the distance $\rho(Y_{n+k}, Y_n)$ tends to zero uniformly in k as $n \to \infty$. Relation (17.4.2) can also be written as $\mathbf{P}(A_\delta) \leq \delta$, where

$$A_\delta := \bigcup_{n\geq n_\delta} \big\{\rho(Y_{n+k}, Y_n) \geq \delta\big\}.$$

Proof of Lemma 17.4.1 By virtue of condition **(B)**, there exists an $N = N_\delta$ such that, for all $k \geq 1$,

$$\mathbf{P}\big(\rho\big(x_0, X_k(x_0)\big) > N\big) \leq \frac{\delta}{4}.$$

Hence

$$\mathbf{P}(A_\delta) \leq \delta/3 + \mathbf{P}\big(A_\delta;\ \rho(x_0, \theta_{n,k}) \leq N\big).$$

The random variable $\theta_{n,k} := U^{-n-k}X_k(x_0)$ has the same distribution as $X_k(x_0)$. Next, by virtue of **(C)**,

$$\rho(Y_{n+k}, Y_n) \leq \rho\big(f_{n+k}\big(x_0, \xi_{-n-k}^{-1}\big), f_n\big(x_0, \xi_{-n}^{-1}\big)\big)$$

$$\le q\big(\xi_{-m}^{-1}\big)\rho\big(f_{n+k-m}\big(x_0,\xi_{-n-k}^{-m-1}\big),f_{n-m}\big(x_0,\xi_{-n}^{-m-1}\big)\big)$$
$$= q\big(\xi_{-m}^{-1}\big)\rho\big(U^{-n-k}X_{n+k-m}(x_0),U^{-n}X_{n-m}(x_0)\big). \qquad (17.4.3)$$

Denote by B_s the set of numbers n of the form $n = lm + s$, $l = 0, 1, 2, \dots$, $0 \le s < m$, and put

$$\lambda_j := \ln q\big(\xi_{-jm}^{-jm+m-1}\big), \quad j = 1, 2, \dots.$$

Then, for $n \in B_s$, we obtain from (17.4.3) and similar relations that

$$\rho(Y_{n+k},Y_n) \le \exp\Bigg\{\sum_{j=1}^{l}\lambda_j\Bigg\}\rho\big(U^{-n-k}X_{k+s}(x_0),U^{-n}X_s(x_0)\big), \qquad (17.4.4)$$

where the last factor (denote it just by ρ) is bounded from above:

$$\rho \le \rho\big(x_0,U^{-n-k}X_{k+s}(x_0)\big) + \rho\big(x_0,U^{-n}X_s(x_0)\big).$$

The random variables $U^{-n}X_j(x_0)$ have the same distribution as $X_j(x_0)$. By virtue of **(B)**, there exists an $N = N_\delta$ such that, for all $j \ge 1$,

$$\mathbf{P}\big(\rho\big(x_0,X_j(x_0)\big) > N\big) \le \frac{\delta}{4m}.$$

Hence, for all n, k and s, we have $\mathbf{P}(\rho > 2N) < \delta/(2m)$, and the right-hand side of (17.4.4) does not exceed $2N\exp\{\sum_{j=1}^{l}\lambda_j\}$ on the complement set $\{\rho \le 2N\}$.

Because $\mathbf{E}\lambda_j \le -m\beta < 0$ and the sequence $\{\lambda_j\}$ is metric transitive, by the ergodic Theorem 16.3.1 we have

$$\sum_{j=1}^{l}\lambda_j < -m\beta l/2$$

for all $l \ge l(\omega)$, where $l(\omega)$ is a proper random variable. Choose l_1 and l_2 so that the inequalities

$$-\frac{m\beta l_1}{2} < \ln\delta - \ln 2N, \qquad \mathbf{P}\big(l(\omega) > l_2\big) < \frac{\delta}{2}$$

hold. Then, putting

$$l_\delta := \max(l_1,l_2), \qquad n_\delta := ml_\delta, \qquad A_\delta^s := \bigcup_{n\ge n_\delta,\ n\in B_s}\big\{\rho(Y_{n+k},Y_n) \ge \delta\big\},$$

we obtain that

$$\mathbf{P}\big(A_\delta^s\big) \le \mathbf{P}(\rho > 2N) + \mathbf{P}\big(A_\delta^s;\rho \le N\big) \le \frac{\delta}{2m} + \mathbf{P}\Bigg(\bigcup_{l\ge l_\delta}\Bigg\{2N\exp\Bigg\{-\sum_{j=0}^{l}\lambda_j\Bigg\} \ge \delta\Bigg\}\Bigg).$$

But the intersection of the events from the term with $\{l_\delta \ge l(\omega)\}$ is empty. Therefore, the former event is a subset of the event $\{l(\omega) > l_\delta\}$, and

$$\mathbf{P}\big(A_\delta^s\big) \le \frac{\delta}{m}, \quad \mathbf{P}(A_\delta) \le \sum_{s=0}^{m-1} \mathbf{P}\big(A_\delta^s\big) \le \delta.$$

The lemma is proved. □

Lemma 17.4.2 (Completeness of $\mathcal{X}$ with respect to convergence in probability) *Let $\mathcal{X}$ be a complete metric space. If a sequence of $\mathcal{X}$-valued random elements η_n is such that, for any $\delta > 0$,*

$$P_n := \sup_{k\ge 0} \mathbf{P}\big(\rho(\eta_{n+k}, \eta_n) > \delta\big) \to 0$$

as $n \to \infty$, then there exists a random element $\eta \in \mathcal{X}$ such that $\eta \xrightarrow{p} \eta$ (that is, $\mathbf{P}(\rho(\eta_n, \eta) > \delta) \to 0$ as $n \to \infty$).

Proof For given ε and δ choose n_k, $k = 0, 1, \ldots$, such that

$$\sup_s \mathbf{P}\big(\rho(\eta_{n_k+s}, \eta_{n_k}) > 2^{-k}\delta\big) < \varepsilon 2^{-k},$$

and, for the sake of brevity, put $\zeta_k := \eta_{n_k}$. Consider the set

$$D := \bigcap_{k=0}^{\infty} D_k, \quad D_k := \big\{\omega\, \rho(\zeta_{k+1}, \zeta_k) \le 2^{-k}\delta\big\}.$$

Then $\mathbf{P}(D) > 1 - 2\varepsilon$ and, for any $\omega \in D$, one has $\rho(\zeta_{k+s}(\omega), \zeta_k(\omega)) < \delta 2^{k-1}$ for all $s \ge 1$. Hence $\zeta_k(\omega)$ is a Cauchy sequence in $\mathcal{X}$ and there exists an $\eta = \eta(\omega) \in \mathcal{X}$ such that $\zeta_k(\omega) \to \eta(\omega)$. Since ε is arbitrary, this means that $\zeta_k \xrightarrow{a.s.} \eta$ as $k \to \infty$, and

$$\begin{aligned}\mathbf{P}\big(\rho(\zeta_0, \eta) > 2\delta\big) &\le \mathbf{P}\left(\bigcup_{k=0}^{\infty} \rho(\zeta_{k+1}, \zeta_k) > 2^{-k}\delta\right)\\ &\le \sum_{k=0}^{\infty} \mathbf{P}\big(\rho(\zeta_{k+1}, \zeta_k) > 2^{-k}\delta\big) \le 2\varepsilon.\end{aligned}$$

Therefore, for any $n \ge n_0$,

$$\mathbf{P}\big(\rho(\eta_n, \eta) > 3\delta\big) \le \mathbf{P}\big(\rho(\eta_n, \eta_{n_0}) > \delta\big) + \mathbf{P}\big(\rho(\zeta_0, \eta) > 2\delta\big) \le 3\varepsilon.$$

Since ε and δ are arbitrary, the lemma is proved. □

Proof of Theorem 17.4.1 From Lemma 17.4.1 it follows that

$$\sup_k \mathbf{P}\big(\rho(Y_{n+k}, Y_n) > \delta\big) \to 0 \quad \text{as } n \to \infty.$$

This means that Y_n is a Cauchy sequence with respect to convergence in probability, and by Lemma 17.4.2 there exists a random variable X^0 such that

$$\begin{aligned} &Y_n \xrightarrow{p} X^0, \\ &U^{-n}X_{n+s}(x_0) = U^s\big(U^{-n-s}X_{n+s}(x_0)\big) = U^sY_{n+s} \to U^sX^0 \equiv X^s. \end{aligned} \tag{17.4.5}$$

By continuity of f,

$$\begin{aligned} U^{-n}X_{n+s+1}(x_0) &= U^{-n}f\big(X_{n+s}(x_0), \xi_{n+s}\big) \\ &= f\big(U^{-n}X_{n+s}(x_0), \xi_s\big) \xrightarrow{p} f\big(X^s, \xi_s\big) = X^{s+1}. \end{aligned}$$

We proved the required convergence for a fixed initial value x_0. For an arbitrary $x \in C_n = \{z : \rho(x_0, z) \le N\}$, one has

$$\rho\big(U^{-n}X_n(x), X^0\big) \le \rho\big(U^{-n}X_n(x), U^{-n}X_n(x_0)\big) + \rho\big(U^{-n}X_n(x_0), X^0\big), \tag{17.4.6}$$

where the first term on the right-hand side converges in probability to 0 uniformly in $x \in C_N$. For $n = lm$ this follows from the inequality (see condition **(C)**)

$$\rho\big(U^{-n}X_n(x), U^{-n}X_n(x_0)\big) \le N \exp\left\{\sum_{j=1}^{l} \lambda_j\right\} \tag{17.4.7}$$

and the above argument. Similar relations hold for $n = lm + s$, $m > s > 0$. This, together with (17.4.5) and (17.4.6), implies that

$$U^{-n}X_{n+s}(x) \xrightarrow{p} X^s = U^sX^0$$

uniformly in $x \in C_N$. This proves the assertion of the theorem in regard to convergence in probability.

We now prove convergence with probability 1. To this end, one should repeat the argument proving Lemma 17.4.1, but bounding $\rho(X^0, U^{-n}X_n(x))$ rather than $\rho(Y_{n+k}, Y_n)$. Assuming for simplicity's sake that $s = 0$ (n is a multiple of m), we get (similarly to (17.4.4)) that, for any x,

$$\rho\big(X^0, U^{-n}X_n(x)\big) \le \rho\big(x, U^{-n}X^0\big) \exp\left\{\sum_{j=1}^{l} \lambda_j\right\}. \tag{17.4.8}$$

The rest of the argument of Lemma 17.4.1 remains unchanged. This implies that, for any $\delta > 0$ and sufficiently large n_δ,

$$\mathbf{P}\left(\bigcup_{n \ge n_\delta} \big\{\rho\big(X^0, U^{-n}X_n(x)\big) > \delta\big\}\right) < \delta.$$

Theorem 17.4.1 is proved. □

Example 17.4.1 (Generalised autoregression) Let $\mathcal{X} = \mathbb{R}$. A generalised autoregression process is defined by the relations

$$X_{n+1} = G\big(\zeta_n F(X_n) + \eta_n\big), \qquad (17.4.9)$$

where F and G are functions mapping $\mathbb{R} \mapsto \mathbb{R}$ and $\xi_n = (\zeta_n, \eta_n)$ is a stationary ergodic driving sequence, so that $\{X_n\}$ is an s.r.s. with the function

$$f(x, y) = G\big(y_1, F(x) + y_2\big), \quad y = (y_1, y_2) \in \mathcal{Y} = \mathbb{R}^2.$$

If the functions F and G are nondecreasing and left continuous, $G(x) \geq 0$ for all $x \in \mathbb{R}$, and the elements ζ_n are nonnegative, then the process (17.4.9) satisfies the condition of Theorem 17.3.1, and therefore $U^{-n+s}X_n(0) \uparrow X^s$ with probability 1 (as $n \to \infty$). To establish convergence to a proper stationary sequence X^s, one has to prove uniform boundedness in probability (in n) of the sequence $X_n(0)$ (see below).

Now we will establish under what conditions the sequence (17.4.9) will satisfy the conditions of Theorem 17.4.1. Suppose that the functions F and G satisfy the Lipschitz condition:

$$\big|G(x_1) - G(x_2)\big| \leq c_G|x_1 - x_2|, \qquad \big|F(x_1) - F(x_2)\big| \leq c_F|x_1 - x_2|.$$

Then

$$\big|f(x_1, \xi_0) - f(x_2, \xi_0)\big| \leq c_G\big|\zeta_0\big(F(x_1) - F(x_2)\big)\big| \leq c_F c_G|\zeta_0||x_1 - x_2|. \quad (17.4.10)$$

Theorem 17.4.2 *Under the above assumptions, the sequence* (17.4.9) *will satisfy condition* **(C)** *if*

$$\ln c_G c_F + \mathbf{E} \ln|\zeta_0| < 0. \qquad (17.4.11)$$

The sequence (17.4.9) *will satisfy condition* **(B)** *if* (17.4.11) *holds and, moreover,*

$$\mathbf{E}\big(\ln|\eta_0|\big)^+ < \infty. \qquad (17.4.12)$$

When (17.4.11) *and* (17.4.12) *hold, the sequence* (17.4.9) *has a stationary majorant, i.e. there exists a stationary sequence M_n (depending on X_0) such that $|X_n| \leq M_n$ for all n.*

Proof That condition **(C)** for $\rho(x_1, x_2) = |x_1 - x_2|$ follows from (17.4.10) is obvious. We prove **(B)**. To do this, we will construct a stationary majorant for $|X_n|$. One could do this using Theorems 17.2.2 and 17.2.2A. In our case, it is simpler to prove it directly, making use of the inequalities

$$\big|G(x)\big| \leq \big|G(0)\big| + c_G|x|, \qquad \big|F(x)\big| \leq \big|F(0)\big| + c_F|x|,$$

where we assume, for simplicity's sake, that $G(0)$ and $F(0)$ are finite. Then

$$\begin{aligned}|X_{n+1}| &\le |G(0)| + c_G|\zeta_n|\cdot|F(X_n)| + c_G|\eta_n| \\ &\le |G(0)| + c_G c_F|\zeta_n|\cdot|X_n| + c_G|\zeta_n|\cdot|F(0)| + c_G|\eta_n| = \beta_n|X(n)| + \gamma_n,\end{aligned}$$

where

$$\beta_n := c_G c_F|\zeta_n| \ge 0, \qquad \gamma_n := |G(0)| + c_G|\zeta_n|\cdot|F(0)| + c_G|\eta_n|$$

$$\mathbf{E}\ln\beta_n < 0, \quad \mathbf{E}(\ln\gamma_n)^+ < \infty.$$

From this we get that, for $X_0 = x$,

$$\begin{aligned}|X_{n+1}| &\le |x|\prod_{j=0}^{n}\beta_j + \sum_{l=0}^{n-1}\Bigl(\prod_{j=n-l}^{n}\beta_j\Bigr)\gamma_{n-l-1} + \gamma_n, \\ U^{-n}|X_{n+1}| &\le |x|\prod_{j=-n}^{0}\beta_j + \sum_{l=0}^{\infty}\Bigl(\prod_{j=-l}^{0}\beta_j\Bigr)\gamma_{-l-1} + \gamma_0.\end{aligned} \tag{17.4.13}$$

Put

$$\alpha_i := \ln\beta_j, \qquad S_l := \sum_{j=-l}^{0}\alpha_j.$$

By the strong law of large numbers, there are only finitely many positive values $S_l - al$, where $2a = \mathbf{E}\alpha_j < 0$. Therefore, for all l except for those with $S_l - al > 0$,

$$\prod_{j=-l}^{0}\beta_j < e^{al}.$$

On the other hand, γ_{-l-1} exceeds the level l only finitely often. This means that the series in (17.4.13) (denote it by R) converges with probability 1. Moreover,

$$S = \sup_{k\ge 0} S_k \ge S_n$$

is a proper random variable. As result, we obtain that, for all n,

$$U^{-n}|X_{n+1}| \le |x|e^S + R + \gamma_0,$$

where all the terms on the right-hand side are proper random variables. The required majorant

$$M_n := U^{n-1}\bigl(|x|e^S + R + \gamma_0\bigr)$$

is constructed. This implies that **(B)** is met. The theorem is proved. □

The assertion of Theorem 17.4.2 can be extended to the multivariable case $\mathcal{X} = \mathbb{R}^d$, $d > 1$, as well (see [6]).

Note that conditions (17.4.11) and (17.4.12) are, in a certain sense, necessary not only for convergence $U^{-n+s} X_n(x) \to X^s$, but also for the boundedness of $X_n(x)$ (or of X^0) only. This fact can be best illustrated in the case when $F(t) \equiv G(t) \equiv t$. In that case, $U^{-n} X_{n+s+1}(x)$ and X^{s+1} admit explicit representations

$$U^{-n} X_{n+s+1}(x) = x \prod_{j=-n}^{s} \zeta_j + \sum_{l=0}^{n+s} \prod_{j=s-l}^{s} \zeta_j \eta_{s-l-1} + \eta_s,$$

$$X^{s+1} = \sum_{l=0}^{\infty} \prod_{j=s-l}^{s} \zeta_j \eta_{s-l-1} + \eta_s.$$

Assume that $\mathbf{E} \ln \zeta \ge 0$, $\eta \equiv 1$, and put

$$s := 0, \qquad z_j := \ln \zeta_j, \qquad Z_l := \sum_{j=-l}^{0} z_j.$$

Then

$$X^1 = 1 + \sum_{l=0}^{\infty} e^{Z_l}, \quad \text{where } \sum_{l=0}^{\infty} \mathrm{I}(Z_l \ge 0) = \infty$$

with probability 1, and consequently $X^1 = \infty$ and $X_n \to \infty$ with probability 1.

If $\mathbf{E}[\ln \eta]^+ = \infty$ and $\zeta = b < 1$ then

$$X^1 = \eta_0 + b \sum_{l=0}^{\infty} \exp\{y_{-l-1} + l \ln b\},$$

where $y_j = \ln \eta_j$; the event $\{y_{-l-1} > -l \ln b\}$ occurs infinitely often with probability 1. This means that $X^1 = \infty$ and $X_n \to \infty$ with probability 1.

Chapter 18
Continuous Time Random Processes

Abstract This chapter presents elements of the general theory of continuous time processes. Section 18.1 introduces the key concepts of random processes, sample paths, cylinder sets and finite-dimensional distributions, the spaces of continuous functions and functions without discontinuities of the second kind, and equivalence of random processes. Section 18.2 presents the fundamental results on regularity of processes: Kolmogorov's theorem on existence of a continuous modification and Kolmogorov–Chentsov's theorem on existence of an equivalent process with trajectories without discontinuities of the second kind. The section also contains discussions of the notions of separability, stochastic continuity and continuity in mean.

18.1 General Definitions

Definition 18.1.1 A *random process*[1] is a family of random variables $\xi(t) = \xi(t, \omega)$ given on a common probability space $\langle \Omega, \mathfrak{F}, \mathbf{P} \rangle$ and depending on a parameter t taking values in some set T.

A random process will be written as $\{\xi(t), t \in T\}$.

The sequences of random variables $\xi_1, \xi_2, \ldots$ considered in the previous sections are random processes for which $T = \{1, 2, 3, \ldots\}$. The same is true of the sums $S_1, S_2, \ldots$ of $\xi_1, \xi_2, \ldots$ Markov chains $\{X_n,\ n = 0, 1, \ldots\}$, martingales $\{X_n;\ n \in \mathbb{N}\}$, stationary and stochastic recursive sequences described in previous chapters are also random processes. The processes for which the set T can be identified with the whole sequence $\{\ldots, -1, 0, 1, \ldots\}$ or a part thereof are usually called *random processes in discrete time*, or *random sequences*.

If T coincides with a certain real interval $T = [a, b]$ (this may be the whole real line $-\infty < t < \infty$ or the half-line $t \ge 0$), then the collection $\{\xi(t),\ t \in T\}$ is said to be a process in *continuous time*.

Simple examples of such objects are renewal processes $\{\eta(t),\ t \ge 0\}$ described in Chap. 10.

[1] As well as the term "random process" one also often uses the terms "stochastic" or "probabilistic" processes.

A.A. Borovkov, *Probability Theory*, Universitext,
DOI 10.1007/978-1-4471-5201-9_18, © Springer-Verlag London 2013

In the present chapter we will be considering *continuous time processes* only. Interpretation of the parameter t as time is, of course, not imperative. It appeared historically because in most problems from the natural sciences which led to the concept of random process the parameter t had the meaning of time, and the value $\xi(t)$ was what one would observe at time t.

The movement of a gas molecule as time passes, the storage level in a water reservoir, oscillations of an airplane's wing etc could be viewed as examples of real world random processes.

The random function

$$\xi(t) = \sum_{k=1}^{\infty} 2^{-k} \xi_k \sin kt, \quad t \in [0, 2\pi],$$

where the ξ_k are independent and identically distributed, is also an example of a random process.

Consider a random process $\{\xi(t),\ t \in T\}$. If $\omega \in \Omega$ is fixed, we obtain a function $\xi(t)$, $t \in T$, which is often called a *sample function, trajectory* or *path* of the process. Thus, the random values here are *functions*. As before, we could consider here a *sample* probability space, which can be constructed for example as follows. Consider the space $\mathcal{X}$ of functions $x(t)$, $t \in T$, to which the trajectories $\xi(t)$ belong. Let, further, $\mathfrak{B}_{\mathcal{X}}^{T}$ be the σ-algebra of subsets of $\mathcal{X}$ generated by the sets of the form

$$C = \{x \in \mathcal{X} : x(t_1) \in B_1, \dots, x(t_n) \in B_n\} \tag{18.1.1}$$

for any n, any $t_1, \dots, t_n$ from T, and any Borel sets $B_1, \dots, B_n$. Sets of this form are called *cylinders*; various finite unions of cylinder sets form an algebra generating $\mathfrak{B}_{\mathcal{X}}^{T}$. If a process $\xi(t, \omega)$ is given, it defines a measurable mapping of $\langle \Omega, \mathfrak{F} \rangle$ into $\langle \mathcal{X}, \mathfrak{B}_{\mathcal{X}}^{T} \rangle$, since clearly $\xi^{-1}(C) = \{\omega : \xi(\cdot, \omega) \in C\} \in \mathfrak{F}$ for any cylinder C, and therefore $\xi^{-1}(B) \in \mathfrak{F}$ for any $B \in \mathfrak{B}_{\mathcal{X}}^{T}$. This mapping induces a distribution $\mathbf{P}_{\xi}$ on $\langle \mathcal{X}, \mathfrak{B}_{\mathcal{X}}^{T} \rangle$ defined by the equalities $\mathbf{P}_{\xi}(B) = \mathbf{P}(\xi^{-1}(B))$. The triplet $\langle \mathcal{X}, \mathfrak{B}_{\mathcal{X}}^{T}, \mathbf{P}_{\xi} \rangle$ is called the *sample probability space*. In that space, an elementary outcome ω is identified with the trajectory of the process, and the measure $\mathbf{P}_{\xi}$ is said to be the *distribution of the process* ξ.

Now if, considering the process $\{\xi(t)\}$, we fix the time epochs $t_1, t_2, \dots, t_n$, we will get a multi-dimensional random variable $(\xi(t_1, \omega), \dots, \xi(t_n, \omega))$. The distributions of such variables are said to be the *finite-dimensional distributions* of the process.

The following function spaces are most often considered as spaces $\mathcal{X}$ in the theory of random processes with continual sets T.

1. The set of all functions on T:

$$\mathcal{X} = \mathbb{R}^T = \prod_{t \in T} \mathbb{R}_t,$$

where $\mathbb{R}_t$ are copies of the real line $(-\infty, \infty)$. This space is usually considered with the σ-algebra $\mathfrak{B}_{\mathbb{R}}^{T}$ of subsets of $\mathbb{R}^T$ generated by cylinders.

2. The space $C(T)$ of continuous functions on T (we will write $C(a,b)$ if $T=[a,b]$). In this space, along with the σ-algebra $\mathfrak{B}_C^T$ generated by cylinder subsets of $C(T)$ (this σ-algebra is smaller that the similar σ-algebra in $\mathbb{R}^T$), one also often considers the σ-algebra $\mathfrak{B}_{C(T)}$ (the Borel σ-algebra) generated by the sets open with respect to the uniform distance

$$\rho(x,y):=\sup_{t\in T}\bigl|y(t)-x(t)\bigr|,\quad x,y\in C(T).$$

It turns out that, in the space $C(T)$, we always have $\mathfrak{B}_{C(T)}=\mathfrak{B}_C^T$ (see, e.g., [14]).

3. The space $D(T)$ of functions having left and right limits $x(t-0)$ and $x(t+0)$ at each point t, the value $x(t)$ being equal either to $x(t-0)$ or to $x(t+0)$. If $T=[a,b]$, it is also assumed that $x(a)=x(a+0)$ and $x(b)=x(b-0)$. This space is often called the *space of functions without discontinuities of the second kind.*[2] The space of functions for which at all other points $x(t)=x(t-0)$ $(x(b)=x(t+0))$ will be denoted by $D_-(T)$ $(D_+(T))$. The space $D_+(T)$ $(D_-(T))$ will be called the space of right-continuous (left-continuous) functions. For example, the trajectories of the renewal processes discussed in Chap. 10 belong to $D_+(0,\infty)$.

In the space $D(T)$ one can also construct the Borel σ-algebra with respect to an appropriate metric, but we will restrict ourselves to using the σ-algebra $\mathfrak{B}_D^T$ of cylindric subsets of $D(T)$.

Now we can formulate the following equivalent definition of a random process. Let $\mathfrak{X}$ be a given function space, and $\mathfrak{G}$ be the σ-algebra of its subsets containing the σ-algebra $\mathfrak{B}_{\mathfrak{X}}^T$ of cylinders.

Definition 18.1.2 A *random process* $\xi(t)=\xi(t,\omega)$ is a measurable (in ω) mapping of $\langle\Omega,\mathfrak{F},\mathbf{P}\rangle$ into $\langle\mathfrak{X},\mathfrak{G},\mathbf{P}_\xi\rangle$ (to each ω one puts into correspondence a function $\xi(t)=\xi(t,\omega)$ so that $\xi^{-1}(G)=\{\omega:\xi(\cdot)\in G\}\in\mathfrak{F}$ for $G\in\mathfrak{G}$). The distribution $\mathbf{P}_\xi$ is said to be the *distribution of the process.*

The condition $\mathfrak{B}_{\mathfrak{X}}^T\subset\mathfrak{G}$ is needed to ensure that the probabilities of cylinder sets and, in particular, the probabilities $\mathbf{P}(\xi(t)\in B)$, $B\in\mathfrak{B}_{\mathfrak{X}}^T$ are correctly defined, which means that $\xi(t)$ are random variables.

So far we have tacitly assumed that the process is *given* and it is known that its trajectories lie in $\mathfrak{X}$. However, this is rarely the case. More often one tries to describe the process $\xi(t)$ in terms of *some characteristics of its distribution*. One could, for example, specify the finite-dimensional distributions of the process. From Kolmogorov's theorem on consistent distributions[3] (see Appendix 2), it follows that

---

[2]A discontinuity of the second kind is associated with either non-fading oscillations of increasing frequency or escape to infinity.

[3]Recall the definition of consistent distributions. Let $\mathbb{R}_t$, $t\in T$, be real lines and $\mathfrak{B}_t$ the σ-algebras of Borel subsets of $\mathbb{R}_t$. Let $T_n=\{t_1,\dots,t_n\}$ be a finite subset of T. The finite-dimensional distribution of $(\xi(t_1,\omega),\dots,\xi(t_n,\omega))$ is the distribution $\mathbf{P}_{T_n}$ on $(\mathbb{R}^{T_n},\mathfrak{B}^{T_n})$, where $\mathbb{R}^{T_n}=\prod_{t\in T_n}\mathbb{R}_t$ and $\mathfrak{B}^{T_n}=\prod_{t\in T_n}\mathfrak{B}_t$. Let two finite subsets T' and T'' of T be given, and $(\mathbb{R}',\mathfrak{B}')$ and $(\mathbb{R}'',\mathfrak{B}'')$ be the respective subspaces of $(\mathbb{R}^T,\mathfrak{B}^T)$. The distributions $\mathbf{P}_{T'}$ and $\mathbf{P}_{T''}$ on $(\mathbb{R}',\mathfrak{B}')$ and $(\mathbb{R}'',\mathfrak{B}'')$

finite-dimensional distributions uniquely specify the distribution $\mathbf{P}_\xi$ of the process on the space $\langle \mathbb{R}^T, \mathfrak{B}_\mathbb{R}^T \rangle$. That theorem can be considered as *the existence theorem for random processes in* $\langle \mathbb{R}^T, \mathfrak{B}_\mathbb{R}^T \rangle$ *with prescribed finite-dimensional distributions*.

The space $\langle \mathbb{R}^T, \mathfrak{B}_\mathbb{R}^T \rangle$ is, however, not quite convenient for studying random processes. The fact is that by no means all relations frequently used in analysis generate events, i.e. the sets which belong to the σ-algebra $\mathfrak{B}_\mathbb{R}^T$ and whose probabilities are defined. Based on the definition, we can be sure that only the elements of the σ-algebra generated by $\{\xi(t) \in B\}$, $t \in T$, B being Borel sets, are events. The set $\{\sup_{t\in T} \xi(t) < c\}$, for instance, does not have to be an event, for we only know its representation in the form $\bigcap_{t\in T}\{\xi(t) < c\}$, which is the intersection of an *uncountable* collection of measurable sets when T is an interval on the real line.

Another inconvenience occurs as well: the distribution $\mathbf{P}_\xi$ on $\langle \mathbb{R}^T, \mathfrak{B}_\mathbb{R}^T \rangle$ does not uniquely specify the properties of the trajectories of $\xi(t)$. The reason is that the space $\mathbb{R}^T$ is very rich, and if we know that $x(\cdot)$ belongs to a set of the form (18.1.1), this gives us no information about the behaviour of $x(t)$ at points t different from $t_1, \dots, t_n$. The same is true of arbitrary sets A from $\mathfrak{B}_\mathbb{R}^T$: roughly speaking, the relation $x(\cdot) \in A$ can determine the values of $x(t)$ at most at a countable set of points. We will see below that even such a set as $\{x(t) \equiv 0\}$ does not belong to $\mathfrak{B}_\mathbb{R}^T$. To specify the behaviour of the *entire* trajectory of the process, it is not sufficient to give a distribution on $\mathfrak{B}_\mathbb{R}^T$—one has to extend this σ-algebra.

Prior to presenting the respective example, we will give the following definition.

Definition 18.1.3 Processes $\xi(t)$ and $\eta(t)$ are said to be *equivalent* (or *stochastically equivalent*) if $\mathbf{P}(\xi(t) = \eta(t)) = 1$ for all $t \in T$. In this case the process η is called a *modification* of ξ.

Finite-dimensional distributions of equivalent process clearly coincide, and therefore the distributions $\mathbf{P}_\xi$ and $\mathbf{P}_\eta$ on $\langle \mathbb{R}^T, \mathfrak{B}_\mathbb{R}^T \rangle$ coincide, too.

Example 18.1.1 Put

$$x_a(t) := \begin{cases} 0 & \text{if } t \neq a, \\ 1 & \text{if } t = a, \end{cases}$$

and complete $\mathfrak{B}_\mathbb{R}^T$ with the elements $x_a(t)$, $a \in [0, 1]$, and the element $x^0(t) \equiv 0$. Let $\gamma \in \mathbf{U}_{0,1}$. Consider two random processes $\xi_0(t)$ and $\xi_1(t)$ defined as follows: $\xi_0(t) \equiv x^0(t)$, $\xi_1(t) = x_\gamma(t)$. Then clearly

$$\mathbf{P}\big(\xi_0(t) = \xi_1(t)\big) = \mathbf{P}(\gamma \neq t) = 1,$$

the processes ξ_0 and ξ_1 are equivalent, and hence their distributions on $\langle \mathbb{R}^T, \mathfrak{B}_\mathbb{R}^T \rangle$ coincide. However, we see that the trajectories of the processes are substantially different.

---

are said to be *consistent* if their projections on the common part of subspaces $\mathbb{R}'$ and $\mathbb{R}''$ (if it exists) coincide.

It is easy to see from the above example that the set of all continuous functions $C(T)$, the set $\{\sup_{t\in[0,1]} x(t) < x\}$, the one-point set $\{x(t) \equiv 0\}$ and many others do not belong to $\mathfrak{B}_{\mathbb{R}}^T$. Indeed, if we assume the contrary—say, that $C(T) \in \mathfrak{B}_{\mathbb{R}}^T$—then we would get from the equivalence of ξ_0 and ξ_1 that $\mathbf{P}(\xi_0 \in C(0,1)) = \mathbf{P}(\xi_1 \in C(0,1))$, while the former of these probabilities is 1 and the latter is 0.

The simplest way of overcoming the above difficulties and inconveniences is to define the processes in the spaces $C(T)$ or $D(T)$ when it is possible. If, for example, $\xi(t) \in C(T)$ and $\eta(t) \in C(T)$, and they are equivalent, then the trajectories of the processes will completely coincide with probability 1, since in that case

$$\bigcap_{\text{rational } t} \big\{\xi(t) = \eta(t)\big\} = \bigcap_{t\in T} \big\{\xi(t) = \eta(t)\big\} = \big\{\xi(t) = \eta(t) \text{ for all } t \in T\big\},$$

where the probability of the event on the left-hand side is defined (this is the probability of the intersection of a countable collection of sets) and equals 1. Similarly, the probabilities, say, of the events

$$\Big\{\sup_{t\in T} \xi(t) < c\Big\} = \bigcap_{t\in T} \big\{\xi(t) < c\big\}$$

are also defined.

The same argument holds for the spaces $D(T)$, because each element $x(\cdot)$ of D is uniquely determined by its values $x(t)$ on a countable everywhere dense set of t values (for example, on the set of rationals).

Now assume that we have somehow established that the original process $\xi(t)$ (let it be given on $\langle \mathbb{R}^T, \mathfrak{B}_{\mathbb{R}}^T \rangle$) has a continuous modification, i.e. an equivalent process $\eta(t)$ such that its trajectories are continuous with probability 1 (or belong to the space $D(T)$). The above means, first of all, that we have somehow extended the σ-algebra $\mathfrak{B}_{\mathbb{R}}^T$—adding, say, the set $C(T)$—and now consider the distribution of ξ on the σ-algebra $\widetilde{\mathfrak{B}}^T = \sigma(\mathfrak{B}_{\mathbb{R}}^T, C(T))$ (otherwise the above would not make sense). But the extension of the distribution of ξ from $\langle \mathbb{R}^T, \mathfrak{B}_{\mathbb{R}}^T \rangle$ to $\langle \mathbb{R}^T, \widetilde{\mathfrak{B}}^T \rangle$ may not be unique. (We saw this in Example 18.1.1; the extension can be given by, say, putting $\mathbf{P}(\xi \in C(T)) = 0$.) What we said above about the process η means that there exists an extension $\mathbf{P}_\eta$ such that $\mathbf{P}_\eta(C(T)) = \mathbf{P}(\eta \in C(T)) = 1$.

Further, it is often better not to deal with the inconvenient space $\langle \mathbb{R}^T, \mathfrak{B}_{\mathbb{R}}^T \rangle$ at all. To avoid it, one can define the distribution of the process η on the restricted space $\langle C(T), \mathfrak{B}_C^T \rangle$. It is clear that

$$\mathfrak{B}_C^T \subset \widetilde{\mathfrak{B}}^T = \sigma\big(\mathfrak{B}_{\mathbb{R}}^T, C(T)\big), \qquad \mathfrak{B}_C^T = \widetilde{\mathfrak{B}}^T \cap C(T)$$

(the former σ-algebra is generated by sets of the form (18.1.1) intersected with $C(T)$). Therefore, considering the distribution of η concentrated on $C(T)$, we can deal with the restriction of the space $\langle \mathbb{R}^T, \widetilde{\mathfrak{B}}^T \rangle$ to $\langle C(T), \mathfrak{B}_C^T \rangle$ and define the probability on the latter as $\mathbf{P}_\eta(A) = \mathbf{P}(\eta \in A)$, $A \in \mathfrak{B}_C^T \subset \widetilde{\mathfrak{B}}^T$. *Thus we have constructed a process η with continuous trajectories which is equivalent to the original process ξ* (if we consider their distributions in $\langle \mathbb{R}^T, \mathfrak{B}_{\mathbb{R}}^T \rangle$).

To realise this construction, one has now to learn how to find from the distribution of a process ξ whether it has a continuous modification η or not.

Before stating and proving the respective theorems, note once again that the above-mentioned difficulties are mainly of a *mathematical character*, i.e. related to the mathematical model of the random process. In real life problems, it is usually clear in advance whether the process under consideration is continuous or not. If it is "physically" continuous, and we want to construct an adequate model, then, of course, of all modifications of the process we have to take the continuous one.

The same argument remains valid if, instead of continuous trajectories, one considers trajectories from $D(T)$. The problem essentially remains the same: the difficulties are eliminated if one can describe the entire trajectory of the process $\xi(\cdot)$ by the values $\xi(t)$ on some countable set of t values. Processes possessing this property will be called *regular*.

18.2 Criteria of Regularity of Processes

First we will find conditions under which a process has a continuous modification. Without loss of generality, we will assume that T is the segment $T = [0, 1]$.

A very simple criterion for the existence of a continuous modification is based on the knowledge of *two-dimensional* distributions of $\xi(t)$ only.

Theorem 18.2.1 (Kolmogorov) *Let $\xi(t)$ be a random process given on $\langle \mathbb{R}^T, \mathfrak{B}_{\mathbb{R}}^T \rangle$ with $T = [0, 1]$. If there exist $a > 0$, $b > 0$ and $c < \infty$ such that, for all t and $t + h$ from the segment* $[0, 1]$,

$$\mathbf{E}\big|\xi(t+h) - \xi(t)\big|^a \le c|h|^{1+b}, \tag{18.2.1}$$

then $\xi(\cdot)$ has a continuous modification.

We will obtain this assertion as a consequence of a more general theorem, of which the conditions are somewhat more difficult to comprehend, but have essentially the same meaning as (18.2.1).

Theorem 18.2.2 *Let for all $t, t + h \in [0, 1]$,*

$$\mathbf{P}\big(\big|\xi(t+h) - \xi(t)\big| > \varepsilon(h)\big) \le q(h),$$

where $\varepsilon(h)$ and $q(h)$ are decreasing even functions of h such that

$$\sum_{n=1}^{\infty} \varepsilon\big(2^{-n}\big) < \infty, \qquad \sum_{n=1}^{\infty} 2^n q\big(2^{-n}\big) < \infty.$$

Then $\xi(\cdot)$ has a continuous modification.

Proof We will make use of approximations of $\xi(t)$ by continuous processes. Put

$$t_{n,r} := r2^{-n}, \quad r = 0, 1, \ldots, 2^n,$$
$$\xi_n(t) := \xi(t_{n,r}) + 2^n(t - t_{n,r})\big[\xi(t_{n,r+1}) - \xi(t_{n,r})\big] \quad \text{for } t \in [t_{n,r}, t_{n,r+1}].$$

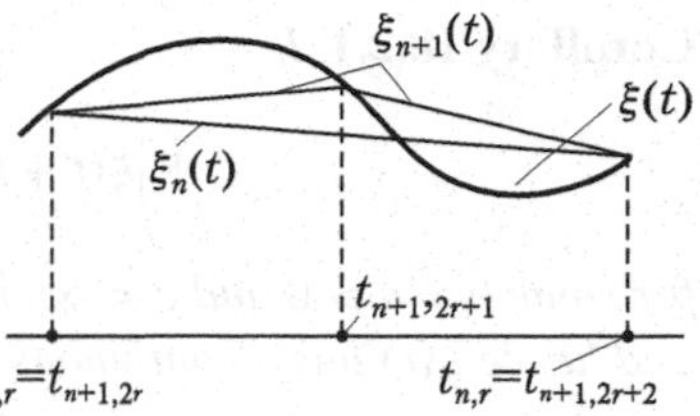

Fig. 18.1 Illustration to the proof of Theorem 18.2.2: construction of piece-wise linear approximations to the process $\xi(t)$

From Fig. 18.1 we see that

$$\big|\xi_{n+1}(t)-\xi_n(t)\big| \le \left|\xi(t_{n+1,2r+1})-\frac{1}{2}\big[\xi(t_{n+1,2r})+\xi(t_{n+1,2r+2})\big]\right| \le \frac{1}{2}(\alpha+\beta),$$

where $\alpha := |\xi(t_{n+1,2r+1})-\xi(t_{n+1,2r})|$, $\beta := |\xi(t_{n+1,2r+1})-\xi(t_{n+1,2r+2})|$. This implies that

$$Z_n := \max_{t\in[t_{n,r},t_{n,r+1}]}\big|\xi_{n+1}(t)-\xi_n(t)\big| \le \frac{1}{2}(\alpha+\beta),$$
$$\mathbf{P}\big(Z_n>\varepsilon\big(2^{-n}\big)\big) \le \mathbf{P}\big(\alpha>\varepsilon\big(2^{-n}\big)\big)+\mathbf{P}\big(\beta>\varepsilon\big(2^{-n}\big)\big) \le 2q\big(2^{-n}\big)$$

(note that since the trajectories of $\xi_n(t)$ are continuous, $\{Z_n>\varepsilon(2^{-n})\}\in\mathfrak{B}_R^T$, which is not the case in the general situation). Since here we have altogether 2^n segments of the form $[t_{n,r},t_{n,r+1}]$, $r=0,1,\ldots,2^n-1$, one has

$$\mathbf{P}\Big(\max_{t\in[0,1]}\big|\xi_{n+1}(t)-\xi_n(t)\big|>\varepsilon(2^{-n})\Big) \le 2^{n+1}q(2^{-n}).$$

Because $\sum_{n=1}^{\infty}2^n q(2^{-n})<\infty$, by the Borel–Cantelli criterion, for almost all ω (i.e. for $\omega\in A$, $\mathbf{P}(A)=1$), there exists an $n(\omega)$ such that, for all $n\ge n(\omega)$,

$$\max_{t\in[0,1]}\big|\xi_{n+1}(t)-\xi_n(t)\big| \equiv \rho(\xi_{n+1},\xi_n)<\varepsilon\big(2^{-n}\big).$$

From this it follows that ξ_n is a Cauchy sequence a.s., since

$$\rho(\xi_n,\xi_m)\le\varepsilon_n := \sum_{n}^{\infty}\varepsilon(2^{-k})\to 0$$

as $n\to\infty$ for all $m>n$, $\omega\in A$. Therefore, for $\omega\in A$, there exists the limit $\eta(t)=\lim_{n\to\infty}\xi_n(t)$, and $|\xi_n(t)-\eta(t)|\le\varepsilon_n$, so that convergence $\xi_n(t)\to\eta(t)$ is uniform. Together with continuity of $\xi_n(t)$ this implies that $\eta(t)$ is also continuous (this argument actually shows that the space $C(0,1)$ is complete).

It remains to verify that ξ and η are equivalent. For $t=t_{n,r}$ one has $\xi_{n+k}(t)=\xi(t)$ for all $k\ge 0$, so that $\eta(t)=\xi(t)$. If $t\ne t_{n,r}$ for all n and r, then there exists a sequence r_n such that $t_{t,r_n}\to t$ and $0<t-t_{t,r_n}<2^{-n}$, and hence

$$\mathbf{P}\big(\big|\xi(t_{t,r_n})-\xi(t)\big|>\varepsilon(t-t_{t,r_n})\big)\le q(t-t_{t,r_n}),$$
$$\mathbf{P}\big(\big|\xi(t_{t,r_n})-\xi(t)\big|>\varepsilon\big(2^{-n}\big)\big)\le q\big(2^{-n}\big).$$

By the Borel–Cantelli criterion this means that $\xi_{n,r_n}\to\xi$ with probability 1. At the same time, by virtue of the continuity of $\eta(t)$ one has $\eta(t_{t,r_n})\to\eta(t)$. Because $\xi(t_{t,r_n})=\eta(t_{t,r_n})$, we have $\xi(t)=\eta(t)$ with probability 1.

The theorem is proved. □

Corollary 18.2.1 *If*

$$\mathbf{E}\left|\xi(t+h)-\xi(t)\right|^a \le \frac{c|h|}{|\log|h||^{1+b}} \tag{18.2.2}$$

for some $b > a > 0$ *and* $c < \infty$, *then the conditions of Theorem* 18.2.2 *are satisfied and hence* $\xi(t)$ *has a continuous modification.*

Condition (18.2.2) will certainly be satisfied if (18.2.1) holds, so that Kolmogorov's theorem is a consequence of Theorem 18.2.2.

Proof of Corollary 18.2.1 Put $\varepsilon(h) := |\log_2|h||^{-\beta}$, $1 < \beta < b/a$. Then

$$\sum_{n=1}^{\infty} \varepsilon\left(2^{-n}\right) = \sum_{n=1}^{\infty} n^{-\beta} < \infty,$$

and from Chebyshev's inequality we have

$$\mathbf{P}\left(\left|\xi(t+a)-\xi(t)\right| > \varepsilon(h)\right) \le \frac{c|h|}{|\log_2|h||^{1+b}}\left(\varepsilon(h)\right)^{-a} = \frac{c|h|}{|\log_2|h||^{1+\delta}} =: q(h),$$

where $\delta = b - a\beta > 0$. It remains to note that

$$\sum_{n=1}^{\infty} 2^n q\left(2^{-n}\right) = \sum_{n=1}^{\infty} \left|\log_2 2^{-n}\right|^{-1-\delta} < \infty.$$

The corollary is proved. □

The criterion for $\xi(t)$ to have a modification belonging to the space $D(T)$ is more complicated to formulate and prove, and is related to weaker conditions imposed on the process. We confine ourselves here to simply stating the following assertion.

Theorem 18.2.3 (Kolmogorov–Chentsov) *If, for some* $\alpha \ge 0$, $\beta \ge 0$, $b > 0$, *and all* t, $h_1 \le t \le 1 - h_2$, $h_1 \ge 0$, $h_2 \ge 0$,

$$\mathbf{E}\left|\xi(t)-\xi(t-h_1)\right|^{\alpha}\left|\xi(t+h_2)-\xi(t)^{\beta}\right| < ch^{1+b}, \quad h = h_1 + h_2, \tag{18.2.3}$$

then there exists a modification of $\xi(t)$ *in* $D(0,1)$.[4]

Condition (18.2.3) admits the following extension:

$$\mathbf{P}\left(\left|\xi(t+h_2)-\xi(t)\right| \cdot \left|\xi(t)-\xi(t-h_1)\right| \ge \varepsilon(h)\right) \le q(h), \tag{18.2.4}$$

where $\varepsilon(h)$ and $q(h)$ have the same meaning as in Theorem 18.2.2. Under condition (18.2.4) the assertion of the theorem remains valid.

The following two examples illustrate, to a certain extent, the character of the conditions of Theorems 18.2.1–18.2.3.

[4] For more details, see, e.g., [9].

Example 18.2.1 Assume that a random process $\xi(t)$ has the form

$$\xi(t) = \sum_{k=1}^{r} \xi_k \varphi_k(t),$$

where $\varphi_k(t)$ satisfy the Hölder condition

$$\left|\varphi_k(t+h) - \varphi_k(t)\right| \le c\,|h|^\alpha,$$

$\alpha > 0$, and $(\xi_1, \dots, \xi_r)$ is an arbitrary random vector such that all $\mathbf{E}|\xi_k|^l$ are finite for some $l > 1/\alpha$. Then the process $\xi(t)$ (which is clearly continuous) satisfies condition (18.2.1). Indeed,

$$\mathbf{E}\left|\xi(t+h) - \xi(t)\right|^l \le c_1 \sum_{k=1}^{r} \mathbf{E}|\xi_k|^l c^l |h|^{\alpha l} \le c_2 |h^{\alpha l}|, \quad al > 1.$$

Example 18.2.2 Let $\gamma \in \mathbf{U}_{0,1}$, $\xi(t) = 0$ for $t < \gamma$, and $\xi(t) = 1$ for $t \ge \gamma$. Then

$$\mathbf{E}\left|\xi(t+h) - \xi(t)\right|^l = \mathbf{P}\big(\gamma \in (t, t+h)\big) = h$$

for any $l > 0$. Here condition (18.2.1) is not satisfied, although $|\xi(t+h) - \xi(t)| \xrightarrow{p} 0$ as $h \to 0$. Condition (18.2.3) is clearly met, for

$$\mathbf{E}\left|\xi(t) - \xi(t-h_1)\right| \cdot \left|\xi(t+h_2) - \xi(t)\right| = 0. \tag{18.2.5}$$

We will get similar results if we take $\xi(t)$ to be the renewal process for a sequence $\gamma_1, \gamma_2, \dots$, where the distribution of γ_j has a density. In that case, instead of (18.2.5) one will obtain the relation

$$\mathbf{E}\left|\xi(t) - \xi(t-h_1)\right| \cdot \left|\xi(t+h_2) - \xi(t)\right| \le c h_1 h_2 \le c h^2.$$

In the general case, when we do not have data for constructing modifications of the process ξ in the spaces $C(T)$ or $D(T)$, one can overcome the difficulties mentioned in Sect. 18.1 with the help of the notion of separability.

Definition 18.2.1 A process $\xi(t)$ is said to be *separable* if there exists a countable set S which is everywhere dense in T and

$$\mathbf{P}\Big(\limsup_{\substack{u\to t\\ u\in S}} \xi(u) \ge \xi(t) \ge \liminf_{\substack{u\to t\\ u\in S}} \xi(u) \text{ for all } t \in T\Big) = 1. \tag{18.2.6}$$

This is equivalent to the property that, for any interval $I \subset T$,

$$\mathbf{P}\Big(\sup_{u\in I\cap S} \xi(u) = \sup_{u\in I} \xi(u);\ \inf_{u\in I\cap S} \xi(u) = \inf_{u\in I} \xi(u)\Big) = 1.$$

It is known (Doob's theorem[5]) that *any random process has a separable modification.*

[5] See [14, 26].

Constructing a separable modification of a process, as well as constructing modifications in spaces $C(T)$ and $D(T)$, means extending the σ-algebra $\mathfrak{B}^T_{\mathbb{R}}$, to which one adds uncountable intersections of the form

$$A = \bigcap_{u \in I} \{\xi(u) \in [a,b]\} = \Big\{\sup_{u\in I} \xi(u) \le b,\ \inf_{u \in I} \xi(u) \ge a\Big\},$$

and extending the measure $\mathbf{P}$ to the extended σ-algebra using the equalities

$$\mathbf{P}(A) = \mathbf{P}\Big(\bigcap_{u \in I \cap S} \{\xi(u) \in [a,b]\}\Big),$$

where in the probability on the right-hand side we already have an element of $\mathfrak{B}^T_{\mathbb{R}}$.

For separable processes, such sets as the set of all nondecreasing functions, the sets $C(T)$, $D(T)$ and so on, are events. Processes from $C(T)$ or $D(T)$ are automatically separable. And vice versa, if a process is separable and admits a continuous modification (modification from $D(T)$) then it will be continuous (belong to $D(T)$) itself. Indeed, if η is a continuous modification of ξ then

$$\mathbf{P}\big(\xi(t) = \eta(t) \text{ for all } t \in S\big) = 1.$$

From this and (18.2.6) we obtain

$$\mathbf{P}\Big(\limsup_{\substack{u\to t\\ u \in S}} \eta(u) \ge \xi(t) \ge \liminf_{\substack{u\to t\\ u \in S}} \eta(u) \text{ for all } t \in T\Big) = 1.$$

Since $\limsup_{u\to t} \eta(u) = \liminf_{u \to t} \eta(u) = \eta(t)$, one has

$$\mathbf{P}\big(\xi(t) = \eta(t) \text{ for all } t \in T\big) = 1.$$

In Example 18.1.1, the process $\xi_1(t)$ is clearly not separable. The process $\xi_0(t)$ is a separable modification of $\xi_1(t)$.

As well as pathwise continuity, there is one more way of characterising the continuity of a random process.

Definition 18.2.2 A random process $\xi(t)$ is said to be *stochastically continuous* if, for all $t \in T$, as $h \to 0$,

$$\xi(t+h) \xrightarrow{p} \xi(t) \qquad \big(\mathbf{P}\big(\big|\xi(t+h) - \xi(t)\big| > \varepsilon\big) \to 0\big).$$

Here we deal with the two-dimensional distributions of $\xi(t)$ only.

It is clear that all processes with continuous trajectories are stochastically continuous. But not only them. The discontinuous processes from Examples 18.1.1 and 18.2.2 are also stochastically continuous. A discontinuous process is not stochastically continuous if, for a (random) discontinuity point τ ($\xi(\tau + 0) \ne \xi(\tau - 0)$), the probability $\mathbf{P}(\tau = t_0)$ is positive for some fixed point t_0.

Definition 18.2.3 A process $\xi(t)$ is said to be *continuous in mean of order* r (*in mean* when $r = 1$; *in mean quadratic* when $r = 2$) if, for all $t \in T$, as $h \to 0$,

$$\xi(t+h) \xrightarrow{(r)} \xi(t) \quad \text{or, which is the same,} \quad \mathbf{E}\big|\xi(t+h) - \xi(t)\big|^r \to 0.$$

The discontinuous process $\xi(t)$ from Example 18.2.2 is continuous in mean of any order. Therefore the continuity in mean and stochastic continuity do not say much about the pathwise properties (they only say that a jump in a neighbourhood of any fixed point t is unlikely). As Kolmogorov's theorem shows, in order to characterise the properties of trajectories, one needs *quantitative* bounds for $\mathbf{E}|\xi(t+h)-\xi(t)|^r$ or for $\mathbf{P}(|\xi(t+h)-\xi(t)|>\varepsilon)$.

Continuity theorems for moments imply that, *for a stochastically continuous process* $\xi(t)$ *and any continuous bounded function* $g(x)$, *the function* $\mathbf{E}g(\xi(t))$ *is continuous.* This assertion remains valid if we replace the boundedness of $g(x)$ with the condition that

$$\sup_t \mathbf{E}\big|g\big(\xi(t)\big)\big|^\alpha < \infty \quad \text{for some } \alpha > 1.$$

The consequent Chaps. 19, 21 and 22 will be devoted to studying random processes which can be given by specifying the explicit form of their finite-dimensional distributions. To this class belong:

1. Processes with independent increments.
2. Markov processes.
3. Gaussian processes.

In Chap. 22 we will also consider some problems of the theory of processes with finite second moments. Chapter 20 contains limit theorems for random processes generated by partial sums of independent random variables.

Chapter 19
Processes with Independent Increments

Abstract Section 19.1 introduces the fundamental concept of infinitely divisible distributions and contains the key theorem on relationship of such processes to processes with independent homogeneous increments. Section 19.2 begins with a definition of the Wiener process based on its finite-dimensional distributions and establishes existence of a continuous modification of the process. It also derives the distribution of the maximum of the Wiener process on a finite interval. The Laws of the Iterated Logarithm for the Wiener process are established in Sect. 19.3. Section 19.4 is devoted to the Poisson processes, while Sect. 19.5 presents a characterisation of the class of processes with independent increments (the Lévy–Khintchin theorem).

19.1 General Properties

Definition 19.1.1 A process $\{\xi(t),\ t \in [a,b]\}$ given on the interval $[a,b]$ is said to be a *process with independent increments* if, for any n and $t_0 < t_1 < \cdots < t_n$, $a \le t_0$, $t_n \le b$, the random variables $\xi(t_0)$, $\xi(t_1) - \xi(t_0), \ldots, \xi(t_n) - \xi(t_{n-1})$ are independent.

A process with independent increments is called *homogeneous* if the distribution of $\xi(t_1) - \xi(t_0)$ is determined by the length of the interval $t_1 - t_0$ only and is independent of t_0.

In what follows, we will everywhere assume for simplicity's sake that $a = 0$, $\xi(0) = 0$ and $b = 1$ or $b = \infty$.

Definition 19.1.2 The distribution of a random variable ξ is called *infinitely divisible* (cf. Sect. 8.8) if, for any n, the variable ξ can be represented as a sum of independent identically distributed random variables: $\xi = \xi_{1,n} + \cdots + \xi_{n,n}$. If $\varphi(\lambda)$ is the ch.f. of ξ, then this is equivalent to the property that $\varphi^{1/n}$ is a ch.f. for any n.

It is clear from the above definitions that, for a homogeneous process with independent increments, the distribution of $\xi(t)$ is infinitely divisible, because $\xi = \xi_{1,n} + \cdots + \xi_{n,n}$, where $\xi_{k,n} = \xi(kt/n) - \xi((k-1)t/n)$ are independent and distributed as $\xi(t/n)$.

A.A. Borovkov, *Probability Theory*, Universitext,
DOI 10.1007/978-1-4471-5201-9_19,

Theorem 19.1.1

(1) *Let* $\{\xi(t),\ t\ge 0\}$ *be a stochastically continuous homogeneous process with independent increments, and let* $\varphi_t(\lambda)=\mathbf{E}e^{i\lambda\xi(t)}$ *be the ch.f. of* $\xi(t)$, $\varphi(\lambda):=\varphi_1(\lambda)$. *Then*

$$\varphi_t(\lambda)=\varphi^t(\lambda), \tag{19.1.1}$$

$\varphi(\lambda)\neq 0$ *for any* λ.

(2) *Let* $\varphi(\lambda)$ *be the ch.f. of an infinitely divisible distribution. Then there exists a random process* $\{\xi(t),\ t\ge 0\}$ *satisfying the conditions of* (1) *and such that*

$$\mathbf{E}e^{i\lambda\xi(1)}=\varphi(\lambda).$$

Note that in the theorem the power $\varphi^t(\lambda)$ of the complex number $\varphi(\lambda)$ is understood as $|\varphi(\lambda)|^t e^{i\alpha(\lambda)t}$, where $\alpha(\lambda)=\arg\varphi(\lambda)$ ($\varphi(\lambda)=|\varphi(\lambda)|e^{i\alpha(\lambda)}$). But $\alpha(\lambda)$ is a multi-valued function, which is defined up to the term $2\pi k$ with integer k. Therefore, for non-integer t, the function $\varphi^t(\lambda)$ will be multi-valued as well. Since any ch.f. is continuous, after crossing the level $2\pi k$ by $\alpha(k)$ (while changing the value of λ from zero, $\alpha(0)=0$), we are to take the "nearest" branch of $\alpha(k)$ so as to ensure continuity of the function $\varphi^t(\lambda)$. For example, for the degenerate distribution I_1 we have $\varphi(\lambda)=e^{i\lambda}$ ($\alpha(\lambda)=\lambda$), so for small $t>0$, $\varepsilon>0$ and for $\lambda=2\pi+\varepsilon$ we are to set $\varphi^t(\lambda)=e^{i(2\pi+\varepsilon)t}$ rather than $\varphi^t(\lambda)=e^{i\varepsilon t}$ (although $\varphi(\lambda)=e^{i\varepsilon}$).

Denote by $\mathcal{L}$ the class of ch.f.s of all infinitely divisible distributions and by $\mathcal{L}_1$ the class of the ch.f.s of the distributions of $\xi(t)$ for stochastically continuous homogeneous processes with independent increments. Then it follows from Theorem 19.1.1 that $\mathcal{L}=\mathcal{L}_1$. The class $\mathcal{L}$ will be characterised in Sect. 19.5.

Proof (1) Let $\xi(t)$ satisfy the conditions of part (1) of the theorem. Then $\xi(t)$ can be represented as a sum of independent increments

$$\xi(t)=\sum_{j=1}^{n}\bigl[\xi(t_j)-\xi(t_{j-1})\bigr],\quad t_0=0,\ t_n=t,\ t_j>t_{j-1}.$$

From this it follows, in particular, that for $t_j=j/n$, $t=1$,

$$\varphi(\lambda)=\bigl[\varphi_{1/n}(\lambda)\bigr]^n,\qquad \varphi_{1/n}(\lambda)=\varphi^{1/n}(\lambda).$$

Raising both sides of the last equality to an integer power k, we obtain that, for any rational $r=k/n$, one has

$$\varphi_{k/n}(\lambda)=\varphi^{k/n}(\lambda),$$

which proves (19.1.1) for $t=r$. Now let t be irrational and $r_n:=\lfloor tn\rfloor/n$. Since $\xi(t)$ is a stochastically continuous process, one has $\xi(r_n)\xrightarrow{p}\xi(t)$ as $n\to\infty$, and hence the corresponding ch.f.s converge: for any λ,

$$\varphi_{r_n}(\lambda)\to\varphi_t(\lambda).$$

But $\varphi_{r_n}(\lambda)=\varphi^{r_n}(\lambda)\to\varphi^t(\lambda)$. Therefore (19.1.1) necessarily holds true.

Further, by stochastic continuity of $\xi(\cdot)$, we have $\varphi_t(\lambda) = \varphi^t(\lambda) \to 1$ as $t \to 0$ for any λ. This implies that $\varphi(\lambda) \neq 0$ for any λ. This completes the proof of the first assertion of the theorem.

(2) Observe first that if $\varphi \in \mathcal{L}$ then, for any $t > 0$, φ^t is again a ch.f. Indeed,

$$\varphi^t(\lambda) = \lim_{n\to\infty} \varphi^{\lfloor tn\rfloor/n}(\lambda),$$

so that $\varphi^t(\lambda)$ is a limit of ch.f.s which is continuous at the point $\lambda = 0$. By the continuity theorem for ch.f.s, this is again a ch.f.

Now we will construct a random process $\xi(t)$ with independent increments by specifying its finite-dimensional distributions. Put

$$0 = t_0 < t_1 < \cdots < t_k, \qquad \Delta_j := \xi(t_j) - \xi(t_{j-1}), \qquad \delta_j := t_j - t_{j-1},$$

and observe that

$$\sum_{j=1}^k \lambda_j \xi(t_j) = \sum_{j=1}^k \lambda_j \sum_{l=1}^j \Delta_l = \sum_{j=1}^k \Delta_j \sum_{l=j}^k \lambda_l.$$

Define the ch.f. of the joint distribution of $\xi(t_1), \dots, \xi(t_k)$ by the equality (postulating independence of Δ_j)

$$\mathbf{E}\exp\left\{i\sum_1^k \lambda_j \xi(t_j)\right\} := \mathbf{E}\exp\left\{i\sum_{j=1}^k \Delta_j \sum_{l=j}^k \lambda_l\right\} = \prod_{j=1}^k \varphi\left(\sum_{l=j}^k \lambda_l\right)^{\delta_j}.$$

Thus, we have used φ to define the finite-dimensional distributions of $\xi(t)$ in $\langle \mathbb{R}^T, \mathfrak{B}_{\mathbb{R}}^T\rangle$ with $T = [0, \infty)$ which, as one can easily see, are consistent. By Kolmogorov's theorem, there exists a distribution of a random process $\xi(t)$ in $\langle \mathbb{R}^T, \mathfrak{B}_{\mathbb{R}}^T\rangle$. That process is by definition a homogeneous processes with independent increments.

To prove stochastic continuity of $\xi(t)$, note that, as $h \to 0$,

$$\mathbf{E}e^{i\lambda(\xi(t+h)-\xi(t))} = \varphi^h(\lambda) \to \varphi_0(\lambda),$$

where

$$\varphi_0(\lambda) = \begin{cases} 1 & \text{if } \varphi(\lambda) \neq 0, \\ 0 & \text{if } \varphi(\lambda) = 0. \end{cases}$$

Thus the limiting function $\varphi_0(\lambda)$ can assume only two values: 0 and 1. But it is bound to be a ch.f. since it is continuous at the point $\lambda = 0$ ($\varphi(\lambda) \neq 0$ in a neighbourhood of the point $\lambda = 0$) and is a limit of ch.f.s. Therefore $\varphi_0(\lambda)$ is continuous, $\varphi_0(\lambda) \equiv 1$, $\varphi^h(\lambda) \to 1$, and

$$\xi(t+h) - \xi(t) \xrightarrow{p} 0 \quad \text{as } h \to 0.$$

The theorem is proved. □

Corollary 19.1.1 *Let the conditions of part* (1) *of Theorem* 19.1.1 *be met. If, for all* t, $\mathbf{E}|\xi(t)| < \infty$ *then*

$$\mathbf{E}\xi(t) = t\mathbf{E}\xi(1).$$

If $\mathbf{E}(\xi(1))^2 < \infty$ *then*

$$\operatorname{Var}\xi(t) = t \operatorname{Var}\xi(1).$$

Proof For the sake of brevity, put $a := \mathbf{E}\xi(1)$. Then, differentiating (19.1.1) in λ at the point $\lambda = 0$, we obtain

$$\begin{aligned}
\mathbf{E}\xi(t) &= -i\varphi_t'(0) = -it\varphi^{t-1}\varphi'(0) = at, \\
\mathbf{E}\xi^2(t) &= -\varphi_t''(0) = -t(t-1)\varphi^{t-2}(0)\big(\varphi'(0)\big)^2 - t\varphi^{t-1}(0)\varphi''(0) \\
&= t(t-1)a^2 + t\mathbf{E}\xi^2(1), \\
\operatorname{Var}\xi(t) &= t\big(\mathbf{E}\xi^2(1) - a^2\big) = t \operatorname{Var}\xi(1).
\end{aligned}$$

The corollary is proved. □

In the next theorem we put, as before, $T = [0, 1]$ or $T = [0, \infty)$.

Theorem 19.1.2 *Homogeneous stochastically continuous processes with independent increments* $\{\xi(t),\ t \in T\}$ *have modifications in the space* $D(T)$, *i.e. the process* $\xi(t)$ *can be given in* $\langle D(T), \mathfrak{B}_D^T\rangle$ *and hence have no discontinuities of the second type.*

Proof To simplify the argument, assume that $\mathbf{E}\xi^2(1)$ exists, or, which is the same, that the second derivative $\varphi''(\lambda)$ exists. Then

$$\begin{aligned}
\mathbf{E}\big(\xi(t) - \xi(t-h)\big)^2 &= \varphi_h''(0) = -h(h-1)\big[\varphi'(0)\big]^2 - h\varphi''(0) \le c|h|, \\
\mathbf{E}\big(|\xi(t+h_2) - \xi(t)|^2 |\xi(t) - \xi(t-h_1)|^2\big) &\le c^2 h_1 h_2 \le c^2(h_1 + h_2),
\end{aligned}$$

and the assertion follows from the second criterion of Theorem 18.2.3. The theorem is proved. □

In the general case, the proof is more complicated: one has to make use of criterion (18.2.4) and bounds for $\mathbf{P}(\xi(t) - \xi(t-h)| > \varepsilon)$.

Now we will consider the two most important processes with independent increments: the so-called Wiener and Poisson processes.

19.2 Wiener Processes. The Properties of Trajectories

Definition 19.2.1 The *Wiener process* is a homogeneous process with independent increments for which the distribution of $\xi(1)$ is normal.

In other words, this is a process for which

$$\varphi(\lambda) = e^{i\lambda\alpha - \sigma^2\lambda^2/2}, \qquad \varphi_t(\lambda) = \varphi^t(\lambda) = e^{i\lambda t\alpha - \sigma^2\lambda^2 t/2}$$

for some α and $\sigma^2 \ge 0$. The second equality means that the increments $\xi(t+u) - \xi(u)$ are normally distributed with parameters $(\alpha t, \sigma^2 t)$. All joint distributions of $\xi(t_1), \dots, \xi(t_n)$ are clearly also normal.

The numbers α and σ are called the *shift* and *diffusion coefficients*, respectively. Introducing the process $\xi_0(t) := (\xi(t) - \alpha t)/\sigma$ which is obtained from $\xi(t)$ by an affine transformation, we obtain that its ch.f. equals

$$\mathbf{E}e^{i\lambda\xi_0(t)} = e^{-i\lambda\alpha t/\sigma}\varphi_t(\lambda/\sigma) = e^{-\lambda^2 t/2}.$$

Such a process with parameters $(0, t)$ is often called the *standard Wiener process*. We consider it in more detail.

Theorem 19.2.1 *The Wiener process has a continuous modification.*

This means, as we know, that the Wiener process $\{\xi(t),\ t \in [0,1]\}$ can be considered as given on the measurable space $\langle C(0,1), \mathfrak{B}_C^{[0,1]}\rangle$ of continuous functions.

Proof We have $\xi(t+h) - \xi(t) \in\!\!\!\!= \mathbf{\Phi}_{0,h}$ and $h^{-1/2}(\xi(t+h) - \xi(t)) \in\!\!\!\!= \mathbf{\Phi}_{0,1}$. Therefore

$$\mathbf{E}\big(\xi(t+h) - \xi(t)\big)^4 = h^2\mathbf{E}\xi(1)^4 = 3h^2.$$

This means that the conditions of Theorem 18.2.1 are satisfied. □

Thus we can assume that $\xi(\cdot) \in C(0,1)$. The standard Wiener process with *continuous* trajectories will be denoted by $\{w(t),\ t \in T\}$.

Now note that the *trajectories of the Wiener process $w(t)$, being continuous, are not differentiable with probability* 1 *at any given point t.*

By virtue of the homogeneity of the process, it suffices to prove its nondifferentiability at the point 0. If, with a positive probability, i.e. on an event set $A \subset \Omega$ with $\mathbf{P}(A) > 0$, there existed the derivative

$$w'(0) = w'(0,\omega) = \lim_{t\to 0}\frac{w(t)}{t},$$

then, on the same event, there would exist the limit

$$\begin{aligned}\lim_{k\to\infty}\frac{w(2^{-k+1}) - w(2^{-k})}{2^{-k}} &= \lim_{k\to\infty}\frac{2w(2^{-k+1})}{2^{-k+1}} - \lim_{k\to\infty}\frac{w(2^{-k})}{2^{-k}}\\ &= 2w'(0) - w'(0) = w'(0).\end{aligned}$$

But this is impossible for the following reason. The independent differences $w(2^{-k+1}) - w(2^{-k})$ have the same distribution as $w(2^{-k})$, and with the positive probability $p = 1 - \Phi(1)$ they exceed the value $\sqrt{2^{-k}}$. That is, the independent events $B_k = \{w(2^{-k+1}) - w(2^{-k}) > \sqrt{2^{-k}}\}$ have the property $\sum_{k=1}^{\infty}\mathbf{P}(B_k) = \infty$. By the Borel–Cantelli criterion, this means that with probability 1 there occur infinitely many events B_k, so that

$$\mathbf{P}\left(\limsup_{k\to\infty}\frac{w(2^{-k+1}) - w(2^{-k})}{\sqrt{2^{-k}}} > 1\right) = 1.$$

In the same way we find that

$$\mathbf{P}\left(\liminf_{k\to\infty}\frac{w(2^{-k+1})-w(2^{-k})}{2^{-k}}<-1\right)=1.$$

This implies that, with probability 1,

$$\limsup_{k\to\infty}\frac{w(2^{-k+1})-w(2^{-k})}{2^{-k}}=\infty,\qquad \liminf_{k\to\infty}\frac{w(2^{-k+1})-w(2^{-k})}{2^{-k}}=-\infty,$$

and therefore the process $w(t)$ is nondifferentiable at any given point t with probability 1.

A stronger assertion also takes place: with probability 1 *there exists no point t at which the trajectory of the process $w(t)$ would have a derivative*. In other words, the Wiener process is *nowhere differentiable* with probability 1. The proof of this fact is much more complicated and lies beyond the scope of the book.

The reader can easily verify that $w(t)$ has, in a certain sense, a parabola property. Namely, for any $c>0$, the process $w^*(t)=c^{-1/2}w(ct)$ is again a Wiener process.

The properties of continuity of trajectories and independence of increments for the Wiener process allow us to find, in an explicit form, the distributions of

$$\overline{w}(t)=\max_{u\in[0,t]}w(u)$$

and of the time of the first passage of a given level which is defined, for a given $x>0$, by

$$\eta(x):=\inf\{t:w(t)\ge x\}=\inf\{t:w(t)=x\}.$$

Theorem 19.2.2

$$\mathbf{P}\big(\overline{w}(t)>x\big)=2\mathbf{P}\big(w(t)>x\big)=2\left(1-\Phi\left(\frac{x}{\sqrt{t}}\right)\right).\tag{19.2.1}$$

The distribution of $\eta(1)$ is stable and has the density

$$\frac{1}{\sqrt{2\pi}\,t^{3/2}}e^{-\frac{1}{2t}},\quad t>0.\tag{19.2.2}$$

Distribution (19.2.1) is sometimes called the *double normal tail law*, while the distribution with density (19.2.2) is called the Lévy distribution (see Sect. 8.8).

Proof Since

$$\{\eta(x)=v\}=\bigcap_{n=1}^{\infty}\{\overline{w}(v-1/n)<x,\,w(v)=x\}\in\mathfrak{F}_v:=\sigma\{w(u);\,u\le v\}$$

and $w(t)-w(v)\stackrel{d}{=}w(t-v)$ for $t>v$ does not depend on $\mathfrak{F}_v$, we have

$$\begin{aligned}\mathbf{P}\big(w(t)>x\big)&=\int_0^t\mathbf{P}\big(\eta(x)\in dv\big)\mathbf{P}\big(w(t-v)>0\big)\\&=\frac{1}{2}\int_0^t\mathbf{P}\big(\eta(x)\in dv\big)=\frac{1}{2}\mathbf{P}\big(\overline{w}(t)\ge x\big).\end{aligned}$$

This implies the first assertion of the theorem.

The same equalities imply that

$$\mathbf{P}\big(\eta(x) < v\big) = \mathbf{P}\big(\overline{w}(v) > x\big) = 2\left(1 - \Phi\left(\frac{x}{\sqrt{v}}\right)\right) = \frac{2}{\sqrt{2\pi}} \int_{x/\sqrt{v}}^{\infty} e^{-s^2/2}\,ds,$$

which yields, for the density f_η of the variable $\eta := \eta(1)$,

$$f_\eta(v) = \frac{e^{-\frac{1}{2v}}}{\sqrt{2\pi}\, v^{3/2}}.$$

In order to prove that this distribution is stable, note that

$$\eta(n) = \eta_1 + \cdots + \eta_n,$$

where η_i are distributed as η and are independent (since the path of $w(t)$ first attains level 1; then level 2, starting at a point with ordinate 1; then level 3, and so on). Using the same argument as above, we obtain that

$$\mathbf{P}\big(\eta(n) < v\big) = \mathbf{P}\big(\overline{w}(v) > n\big) = \mathbf{P}\big(\overline{w}\big(vn^{-2}\big) > 1\big) = \mathbf{P}\big(\eta < vn^{-2}\big),$$

so the distributions of η and $\eta(n)$ coincide up to a scale transformation. This implies the stability of the distribution of η (see Sect. 8.8). Since $\eta \ge 0$ and $\mathbf{P}(\eta > x) \sim \sqrt{\frac{2}{\pi x}}$ as $x \to \infty$, we obtain that it is, up to a scale transformation, the distribution $\mathbf{F}_{1/2,1}$ with parameters $\beta = 1/2$, $\rho = 1$ (cf. Sect. 8.8). The theorem is proved. □

19.3 The Laws of the Iterated Logarithm

Using an argument similar to that employed at the end of the previous section, one can establish a much stronger assertion: the trajectory of $w(t)$ in the neighbourhood of the point $t = 0$, graphically speaking, "completely shades" the interior of the domain bounded by the two curves

$$y = \pm\sqrt{2t \ln\ln\frac{1}{t}}.$$

The exterior of this domain remains untouched. This is the so-called *law of the iterated logarithm.*

Theorem 19.3.1

$$\mathbf{P}\left(\limsup_{t\to 0} \frac{w(t)}{\sqrt{2t\ln\ln\frac{1}{t}}} = 1\right) = 1,$$

$$\mathbf{P}\left(\liminf_{t\to 0} \frac{w(t)}{\sqrt{2t\ln\ln\frac{1}{t}}} = -1\right) = 1.$$

Thus, if we consider the sequence of random variables $w(t_n)$, $t_n \downarrow 0$, then, for any $\varepsilon > 0$,

$$(1 \pm \varepsilon)\sqrt{2t_n \ln\ln \frac{1}{t_n}}$$

will be upper and lower sequences, respectively, for that sequence.

For processes, we could introduce in a natural way the notions of *upper* and *lower functions*. If, for example, a process $\xi(t)$ belongs to $C(0,\infty)$ or $D(0,\infty)$ (or is separable on $(0,\infty)$), then the respective definition for the case $t \to \infty$ has the following form.

Definition 19.3.1 A function $a(t)$ is said to be *upper* (*lower*) for the process $\xi(t)$ if, for some sequence $t_n \uparrow \infty$, the events $A_n = \{\sup_{u \ge t_n}(\xi(t) - a(t)) > 0\}$ occur finitely (infinitely) often with probability 1.

Along with Theorem 19.3.1, we will obtain here the conventional law of the iterated logarithm. The proofs of the both assertions are essentially identical. We will prove the latter and derive the former as a consequence.

Theorem 19.3.2 (The Law of the Iterated Logarithm)

$$\mathbf{P}\biggl(\limsup_{t\to\infty} \frac{w(t)}{\sqrt{2t\ln\ln t}} = 1\biggr) = 1,$$

$$\mathbf{P}\biggl(\liminf_{t\to\infty} \frac{w(t)}{\sqrt{2t\ln\ln t}} = -1\biggr) = 1.$$

Thus, for any $\varepsilon > 0$, the functions $(1 \pm \varepsilon)\sqrt{2t\ln\ln t}$ are, respectively, upper and lower for $w(t)$ as $t \to \infty$.

Proof of Theorem 19.3.2 First observe that, by L'Hospital's rule,

$$\begin{aligned}\mathbf{P}\bigl(w(t) > x\bigr) &= \frac{1}{\sqrt{2\pi t}} \int_x^\infty e^{-u^2/2t}\,du \\ &= \frac{1}{\sqrt{2\pi t}} \int_{x/\sqrt{t}}^\infty e^{-u^2/2t}\,du \sim \frac{\sqrt{t}}{\sqrt{2\pi}\,x} e^{-x^2/2t} \end{aligned} \tag{19.3.1}$$

as $x/\sqrt{t} \to \infty$.

Let $a > 1$ and $x_k := \sqrt{2a^k \ln\ln a^k}$. We have to show that, for any $\varepsilon > 0$,

$$\mathbf{P}\biggl(\limsup_{t\to\infty} \frac{w(t)}{\sqrt{2t\ln\ln t}} < 1 + \varepsilon\biggr) = 1, \tag{19.3.2}$$

i.e. that, with probability 1, for all sufficiently large t,

$$w(t) < (1 + \varepsilon)\sqrt{2t\ln\ln t}.$$

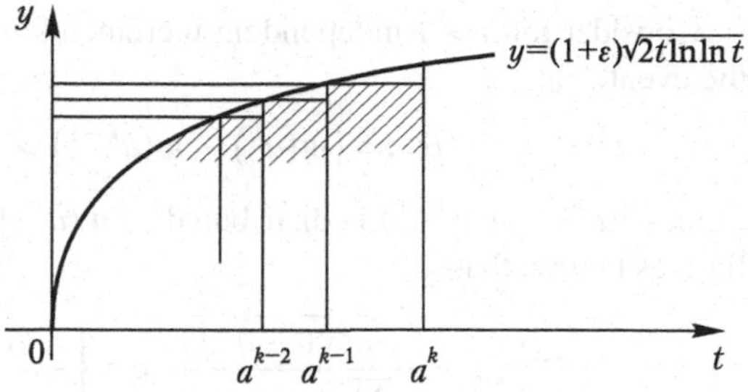

Fig. 19.1 Illustration to the proof of Theorem 19.3.2: replacing the curvilinear boundary with a step function

To this end it suffices to establish that, with probability 1, there occur only finitely many events

$$B_k := \Big\{ \sup_{a^{k-1}<u\le a^k} w(u) > (1+\varepsilon)x_{k-1} \Big\}.$$

Consider the events

$$A_k = \Big\{ \sup_{u\le a^k} w(u) > (1+\varepsilon)x_{k-1} \Big\} \supset B_k$$

(see Fig. 19.1). Because $x_k/\sqrt{a^k} \to \infty$ as $k\to\infty$, by Theorem 19.2.2 one has

$$\begin{aligned}
\mathbf{P}(A_k) &= 2\mathbf{P}\big(w\big(a^k\big) > (1+\varepsilon)x_{k-1}\big) \\
&\sim \sqrt{\frac{2}{\pi}} \frac{\sqrt{a^k}}{(1+\varepsilon)x_{k-1}} \exp\bigg\{ -\frac{2(1+\varepsilon)a^{k-1}\ln\ln a^{k-1}}{2a^k} \bigg\} \\
&= \frac{1}{(1+\varepsilon)} \sqrt{\frac{1}{\pi a \ln\ln a^{k-1}}} \frac{1}{(\ln a^{k-1})^{(1+\varepsilon)^2/a}} \\
&= c(a,\varepsilon) \frac{1}{\sqrt{(\ln(k-1)+\ln\ln a)}(k-1)^{(1+\varepsilon)^2/a}}.
\end{aligned}$$

Put $a := 1+\varepsilon > 1$. Then clearly

$$\mathbf{P}(A_k) \sim \frac{c(\varepsilon)}{k^{1+\varepsilon}\sqrt{\ln k}}$$

as $k \to \infty$.

In the above formulas, $c(a,\varepsilon)$ and $c(\varepsilon)$ are some constants depending on the indicated parameters. The obtained relation implies that $\sum_{k=1}^{\infty} \mathbf{P}(A_k) < \infty$ and hence $\sum_{k=1}^{\infty} \mathbf{P}(B_k) < \infty$ (for $B_k \subset A_k$), so that by the Borel–Cantelli criterion (Theorem 11.1.1) with probability 1 the events B_k occur only finitely often.

We now prove that, for an arbitrary $\varepsilon > 0$,

$$\mathbf{P}\bigg(\limsup_{t\to\infty} \frac{w(t)}{\sqrt{2t\ln\ln t}} > 1-\varepsilon \bigg) = 1. \qquad (19.3.3)$$

It is evident that, together with (19.3.2), this will mean that the first assertion of the theorem is true.

Consider for $a > 1$ independent increments $w(a^k) - w(a^{k-1})$ and denote by B_k the event

$$B_k := \{w(a^k) - w(a^{k-1}) > (1 - \varepsilon/2) r x_k\}.$$

Since $w(a^k) - w(a^{k-1})$ is distributed as $w(a^k(1 - a^{-1}))$, by virtue of (19.3.1) we find, as before, that

$$\begin{aligned}\mathbf{P}(B_k) &\sim \frac{\sqrt{a^k(1-a^{-1})}}{\sqrt{2\pi}(1-\varepsilon/2)x_k} \exp\left\{-\frac{(1-\varepsilon/2)^2 2a^k \ln\ln a^{-k}}{2a^k(1-a^{-1})}\right\} \\ &\sim \frac{c_1(a,\varepsilon)}{\sqrt{\ln k}} k^{-(1-\varepsilon/2)^2/(1-a^{-1})}.\end{aligned}$$

This implies that, for $a \geq 2/\varepsilon$, the series $\sum_{k=1}^{\infty} \mathbf{P}(B_k)$ diverges, and hence by the Borel–Cantelli criterion the events B_k occur infinitely often, with probability 1.

Further, by the symmetry of the process $w(t)$, it follows from relation (19.3.2) that, for all k large enough and any $\delta > 0$,

$$w(a^k) > -(1+\delta)x_k.$$

Together with the preceding argument this shows that the event

$$w(a^{k-1}) + [w(a^k) - w(a^{k-1})] = w(a^k) > -(1+\delta)x_{k-1} + (1-\varepsilon/2)x_k$$

will occur infinitely often. But the right hand-side of the above inequality can be made greater than $(1-\varepsilon)x_k$ by choosing an appropriate a. Indeed,

$$-(1+\delta)x_{k-1} + \frac{\varepsilon}{2}x_k > 0$$

once

$$(1+\delta)\sqrt{\frac{\ln\ln a^{k-1}}{a \ln\ln a^k}} < \frac{\varepsilon}{2},$$

which, in turn, can easily be achieved by taking a large enough. Thus relation (19.3.3) is proved.

The second assertion of the theorem clearly follows from the first by virtue of the symmetry of the distribution of $w(t)$. □

Now we can obtain as a consequence the local law of the iterated logarithm for the case where $t \to 0$.

Proof of Theorem 19.3.1 Consider the process $\{W(u) := uw(1/u),\ u \geq 0\}$, where we put $W(0) := 0$. The remarkable fact is that the *process* $\{W(u),\ u \geq 0\}$ *is also the standard Wiener process*. Indeed, for $t > u$,

$$\begin{aligned}\mathbf{E}\exp\{i\lambda(W(t) - W(u))\} &= \mathbf{E}\exp\left\{i\lambda\left[tw\left(\frac{1}{t}\right) - uw\left(\frac{1}{u}\right)\right]\right\} \\ &= \mathbf{E}\exp\left(i\lambda\left[w\left(\frac{1}{t}\right)(t-u) - u\left(w\left(\frac{1}{u}\right) - w\left(\frac{1}{t}\right)\right)\right]\right)\end{aligned}$$

$$= \exp\left\{-\frac{\lambda^2}{2}(t-u)^2\frac{1}{t} - \frac{\lambda^2 u^2}{2}\left(\frac{1}{u} - \frac{1}{t}\right)\right\}$$
$$= \exp\left\{-\frac{\lambda^2}{2}(t-u)\right\}.$$

The independence of increments is easiest to prove by establishing their noncorrelatedness. Indeed,

$$\mathbf{E}\big[W(u)\big(W(t) - W(u)\big)\big] = \mathbf{E}\left[uw\left(\frac{1}{u}\right)\left(tw\left(\frac{1}{t}\right) - uw\left(\frac{1}{u}\right)\right)\right]$$
$$= \mathbf{E}\left[uw\left(\frac{1}{t}\right)tw\left(\frac{1}{t}\right) - u^2 w^2\left(\frac{1}{u}\right)\right] = u - u = 0.$$

To complete the proof of the theorem, it remains to observe that

$$\limsup_{t\to\infty} \frac{w(t)}{\sqrt{2t\ln\ln t}} = \limsup_{u\to 0} \frac{uw(1/u)}{\sqrt{2u\ln\ln\frac{1}{u}}} = \limsup_{u\to 0} \frac{W(u)}{\sqrt{2u\ln\ln\frac{1}{u}}}.$$

The theorem is proved. □

We could also prove the theorem by repeating the argument from the proof of Theorem 19.3.2 with $a < 1$.

In conclusion we note that Wiener processes play an important role in many theoretical probabilistic considerations and serve as models for describing various real-life processes. For example, they provide a good model for the movement of a diffusing particle. In this connection, the Wiener processes are also often called *Brownian motion processes*.

Wiener processes prove to be, in a certain sense, the limiting processes for random polygons constructed on the vertices $(k/n, S_k/\sqrt{n})$, where S_k are sums of random variables ξ_j with $\mathbf{E}\xi_j = 0$ and $\mathrm{Var}(\xi_j) = 1$. We will discuss this in more detail in Chap. 20. The concept of the stochastic integral and many other constructions and results are also closely related to the Wiener process.

19.4 The Poisson Process

Definition 19.4.1 A homogeneous process $\xi(t)$ with independent increments is said to be the *Poisson process* if $\xi(t) - \xi(0)$ has the Poisson distribution.

For simplicity's sake put $\xi(0) = 0$. If $\xi(1) \in \Pi_\mu$, then

$$\varphi(\lambda) := \mathbf{E}e^{i\lambda\xi(1)} = \exp\big\{\mu\big(e^{i\lambda} - 1\big)\big\}$$

and, as we know,

$$\varphi_t(\lambda) = \mathbf{E}e^{i\lambda\xi(t)} = \varphi^t(\lambda) = \exp\big\{\mu t\big(e^{i\lambda} - 1\big)\big\},$$

so that $\xi(t) \Subset \Pi_{\mu t}$. We consider the properties of the Poisson process. First of all, for each t, $\xi(t)$ takes only integer values $0, 1, 2, \ldots$. Divide the interval $[0, t)$ into segments $[0, t_1), [t_1, t_2), \ldots, [t_{n-1}, t_n)$ of lengths $\Delta_i = t_i - t_{i-1}$, $i = 1, \ldots, n$. For small Δ_i the distributions of the increments $\xi(t_i) - \xi(t_{i-1})$ will have the property that

$$\begin{aligned}
&\mathbf{P}\big(\xi(t) - \xi(t_{i-1}) = 0\big) = \mathbf{P}\big(\xi(\Delta_i) = 0\big) = e^{-mu\Delta_i} = 1 - \mu\Delta_i + O\big(\Delta_i^2\big),\\
&\mathbf{P}\big(\xi(t) - \xi(t_{i-1}) = 1\big) = \mu\Delta_i e^{-\mu\Delta_i} = \mu\Delta_i + O\big(\Delta_i^2\big),\\
&\mathbf{P}\big(\xi(t) - \xi(t_{i-1}) \ge 2\big) = O\big(\Delta_i^2\big).
\end{aligned} \tag{19.4.1}$$

Consider "embedded" rational partitions $\mathcal{R}(n) = \{t_1, \ldots, t_n\}$ of the interval $[0, t]$ such that $\mathcal{R}(n) \subset \mathcal{R}(n+1)$ and $\bigcup \mathcal{R}(n) = \mathcal{R}_1$ is the set of all rationals in $[0, t]$.

Note the following three properties.

(1) Let $\nu(n)$ be the number of intervals in the partition $\mathcal{R}(n)$ on which the increments of the process ξ are non-zero. For each ω, $\nu(n)$ is non-decreasing as $n \to \infty$. Furthermore, the number $\nu(n)$ can be represented as a sum of independent random variables which are equal to 1 if there is an increment on the i-th interval and 0 otherwise. Therefore, by (19.4.1)

$$\begin{aligned}
\mathbf{P}\big(\nu(n) \ne \xi(t)\big) &= \mathbf{P}\Bigg(\bigcup_{t_i \in \mathcal{R}(n)} \big\{\xi(t_i) - \xi(t_{i-1}) \ge 2\big\} \cup \big\{\xi(t) - \xi(t_n) \ge 1\big\}\Bigg)\\
&= O\Bigg(\sum_{j=1}^{n} \Delta_j^2 + (t - t_n)\Bigg),
\end{aligned}$$

where $\sum_{j=1}^{n} \Delta_j^2 \le t \max \Delta_j \to 0$ as $n \to \infty$, so that a.s.

$$\nu(n) \uparrow \xi(t) \Subset \Pi_{\mu t}$$

as the partitions refine.

(2) Because the maximum length of the intervals Δ_j tends to 0 as $n \to \infty$, the total length of the intervals containing jumps converges to 0.

Therefore, taking the unions of the remaining adjacent intervals Δ_j (i.e. where there are no increments of ξ), for each ω we obtain in the limit, as $n \to \infty$, $\xi(t) + 1$ intervals $(0, T_1), (T_1, T_2), \ldots, (T_\nu, t)$ on which the increments of ξ are null.

(3) Finally, by (19.4.1) the probability that at least one of the increments on the intervals Δ_j exceeds one is $\sum_j O(\Delta_j^2) = o(1)$ as $n \to \infty$, so that, with probability 1, the jumps at the points T_k are equal to 1.

Thus we have shown that, on the segment $[0, t]$, for each ω there exists a finite number $\xi(t)$ of points $T_1, \ldots, T_{\xi(t)}$ such that $\xi(u)$ takes at the rational points of the intervals (T_k, T_{k+1}) one and the same constant value equal to k. This means that one can extend the trajectories of the process $\xi(u)$, say, by continuity from the right so that $\xi(u) = k$ for all $u \in [T_k, T_{k+1})$.

Thus, for the original process $\xi(t)$ we have constructed a *modification* $\overline{\xi}(t)$ *with trajectories in* $D_+(T)$. The equivalence of $\overline{\xi}$ and ξ follows from the very construction since, by virtue of (1),

$$\mathbf{P}\big(\overline{\xi}(t) = \xi(t)\big) = \mathbf{P}\Big(\lim_{n\to\infty} \nu(n) = \xi(t)\Big) = 1.$$

One usually considers just such right (or left) continuous modifications of the Poisson process. We have already dealt with processes of this kind in Chap. 10 where more general objects—renewal processes—were defined from scratch using trajectories. That the Poisson process is a renewal process is seen from the following considerations. It is easy to establish from relations (19.4.1) that the distributions of the random variables $T_1, T_2 - T_1, T_3 - T_2, \ldots$ coincide and that these variables are independent. Indeed, the difference $T_j - T_{j-1}$, $j \ge 1$, $T_0 = 0$, can be approximated by the sum $(\gamma_j - \gamma_{j-1})\Delta$ of the lengths of identical intervals of size $\Delta_i = \Delta$, where γ_j is the number of the interval in which the j-th non-zero increment of ξ occurred. Since the process $\xi(t)$ is homogeneous with independent increments, we have

$$\mathbf{P}\big((\gamma_j - \gamma_{j-1})\Delta > u\big) = \mathbf{P}\left(\gamma_1 > \frac{u}{\Delta}\right) = \big(e^{-\mu\Delta}\big)^{\lfloor u/\Delta \rfloor} \to e^{-\mu u},$$

$$\mathbf{P}\big((\gamma_j - \gamma_{j-1})\Delta > u\big) \to \mathbf{P}(T_j - T_{j-1} > u)$$

as $\Delta \to 0$. Hence the variables $\tau_j := T_j - T_{j-1}$, $j = 1, 2, 3, \ldots$, have the exponential distribution, and the value $\xi(t) + 1$ can be considered as the first crossing time of the level t by the sums T_j:

$$\xi(t) = \max\{k : T_k \le t\}, \qquad \xi(t) + 1 = \min\{k : T_k > t\}.$$

Thus we obtain that the Poisson process $\xi(t)$ coincides with the renewal process $\eta(t)$ (see Chap. 10) for exponentially distributed variables $\tau_1, \tau_2, \ldots$ with $\mathbf{P}(\tau_j > u) = e^{-\mu u}$.

The above and the properties of the Poisson process also imply the following remarkable property of exponentially distributed random variables. The numbers of jump points (i.e. sums T_k) which fall into disjoint time intervals δ_j are independent, these numbers being distributed according to the Poisson laws with parameters $\mu\delta_j$.

Using the last fact, one can construct a more general model of a pure jump homogeneous process with independent increments. Consider an arbitrary sequence of independent identically distributed random variables $\zeta_1, \zeta_2, \ldots$ that have a ch.f. $\beta(\lambda)$ and are independent of the σ-algebra generated by the process $\xi(t)$. Construct now a new process $\zeta(t)$ as follows. To each ω we put into correspondence a new trajectory obtained from the trajectory $\xi(t)$ by replacing the first unit jump with the variable ζ_1, the second one with the variable ζ_2, and so on. It is easy to see that $\zeta(t)$ will also be a process with independent increments. The value $\zeta(t)$ will be equal to the sum

$$\zeta(t) = \zeta_1 + \cdots + \zeta_{\xi(t)} \tag{19.4.2}$$

of the random number $\xi(t)$ of random variables $\zeta_1, \zeta_2, \ldots$, where $\xi(t)$ is independent of $\{\zeta_k\}$ by construction.

Hence, by the total probability formula,

$$\begin{aligned}\mathbf{E}e^{i\lambda\zeta(t)} &= \sum_{k=0}^{\infty} \mathbf{P}\big(\xi(t) = k\big)\mathbf{E}e^{i\lambda(\zeta_1 + \cdots + \zeta_k)} \\ &= \sum_{k=0}^{\infty} \frac{(\mu t)^k}{k!} e^{-\mu t}\big(\beta(\lambda)\big)^k = e^{-\mu t + \mu t\beta(\lambda)} = e^{\mu t(\beta(\lambda)-1)}.\end{aligned} \tag{19.4.3}$$

Definition 19.4.2 The process $\zeta(t)$ defined by formula (6) or ch.f. (7) is called a *compound Poisson process*. It is evidently a special case of the generalised renewal process (see Sect. 10.6).

As we have already noted, it is again a homogeneous process with independent increments. In formula (19.4.3), the parameter μ determines the jumps' intensity in the process $\zeta(t)$, while the ch.f. $\beta(\lambda)$ specifies their distribution. If we add a constant "drift" qt to the process $\zeta(t)$, then $\widetilde{\zeta}(t) = \zeta(t) + qt$ will clearly also be a homogeneous process with independent increments having the ch.f. $\mathbf{E}e^{i\lambda\widetilde{\zeta}(t)} = e^{t(i\lambda q+\mu(\beta(\lambda)-1))}$.

Finally, if a Wiener process $w(t)$ with zero drift and diffusion coefficient σ is given on the same probability space and is independent of $\widetilde{\zeta}(t)$, and to each ω we put into correspondence a trajectory of $\widetilde{\zeta}(t) + w(t)$, we again obtain a process with independent increments, with ch.f. $\exp\{t(i\lambda q + \mu(\beta(\lambda) - 1) - \lambda^2\sigma^2/2)\}$.

One should note, however, that these constructions by no means exhaust the whole class of processes with independent increments (and therefore the class of infinitely divisible distributions).

A description of the entire class will be given in the next section.

The Poisson processes, as well as Wiener processes, are often used as mathematical models in various applications. For example, the process of counts of cosmic particles of certain energy registered by a sensor in a given volume, or of collisions of elementary particles in an accelerator are described by the Poisson process. The same is true of the process of incoming telephone calls at a switchboard and many other processes.

Due to representation (19.4.2), the study of compound Poisson processes reduces, in many aspects, to the study of the properties of sums of independent random variables.

19.5 Description of the Class of Processes with Independent Increments

We saw in Theorem 19.1.1 that, to describe the class of distributions of stochastically continuous processes with independent increments, it suffices to describe the class of all infinitely divisible distributions. Let, as before, $\mathfrak{L}$ be the class of the ch.f.s of infinitely divisible distributions.

Lemma 19.5.1 *The class $\mathfrak{L}$ is closed with respect to the operations of multiplication and passage to the limit (when the limit is again a ch.f.).*

Proof (1) Let $\varphi_1 \in \mathfrak{L}$ and $\varphi_2 \in \mathfrak{L}$. Then $\varphi_1\varphi_2 = (\varphi_1^{1/n} \cdot \varphi_2^{1/n})^n$, where $\varphi_1^{1/n} \cdot \varphi_2^{1/n}$ is a ch.f.

(2) Let $\varphi_n \in \mathfrak{L}$, $\varphi_n \to \varphi$, and φ be a ch.f. Then, for any m, $\varphi_n^{1/m} \to \varphi^{1/m}$ as $n \to \infty$, where $\varphi^{1/m}$ is continuous at zero and hence is a ch.f. The lemma is proved. □

Denote by $\mathfrak{L}_\Pi \subset \mathfrak{L}$ the class of ch.f.s whose logarithms have the form

$$\ln\varphi(\lambda) = i\lambda q + \sum_k c_k\left(e^{i\lambda b_k} - 1\right), \quad c_k \ge 0, \quad \sum_k c_k < \infty.$$

We will call this the *Poisson class*. We already know that it corresponds to compound Poisson processes with drift q and intensities c_k of jumps of size b_k (note that $\sum_k c_k(e^{i\lambda b_k} - 1) = (\sum_k c_k)\mathbf{E}(e^{i\lambda\xi} - 1)$, where ζ assumes the values b_k with probabilities $c_k/\sum_j c_j$).

Lemma 19.5.2 *A ch.f. φ belongs to $\mathfrak{L}$ if and only if $\varphi = \lim_{n\to\infty}\varphi_n$, $\varphi_n \in \mathfrak{L}_\Pi$.*

Proof Sufficiency. Let

$$\ln\varphi_n = \sum_k \left(i\lambda q_{k,n} + c_{k,n}\left(e^{i\lambda b_{k,n}} - 1\right)\right),$$

and $\varphi = \lim\varphi_n$ be a ch.f. It is evident that $\varphi_n^{1/m} \in \mathfrak{L}_\Pi \subset \mathfrak{L}$ and $\varphi_n^{1/m} \to \varphi^{1/m}$. Therefore $\varphi^{1/m}$, being a limit of a sequence of ch.f.s which is continuous at zero, is a ch.f. itself, so that $\varphi \in \mathfrak{L}$.

Necessity. Let $\varphi \in \mathfrak{L}$. Then $\varphi(\lambda) \neq 0$ and there exists $\beta := \ln\varphi$ with $n(\varphi^{1/n} - 1) \to \beta$, and

$$\varphi^{1/n} - 1 = \int \left(e^{i\lambda x} - 1\right) dF_n(x).$$

The integral of the continuous function on the right-hand side can be viewed as a Riemann–Stieltjes integral. This means that for F_n there exists a partition of the real axis into intervals Δ_{nk} such that, for $x_{nk} \in \Delta_{nk}$ and $r_n < cn^{-2}$,

$$\int \left(e^{i\lambda x} - 1\right) dF_n(x) = \sum_k \int \left(e^{i\lambda x} - 1\right) P_n(\Delta_{nk}) + r_n$$

($P_n(\Delta)$ is the probability of hitting the interval Δ corresponding to F_n). We obtain

$$\beta = \lim n\left(\varphi^{1/n} - 1\right) = \lim_{n\to\infty}\left[n\sum_k \left(e^{i\lambda x_{nk}} - 1\right)P_n(\Delta_{nk})\right].$$

The lemma is proved. □

Theorem 19.5.1 (Lévy–Khintchin) *A ch.f. φ belongs to $\mathfrak{L}$ if and only if the function $\beta := \ln\varphi$ admits a representation of the form*

$$\beta = \beta(\lambda; a, \Psi) = i\lambda q + \int \left(e^{i\lambda x} - 1 - \frac{i\lambda x}{1+x^2}\right)\frac{1+x^2}{x^2}\, d\Psi(x), \qquad (19.5.1)$$

where Ψ is a non-decreasing function of bounded variation (i.e., a distribution function up to a constant factor), the integrand being assumed equal to $-\lambda^2/2$ at the point $x = 0$ (by continuity).

Proof Assume that β has the form (19.5.1). Then $\beta(\lambda)$ is a continuous function, since it is (up to a continuous additive term $i\lambda a$) a uniformly convergent integral of a continuous bounded function. Further, let $x_{nk} \neq 0$, $k = 1, \dots, n$, be points of refining partitions of intervals $[-\sqrt{n}, \sqrt{n})$. Then $\beta^0(\lambda) = \beta(\lambda) - i\lambda q$ can be represented as $\beta^0 = \lim \beta_n$ with

$$\beta_n(\lambda) := \sum_{k=1}^{n} \bigl[i\lambda q_{kn} + c_{kn}\bigl(e^{i\lambda b_{kn}} - 1\bigr)\bigr] \in \mathcal{L}_\Pi,$$

where, under a natural notational convention, one should put

$$c_{kn} = \frac{1 + x_{kn}^2}{x_{kn}^2} \Psi\bigl([x_{kn}, x_{k+1,n})\bigr), \qquad q_{kn} = \frac{1}{x_{kn}} \Psi\bigl([x_{kn}, x_{k+1,n})\bigr), \qquad b_{kn} = x_{kn},$$

Ψ being used to denote the measure $\Psi(A) = \int_A d\Psi(x)$. We obtain that φ is a limit of the sequence of ch.f.s $\varphi_n \in \mathcal{L}_\Pi$. It remains to make use of Lemma 19.5.2.

Now let $\varphi \in \mathcal{L}$. Then

$$\begin{aligned}
\beta &= \lim n\bigl(\varphi^{1/n} - 1\bigr) = \lim \int \bigl(e^{i\lambda x} - 1\bigr) n\, dF_n(x) \\
&= \lim \Biggl[i\lambda \int \frac{nx}{1+x^2}\, dF_n(x) \\
&\quad + \int \biggl(e^{i\lambda x} - 1 - \frac{i\lambda x}{1+x^2}\biggr) \frac{1+x^2}{x^2} \frac{nx^2}{1+x^2}\, dF_n(x) \Biggr].
\end{aligned} \tag{19.5.2}$$

If we put

$$q_n := \int \frac{nx}{1+x^2}\, dF_n(x), \qquad \Psi_n(x) := \frac{nx^2}{1+x^2}\, dF_n(x), \tag{19.5.3}$$

then on the right-hand side of (19.5.2) we will have $\lim \beta_n$, $\beta_n = \beta(\lambda; q_n, \Psi_n)$.

Now assume for a moment that the following continuity theorem holds for functions from $\mathcal{L}$.

Lemma 19.5.3 *If $\beta_n = \beta(\lambda; q_n, \Psi_n) \to \beta$ and β is continuous at the point $\lambda = 0$, then $\beta(\lambda)$ has the form $\beta(\lambda; q, \Psi)$, $q_n \to q$ and $\Psi_n \Rightarrow \Psi$.*

The symbol $\Rightarrow$ in the lemma means convergence at the points of continuity of the limiting function (as in the case of distribution functions) and that $\Psi_n(\pm\infty) \to \Psi(\pm\infty)$.

If the lemma is true, the required assertion of the theorem will follow in an obvious way from (19.5.2) and (19.5.3). It remains to prove the lemma.

Proof of Lemma 19.5.3 Observe first that the correspondence $\beta(\lambda; q, \Psi) \leftrightarrow (q, \Psi)$ is one-to-one. Since in one direction it is obvious, we only have to verify that β uniquely determines q and Ψ. To each β we put into correspondence the function

$$\begin{aligned}\gamma(\lambda) &= \int_0^1 \left[\beta(\lambda) - \frac{1}{2}\big(\beta(\lambda+h) - \beta(\lambda-h)\big)\right] dh \\ &= \int_0^1 \int \left(e^{i\lambda x} - \frac{1}{2}\big(e^{i(\lambda+h)x} - e^{i(\lambda-h)x}\big)\right) \frac{1+x^2}{x^2} \, d\Psi(x)\, dh,\end{aligned}$$

where

$$\frac{1}{2}\big(e^{i(\lambda+h)x} - e^{i(\lambda-h)x}\big) = e^{i\lambda x} \cos hx,$$

$$\int_0^1 e^{i\lambda x}(1 - \cos hx)\, dh = e^{i\lambda x}\left(1 - \frac{\sin x}{x}\right),$$

$$0 < c_1 < \left(1 - \frac{\sin x}{x}\right)\frac{1+x^2}{x^2} < c_2 < \infty.$$

Therefore

$$\gamma(\lambda) = \int e^{i\lambda x}\, d\Gamma(x),$$

where

$$\Gamma(x) = \int_{-\infty}^x \left(1 - \frac{\sin u}{u}\right)\frac{1+u^2}{u^2}\, d\Psi(u)$$

is (up to a constant multiplier) a distribution function, for which $\gamma(\lambda)$ plays the role of its ch.f. Clearly,

$$\Psi(x) = \int_{-\infty}^x \frac{1+u^2}{u^2}\left(1 - \frac{\sin u}{u}\right)^{-1} d\Gamma(u),$$

so that we obtained a chain of univalent correspondences $\beta \to \gamma \to \Gamma \to \Psi$ which proves the assertion.

We return to the proof of Lemma 19.5.3. Because $e^{\beta_n} \to e^{\beta}$, e^{β_n} is a ch.f., and e^{β} is continuous at the point $\lambda = 0$, we see that e^{β} is a ch.f. and hence a continuous function. This means that the convergence $\varphi_n \to \varphi$ is uniform on any interval,

$$\begin{aligned}\gamma_n(\lambda) &= \int_0^1 \left[\beta_n(\lambda) - \frac{1}{2}\big(\beta_n(\lambda+h) + \beta_n(\lambda-h)\big)\right] dh \\ &\to \int_0^1 \left[\beta(\lambda) - \frac{1}{1}\big(\beta(\lambda+h) + \beta(\lambda-h)\big)\right] dh =: \gamma(\lambda),\end{aligned}$$

and the function $\gamma(u)$ is continuous. By the continuity theorem for ch.f.s, this means that $\gamma(u)$ is a ch.f. (of a finite measure Γ), $\Gamma_n \Rightarrow \Gamma$ (where Γ_n is the preimage of γ_n), $\Psi_n \Rightarrow \Psi$, and $q_n \to q$. Thus we establish that

$$\begin{aligned}\beta = \lim \beta_n &= \lim\left[i\lambda q_n + \int \left(e^{i\lambda x} - 1 - \frac{i\lambda x}{1+x^2}\right) d\Psi_n(x)\right] \\ &= i\lambda q + \int \left(e^{i\lambda x} - 1 - \frac{i\lambda x}{1+x^2}\right) d\Psi(x) = \beta(\lambda; q, \Psi).\end{aligned}$$

Lemma 19.5.3 is proved. □

Theorem 19.5.1 is proved. □

Now we will make several remarks in regard to the structure of the process $\xi(t)$ and its relationship to representation (19.5.1). The function Ψ in (19.5.1) corresponds to the so-called *spectral measure of the process* $\xi(t)$ (recall that we agreed to use the same symbol Ψ for the measure itself: $\Psi(A) = \int_A d\Psi(x)$). It can be represented in the form $\mu\Psi_1(x)$, where $\mu = \Psi(\infty) - \Psi(-\infty)$ and $\Psi_1(x)$ is a distribution function.

(1) The spectral measure of the Wiener process is concentrated at the point 0. If $\Psi(\{0\}) = \sigma^2$, then $\xi(1) \in \mathbf{\Phi}_{q,\sigma^2}$.

(2) The spectral measure Ψ of a compound Poisson process has the property

$$\int |x|^{-2}\, d\Psi(x) < \infty.$$

In that case

$$G(x) = \int_{-\infty}^{x} \frac{1+u^2}{u^2}\, d\Psi(u)$$

possesses the properties of a distribution function, and $\psi(\lambda; q, \Psi)$ may be written in the form

$$i\lambda q_1 + \int \left(e^{i\lambda x} - 1\right) dG(x),$$

where

$$q_1 = q - \int x^{-1}\, d\Psi(x).$$

(3) Consider now the general case, but under the condition that $\Psi(\{0\}) = 0$. As we know, the function ψ can be approximated for small Δ by expressions of the form (we put $\Delta_k = [(k-1)\Delta, k\Delta)$)

$$i\lambda q + \sum_{\substack{k=-\infty \\ k\neq 0}}^{\infty} \left[-\frac{i\lambda}{k\Delta}\Psi(\Delta_k) + \left(e^{i\lambda k\Delta} - 1\right) \frac{1+(k\Delta)^2}{(k\Delta)^2}\Psi(\Delta_k) \right],$$

which corresponds to the sum of Poisson processes with jumps of sizes $k\Delta$ of the respective intensities

$$\frac{1+(k\Delta)^2}{(k\Delta)^2}\Psi(\Delta_k).$$

If, say,

$$\int_{+0}^{\infty} \frac{d\Psi(x)}{x^2} = \infty,$$

then for any $\varepsilon > 0$ the total intensity of these processes with jumps from the interval $(0, \varepsilon)$ will increase to infinity as $\Delta \to 0$. This means that, with probability 1, on any time interval δ there will be at least one jump of size smaller than any given $\varepsilon > 0$, so that the trajectories of $\xi(t)$ will be everywhere discontinuous. To "compensate" these jumps, a drift of size $\Psi(\Delta_k)/k\Delta$ is added, the "total value" of such drifts being possibly unbounded (if $\int_{+0}^{\infty} x^{-1} d\Psi(x) = \infty$).

(4) For stable processes (see Sect. 8.8) the functions $\Psi(x)$ have power "branches", smooth on the half-axes, possessing the property $c_1\Psi'(x) = \Psi'(c_2 x)$ for appropriate c_1 and c_2.

Chapter 20
Functional Limit Theorems

Abstract The chapter begins with Sect. 20.1 presenting the classical Functional Central Limit Theorem in the triangular array scheme. It establishes not only convergence of the distributions of the scaled trajectories of random walks to that of the Wiener process, but also convergence rates for Lipshchitz sets and distribution functions of Lipshchitz functionals in the case of finite third moments when the Lyapunov condition is met. Section 20.2 uses the Law of the Iterated Logarithm for the Wiener process to establish such a low for the trajectory of a random walk with independent non-identically distributed jumps. Section 20.3 is devoted to proving convergence to the Poisson process of the processes of cumulative sums of independent random indicators with low success probabilities and also that of the so-called thinning renewal processes.

20.1 Convergence to the Wiener Process

We have already pointed out in Sect. 19.2 that the Wiener processes are, in a certain sense, limiting to random polygons with vertices at the points $(k/n, S_k/\sqrt{n})$, where $S_k = \xi_1 + \cdots + \xi_k$ are partial sums of independent identically distributed random variables $\xi_1, \xi_2, \ldots$ with zero means and finite variances. Now we will give a more precise and general meaning to this statement.

Let

$$\xi_{1,n}, \ldots, \xi_{n,n} \tag{20.1.1}$$

be independent random variables in the triangular array scheme (see Sects. 8.3, 8.4),

$$\zeta_{k,n} := \sum_{j=1}^{k} \xi_{j,n}, \qquad \mathbf{E}\xi_{k,n} = 0, \qquad \mathbf{E}\xi_{k,n}^2 = \sigma_{k,n}^2,$$

that have finite third moments $\mathbf{E}|\xi_{k,n}|^3 = \mu_{k,n} < \infty$.

We will assume without loss of generality (see Sect. 8.4) that

$$\mathrm{Var}(\zeta_{n,n}) = \sum_{j=1}^{n} \sigma_{j,n}^2 = 1.$$

A.A. Borovkov, *Probability Theory*, Universitext,
DOI 10.1007/978-1-4471-5201-9_20, © Springer-Verlag London 2013

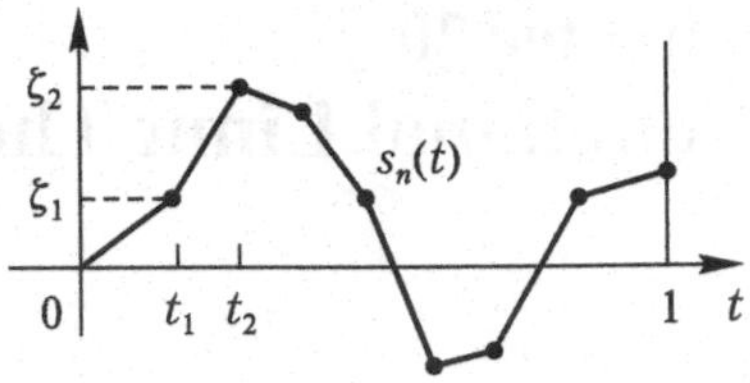

Fig. 20.1 The random polygon $s_n(t)$ constructed from the random walk $\zeta_0, \zeta_1, \zeta_2, \ldots$

Put

$$t_{k,n} := \sum_{j=1}^{k} \sigma_{j,n}^2,$$

so that $t_{0,n} = 0$, $t_{n,n} = 1$, and consider a random polygon with vertices at the points (t_k, ζ_k), where we suppress the second subscript n for brevity's sake: $t_k = t_{k,n}$, $\zeta_k = \zeta_{k,n}$.

We obtain a random process on $[0, 1]$ with continuous trajectories, which will be denoted by $s_n = s_n(t)$ (see Fig. 20.1). The functional limit theorem (or invariance principle; the motivation behind this second name will be commented on below) states that for any functional f given on the space $C(0, 1)$ and continuous in the uniform metric, the distribution of $f(s_n)$ converges weakly to that of $f(w)$ as $n \to \infty$:

$$f(s_n) \Longrightarrow f(w), \tag{20.1.2}$$

where $w = w(t)$ is the standard Wiener process. The conventional central limit theorem is a special case of this statement (one should take $f(x)$ to be $x(1)$).

The above assertion is equivalent to each of the following two statements:

1. For any bounded continuous functional f,

$$\mathbf{E}f(s_n) \to \mathbf{E}f(w), \quad n \to \infty. \tag{20.1.3}$$

2. For any set G from the σ-algebra $\mathfrak{B}_{C(0,1)}$ of Borel sets in the space $C(0, 1)$ ($\mathfrak{B}_{C(0,1)}$ is generated by open balls in the metric space $C(0, 1)$ endowed with the uniform distance ρ; as we already noted, $\mathfrak{B}_{C(0,1)} = \mathfrak{B}_C^{[0,1]}$) such that $\mathbf{P}(w \in \partial G) = 0$, where ∂G is the boundary of the set G, one has

$$\mathbf{P}(s_n \in G) \to \mathbf{P}(w \in G), \quad n \to \infty. \tag{20.1.4}$$

Relations (20.1.3) and (20.1.4) are equivalent definitions of weak convergence of the distributions $\mathbf{P}_n$ of the processes s_n to the distribution $\mathbf{W}$ of the Wiener process w in the space $\langle C(0, 1), \mathfrak{B}_{C(0,1)} \rangle$. More details can be found in Appendix 3 and in [1] and [14].

The main results of the present section are the following theorems.

As before, put $L_3 := \sum_{k=1}^{n} \mu_{k,n}$.

Theorem 20.1.1 *Let* $L_3 \to 0$ *as* $n \to \infty$ (the Lyapunov condition). *Then the convergence relations* (20.1.2)–(20.1.4) *hold true.*

Remark 20.1.1 The condition $L_3 \to 0$ can be relaxed here to the Lindeberg condition. In this version the above convergence theorem is known under the name of the *Donsker–Prokhorov invariance principle.*

Along with Theorem 20.1.1 we will obtain a more precise assertion.

Definition 20.1.1 A set G is said to be *Lipschitz* if $\mathbf{W}(G^{(\varepsilon)}) - \mathbf{W}(G) \le c\varepsilon$ for some $c < \infty$, where $G^{(\varepsilon)}$ is the ε-neighbourhood of G and $\mathbf{W}$ is the measure corresponding to the Wiener process.

In the sequel we will denote by the letter c (with or without subscripts) absolute constants, possibly having different values.

Theorem 20.1.2 *If G is a Lipschitz set, then*

$$\bigl|\mathbf{P}(s_n \in G) - \mathbf{P}(w \in G)\bigr| < cL_3^{1/4}. \tag{20.1.5}$$

In the case when $\xi_{k,n} = \xi_k/\sqrt{n}$, where the ξ_k do not depend on n and are identically distributed with $\mathbf{E}\xi_k = 0$ and $\operatorname{Var}(\xi_k) = 1$, the right-hand side of (20.1.5) becomes $cn^{-1/8}$.

A similar bound can be obtained for functionals. A functional on $C(0, 1)$ is said to be *Lipschitz* if the following two conditions are met:

(1) $|f(x) - f(y)| < c\rho(x, y)$;
(2) the distribution of $f(w)$ has a bounded density.

Corollary 20.1.1 *If f is a Lipschitz functional, then $G_v := \{f(x) < v\}$ is a Lipschitz set (with one and the same constant for all v), so that by Theorem* 20.1.2

$$\sup_v \bigl|\mathbf{P}\bigl(f(w) < v\bigr) - \mathbf{P}\bigl(f(s_n) < v\bigr)\bigr| \le cL_3^{1/4}.$$

The above theorems are consequences of Theorem 20.1.3 to be stated below. Let

$$\eta_{1,n}, \ldots, \eta_{n,n} \tag{20.1.6}$$

be any other sequence of independent identically distributed random variables in the triangular array scheme with the same two first moments $\mathbf{E}\eta_{k,n} = 0$, $\mathbf{E}\eta_{k,n}^2 = \sigma_{k,n}^4$, and finite third moments. Denote by $F_{k,n}$ and $\Phi_{k,n}$ the distribution functions of $\xi_{k,n}$ and $\eta_{k,n}$, respectively, and put

$$\nu_{k,n} := \mathbf{E}|\eta_{k,n}|^3 < \infty, \qquad N_3 := \sum_{k=1}^{n} \nu_{k,n},$$

$$\mu_{k,n}^0 := \int |x|^3 \bigl|d\bigl(F_{k,n}(x) - \Phi_{k,n}(x)\bigr)\bigr| \le \mu_{k,n} + \nu_{k,n},$$

$$L_3^0 := \sum_{k=1}^{n} \mu_{k,n}^0 \le L_3 + N_3.$$

Denote by $s'_n(t)$ the random process constructed in the same way as $s_n(t)$ but using the sequence $\{\eta_{k,n}\}$.

Theorem 20.1.3 *For any $A \in \mathfrak{B}_{C(0,1)}$ and any $\varepsilon > 0$,*

$$\mathbf{P}(s_n \in A) \le \mathbf{P}\big(s'_n \in A^{(2\varepsilon)}\big) + \frac{cL_3^0}{\varepsilon^3}.$$

In order to prove Theorem 20.1.3, we will first obtain its finite-dimensional analogue. Denote by ζ and η the vectors $\zeta = (\zeta_1, \dots, \zeta_n)$ and $\eta = (\eta_1, \dots, \eta_2)$ respectively, where $\zeta_k := \sum_{j=1}^{k} \zeta_{j,n}$ and $\eta_k := \sum_{j=1}^{k} \eta_{j,n}$, and by $B^{(\varepsilon)}$ the ε-neighbourhood of a set $B \in \mathbb{R}^n$:

$$B^{(\varepsilon)} := \bigcup_{\substack{x\in B\\ |v|\le\varepsilon}} (x+v),$$

where $x = (x_1, \dots, x_n)$, $v = (v_1, \dots, v_n)$, and $|v| = \max_k |v_k|$.

Lemma 20.1.1 *Let B be an arbitrary Borel subset of $\mathbb{R}^n$. Then, for any $\varepsilon > 0$,*

$$\mathbf{P}(\zeta \in B) \le \mathbf{P}\big(\eta \in B^{(2\varepsilon)}\big) + \frac{cL_3^0}{\varepsilon^3}.$$

Proof[1] Introduce a collection of nested neighbourhoods

$$B^{(\varepsilon)}(k) := \bigcup_{\substack{x\in B\\ |v|\le\varepsilon}} (x_1, \dots, x_k, x_{k+1} + v_{k+1}, \dots, x_n + v_n), \qquad k = 0, \dots, n,$$

$$B := B^{(\varepsilon)}(n) \subset B^{(\varepsilon)}(n-1) \subset \dots \subset B^{(\varepsilon)}(1) \subset B^{(\varepsilon)}(0) = B^{(\varepsilon)}$$

and denote by e_k the vector $(0, \dots, 0, 1, 0, \dots, 0)$, where 1 stands in the k-th position. It is obvious that if $x \in B^{(\varepsilon)}(k)$, then

$$x + \mathsf{e}_k v_k \in B^{(\varepsilon)}(k-1) \quad \text{if } |v_k| \le \varepsilon. \tag{20.1.7}$$

Further, together with arrays (20.1.1) and (20.1.6), consider the collection of "transitional" arrays

$$\xi_{1,n}, \dots, \xi_{k,n}, \eta_{k+1,n}, \dots, \eta_{n,n}, \quad k = 0, \dots, n. \tag{20.1.8}$$

Denote by $\zeta(k) = (\zeta_1(k), \dots, \zeta_n(k))$ the vectors formed by the cumulative sums of random variables from the k-th row (20.1.8), so that

$$\zeta_j(k) = \begin{cases} \zeta_j & \text{for } j \le k, \\ \zeta_k + \eta_{k+1,n} + \dots + \eta_{j,n} & \text{for } j > k. \end{cases}$$

To continue the proof of Lemma 20.1.1 we need the following.

[1] The extension of the approach to proving the central limit theorem used in Sect. 8.5, which is used in this demonstration, was suggested by A.V. Sakhanenko.

Lemma 20.1.2 *For any random variable δ such that $\mathbf{P}(|\delta| \le \varepsilon) = 1$, one has*

$$\mathbf{P}(\zeta \in B) \le \mathbf{P}\big(\eta \in B^{(2\varepsilon)}\big) + \sum_{k=1}^{n} \Delta_k, \tag{20.1.9}$$

where

$$\Delta_k = \mathbf{P}\big(\zeta(k) + \delta \mathrm{e}(k-1) \in B^{(\varepsilon)}(k-1)\big) - \mathbf{P}\big(\zeta(k-1) + \delta \mathrm{e}(k-1) \in B^{(\varepsilon)}(k-1)\big),$$

$$\mathrm{e}(r) = \sum_{j=r+1}^{n} \mathrm{e}_j = (0, \dots, 0, 1, \dots, 1).$$

Proof Indeed, by virtue of (20.1.7),

$$\begin{aligned}\mathbf{P}(\zeta \in B) &= \mathbf{P}\big(\zeta(n) \in B^{(\varepsilon)}(n)\big) \le \mathbf{P}\big(\zeta(n) + \mathrm{e}(n-1)\delta \in B^{(\varepsilon)}(n-1)\big)\\ &\equiv \mathbf{P}\big(\zeta(n-1) + \mathrm{e}(n-1)\delta \in B^{(\varepsilon)}(n-1)\big) + \Delta_n.\end{aligned}$$

Reapplying the same calculations to the right-hand side, we obtain that

$$\begin{aligned}&\mathbf{P}\big(\zeta(n-1) + \mathrm{e}(n-1)\delta \in B^{(\varepsilon)}(n-1)\big)\\ &\quad\le \mathbf{P}\big(\zeta(n-1) + \mathrm{e}(n-1)\delta + \mathrm{e}_{n-1}\delta \in B^{(\varepsilon)}(n-2)\big)\\ &\quad= \mathbf{P}\big(\zeta(n-1) + \mathrm{e}(n-2)\delta \in B^{(\varepsilon)}(n-2)\big)\\ &\quad\equiv \mathbf{P}\big(\zeta(n-2) + \mathrm{e}(n-2)\delta \in B^{(\varepsilon)}(n-2)\big) + \Delta_{n-1}\end{aligned}$$

....

$$\begin{aligned}&\mathbf{P}\big(\zeta(1) + \mathrm{e}(1)\delta \in B^{(\varepsilon)}(1)\big) \le \mathbf{P}\big(\zeta(1) + \mathrm{e}(1)\delta + \mathrm{e}_1\delta \in B^{(\varepsilon)}(0)\big)\\ &\quad= \mathbf{P}\big(\zeta(1) + \mathrm{e}(0)\delta \in B^{(\varepsilon)}(0)\big) \equiv \mathbf{P}\big(\zeta(0) + \mathrm{e}(0)\delta \in B^{(\varepsilon)}(0)\big) + \Delta_1.\end{aligned}$$

Since $\zeta(0) = \eta$ and $\mathbf{P}(\eta + \mathrm{e}(0)\delta \in B^{(\varepsilon)}) \le \mathbf{P}(\eta \in B^{(2\varepsilon)})$, inequality (20.1.9) is proved. Lemma 20.1.2 is proved. □

To obtain Lemma 20.1.1, we now have to estimate Δ_k. It will be convenient to consider, along with (20.1.8), the sequences

$$\xi_{1,n}, \dots, \xi_{k-1,n}, y, \eta_{k+1,n}, \dots, \eta_{n,n}$$

and denote by $\zeta(k, y) = (\zeta_1(k, y), \dots, \zeta_n(k, y))$ the respective vectors of cumulative sums, so that

$$\begin{aligned}\zeta(k, \xi_{k,n}) &= \zeta(k) = \zeta(k, 0) + \xi_{k,n}\mathrm{e}_{(k-1)},\\ \zeta(k, \eta_{k,n}) &= \zeta(k-1) = \zeta(k, 0) + \eta_{k,n}\mathrm{e}_{(k-1)}.\end{aligned}$$

Then Δ_k can be written in the form

$$\begin{aligned}\Delta_k &= \mathbf{P}\big(\zeta(k, 0) + (\delta + \xi_{k,n})\mathrm{e}(k-1) \in B^{(\varepsilon)}(k-1)\big)\\ &\quad- \mathbf{P}\big(\zeta(k, 0) + (\delta + \eta_{k,n})\mathrm{e}(k-1) \in B^{(\varepsilon)}(k-1)\big).\end{aligned} \tag{20.1.10}$$

Take δ to be a random variable independent of ζ and η. Then it will be convenient to use conditional expectation to estimate the probabilities participating in (20.1.10), because, for instance, in the equality

$$\mathbf{P}\big(\zeta(k,0)+(\delta+\xi_{k,n})\mathrm{e}(k-1)\in B^{(\varepsilon)}(k-1)\big)$$
$$=\mathbf{EP}\big((\delta+\xi_{k,n})\mathrm{e}(k-1)\in B^{(\varepsilon)}(k-1)-\zeta(k,0)\;\big|\;\zeta(k,0)\big) \quad (20.1.11)$$

the set $C=B^{(\varepsilon)}(k-1)-\zeta(k,0)$ may be assumed fixed (see the properties of conditional expectations; here δ and $\xi_{k,n}$ are independent of $\zeta(k,0)$). Denote by D the set of all ys for which $y\,\mathrm{e}(k-1)\in C$. We have to bound the difference

$$\mathbf{P}(\delta+\xi_{k,n}\in D)-\mathbf{P}(\delta+\eta_{k,n}\in D). \quad (20.1.12)$$

We make use of Lemma 8.5.1. To transform (20.1.12) to a form convenient for applying the lemma, take δ to be a random variable having a thrice continuously differentiable density $g(t)$ and put for brevity $\xi_{k,n}=\xi$ and $\eta_{k,n}=\eta$. Then $\delta+\xi$ will have a density equal to

$$\int dF_\xi(t)g(y-t)=\mathbf{E}g(y-\xi),$$

so that

$$\mathbf{P}(\delta+\xi\in D)=\int_D \mathbf{E}g(y-\xi)\,dy=\mathbf{E}\int_D g(y-\xi)\,dy.$$

Now putting

$$h(x):=\int_D g(y-x)\,dy,$$

we have

$$\mathbf{P}(\delta+\xi\in D)=\mathbf{E}h(\xi),$$

where h is a thrice continuously differentiable function,

$$\big|h'''(x)\big|\le\int\big|g'''(y)\big|\,dy=:h_3.$$

Applying now Lemma 8.5.1 we obtain that

$$\big|\mathbf{P}(\delta+\xi\in D)-\mathbf{P}(\delta+\eta\in D)\big|=\big|\mathbf{E}\big(h(\xi)-h(\eta)\big)\big|\le\frac{h_3}{6}\mu^0_{k,n},$$

$$\mu^0_{k,n}=\int|x^3|\big|d\big(F_{k,n}(x)-\Phi_{k,n}(x)\big)\big|.$$

Because the right-hand side here does not depend on $\xi(k,0)$ and D in any way, we get, returning to (20.1.10) and (20.1.11), the estimate

$$|\Delta_k|\le\frac{h_3}{6}\mu^0_{n,k}. \quad (20.1.13)$$

Now let $g_1(x)$ be a smooth density concentrated on $[-1,1]$. Then, putting

$$g(x):=g_1\left(\frac{x}{\varepsilon}\right)\frac{1}{\varepsilon},$$

we obtain that

$$h_3 = \int \frac{1}{\varepsilon^4}\left|g_1'''\left(\frac{y}{\varepsilon}\right)\right| dy = \frac{1}{\varepsilon^3}\int \left|g_1'''(y)\right| dy = \frac{c_1}{\varepsilon^3}, \quad c_1 = \text{const.} \qquad (20.1.14)$$

The assertion of Lemma 20.1.1 now follows from (20.1.9), (20.1.13) and (20.1.14). □

Proof of Theorem 20.1.3 This theorem is a consequence of Lemma 20.1.1. Indeed, let $B \in \mathbb{R}^n$ be such that the events $\{s_n \in A\}$ and $\{\zeta \in B\}$ are equivalent (s_n is completely determined by ζ). Then clearly $\{s_n \in A^{(\varepsilon)}\} = \{\zeta \in B^{(\varepsilon)}\}$ and the assertion of Theorem 20.1.3 repeats that of Lemma 20.1.1. Theorem 20.1.3 is proved. □

Proof of Theorem 20.1.1 Let $w(t)$ be the standard Wiener process. Put $\eta_{k,n} := w(t_{k,n}) - w(t_{k-1,n})$. Then the sequence $\eta_{1,n}, \ldots, \eta_{n,n}$ satisfies all the required conditions, for

$$\mathbf{E}\eta_{k,n} = 0, \qquad \mathbf{E}\eta_{k,n}^2 = \sigma_{k,n}^2, \qquad \nu_{k,n} = \mathbf{E}|\xi_{k,n}|^3 = c_3\sigma_{k,n}^3 < \infty.$$

Note also that

$$\sigma_{k,n}^3 = \left(\mathbf{E}|\xi_{k,n}|^2\right)^{3/2} \le \mathbf{E}|\xi_{k,n}|^3 = \mu_{k,n},$$

so that

$$N_3 = \sum_{k=1}^{n} \nu_{k,n} = c_3 \sum_{k=1}^{n} \sigma_{k,n}^3 \le c_3 L_3 \to 0.$$

We will need the following

Lemma 20.1.3 $\mathbf{P}(\rho(w, s_n') > \varepsilon) \le cN_3/\varepsilon^3$.

Proof The event $\{\rho(w, s_n') > \varepsilon\}$ is equal to $\bigcup_k A_k$, where

$$A_k := \left\{\sup_{t\in I_k}\left|w(t) - s'(t)\right| > \varepsilon\right\} \subset \left\{\sup_{t\in I_k}\left|w(t)\right| > \frac{\varepsilon}{2}\right\}, \qquad I_k := [t_{k-1}, t_k].$$

Therefore, recalling that $t_k - t_{k-1} = \sigma_{k,n}^2$ and $w(t) \stackrel{d}{=} \sigma w(t/\sigma^2)$, we have

$$\mathbf{P}(A_k) \le \mathbf{P}\left(\sup_{t\in[0,1)}\left|w(t)\right| > \frac{\varepsilon}{2\sigma_{k,n}}\right) \le 2\left(1 - \Phi\left(\frac{\varepsilon}{2\sigma_{k,n}}\right)\right).$$

The function $(1 - \Phi(t))$ vanishes as $t \to \infty$ much faster than t^{-3}. Hence

$$2\left(1 - \Phi\left(\frac{\varepsilon}{2\sigma_{k,n}}\right)\right) \le c\frac{\sigma_{k,n}^3}{\varepsilon^3}, \qquad \mathbf{P}\left(\bigcup_k A_k\right) \le \frac{cN_3}{\varepsilon^3}.$$

Lemma 20.1.3 is proved. □

We see from the proof that the bound stated by Lemma 20.1.3 is rather crude.

We return to the proof of Theorem 20.1.1. Because

$$\mathbf{P}(s'_n \in G) = \mathbf{P}\big(s'_n \in G, \rho(w, s'_n) \le \varepsilon\big) + \mathbf{P}\big(s'_n \in G, \rho(w, s'_n) > \varepsilon\big),$$

we have

$$\mathbf{P}(s'_n \in G) \le \mathbf{P}\big(w \in G^{(\varepsilon)}\big) + \frac{cN_3}{\varepsilon^3} \tag{20.1.15}$$

and, by Theorem 20.1.3,

$$\mathbf{P}(s_n \in G) \le \mathbf{P}\big(w \in G^{(3\varepsilon)}\big) + \frac{c(L_3^0 + N_3)}{\varepsilon^3}.$$

Now we prove the converse inequality. Introduce the set $G^{(-\varepsilon)} := G - (\partial G)^{(\varepsilon)}$. Then $[G^{(-\varepsilon)}]^{(\varepsilon)} =: G^0 \subset G$. Swapping s_n and s'_n in Theorem 20.1.3 and applying the latter to the set $G^{(2\varepsilon)}$, we obtain

$$\mathbf{P}\big(s_n \in G^0\big) \ge \mathbf{P}\big(s'_n \in G^{(-2\varepsilon)}\big) - \frac{c(L_3^0 + N_3)}{\varepsilon^3}. \tag{20.1.16}$$

Swapping w and s'_n in (20.1.15) and applying that relation to $G^{(-3\varepsilon)}$, we find that

$$\mathbf{P}\big(s'_n \in G^{(-2\varepsilon)}\big) \ge \mathbf{P}\big(w \in G^{(-3\varepsilon)}\big) - \frac{cN_3}{\varepsilon^3}.$$

This and (20.1.16) imply that

$$\mathbf{P}(s_n \in G) \ge \mathbf{P}\big(s_n \in G^0\big) \ge \mathbf{P}\big(w \in G^{(-3\varepsilon)}\big) - \frac{c(L_3^0 + N_3)}{\varepsilon^3}.$$

Setting

$$\mathbf{P}\big(w \in G^{(\varepsilon)}\big) - \mathbf{P}(w \in G) = \mathbf{W}\big(G^{(\varepsilon)}\big) - \mathbf{W}(G) =: \mathbf{W}_G(\varepsilon)$$

and taking into account that $N_3 \le cL_3$ and $L_3^0 \le L_3 + N_3$, we will obtain that

$$-\mathbf{W}_G(-3\varepsilon) + \frac{cL_3}{\varepsilon^3} \le \mathbf{P}(s_n \in G) - \mathbf{W}(G) \le \mathbf{W}_G(3\varepsilon) + \frac{cL_3}{\varepsilon^3}. \tag{20.1.17}$$

If $\mathbf{W}(\partial G) = 0$ then clearly

$$\mathbf{W}\big(G^{(3\varepsilon)}\big) - \mathbf{W}\big(G^{(-3\varepsilon)}\big) \to 0$$

as $\varepsilon \to 0$, and $\mathbf{W}_G(\pm 3\varepsilon) \to 0$. From this and (20.1.17) it is easy to derive that

$$\mathbf{P}(s_n \in G) \to \mathbf{P}(w \in G), \quad n \to \infty.$$

Convergence $f(s_n) \Longrightarrow f(w)$ for continuous functionals follows from (20.1.4), since if v is a point of continuity of the distribution of $f(w)$ then the set $G_v = \{x \in C(0, 1) : f(x) < v\}$ has the property

$$\mathbf{W}(\partial G_v) = \mathbf{P}\big(f(w) = v\big) = 0$$

and therefore

$$\mathbf{P}\big(f(s_n) < v\big) \to \mathbf{P}\big(f(w) \in v\big).$$

Theorem 20.1.1 is proved. □

Proof of Theorem 20.1.2 If G is a Lipschitz set, then

$$\left|\Delta \mathbf{W}_G(\pm 3\varepsilon)\right| < c\varepsilon,$$

and by (20.1.17)

$$\left|\mathbf{P}(s_n \in G) - \mathbf{W}(G)\right| < c\left(\varepsilon + \frac{L_3}{\varepsilon^3}\right).$$

Putting $\varepsilon := L_3^{1/4}$ we obtain the required assertion. Theorem 20.1.2 is proved. □

The reason for the name "*invariance principle*" used to refer to the main assertions of this section is best illustrated by Theorem 20.1.3. By virtue of the theorem, one can approximate the value of $\mathbf{P}(s_n \in A)$ by $\mathbf{P}(s_n' \in A)$ for any other sequence (20.1.6) having the same first two moments as (20.1.1). In that sense, the asymptotics of $\mathbf{P}(s_n \in A)$ are invariant with respect to particular distributions of the underlying sequences with fixed first two moments. For example, the calculation of $\mathbf{P}(s_n \in G)$ or $\mathbf{P}(w(t) \in G)$ can be replaced with that of $\mathbf{P}(s_n' \in G)$ for a Bernoulli sequence, which is convenient for various numerical methods. On the other hand, the probabilities $\mathbf{P}(w \in G)$ for a whole class of regions G were found in explicit form (see e.g. [32]). We know, for example, that $\mathbf{P}(\sup_{t\in[0,1]} w(t) > y) = 2(1 - \Phi(y))$. (This implies, in particular, that $G = \{x \in C(0,1) \sup_{t\in[0,1]} x(t) > y\}$ is a Lipschitz set.) Hence for the distribution of the maximum $\overline{S}_n = \max_{k\le n} S_k$ of the sums $S_k = \sum_{j=1}^k \xi_j$, when $\mathbf{E}\xi_k = 0$ and $\operatorname{Var}\xi_k = \sigma^2$, we have

$$\mathbf{P}(\overline{S}_n > x\sigma\sqrt{n}) \to 2\bigl(1 - \Phi(x)\bigr), \quad n \to \infty,$$

and one can use this relation for the approximate calculation of the distribution of $\overline{S}_n$ which is, as we saw in Chap. 12, of substantial interest in applications.

In the same way we can approximate the joint distribution of S_n, $\overline{S}_n$, and $\underline{S}_n := \min_{k\le n} S_k$ (i.e. the probabilities of the form $\mathbf{P}(\overline{S}_n < x\sqrt{n},\ \underline{S}_n > y\sqrt{n},\ S_n \in B)$) using the respective formulas for the Wiener process given in Skorokhod (1991).

Remark 20.1.2 In conclusion of this section note that all the above assertions will remain true if, instead of $s_n(t)$, we consider in them the step function $s_n^*(t) = \zeta_{k,n}$ for $t \in [t_k, t_{k+1})$. One can verify this by repeating anew all the arguments for s^*. Another way to obtain, say, Theorems 20.1.1 and 20.1.2 for s_n^* is to make use of the already obtained results and bound the distance $\rho(s_n, s_n^*)$. Because

$$\bigl\{\rho\bigl(s_n, s_n^*\bigr) > \varepsilon\bigr\} \subset \bigcup_{k=1}^{n} \bigl\{|\xi_{k,n}| > \varepsilon\bigr\},$$

one has

$$\mathbf{P}\bigl(\rho\bigl(s_n, s_n^*\bigr) > \varepsilon\bigr) \le \sum_{k=1}^{n} \mathbf{P}\bigl(|\xi_{k,n}| > \varepsilon\bigr) \le \sum_{k=1}^{n} \frac{\mu_{k,n}}{\varepsilon^3} = \frac{L_3}{\varepsilon^3}.$$

Recall that a similar bound was obtained for $\rho(s_n', w)$, and this allowed us to replace, where it was needed, the process s_n' with w. Therefore, using the same argument, one can replace s_n with s_n^*. In that case, we can consider convergence of the distributions of functionals $f(s_n^*)$ defined on $D(0,1)$ (and continuous in the uniform metric ρ). Sometimes the use of s_n^* is more convenient than that of s_n. This is the case, for example, when one has to find the limiting distribution of

$$\sum_{k=1}^{n} g(\zeta_{k,n}) = n \int g\big(s_n^*(t)\big)\,dt$$

($\xi_{k,n}$ are identically distributed). It follows from the above representation that

$$\frac{1}{n}\sum_{k=1}^{n} g(\zeta_{k,n}) \Longrightarrow \int g\big(w(t)\big)\,dt, \quad n\to\infty.$$

20.2 The Law of the Iterated Logarithm

Let $\xi_1, \xi_2, \dots$ be a sequence of independent random variables,

$$\mathbf{E}\xi_k = 0, \qquad \mathbf{E}\xi_k^2 = \sigma_k^2, \qquad \mathbf{E}|\xi_k|^3 = \mu_k,$$

$$S_n = \sum_{k=1}^{n} \xi_k, \qquad B_n^2 = \sum_{k=1}^{n} \sigma_k^2, \qquad M_n = \sum_{k=1}^{n} \mu_k.$$

In this notation, the Lyapunov ratio is equal to

$$L_3 = L_{3,n} = \frac{M_n}{B_n^3}.$$

In the present section, we will show that the law of the iterated logarithm for the Wiener process and Theorem 20.1.2 imply the following.

Theorem 20.2.1 (The law of the iterated logarithm for sums of random variables) *If $B_n \to \infty$ as $n \to \infty$ and $L_{3,n} < c/\ln B_n$ for some $c < \infty$, then*

$$\mathbf{P}\left(\overline{\lim_{n\to\infty}} \frac{S_n}{B_n\sqrt{2\ln\ln B_n}} = 1\right) = 1, \tag{20.2.1}$$

$$\mathbf{P}\left(\underline{\lim_{n\to\infty}} \frac{S_n}{B_n\sqrt{2\ln\ln B_n}} = -1\right) = 1. \tag{20.2.2}$$

Thus all the sequences which lie above

$$(1+\varepsilon)B_n\sqrt{2\ln\ln B_n}$$

will be upper for the sequence of sums S_n, while all the sequences below

$$(1-\varepsilon)B_n\sqrt{2\ln\ln B_n}$$

will be lower.

The conditions of the theorem will clearly be satisfied for identically distributed ξ_k, for in that case $B_n^2 = \sigma_1^2 n$, $L_{3,n} = \mu_1/(\sigma_1^3\sqrt{n})$.

Proof We turn to the proof of the law of the iterated logarithm in Theorem 19.3.2 and apply it to the sequence S_n. We will not need to introduce any essential changes. One just has to consider S_{n_k} instead of $w(a^k)$, where $n_k = \min\{n : B_n^2 \ge a^k\}$, and replace a^k with $B_{n_k}^2$ where it is needed. By the Lyapunov condition, $\max_{k\le n}\sigma_k^2 = o(B_n^2)$, so that $B_{n_k-1}^2 \sim B_{n_k}^2 \sim a^k$ as $k\to\infty$.

The key point in the proof of Theorem 19.3.2 is the proof of convergence (for any $\varepsilon > 0$) of the series

$$\sum_k \mathbf{P}\Big(\sup_{u\le a^k} w(u) > (1+\varepsilon)x_{k-1}\Big) \tag{20.2.3}$$

and divergence of the series

$$\sum_k \mathbf{P}\bigg(w\big(a^k\big) - w\big(a^{k-1}\big) > \Big(1-\frac{\varepsilon}{2}\Big)x_k\bigg), \tag{20.2.4}$$

where

$$x_k = \sqrt{2a^k \ln\ln a^k}, \qquad w\big(a^k\big) - w\big(a^{k-1}\big) \stackrel{p}{=} w\big(a^k\big(1-a^{-1}\big)\big).$$

In our case, if one follows the same argument, one has to prove the convergence of the series

$$\sum_k \mathbf{P}\big(\overline{S}_{n_k} > (1+\varepsilon)\, y_{k-1}\big) \tag{20.2.5}$$

and divergence of the series

$$\sum_k \mathbf{P}\bigg(S_{n_k} - S_{n_{k-1}} > \Big(1-\frac{\varepsilon}{2}\Big)y_k\bigg), \tag{20.2.6}$$

where $y_k = \sqrt{2B_{n_k}^2 \ln\ln B_{n_k}^2} \sim x_k$. But the asymptotic behaviour of the probabilities of the events in (20.2.3), (20.2.5) and (20.2.4), (20.2.6) under the conditions $L_{3,n} < c/\ln B_n$ will essentially be the same. To establish this, we will make use of the inequality

$$\bigg|\mathbf{P}\Big(\frac{s_n}{B_n} \in G\Big) - \mathbf{P}\big(w \in G^{(\pm\delta)}\big)\bigg| < \frac{cL_{3,n}}{\delta^3}, \tag{20.2.7}$$

which follows from the proof of Theorem 20.1.3. By this inequality,

$$\begin{aligned}\mathbf{P}\Big(\frac{\overline{S}_n}{B_n} > (1+3\varepsilon)x\Big) &\le \mathbf{P}\Big(\sup_{u\le 1} w(u) > (1+2\varepsilon)\,x\Big) + \frac{cL_{3,n}}{(\varepsilon x)^3}\\ &= \mathbf{P}\Big(\sup_{u\le B_n^2} w(u) > (1+2\varepsilon)x B_n\Big) + \frac{cL_{3,n}}{(\varepsilon x)^3}.\end{aligned}$$

Therefore (see (20.2.5)), putting $n := n_k$ and $x := y_{k-1}/B_n$, we obtain

$$\mathbf{P}\big(\overline{S}_{n_k} > (1+3\varepsilon)y_{k-1}\big) \le \mathbf{P}\Big(\sup_{u\le B_{n_k}^2} w(u) > (1+2\varepsilon)\,y_{k-1}\Big) + \frac{cL_{3,n_k}}{\varepsilon^3(\ln\ln B_{n_k}^2)^{3/2}}.$$

Here

$$y_{k-1} \sim x_{k-1}, \qquad B_{n_k}^2 \ge a^k, \qquad \ln\ln B_{n_k}^2 \sim \ln\ln a^k \sim \ln k,$$
$$L_{3,n_k} \le \frac{c}{\ln B_{n_k}} \sim \frac{c}{\ln a^k} \sim \frac{c_1}{k}.$$

Consequently, for all sufficiently large k (recall that the letter c denotes different constants),

$$\mathbf{P}\big(\overline{S}_{n_k} > (1+3\varepsilon)y_{k-1}\big) \le \mathbf{P}\Big(\sup_{u\le a^k} w(u) > (1+\varepsilon)x_{k-1}\Big) + \frac{c}{\varepsilon^3 k(\ln k)^{3/2}}.$$

Since

$$\sum_{k=1}^{\infty} \frac{1}{k(\ln k)^{3/2}} < \infty, \tag{20.2.8}$$

the above inequality means that the convergence of series (20.2.3) implies that of series (20.2.5). The first part of the theorem is proved.

The second part is proved in a similar way. Consider series (20.2.6). By (20.2.7),

$$\begin{aligned}
&\mathbf{P}\big(S_{n_k} - S_{n_{k-1}} > (1-3\varepsilon)y_k\big) \\
&\quad = \mathbf{P}\left(s_{n_k}(1) - s_{n_k}(r_k) > (1-3\varepsilon)\frac{y_k}{B_{n_k}}\right) \\
&\quad \ge \mathbf{P}\left(w(1) - w(r_k) > (1-2\varepsilon)\frac{y_k}{B_{n_k}}\right) - \frac{cL_{3,n_k}}{\varepsilon^3}\big(\ln\ln B_{n_k}^2\big)^{-3/2},
\end{aligned} \tag{20.2.9}$$

where $r_k = B_{n_{k-1}}^2 B_{n_k}^{-2} \to a^{-1}$ due to the fact that

$$B_{n_k}^2 = a^k + \theta_k\sigma_{n_k}^2, \quad 0 \le \theta_k \le 1, \quad \sigma_{n_k}^2 = o\big(B_{n_k}^2\big).$$

The first term on the right-hand side of (20.2.9) is equal to

$$\begin{aligned}
\mathbf{P}\left(w(1-r_k) > (1-2\varepsilon)\frac{y_k}{B_{n_k}}\right) &\ge \mathbf{P}\big(w\big(a^k(1-r_k)\big) > (1-\varepsilon)x_k\big) \\
&= \mathbf{P}\left(w\big(a^k\big(1-a^{-1}\big)\big) > (1-\varepsilon)x_k\sqrt{\frac{1-a^{-1}}{1-r_k}}\right) \\
&\ge \mathbf{P}\left(w\big(a^k\big) - w\big(a^{k-1}\big) > \left(1-\frac{\varepsilon}{2}\right)x_k\right).
\end{aligned}$$

As before, the series consisting of the second terms on the right-hand side of (20.2.9) converges by virtue of (20.2.8). Therefore the established inequalities mean that the divergence of series (20.2.4) implies that of series (20.2.6). The theorem is proved. □

Now we will present an example that we need to complete the proof in Remark 4.4.1.

Example 20.2.1 Let ζ_k be independent and identically distributed, $\mathbf{E}\zeta_k = 0$, $\mathbf{E}\zeta_k^2 = 1$, $\mathbf{E}|\zeta_k|^3 = \mu < \infty$ and $\xi_k = \sqrt{2k}\,\zeta_k$. Here we have $B_n^2 = n^2(1 + 1/n)$. In Remark 4.4.1 we used the assertion that (in a somewhat different notation)

$$\mathbf{P}\left(\bigcup_{n=1}^{\infty}\{S_n < -n\}\right) = 1$$

or, which is the same (as the sign of S_n is inessential),

$$\mathbf{P}\left(\bigcup_{n=1}^{\infty}\{S_n > n\}\right) = \mathbf{P}\left(\bigcup_{n=1}^{\infty}\left\{S_n > B_n\left(1 + O\left(\frac{1}{n}\right)\right)\right\}\right) = 1. \tag{20.2.10}$$

To verify it, we will show that any sequence of the form $B_n' = B_n(1+O(1/n))$ is lower for $\{S_k\}$. In our case,

$$M_n = \sum_{k=1}^{n}(2k)^{3/2}\mu \sim cn^{5/2}, \qquad L_{3.n} = \frac{M_n}{B_n^3} \sim cn^{-1/2} \ll \frac{1}{\ln B_n}.$$

This means that the conditions of Theorem 20.2.1 are met, and hence any sequence which lies lower than $(1-\varepsilon)n\sqrt{2\ln\ln n}$ (in particular, the sequence $B_n' = n$) is lower for $\{S_k\}$. This proves (20.2.10). □

Let us return to Theorem 20.2.1. As we saw in Sect. 19.3, the proof of the law of the iterated logarithm is based on the asymptotics (the rate of decrease) of the function $1 - \Phi(x)$ as $x \to \infty$. Therefore, the conditions for the law of the iterated logarithm for the sums S_n are related to the width of the range of x values for which the probabilities

$$\mathbf{P}_{n\pm}(x) := \mathbf{P}\left(\pm\frac{S_n}{B_n} > x\right)$$

are approximated by the normal law (i.e. by the function $1 - \Phi(x)$). Here we encounter the problem of large deviations (see Chap. 9). If

$$\frac{\mathbf{P}_{n\pm}(x)}{1 - \Phi(x)} \to 1 \tag{20.2.11}$$

as $n \to \infty$ for all

$$x \le \sqrt{2\ln\ln B_n}(1-\varepsilon) \tag{20.2.12}$$

and some $\varepsilon > 0$ then the proof of the law of the iterated logarithm for the Wiener process given in Sect. 19.3 can easily be extended to the sums S_n/B_n (to estimate $\mathbf{P}(\overline{S}_n/B_n > x)$ one has to use the Kolmogorov inequality; see Corollary 11.2.1).

One way to establish (20.2.11) and (20.2.12) is to use estimates for the rate of convergence in the central limit theorem. This approach was employed in the proof of Theorem 20.2.1, where we used Theorem 20.1.3. However, to ensure that

(20.2.11) and (20.2.12) hold one can use weaker assertions than Theorem 20.1.3. To some extent, this fact is illustrated by the following assertion (see [32]):

Theorem 20.2.2 *If $B_n \to \infty$ and $B_{n+1}/B_n \to 1$ as $n \to \infty$, and*

$$\sup_x \left| \mathbf{P}\left(\frac{S_n}{B_n} < x\right) - \Phi(x) \right| \le c(\ln B_n)^{-1-\delta}$$

for some $\delta > 0$ and $c < \infty$, then the law of the iterated logarithm holds.

If $\xi_k \stackrel{d}{=} \xi$ are identically distributed then Theorem 20.2.1 implies that the law of the iterated logarithm is valid whenever $\mathbf{E}|\xi|^3$ exists. In fact, however, for identically distributed ξ_k, the law of the iterated logarithm always holds in the case of a finite second moment, without any additional conditions.

Theorem 20.2.3 (Hartman–Wintner, [32]) *If the ξ_k are identically distributed, $\mathbf{E}\xi_k = 0$, and $\mathbf{E}\xi_k^2 = 1$, then* (20.2.1) *and* (20.2.2) *hold with B_n^2 replaced with n. Every point from the segment $[-1, 1]$ is a limiting one for the sequence*

$$\frac{S_n}{\sqrt{2n \ln\ln n}}, \qquad n \ge 1.$$

The last assertion of the theorem means that, for each $t \in [-1, 1]$ and any $\varepsilon > 0$, the interval $(t - \varepsilon, t + \varepsilon)$ contains, with probability 1, infinitely many elements of the sequence

$$\frac{S_n}{\sqrt{2n \ln\ln n}}.$$

20.3 Convergence to the Poisson Process

20.3.1 Convergence of the Processes of Cumulative Sums

The theorems of Sects. 20.1 and 20.2 show that the Wiener process describes rather well the evolution of the cumulative sums when summing "conventional" random variables $\xi_{k,n}$ satisfying the Lyapunov condition. It turns out that the Poisson process describes in a similar way the evolution of the cumulative sums when the random variables $\xi_{k,n}$ correspond to the occurrence of rare events.

As in Sect. 5.4, first we will not consider the triangular array scheme, but obtain precise inequalities describing the proximity of the processes under study. Consider independent random variables $\xi_1, \dots, \xi_n$ with Bernoulli distributions:

$$\mathbf{P}(\xi_k = 1) = p_k, \qquad \mathbf{P}(\xi_k = 0) = 1 - p_k, \qquad \sum_{k=1}^{n} p_k = \mu.$$

We will assume that $\overline{p} := \max_{k \le n} p_k$ is small and the number μ is "comparable with 1". Put

$$q_0 := 0, \qquad q_k := \frac{p_k}{\mu}, \qquad Q_k := \sum_{j=0}^{k} q_j, \quad k \ge 0,$$

and form a random function $s_n(t)$ on [0, 1] in the following way. Put $s_n(0) := 0$,

$$s_n(t) := S_k = \sum_{j=1}^{k} \xi_j \quad \text{for } t \in (Q_{k-1}, Q_k],\ k = 1, \dots, n.$$

Here it is more convenient to use a step function rather than a continuous trajectory $s_n(t)$ (cf. Remark 20.1.2). The assertions to be obtained in this section are similar to the invariance principle and state that the process $s_n(t)$ converges in a certain sense to the Poisson process $\xi(t)$ with intensity μ on [0, 1]. This convergence could of course be treated as weak convergence of distributions in the metric space $D(0, 1)$. But in the framework of the present book, it is apparently inexpedient for at least two reasons:

1. To do that, we would have to introduce a metric in $D(0, 1)$ and study its properties, which is somewhat complicated by itself.

2. The trajectories of the processes $s_n(t)$ and $\xi(t)$ are of a simple form, and characterising their closeness can be done in a simpler and more precise way without using more general concepts. Indeed, as we saw, the trajectory of $\xi(t)$ on [0, 1] is completely determined by the collection of random variables $(\pi(1); T_1, \dots, T_{\pi(1)})$, where T_k is the epoch of the k-th jump of the process, $T_{k+1} - T_k \in\!\!\!= \mathbf{\Gamma}_\mu$. A similar characterisation is valid for the trajectories of $s_n(t)$: they are determined by the vector $(s_n(1), \theta_1, \dots, \theta_{s_n(1)})$, where $\theta_k = Q_{\gamma_k}$, $\gamma_1, \gamma_2, \dots$ are the values j for which $\xi_j = 1$. We will say that the distributions of $s_n(t)$ and $\pi(t)$ are close to each other if the distributions of the above vectors are close. This convention will correspond to convergence of the processes in a rather strong and natural sense.

It is not hard to see from what we said before about the Poisson processes (see Sect. 19.4) that the introduced convergence of the distributions of the jump points of the process $s_n(t)$ is equivalent to convergence of the finite-dimensional distributions of $s_n(t)$ to those of $\pi(t)$ (we know that the trajectories of $s_n(t)$ are step functions).

Theorem 20.3.1 *The processes $s_n(t)$ and $\pi(t)$ can be constructed on a common probability space so that*

$$\mathbf{P}\big(s_n(1) = \pi(1);\ \theta_k - q_{\gamma_k} \le T_k < \theta_k,\ k = 1, \dots, \pi(1)\big) \ge 1 - \sum_{j=1}^{n} p_j^2. \tag{20.3.1}$$

Since $\sum_{j=1}^{n} p_j^2 \le \mu\overline{p}$, the smallness of $\overline{p}$ means that, with probability close to 1, the values of $s_n(1)$ and $\pi(1)$ coincide (cf. Theorem 5.4.2) and the positions of the respective points of jumps of the processes $s_n(t)$ and $\pi(t)$ do not differ much.

Put $\overline{q}=\overline{p}/\mu$ and, for a fixed $k\ge 1$, denote by $B^{(\varepsilon)}$ the ε-neighbourhood of the orthant set $B:=\{(x_1,\dots,x_k):x_j<v_j,\ j\le k\}$ for some $v_j>0$. Theorem 20.3.1 implies the following.

Corollary 20.3.1 *For any* $k=1,\dots,n$,

$$\mathbf{P}\bigl(s_n(1)=k,(\theta_1,\dots,\theta_k)\in B\bigr)\le \mathbf{P}\bigl(\pi(1)=k,(T_1,\dots,T_k)\in B\bigr)+\sum_{j=1}^n p_j^2;$$

$$\mathbf{P}\bigl(\pi(1)=k,(T_1,\dots,T_k)\in B\bigr)\le \mathbf{P}\bigl(s_n(1)=k,(\theta_1,\dots,\theta_k)\in B^{(\overline{q})}\bigr)+\sum_{j=1}^n p_j^2.$$

Proof Let A_n denote the event appearing on the left-hand side of (20.3.1),

$$D_n:=\bigl\{s_n(1)=k,\ (\theta_1,\dots,\theta_k)\in B\bigr\},$$
$$C_n:=\bigl\{\pi(1)=k,\ (T_1,\dots,T_k)\in B\bigr\}.$$

Then, by virtue of (20.3.1),

$$\begin{aligned}\mathbf{P}(D_n)&\le \mathbf{P}(D_nA_n)+\sum_{j=1}^n p_j^2\\&\le \mathbf{P}\bigl(D_n,\pi(1)=k,\ (T_1,\dots,T_k)\in B\bigr)+\sum_{j=1}^n p_j^2\\&\le \mathbf{P}\bigl(\pi(1)=k,\ (T_1,\dots,T_k)\in B\bigr)+\sum_{j=1}^n p_j^2.\end{aligned}$$

The converse inequality is established similarly. The corollary is proved. □

Proof of Theorem 20.3.1 Let $\eta_k:=\pi(Q_k)-\pi(Q_{k-1})$, $k=1,\dots,n$. The theorem will be proved if we construct $\{\xi_k\}$ and $\{\eta_k\}$ on a common probability space so that

$$\mathbf{P}\left(\bigcup_{k=1}^n\{\xi_k\ne\eta_k\}\right)\le\sum_{j=1}^n p_j^2. \tag{20.3.2}$$

A construction leading to (20.3.2) has essentially already been used in Theorem 5.4.2. The required construction will be obtained if we consider independent random variables $\omega_1,\dots,\omega_n$; $\omega_k\Subset\mathbf{U}_{0,1}$, and put

$$\xi_k:=\begin{cases}0 & \text{if } \omega_k<1-p_k,\\ 1 & \text{if } \omega_k\ge 1-p_k,\end{cases}\qquad \eta_k:=\begin{cases}0 & \text{if } \omega_k<e^{-p_k}=:\pi_{0,k},\\ j\ge 1 & \text{if } \omega_k\in[\pi_{j-1,k},\pi_{j,k}),\end{cases}$$

where $\pi_{j,k}=\mathbf{\Pi}_{p_k}([0,j))$, $j=0,1,\dots$. Then $\eta_k\Subset\mathbf{\Pi}_{p_k}$, $\sum_{k=1}^n\eta_k\Subset\mathbf{\Pi}_\mu$,

$$\{\xi_k\ne\eta_k\}=\bigl\{\omega_k\in[1-p_k,e^{-p_k})\cup\bigl[e^{-p_k}+p_ke^{-p_k},1\bigr]\bigr\}.$$

Therefore,

$$\mathbf{P}(\xi_k \neq \eta_k) \le p_k^2$$

and we get (20.3.2). The theorem is proved. □

If we now consider the triangular array scheme $\xi_{1,n}, \ldots, \xi_{n,n}$, for which

$$\mathbf{P}(\xi_{k,n} = 1) = p_{k,n}, \qquad \mathbf{P}(\xi_{k,n} = 0) = 1 - p_{k,n},$$

$$\sum_{k=1}^{n} p_{k,n} =: \mu_n \to \mu, \qquad \overline{p}_n = \max_{k \le n} p_{k,n} \to 0$$

as $n \to \infty$, then Theorem 20.3.1 easily implies *convergence of the finite-dimensional distributions of the processes* $s_n(t)$ *to* $\pi(t)$, where $s_n(t)$ is constructed as before and $\pi(t)$ is the Poisson process with parameter μ. Consider, for example, the two-dimensional distributions $\mathbf{P}(s_n(t) \ge j,\ s_n(1) = k)$ for $t \in (0, 1)$, $j \le k$. In the notation of Theorem 20.3.1 (to be precise, we have to add the subscript n where appropriate; e.g., the Poisson processes with parameters μ_n and μ will be denoted by $\pi_n(t)$ and $\pi(t)$, respectively), we obtain

$$\mathbf{P}\big(s_n(t) \ge j,\ s_n(1) = k\big) = \mathbf{P}\big(s_n(1) = k,\ \theta_j < t\big).$$

By Corollary 20.3.1 the right-hand side does not exceed

$$\mathbf{P}\big(\pi_n(1) = k,\ T_j < t\big) + \sum_{l=1}^{n} p_{l,n}^2,$$

where, as is easy to see,

$$\begin{aligned}
\mathbf{P}\big(\pi_n(1) = k,\ T_j < t\big) &= \mathbf{P}\big(\pi_n(1) = k,\ \pi_n(t) \ge j\big) \\
&= \sum_{l=j}^{k} \mathbf{P}\big(\pi_n(t) = l\big)\mathbf{P}\big(\pi_n(1-t) = k - l\big) \\
&\to \mathbf{P}\big(\pi(1) = k,\ \pi(t) \ge j\big)
\end{aligned}$$

as $n \to \infty$, so that

$$\mathbf{P}\big(s_n(t) \ge j,\ s_n(1) = k\big) \le \mathbf{P}\big(\pi(t) \ge j,\ \pi(1) = k\big) + o(1).$$

The converse inequality is established in a similar way (by using the convergence $\overline{q}_n \to 0$ as $n \to \infty$). The required convergence of the finite-dimensional distributions is proved. □

20.3.2 Convergence of Sums of Thinning Renewal Processes

The Poisson process can appear as a limiting one in a somewhat different set-up—as a limit for the sum of a large number of homogeneous "slow" renewal processes.

We formulate the setting of the problem more precisely. Let $\eta_i(t)$, $i = 1, 2, \dots, n$, be mutually independent *arbitrary homogeneous renewal processes* in the "triangular array scheme" (i.e. they depend on n) generated by sequences $\{\tau_k^{(i)}\}_{k=1}^{\infty}$ for which (see Chap. 10; $\tau_k^{(i)} \stackrel{d}{=} \tau^{(i)}$ for $k \ge 2$)

$$\mathbf{E}\eta_i(t) = \frac{t}{a_i}, \qquad a_i := a_{i,n} = \mathbf{E}\tau^{(i)} \to \infty, \qquad \sum_{i=1}^{n} \frac{1}{a_i} \to \mu$$

for a fixed μ, and

$$F_i(t) := \mathbf{P}\big(\tau^{(i)} < t\big) \le r_{t,n} \to 0$$

and for any fixed t as $n \to \infty$, where $r_{t,n}$ does not depend on i.

Theorem 20.3.2 *Under the above conditions, the finite-dimensional distributions of the process*

$$\zeta_n(t) := \sum_{i=1}^{n} \eta_i(t)$$

converge as $n \to \infty$ to those of the Poisson process $\pi(t)$ with the parameter μ: for any $l \ge 1$, $0 \le k_1 \le k_2 \le \cdots \le k_l$,

$$\mathbf{P}\big(\zeta_n(t_1) = k_1, \dots, \zeta_n(t_l) = k_l\big) \to \mathbf{P}\big(\pi(t_1) = k_1, \dots, \pi(t_l) = k_l\big).$$

(On convergence to the Poisson process, see the remark preceding Theorem 20.3.1.)

Proof First we will prove convergence of the distributions of the increments

$$\zeta_n(t + u) - \zeta_n(u)$$

to the Poisson distribution with parameter μt. Put $\Delta_i := \eta_i(t + u) - \eta_i(u)$, $p_i := t/a_i$. We have ($\chi_i(u)$ is the excess for the process η_i; see Sects. 10.2, 10.4)

$$\begin{aligned}
\mathbf{E}\Delta_i &= p_i, \\
\mathbf{P}(\Delta_i \ge l) &\le \mathbf{P}\big(\chi_i(u) < t\big)\big[\mathbf{P}\big(\xi_2^{(1)}\big) < t\big]^{l-1} \\
&\le \frac{1}{a_i} \int_0^t \mathbf{P}\big(\xi_2^{(1)} > z\big)\, dz \cdot F_i(t)^{l-1} \le \frac{t}{a_i} (r_{t,n})^{l-1} = p_i r_{t,n}^{l-1}.
\end{aligned}$$

This implies that

$$\begin{gathered}
\mathbf{E}\Delta_i = p_i = \sum_{1}^{\infty} l\mathbf{P}(\Delta_i = l) = \mathbf{P}(\delta_i = 1) + o(p_i), \\
\mathbf{P}(\Delta_i = 1) = p_i + o(p_i), \qquad \mathbf{P}(\Delta_i = 0) = 1 - p_i + o(p_i).
\end{gathered} \tag{20.3.3}$$

Therefore the conditions of Corollary 5.4.2 are met, which implies that

$$\zeta_n(t + u) - \zeta_n(u) = \sum_{i=1}^{n} \Delta_i \Leftrightarrow \mathbf{\Pi}_{\mu t}. \tag{20.3.4}$$

It remains to prove the asymptotic independence of the increments. For simplicity's sake, consider only two increments, on the intervals $(u, 0)$ and $(u, u+t)$, and assume that $\zeta_n(u) = k$. Moreover, suppose that the following event A occurred: the renewals occurred in the processes with numbers $i_1, \ldots, i_k$. It suffices to verify that, given this condition, (20.3.4) will still remain true. Let B be the event that there again were renewals on the interval $(u, u+t)$ in the processes with the numbers $i_1, \ldots, i_k$. Evidently,

$$\mathbf{P}(B \mid A) \le \sum_{l=1}^{k} \mathbf{P}\bigl(\tau^{(i_l)} < t+u\bigr) \le k r_{t+u,n} \to 0.$$

Thus the contribution of the processes η_{i_l}, $l = 1, \ldots, k$, to the sum (20.3.4) given condition A is negligibly small. Consider the remaining $n-k$ processes. For them,

$$\begin{aligned}\mathbf{P}(\Delta_i \ge 1 \mid A) &= \frac{\mathbf{P}(\chi_i(0) \in (u, u+t))}{\mathbf{P}(\chi_i(0) > u)} \\ &= \frac{1}{a_i} \int_u^{u+t} \bigl(1 - F_i(z)\bigr)\, da \left[1 - \frac{1}{a_i} \int_0^u \bigl(1 - F_i(z)\bigr)\, dz\right]^{-1} \\ &= p_i + o(p_i). \end{aligned} \tag{20.3.5}$$

Since relation (20.3.3) remains true for conditional distributions of Δ_i (given A and for $i \ne i_l$, $l = 1, \ldots, k$), we obtain, similarly to the above argument (using now instead of the equality $\sum_{i=1}^{\infty} l\mathbf{P}(\Delta_i = l) = p_i$ the relation $\sum_{i=1}^{\infty} l\mathbf{P}(\Delta_i = l \mid A) = p_i + o(p_i)$ which follows from (20.3.5)) that

$$\mathbf{P}(\Delta_i = 1 \mid A) = p_i + o(p_i), \qquad \mathbf{P}(\Delta_i = 0 \mid A) = 1 - p_i + o(p_i).$$

It remains to once again make use of Corollary 5.4.2. □

Chapter 21
Markov Processes

Abstract This chapter presents the fundamentals of the theory of general Markov processes in continuous time. Section 21.1 contains the definitions and a discussion of the Markov property and transition functions, and derives the Chapman–Kolmogorov equation. Section 21.2 studies Markov processes in countable state spaces, deriving systems of backward and forward differential equations for transition probabilities. It also establishes the ergodic theorem and contains examples illustrating the presented theory. Section 21.3 deals with continuous time branching processes. Then the elements of the general theory of semi-Markov processes are presented in Sect. 21.4, including the ergodic theorem and some other related results for such processes. Section 21.5 discusses the so-called regenerative processes, establishing their ergodicity and the Laws of Large Numbers and Central Limit Theorem for integrals of functions of their trajectories. Section 21.6 is devoted to diffusion processes. It begins with the classical definition of diffusion, derives the forward and backward Kolmogorov equations for the transition probability function of a diffusion process, and gives a couple of examples of using the equations to compute important characteristics of the respective processes.

21.1 Definitions and General Properties

Markov processes in discrete time (Markov chains) were considered in Chap. 13. Recall that their main property was independence of the "future" of the process of its "past" given its "present" is fixed. The same principle underlies the definition of Markov processes in the general case.

21.1.1 Definition and Basic Properties

Let $\langle \Omega, \mathfrak{F}, \mathbf{P} \rangle$ be a probability space and $\{\xi(t) = \xi(t, \omega),\ t \geq 0\}$ a random process given on it. Set

$$\mathfrak{F}_1 := \sigma\bigl(\xi(u);\ u \leq t\bigr), \qquad \mathfrak{F}_{[t,\infty)} := \sigma\bigl(\xi(u);\ u \geq t\bigr),$$

A.A. Borovkov, *Probability Theory*, Universitext,
DOI 10.1007/978-1-4471-5201-9_21, © Springer-Verlag London 2013

so that the variable $\xi(u)$ is $\mathfrak{F}_t$-measurable for $u \le t$ and $\mathfrak{F}_{[t,\infty)}$-measurable for $u \ge t$. The σ-algebra $\sigma(\mathfrak{F}_t, \mathfrak{F}_{[t,\infty)})$ is generated by the variables $\xi(u)$ for all u and may coincide with $\mathfrak{F}$ in the case of the sample probability space.

Definition 21.1.1 We say that $\xi(t)$ is a *Markov process* if, for any t, $A \in \mathfrak{F}_t$, and $B \in \mathfrak{F}_{[t,\infty)}$, we have

$$\mathbf{P}\big(AB\big|\xi(t)\big) = \mathbf{P}\big(A\big|\xi(t)\big)\mathbf{P}\big(B\big|\xi(t)\big). \tag{21.1.1}$$

This expresses precisely the fact that the future is independent of the past when the present is fixed (conditional independence of $\mathfrak{F}_t$ and $\mathfrak{F}_{[t,\infty)}$ given $\xi(t)$).

We will now show that the above definition is equivalent to the following.

Definition 21.1.2 We say that $\xi(t)$ is a *Markov process* if, for any bounded $\mathfrak{F}_{[t,\infty)}$-measurable random variable η,

$$\mathbf{E}(\eta|\mathfrak{F}_t) = \mathbf{E}\big(\eta\big|\xi(t)\big). \tag{21.1.2}$$

It suffices to take η to be functions of the form $\eta = f(\xi(s))$ for $s \ge t$.

Proof of the equivalence Let (21.1.1) hold. By the monotone convergence theorem it suffices to prove (21.1.2) for simple functions η. To this end it suffices, in turn, to prove (21.1.2) for $\eta = \mathrm{I}_B$, the indicator of the set $B \in \mathfrak{F}_{[t,\infty)}$. Let $A \in \mathfrak{F}_t$. Then, by (21.1.1),

$$\begin{aligned}\mathbf{P}(AB) &= \mathbf{E}\mathbf{P}\big(AB\big|\xi(t)\big) = \mathbf{E}\big[\mathbf{P}\big(A\big|\xi(t)\big)\mathbf{P}\big(B\big|\xi(t)\big)\big] \\ &= \mathbf{E}\mathbf{E}\big[\mathrm{I}_A\mathbf{P}\big(B\big|\xi(t)\big)\big|\xi(t)\big] = \mathbf{E}\big[\mathrm{I}_A\mathbf{P}\big(B\big|\xi(t)\big)\big].\end{aligned} \tag{21.1.3}$$

On the other hand,

$$\mathbf{P}(AB) = \mathbf{E}[\mathrm{I}_A\mathrm{I}_B] = \mathbf{E}\big[\mathrm{I}_A\mathbf{P}(B|\mathfrak{F}_t)\big]. \tag{21.1.4}$$

Because (21.1.3) and (21.1.4) hold for any $A \in \mathfrak{F}_t$, this means that $\mathbf{P}(B|\mathfrak{F}_t) = \mathbf{P}(B|\xi(t))$.

Conversely, let (21.1.2) hold. Then, for $A \in \mathfrak{F}_t$ and $B \in \mathfrak{F}_{[t,\infty)}$, we have

$$\begin{aligned}\mathbf{P}\big(AB\big|\xi(t)\big) &= \mathbf{E}\big[\mathbf{E}(\mathrm{I}_A\mathrm{I}_B|\mathfrak{F}_t)\big|\xi(t)\big] = \mathbf{E}\big[\mathrm{I}_A\mathbf{E}(\mathrm{I}_B|\mathfrak{F}_t)\big|\xi(t)\big] \\ &= \mathbf{E}\big[\mathrm{I}_A\mathbf{E}\big(\mathrm{I}_B\big|\xi(t)\big)\big|\xi(t)\big] = \mathbf{P}\big(B\big|\xi(t)\big)\mathbf{P}\big(A\big|\xi(t)\big).\end{aligned}$$

□

It remains to verify that it suffices to take $\eta = f(\xi(s))$, $s \ge t$, in (21.1.2). In order to do this, we need one more equivalent definition of a Markov process.

Definition 21.1.3 We say that $\xi(t)$ is a *Markov process* if, for any bounded function f and any $t_1 < t_2 < \cdots < t_n \le t$,

$$\mathbf{E}\big(f\big(\xi(t)\big)\big|\xi(t_1), \ldots, \xi(t_n)\big) = \mathbf{E}\big(f\big(\xi(t)\big|\xi(t_n)\big)\big). \tag{21.1.5}$$

Proof of the equivalence Relation (21.1.5) follows in an obvious way from (21.1.2). Now assume that (21.1.5) holds. Then, for any $A \in \sigma(\xi(t_1), \dots, \xi(t_n))$,

$$\mathbf{E}(f(\xi(t)); A) = \mathbf{E}[\mathbf{E}(f(\xi(t)) | \xi(t_n)); A]. \tag{21.1.6}$$

Both parts of (21.1.6) are measures coinciding on the algebra of cylinder sets. Therefore, by the theorem on uniqueness of extension of a measure, they coincide on the σ-algebra generated by these sets, i.e. on $\mathfrak{F}_{t_n}$. In other words, (21.1.6) holds for any $A \in \mathfrak{F}_{t_n}$, which is equivalent to the equality

$$\mathbf{E}[f(\xi(t)) | \mathfrak{F}_{t_n}] = \mathbf{E}[f(\xi(t)) | \xi(t_n)]$$

for any $t_n \le t$. Relation (21.1.2) for $\eta = f(\xi(t))$ is proved. □

We now prove that in (21.1.2) it suffices to take $\eta = f(\xi(s))$, $s \ge t$. Let $t \le u_1 < \dots < u_n$. We prove that then (21.1.2) is true for

$$\eta = \prod_{i=1}^{n} f_i(\xi(u_i)). \tag{21.1.7}$$

We will make use of induction and assume that equality (21.1.2) holds for the functions

$$\gamma = \prod_{i=1}^{n-1} f_i(\xi(u_i))$$

(for $n = 1$ relation (21.1.2) is true). Then, putting $g(u_{n-1}) := \mathbf{E}[f_n(\xi(u_n)) | \xi(u_{n-1})]$, we obtain

$$\begin{aligned} \mathbf{E}(\eta | \mathfrak{F}_t) &= \mathbf{E}[\mathbf{E}(\eta | \mathfrak{F}_{u_{n-1}}) | \mathfrak{F}_t] = \mathbf{E}[\gamma \mathbf{E}(f_n(\xi(u_n)) | \mathfrak{F}_{u_{n-1}}) | \mathfrak{F}_t] \\ &= \mathbf{E}[\gamma \mathbf{E}(f_n(\xi(u_n)) | \xi(u_{n-1})) | \mathfrak{F}_t] = \mathbf{E}[\gamma g(\xi(u_{n-1})) | \mathfrak{F}_t]. \end{aligned}$$

By the induction hypothesis this implies that

$$\mathbf{E}(\eta | \mathfrak{F}_t) = \mathbf{E}[\gamma g(\xi(u_{n-1})) | \xi(t)]$$

and, therefore, that $\mathbf{E}(\eta | \mathfrak{F}_t)$ is $\sigma(\xi(t))$-measurable and

$$\mathbf{E}(\eta | \xi(t)) = \mathbf{E}(\mathbf{E}(\eta | \mathfrak{F}_t) | \xi(t)) = \mathbf{E}(\eta | \mathfrak{F}_t).$$

We proved that (21.1.2) holds for $\sigma(\xi(u_1), \dots, \xi(u_n))$-measurable functions of the form (21.1.7). By passing to the limit we establish first that (21.1.2) holds for simple functions, and then that it holds for any $\mathfrak{F}_{[t,\infty)}$-measurable functions. □

21.1.2 Transition Probability

We saw that, for a Markov process $\xi(t)$, the conditional probability

$$\mathbf{P}(\xi(t) \in B | \mathfrak{F}_s) = \mathbf{P}(\xi(t) \in B | \xi(s)) \quad \text{for } t > s$$

is a Borel function of $\xi(s)$ which we will denote by

$$P\big(s, \xi(s); t, B\big) := \mathbf{P}\big(\xi(t) \in B \big| \xi(s)\big).$$

One can say that $P(s, x; t, B)$ as a function of B and x is the conditional distribution (see Sect. 4.9) of $\xi(t)$ given that $\xi(s) = x$. By the Markov property, it satisfies the relation $(s < u < t)$

$$P(s, x; t, B) = \int P(s, x; u, dy) P(u, y; t, B), \tag{21.1.8}$$

which follows from the equality

$$\begin{aligned}&\mathbf{P}\big(\xi(t) \in B \big| \xi(s) = x\big)\\ &\quad = \mathbf{E}\big[\mathbf{P}\big(\xi(t) \in B \big| \mathfrak{F}_u\big) \big| \xi(s) = x\big] = \mathbf{E}\big[P\big(u, \xi(u); t, B\big) \big| \xi(s) = x\big].\end{aligned}$$

Equation (21.1.8) is called the *Chapman–Kolmogorov equation*.

The function $P(s, x; t, B)$ can be used in an analytic definition of a Markov process. First we need to clarify what properties a function $P_{x,B}(s, t)$ should possess in order that there exists a Markov process $\xi(t)$ for which

$$P_{x,B}(s, t) = P(s, x; t, B).$$

Let $\langle \mathcal{X}, \mathfrak{B}_{\mathcal{X}} \rangle$ be a measurable space.

Definition 21.1.4 A function $P_{x,B}(s, t)$ is said to be a *transition function on* $\langle \mathcal{X}, \mathfrak{B}_{\mathcal{X}} \rangle$ if it satisfies the following conditions:

(1) As a function of B, $P_{x,B}(s, t)$ is a probability distribution for each $s \le t$, $x \in \mathcal{X}$.
(2) $P_{x,B}(s, t)$ is measurable in x for each $s \le t$ and $B \in \mathfrak{B}_{\mathcal{X}}$.
(3) For $0 \le s < u < t$ and all x and B,

$$P_{x,B}(s, t) = \int P_{x,dy}(s, u) P_{y,B}(u, t)$$

(the Chapman–Kolmogorov equation).
(4) $P_{x,B}(s, t) = \mathrm{I}_B(x)$ for $s = t$.

Here properties (1) and (2) ensure that $P_{x,B}(s, t)$ can be a conditional distribution (cf. Sect. 4.9).

Now define, with the help of $P_{x,B}(s, t)$, the finite-dimensional distributions of a process $\xi(t)$ with the initial condition $\xi(0) = a$ by the formula

$$\begin{aligned}&\mathbf{P}\big(\xi(t) \in dy_1, \ldots, \xi(t_n) \in dy_n\big)\\ &\quad = P_{a,dy_1}(0, t_1) P_{y_1,dy_2}(t_1, t_2) \cdots P_{y_{n-1},dy_n}(t_{n-1}, t_n).\end{aligned} \tag{21.1.9}$$

By virtue of properties (3) and (4), these distributions are consistent and therefore by the Kolmogorov theorem define a process $\xi(t)$ in $\langle \mathbb{R}^T, \mathfrak{B}_R^T \rangle$, where $T = [0, \infty)$.

By formula (21.1.9) and rule (21.1.5),

$$\begin{aligned}&\mathbf{P}\big(\xi(t_n)\in B_n\big|\big(\xi(t_1),\dots,\xi(t_{n-1})\big)=(y_1,\dots,y_{n-1})\big)\\&\quad=P_{y_{n-1},B_n}(t_{n-1},t_n)=\mathbf{P}\big(\xi(t_n)\in B_n\big|\xi(t_{n-1})=y_{n-1}\big)\\&\quad=P(t_{n-1},y_{n-1};t_n,B_n).\end{aligned}$$

We could also verify this equality in a more formal way using the fact that the integrals of both sides over the set $\{\xi(t_1)\in B_1,\dots,\xi(t_{n-1})\in B_{n-1}\}$ coincide.

Thus, by virtue of Definition 21.1.3, we have constructed a Markov process $\xi(t)$ for which

$$P(s,x;t,B)=P_{x,B}(s,t).$$

This function will also be called the *transition function* (or *transition probability*) *of the process* $\xi(t)$.

Definition 21.1.5 A Markov process $\xi(t)$ is said to be *homogeneous* if $P(s,x;t,B)$, as a function of s and t, depends on the difference $t-s$ only:

$$P(s,x;t,B)=P(t-s;x,B).$$

This is the probability of transition during a time interval of length $t-s$ from x to B. If

$$P(u;t,B)=\int_B p(u;t,y)\,dy$$

then the function $p(u;x,y)$ is said to be a *transition density*.

It is not hard to see that the Wiener and Poisson processes are both homogeneous Markov processes. For example, for the Wiener process,

$$P(u;x,y)=\frac{1}{\sqrt{2\pi u}}e^{-(x-y)^2/2u}.$$

21.2 Markov Processes with Countable State Spaces. Examples

21.2.1 Basic Properties of the Process

Assume without loss of generality that the "discrete state space" $\mathcal{X}$ coincides with the set of integers $\{0,1,2,\dots\}$. For simplicity's sake we will only consider homogeneous Markov processes.

The transition function of such a process is determined by the collection of functions $P(t;i,j)=p_{ij}(t)$ which form a stochastic matrix $P(t)=\|p_{ij}(t)\|$ (with $p_{ij}(t)\ge 0$, $\sum_j p_{ij}(t)=1$). Chapman–Kolmogorov's equation now takes the form

$$p_{ij}(t+s)=\sum_k p_{ik}(t)p_{kj}(s),$$

or, which is the same, in the matrix form,

$$P(t+s) = P(t)P(s) = P(s)P(t). \qquad (21.2.1)$$

In what follows, we consider only *stochastically continuous processes* for which $\xi(t+s) \xrightarrow{p} \xi(t)$ as $s \to 0$, which is equivalent in the case under consideration to each of the following three relations:

$$\mathbf{P}\big(\xi(t+s) \neq \xi(t)\big) \to 0, \qquad P(t+s) \to P(t), \qquad P(s) \to P(0) \equiv E \quad (21.2.2)$$

as $s \to 0$ (component-wise; E is the unit matrix).

We will also assume that convergence in (21.2.2) is *uniform* (for a finite $\mathcal{X}$ this is always the case).

According to the separability requirement, we will assume that $\xi(t)$ cannot change its state in "zero time" more than once (thus excluding the effects illustrated in Example 18.1.1, i.e. assuming that if $\xi(t) = j$ then, with probability 1, $\xi(t+s) = j$ for $s \in [0, \tau)$, $\tau = \tau(\omega) > 0$). In that case, the trajectories of the processes will be piece-wise constant (right-continuous for definiteness), i.e. the time axis is divided into half-intervals $[0, \tau_1), [\tau_1, \tau_1 + \tau_2), \ldots$, on which $\xi(t)$ is constant. Put

$$q_j(t) := \mathbf{P}\big(\xi(u) = j,\ 0 \le u < t \big| \xi(0) = j\big) = \mathbf{P}(\tau_1 \ge t).$$

Theorem 21.2.1 *Under the above assumptions (stochastic continuity and separability),*

$$q_i(t) = e^{-q_i t},$$

where $q_i < \infty$; moreover, $q_i > 0$ if $p_{ii}(t) \not\equiv 1$. There exist the limits

$$\lim_{t\to 0} \frac{1 - p_{ii}(t)}{t} = q_i, \qquad \lim_{t\to 0} \frac{p_{ij}(t)}{t} = q_{ij}, \quad i \neq j, \qquad (21.2.3)$$

where $\sum_{j:j\neq i} q_{ij} = q_i$.

Proof By the Markov property,

$$q_i(t+s) = q_i(t) q_i(s),$$

and $q_i(t) \downarrow$. Therefore there exists a unique solution $q_i(t) = e^{-q_i t}$ of this equation, where $q_i < \infty$, since $\mathbf{P}(\tau_1 > 0) = 1$ and $q_i > 0$, because $q_i(t) < 1$ when $p_{ii}(t) \not\equiv 1$.

Let further $0 < t_0 < t_1 \cdots < t_n < t$. Since the events

$$\big\{\xi(u) = i \text{ for } u \le t_r, \xi(t_{r+1}) = j\big\}, \quad r = 0, \ldots, n-1;\ j \neq i,$$

are disjoint,

$$p_{ii}(t) = q_i(t) + \sum_{r=0}^{n-1} \sum_{j:j\neq i} q_i(t_r) p_{ij}(t_{r+1} - t_r) p_{ji}(t - t_{r+1}). \qquad (21.2.4)$$

Here, by condition (21.2.2), $p_{ji}(t - t_{r+1}) < \varepsilon_t$ for all $j \neq i$, and $\varepsilon_t \to 0$ as $t \to 0$, so that the sum in (21.2.4) does not exceed

$$\varepsilon_t \sum_{r=0}^{n-1} \sum_{j:j\neq i} q_i(t_r) p_{ij}(t_{r+1} - t_r) = \varepsilon_t \mathbf{P}\left(\bigcup_{r=1}^{n} \{\xi(t_r) \neq i\} \big| \xi(0) = i \right) < \varepsilon_t \big(1 - q_i(t)\big),$$

$$p_{ii}(t) \le q_i(t) + \varepsilon_t \big(1 - q_i(t)\big).$$

Together with the obvious inequality $p_{ii}(t) \ge q_i(t)$ this gives

$$1 - q_i(t) \ge 1 - p_{ii}(t) \ge \big(1 - q_i(t)\big)(1 + \varepsilon_t)$$

(i.e. the asymptotic behaviour of $1 - q_i(t)$ and $1 - p_{ii}(t)$ as $t \to \infty$ is identical). This implies the second assertion of the theorem (i.e., the first relation in (21.2.3)).

Now let $t_r := rt/n$. Consider the transition probabilities

$$\begin{aligned} p_{ij}(t) &\ge \sum_{r=0}^{n-1} q_i(t_r) p_{ij}(t/n) q_j(t - t_{r+1}) \\ &\ge (1 - \varepsilon_t) p_{ij}(t/n) \sum_{r=0}^{n-1} e^{-q_i rt/n} \ge (1 - \varepsilon_t) p_{ij}(t/n) \frac{(1 - e^{-q_i t})n}{q_i t}. \end{aligned}$$

This implies that

$$p_{ij}(t) \ge (1 - \varepsilon_t) \left(\frac{1 - e^{-q_i t}}{q_i} \right) \limsup_{\delta \to 0} \frac{p_{ij}(\delta)}{\delta},$$

and that the upper limit on the right-hand side is bounded. Passing to the limit as $t \to 0$, we obtain

$$\liminf_{t \to 0} \frac{p_{ij}(t)}{t} \ge \limsup_{\delta \to 0} \frac{p_{ij}(\delta)}{\delta}.$$

Since $\sum_{j:j\neq i} p_{ij}(t) = 1 - p_{ii}(t)$, we have $\sum_{j:j\neq i} q_{ij} = q_i$. The theorem is proved. □

The theorem shows that the quantities

$$p_{ij} = \frac{q_{ij}}{q_i}, \quad j \neq i, \qquad p_{ii} = 0$$

form a stochastic matrix and give the probabilities of transition from i to j during an infinitesimal time interval Δ given the process $\xi(t)$ left the state i during that time interval:

$$\mathbf{P}\big(\xi(t + \Delta) = j \big| \xi(t) = i,\ \xi(t + \Delta) \neq i\big) = \frac{p_{ij}(\Delta)}{1 - p_{ii}(\Delta)} \to \frac{q_{ij}}{q_i}$$

as $\Delta \to 0$.

Thus the evolution of $\xi(t)$ can be thought of as follows. If $\xi(0) = X_0$, then $\xi(t)$ stays at X_0 for a random time $\tau_1 \in \boldsymbol{\Gamma}_{q_{X_0}}$. Then $\xi(t)$ passes to a state X_1 with probability $p_{X_0 X_1}$. Further, $\xi(t) = X_1$ over the time interval $[\tau_1, \tau_1 + \tau_2)$, $\tau_2 \in \boldsymbol{\Gamma}_{q_{X_1}}$, after which the system changes its state to X_2 and so on. It is clear that $X_0, X_1, \ldots$ is a homogeneous Markov chain with the transition matrix $\|p_{ij}\|$. Therefore the further study of $\xi(t)$ can be reduced in many aspects to that of the Markov chain $\{X_n;\ n \ge 0\}$, which was carried out in detail in Chap. 13.

We see that the evolution of $\xi(t)$ is completely specified by the quantities q_{ij} and q_i forming the matrix

$$Q = \|q_{ij}\| = \lim_{t\to 0} \frac{P(t) - P(0)}{t}, \tag{21.2.5}$$

where we put $q_{ii} := -q_i$, so that $\sum_j q_{ij} = 0$. We can also justify this claim using an analytical approach. To simplify the technical side of the exposition, we will assume, where it is needed, that the entries of the matrix Q are bounded and convergence in (21.2.3) is uniform in i.

Denote by e^A the matrix-valued function

$$e^A = E + \sum_{k=1}^{\infty} \frac{1}{k!} A^k.$$

Theorem 21.2.2 *The transition probabilities $p_{ij}(t)$ satisfy the systems of differential equations*

$$P'(t) = P(t)Q, \tag{21.2.6}$$

$$P'(t) = QP(t). \tag{21.2.7}$$

Each of the systems (21.2.6) *and* (21.2.7) *has a unique solution*

$$P(t) = e^{Qt}.$$

It is clear that the solution can be obtained immediately by formally integrating equation (21.2.6).

Proof By virtue of (21.2.1), (21.2.2) and (21.2.5),

$$P'(t) = \lim_{s\to 0} \frac{P(t+s) - P(t)}{s} = \lim_{s\to 0} P(t)\frac{P(s) - E}{s} = P(t)Q. \tag{21.2.8}$$

In the same way we obtain, from the equality

$$P(t+s) - P(t) = \big(P(s) - E\big)P(t),$$

the second equation in (21.2.7). The passage to the limit is justified by the assumptions we made.

Further, it follows from (21.2.6) that the function $P(t)$ is infinitely differentiable, and

$$P^{(k)}(t) = P(t)Q^k,$$

$$P(t) - P(0) = \sum_{k=1}^{\infty} P^{(k)}(0)\frac{t^k}{k!} = \sum_{k=1}^{\infty} \frac{Q^k t^k}{k!},$$

$$P(t) = P(0)e^{Qt}.$$

The theorem is proved. □

Because of the derivation method, (21.2.6) is called the *backward Kolmogorov equation*, and (21.2.7) is known as the *forward Kolmogorov equation* (the time increment is taken *after* or *before* the basic time interval).

The difference between these equations becomes even more graphical in the case of inhomogeneous Markov processes, when the transition probabilities

$$\mathbf{P}\big(\xi(t)=j\,|\,\xi(s)=i\big)=p_{ij}(s,t),\quad s\le t,$$

depend on two time arguments: s and t. In that case, (21.2.1) becomes the equality $P(s,t+u)=P(s,t)P(t,t+u)$, and the backward and forward equations have the form

$$\frac{\partial P(s,t)}{\partial s}=P(s,t)Q(s),\qquad \frac{\partial P(s,t)}{\partial t}=Q(t)P(s,t),$$

respectively, where

$$Q(t)=\lim_{u\to 0}\frac{P(t,t+u)-E}{u}.$$

The reader can derive these relations independently.

What are the general conditions for existence of a stationary limiting distribution? We can use here an approach similar to that employed in Chap. 13.

Let $\xi^{(i)}(t)$ be a process with the initial value $\xi^{(i)}(0)=i$ and right-continuous trajectories. For a given i_0, put

$$\nu^{(i)}:=\min\big\{t\ge 0:\xi^{(i)}(t)=i_0\big\}=:\nu_0,$$
$$\nu_k:=\min\big\{t\ge \nu_{k-1}+1:\xi^{(i)}(t)=i_0\big\},\quad k=1,2,\ldots.$$

Here in the second formula we consider the values $t\ge\nu_{k-1}+1$, since for $t\ge\nu_{k-1}$ we would have $\nu_k\equiv\nu_{k-1}$. Clearly, $\mathbf{P}(\nu_k-\nu_{k-1}=1)>0$, and $\mathbf{P}(\nu_k-\nu_{k-1}\in(t,t+h))>0$ for any $t\ge 1$ and $h>0$ provided that $p_{i_0i_0}(t)\not\equiv 1$.

Note also that the variables ν_k, $k=0,1,\ldots$, are not defined for all elementary outcomes. We put $\nu_0=\infty$ if $\xi^{(i)}(t)\ne i_0$ for all $t\ge 0$. A similar convention is used for ν_k, $k\ge 1$. The following ergodic theorem holds.

Theorem 21.2.3 *Let there exist a state i_0 such that $\mathbf{E}\nu_1<\infty$ and $\mathbf{P}(\nu^{(i)}<\infty)=1$ for all $i\in\mathfrak{X}_0\subset\mathfrak{X}$. Then there exist the limits*

$$\lim_{t\to\infty}p_{ij}(t)=p_j \tag{21.2.9}$$

which are independent of $i\in\mathfrak{X}_0$.

Proof As was the case for Markov chains, the epochs $\nu_1,\nu_2,\ldots$ divide the time axis into independent cycles of the same nature, each of them being completed when the system returns for the first time (after one time unit) to the state i_0. Consider the renewal process generated by the sums ν_k, $k=0,1,\ldots$, of independent random variables ν_0, $\nu_k-\nu_{k-1}$, $k=1,2,\ldots$. Let

$$\eta(t):=\min\{k:\nu_k>t\},\qquad \gamma(t):=t-\nu_{\eta(t)-1},\qquad H(t):=\sum_{k=0}^{\infty}\mathbf{P}(\nu_k<t).$$

The event $A_{dv} := \{\gamma(t) \in [v, v+dv)\}$ can be represented as the intersection of the events

$$B_{dv} := \bigcup_{k\geq 0} \{\nu_k \in (t-v-dv, t-v]\} \in \mathfrak{F}_{1-v}$$

and $C_v := \{\xi(u) \neq i_0 \text{ for } u \in [t-v+1, t]\} \in \mathfrak{F}_{[t-v,\infty)}$. We have

$$\begin{aligned} p_{ij}(t) &= \int_0^t \mathbf{P}\big(\xi^{(i)}(t) = j,\ \gamma(t) \in [v, v+dv)\big) = \int_0^t \mathbf{P}\big(\xi^{(i)} = j,\ B_{dv}C_v\big) \\ &= \int_0^t \mathbf{E}\big[\mathrm{I}_{B_{dv}} \mathbf{P}\big(\xi^{(i)}(t) = j,\ C_v \big| \mathfrak{F}_{t-v}\big)\big] \\ &= \int_0^t \mathbf{E}\big[\mathrm{I}_{B_{dv}} \mathbf{P}\big(\xi^{(i)}(t) = j, C_v \big| \xi(t-v)\big)\big]. \end{aligned}$$

On the set B_{dv}, one has $\xi(t-v) = i_0$, and hence the probability inside the last integral is equal to

$$\mathbf{P}\big(\xi^{(i_0)}(v) = j,\ \xi(u) \neq i_0 \text{ for } u \in [1, v]\big) =: g(v)$$

and is independent of t and i. Since $\mathbf{P}(B_{dv}) = dH(t-v)$, one has

$$p_{ij}(t) = \int_0^t g(v)\mathbf{P}(B_{dv}) = \int_0^t g(v)\,dH(t-v).$$

By the key renewal theorem, as $t \to \infty$, this integral converges to

$$\frac{1}{\mathbf{E}\nu_1} \int_0^\infty g(v)\,dv.$$

The existence of the last integral follows from the inequality $g(v) \leq \mathbf{P}(\nu_1 > v)$. The theorem is proved. □

Theorem 21.2.4 *If the stationary distribution*

$$P = \lim_{t\to\infty} P(t)$$

exists with all the rows of the matrix P being identical, then it is a unique solution of the equation

$$PQ = 0. \tag{21.2.10}$$

It is evident that Eq. (21.2.10) is obtained by setting $P'(t) = 0$ in (21.2.6). Equation (21.2.7) gives the trivial equality $QP = 0$.

Proof Equation (21.2.10) is obtained by passing to the limit in (21.2.8) first as $t \to \infty$ and then as $s \to 0$. Now assume that P_1 is a solution of (21.2.10), i.e. $P_1 Q = 0$. Then $P_1 P(t) = P_1$ for $t < 1$, since

$$P_1\big(P(t) - P(0)\big) = P_1 \sum_{k=1}^{\infty} \frac{Q^k t^k}{k!} = 0.$$

Further, $P_1 = P_1 P^k(t) = P_1 P(kt)$, $P(kt) \to P$ as $k \to \infty$, and hence $P_1 = P_1 P = P$. The theorem is proved. □

Now consider a Markov chain $\{X_n\}$ in discrete time with transition probabilities $p_{ij} = q_{ij}/q_i$, $i \neq j$, $p_{ii} = 0$. Suppose that this chain is ergodic (see Theorem 13.4.1). Then its stationary probabilities $\{\pi_j\}$ satisfy Eqs. (13.4.2). Now note that Eq. (21.2.10) can be written in the form

$$p_j q_j = \sum_k p_k q_k p_{kj}$$

which has an obvious solution $p_j = c\pi_j/q_j$, $c = \text{const}$. Therefore, if

$$\sum \frac{\pi_j}{q_j} < \infty \tag{21.2.11}$$

then there exists a solution to (21.2.10) given by

$$p_j = \frac{\pi_j}{q_j} \left(\sum \frac{\pi_j}{q_j} \right)^{-1}. \tag{21.2.12}$$

In Sects. 21.4 and 21.5 we will derive the ergodic theorem for processes of a more general form than the one in the present section. That theorem will imply, in particular, that ergodicity of $\{X_n\}$ and convergence (21.2.11) imply (21.2.9). Recall that, for ergodicity of $\{X_n\}$, it suffices, in turn, that Eqs. (13.4.2) have a solution $\{\pi_j\}$. Thus the existence of solution (21.2.12) implies the ergodicity of $\xi(t)$.

21.2.2 Examples

Example 21.2.1 The *Poisson process* $\xi(t)$ with parameter λ is a Markov process for which $q_i = \lambda$, $q_{i,i+1} = \lambda$, and $p_{i,i+1} = 1$, $i = 1, 0, \ldots$. For this process, the stationary distribution $p = (p_0, p_1, \ldots)$ does not exist (each trajectory goes to infinity).

Example 21.2.2 Birth-and-death processes. These are processes for which, for $i \geq 1$,

$$p_{ij}(\Delta) = \begin{cases} \lambda_i \Delta + o(\Delta) & \text{for } j = i+1, \\ \mu_i \Delta + o(\Delta) & \text{for } j = i-1, \\ o(\Delta) & \text{for } |j-i| \geq 2, \end{cases}$$

so that

$$p_{ij} = \begin{cases} \frac{\lambda_i}{\lambda_i + \mu_i} & \text{for } j = i+1, \\ \frac{\mu_i}{\lambda_i + \mu_i} & \text{for } j = i-1 \end{cases}$$

are probabilities of birth and death, respectively, of a particle in a certain population given that the population consisted of i particles and changed its composition. For

$i=0$ one should put $\mu_0 := 0$. Establishing conditions for the existence of a stationary regime is a rather difficult problem (related mainly to finding conditions under which the trajectory escapes to infinity). If the stationary regime exists, then according to Theorem 21.2.4 the stationary probabilities p_j can be uniquely determined from the recursive relations (see Eq. (21.2.10), in our case $q_{ii} = -q_i = -(\lambda_i + \mu_i)$)

$$\begin{aligned} -p_0\lambda_0 + p_1\mu_1 &= 0, \\ p_0\lambda_0 - p_1(\lambda_1 + \mu_1) + p_2\mu_2 &= 0, \\ &\dots\dots \\ p_{k-1}\lambda_{k-1} - p_k(\lambda_k + \mu_k) + p_{k+1}\mu_{k+1} &= 0, \\ &\dots\dots \end{aligned} \tag{21.2.13}$$

and condition $\sum p_j = 1$.

Example 21.2.3 The telephone lines problem from queueing theory. Suppose we are given a system consisting of infinitely many communication channels which are used for telephone conversations. The probability that, for a busy channel, the transmitted conversation terminates during a small time interval $(t, t+\Delta)$ is equal to $\lambda\Delta + o(\Delta)$. The probability that a request for a new conversation (a new call) arrives during the same time interval is $\mu\Delta + o(\Delta)$. Thus the "arrival flow" of calls is nothing else but the Poisson process with parameter λ, and the number $\xi(t)$ of busy channels at time t is the value of the birth-and-death process for which $\lambda_i = \lambda$ and $\mu_i = i\mu$.

In that case, it is not hard to verify with the help of Theorem 21.2.3 that there always exists a stationary limiting distribution, for which Eqs. (21.2.13) have the form

$$\begin{aligned} \lambda p_0 &= \mu p_1, \\ &\dots\dots \\ (\lambda + \mu k)p_k &= \lambda p_{k-1} + (k+1)\mu p_{k+1}, \\ &\dots\dots \end{aligned} \tag{21.2.14}$$

From this we get that

$$p_1 = p_0\frac{\lambda}{\mu}, \qquad p_2 = \frac{p_0}{2}\left(\frac{\lambda}{\mu}\right)^2, \qquad \dots, \qquad p_k = \left(\frac{\lambda}{\mu}\right)^k \frac{p_0}{k!}, \tag{21.2.15}$$

so that $p_0 = e^{-\lambda/\mu}$, and the limiting distribution will be the Poisson law with parameter λ/μ.

If the number of channels n is finite, the calls which find all the lines busy will be rejected, and in (21.2.13) one has to put $\lambda_n = 0$, $p_{n+1} = p_{n+2} = \cdots = 0$. In that case, the last equation in (21.2.14) will have the form $\mu n p_n = \lambda p_{n-1}$. Since the formulas (21.2.15) will remain true for $k \le n$, we obtain the so-called Erlang formulas for the stationary distribution:

$$p_k = \left(\frac{\lambda}{\mu}\right)^k \frac{1}{k!}\left[\sum_{j=0}^{n} \frac{1}{j!}\left(\frac{\lambda}{\mu}\right)^j\right]^{-1}$$

(the truncated Poisson distribution).

The next example will be considered in a separate section.

21.3 Branching Processes

The essence of the mathematical model describing a branching process remains roughly the same as in Sect. 7.7.2. A continuous time branching process can be defined as follows. Let $\xi^{(i)}(t)$ denote the number of particles at time t with the initial condition $\xi^{(i)}(0) = i$. Each particle, independently of all others, splits during the time interval $(t, t+\Delta)$ with probability $\mu\Delta + o(\Delta)$ into a random number $\eta \neq 1$ of particles (if $\eta = 0$, we say that the particle dies). Thus,

$$\xi^{(i)}(t) = \xi_1^{(1)}(t) + \cdots + \xi_i^{(1)}(t), \tag{21.3.1}$$

where $\xi_k^{(1)}(t)$ are independent and distributed as $\xi^{(1)}(t)$. Moreover,

$$\begin{aligned} p_{ij}(\Delta) &= i\mu\Delta h_{j-i+1} + o(\Delta), \quad j \neq i; \qquad h_k = \mathbf{P}(\eta = k); \quad h_1 = 0; \\ p_{ii}(\Delta) &= 1 - i\mu\Delta + o(\Delta), \end{aligned} \tag{21.3.2}$$

so that here $q_{ij} = i\mu h_{j-i+1}$, $q_{ii} = -i\mu$.

By formula (21.3.2), $i\mu\Delta$ is the principal part of the probability that at least one particle will split. Clearly, the state 0 is absorbing. It will not be absorbing any more if one considers processes with *immigration* when a Poisson process (with intensity λ) of "outside" particles is added to the process $\xi^{(i)}(t)$. Then

$$\begin{aligned} p_{ij}(\Delta) &= i\mu\Delta h_{j-i+1} + o(\Delta) \quad \text{for } j - i \neq 0, 1, \\ p_{i,i+1}(\Delta) &= \Delta(i\mu h_2 + \lambda) + o(\Delta). \end{aligned}$$

We return to the branching process (21.3.1), (21.3.2). By (21.3.1) we have

$$r^{(i)}(t,z) := \mathbf{E} z^{\xi^{(i)}(t)} = \left[\mathbf{E} z^{\xi^{(1)}(t)}\right]^i = r^i(t,z) = \sum_{k=0}^{\infty} z^k p_{ik}(t),$$

where

$$r(t,z) := \mathbf{E} z^{\xi^{(1)}(t)} = \sum_{k=0}^{\infty} z^k p_{1k}(t). \tag{21.3.3}$$

Equation (21.2.7) implies

$$p'_{1k}(t) = \sum_{l=0}^{\infty} q_{1l} p_{lk}(t).$$

Therefore, differentiating (21.3.3) with respect to t, we find that

$$\begin{aligned} r'_t(t,z) &= \sum_{k=0}^{\infty} z^k p'_{1k}(t) = \sum_{k=0}^{\infty} \sum_{l=0}^{\infty} q_{1l} p_{lk}(t) z^k \\ &= \sum_{l=0}^{\infty} q_{1l} \sum_{k=0}^{\infty} z^k p_{lk}(t) = \sum_{l=0}^{\infty} q_{1l} r^l(t,z). \end{aligned} \tag{21.3.4}$$

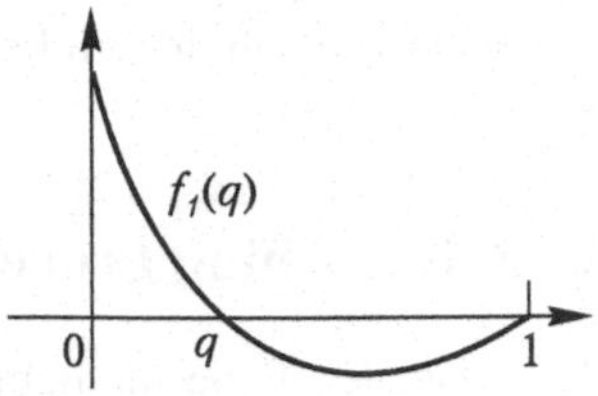

Fig. 21.1 The form of the plot of the function f_1. The smaller root of the equation $f_1(q) = q$ gives the probability of the eventual extinction of the branching process

But $q_{1l} = \mu p_l$ for $l \neq 1$, $q_{11} = -\mu$, and putting

$$f(s) := \sum_{l=0}^{\infty} q_{1l} s^l = \mu\big(\mathbf{E}s^\eta - s\big) = \mu\left(\sum_{l=0}^{\infty} h_l s^l - s\right),$$

we can write (21.3.4) in the form

$$r'_t(t, z) = f\big(r(t, z)\big).$$

We have obtained a differential equation for $r = r(t, z)$ (equivalent to (21.3.2)) which is more convenient to write in the form

$$\frac{dr}{f(r)} = dt, \qquad t = \int_{r(0,z)}^{r(t,z)} \frac{dy}{f(y)} = \int_{z}^{r(t,z)} \frac{dy}{f(y)}.$$

Consider the behaviour of the function $f_1(y) = \mathbf{E}y^\eta - y$ on $[0, 1]$. Clearly, $f_1(0) = \mathbf{P}(\eta = 0)$, $f_1(1) = 0$, and

$$f_1'(1) = \mathbf{E}\eta - 1, \qquad f_1''(y) = \mathbf{E}\eta(\eta - 1)y^{\eta-2} > 0.$$

Consequently, the function $f_1(y)$ is convex and has no zeros in $(0, 1)$ if $\mathbf{E}\eta \le 1$. When $\mathbf{E}\eta > 1$, there exists a point $q \in (0, 1)$ such that $f_1(q) = 0$, $f_1'(q) < 0$ (see Fig. 21.1), and $f_1(y) = (y - q)f_1'(q) + O((y - q)^2)$ in the vicinity of this point.

Thus if $\mathbf{E}\eta > 1$, $z < q$ and $r \uparrow q$, then, by virtue of the representation

$$\frac{1}{f_1(y)} = \frac{1}{(y - q)f_1'(q)} + O(1),$$

we obtain

$$t = \int_z^r \frac{dy}{f(y)} = \frac{1}{\mu f_1'(q)} \ln\left(\frac{r - q}{z - q}\right) + O(1).$$

This implies that, as $t \to \infty$,

$$\begin{aligned} r(t, z) - q &= (z - q)e^{\mu t f_1'(q) + O(1)} \sim (z - q)e^{\mu t f_1'(q)}, \\ r(t, z) &= q + O\big(e^{-\alpha t}\big), \quad \alpha = -\mu f_1'(q) > 0. \end{aligned} \tag{21.3.5}$$

In particular, the extinction probability

$$p_{10}(t) = r(t, 0) = q + O\big(e^{-\alpha t}\big)$$

converges exponentially fast to q, $p_{10}(\infty) = q$. Comparing our results with those from Sect. 7.7, the reader can see that the extinction probability for a discrete time branching process had the same value (we could also come to this conclusion directly). Since $p_{k0}(t) = [p_{10}(t)]^k$, one has $p_{k0}(\infty) = q^k$.

It follows from (21.3.5) that the remaining "probability mass" of the distribution of $\xi(t)$ quickly moves to infinity as $t \to \infty$.

If $\mathbf{E}\eta < 1$, the above argument remains valid with q replaced with 1, so that the extinction probability is $p_{10}(\infty) = p_{k0}(\infty) = 1$.

If $\mathbf{E}\eta = 1$, then

$$f_1(y) = \frac{(y-1)^2}{2} f_1''(1) + O\big((y-1)^3\big),$$

$$t = \int_z^r \frac{dy}{f(y)} \sim -\frac{2}{\mu f_1''(1)} \cdot \frac{1}{r-1}, \qquad r(t,z) - 1 \sim -\frac{2}{\mu t f_1''(1)}.$$

Thus the extinction probability $r(t,0) = p_{10}(t)$ also tends to 1 in this case.

21.4 Semi-Markov Processes

21.4.1 Semi-Markov Processes on the States of a Chain

Semi-Markov processes can be described as follows. Let an aperiodic discrete time irreducible Markov chain $\{X_n\}$ with the state space $\mathcal{X} = \{0, 1, 2, \ldots\}$ be given. To each state i we put into correspondence the distribution $F_i(t)$ of a positive random variable $\zeta^{(i)}$:

$$F_i(t) = \mathbf{P}\big(\zeta^{(i)} < t\big).$$

Consider independent of the chain $\{X_n\}$ and of each other the sequences $\zeta_1^{(i)}$, $\zeta_2^{(i)}, \ldots; \zeta_j^{(i)} \stackrel{d}{=} \zeta^{(i)}$, of independent random variables with the distribution F_i. Let, moreover, the distribution of the initial random vector (X_0, ζ_0), $X_0 \in \mathcal{X}$, $\zeta_0 \ge 0$, be given. The evolution of the semi-Markov process $\xi(u)$ is described as follows:

$$\begin{aligned}
&\xi(u) = X_0 \quad \text{for } 0 \le u < \zeta_0,\\
&\xi(u) = X_1 \quad \text{for } \zeta_0 \le u < \zeta_0 + \zeta_1^{(X_1)},\\
&\xi(u) = X_2 \quad \text{for } \zeta_0 + \zeta_1^{(X_1)} \le u < \zeta_0 + \zeta_1^{(X_1)} + \zeta_2^{(X_2)},\\
&\ldots,\\
&\xi(u) = X_n \quad \text{for } Z_{n-1} \le u < Z_n,\ Z_n = \zeta_0 + \zeta_1^{(X_1)} + \cdots + \zeta_n^{(X_n)},
\end{aligned} \tag{21.4.1}$$

and so on. Thus, upon entering state $X_n = j$, the trajectory of $\xi(u)$ remains in that state for a random time $\zeta_n^{(X_n)} = \zeta_n^{(j)}$, then switches to state X_{n+1} and so on. It is evident that such a process is, generally speaking, not Markovian. It will be a Markov process only if

$$1 - F_i(t) = e^{-q_i t}, \quad q_i > 0,$$

and will then coincide with the process described in Sect. 21.2.

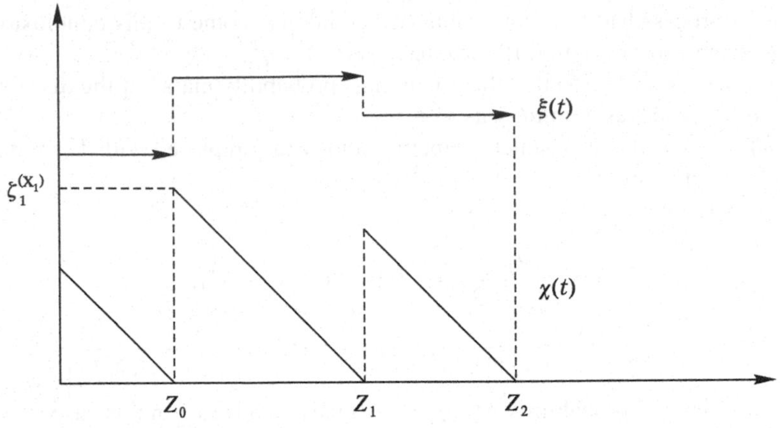

Fig. 21.2 The trajectories of the semi-Markov process $\xi(t)$ and of the residual sojourn time process $\chi(t)$

If the distribution F_i is not exponential, then, given the value $\xi(t) = i$, the time between t and the next jump epoch will depend on the epoch of the preceding jump of $\xi(\cdot)$, because

$$\mathbf{P}\big(\zeta^{(i)} > v + u \big| \zeta^{(i)} > v\big) = \frac{1 - F_i(v+u)}{1 - F_i(v)}$$

for non-exponential F_i depends on v. It is this property that means that the process is non-Markovian, for fixing the "present" (i.e. the value of $\xi(t)$) does not make the "future" of the process $\xi(u)$ independent of the "past" (i.e. of the trajectory of $\xi(u)$ for $u < t$).

The process $\xi(t)$ can be "complemented" to a Markov one by adding to it the component $\chi(t)$ of which the value gives the time u for which the trajectory $\xi(t+u)$, $u \ge 0$, will remain in the current state $\xi(t)$. In other words, $\chi(t)$ is the excess of level t for the random walk $Z_0, Z_1, \ldots$ (see Fig. 21.2):

$$\chi(t) = Z_{\nu(t)+1} - t, \qquad \nu(t) = \max\{k : Z_k \le t\}.$$

The process $\chi(t)$ is Markovian and has "saw-like" trajectories deterministic inside the intervals (Z_k, Z_{k+1}). The process $X(t) = (\xi(t), \chi(t))$ is obviously Markovian, since the value of $X(t)$ uniquely determines the law of evolution of the process $X(t+u)$ for $u \ge 0$ whatever the "history" $X(v)$, $v < t$, is. Similarly, we could consider the Markov process $Y(t) = (\xi(t), \gamma(t))$, where $\gamma(t)$ is the defect of level t for the walk $Z_0, Z_1, \ldots$:

$$\gamma(t) = t - Z_{\nu(t)}.$$

21.4.2 The Ergodic Theorem

In the sequel, we will distinguish between the following two cases.

(A) The *arithmetic* case when the possible values of $\zeta^{(i)}$, $i=0,1,\ldots$, are multiples of a certain value h which can be assumed without loss of generality to be equal to 1. In that case we will also assume that the g.c.d. of the possible values of the sums of the variables $\zeta^{(i)}$ is also equal to $h=1$. This is clearly equivalent to assuming that the g.c.d. of the possible values of recurrence times $\theta^{(i)}$ of $\xi(t)$ to the state i is equal to 1 for any fixed i.

(NA) The *non-arithmetic* case, when condition (A) does not hold.

Put $a_i := \mathbf{E}\zeta^{(i)}$.

Theorem 21.4.1 *Let the Markov chain $\{X_n\}$ be ergodic (satisfy the conditions of Theorem* 13.4.1) *and $\{\pi_j\}$ be the stationary distribution of that chain. Then, in the non-arithmetic case* (NA), *for any initial distribution (ζ_0, X_0) there exists the limit*

$$\lim_{t\to\infty} \mathbf{P}\big(\xi(t)=i,\ \chi(t)>v\big)=\frac{\pi_i}{\sum \pi_j a_j}\int_v^\infty \mathbf{P}\big(\zeta^{(i)}>u\big)\,du. \tag{21.4.2}$$

In the arithmetic case (A), (21.4.2) *holds for integer-valued v (the integral becomes a sum in that case). It follows from* (21.4.2) *that the following limit exists*

$$\lim_{t\to\infty} \mathbf{P}\big(\xi(t)=i\big)=\frac{\pi_i a_i}{\sum \pi_j a_j}.$$

Proof For definiteness we restrict ourselves to the non-arithmetic case (NA). In Sect. 13.4 we considered the times $\tau^{(i)}$ between consecutive visits of $\{X_n\}$ to state i. These times could be called "embedded", as well as the chain $\{X_n\}$ itself in regard to the process $\xi(t)$. Along with the times $\tau^{(i)}$, we will need the "real" times $\theta^{(i)}$ between the visits of the process $\xi(t)$ to the state i. Let, for instance, $X_1=1$. Then

$$\theta^{(1)} \stackrel{d}{=} \zeta_1^{(X_1)}+\zeta_2^{(X_2)}+\cdots+\zeta_\tau^{(X_\tau)},$$

where $\tau=\tau^{(1)}$. For definiteness and to reduce notation, we fix for the moment the value $i=1$ and put $\theta^{(1)}=:\theta$. Let first

$$\zeta_0 \stackrel{d}{=} \zeta^{(1)}, \quad X_0=1. \tag{21.4.3}$$

Then the whole trajectory of the process $X(t)$ for $t\ge 0$ will be divided into identically distributed independent cycles by the epochs when the process hits the state $\xi(t)=1$. We denote the lengths of these cycles by $\theta_1,\theta_2\ldots$; they are independent and identically distributed. We show that

$$\mathbf{E}\theta=\frac{1}{\pi_1}\sum a_j\pi_j. \tag{21.4.4}$$

Denote by $\theta(n)$ the "real" time spent on n transitions of the governing chain $\{X_n\}$. Then

$$\theta_1+\cdots+\theta_{\eta(n)-1}\le\theta(n)\le\theta_1+\cdots+\theta_{\eta(n)}, \tag{21.4.5}$$

where $\eta(n):=\min\{k: T_k>n\}$, $T_k=\sum_{j=1}^k \tau_j$, τ_j are independent and distributed as τ. We prove that, as $n\to\infty$,

$$\mathbf{E}\theta(n)\sim n\pi_1\mathbf{E}\theta. \tag{21.4.6}$$

By Wald's identity and (21.4.5),

$$\mathbf{E}\theta(n) \le \mathbf{E}\theta\mathbf{E}\eta(n), \tag{21.4.7}$$

where $\mathbf{E}\eta(n) \sim n/\mathbf{E}\tau = n\pi_1$.

Now we bound from below the expectation $\mathbf{E}\theta(n)$. Put $m := \lfloor n\pi_1 - \varepsilon n \rfloor$, $\Theta_n := \sum_{j=1}^{n} \theta_j$. Then

$$\begin{aligned}\mathbf{E}\theta(n) &\ge \mathbf{E}\big(\theta(n);\ \eta(n) > m\big)\\ &\ge \mathbf{E}\big(\Theta_m;\ \eta(n) > m\big) = m\mathbf{E}\theta - \mathbf{E}\big(\Theta_m;\ \eta(n) \le m\big).\end{aligned} \tag{21.4.8}$$

Here the random variable $\Theta_m/m \ge 0$ possesses the properties

$$\Theta_m/m \xrightarrow{p} \mathbf{E}\theta \quad \text{as } m \to \infty, \qquad \mathbf{E}(\Theta_m/m) = \mathbf{E}\theta.$$

Therefore it satisfies the conditions of part 4 of Lemma 6.1.1 and is uniformly integrable. This, in turn, by Lemma 6.1.2 and convergence $\mathbf{P}(\eta(n) \le m) \to 0$ means that the last term on the right-hand side of (21.4.8) is $o(m)$. By virtue of (21.4.8), since $\varepsilon > 0$ is arbitrary, we obtain that

$$\liminf_{n\to\infty} n^{-1}\mathbf{E}\theta(n) \ge \pi_1\mathbf{E}\theta.$$

This together with (21.4.7) proves (21.4.6).

Now we will calculate the value of $\mathbf{E}\theta(n)$ using another approach. The variable $\theta(n)$ admits the representation

$$\theta(n) = \sum_j \big(\zeta_1^{(j)} + \cdots + \zeta_{N(j,n)}^{(j)}\big),$$

where $N(j,n)$ is the number of visits of the trajectory of $\{X_k\}$ to the state j during the first n steps. Since $\{\zeta_k^{(j)}\}_{k=1}^{\infty}$ and $N(j,n)$ are independent for each j, we have

$$\mathbf{E}\theta(n) = \sum_j a_j \mathbf{E}N(j,n), \qquad \mathbf{E}N(j,n) = \sum_{k=1}^{n} p_{1j}(k).$$

Because $p_{1j}(k) \to \pi_j$ as $k \to \infty$, one has

$$\lim_{n\to\infty} n^{-1}\mathbf{E}N(j,n) = \pi_j.$$

Moreover,

$$\pi_j = \sum \pi_l p_{lj}(k) \ge \pi_1 p_{1j}(k)$$

and, therefore,

$$p_{1j}(k) \le \pi_j/\pi_1.$$

Hence

$$n^{-1}\mathbf{E}N(j,n) \le \pi_j/\pi_1,$$

and in the case when $\sum_j a_j \pi_j < \infty$, the series $\sum_j a_j n^{-1} \mathbf{E} N(j,n)$ converges uniformly in n. Consequently, the following limit exists

$$\lim_{n\to\infty} n^{-1} \mathbf{E}\theta(n) = \sum_j a_j \pi_j.$$

Comparing this with (21.4.6) we obtain (21.4.4). If $\mathbf{E}\theta = \infty$ then clearly $\mathbf{E}\theta(n) = \infty$ and $\sum_j a_j \pi_j = \infty$, and vice versa, if $\sum_j a_j \pi_j = \infty$ then $\mathbf{E}\theta = \infty$.

Consider now the random walk $\{\Theta_k\}$. To the k-th cycle there correspond T_k transitions. Therefore, by the total probability formula,

$$\mathbf{P}\big(\xi(t) = 1, \chi(t) > v\big) = \sum_{k=1}^{\infty} \int_0^t \mathbf{P}\big(\Theta_k \in du, \zeta^{(1)}_{T_k+1} > t - u + v\big),$$

where $\zeta^{(1)}_{T_k+1}$ is independent of Θ_k and distributed as $\zeta^{(1)}$ (see Lemma 11.2.1 or the strong Markov property). Therefore, denoting by $H_\theta(u) := \sum_{k=1}^{\infty} \mathbf{P}(\Theta_k < u)$ the renewal function for the sequence $\{\Theta_k\}$, we obtain for the non-arithmetic case (NA), by virtue of the renewal theorem (see Theorem 10.4.1 and (10.4.2)), that, as $t \to \infty$,

$$\begin{aligned}
&\mathbf{P}\big(\xi(t) = 1, \chi(t) > v\big) \\
&\quad = \int_0^t dH_\theta(u) \mathbf{P}\big(\zeta^{(1)} > t - u + v\big) \\
&\quad \to \frac{1}{\mathbf{E}\theta} \int_0^\infty \mathbf{P}\big(\zeta^{(1)} > u + v\big)\, dv = \frac{1}{\mathbf{E}\theta} \int_v^\infty \mathbf{P}\big(\zeta^{(1)} > u\big)\, du. \qquad (21.4.9)
\end{aligned}$$

We have proved assertion (21.4.2) for $i = 1$ and initial conditions (21.4.3). The transition to arbitrary initial conditions is quite obvious and is done in exactly the same way as in the proof of the ergodic theorems of Chap. 13.

If $\sum a_i \pi_i = \infty$ then, as we have already observed, $\mathbf{E}\theta = \infty$ and, by the renewal theorem and (21.4.9), one has $\mathbf{P}(\xi(t) = 1,\ \chi(t) > v) \to 0$ as $t \to \infty$. It remains to note that instead of $i = 1$ we can fix any other value of i. The theorem is proved. □

In the same way we could also prove that

$$\lim_{t\to\infty} \mathbf{P}\big(\xi(t) = i,\ \gamma(t) > v\big) = \frac{\pi_i}{\sum a_j \pi_j} \int_v^\infty \mathbf{P}\big(\zeta^{(i)} > y\big)\, dy,$$

$$\lim_{t\to\infty} \mathbf{P}\big(\xi(t) = i,\ \chi(t) > u,\ \gamma(t) > v\big) = \frac{\pi_i}{\sum a_j \pi_j} \int_{u+v}^\infty \mathbf{P}\big(\zeta^{(i)} > y\big)\, dy$$

(see Theorem 10.4.3).

21.4.3 Semi-Markov Processes on Chain Transitions

Along with the semi-Markov processes $\xi(t)$ described at the beginning of the present section, one sometimes considers semi-Markov processes "given on the transitions" of the chain $\{X_n\}$. In that case, the distributions F_{ij} of random variables $\zeta^{(ij)} > 0$ are given and, similarly to (21.4.1), for the initial condition (X_0, X_1, ζ_0) one puts

$$\begin{aligned} \xi(u) &:= (X_0, X_1) \quad \text{for } 0 \le u < \zeta_0 \\ \xi(u) &:= (X_1, X_2) \quad \text{for } \zeta_0 \le u < \zeta_0 + \zeta_1^{(X_0,X_1)} \\ \xi(u) &:= (X_2, X_3) \quad \text{for } \zeta_0 + \zeta_1^{(X_0,X_1)} \le u < \zeta_0 + \zeta_1^{(X_0,X_1)} + \zeta_2^{(X_1,X_2)}, \end{aligned} \tag{21.4.10}$$

and so on. Although at first glance this is a very general model, it can be completely reduced to the semi-Markov processes (21.4.1). To that end, one has to notice that the "two-dimensional" sequence $Y_n = (X_n, X_{n+1})$, $n = 0, 1, \ldots$, also forms a Markov chain. Its transition probabilities have the form

$$p_{(ij)(kl)} = \begin{cases} p_{jl} & \text{for } k = j, \\ 0 & \text{for } k \ne j, \end{cases}$$

$$p_{(ij)(kl)}(n) = p_{jk}(n) p_{kl} \quad \text{for } n > 1,$$

so that if the chain $\{X_n\}$ is ergodic, then $\{Y_n\}$ is also ergodic and

$$p_{(ij)(kl)}(n) \to \pi_k p_{kl}.$$

This enables one to restate Theorem 21.4.1 easily for the semi-Markov processes (21.4.10) given on the transitions of the Markov chain $\{X_n\}$, since the process (21.4.10) will be an ordinary semi-Markov process given on the chain $\{Y_n\}$.

Corollary 21.4.1 *If the chain $\{X_n\}$ is ergodic then, in the non-arithmetic case,*

$$\lim_{t\to\infty} \mathbf{P}\big(\xi(t) = (i, j),\ \chi(t) > v\big) = \frac{\pi_i p_{ij}}{\sum_{k,l} a_{kl} \pi_k p_{kl}} \int_v^\infty \mathbf{P}\big(\zeta^{(ij)} > u\big)\, du, \quad a_{kl} = \mathbf{E}\zeta^{(kl)}.$$

In the arithmetic case v must be a multiple of the lattice span.

We will make one more remark which could be helpful when studying semi-Markov processes and which concerns the so-called semi-Markov renewal functions $H_{ij}(t)$. Denote by $T_{ij}(n)$ the epoch (in the "real time") of the n-th jump of the process $\xi(t)$ from state i to j. Put

$$H_{ij}(t) := \sum_{n=1}^{\infty} \mathbf{P}\big(T_{ij}(n) < t\big).$$

If $\nu_{ij}(t)$ is the number of jumps from state i to j during the time interval $[0, t)$, then clearly $H_{ij}(t) = \mathbf{E}\nu_{ij}(t)$.

Set $\Delta f(t) := f(t + \Delta) - f(t)$, $\Delta > 0$.

Corollary 21.4.2 *In the non-arithmetic case,*

$$\lim_{t\to\infty} \Delta H_{ij}(t) = \frac{\pi_i p_{ij} \Delta}{\sum_l a_l \pi_l}. \tag{21.4.11}$$

In the arithmetic case v must be a multiple of the lattice span.

Proof Denote by $\nu_{ij}^{(k)}(u)$ the number of transitions of the process $\xi(t)$ from i to j during the time interval $(0, u)$ given the initial condition $(k, 0)$. Then, by the total probability formula,

$$\mathbf{E}\Delta\nu_{ij}(t) = \int_0^\Delta \sum_{k=0}^\infty \mathbf{P}\big(\xi(t) = k,\ \chi(t) \in du\big)\mathbf{E}\nu_{ij}^{(k)}(\Delta - u).$$

Since $\nu_{ij}^{(k)}(u) \le \nu_{ij}^{(i)}(u)$, by Theorem 21.4.1 one has

$$h_{ij}(\Delta) := \lim_{t\to\infty} \mathbf{E}\Delta\nu_{ij}(t) = \frac{1}{\sum_l a_l\pi_l} \sum_{k=0}^\infty \pi_k \int_0^\Delta \mathbf{P}\big(\zeta^{(k)} > u\big)\mathbf{E}\nu_{ij}^{(k)}(\Delta - u)\,du. \tag{21.4.12}$$

Further,

$$\mathbf{P}\big(\zeta^{(i)} < \Delta - u\big) \le F_i(\Delta) \to 0$$

as $\Delta \to 0$, and

$$\begin{aligned}
&\mathbf{P}\big(\nu_{ij}^{(k)}(\Delta - u) = s\big) \le \big(p_{ij}F_i(\Delta)\big)^s, \quad k \ne i,\\
&\mathbf{P}\big(\nu_{ij}^{(i)}(\Delta - u) = s + 1\big) \le \big(p_{ij}F_i(\Delta)\big)^s, \quad s \ge 1,\\
&\mathbf{P}\big(\nu_{ij}^{(i)}(\Delta - u) = 1\big) = p_{ij} + o\big(F_i(\Delta)\big).
\end{aligned}$$

It follows from the aforesaid that

$$\mathbf{E}\nu_{ij}^{(k)}(\Delta - u) = o\big(F_i(\Delta)\big), \qquad \mathbf{E}\nu_{ij}^{(i)}(\Delta - u) = p_{ij} + o\big(F_i(\Delta)\big).$$

Therefore,

$$h_{ij}(\Delta) = \frac{\pi_i p_{ij}\Delta}{\sum_l a_l\pi_l} + o(\Delta). \tag{21.4.13}$$

Further, from the equality

$$H_{ij}(t + 2\Delta) - H_{ij}(t) = \Delta H_{ij}(t) + \Delta H_{ij}(t + \Delta)$$

we obtain that $h_{ij}(2\Delta) = 2h_{ij}(\Delta)$, which means that $h_{ij}(\Delta)$ is linear. Together with (21.4.13) this proves (21.4.11). The corollary is proved. □

The class of processes for which one can prove ergodicity using the same methods as the one used for semi-Markov processes and also in Chap. 13, can be somewhat extended. For this broader class of processes we will prove in the next section the ergodic theorem, and also the laws of large numbers and the central limit theorem for integrals of such processes.

21.5 Regenerative Processes

21.5.1 Regenerative Processes. The Ergodic Theorem

Let $X(t)$ and $X_0(t)$; $t \ge 0$, be processes given in the space $D(0,\infty)$ of functions without discontinuities of the second type (the state space of these processes could be any metric space, not necessarily the real line). The process $X(t)$ is said to be *regenerative* if it possesses the following properties:

(1) There exists a state x_0 which is visited by the process X with probability 1. After each such visit, the evolution of the process starts anew as if it were the original process $X(t)$ starting at the state $X(0)=x_0$. We will denote this new process by $X_0(t)$ where $X_0(0)=x_0$. To state this property more precisely, we introduce the time τ_0 of the first visit to x_0 by X:

$$\tau_0 := \inf\bigl\{t \ge 0 : X(t) = x_0\bigr\}.$$

However, it is not clear from this definition whether τ_0 is a random variable. For definiteness, assume that the process X is such that for τ_0 one has

$$\{\tau_0 > t\} = \bigcup_n \bigcap_{t_k \in S} \bigl\{\bigl|X(t_k) - x_0\bigr| > 1/n\bigr\},$$

where S is a countable set everywhere dense in $[0,t]$. In that case the set $\{\tau_0 > t\}$ is clearly an event and τ_0 is a random variable. The above stated property means that τ_0 is a proper random variable: $\mathbf{P}(\tau_0 < \infty) = 1$, and that the distribution of $X(\tau_0 + u)$, $u \ge 0$, coincides with that of $X_0(u)$, $u \ge 0$, whatever the "history" of the process $X(t)$, $t \le \tau_0$.

(2) The recurrence time τ of the state x_0 has finite expectation $\mathbf{E}\tau < \infty$, $\tau := \inf\{t : X_0(t) = x_0\}$.

The aforesaid means that the evolution of the process is split into independent identically distributed cycles by its visits to the state x_0. The visit times to x_0 are called *regeneration times*. The behaviour of the process inside the cycles may be arbitrary, and no further conditions, including Markovity, are imposed.

We introduce the so-called "taboo probability"

$$P(t,B) := \mathbf{P}\bigl(X_0(t) \in B,\ \tau > t\bigr).$$

We will assume that, as a function of t, $P(t,B)$ is measurable and Riemann integrable.

Theorem 21.5.1 *Let $X(t)$ be a regenerative process and the random variable τ be non-lattice. Then, for any Borel set B, as $t \to \infty$,*

$$\mathbf{P}\bigl(X(t) \in B\bigr) \to \pi(B) = \frac{1}{\mathbf{E}\tau} \int_0^\infty \mathbf{P}(u,B)\,du.$$

If τ is a lattice variable (which is the case for processes $X(t)$ in discrete time), the assertion holds true with the following obvious changes: $t \to \infty$ along the lattice and the integral is replaced with a sum.

Proof Let $T_0 := 0$, $T_k := \tau_1 + \cdots + \tau_k$ be the epoch of the k-th regeneration of the process $X_0(t)$, and

$$H(u) := \sum_{k=0}^{\infty} \mathbf{P}(\tau_k < u)$$

($\tau_k \stackrel{d}{=} \tau$ are independent). Then, using the total probability formula and the key renewal theorem, we obtain, as $t \to \infty$,

$$\begin{aligned}\mathbf{P}\big(X_0(t) \in B\big) &= \sum_{k=0}^{\infty} \int_0^t \mathbf{P}(T_k \in du)\, P(t-u, B) \\ &= \int_0^t dH(u)\, P(t-u, B) \to \frac{1}{\mathbf{E}\tau} \int_0^{\infty} P(u, B)\, du = \pi(B).\end{aligned}$$

For the process $X(t)$ one gets

$$\mathbf{P}\big(X(t) \in B\big) = \int_0^t \mathbf{P}(t_0 \in du)\mathbf{P}\big(X_0(t-u) \in B\big) \to \pi(B).$$

The theorem is proved. □

21.5.2 The Laws of Large Numbers and Central Limit Theorem for Integrals of Regenerative Processes

Consider a measurable mapping $f : \mathcal{X} \to \mathbb{R}$ of the state space $\mathcal{X}$ of a process $X(t)$ to the real line $\mathbb{R}$. As in Sect. 21.4.2, for the sake of simplicity, we can assume that $\mathcal{X} = \mathbb{R}$ and the trajectories of $X(t)$ lie in the space $D(0, \infty)$ of functions without discontinuities of the second kind. In this case the paths $f(X(u))$, $u \ge 0$, will be measurable functions, for which the integral

$$S(t) = \int_0^t f\big(X(u)\big)\, du$$

is well defined. For such integrals we have the following law of large numbers. Set

$$\zeta := \int_0^{\tau} f\big(X_0(u)\big)\, du, \qquad a := \mathbf{E}\tau.$$

Theorem 21.5.2 *Let the conditions of Theorem* 21.5.1 *be satisfied and there exist* $a_\zeta := \mathbf{E}\zeta$. *Then, as* $t \to \infty$,

$$\frac{S(t)}{t} \xrightarrow{p} \frac{a_\zeta}{a}.$$

For conditions of existence of $\mathbf{E}\zeta$, see Theorem 21.5.4 below.

Proof The proof of the theorem largely repeats that of the similar assertion (Theorem 13.8.1) for sums of random variables defined on a Markov chain. Divide the domain $u \ge 0$ into half-intervals

$$(0, T_0], \quad (T_{k-1}, T_k], \quad k \ge 1, \quad T_0 = \tau_0,$$

where T_k are the epochs of hitting the state x_0 by the process $X(t)$, $\tau_k = T_k - T_{k-1}$ for $k \ge 1$ are independent and distributed as τ. Then the random variables

$$\zeta_k = \int_{T_{k-1}}^{T_k} f\bigl(X(u)\bigr)\,du, \quad k \ge 1$$

are independent, distributed as ζ, and have finite expectation a_ζ. The integral $S(t)$ can be represented as

$$S(t) = z_0 + \sum_{k=1}^{\nu(t)} \zeta_k + z_t,$$

where

$$\nu(t) := \max\{k : T_k \le t\}, \qquad z_0 := \int_0^{T_0} f\bigl(X(u)\bigr)\,du, \qquad z_t := \int_{T_{\nu(t)}}^{t} f\bigl(X(u)\bigr)\,du.$$

Since τ_0 is a proper random variable, z_0 is a proper random variable as well, and hence $z_0/t \xrightarrow{a.s.} 0$ as $t \to \infty$. Further,

$$z_t \stackrel{d}{=} \int_0^{\gamma(t)} f\bigl(X_0(u)\bigr)\,du,$$

where $\gamma(t) = t - T_{\nu(t)}$ has a proper limiting distribution as $t \to \infty$ (see Chap. 10), so $z_t/t \xrightarrow{p} 0$ as $t \to \infty$. The sum $S_{\nu(t)} = \sum_{k=1}^{\nu(t)} \zeta_k$ is nothing else but the generalised renewal process studied in Chaps. 10 and 11. By Theorem 11.5.2, as $t \to \infty$,

$$\frac{S_{\nu(t)}}{t} \xrightarrow{p} \frac{a_\zeta}{a}.$$

The theorem is proved. □

In order to prove the strong law of large numbers we need a somewhat more restrictive condition than that in Theorem 21.5.2. Put

$$\zeta^* := \int_0^{\tau} \bigl|f\bigl(X_0(u)\bigr)\bigr|\,du.$$

Theorem 21.5.3 *Let the conditions of Theorem* 21.5.1 *be satisfied and* $\mathbf{E}\zeta^* < \infty$. *Then*

$$\frac{S(t)}{t} \xrightarrow{a.s.} \frac{a_\zeta}{a}.$$

The proof essentially repeats (as was the case for Theorem 21.5.2) that of the law of large numbers for sums of random variables defined on a Markov chain (see Theorem 13.8.3). One only needs to use, instead of (13.8.18), the relation

$$\sup_{T_k \le u \le T_{k+1}} \left| \int_{T_k}^{u} f\big(X(v)\big)\, dv \right| \le \zeta_k^* = \int_{T_k}^{T_{k+1}} \big| f\big(X(v)\big)\big|\, dv$$

and the fact that $\mathbf{E}\,\zeta_k^* < \infty$. The theorem is proved. □

Here an analogue of Theorem 13.8.2, in which the conditions of existence of $\mathbf{E}\,\zeta^*$ and $\mathbf{E}\zeta$ are elucidated, is the following.

Theorem 21.5.4 (Generalisation of Wald's identity) *Let the conditions of Theorem* 21.5.1 *be met and there exist*

$$\mathbf{E}\big|f\big(X(\infty)\big)\big| = \int \big|f(x)\big|\pi(dx),$$

where $X(\infty)$ is a random variable with the stationary distribution π. Then there exist

$$\mathbf{E}\zeta^* = \mathbf{E}\tau\mathbf{E}\big|f\big(X(\infty)\big)\big|, \qquad \mathbf{E}\zeta = \mathbf{E}\tau\mathbf{E}f\big(X(\infty)\big).$$

The proof of Theorem 21.5.4 repeats, with obvious changes, that of Theorem 13.8.2. □

Theorem 21.5.5 (The central limit theorem) *Let the conditions of Theorem* 21.5.1 *be met and* $\mathbf{E}\tau^2 < \infty$, $\mathbf{E}\zeta^2 < \infty$. *Then*

$$\frac{S(t) - rt}{d\sqrt{t/a}} \Subset\!\Rightarrow \Phi_{0,1}, \quad t \to \infty,$$

where $r = a_\zeta/a$, $d^2 = \mathbf{D}(\zeta - r\tau)$.

The proof, as in the case of Theorems 21.5.2–21.5.4, repeats, up to evident changes, that of Theorem 13.8.4. □

Here an analogue of Theorem 13.8.5 (on the conditions of existence of variance and on an identity for $a^{-1}d^2$) looks more complicated than under the conditions of Sect. 13.8 and is omitted.

21.6 Diffusion Processes

Now we will consider an important class of Markov processes with continuous trajectories.

Definition 21.6.1 A homogeneous Markov process $\xi(t)$ with state space $\langle \mathbb{R}, \mathfrak{B}\rangle$ and the transition function $P(t,x,B)$ is said to be a *diffusion process* if, for some finite functions $a(x)$ and $b^2(x) > 0$,

(1) $\lim_{\Delta\to 0}\frac{1}{\Delta}\int(y-x)P(\Delta,x,dy)=a(x)$,
(2) $\lim_{\Delta\to 0}\frac{1}{\Delta}\int(y-x)^2P(\Delta,x,dy)=b^2(x)$,
(3) for some $\delta>0$ and $c<\infty$,

$$\int|y-x|^{2+\delta}P(\Delta,x,dy)<c\Delta^{1+\delta/2}.$$

Put $\Delta\xi(t):=\xi(t+\Delta)-\xi(t)$. Then the above conditions can be written in the form:

$$\begin{aligned}&\mathbf{E}\big[\Delta\xi(t)\big|\xi(t)=x\big]\sim a(x)\Delta,\\&\mathbf{E}\big[\big(\Delta\xi(t)\big)^2\big|\xi(t)=x\big]\sim b^2(x)\Delta,\\&\mathbf{E}\big[\big|\Delta\xi(t)\big|^{2+\delta}\big|\xi(t)=x\big]<c\Delta^{1+\delta/2}\quad\text{as }\Delta\to 0.\end{aligned}$$

The coefficients $a(x)$ and $b(x)$ are called the *shift* and *diffusion coefficients*, respectively. Condition (3) is an analogue of the Lyapunov condition. It could be replaced with a Lindeberg type condition:

(3a) $\mathbf{E}[(\Delta\xi(t))^2;\ |\Delta\xi(t)|>\varepsilon]=o(\Delta)$ for any $\varepsilon>0$ as $\Delta\to 0$.

It follows immediately from condition (3) and the Kolmogorov theorem that a diffusion process $\xi(t)$ can be thought of as a process with continuous trajectories.

The standard Wiener process $w(t)$ is a diffusion process, since in that case

$$P(t;x,B)=\frac{1}{\sqrt{2\pi t}}\int_B e^{-(x-y)^2/(2t)}\,dy,$$
$$\mathbf{E}\Delta w(t)=0,\qquad \mathbf{E}\big[\Delta w(t)\big]^2=\Delta,\qquad \mathbf{E}\big[\Delta w(t)\big]^4=3\Delta^2.$$

Therefore the Wiener process has zero shift and a constant diffusion coefficient. Clearly, the process $w(t)+at$ will have shift a and the same diffusion coefficient.

We saw in Sect. 21.2 that the "local" characteristic Q of a Markov process $\xi(t)$ with a discrete state space $\mathcal{X}$ specifies uniquely the evolution law of the process. A similar situation takes place for diffusion processes: the *distribution of the process is determined uniquely by the coefficients $a(x)$ and $b(x)$*. The way to establishing this fact again lies via the Chapman–Kolmogorov equation.

Theorem 21.6.1 *If the transition probability $P(t;x,B)$ of a diffusion process is twice continuously differentiable with respect to x, then $P(t;x,B)$ is differentiable with respect to t and satisfies the equation*

$$\frac{\partial P}{\partial t}=a\frac{\partial P}{\partial x}+\frac{b^2}{2}\frac{\partial^2 P}{\partial x^2}\tag{21.6.1}$$

with the initial condition

$$P(0;x,B)=\mathrm{I}_B(x).\tag{21.6.2}$$

Remark 21.6.1 The conditions of the theorem on smoothness of the transition function P can actually be proved under the assumption that a and b are continuous, $b\ge b_0>0$, $|a|\le c(|x|+1)$ and $b^2\le c(|x|+1)$.

Proof of Theorem 21.6.1 For brevity's sake denote by P'_t, P'_x, and P''_x the partial derivatives $\frac{\partial P}{\partial t}$, $\frac{\partial P}{\partial x}$ and $\frac{\partial^2 P}{\partial x^2}$, respectively, and make use of the relation

$$\begin{aligned} &P(t; y, B) - P(t; x, B) \\ &\quad = (y-x)P'_x + \frac{(y-x)^2}{2}P''_x + \frac{(y-x)^2}{2}\big[P''_x(t; y_x, B) - P''_x(t; x, B)\big], \\ &y_x \in (x, y). \end{aligned} \tag{21.6.3}$$

Then by the Chapman–Kolmogorov equation

$$\begin{aligned} P(t+\Delta; x, B) - P(t; x, B) &= \int P(\Delta; x, dy)\big[P(t; y, B) - P(t; x, B)\big] \\ &= a(x)P'_x\Delta + \frac{b^2(x)}{2}P''_x\Delta + o(\Delta) + R, \end{aligned} \tag{21.6.4}$$

where

$$R = \int \frac{(y-x)^2}{2}\big[P''_x(t; y_x, B) - P''_x(t; x, B)\big]P(\Delta; x, dy) = \int_{|y-x|\le\varepsilon} + \int_{|y-x|>\varepsilon}.$$

The first integral, by virtue of the continuity of P''_x, does not exceed

$$\delta(\varepsilon)\left[\frac{b^2(x)}{2}\Delta + o(\Delta)\right],$$

where $\delta(\varepsilon) \to 0$ as $\varepsilon \to 0$; the second integral is $o(\Delta)$ by condition (3a). Since ε is arbitrary, one has $R = o(\Delta)$ and it follows from the above that

$$P'_t = \lim_{\Delta\to 0} \frac{P(t+\Delta; x, B) - P(t; x, B)}{\Delta} = a(x)P'_x + \frac{b^2(x)}{2}P''_x.$$

This proves (21.6.1). The theorem is proved. □

It is known from the theory of differential equations that, under wide assumptions about the coefficients a and b and for $B = (-\infty, z)$, the Cauchy problem (21.6.1)–(21.6.2) has a unique solution P which is infinitely many times differentiable with respect to t, x and z. From this it follows that $P(t; x, B)$ has a density $p(t; x, z)$ which is the fundamental solution of (21.6.1).

It is also not difficult to derive from Theorem 21.6.1 that, along with $P(t; x, B)$, the function

$$u(t, x) = \int g(z)P(t; x, dz) = \mathbf{E}\big[g\big(\xi^{(x)}(t)\big)\big]$$

will also satisfy Eq. (21.6.1) for any smooth function g with a compact support, $\xi^{(x)}(t)$ being the diffusion process with the initial value $\xi^{(x)}(0) = x$.

In the proof of Theorem 21.6.1 we considered (see (21.6.4)) the time increment Δ *preceding* the main time interval. In this connection Eqs. (21.6.1) are called *backward* Kolmogorov equations. *Forward* equations can be derived in a similar way.

Theorem 21.6.2 (Forward Kolmogorov equations) *Let the transition density $p(t; x, y)$ be such that the derivatives*

$$\frac{\partial}{\partial y}\big[a(y)p(t; x, y)\big] \quad \text{and} \quad \frac{\partial^2}{\partial y^2}\big[b^2(y)p(t; x, y)\big]$$

exist and are continuous. Then $p(t, x, y)$ satisfies the equation

$$Dp := \frac{\partial p}{\partial t} + \frac{\partial}{\partial y}\big[a(y)p(t; x, y)\big] - \frac{1}{2}\frac{\partial^2}{\partial y^2}\big[b^2(y)p(t; x, y)\big] = 0. \qquad (21.6.5)$$

Proof Let $g(y)$ be a smooth function with a bounded support,

$$u(t, x) := \mathbf{E}g\big(\xi^{(x)}(t)\big) = \int g(y)p(x; t, y)\,dy.$$

Then

$$u(t + \Delta, x) - u(t, x)$$
$$= \int p(t; x, z)\left[p(\Delta; z, y)g(y)\,dy - \int p(\Delta, z, y)g(z)\,dy \right] dz. \qquad (21.6.6)$$

Expanding the difference $g(y) - g(z)$ into a series, we obtain in the same way as in the proof of Theorem 21.4.1 that, by virtue of properties (1)–(3), the expression in the brackets is

$$\left[a(z)g'(z) + \frac{b^2(z)}{2}g''(z) \right]\Delta + o(\Delta).$$

This implies that there exists the derivative

$$\frac{\partial u}{\partial t} = \int p(t; x, z)\left[a(z)g'(z)\,dz + \frac{1}{2}\frac{b^2(z)}{2}g''(z) \right] dz.$$

Integrating by parts we get

$$\frac{\partial u}{\partial t} = \int \left\{ -\frac{\partial}{\partial z}\big[a(z)p(t; x, z)\big] + \frac{1}{2}\frac{\partial}{\partial z^2}\big[b^2(z)p(t; x, z)\big] \right\} g(z)\,dz = 0$$

or, which is the same,

$$\int Dp(t; x, z)g(z)\,dz = 0.$$

Since g is arbitrary, (21.6.5) follows. The theorem is proved. □

As in the case of discrete $\mathcal{X}$, the difference between the forward and backward Kolmogorov equations becomes more graphical for non-homogeneous diffusion processes, when the transition probabilities $P(s, x; t, B)$ depend on two time variables, while a and b in conditions (1)–(3) are functions of s and x. Then the backward Kolmogorov equation (for densities) will relate the derivatives of the transition densities $p(s, x; t, y)$ with respect to the first two variables, while the forward equation will hold for the derivatives with respect to the last two variables.

We return to homogeneous diffusion processes. One can study conditions ensuring the existence of the limiting stationary distribution of $\xi^{(x)}(t)$ as $t \to \infty$ which is independent of x using the same approach as in Sect. 21.2. Theorem 21.2.3 will remain valid (one simply has to replace i_0 in it with x_0, in agreement with the notation of the present section). The proof of Theorem 21.2.3 also remains valid, but will need a somewhat more precise argument (in the new situation, on the event B_{dv} one has $\xi(t-v) \in dx_0$ instead of $\xi(t-v) = x_0$).

If the stationary distribution density

$$\lim_{t\to\infty} p(t; x, y) = p(y) \tag{21.6.7}$$

exists, how could one find it? Since the dependence of $p(t; x, y)$ of t and x vanishes as $t \to \infty$, the backward Kolmogorov equations turn into the identity $0 = 0$ as $t \to \infty$. Turning to the forward equations and passing in (21.6.6) to the limit first as $t \to \infty$ and then as $\Delta \to 0$, we come, using the same argument as in the proof of Theorem 21.2.3, to the following conclusion.

Corollary 21.6.1 *If* (21.6.7) *and the conditions of Theorem* 21.6.2 *hold, then the stationary density* $p(y)$ *satisfies the equation*

$$-\big[a(y)p(y)\big]' + \frac{1}{2}\big[b^2(y)p(y)\big]'' = 0$$

(*which is obtained from* (21.6.5) *if we put* $\frac{\partial p}{\partial t} = 0$).

Example 21.6.1 The Ornstein–Uhlenbeck process

$$\xi^{(x)}(t) = xe^{at} + \sigma e^{at} w\left(\frac{1 - e^{-2at}}{2a}\right),$$

where $w(u)$ is the standard Wiener process, is a homogeneous diffusion process with the transition density

$$p(t; x, y) = \frac{1}{\sqrt{2\pi}\sigma(t)} \exp\left\{-\frac{(y - xe^{at})^2}{2\sigma^2(t)}\right\}, \qquad \sigma^2(t) = \frac{\sigma^2}{2a}\left(e^{2at} - 1\right). \tag{21.6.8}$$

We leave it to the reader to verify that this process has coefficients $a(x) = ax$, $b(x) = \sigma = \text{const}$, and that function (21.6.8) satisfies the forward and backward equations. For $a < 0$, there exists a stationary process (the definition is given in the next chapter)

$$\xi(t) = \sigma e^{at} w\left(\frac{e^{-2at}}{2a}\right),$$

of which the density (which does not depend on t) is equal to

$$p(y) = \lim_{t\to\infty} p(x; t, y) = \frac{1}{\sqrt{2\pi}\sigma(\infty)} \exp\left\{-\frac{y^2}{2\sigma^2(\infty)}\right\}, \qquad \sigma(\infty) = -\frac{\sigma^2}{2a}.$$

In conclusion of this section we will consider the problem, important for various applications, of finding the probability that the trajectory of a diffusion process will not leave a given strip. For simplicity's sake we confine ourselves to considering this problem for the Wiener process. Let $c > 0$ and $d < 0$.

Put

$$U(t; x, B) := \mathbf{P}\big(w^{(x)}(u) \in (d, c) \text{ for all } u \in [0, t];\ w^{(x)}(t) \in B\big)$$
$$= \mathbf{P}\Big(\sup_{u \le t} w^{(x)}(u) < c,\ \inf_{u \le t} w^{(x)}(u) > d,\ w^{(x)}(t) \in B\Big).$$

Leaving out the verification of the fact that the function U is twice continuously differentiable, we will only prove the following proposition.

Theorem 21.6.3 *The function U satisfies Eq.* (21.6.1) *with the initial condition*

$$U(0; x, B) = \mathrm{I}_B(x) \tag{21.6.9}$$

and boundary conditions

$$U(t; c, B) = U(t; d, B) = 0. \tag{21.6.10}$$

Proof First of all note that the function $U(t; x, B)$ for $x \in (d, c)$ satisfies conditions (1)–(3) imposed on the transition function $P(t; x, B)$. Indeed, consider, for instance, property (1).

We have to verify that

$$\int_d^c (y - x) U(\Delta; x, dy) = \Delta a(x) + o(\Delta) \tag{21.6.11}$$

(with $a(x) = 0$ in our case). But $U(t, x, B) = P(t; x, B) - V(t; x, B)$, where

$$V(t; x, B) = \mathbf{P}\Big(\Big\{\sup_{u \le t} w^{(x)}(u) \ge c \text{ or } \inf_{u \le t} w^{(x)}(u) \le d\Big\} \cap \big\{w^{(x)}(t) \in B\big\}\Big),$$

and

$$\int_d^c (y - x) V(\Delta; x, dy)$$
$$\le \max(c, -d)\Big[\mathbf{P}\Big(\sup_{u \le \Delta} w^{(x)}(u) \ge c\Big) + \mathbf{P}\Big(\inf_{u \le \Delta} w^{(x)}(u) \le d\Big)\Big].$$

The first probability in the brackets is given, as we know (see (20.2.1) and Theorem 19.2.2), by the value

$$2\mathbf{P}\big(w^{(x)}(\Delta) > c\big) = 2\mathbf{P}\Big(w(1) > \frac{c - x}{\sqrt{\Delta}}\Big) \sim \frac{2}{\sqrt{2\pi}\, z} e^{-z^2/2}, \quad z = \frac{c - x}{\sqrt{\Delta}}.$$

For any $x < c$ and $k > 0$, it is $o(\Delta^k)$. The same holds for the second probability. Therefore (21.6.11) is proved. In the same way one can verify properties (2) and (3).

Further, because by the total probability formula, for $x \in (d, c)$,

$$U(t + \Delta; x, B) = \int_d^c U(\Delta; x, dy) U(t; y, B),$$

using an expansion of the form (21.6.3) for the function U, we obtain in the same way as in (21.6.4) that

$$U(t+\Delta;x,B)-U(t;x,B)=\int U(\Delta;x,dy)\big[U(t;y,B)-U(t;x,B)\big]$$
$$=a(x)\frac{\partial U}{\partial x}\Delta+\frac{b^2(x)}{2}\frac{\partial^2 U}{\partial x^2}\Delta+o(\Delta).$$

This implies that $\frac{\partial U}{\partial t}$ exists and that Eq. (21.6.1) holds for the function U.

That the boundary and initial conditions are met is obvious. The theorem is proved. □

The reader can verify that the function

$$u(t;x,y):=\frac{\partial}{\partial y}U\big(t;x,(-\infty,y)\big),\quad y\in(d,c),$$

playing the role of the fundamental solution to the boundary problem (21.6.9)–(21.6.10) (the function u satisfies (21.6.1) with the boundary conditions (21.6.10) and the initial conditions degenerating into the δ-function), is equal to

$$u(t;x,y)=\frac{1}{\sqrt{2\pi t}}\Bigg[\sum_{k=-\infty}^{\infty}\exp\bigg\{-\frac{[y+2k(c-d)]^2}{2t}\bigg\}$$
$$-\sum_{k=0}^{\infty}\exp\bigg\{-\frac{[y-2c-2k(c-d)]^2}{2t}\bigg\}$$
$$-\sum_{k=0}^{\infty}\exp\bigg\{-\frac{[y-2d-2k(c-d)]^2}{2t}\bigg\}\Bigg].$$

This expression can also be obtained directly from probabilistic considerations (see, e.g., [32]).

Chapter 22
Processes with Finite Second Moments. Gaussian Processes

Abstract The chapter is devoted to the classical "second-order theory" of time-homogeneous processes with finite second moments. Section 22.1 explores the relationships between the covariance function properties and those of the process itself and proves the ergodic theorem (in quadratic mean) for processes with covariance functions vanishing at the infinity. Section 22.2 is devoted to the special case of Gaussian processes, while Sect. 22.3 solves the best linear prediction problem.

22.1 Processes with Finite Second Moments

Let $\{\xi(t),\ -\infty < t < \infty\}$ be a random process for which there exist the moments $a(t) = \mathbf{E}\xi(t)$ and $R(t, u) = \mathbf{E}\xi(t)\xi(u)$. Since it is always possible to study the process $\xi(t) - a(t)$ instead of $\xi(t)$, we can assume without loss of generality that $a(t) \equiv 0$.

Definition 22.1.1 The function $R(t, u)$ is said to be the *covariance function* of the process $\xi(t)$.

Definition 22.1.2 A function $R(t, u)$ is said to be *nonnegative* (*positive*) *definite* if, for any $k;\ u_1, \ldots, u_k;\ a_1, \ldots, a_k \neq 0$,

$$\sum_{i,j} a_i a_j R(u_i, u_j) \geq 0 \quad (> 0).$$

It is evident that the covariance function $R(t, u)$ is nonnegative definite, because

$$\sum_{i,j} a_i a_j R(u_i, u_j) = \mathbf{E}\Bigl(\sum_{i,j} a_j \xi(u_i)\Bigr)^2 \geq 0.$$

Definition 22.1.3 A process $\xi(t)$ is said to be *unpredictable* if no linear combination of the variables $\xi(u_1), \ldots, \xi(u_k)$ is zero with probability 1, i.e. if there exist no $u_1, \ldots, u_k;\ a_1, \ldots, a_k$ such that

$$\mathbf{P}\Bigl(\sum_i a_i \xi(u_i) = 0\Bigr) = 1.$$

DOI 10.1007/978-1-4471-5201-9_22,

If $R(t,u)$ is the covariance function of an unpredictable process, then $R(t,u)$ is positive definite. We will see below that the converse assertion is also true in a certain sense.

Unpredictability means that we cannot represent $\xi(t_k)$ as a linear combination of $\xi(t_j)$, $j<k$.

Example 22.1.1 The process $\xi(t)=\sum_{k=1}^{N}\xi_k g_k(t)$, where $g_k(t)$ are linearly independent and ξ_k are independent, is not unpredictable, because from $\xi(t_1),\dots,\xi(t_N)$ we can determine the values $\xi(t)$ for all other t.

Consider the Hilbert space L_2 of all random variables η on $\langle\Omega,\mathfrak{F},\mathbf{P}\rangle$ having finite second moments, $\mathbf{E}\eta=0$, endowed with the inner product $(\eta_1,\eta_2)=\mathbf{E}\eta_1\eta_2$ corresponding to the distance $\|\eta_1-\eta_2\|=[\mathbf{E}(\eta_1-\eta_2)^2]^{1/2}$. Convergence in L_2 is obviously convergence in mean quadratic.

A random process $\xi(t)$ may be thought of as a curve in L_2.

Definition 22.1.4 A random process $\xi(t)$ is said to be *wide sense stationary* if the function $R(t,u)=:R(t-u)$ depends on the difference $t-u$ only. The function $R(s)$ is called nonnegative (positive) definite if the function $R(t,t+s)$ is of the respective type. For brevity, we will often call wide sense stationary processes simply stationary.

For the Wiener process, $R(t,u)=\mathbf{E}w(t)w(u)=\min(t,u)$, so that $w(t)$ cannot be stationary. But the process $\xi(t)=w(t+1)-w(t)$ will already be stationary.

It is obvious that, for a stationary process, the function $R(s)$ is even and $\mathbf{E}\xi^2(t)=R(0)=\text{const}$. For simplicity's sake, put $R(0)=1$. Then, by the Cauchy–Bunjakovsky inequality,

$$\left|R(s)\right|=\left|\mathbf{E}\xi(t)\xi(t+s)\right|\le\left[\mathbf{E}\xi^2(t)\mathbf{E}\xi^2(t+s)\right]^{1/2}=R(0)=1.$$

Theorem 22.1.1

(1) *A process $\xi(t)$ is continuous in mean quadratic ($\xi(t+\Delta)\xrightarrow{(2)}\xi(t)$ as $\Delta\to 0$) if and only if the function $R(u)$ is continuous at zero.*
(2) *If the function $R(u)$ is continuous at zero, then it is continuous everywhere.*

Proof

$$(1)\quad \left\|\xi(t+\Delta)-\xi(t)\right\|^2=\mathbf{E}\big(\xi(t+\Delta)-\xi(t)\big)^2=2R(0)-2R(\Delta).$$

$$\begin{aligned}(2)\qquad R(t+\Delta)-R(t)&=\mathbf{E}\big(\xi(t+\Delta)\xi(0)-\xi(t)\xi(0)\big)\\&=\big(\xi(0),\xi(t+\Delta)-\xi(t)\big)\le\left\|\xi(t+\Delta)-\xi(t)\right\|\\&=\sqrt{2\big(R(0)-R(\Delta)\big)}.\end{aligned}\qquad(22.1.1)$$

The theorem is proved. □

A process $\xi(t)$ continuous in mean quadratic will be stochastically continuous, as we can see from Chaps. 6 and 18. The continuity in mean quadratic does not, however, imply path-wise continuity. The reader can verify this by considering the example of the process

$$\xi(t) = \eta(t+1) - \eta(t) - 1,$$

where $\eta(t)$ is the Poisson process with parameter 1. For that process, the covariance function

$$R(t) = \begin{cases} 0 & \text{for } t \ge 1, \\ 1-t & \text{for } 0 \le t \le 1 \end{cases}$$

is continuous, although the trajectories of $\xi(t)$ are not. If

$$\bigl|R(\Delta) - R(0)\bigr| < c\Delta^{1+\varepsilon} \tag{22.1.2}$$

for some $\varepsilon > 0$ then, by the Kolmogorov theorem (see Theorem 18.2.1), $\xi(t)$ has a continuous modification. From this it follows, in particular, that if $R(t)$ is twice differentiable at the point $t = 0$, then the trajectories of $\xi(t)$ may be assumed continuous. Indeed, in that case, since $R(t)$ is even, one has

$$R'(0) = 0 \quad \text{and} \quad R(\Delta) - R(0) \sim \frac{1}{2} R''(0)\Delta^2.$$

As a whole, the smoother the covariance function is at zero, the smoother the trajectories of $\xi(t)$ are.

Assume that the trajectories of $\xi(t)$ are measurable (for example, belong to the space D).

Theorem 22.1.2 (The simplest ergodic theorem) *If*

$$R(s) \to 0 \quad \textit{as } s \to \infty, \tag{22.1.3}$$

then

$$\zeta_T := \frac{1}{T} \int_0^T \xi(t)\,dt \xrightarrow{(2)} 0.$$

Proof Clearly,

$$\|\zeta_T\|^2 = \frac{1}{T^2} \int_0^T \int_0^T R(t-u)\,dt\,du.$$

Since $R(s)$ is even,

$$J := \int_0^T \int_0^T R(t-u)\,dt\,du = 2 \int_0^T \int_u^T R(t-u)\,dt\,du.$$

Making the orthogonal change of variables $v = (t-u)/\sqrt{2}$, $s = (t+u)/\sqrt{2}$, we obtain

$$J \le 2 \int_{s=0}^{T/\sqrt{2}} \int_{v=0}^{T/\sqrt{2}} R(v\sqrt{2})\,dv\,ds \le 2T \int_0^T R(v)\,dv,$$

$$\|\zeta_T\|^2 \le \frac{2}{T} \int_0^T R(v)\,dv \to 0.$$

The theorem is proved. □

Example 22.1.2 The stationary white noise process $\xi(t)$ is defined as a process with independent values, i.e. a process such that, for any $t_1, \dots, t_n$, the variables $\xi(t_1), \dots, \xi(t_n)$ are independent. For such a process,

$$R(t) = \begin{cases} 1 & \text{for } t = 0, \\ 0 & \text{for } t \neq 0, \end{cases}$$

and thus condition (22.1.3) is met. However, one cannot apply Theorem 22.1.2 here, for the trajectories of $\xi(t)$ will be non-measurable with probability 1 (for example, the set $B = \{t : \xi(t) > 0\}$ is non-measurable with probability 1).

Definition 22.1.5 A process $\xi(t)$ is said to be *strict sense stationary* if, for any $t_1, \dots, t_k$, the distribution of $(\xi(t_1 + u), \xi(t_2 + u), \dots, \xi(t_k + u))$ is independent of u.

It is obvious that if $\xi(t)$ is a strict sense stationary process then

$$\mathbf{E}\xi(t)\xi(u) = \mathbf{E}\xi(t - u)\xi(0) = R(t - u),$$

and $\xi(t)$ will be wide sense stationary. The converse is, of course, not true. However, there exists a class of processes for which both concepts of stationarity coincide.

22.2 Gaussian Processes

Definition 22.2.1 A process $\xi(t)$ is said to be *Gaussian* if its finite-dimensional distributions are normal.

We again assume that $\mathbf{E}\xi(t) = 0$ and $R(t, u) = \mathbf{E}\xi(t)\xi(u)$.

The finite-dimensional distributions are completely determined by the ch.f.s ($\lambda = (\lambda_1, \dots, \lambda_k)$, $\xi = (\xi(t_1), \dots, \xi(t_k))$)

$$\mathbf{E}e^{i(\lambda,\xi)} = \mathbf{E}e^{i\sum_j \lambda_j \xi(t_j)} = e^{-\frac{1}{2}\lambda R \lambda^T},$$

where $R = \|R(t_i, t_j)\|$ and the superscript T stands for transposition, so that

$$\lambda R \lambda^T = \sum_{i,j} \lambda_i \lambda_j R(t_i, t_j).$$

Thus for a Gaussian process the finite-dimensional distributions are completely determined by the covariance function $R(t, u)$.

We saw that for an unpredictable process $\xi(t)$, the function $R(t, u)$ is positive definite. A converse assertion may be stated in the following form.

Theorem 22.2.1 *If the function $R(t, u)$ is positive definite, then there exists an unpredictable Gaussian process with the covariance function $R(t, u)$.*

Proof For arbitrary $t_1, \dots, t_k$, define the finite-dimensional distribution of the vector $\xi(t_1), \dots, \xi(t_k)$ via the density

$$p_{t_1,\dots,t_k}(x_1, \dots, x_k) = \frac{\sqrt{|A|}}{(2\pi)^{k/2}} \exp\left\{-\frac{1}{2} x A x^T\right\},$$

where A is the matrix inverse to the covariance matrix $R = \|R(t_i, t_j)\|$ (see Sect. 7.6) and $|A|$ is the determinant of A. These distributions will clearly be consistent, because the covariance matrices are consistent (the matrix for $\xi(t_1), \dots, \xi(t_{k-1})$ is a submatrix of R). It remains to make use of the Kolmogorov theorem. The theorem is proved. □

Example 22.2.1 Let $w(t)$ be the standard Wiener process. The process

$$w^0(t) = w(t) - t w(1), \quad t \in [0, 1],$$

is called the *Brownian bridge* (its "ends are fixed": $w^0(0) = w^0(1) = 0$). The covariance function of $w^0(t)$ is equal to

$$R(t, u) = \mathbf{E}\big(w(t) - t w(1)\big)\big(w(u) - u w(1)\big) = t(1 - u)$$

for $u \ge t$.

A Gaussian wide sense stationary process $\xi(t)$ is strict sense stationary. This immediately follows from the fact that for $R(t, u) = R(t - u)$ the finite-dimensional distributions of $\xi(t)$ become invariant with respect to time shift:

$$p_{t_1,\dots,t_k}(x_1, \dots, x_k) = p_{t_1+u,\dots,t_k+u}(x_1, \dots, x_k)$$

since $\|R(t_i + u, t_j + u)\| = \|R(t_i, t_j)\|$.

If $\xi(t)$ is a Gaussian process, then conditions ensuring the smoothness of its trajectories can be substantially relaxed in comparison with (22.1.2).

Let for simplicity's sake the Gaussian process $\xi(t)$ be stationary.

Theorem 22.2.2 *If, for $h < 1$,*

$$\big|R(h) - R(0)\big| < c\left(\log \frac{1}{h}\right)^{-\alpha}, \quad \alpha > 3,\ c < \infty,$$

then the trajectories of $\xi(t)$ can be assumed continuous.

Proof We make use of Theorem 18.2.2 and put $\varepsilon(h) = (\log \frac{1}{h})^{-\beta}$ for $1 < \beta < (\alpha - 1)/2$ (we take logarithms to the base 2). Then

$$\sum_{n=1}^{\infty} \varepsilon(2^{-n}) = \sum_{n=1}^{\infty} n^{-\beta} < \infty,$$

and, by (22.1.1),

$$\begin{aligned}\mathbf{P}\big(\big|\xi(t+h)-\xi(t)\big|>\varepsilon(h)\big)&=2\left[1-\Phi\left(\frac{\varepsilon(h)}{\sqrt{2(1-R(h))}}\right)\right]\\&\le 2\left[1-\Phi\left(c\varepsilon(h)\left(\log\frac{1}{h}\right)^{\alpha/2}\right)\right]=2\left[1-\Phi\left(c\left(\log\frac{1}{h}\right)^{\alpha/2-\beta}\right)\right].\end{aligned}\tag{22.2.1}$$

Since the argument of Φ increases unboundedly as $h\to 0$, $\gamma=\alpha-2\beta>1$, and by (19.3.1)

$$1-\Phi(x)\sim\frac{1}{\sqrt{2\pi}x}e^{-x^2/2}\quad\text{as } x\to\infty,$$

we see that the right-hand side of (22.2.1) does not exceed

$$q(h):=c_1\left(\log\frac{1}{h}\right)^{\beta-\alpha/2}\exp\left\{-c_2\left(\log\frac{1}{h}\right)^{\alpha-2\beta}\right\},$$

so that

$$\sum_{n=1}^{\infty}2^nq\left(2^{-n}\right)=c_1\sum_{n=1}^{\infty}n^{-\gamma/2}\exp\left\{-c_2n^{\gamma}+n\ln 2\right\}<\infty,$$

because $c_2>0$ and $\gamma>1$. The conditions of Theorem 18.2.2 are met, and so Theorem 22.2.2 is proved. □

22.3 Prediction Problem

Suppose the distribution of a process $\xi(t)$ is known, and one is given the trajectory of $\xi(t)$ on a set $B\subset(-\infty,t]$, B being either an interval or a finite collection of points. What could be said about the value $\xi(t+u)$? Our aim will be to find a random variable ζ, which is $\mathfrak{F}_B=\sigma(\xi(v),\ v\in B)$-measurable (and called a *prediction*) and such that $\mathbf{E}(\xi(t+u)-\zeta)^2$ assumes the smallest possible value. The answer to that problem is actually known (see Sect. 4.8):

$$\zeta=\mathbf{E}\big(\xi(t+u)\big|\mathfrak{F}_B\big).$$

Let $\xi(t)$ be a Gaussian process, $B=\{t_1,\dots,t_k\}$, $t_1<t_2<\dots<t_k<t_0=t+u$, $A=(\sigma^2)^{-1}=\|a_{ij}\|$ and $\sigma^2=\|\mathbf{E}\xi(t_i)\xi(t_j)\|_{i,j=1,\dots,k,0}$. Then the distribution of the vector $(\xi(t_1),\dots,\xi(t_0))$ has the density

$$f(x_1,\dots,x_k,x_0)=\frac{\sqrt{|A|}}{(2\pi)^{(k+1)/2}}\exp\left\{-\frac{1}{2}\sum_{i,j}x_ix_ja_{ij}\right\},$$

and the conditional distribution of $\xi(t_0)$ given $\xi(t_1),\dots,\xi(t_k)$ has density equal to the ratio

$$\frac{f(x_1,\dots,x_k,x_0)}{\int_{-\infty}^{\infty}f(x_1,\dots,x_k,x_0)\,dx_0}.$$

The exponential part of this ratio has the form

$$\exp\left\{-\frac{a_{00}x_0^2}{2} - \sum_{j=1}^{k} x_0 x_j a_{j0}\right\}.$$

This means that the conditional distribution under consideration is the normal law $\mathbf{\Phi}_{\alpha,d^2}$, where

$$\alpha = -\sum_j \frac{x_j a_{j0}}{a_{00}}, \qquad d^2 = \frac{1}{a_{00}}.$$

Thus, in our case the best prediction ζ is equal to

$$\zeta = -\sum_{j=1}^{k} \frac{\xi(t_j)a_{0j}}{a_{00}}.$$

The mean quadratic error of this prediction equals $\sqrt{1/a_{00}}$.

We have obtained a *linear prediction*. In the general case, the linearity property is usually violated.

Consider now the problem of the best linear prediction in the case of an arbitrary process $\xi(t)$ with finite second moments. For simplicity's sake we assume again that $B = \{t_1, \dots, t_k\}$.

Denote by $H(\xi)$ the subspace of L_2 generated by the random variables $\xi(t)$, $-\infty < t < \infty$, and by $H_B(\xi)$ the subspace of $H(\xi)$ generated (or spanned by) $\xi(t_1), \dots, \xi(t_k)$. Elements of $H_B(\xi)$ have the form

$$\sum_{j=1}^{k} a_j \xi(t_j).$$

The existence and the form of the best linear prediction in this case are established by the following assertion.

Theorem 22.3.1 *There exists a unique point $\zeta \in H_B(\xi)$ (the projection of $\xi(t+u)$ onto $H_B(\xi)$, see Fig. 22.1) such that*

$$\xi(t+u) - \zeta \perp H_B(\xi). \tag{22.3.1}$$

Relation (22.3.1) *is equivalent to*

$$\big\|\xi(t+u) - \zeta\big\| = \min_{\theta \in H_B(\xi)} \big\|\xi(t+u) - \theta\big\|. \tag{22.3.2}$$

Explicit formulas for the coefficients a_j in the representation $\zeta = \sum a_j \xi(t_j)$ are given in the proof.

Proof Relation (22.3.1) is equivalent to the equations

$$\big(\xi(t+u) - \zeta, \xi(t_j)\big) = 0, \quad j = 1, \dots, k.$$

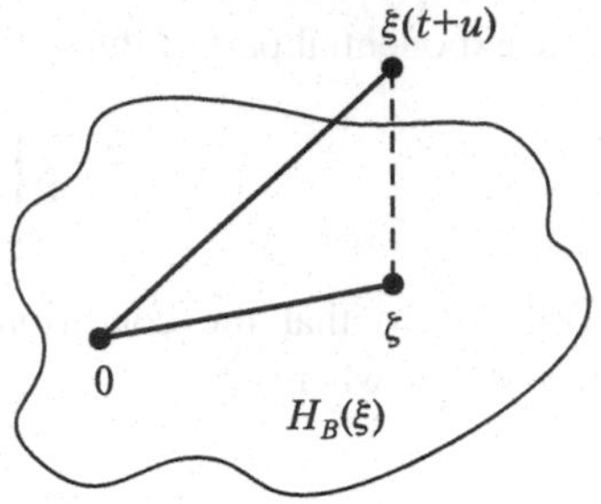

Fig. 22.1 Illustration to Theorem 22.3.1: the point ζ is the projection of $\xi(t+u)$ onto $H_B(\xi)$

Substituting here

$$\zeta = \sum_{l=1}^{k} a_l \xi(t_l) \in H_B(\xi),$$

we obtain

$$R(t+u, t_j) = \sum_{l=1}^{k} a_l R(t_j, t_l), \quad j = 1, \dots, k, \tag{22.3.3}$$

or, in vector form, $R_{t+u} = aR$, where

$$a = (a_1, \dots, a_k),$$
$$R_{t+u} = \big(R(t+u, t_1), \dots, R(t+u, t_k)\big), \quad R = \big\| R(t_i, t_j) \big\|.$$

If the process $\xi(t)$ is unpredictable, then the matrix R is non-degenerate and Eq. (22.3.3) has a unique solution:

$$a = R_{t+u} R^{-1}. \tag{22.3.4}$$

If $\xi(t)$ is not unpredictable, then either R^{-1} still exists and then (22.3.4) holds, or R is degenerate. In that case, one has to choose from the collection $\xi(t_1), \dots, \xi(t_k)$ only $l < k$ linearly independent elements for which all the above remains true after replacing k with l.

The equivalence of (22.3.1) and (22.3.2) follows from the following considerations. Let θ be any other element of $H_B(\xi)$. Then

$$\eta := \theta - \zeta \in H_B(\xi), \qquad \eta \perp \xi(t+u) - \zeta,$$

so that

$$\big\|\xi(t+u) - \theta\big\| = \big\|\xi(t+u) - \zeta\big\| + \|\eta\| \ge \big\|\xi(t+u) - \zeta\big\|.$$

The theorem is proved. □

Remark 22.3.1 It can happen (in the case where the process $\xi(t)$ is not unpredictable) that $\xi(t+u) \in H_B(\xi)$. Then the error of the prediction ζ will be equal to zero.

Appendix 1
Extension of a Probability Measure

In this appendix we will prove Carathéodory's theorem, which was used in Sect. 2.1.

Let $\mathcal{A}$ be an algebra of subsets of Ω on which a probability measure $\mathbf{P}$, i.e., a real-valued function satisfying conditions P1–P3 of Chap. 2, is given. Let $\mathcal{P}$ denote the class of all subsets of Ω. For any $A \in \mathcal{P}$, there always exists a sequence $\{A_n\}_{n=1}^{\infty}$ of disjoint sets from $\mathcal{A}$ such that $\bigcup_{n=1}^{\infty} A_n \supset A$ (it suffices to take $A_1 = \Omega$ and $A_n = \varnothing$, $n \ge 2$). Denote by $\gamma(A)$ the class of all such sequences and introduce on $\mathcal{P}$ the real-valued function

$$\mathbf{P}^*(A) := \inf\left\{\sum_{n=1}^{\infty} \mathbf{P}(A_n);\ \{A_n\} \in \gamma(A)\right\}.$$

This function (the outer measure on $\mathcal{P}$ induced by the measure $\mathbf{P}$ on $\mathcal{A}$) has the following properties:

(1) $\mathbf{P}^*(A) \le \mathbf{P}^*(B) \le 1$ if $A \subset B$.
(2) $\mathbf{P}^*(\bigcup_{n=1}^{\infty} A_n) = \sum_{n=1}^{\infty} \mathbf{P}(A_n)$ if the sets $A_n \in \mathcal{A}$, $n = 1, 2, \dots$, are disjoint.
(3) $\mathbf{P}^*(\bigcup_{n=1}^{\infty} A_n) \le \sum_{n=1}^{\infty} \mathbf{P}^*(A_n)$ for any $A_1, A_2, \ldots \in \mathcal{P}$.

Property (1) is obvious. Property (2) is established by the following argument. Let $\{B_n\}$ be any sequence from $\gamma(A)$, where $A = \bigcup_{n=1}^{\infty} A_n$. Since $\bigcup_{m=1}^{\infty} A_n B_m = A_n \in \mathcal{A}$, one has $\mathbf{P}(A_n) = \sum_{m=1}^{\infty} \mathbf{P}(A_n B_m)$. Therefore,

$$\sum_{n=1}^{\infty} \mathbf{P}(A_n) = \sum_{n}\sum_{m} \mathbf{P}(A_n B_m) = \sum_{m=1}^{\infty}\sum_{n=1}^{\infty} \mathbf{P}(A_n B_m).$$

But, for each $N < \infty$,

$$\sum_{n=1}^{N} \mathbf{P}(A_n B_m) \le \mathbf{P}(B_m).$$

A.A. Borovkov, *Probability Theory*, Universitext,
DOI 10.1007/978-1-4471-5201-9, © Springer-Verlag London 2013

Hence this equality holds for $N=\infty$ as well and, for any sequence $\{B_m\}_{m=1}^{\infty}\in\gamma(A)$,

$$\sum_{n=1}^{N}\mathbf{P}(A_n)\le\mathbf{P}(B_m).$$

This implies that $\mathbf{P}^*(A)\ge\sum_{n=1}^{\infty}\mathbf{P}(A_n)$. Because the converse inequality is obvious, we have $\mathbf{P}^*(A)=\sum_{n=1}^{\infty}\mathbf{P}(A_n)$.

Proof of property (3) Consider, for some $\varepsilon>0$, sequences $\{A_{nk}\}_{k=1}^{\infty}\in\gamma(A_n)$ such that

$$\sum_{k=1}^{\infty}\mathbf{P}(A_{nk})\le\mathbf{P}^*(A_n)+\frac{\varepsilon}{2^n}.$$

The sequence of sets $\{A_{nk}\}_{n,k=1}^{\infty}$ clearly contains $\bigcup A_n$ and therefore

$$\mathbf{P}^*\Big(\bigcup A_n\Big)\le\sum_n\sum_k\mathbf{P}(A_{nk})\le\sum_{n=1}^{\infty}\mathbf{P}^*(A_n)+\varepsilon.$$

Since ε is arbitrary, property (3) is proved. □

Introduce now the binary operation of *symmetric difference* $\oplus$ on arbitrary sets A and B from $\mathcal{P}$ by means of the equality

$$A\oplus B:=A\overline{B}\cup\overline{A}B.$$

It is not hard to see that

$$A\oplus B=B\oplus A=\overline{A}\oplus\overline{B}\subset A\cup B,\qquad A\oplus A=\varnothing,$$

$$A\oplus\varnothing=A,\qquad (A\oplus B)\oplus C=A\oplus(B\oplus C).$$

With the help of this operation and the function $\mathbf{P}^*$, we introduce on $\mathcal{P}$ a distance ρ by putting, for any $A,B\in\mathcal{P}$,

$$\rho(A,B):=\mathbf{P}^*(A\oplus B).$$

This construction is quite similar to the one used in Sect. 3.4 (we considered there the distance $d(A,B)=\mathbf{P}(A\oplus B)$ between measurable sets A and B). The properties of the distance ρ are the same as in (3.4.2). We will need the following properties:

(1) $\rho(A,B)=\rho(B,A)\ge 0$, $\rho(A,A)=0$,
(2) $\rho(\overline{A},\overline{B})=\rho(A,B)$,
(3) $\rho(AB,CD)\le\rho(A,C)+\rho(B,D)$,
(4) $\rho(\bigcup A_k,\bigcup B_k)\le\sum_k\rho(A_k,B_k)$.

We also note that

(5) $|\mathbf{P}^*(A) - \mathbf{P}^*(B)| \le \rho(A, B)$, and therefore $\mathbf{P}^*(\cdot)$ is a uniformly continuous function with respect to ρ.

Properties (1)–(3) were listed in (3.4.2); in the present context, they are proved in exactly the same way based on the properties of the measure $\mathbf{P}^*$. Property (4) follows from property (3) of the measure $\mathbf{P}^*$ and the relation (we put here $A = \bigcup A_n$ and $B = \bigcup B_n$)

$$A \oplus B \subset \bigcup (A_n \oplus B_n),$$

because

$$\begin{aligned} A \oplus B &= \Big[\Big(\bigcup A_n\Big) \cap \Big(\bigcap \overline{B}_n\Big)\Big] \cup \Big[\Big(\bigcap \overline{A}_n\Big) \cap \Big(\bigcup B_n\Big)\Big] \\ &\subset \Big[\bigcup A_n \overline{B}_n\Big] \cup \Big[\bigcup B_n \overline{A}_n\Big] = \bigcup (A_n \overline{B}_n \cup \overline{A}_n B_n) = \bigcup (A_n \oplus B_n). \end{aligned}$$

Property (5) follows from the fact that

$$A \subset B \cup (A \oplus B), \qquad B \subset A \cup (A \oplus B) \tag{A1.1}$$

and therefore

$$\mathbf{P}^*(A) - \mathbf{P}^*(B) \le \mathbf{P}^*(A \oplus B) = \rho(A, B),$$
$$\mathbf{P}^*(B) - \mathbf{P}^*(A) \le \mathbf{P}^*(A \oplus B) = \rho(A, B).$$

Similarly to the terminology adopted in Sect. 3.4 we call a set $A \in \mathcal{P}$ *approximable* if there exists a sequence $A_n \in \mathcal{A}$ for which $\rho(A, A_n) \to 0$. The totality of all approximable sets we denote by $\mathfrak{A}$. This is clearly the closure of $\mathcal{A}$ with respect to ρ.

Lemma A1.1 *$\mathfrak{A}$ is a σ-algebra.*

Proof We verify that $\mathfrak{A}$ satisfies properties A1, A2′ and A3 of σ-algebras of Chap. 2. Property A1: $\Omega \in \mathfrak{A}$ is obvious, for $\mathcal{A} \in \mathfrak{A}$. Property A3 ($\overline{A} \in \mathfrak{A}$ if $A \in \mathfrak{A}$) follows from the fact that, for $A \in \mathfrak{A}$, there exist $A_n \in \mathcal{A}$ such that, as $n \to \infty$,

$$\rho(A, A_n) \to 0, \quad \rho(\overline{A}, \overline{A}_n) = \rho(A, A_n) \to 0.$$

Finally, consider property A2′. We show first that if $A_n \in \mathcal{A}$, then $A = \bigcup A_n \in \mathfrak{A}$. Indeed, we can assume without loss of generality that the A_n are disjoint. Then, by virtue of the properties of the measure $\mathbf{P}^*$, for any $\varepsilon > 0$,

$$\sum \mathbf{P}(A_k) \le \mathbf{P}^*(\Omega) = 1,$$

$$\rho\Big(A, \bigcup_{k=1}^{n} A_k\Big) = \mathbf{P}^*\Big(\bigcup_{k=n+1}^{\infty} A_k\Big) = \sum_{k=n+1}^{\infty} \mathbf{P}(A_k) < \varepsilon$$

for n large enough.

Now let $A_n \in \mathfrak{A}$. We have to show that

$$A = \bigcup_{n=1}^{\infty} A_n \in \mathfrak{A}.$$

Let $\{B_n\}$ be a sequence of sets from $\mathcal{A}$ such that $\rho(A_n, B_n) < \varepsilon/2^n$. Then one has $B = \bigcup B_n \in \mathfrak{A}$ and, by property (4) of the distance ρ,

$$\rho(A, B) \le \sum_{n=1}^{\infty} \rho(A_n, B_n) < \varepsilon.$$

The lemma is proved. □

Now we can prove the main assertion.[1]

Theorem A1.1 *The probability* $\mathbf{P}$ *can be extended from the algebra* $\mathcal{A}$ *to some probability* $\overline{\mathbf{P}}$ *given on the* σ*-algebra* $\mathfrak{A}$.

Proof For $A \in \mathfrak{A}$, put

$$\overline{\mathbf{P}}(A) := \mathbf{P}^*(A).$$

It is evident that $\overline{\mathbf{P}}(A) = \mathbf{P}(A)$ for $A \in \mathcal{A}$, and $\overline{\mathbf{P}}(\Omega) = 1$. To verify that $\overline{\mathbf{P}}$ is a probability we just have to prove the countable additivity of $\overline{\mathbf{P}}$. We first prove the finite additivity. It suffices to prove it for two sets:

$$\mathbf{P}^*(A \cup B) = \mathbf{P}^*(A) + \mathbf{P}^*(B), \tag{A1.2}$$

where $A, B \in \mathfrak{A}$ and $A \cap B = \varnothing$. Let $A_n \in \mathcal{A}$ and $B_n \in \mathcal{A}$ be such that $\rho(A, A_n) \to 0$ and $\rho(B, B_n) \to 0$ as $n \to \infty$. Then

$$\big|\mathbf{P}^*(A \cup B) - \mathbf{P}^*(A_n \cup B_n)\big| \le \rho(A \cup B, A_n \cup B_n) \le \rho(A, A_n) + \rho(B, B_n) \to 0,$$

$$\mathbf{P}^*(A_n \cup B_n) = \mathbf{P}(A_n \cup B_n) = \mathbf{P}(A_n) + \mathbf{P}(B_n) - \mathbf{P}(A_n B_n). \tag{A1.3}$$

Here

$$\mathbf{P}(A_n) \to \mathbf{P}^*(A), \qquad \mathbf{P}(B_n) \to \mathbf{P}^*(B),$$

[1] The theorem on the extension of a measure to the minimum σ-algebra containing $\mathcal{A}$ was obtained by C. Carathéodory. The metrisation of normed Boolean algebras $\mathcal{A}$ by the distance $\rho(A, B) = \mathbf{P}(A \oplus B)$ was used by many authors (see, e.g., the talk by A.N. Kolmogorov at the 6th Polish Mathematical Congress in 1948 and Halmos [19]).

It was L.Ya. Savel'ev who suggested the use of the continuity properties of the measure with respect to the distance $\rho(A, B) = \mathbf{P}^*(A \oplus B)$ in order to extend it.

$$\begin{aligned}\mathbf{P}(A_n B_n) &\le \mathbf{P}^*(A_n B) + \mathbf{P}^*(B_n \overline{B}) \\ &\le \mathbf{P}^*(A_n \overline{A}) + \mathbf{P}^*(B_n \overline{B}) \le \rho(A, A_n) + \rho(B, B_n) \to 0.\end{aligned}$$

Hence (A1.3) implies (A1.2).

We now prove countable additivity. Let $A_n \in \mathfrak{A}$ be disjoint. Then, putting

$$A = \bigcup_{n=1}^{\infty} A_n,$$

we obtain from the finite additivity of $\overline{\mathbf{P}}$ that

$$\overline{\mathbf{P}}(A) = \sum_{k=1}^{n} \overline{\mathbf{P}}(A_k) + \overline{\mathbf{P}}\left(\bigcup_{k=n+1}^{\infty} A_k \right).$$

Therefore

$$\overline{\mathbf{P}}(A) \ge \sum_{k=1}^{\infty} \overline{\mathbf{P}}(A_k).$$

On the other hand,

$$\overline{\mathbf{P}}(A) = \mathbf{P}^*(A) \le \sum_{k=1}^{\infty} \mathbf{P}^*(A_k) = \sum_{k=1}^{\infty} \overline{\mathbf{P}}(A_k).$$

The theorem is proved. □

Theorem A1.2 *The extension of the probability* **P** *from the algebra* $\mathcal{A}$ *to the* σ-*algebra* $\mathfrak{A}$ *is unique.*

Proof Assume that there exists another probability $\mathbf{P}_1$ on $\mathfrak{A}$, which coincides with **P** on $\mathcal{A}$ and is such that, for some $A \in \mathfrak{A}$,

$$\mathbf{P}_1(A) \ne \overline{\mathbf{P}}(A).$$

Suppose first that $\varepsilon = \mathbf{P}_1(A) - \overline{\mathbf{P}}(A) > 0$. Consider a sequence $\{B_n\} \in \gamma(A)$ such that

$$\sum_{n=1}^{\infty} \mathbf{P}(B_n) - \overline{\mathbf{P}}(A) < \frac{\varepsilon}{2}.$$

Then

$$\mathbf{P}_1(A) = \overline{\mathbf{P}}(A) + \varepsilon \ge \sum_{n=1}^{\infty} \mathbf{P}(B_n) + \varepsilon/2$$

which contradicts the assumption that $A \subset \bigcup_{n=1}^{\infty} B_n$. Therefore

$$\mathbf{P}_1(A) \le \overline{\mathbf{P}}(A), \quad A \in \mathfrak{A}.$$

Since $\overline{\mathbf{P}}$ is ρ-continuous at the point $\varnothing$, it follows that $\mathbf{P}_1$ is also ρ-continuous at the point $\varnothing$, and hence at any "point" $A \in \mathfrak{A}$. Indeed, by virtue of (A1.1),

$$\left|\mathbf{P}_1(A) - \mathbf{P}_1(B)\right| \le \mathbf{P}_1(A \oplus B) \le \overline{\mathbf{P}}(A \oplus B) \to 0$$

if only $\rho(A, B) = \overline{\mathbf{P}}(A \oplus B) \to 0$. Hence, for $A \in \mathfrak{A}$,

$$\overline{\mathbf{P}}(A) = \lim_{\substack{B \to A \\ B \in \mathfrak{A}}} \mathbf{P}(B) = \lim_{\substack{B \to A \\ B \in \mathfrak{A}}} \mathbf{P}_1(B) = \mathbf{P}_1(A).$$

The theorem is proved. □

Let $\mathfrak{A}^* = \sigma(\mathcal{A})$ be the σ-algebra generated by $\mathcal{A}$. Since $\mathcal{A} \subset \mathfrak{A}$, we have $\mathfrak{A}^* \in \mathfrak{A}$, and the next statement follows in an obvious way from the above assertions.

Corollary A1.1 *The probability* $\mathbf{P}$ *can be uniquely extended from the algebra* $\mathcal{A}$ *to the* σ*-algebra* $\mathfrak{A}^*$ *generated by* $\mathcal{A}$.

Remark A1.1 The σ-algebra $\mathfrak{A}$ defined above as the closure of the algebra $\mathcal{A}$ with respect to the introduced distance ρ is in many cases wider than the σ-algebra $\mathfrak{A}^* = \sigma(\mathcal{A})$ generated by $\mathcal{A}$. This fact is closely related to the concept of the *completion* of a measure. To explain the concept, we assume from the very beginning that $\mathcal{A} = \mathfrak{F}$ is a σ-algebra. Then the measure $\overline{\mathbf{P}}$ can be constructed in a rather simple way. To do this we extend the measure $\mathbf{P}$ from $\langle \Omega, \mathfrak{F} \rangle$ to a σ-algebra which is wider than $\mathfrak{F}$ and is constructed as follows. We will say that a subset N of Ω belongs to the class $\mathcal{N}$ if there exists an $A = A(N) \in \mathfrak{F}$ such that $N \subset A$ and $\mathbf{P}(A) = 0$. It is not hard to see that the class of all sets of the form $B \cup N$, where $B \in \mathfrak{F}$ and $N \in \mathcal{N}$, also forms a σ-algebra. Denote it by $\mathfrak{F}_{\mathcal{N}}$. Putting $\mathbf{P}(B \cup N) := \mathbf{P}(B)$ we obtain an extension of $\mathbf{P}$ to $\langle \Omega, \mathfrak{F}_{\mathcal{N}} \rangle$. Such a measure is said to be *complete*, and the above operation itself is called the *completion of the measure* $\mathbf{P}$.

Now we can say that the measure $\overline{\mathbf{P}}$ constructed in Theorem A1.1 is complete, and the σ-algebra $\mathfrak{A}$ coincides with $\mathfrak{F}_{\mathcal{N}}$.

If, for example, $\Omega = [0, 1]$ and $\mathcal{A}$ is the algebra generated by the intervals, then $\mathfrak{A}^* = \sigma(\mathcal{A})$ will, as we already know, be the Borel σ-algebra, and $\mathfrak{A}$ will be the Lebesgue extension of $\mathfrak{A}^*$ consisting of all "Lebesgue measurable" sets.

Appendix 2
Kolmogorov's Theorem on Consistent Distributions

In this appendix we will prove the Kolmogorov theorem asserting that consistent distributions define a unique probability measure such that the consistent distributions are its projections. We used this theorem in Sect. 5.5 and in some other places, where distributions on infinite-dimensional spaces were considered.

Let T be an index set and, for each $t \in T$, $\mathbb{R}_t$ be the real line $(-\infty, \infty)$. Let $N \in T$ be a finite subset of T. Then the product space

$$\prod_{t\in T} \mathbb{R}_t = \mathbb{R}^N$$

is a Euclidean space of dimension equal to the number n of elements in N, spanned on n axes of the space

$$\mathbb{R}^T = \prod_{t\in T} \mathbb{R}_t.$$

Assume that, for any finite subset $N \subset T$, a probability measure $\mathbf{P}_N$ is given on $\langle \mathbb{R}^N, \mathfrak{B}^N \rangle$, where $\mathfrak{B}^N$ is the σ-algebra of Borel subsets of $\mathbb{R}^N$. Thereby a family of measures is given on $\mathbb{R}^T$. The family is said to be *consistent* if, for any $L \subset N$ and any Borel set B from $\mathbb{R}^L$,

$$\mathbf{P}_L(B) = \mathbf{P}_N\big(B \times \mathbb{R}^{N-L}\big).$$

The measure $\mathbf{P}_L$ is said to be the *projection* of $\mathbf{P}_N$ onto $\mathbb{R}^L$. A set from $\mathbb{R}^T$ that can be represented in the form $B \times \mathbb{R}^{T-N}$, where $B \in \mathfrak{B}^N$ and N is a finite set, is called a *cylinder set* in $\mathbb{R}^T$. The set B is said to be the *base* of the cylinder.

Denote by $\mathfrak{B}^T$ the σ-algebra of sets from $\mathbb{R}^T$ generated by all cylinder sets.

Theorem A2.1 (Kolmogorov) *If a consistent family of probability measures is given on $\mathbb{R}^T$, then there exists a unique probability measure $\mathbf{P}$ on $\langle \mathbb{R}^T, \mathfrak{B}^T \rangle$ such that, for any N, the measure $\mathbf{P}_N$ coincides with the projection of $\mathbf{P}$ onto $\mathbb{R}^N$.*

A.A. Borovkov, *Probability Theory*, Universitext,
DOI 10.1007/978-1-4471-5201-9, © Springer-Verlag London 2013

Proof The cylinder subsets of $\mathbb{R}^T$ form an algebra. We show that, for $B \in \mathfrak{B}^N$, the relations

$$\mathbf{P}\big(B \times \mathbb{R}^{T-N}\big) = \mathbf{P}_N(B) \tag{A2.1}$$

define a measure on this algebra. First of all, by consistency of the measures $\mathbf{P}_N$, this definition of probability on cylinder sets is consistent (we mean the cases when $B = B_1 \times \mathbb{R}^{N-L}$ for $B_1 \in \mathfrak{B}^L$; then the left-hand side of (A2.1) will also be equal to $\mathbf{P}(B_1 \times \mathbb{R}^{T-L})$). Further, the thus defined probability is additive. Indeed, let $B_1 \times \mathbb{R}^{T-N_1}$ and $B_2 \times \mathbb{R}^{T-N_2}$ be two disjoint cylinder sets. Then, putting $N = N_1 \cup N_2$, we will have

$$\begin{aligned}
&\mathbf{P}\big(\big(B_1 \times \mathbb{R}^{T-N_1}\big) \cup \big(B_2 \times \mathbb{R}^{T-N_2}\big)\big) \\
&\quad = \mathbf{P}\big(\big\{\big(B_1 \times \mathbb{R}^{N-N_1}\big) \cup \big(B_2 \times \mathbb{R}^{N-N_2}\big)\big\} \times \mathbb{R}^{T-N}\big) \\
&\quad = \mathbf{P}_N\big(\big\{\big(B_1 \times \mathbb{R}^{N-N_1}\big) \cup \big(B_2 \times \mathbb{R}^{N-N_2}\big)\big\}\big) \\
&\quad = \mathbf{P}_N\big(B_1 \times \mathbb{R}^{N-N_1}\big) + \mathbf{P}_N\big(B_2 \times \mathbb{R}^{N-N_2}\big) \\
&\quad = \mathbf{P}\big(B_1 \times \mathbb{R}^{T-N_1}\big) + \mathbf{P}\big(B_2 \times \mathbb{R}^{T-N_2}\big).
\end{aligned}$$

To verify that $\mathbf{P}$ is countably additive, we make use of the equivalence of properties P3 and P3′ (see Chap. 2). By this equivalence, it suffices to show that if $\mathcal{B}_n$, $n = 1, 2, \dots$, is a decreasing sequence of cylinder sets and, for some $\varepsilon > 0$, we have $\mathbf{P}(\mathcal{B}) > \varepsilon$, $n = 1, 2, \dots$, then $\mathcal{B} = \bigcap_{n=1}^{\infty} \mathcal{B}_n$ is not empty. Since the $\mathcal{B}_n$ are enclosed in all the preceding sets, in the representation $\mathcal{B}_n = B_n \times \mathbb{R}^{T-N_n}$ one has $N_n \subset N_{n+1}$ and $B_{n+1} \cap \mathbb{R}^{N_n} \subset B_n$. Without loss of generality, we will assume that the number of elements in the set $N_n = \{t_1, \dots, t_n\}$ is equal to n, and denote by x_i (with various superscripts) the coordinates in the space $\mathbb{R}_{t_i}$.

Thus, let

$$\mathbf{P}(\mathcal{B}_n) = \mathbf{P}_{N_n}(B_n) \ge \varepsilon > 0.$$

We prove that the intersection

$$\mathcal{B} = \bigcap_{n=1}^{\infty} \mathcal{B}_n$$

is non-empty. For any Borel set $B_n \subset \mathbb{R}^{N_n}$, there exists a compactum K_n such that

$$K_n \subset B_n, \quad \mathbf{P}_{N_n}(B_n - K_n) < \frac{\varepsilon}{2^{n+1}}.$$

Setting $\mathcal{K}_n := K_n \times \mathbb{R}^{T-N_n}$, we obtain

$$\mathbf{P}(\mathcal{B}_n - \mathcal{K}_n) = \mathbf{P}_{N_n}(B_n - K_n) < \frac{\varepsilon}{2^{n+1}}.$$

Introduce the sets $\mathcal{D}_n := \bigcap_{k=1}^n \mathcal{K}_k$. It is easy to see that $\mathcal{D}_n \subset \mathcal{B}_n$ are also cylinders. Because

$$\mathcal{B}_n - \bigcap_{k=1}^n \mathcal{K}_k \subset \bigcap_{k=1}^n (\mathcal{B}_k - \mathcal{K}_k),$$

we have

$$\mathbf{P}(\mathcal{B}_n - \mathcal{D}_n) \le \mathbf{P}\left(\bigcap_{k=1}^n (\mathcal{B}_k - \mathcal{K}_k) \right) \le \sum_{k=1}^n \mathbf{P}(\mathcal{B}_k - \mathcal{K}_k) \le \frac{\varepsilon}{2};$$

$$\mathbf{P}(\mathcal{D}_n) \ge \mathbf{P}(\mathcal{B}_n) - \frac{\varepsilon}{2} \ge \frac{\varepsilon}{2}.$$

It follows that $\mathcal{D}_n$ is a decreasing sequence of non-empty cylinder sets. Denote by $X^n = (x_1^n, x_2^n, \ldots, x_n^n)$ an arbitrary point of the base

$$D_n = \bigcap_{k=1}^n K_k \times \mathbb{R}^{N_n - N_k}$$

of the cylinder $\mathcal{D}_n$. The point specifies a cylinder subset $\mathcal{X}$ of $\mathbb{R}^T$. Since the sets $\mathcal{D}_n$ decrease, we have $(x_1^{n+r}, x_2^{n+r}, \ldots, x_n^{n+r}) \in K_n$ for any $r \ge 0$. By compactness of K_n, we can choose a subsequence n_{1k} such that $x_1^{n_{1k}} \to x_1$ as $k \to \infty$. From this subsequence, one can choose a subsequence n_{2k} such that $x_2^{n_{2k}} \to x_2$, and so on.

Now consider the diagonal sequence of the points (or, more precisely, cylinder sets) $X^{n_{kk}} = (x_1^{n_{kk}}, x_2^{n_{kk}}, \ldots, x_{n_{kk}}^{n_{kk}})$. It is clear that

$$X^{n_{kk}} \to X = (x_1, x_2, \ldots)$$

(component-wise) as $k \to \infty$, and that

$$\left(x_1^{n_{kk}}, x_2^{n_{kk}}, \ldots, x_m^{n_{kk}}\right) \to (x_1, \ldots, x_m) \in K_m$$

for any m. This means that, for the set $\mathcal{X}$ corresponding to the point X, one has

$$\mathcal{X} := \left\{ y(t) \in \mathbb{R}^T : y(t_1) = x_1, y(t_2) = x_2, \ldots \right\} \subset K_m \subset \mathcal{B}_m$$

for any m, and therefore

$$\mathcal{X} \subset \bigcap_{m=1}^\infty \mathcal{B}_m.$$

Thus $\mathcal{B}$ is non-empty, and the countable additivity of $\mathbf{P}$ on the algebra of cylinder sets is proved. Hence $\mathbf{P}$ is a measure, and it remains to make use of the theorem on the extension of a measure from an algebra to the σ-algebra generated by that algebra.

The theorem is proved. □

Appendix 3
Elements of Measure Theory and Integration

In this appendix, the properties of integrals with respect to a measure are presented in more detail than in Chaps. 4 and 6. We also prove the basic theorems on decomposition of measure and on convergence of sequences of measures.

3.1 Measure Spaces

Let $\langle \Omega, \mathfrak{F} \rangle$ be a measurable space. We will say that a *measure space* $\langle \Omega, \mathfrak{F}, \mu \rangle$ is given if μ is a nonnegative countably additive set function on $\mathfrak{F}$, i.e. a function having the following properties:

(1) $\mu(\bigcup_j A_j) = \sum_j \mu(A_j)$ for any countable collection of disjoint sets $A_j \in \mathfrak{F}$ (σ-additivity);
(2) $\mu(A) \ge 0$ for any $A \in \mathfrak{F}$;
(3) $\mu(\varnothing) = 0$, where $\varnothing$ is the empty set.

The value $\mu(A)$ is called the *measure* of the set A. We will only consider *finite* and σ*-finite* measures. In the former case one assumes that $\mu(\Omega) < \infty$. In the latter case there exists a partition of Ω into countably many sets A_j such that $\mu(A_j) < \infty$.

A probability space is an example of a space with a finite (unit) measure. The space $\langle \mathbb{R}, \mathfrak{B}, \mu \rangle$, where $\mathbb{R}$ is the real line, $\mathfrak{B}$ is the σ-algebra of Borel sets, and μ is the Lebesgue measure, is an example of a space with a σ-finite measure.

We can also consider such set functions $\mu(A)$ that satisfy conditions (1) and (3) only, but are not necessarily nonnegative. Such functions are called *signed measures*. Any finite signed measure (i.e., such that $\sup_A \mu(A) < \infty$ and $\inf_A \mu(A) > -\infty$) can be represented as a difference of two nonnegative measures (the Hahn decomposition theorem, see Sect. 3.5 of the present appendix). We will need signed measures in Sect. 3.5 only. Everywhere else, unless otherwise specified, by measures we will understand set functions possessing properties (1)–(3).

In the same manner as when establishing the simplest properties of probability, one easily establishes the following properties of measures:

A.A. Borovkov, *Probability Theory*, Universitext,
DOI 10.1007/978-1-4471-5201-9, © Springer-Verlag London 2013

(1) $\mu(A) \le \mu(B)$ if $A \subset B$,
(2) $\mu(\bigcup_j A_j) \le \sum_j \mu(A_j)$ for any A_j,
(3) if $A_n \subset A_{n+1}$ and $\bigcup_n A_n = A$ then $\mu(A_n) \to \mu(A)$, or, which is the same,
(3′) if $A_n \supset A_{n+1}$, $\bigcap_n A_n = A$, and $\mu(A_1) < \infty$ then $\mu(A_n) \to \mu(A)$.

Consider further measurable functions on $\langle \Omega, \mathfrak{F} \rangle$, i.e., functions $\xi(\omega)$ having the property $\{\omega : \xi(\omega) \in B\} \in \mathfrak{F}$ for any Borel subset B of the real line.

The notions of *convergence in measure* and *convergence almost everywhere* are introduced similarly to the case of probability measure.

We will say that a sequence of measurable functions ξ_n *converges to* ξ *almost everywhere* (a.e.): $\xi_n \xrightarrow{a.e.} \xi$ as $n \to \infty$ if $\xi_n(\omega) \to \xi(\omega)$ for all ω except from a set of measure 0.

We will say that the ξ_n *converge to* ξ *in measure*: $\xi_n \xrightarrow{\mu} \xi$ if, for any $\varepsilon > 0$, as $n \to \infty$,

$$\mu(\{|\xi_n - \xi| > \varepsilon\}) \to 0.$$

Now we turn to the construction of integrals and the study of their properties. First we consider *finite measures* assuming them without loss of generality to be *probability measures*. In that case we will write $\mathbf{P}(A)$ instead of $\mu(A)$. We will turn to integrals with respect to arbitrary measures in Sect. 3.4.

3.2 The Integral with Respect to a Probability Measure

3.2.1 The Integrals of a Simple Function

A measurable function $\xi(\omega)$ is said to be *simple* if its range is finite. The *indicator* of a set $F \in \mathfrak{F}$ is the simple function

$$\mathrm{I}_F(\omega) = \begin{cases} 1, & \text{if } \omega \in F, \\ 0, & \text{if } \omega \notin F. \end{cases}$$

Clearly, any simple function $\xi(\omega)$ can be written in the form

$$\xi(\omega) = \sum_{k=1}^{n} x_k \mathrm{I}_{F_k}(\omega),$$

where x_k, $k = 1, 2, \dots, n$, are values assumed by ξ, and $F_k = \{\omega : \xi(\omega) = x_k\}$. The sets $F_k \in \mathfrak{F}$ are disjoint, and $\bigcup_{k=1}^{n} F_k = \Omega$. The *integral* of the simple function $\xi(\omega)$ with respect to a measure $\mathbf{P}$ is defined as the quantity

$$\int \xi \, d\mathbf{P} = \int \xi(\omega) \, d\mathbf{P}(\omega) = \sum_{k=1}^{n} x_k \mathbf{P}(F_k) = \mathbf{E}\xi.$$

The *integral* of the simple function $\xi(\omega)$ *over a set* $A \in \mathfrak{F}$ is defined as

$$\int_A \xi \, d\mathbf{P} = \int \xi(\omega) \mathrm{I}_A(\omega) \, d\mathbf{P}(\omega).$$

That these definitions are consistent (the partitions into sets F_k may be different) can be verified in an obvious way.

3.2.2 The Integrals of an Arbitrary Function

Lemma A3.2.1 *Let* $\xi(\omega) > 0$. *There exists a sequence* $\xi_n(\omega)$ *of simple functions such that* $\xi_n(\omega) \uparrow \xi(\omega)$ *as* $n \to \infty$ *for all* $\omega \in \Omega$.

Proof Partition the segment $[0, n]$ into $n2^n$ equal intervals. Let

$$x_0 = 0, \quad x_1 = 2^{-n}, \quad \ldots, \quad x_{n2^n} = n,$$

denote the partition points, so that $x_{i+1} - x_i = 2^{-n}$. Put

$$F_i := \{\omega : x_i \le \xi(\omega) < x_{i+1}\}, \quad i = 1, 2, \ldots, n2^n - 1;$$

$$F_0 := \{0 \le \xi(\omega) < x_1\} \cup \{\xi(\omega) \ge n\}, \qquad \xi_n(\omega) := \sum_{i=0}^{n2^n-1} x_i \mathrm{I}_{F_i}(\omega) \le \xi(\omega).$$

The function $\xi_n(\omega)$ is clearly simple, $\xi_n(\omega) \le \xi_{n+1}(\omega) \le \xi(\omega)$ for all ω, and has the property that if $n > \xi(\omega)$ at a point $\omega \in \Omega$ then

$$0 \le \xi(\omega) - \xi_n(\omega) \le \frac{1}{2^n}.$$

The lemma is proved. □

Lemma A3.2.2 *Let* $\xi_n \uparrow \xi \ge 0$ *and* $\eta_n \uparrow \xi \ge 0$ *be sequences of simple functions. Then*

$$\lim_{n\to\infty} \int \xi_n \, d\mathbf{P} = \lim_{n\to\infty} \int \eta_n \, d\mathbf{P}.$$

Proof We verify that, for any m,

$$\int \xi_m \, d\mathbf{P} \le \lim_{n\to\infty} \int \eta_n \, d\mathbf{P}.$$

The function ξ_n is simple. Therefore it is bounded by some constant: $\xi_m \le c_m$. Hence, for any integer n and $\varepsilon > 0$,

$$\xi_m - \eta_n \le c_m \cdot \mathrm{I}_{\{\xi_m \ge \eta_n + \varepsilon\}} + \varepsilon.$$

This implies that

$$\mathbf{E}\xi_m \le c_m \mathbf{P}\{\xi_m \ge \eta_n + \varepsilon\} + \varepsilon + \mathbf{E}\eta_n.$$

The probability on the right-hand side vanishes as $n \to \infty$:

$$\mathbf{P}\{\xi_m \ge \eta_n + \varepsilon\} \le \mathbf{P}\{\xi \ge \eta_n + \varepsilon\} \to 0,$$

because η_n converges almost surely (and hence in probability) to ξ. Therefore $\mathbf{E}\xi_m \le \varepsilon + \lim_{n\to\infty} \mathbf{E}\eta_n$. Since ε is arbitrary,

$$\lim_{n\to\infty} \mathbf{E}\xi_n \le \lim_{n\to\infty} \mathbf{E}\eta_n.$$

Swapping $\{\xi_n\}$ and $\{\eta_n\}$, we obtain the converse inequality.

The lemma is proved. □

The assertions of Lemmas A3.2.1 and A3.2.2 make the following definitions consistent.

The *integral of a nonnegative measurable function* $\xi(\omega)$ (with respect to measure $\mathbf{P}$) is the quantity

$$\int \xi\, d\mathbf{P} = \lim_{n\to\infty} \int \xi_n\, d\mathbf{P}, \qquad \text{(A3.2.1)}$$

where ξ_n is a sequence of simple functions such that $\xi_n \uparrow \xi$ as $n \to \infty$.

The integral $\int \xi\, d\mathbf{P}$ will also be denoted by $\mathbf{E}\xi$. We will say that the integral $\int \xi\, d\mathbf{P}$ *exists* and ξ is *integrable* if $\mathbf{E}\xi < \infty$.

The *integral of an arbitrary function* (assuming values of both signs) $\xi(\omega)$ (with respect to measure $\mathbf{P}$) is the quantity

$$\mathbf{E}\xi = \mathbf{E}\xi^+ - \mathbf{E}\xi^-, \qquad \xi^\pm := \max(0, \pm\xi),$$

which is defined when at least one of the values $\mathbf{E}\xi^\pm$ is finite. Otherwise $\mathbf{E}\xi$ is undefined. The integral $\mathbf{E}\xi$ *exists if and only if* $\mathbf{E}|\xi| < \infty$ *exists* (for $|\xi| = \xi^+ + \xi^-$). If $\mathbf{E}\xi$ exists then

$$\mathbf{E}(\xi; A) := \int_A \xi\, d\mathbf{P} = \mathbf{E}\xi \mathrm{I}_A$$

exists for any $A \in \mathfrak{F}$ as well.

Lemma A3.2.3 *If* $\mathbf{E}\xi$ *exists and* $B_n \in \mathfrak{F}$ *is a sequence of sets such that* $\mathbf{P}(B_n) \to 0$ *as* $n \to \infty$, *then*

$$\mathbf{E}(\xi; B_n) \to 0.$$

Proof For any sequence $|\xi_m| \uparrow |\xi|$ of simple functions and $A_m := \{|\xi| \le m\}$ one has

$$\mathbf{E}|\xi| \ge \lim_{m\to\infty} \mathbf{E}|\xi|\mathrm{I}_{A_m} \ge \lim_{m\to\infty} \mathbf{E}|\xi_m|\mathrm{I}_{A_m} = \mathbf{E}|\xi|,$$

since $|\xi_m|\mathrm{I}_{A_m} \uparrow |\xi|$. This implies that

$$\mathbf{E}|\xi| = \lim_{m\to\infty} \mathbf{E}|\xi|\mathrm{I}_{A_m} = \lim_{m\to\infty} \mathbf{E}\big(|\xi|;\ |\xi| \le m\big),$$

and hence, for any $\varepsilon > 0$, there exists an $m(\varepsilon)$ such that

$$\mathbf{E}|\xi| - \mathbf{E}\big(|\xi|;\ |\xi| \le m\big) < \varepsilon$$

for $m > m(\varepsilon)$. Consequently, for such m, one has

$$\mathbf{E}\big(|\xi|;\ B_n\big) = \mathbf{E}\big(|\xi|;\ \{|\xi| \le m\}B_n\big) + \mathbf{E}\big(|\xi|;\ \{|\xi| > m\}B_n\big) \le m\mathbf{P}(B_n) + \varepsilon,$$

and hence

$$\limsup_{n\to\infty} \mathbf{E}\big(|\xi|;\ B_n\big) \le \varepsilon.$$

The lemma is proved. □

Note that Lemma 6.1.2 somewhat extends Lemma A3.2.3.

Corollary A3.2.1 *If* $\mathbf{E}\xi$ *is well-defined (the values* $\pm\infty$ *not being excluded) and* $B_n \in \mathfrak{F}$ *is a sequence of sets such that* $\mathbf{P}(B_n) \to 1$ *as* $n \to \infty$, *then*

$$\mathbf{E}(\xi;\ B_n) \to \mathbf{E}\xi.$$

Proof If $\mathbf{E}\xi$ exists then the required assertion follows from Lemma A3.2.3.

Now let $\mathbf{E}\xi = \infty$. Then $\mathbf{E}\xi^- < \infty$ and $\mathbf{E}\xi^+ = \infty$, where $\xi^\pm = \max(0, \pm\xi)$. It follows that $\mathbf{E}(\xi^-;\ B_n) \to \mathbf{E}\xi^-$ as $n \to \infty$. We show that

$$\mathbf{E}\big(\xi^+;\ B_n\big) \to \infty. \tag{A3.2.2}$$

Let $A_k := \{\xi \in [2^{k-1}, 2^k)\}$, $k = 1, 2, \ldots$; $p_k := \mathbf{P}(A_k)$. We can assume without loss of generality that all $p_k > 0$ (if this is not the case we can consider a subsequence k_j such that all $p_{k_j} > 0$). Since $\mathbf{E}\xi^+ \le 1 + \sum_{k=1}^\infty 2^k p_k$, we have $\sum_{k=1}^\infty 2^k p_k = \infty$. For a given $N > 1$, choose n large enough such that $\mathbf{P}(B_n A_k) > p_k/2$ for all $k \le N$. Then

$$\mathbf{E}\big(\xi^+;\ B_n\big) \ge \sum_{k=1}^{N} 2^{k-2} p_k,$$

where the right-hand side can be made arbitrarily large by an appropriate choice of N. This proves (A3.2.2). Since $\xi = \xi^+ - \xi^-$, the required convergence is proved.

The case $\mathbf{E}\xi = -\infty$ can be dealt with in the same way. The corollary is proved. □

3.2.3 Properties of Integrals

I1. *If sets* $A_j \in \mathfrak{F}$ *are disjoint and* $\bigcup_j A_j = \Omega$ *then*

$$\int \xi\, d\mathbf{P} = \sum_j \int_{A_j} \xi\, d\mathbf{P}. \tag{A3.2.3}$$

Proof It suffices to prove this relation for $\xi(\omega) \geq 0$. For simple functions equality (A3.2.3) is obvious, because

$$\int \xi\, d\mathbf{P} = \sum_k x_k \mathbf{P}(\xi = x_k) = \sum_j \sum_k x_k \mathbf{P}(\xi = x_k;\, A_j).$$

In the general case, using definition (A3.2.1) one gets

$$\int \xi\, d\mathbf{P} = \lim_{n\to\infty} \int \xi_n\, d\mathbf{P} = \lim_{n\to\infty} \sum_j \int_{A_j} \xi_n\, d\mathbf{P}$$

$$= \sum_j \lim_{n\to\infty} \int_{A_j} \xi_n\, d\mathbf{P} = \sum_j \int_{A_j} \xi\, d\mathbf{P}. \tag{A3.2.4}$$

Swapping summation and passage to the limit is justified here, for by Lemma A3.2.3

$$\sum_{j=N}^{\infty} \int_{A_j} \xi_n\, d\mathbf{P} = \mathbf{E}\left(\xi_n;\, \bigcup_{j=N}^{\infty} A_j\right) \leq \mathbf{E}\left(\xi;\, \bigcup_{j=N}^{\infty} A_j\right) \to 0$$

as $N \to \infty$ uniformly in n. □

I2.

$$\int (\xi + \eta)\, d\mathbf{P} = \int \xi\, d\mathbf{P} + \int \eta\, d\mathbf{P}.$$

Proof For simple functions this property is obvious. Hence, for $\xi \geq 0$ and $\eta \geq 0$, this property follows from the additivity of the limit.

In the general case we have ($\xi^{\pm}$ and $\eta^{\pm}$ are defined here as before)

$$\int (\xi + \eta)\, d\mathbf{P} = \int (\xi^+ + \eta^+)\, d\mathbf{P} - \int (\xi^- + \eta^-)\, d\mathbf{P}$$

$$= \int \xi^+\, d\mathbf{P} - \int \xi^-\, d\mathbf{P} + \int \eta^+\, d\mathbf{P} - \int \eta^-\, d\mathbf{P} = \int \xi\, d\mathbf{P} + \int \eta\, d\mathbf{P}.$$ □

I3. *If* c *is an arbitrary constant, then*

$$\int c\xi\, d\mathbf{P} = c \int \xi\, d\mathbf{P}.$$

I4. *If* $\xi \le \eta$, *then* $\int \xi \, d\mathbf{P} \le \int \eta \, d\mathbf{P}$.

The proof of properties I3 and I4 is obvious. Since

$$\int \xi \, d\mathbf{P} = \mathbf{E}\xi,$$

we can write down properties I1–I4 in terms of expectations as follows:

I1. $\mathbf{E}\xi = \sum_j \mathbf{E}(\xi; A_j)$ *if* A_j *are disjoint and* $\bigcup_j A_j = \Omega$.
I2. $\mathbf{E}(\xi + \eta) = \mathbf{E}\xi + \mathbf{E}\eta$.
I3. $\mathbf{E}a\xi = a\mathbf{E}\xi$.
I4. $\mathbf{E}\xi \le \mathbf{E}\eta$, if $\xi \le \eta$.

Note also the following properties of integrals which easily follow from I1–I4.

I5. $|\mathbf{E}\xi| \le \mathbf{E}|\xi|$.
I6. *If* $c_1 \le \xi \le c_2$, *then* $c_1 \le \mathbf{E}\xi \le c_2$.
I7. *If* $\xi \ge 0$ *and* $\mathbf{E}\xi = 0$, *then* $\mathbf{P}(\xi = 0) = 1$.

This property follows from the Chebyshev inequality: $\mathbf{P}(\xi \ge \varepsilon) \le \mathbf{E}\xi/\varepsilon = 0$ for any $\varepsilon > 0$.

I8. *If* $\mathbf{P}(\xi = \eta) = 1$ *and* $\mathbf{E}\xi$ *exists then* $\mathbf{E}\xi = \mathbf{E}\eta$.

Indeed,

$$\mathbf{E}\eta = \lim_{n\to\infty} \mathbf{E}\big(\eta; |\eta| \le n\big) = \lim_{n\to\infty} \mathbf{E}\big(\xi; |\xi| \le n\big) = \mathbf{E}\xi.$$

3.3 Further Properties of Integrals

3.3.1 Convergence Theorems

A number of convergence theorems were proved in Sect. 6.1. One of them was the dominated convergence theorem (Corollary 6.1.3):

If $\xi_n \xrightarrow{p} \xi$ *as* $n \to \infty$ *and* $|\xi_n| \le \eta$, $\mathbf{E}\eta < \infty$, *then the expectation* $\mathbf{E}\xi$ *exists and* $\mathbf{E}\xi_n \to \mathbf{E}\xi$.

Now we will present some further useful assertions concerning convergence of integrals.

Theorem A3.3.1 (Monotone convergence) *If* $0 \le \xi_n \uparrow \xi$, *then* $\mathbf{E}\xi = \lim_{n\to\infty} \mathbf{E}\xi_n$.

Proof In addition to Corollary 6.1.3, here we only need to prove that $\mathbf{E}\xi_n \to \infty$ if $\mathbf{E}\xi = \infty$. Put $\xi_n^N := \min(\xi_n, N)$ and $\xi^N := \min(\xi, N)$. Then clearly $\xi_n^N \uparrow \xi^N$ as $n \to \infty$, and $\mathbf{E}\xi_n^N \uparrow \mathbf{E}\xi^N$. Therefore the value $\mathbf{E}\xi_n^N \le \mathbf{E}\xi_n$ can be made arbitrarily large by choosing appropriate n and N. The theorem is proved. □

These theorems can be generalised in the following way. To make the extension of the convergence theorems to the case of integrals with respect to signed measures in Sect. 3.4 more convenient, we will now write $\mathbf{E}\xi$ in the form of the integral $\int \xi \, d\mathbf{P}$.

Theorem A3.3.2 (Fatou–Lebesgue) *Let η and ζ be integrable. If $\xi_n \le \eta$ then*

$$\limsup_{n\to\infty} \int \xi_n \, d\mathbf{P} \le \int \limsup_{n\to\infty} \xi_n \, d\mathbf{P}. \tag{A3.3.1}$$

If $\xi_n \ge \zeta$ then

$$\liminf_{n\to\infty} \int \xi_n \, d\mathbf{P} \ge \int \liminf_{n\to\infty} \xi_n \, d\mathbf{P}. \tag{A3.3.2}$$

If $\xi_n \uparrow \xi$ and $\xi_n \ge \zeta$, or $\xi_n \xrightarrow{a.e.} \xi$ and $\zeta \le \xi_n \le \eta$, then

$$\lim_{n\to\infty} \int \xi_n \, d\mathbf{P} = \int \xi \, d\mathbf{P}. \tag{A3.3.3}$$

Proof We prove for instance (A3.3.2). Assume without loss of generality that $\zeta \equiv 0$. In this case, as $n \to \infty$,

$$\xi \ge \eta_n := \inf_{k\ge n} \xi_k \uparrow \liminf_{k\to\infty} \xi_k, \qquad \eta_n \ge 0,$$

and by the monotone convergence theorem

$$\liminf_{n\to\infty} \int \xi_n \, d\mathbf{P} \ge \lim_{n\to\infty} \int \eta_n \, d\mathbf{P} = \int \liminf_{n\to\infty} \xi_n \, d\mathbf{P}.$$

Applying (A3.3.2) to the sequence $\eta - \xi_n$ we obtain (A3.3.1); (A3.3.3) follows from the previous theorems. The theorem is proved. □

3.3.2 Connection to Integration with Respect to a Measure on the Real Line

Let $g(x)$ be a Borel function given on the real line $\mathbb{R}$ (if $\mathfrak{B}$ is the σ-algebra of Borel sets on the line and $B \in \mathfrak{B}$, then $\{x : g(x) \in B\} \in \mathfrak{B}$). If ξ is a random variable then $\eta := g(\xi(\omega))$ will clearly also be a random variable. As we saw in Sect. 3.2, a random variable ξ induces the probability space $\langle \mathbb{R}, \mathfrak{B}, \mathbf{F}_\xi \rangle$ with measure $\mathbf{F}_\xi$ on the line such that $\mathbf{F}_\xi(B) = \mathbf{P}(\xi \in B)$. Therefore one can speak about integrals with respect to that measure.

Theorem A3.3.3 *If $\eta = g(\xi(\omega))$ and $\mathbf{E}\eta$ exists, then*

$$\mathbf{E}\eta = \int_\Omega \eta \, d\mathbf{P} = \int_{\mathbb{R}} g(x) \mathbf{F}_\xi(dx)$$

(on the right-hand side we used a somewhat different notation for $\int g \, d\mathbf{F}_\xi$).

Proof Let first $g(x) = \mathrm{I}_B(x)$ be the indicator of a set $B \in \mathfrak{B}$. Then $\eta = g(\xi(\omega)) = \mathrm{I}_{\{\xi \in B\}}(\omega)$ and $\mathbf{E}\eta = \mathbf{P}(\xi \in B)$. Therefore

$$\int g(x)\mathbf{F}_\xi(dx) = \int \mathrm{I}_B(x)\mathbf{F}_\xi(dx) = \mathbf{F}_\xi(B) = \mathbf{P}(\xi \in B) = \mathbf{E}\eta.$$

Using the properties of the integral it is easy to establish that the assertion of the theorem holds for simple functions g. Passing to the limit extends that assertion to bounded functions. Now let $g \geq 0$. If the function $g(\xi)\mathrm{I}_B(\xi) = \eta(\omega)\mathrm{I}_{\{\xi \in B\}}(\omega)$ is bounded, then

$$\int_B g(x)\,\mathbf{F}_\xi(dx) = \mathbf{E}(\eta;\, \xi \in B).$$

Therefore

$$\int_{\{g \leq n\}} g\,d\mathbf{F}_\xi = \mathbf{E}(\eta;\, \eta \leq n).$$

Passing to the limit as $n \to \infty$ we get the assertion of the theorem. Considering the case when g takes values of both signs does not create any difficulties. The theorem is proved. □

Introducing the notation

$$F_\xi(x) = \mathbf{P}(\xi < x),$$

we can also consider, along with the integral just discussed,

$$\int_{\mathbb{R}} g(x)\mathbf{F}_\xi(dx), \tag{A3.3.4}$$

the Riemann–Stieltjes integral

$$\int g(x)\,dF_\xi(x), \tag{A3.3.5}$$

the definition of which was given in Sect. 3.6. It was also shown there that, for *continuous* functions $g(x)$, these integrals coincide. Moreover, we discussed in Sect. 3.6 some other conditions for these integrals to coincide.

Also recall that if

$$F_\xi(x) = \int_{-\infty}^{x} f_\xi(t)\,dt$$

and the functions $g(x)$ and $f_\xi(x)$ are Riemann integrable, then integrals (A3.3.4) and (A3.3.5) coincide with the Riemann integral

$$\int g(x) f_\xi(x)\,dx.$$

3.3.3 Product Measures and Iterated Integrals

Consider a two-dimensional random variable $\zeta = (\xi, \eta)$ given on $\langle \Omega, \mathfrak{F}, \mathbf{P} \rangle$. The random variables ξ and η induce a sample probability space $\langle \mathbb{R}^2, \mathfrak{B}^2, \mathbf{F}_{\xi,\eta} \rangle$ with the measure $\mathbf{F}_{\xi,\eta}$ given on elements of the σ-algebra $\mathfrak{B}^2$ of Borel sets on the plane (the σ-algebra generated by rectangles) and such that

$$\mathbf{F}_{\xi,\eta}(A \times B) = \mathbf{P}(\xi \in A, \eta \in B).$$

Here $A \times B$ is the set of points (x, y) for which $x \in A$ and $y \in B$. If $g(x, y)$ is a Borel function ($\{(x, y) : g(x, y) \in B\} \in \mathfrak{B}^2$ for each $B \in \mathfrak{B}$), then it easily follows from the above that

$$\mathbf{E}g(\xi, \eta) = \int_{\mathbb{R}^2} g(x, y)\mathbf{F}_{\xi,\eta}(dx\,dy), \tag{A3.3.6}$$

since both integrals are equal to $\int_{\mathbb{R}} x\mathbf{F}_\theta(dx)$ for $\theta = g(\xi, \eta)$.

Now let ξ and η be independent random variables, i.e.

$$\mathbf{P}(\xi \in A,\ \eta \in B) = \mathbf{P}(\xi \in A)\mathbf{P}(\eta \in B)$$

for any $A, B \in \mathfrak{B}$.

Theorem A3.3.4 (Fubini's theorem on iterated integrals) *If $g(x, y) \geq 0$ is a Borel function and ξ and η are independent, then*

$$\mathbf{E}g(\xi, \eta) = \mathbf{E}\big[\mathbf{E}g(x, \eta)|_{x=\xi}\big].$$

For arbitrary Borel functions $g(x, y)$ the above equality holds if $\mathbf{E}g(\xi, \eta)$ exists.

This very assertion we stated in Chap. 3 in the form

$$\int g(x, y)\mathbf{F}_{\xi,\eta}(dx\,dy) = \int \left[\int g(x, y)\mathbf{F}_\eta(dy)\right]\mathbf{F}_\xi(dx). \tag{A3.3.7}$$

We will need the following.

Lemma A3.3.1 1. *The section*

$$B_x := \{y : (x, y) \in B\}$$

of any set $B \in \mathfrak{B}^2$ is measurable: $B_x \in \mathfrak{B}$.

2. *The section $g_x(y) = g(x, y)$ of any Borel function g ($\mathfrak{B}^2$-measurable) is a Borel function.*

3. *The integral*

$$\int g(x, y)\mathbf{F}_\eta(dy) \tag{A3.3.8}$$

of a Borel function g is a Borel function of x.

Proof 1. Let $\mathcal{K}_1$ be the class of all sets from $\mathfrak{B}^2$ of which all x-sections are measurable. It is evident that $\mathcal{K}_1$ contains all rectangles $B = B_{(1)} \times B_{(2)}$, where $B_{(1)} \in \mathfrak{B}$ and $B_{(2)} \in \mathfrak{B}$. Moreover, $\mathcal{K}_1$ is a σ-algebra. Indeed, consider for example the set $B = \bigcup_k B^{(k)}$ where $B^{(k)} \in \mathcal{K}_1$. The operation $\bigcup$ on the sets $B^{(k)}$ leads to the same operation on their sections, so that $B_x = \bigcup_k B_x^{(k)} \in \mathfrak{B}$. For the other operations ($\bigcap$ and taking complements) the situation is similar. Thus, $\mathcal{K}_1$ is a σ-algebra containing all rectangles. This means that $\mathfrak{B}^2 \subset \mathcal{K}_1$.

2. For $B \in \mathfrak{B}$, one has

$$\begin{aligned} g_x^{-1}(B) &= \{y : g_x(y) \in B\} = \{y : g(x, y) \in B\} \\ &= \{y : (x, y) \in g^{-1}(B)\} = [g^{-1}(B)]_x \in \mathfrak{B}. \end{aligned}$$

3. Integral (A3.3.8) is, as a function of x, the result of passing to the limit in a sequence of measurable functions, and hence is measurable itself. The lemma is proved. □

Proof of Theorem A3.3.4 First we prove (A3.3.7) in the case where $g(x, y) = \mathrm{I}_B(x, y)$, so that the theorem turns into the formula for consecutive computation of the measure of the set $B \in \mathfrak{B}^2$:

$$\mathbf{P}\big((\xi, \eta) \in B\big) = \int \mathbf{F}_\eta\big((x, y) \in B\big)\mathbf{F}_\xi(dx) = \int \mathbf{F}_\eta(B_x)\mathbf{F}_\xi(dx). \qquad \text{(A3.3.9)}$$

We introduce the set function

$$\mathbf{Q}(B) := \int \mathbf{F}_\eta(B_x)\mathbf{F}_\xi(dx).$$

Clearly, $\mathbf{Q}(B) \ge 0$ and $\mathbf{Q}(\varnothing) = 0$. Further, if $B = \bigcup_k B^{(k)}$ and $B^{(k)}$ are disjoint, then $B_x = \bigcup_k B_x^{(k)}$ and $B_x^{(k)}$ are also disjoint, and

$$\mathbf{Q}(B) = \int \mathbf{F}_\eta\Big(\bigcup_k B_x^{(k)}\Big)\mathbf{F}_\xi(dx) = \sum_k \int \mathbf{F}_\eta\big(B_x^{(k)}\big)\mathbf{F}_\xi(dx) = \sum_k \mathbf{Q}\big(B^{(k)}\big).$$

This means that $\mathbf{Q}(B)$ is a measure.

The measure $\mathbf{Q}(B)$ coincides with $\mathbf{F}_{\xi,\eta}(B) = \mathbf{P}((\xi, \eta) \in B)$ on rectangles $B = B_{(1)} \times B_{(2)}$. Indeed, for rectangles,

$$B_x = \begin{cases} B_{(2)} & \text{for } x \in B_{(1)}, \\ \varnothing & \text{for } x \notin B_{(1)}, \end{cases}$$

and

$$\begin{aligned} \mathbf{P}\big((\xi, \eta) \in B\big) &= \mathbf{F}_\xi(B_{(1)})\mathbf{F}_\eta(B_{(2)}) \\ &= \int_{B_{(1)}} \mathbf{F}_\eta(B_{(2)})\mathbf{F}_\xi(dx) = \int \mathbf{F}_\eta(B_x)\mathbf{F}_\xi(dx) = \mathbf{Q}(B). \end{aligned}$$

This means that the measures $\mathbf{Q}$ and $\mathbf{F}_{\xi,\eta}$ coincide on the algebra generated by rectangles. By the measure extension theorem we obtain that $\mathbf{Q} = \mathbf{F}_{\xi,\eta}$.

We have proved (A3.3.9). This implies that Fubini's theorem holds for simple functions $g_N = \sum_{j=1}^{N} c_j \mathrm{I}_{A_j}$, because

$$\begin{aligned}\mathbf{E}g_N(\xi,\eta) &= \sum_{j=1}^{N} c_j \mathbf{E}\mathrm{I}_{A_j}(\xi,\eta) \\ &= \sum_{j=1}^{N} c_j \int \mathbf{E}\mathrm{I}_{A_j}(x,\eta)\mathbf{F}_\xi(dx) = \int \mathbf{E}g_N(x,\eta)\mathbf{F}_\xi(dx)\end{aligned} \quad \text{(A3.3.10)}$$

Now if $g \ge 0$ is an arbitrary Borel function then there exists a sequence of simple functions $g_N \uparrow g$ and, as in (A3.2.1), it remains to pass to the limit:

$$\begin{aligned}\mathbf{E}g(\xi,\eta) &= \lim_{N\to\infty} \mathbf{E}g_N(\xi,\eta) \\ &= \lim_{N\to\infty} \int \mathbf{E}g_N(\xi,\eta)\mathbf{F}_\xi(dx) = \int \mathbf{E}g_N(\xi,\eta)\mathbf{F}_\xi(dx).\end{aligned}$$

For an arbitrary function g one has to use the representation $g = g^+ - g^-$, $g^+ \ge 0$, $g^- \ge 0$. The theorem is proved. □

Remark A3.1 We see from the proof of the theorem that the random variables ξ and η do not need to be scalar. The assertion remains true in a more general form (see property 5A in Sect. 4.8) and, in particular, for vector-valued ξ and η.

3.4 The Integral with Respect to an Arbitrary Measure

If μ is a finite measure on $\langle\Omega,\mathfrak{F}\rangle$, $\mu(\Omega) < \infty$, then the definition of the integral $\int \xi\, d\mu$ with respect to the measure μ does not differ from that of the integral with respect to a probability measure (one could just put $\int_A \xi\, d\mu = \mu(\Omega)\int_A \xi\, d\mathbf{P}$, where $\mathbf{P}(B) = \mu(B)/\mu(\Omega)$ is a probability distribution). If μ is σ-finite and $\mu(\Omega) = \infty$, then the situation is somewhat more complicated, although it can still be reduced to the already used constructions. First we will make several preliminary remarks.

Let $\langle\Omega,\mathfrak{F},\mathbf{P}\rangle$ be a probability space and $f = f(\omega) \ge 0$ an a.e. finite nonnegative measurable function (i.e., a random variable). Consider the set function

$$\mu(A) := \int_A f\, d\mathbf{P}. \quad \text{(A3.4.1)}$$

If f is integrable ($\mu(\Omega) < \infty$) then $\mu(A)$ is a finite σ-additive measure (see property I1) satisfying conditions (1)–(3) of Sect. 3.1 of the present appendix. In other words, μ is a finite measure on $\langle \Omega, \mathfrak{F} \rangle$. But if f is not integrable, then μ is a σ-finite measure, which immediately follows from the representation

$$\mu(A) = \sum_{k=1}^{\infty} \int_{A \cap \{k-1 \le f < k\}} f \, d\mathbf{P}$$

(the integrals in the sum that are equal to $\int_A f \mathrm{I}_{(k-1 \le f < k)} \, d\mathbf{P}$ are clearly finite measures).

Thus, the integral of the form (A3.4.1) is a measure for any distribution $\mathbf{P}$ and function $f \ge 0$. It turns out that the following assertion, converse in a certain sense to the above, also holds.

Lemma A3.4.1 *For any measure* μ *on* $\langle \Omega, \mathfrak{F} \rangle$, *there exists a distribution* $\mathbf{P}$ *on that space and a measurable function* $f > 0$ *such that representation* (A3.4.1) *holds.*

Thus, any measure can be represented as an integral with respect to a probability measure (i.e., in the form $\mathbf{E}(f; A)$ for the respective function f and distribution $\mathbf{P}$).

Proof Let μ be a σ-finite measure on $\langle \Omega, \mathfrak{F} \rangle$, and sets $B_j \in \mathfrak{F}$, $j = 1, 2, \ldots$, possess the properties $\bigcup_{j=1}^{\infty} B_j = \Omega$, $B_i B_j = \varnothing$ for $i \ne j$, and $\mu(B_j) < \infty$. Put

$$\mathbf{P}(A) := \sum_{k=1}^{\infty} \frac{\mu(AB_k)}{2^k \mu(B_k)}. \tag{A3.4.2}$$

Obviously, $\mathbf{P}(\Omega) = 1$ and $\mathbf{P}$ is a measure. Further, if $A \subset B_k$ then

$$\mu(A) = 2^k \mu(B_k) \mathbf{P}(A).$$

This means that we should put $f(\omega) := 2^k \mu(B_k)$ for $\omega \in B_k$. Then the set function

$$\lambda(A) := \int_A f \, d\mathbf{P} = \int_\Omega f \mathrm{I}_A \, d\mathbf{P}$$

will coincide with $\mu(A)$:

$$\begin{aligned} \lambda(A) &= \sum_{k=1}^{\infty} 2^k \mu(B_k) \mathbf{P}(AB_k) \\ &= \sum_{k=1}^{\infty} 2^k \mu(B_k) \sum_{j=1}^{\infty} \frac{\mu(AB_k B_j)}{2^j \mu(B_j)} = \sum_{k=1}^{\infty} \mu(AB_k) = \mu(A). \end{aligned}$$

The lemma is proved. □

Besides the required assertion, we also obtain that in representation (A3.4.1) the range of values of the function f can be assumed without loss of generality to be countable.

The function f for which equality (A3.4.1) holds is called the *density* of the measure $\boldsymbol{\mu}$ with respect to $\mathbf{P}$ (or *Radon–Nikodym derivative* of the measure $\boldsymbol{\mu}$ with respect to $\mathbf{P}$) and is denoted by $d\boldsymbol{\mu}/d\mathbf{P}$. It is evident that alteration of the function $f = d\boldsymbol{\mu}/d\mathbf{P}$ on a set of zero $\mathbf{P}$-measure leaves the equality (A3.4.1) unchanged.

Now let $\boldsymbol{\mu}$ and $\mathbf{P}$ be two *given arbitrary measures*. The question of under what conditions these two measures $\boldsymbol{\mu}$ and $\mathbf{P}$ could be related by (A3.4.1) and whether the function f is determined uniquely thereby (up to values on a set of zero $\mathbf{P}$-measure) is rather important for probability theory. (We stress that, in the preceding considerations, the measure $\mathbf{P}$ was constructed in a special way from the measure $\boldsymbol{\mu}$, or vice versa.) Answers to these questions are given by the Radon–Nikodym theorem to be discussed in the next section.

Now, using the simple assertion of Lemma A3.4.1 we have just proved, we will give the definition of the integral with respect to an arbitrary measure $\boldsymbol{\mu}$.

Let $\boldsymbol{\mu}$ be an arbitrary σ-finite measure on $\langle \Omega, \mathfrak{F} \rangle$ and $\xi \geq 0$ a $\mathfrak{F}$-measurable function.

The *integral* $\int_A \xi \, d\boldsymbol{\mu}$ over a set $A \in \mathfrak{F}$ of the function $\xi \geq 0$ with respect to the measure $\boldsymbol{\mu}$ is the integral

$$\int_A \xi \, d\boldsymbol{\mu} = \int_A \left(\xi \frac{d\boldsymbol{\mu}}{d\mathbf{P}} \right) d\mathbf{P} \tag{A3.4.3}$$

with respect to any distribution $\mathbf{P}$ satisfying equality (A3.4.1) (for example, with respect to measure (A3.4.2)).

This definition is *consistent* because it does not depend on the choice of $\mathbf{P}$. Indeed, for simple functions ξ ($\xi(\omega) = x_k$ for $\omega \in F_k$),

$$\int_A \xi \, d\boldsymbol{\mu} = \sum_k x_k \int_A \frac{d\boldsymbol{\mu}}{d\mathbf{P}} \mathrm{I}_{B_k} \, d\mathbf{P} = \sum_k x_k \int_{AB_k} \frac{d\boldsymbol{\mu}}{d\mathbf{P}} \, d\mathbf{P} = \sum_k x_k \boldsymbol{\mu}(AB_k).$$

If now $\xi \geq 0$ is an arbitrary function, then by the monotone convergence theorem the integral $\int_A \xi \, d\boldsymbol{\mu}$ is equal to

$$\lim_{n \to \infty} \int_A \xi^{(n)} \frac{d\boldsymbol{\mu}}{d\mathbf{P}} \, d\mathbf{P} = \lim_{n \to \infty} \int_A \xi^{(n)} d\boldsymbol{\mu},$$

where $\xi^{(n)} \uparrow \xi$ is a sequence of simple functions which converge monotonically to ξ (see Lemma A3.2.1). In both cases, the result does not depend on the choice of $\mathbf{P}$.

The integral

$$\int_A \xi \, d\boldsymbol{\mu}$$

of an arbitrary measurable function ξ is defined by

$$\int_A \xi \, d\boldsymbol{\mu} = \int_A \xi^+ d\boldsymbol{\mu} - \int_A \xi^- d\boldsymbol{\mu},$$

when both expressions on the right-hand side are finite. (In that case one says that the integral $\int_A \xi\, d\mu$ exists.) Here, as before, $\xi^+ = \max(0, \xi) \geq 0$ and $\xi^- = \max(0, -\xi) \geq 0$, so that $\xi = \xi^+ - \xi^-$.

Thus we see that the above definition of the integral with respect to an arbitrary measure is essentially equivalent to the construction used in Sect. 3.2 of the present appendix. However, the definition in the form (A3.4.3) saves us from the necessity of repeating what we have already done (and now in a more complex setting) and enables one to transfer all the properties of the integrals $\int \xi\, d\mathbf{P}$ to the general case. We will list the basic properties preserving the existing numeration.

I1. $\int \xi\, d\mu = \sum_j \int_{A_j} \xi\, d\mu$ if A_j are disjoint and $\bigcup_j A_j = \Omega$.
I2. $\int (\xi + \eta)\, d\mu = \int \xi\, d\mu + \int \eta\, d\mu$.
I3. $\int a\xi\, d\mu = a \int \xi\, d\mu$.
I4. $\xi\, d\mu \leq \int \eta\, d\mu$ if $\xi \leq \eta$.
I5. $|\int \xi\, d\mu| \leq \int |\xi|\, d\mu$.
I6. If $c_1 \leq \xi(\omega) \leq c_2$ for $\omega \in A$, then $c_1 \mu(A) \leq \int_A \xi\, d\mu \leq c_2 \mu(A)$.
I7. If $\xi \geq 0$ and $\int \xi\, d\mu = 0$, then $\mu(\xi > 0) = 0$.
I8. If $\mu(\xi \neq \eta) = 0$, then $\int \xi\, d\mu = \int \eta\, d\mu$.

It is clear that all the convergence theorems remain valid as well.

Theorem A3.4.1 (The dominated convergence theorem) *Let $|\xi_n| \leq \eta$ and $\int \eta\, d\mu$ exist. If $\xi_n \xrightarrow{\mu}$ or $\xi_n \to \xi$ a.e. as $n \to \infty$ then*

$$\int \xi_n\, d\mu \to \int \xi\, d\mu.$$

Theorem A3.4.2 (The monotone convergence theorem) *If $0 \leq \xi_n \uparrow \xi$ as $n \to \infty$ then*

$$\int \xi_n\, d\mu \to \int \xi\, d\mu.$$

Theorem A3.4.3 (Fatou–Lebesgue) *The statement and proof of this theorem is obtained from those of Theorem* A3.3.2 *by replacing* $\mathbf{P}$ *with* μ.

In conclusion we note that if $\Omega = \mathbb{R} = (-\infty, \infty)$, $\mathfrak{F} = \mathfrak{B}$ is the σ-algebra of Borel sets, μ is the Lebesgue measure, and the function $g(x)$ is continuous, then the integral $\int_{[a,b]} g(x)\, d\mu(x)$ coincides with the Riemann integral $\int_a^b g(x)\, dx$. This follows from the preceding remarks in part 2 of Sect. 3.3 of this appendix.

3.5 The Lebesgue Decomposition Theorem and the Radon–Nikodym Theorem

We return to a question that has already been asked in the previous section. Under what conditions on measures μ and λ given on $\langle \Omega, \mathfrak{F} \rangle$ can the measure μ be represented as

$$\mu(A) = \int_A f\, d\lambda?$$

We do not assume here that λ is a probability measure.

Definition A3.5.1 A measure μ is said to be *absolutely continuous* with respect to a measure λ (we write $\mu \prec \lambda$) if, for any A such that $\lambda(A) = 0$, one has $\mu(A) = 0$.

Definition A3.5.2 A set N_μ is said to be a *support*[1] of measure μ if $\mu(\Omega - N_\mu) = 0$.

Definition A3.5.2 specifies a rather wide class of sets which can be called the support of the measure μ when μ is concentrated on a part of the space Ω. If $\Omega = \mathbb{R}$ is the real line (and in some other cases as well), one can use another definition which specifies a unique set for each measure. Consider the collection of all intervals $(a, b) \subset \mathbb{R}$ with rational endpoints a and b. This collection is countable. Remove from $\Omega = \mathbb{R}$ all such intervals for which $\mu((a, b)) = 0$. The remaining set (which is measurable) is called the *support of the measure* μ.

Definition A3.5.3 One says that a measure μ is *singular*with respect to λ if there exists a support N_λ of the measure λ such that $\mu(N_\lambda) = 0$. Or, which is the same, if there exists a support N_μ of the measure μ such that $\lambda(N_\mu) = 0$.

The last definition, in contrast to Definition A3.5.1, is symmetric, so one can speak about *mutually singular measures* μ *and* λ (this relation is often written as $\mu \perp \lambda$).

Theorem A3.5.1 (Radon–Nikodym) *A necessary and sufficient condition for the absolute continuity* $\mu \prec \lambda$ *is that there exists a function* f *unique up to* λ*-equivalence (i.e., up to values on a set of zero* λ*-measure) such that*[2]

$$\mu(A) = \int_A f\, d\lambda.$$

As we have already noted, the function f is called the *Radon–Nikodym derivative* $d\mu/d\lambda$ of the measure μ with respect to λ (or *density* of μ with respect to λ).

Since sufficiency in the assertion of the theorem is obvious, we will obtain the Radon–Nikodym theorem as a consequence of the following Lebesgue decomposition theorem.

---

[1]The conventional definition of support refers to the case when Ω is a topological space. Then the support of μ is the smallest closed set such that its complement is of μ-measure zero.

[2]This equality is sometimes adopted as a definition of absolute continuity.

Theorem A3.5.2 (Lebesgue) *Let μ and λ be two σ-finite measures given on $\langle \Omega, \mathfrak{F} \rangle$. There exists a unique decomposition of the measure μ into two components μ_a and μ_s such that*

$$\mu_a \prec \lambda, \qquad \mu_s \perp \lambda.$$

Moreover, there exists a function f, unique up to λ-equivalence, such that

$$\mu_a(A) = \int_A f \, d\lambda.$$

It is obvious that if $\mu \prec \lambda$ then $\mu_s = 0$, and the Lebesgue theorem then implies the Radon–Nikodym theorem.

Proof Since μ and λ are σ-finite, there exist increasing sequences of sets Ω_n^μ and Ω_n^λ such that

$$\mu(\Omega_n^\mu) < \infty, \qquad \lambda(\Omega_n^\lambda) < \infty, \qquad \bigcup_n \Omega_n^\mu = \Omega, \qquad \bigcup_n \Omega_n^\lambda = \Omega.$$

Putting $\Omega_n := \Omega_n^\mu \cap \Omega_n^\lambda$ we obtain a sequence of sets increasing to Ω for which

$$\mu(\Omega_n) < \infty, \qquad \lambda(\Omega_n) < \infty.$$

If we prove the decomposition theorem for restrictions of the measures μ and λ to $\langle B_n, \mathfrak{F}_n \rangle$, where $B_n = \Omega_{n+1} - \Omega_n$ and $\mathfrak{F}_n$ is formed by sets $B_n A$, $A \in \mathfrak{F}$, we will thereby prove it for the whole Ω. It will suffice to take μ_a and μ_s to be the sums of the respective components for each of the restrictions. This remark means that we can consider the case of finite measures only.

Thus let μ and λ be finite measures.

(a) Let $\mathcal{F}$ be the class of functions $f \geq 0$ such that

$$\int_A f \, d\lambda \leq \mu(A) \quad \text{for all } A \in \mathfrak{F} \tag{A3.5.1}$$

(the class $\mathcal{F}$ is non-empty, for the function $f = 0$ belongs to $\mathcal{F}$). Set

$$\alpha := \sup_{f \in \mathcal{F}} \int_\Omega f \, d\lambda \leq \mu(\Omega) < \infty$$

and choose a sequence f_n such that, as $n \to \infty$,

$$\int f_n \, d\lambda \to \alpha.$$

Put $\widehat{f_n} := \max(f_1, \ldots, f_n)$. Then clearly $\widehat{f_n} \uparrow \widehat{f} := \sup f_n$ and by the monotone convergence theorem

$$\int_A \widehat{f_n} \, d\lambda \to \int_A \widehat{f} \, d\lambda. \tag{A3.5.2}$$

We now show that $\widehat{f} \in \mathcal{F}$, i.e., that (A3.5.1) holds for $\widehat{f}$. To this end, it suffices by virtue of (A3.5.2) to notice that $\widehat{f}_n \in \mathcal{F}$. Let A_k, $k = 1, \ldots, n$ be disjoint sets on which $\widehat{f}_n = f_k$. Then $\bigcup_{k=1}^n A_k = \Omega$ and

$$\int_A \widehat{f}_n \, d\lambda = \sum_{k=1}^n \int_{AA_k} f_k \, d\lambda \le \sum_{k=1}^n \mu(AA_k) = \mu(A).$$

Thus, for the "maximum" element f' of $\mathcal{F}$, (A3.5.1) also holds.

(b) Putting

$$\mu_a(A) := \int_A \widehat{f} \, d\lambda, \qquad \mu_s = \mu - \mu_a \tag{A3.5.3}$$

we prove that μ_s is singular with respect to λ. We will need the following assertion about the decomposition of an arbitrary signed measure (for the definition, see Sect. 3.3.1 of this appendix).

Theorem A3.5.3 (The Hahn theorem on decomposition of a measure) *For any signed finite measure γ, there exist disjoint sets $D^+ \in \mathfrak{F}$ and $D^- \in \mathfrak{F}$ such that, for any $A \in \mathfrak{F}$,*

$$\gamma(AD^+) \ge 0, \qquad \gamma(AD^-) \le 0.$$

Proof We first prove that there exists a set $D \in \mathfrak{F}$ on which $\gamma(A)$ attains its upper bound.

Let $B_n \in \mathfrak{F}$ be a sequence such that $\gamma(B_n) \to \Gamma = \sup_A \gamma(A)$ as $n \to \infty$. Put $B := \bigcup_k B_k$ and consider, for a fixed n, the decomposition of T into 2^n sets $B_{n,m}$, $m = 1, \ldots, 2^n$, of the form $\bigcap_{k=1}^n B_k'$, where $B_k' = B_k$ or $B - B_k$, $k \le n$. For $n < N$, each $B_{n,m}$ is a finite union of sets $B_{N,M}$, $1 \le M \le 2^N$. Denote by D_n the sum of all $B_{n,m}$ for which $\gamma(B_{n,m}) \ge 0$. Then $\gamma(B_n) \le \gamma(D_n)$.

On the other hand, for $N > n$, each $B_{N,M}$ either belongs to D_n or is disjoint with it. Therefore

$$\gamma(D_n) \le \gamma(D_n + D_{n+1} + \cdots + D_N).$$

This implies that, for the set $D = \lim \bigcup_{k=n}^\infty D_k$, one has $\gamma(B_n) \le \gamma(D)$, $\Gamma \le \gamma(D)$. Recalling the definition of Γ, we obtain that $\gamma(D) = \Gamma$.

Thus we have proved the existence of a set D on which $\gamma(D)$ attains its maximum. We now show that, for any $A \in \mathfrak{F}$, one has $\gamma(AD) \ge 0$ and $\gamma(A\overline{D}) \le 0$, where $\overline{D} = \Omega - D$. Indeed, assuming, for instance, that $\gamma(AD) < 0$, we come to a contradiction, for in that case

$$\gamma(D - AD) = \gamma(D) - \gamma(AD) > \gamma(D).$$

Similarly, assuming that $\gamma(A\overline{D}) > 0$, we would get

$$\gamma(D + A\overline{D}) = \gamma(D) + \gamma(A\overline{D}) > \gamma(D).$$

It remains to put $D^+ := D$, $D^- := \overline{D}$. The theorem is proved. □

Corollary A3.5.1 *Any finite signed measure γ can be represented as $\gamma = \gamma^+ - \gamma^-$, where $\gamma^\pm$ are finite nonnegative measures.*

To prove the corollary, it suffices to put

$$\gamma^\pm(A) := \pm\gamma\big(A \cap D^\pm\big),$$

where $D^\pm$ are the sets from the Hahn decomposition theorem. □

We return to the proof of the fact that the measure μ_s in equality (A3.5.3) is singular. Let D_n^+ be the set in the Hahn decomposition of the signed measure

$$\nu_n = \mu_s - \frac{1}{n}\lambda.$$

Put $N = \bigcap_n D_n^-$. Then $\overline{N} = \bigcup_n D_n^+$ and, for all n and $A \in \mathfrak{F}$,

$$0 \le \mu_s(AN) \le \frac{1}{n}\lambda(AN).$$

From here, letting $n \to \infty$, we obtain $\mu_s(AN) = 0$ and hence $\mu_s(A) = \mu_s(A\overline{N})$. That is, the set $\overline{N}$ is a support of the measure μ_s.

Further, because

$$\mu_a(A) = \mu(A) - \mu_s(A\overline{N}) \le \mu(A) - \mu_s\big(AD_n^+\big),$$

we have

$$\int_A \Big(\widehat{f} + \frac{1}{n}\mathrm{I}_{D_n^+}\Big)\, d\lambda = \mu_a(A) + \frac{1}{n}\lambda\big(AD_n^+\big) \le \mu(A) - \nu_n\big(AD_n^+\big) \le \mu(A).$$

This means that $\widehat{f} + \frac{1}{n}\mathrm{I}_{D_n^+} \in \mathfrak{F}$ and hence

$$\alpha \ge \int \Big(\widehat{f} + \frac{1}{n}\mathrm{I}_{D_n^+}\Big)\, d\lambda = \alpha + \frac{1}{n}\lambda\big(D_n^+\big).$$

This implies $\lambda(D_n^+) = 0$ and $\lambda(\overline{N}) = 0$, so that μ_s is singular with respect to λ since $\overline{N}$ is a support of μ_s. □

Uniqueness of the decomposition $\mu = \mu_a + \mu_s$ can be established as follows. Assume that $\mu = \mu_a' + \mu_s'$ is another decomposition. Then $\gamma := \mu_a' - \mu_a = \mu_s - \mu_s'$. By singularity, there exist sets N and N' such that $\mu_s(\overline{N}) = 0$, $\lambda(N) = 0$, $\mu_s'(\overline{N}') = 0$, and $\lambda(N') = 0$. Clearly, $\lambda(D) = 0$, where $D = N \cup N'$. If we assumed that $\gamma = \mu_a' - \mu_a = \mu_s - \mu_s' \ne 0$, then there would exist an $A \in \mathfrak{F}$ such that $\gamma(A) \ne 0$. Therefore, either $\gamma(AD) \ne 0$ or $\gamma(A\overline{D}) \ne 0$. However, the former is impossible, for $\lambda(D) = 0$ implies $\mu_a'(D) = \mu_a(D) = 0$. The latter is also impossible, since $\overline{D} = \overline{N}\,\overline{N}'$ and hence $\mu_s(\overline{D}) = \mu_s'(\overline{D}) = 0$.

Uniqueness of the function f (up to $\boldsymbol{\lambda}$-equivalence) follows from the observation that the equalities

$$\mu_a(A) = \int_A f\, d\boldsymbol{\lambda} = \int_A f' d\boldsymbol{\lambda}, \quad \int_A (f - f')\, d\boldsymbol{\lambda} = 0$$

for all A imply the equality $f - f' = 0$ a.e. Assuming, say, that $\boldsymbol{\lambda}(A) > 0$ for $A = \{\omega : f - f' > \varepsilon\}$ would yield for such A the relation $\int_A (f - f')\, d\boldsymbol{\lambda} > 0$. The theorem is proved. □

One of the most important applications of the Radon–Nikodym theorem is the proof of *existence and uniqueness of conditional expectations.*

Proof Let $\mathfrak{F}_0$ be a σ-subalgebra of $\mathfrak{F}$ and ξ a random variable on $\langle \Omega, \mathfrak{F}, \mathbf{P} \rangle$ such that $\mathbf{E}\xi$ exists. In Sect. 4.8 we defined the conditional expectation $\mathbf{E}(\xi \mid \mathfrak{F}_0)$ of the variable ξ given $\mathfrak{F}_0$ as an $\mathfrak{F}_0$-measurable random variable η for which

$$\mathbf{E}(\eta; B) = \mathbf{E}(\xi; B) \tag{A3.5.4}$$

for any $B \in \mathfrak{F}$. We can assume without loss of generality that $\xi \geq 0$ (an arbitrary function ξ can be represented as a difference of two positive functions). Then the right-hand side of (A3.5.4) will be a measure on $\langle \Omega, \mathfrak{F}_0 \rangle$. Since $\mathbf{E}(\xi; B) = 0$ if $\mathbf{P}(B) = 0$, this measure will be absolutely continuous with respect to $\mathbf{P}$. This implies, by the Radon–Nikodym theorem, the existence of a unique (up to $\mathbf{P}$-equivalence) measurable function η on $\langle \Omega, \mathfrak{F}_0 \rangle$ such that, for any $B \in \mathfrak{F}_0$,

$$\mathbf{E}(\xi; B) = \int_B \eta\, d\mathbf{P}.$$

This relation is clearly equivalent to (A3.5.4). It establishes the required existence and uniqueness of the conditional expectation. □

Another consequence of the assertions proved in the present section was mentioned in Sect. 3.6 and is related to the Lebesgue theorem stating that any distribution $\mathbf{P}$ on the real line $\mathbb{R} = (-\infty, \infty)$ (or the respective distribution function) has a unique representation as a sum of the three components $\mathbf{P} = \mathbf{P}_a + \mathbf{P}_s + \mathbf{P}_\partial$, where the component $\mathbf{P}_a$ is absolutely continuous with respect to Lebesgue measure:

$$\mathbf{P}_a(A) = \int_A f(x)\, dx;$$

$\mathbf{P}_\partial$ is the discrete component concentrated on an at most countable set of points $x_1, x_2, \ldots$ such that $\mathbf{P}(\{x_k\}) > 0$, and the component $\mathbf{P}_s$ has a support of Lebesgue measure zero and a continuous distribution function. This is an immediate consequence of the Lebesgue decomposition theorem. One just has to extract the discrete part from the singular (with respect to Lebesgue measure $\boldsymbol{\lambda}$) component of $\mathbf{P}$, first removing all the points x for which $\mathbf{P}(\{x\}) \geq 1/2$, then all points x for which

$\mathbf{P}(\{x\}) \geq 1/3$, and so on. It is clear that in this way we will get at most a countable set of xs, and that this process determines uniquely the discrete component $\mathbf{P}_{\partial}$.

All the aforesaid clearly also applies to distributions in n-dimensional Euclidean spaces $\mathbb{R}^n$.

3.6 Weak Convergence and Convergence in Total Variation of Distributions in Arbitrary Spaces

3.6.1 Weak Convergence

In Sects. 6.2 and 7.6 we studied weak convergence of distributions of random variables and vectors, i.e., weak convergence of distributions in $\mathbb{R}^k$, $k \geq 1$. Now we want to introduce the notion of weak convergence in more general spaces $\mathcal{X}$. As the definitions given in Sect. 6.2 show, we will need continuous functions $f(x)$ on $\mathcal{X}$. This is possible only if the space $\mathcal{X}$ is endowed with a topology. For simplicity's sake, we restrict ourselves to the case where the space $\mathcal{X}$ is endowed with a metric $\rho(x, y)$. Thus, assume we are given a measurable space $\langle \mathcal{X}, \mathfrak{B} \rangle$ with a metric ρ which is "consistent" with the σ-algebra $\mathfrak{B}$, i.e., all open (with respect to the metric ρ) sets from $\mathcal{X}$ belong to $\mathfrak{B}$ (cf. Sect. 16.1), so that any continuous (with respect to ρ) functional will be $\mathfrak{B}$-measurable. This means that if a distribution $\mathbf{Q}$ is given on $\langle \mathcal{X}, \mathfrak{B} \rangle$ (i.e., a probability space $\langle \mathcal{X}, \mathfrak{B}, \mathbf{Q} \rangle$ is given), then $\{x : f(x) < t\} \in \mathfrak{B}$ for any t, and the probabilities of these sets are defined.

Now let $\langle \Omega, \mathfrak{F}, \mathbf{P} \rangle$ be the basic probability space. A measurable mapping $\xi = \xi(\omega)$ of the space $\langle \Omega, \mathfrak{F} \rangle$ to $\langle \mathcal{X}, \mathfrak{B} \rangle$ is called an *$\mathcal{X}$-valued random element.* If $\langle \Omega, \mathfrak{F} \rangle = \langle \mathcal{X}, \mathfrak{B} \rangle$, the mapping ξ may be the identity mapping. The space $\langle \mathcal{X}, \mathfrak{B} \rangle$ is said to be the *sample* or *state space* of the random element ξ. When a functional f is continuous, $f(\xi)$ is a random variable in $\langle \mathbb{R}, \mathfrak{B} \rangle$.

Definition A3.6.1 Let a distribution $\mathbf{P}$ and a sequence of distributions $\mathbf{P}_n$ be given on the space $\langle \mathcal{X}, \mathfrak{B} \rangle$. The sequence $\mathbf{P}_n$ is said to *converge weakly* to $\mathbf{P}$: $\mathbf{P}_n \Rightarrow \mathbf{P}$ as $n \to \infty$ if, for any bounded continuous functional f ($f \in C_b(\mathcal{X})$),

$$\int f(x)\, d\mathbf{P}_n(x) \to \int f(x)\, d\mathbf{P}(x). \tag{A3.6.1}$$

If ξ_n and ξ are random elements having the distributions $\mathbf{P}_n$ and $\mathbf{P}$, respectively, then (A3.6.1) is equivalent to

$$\mathbf{E} f(\xi_n) \to \mathbf{E} f(\xi). \tag{A3.6.2}$$

This, in turn, for any continuous functional f ($f \in C(\mathcal{X})$), is equivalent to

$$f(\xi_n) \Rightarrow f(\xi). \tag{A3.6.3}$$

Indeed, (A3.6.3) means that, for any bounded continuous function g ($g \in C_b(\mathbb{R})$),

$$\mathbf{E}g\big(f(\xi_n)\big) \to \mathbf{E}g\big(f(\xi)\big), \tag{A3.6.4}$$

which is equivalent to (A3.6.2).

If $\mathcal{X} = \mathcal{X}(T)$ is the space of real-valued functions $x(t)$, $t \in T$, given on a parametric set T, and a measurable mapping $\xi(\omega)$ of the basic probability space $\langle \Omega, \mathfrak{F}, \mathbf{P} \rangle$ into $\langle \mathcal{X}, \mathfrak{B} \rangle$ is given, then the random element $\xi(\omega) = \xi(\omega, t)$ will be a *random process* (see Sect. 18.1) if $\{x : x(t) < u\} \in \mathfrak{B}$ for all t, u. In that case (A3.6.1)–(A3.6.4) will refer to the weak convergence of the distributions of random processes which has already been studied in Chap. 20.

In the metric space $\mathcal{X}$, for any $A \in \mathcal{X}$, one can define its boundary

$$\partial A = [A] - (A),$$

where $[A]$ is the closure of A, (A) being its interior ($(A) = \mathcal{X} - [\overline{A}]$, where $\overline{A}$ is the complement of A).

Definition A3.6.2 A set A is said to be a *continuity set* of the distribution $\mathbf{P}$ (or $\mathbf{P}$-continuous set) if $\mathbf{P}(\partial A) = 0$. We will denote the class of all $\mathbf{P}$-continuous sets by $\mathfrak{D}_P$.

The following criterion of weak convergence of distributions holds true.

Theorem A3.6.1 *The following four conditions are equivalent*:

(i) $\mathbf{P}_n \Rightarrow \mathbf{P}$,
(ii) $\lim_{n\to\infty} \mathbf{P}_n(A) = \mathbf{P}(A)$ *for all* $A \in \mathfrak{D}_P$,
(iii) $\limsup_{n\to\infty} \mathbf{P}_n(F) \le \mathbf{P}(F)$ *for all closed* $F \subset \mathcal{X}$,
(iv) $\liminf_{n\to\infty} \mathbf{P}_n(G) \ge \mathbf{P}(G)$ *for all open* $G \subset \mathcal{X}$.

Observe that if $\mathbf{P}_n \Rightarrow \mathbf{P}$, then convergence (A3.6.1)–(A3.6.3) takes place for a wider class of functionals than $C_b(\mathcal{X})$ ($C(\mathcal{X})$), namely, for the so-called $\mathbf{P}$-continuous functionals (or functionals continuous with $\mathbf{P}$-probability 1). We will call so the functionals f for which $f(x_n) \to f(x)$ as $\rho(x_n, x) \to 0$ not for all $x \in \mathcal{X}$, but only for $x \in A$, $\mathbf{P}(A) = 1$. The class of $\mathbf{P}$-continuous functionals will be denoted by $C_P(\mathcal{X})$.

The classes $\mathfrak{D}_P$ and $C_P(\mathcal{X})$, and also the classes of all closed and open sets participating in Theorem A3.6.1, are very wide which makes verifying the conditions of Theorem A3.6.1 rather difficult and cumbersome. These classes can be substantially restricted if we consider not arbitrary but only *relatively compact* sequences $\mathbf{P}_n$ (from any subsequence $\mathbf{P}'_n$ one can choose a convergent subsequence; this approach was already used in Sect. 6.3).

Definition A3.6.3 A class $\mathfrak{D}$ of sets from $\mathfrak{B}$ is said to *determine the measure* $\mathbf{P}$ if, for a measure $\mathbf{Q}$, the equalities $\mathbf{P}(A) = \mathbf{Q}(A)$ for all $A \in \mathfrak{D}\mathfrak{D}_P$ imply $\mathbf{Q} = \mathbf{P}$.

A class $\mathfrak{D}$ determines the measure $\mathbf{P}$ if $\mathfrak{D}$ is an algebra and $\sigma(\mathfrak{D}\mathfrak{D}_P) = \mathfrak{B}_\mathfrak{X}$ (condition $\sigma(\mathfrak{D}) = \mathfrak{B}_\mathfrak{X}$ is insufficient (see e.g. [4])).

In a similar way we introduce the class $\mathfrak{F}$ of functionals f determining the distribution $\mathbf{P}$ of a random element $\xi = \xi^P$: for any $\mathbf{Q}$, the coincidence of the distributions of $f(\xi^P)$ and $f(\xi^Q)$ for all $f \in \mathfrak{F}C_P(X)$ implies $\mathbf{P} = \mathbf{Q}$.

Theorem A3.6.2 *A necessary and sufficient condition for convergence* $\mathbf{P}_n \Rightarrow \mathbf{P}$ *is that*:

(1) *the sequence* $\mathbf{P}_n$ *is relatively compact; and*
(2) *there exists a class of sets* $\mathfrak{D} \subset \mathfrak{B}_\mathfrak{X}$ *determining the measure* $\mathbf{P}$ *and such that* $\mathbf{P}_n(A) \to \mathbf{P}(A)$ *for any* $A \in \mathfrak{D}\mathfrak{D}_P$.

An alternative to condition (2) is the existence of a class of functionals $\mathfrak{F}$ which determines the measure $\mathbf{P}$ and is such that $\mathbf{P}(f(\xi_n) < t) \Rightarrow \mathbf{P}(f(\xi) < t)$ for all $f \in \mathfrak{F}C_P(X)$.

The following notion of tightness plays an important role in establishing the compactness of $\{\mathbf{P}_n\}$.

Definition A3.6.4 A family of distributions $\{\mathbf{P}_n\}$ on $\langle \mathfrak{X}, \mathfrak{B} \rangle$ is said to be *tight* if, for any $\varepsilon > 0$, there exists a compact set $K = K_\varepsilon \subset \mathfrak{X}$ such that $\mathbf{P}_n(K) > 1 - \varepsilon$ for all n.

Theorem A3.6.3 (Prokhorov) *If* $\{\mathbf{P}_n\}$ *is a tight family of distributions then it is relatively compact. If* $\mathfrak{X}$ *is a complete separable space, the converse assertion is also true.*

Since, for many functional spaces (in particular, for spaces $C(0, T)$ and $D(0, T)$), there exist simple explicit criteria for compactness of sets, one can now establish conditions ensuring convergence $\mathbf{P}_n \Rightarrow \mathbf{P}$ in these spaces. It is well known, for example, that in the above-mentioned spaces compacta are, roughly speaking, of the form $\{x : \omega_\Delta(x) \le \varepsilon(\Delta)\}$, where $\omega_\Delta(x)$ is the so-called "modulus of continuity" (in the space C or D, respectively) of the element x, and $\varepsilon(\Delta) \ge 0$ is an arbitrary function vanishing as $\Delta \downarrow 0$.

The proofs of Theorems A3.6.1–A3.6.3 can be found, for example, in [1]. We do not present them here as they are somewhat beyond the scope of this book and, on the other hand, the theorems themselves are not used in the body of the text. We use only the special cases of these theorems given in Sects. 6.2 and 6.3.

The invariance principle of Sect. 20.1 is a theorem about weak convergence of distributions in the space $C(0, 1)$. In order to use Theorems A3.6.2 and A3.6.3 to prove this result, one has to choose the class $\mathfrak{D}$ to be the class of cylinder sets. Convergence of $\mathbf{P}_n$ to $\mathbf{P}$ on sets from this class $\mathfrak{D}$ is the convergence of finite-dimensional distributions of processes $s_n(t)$ generated by sums of random variables (see Sect. 20.1). Since the increments of $s_n(t)$ are essentially independent, the demonstration of that part of the theorem reduces to proving asymptotic normality of these increments, which follows immediately from the central limit theorem.

The condition of compactness of the family of distributions in $C(0, 1)$ requires, according to Theorem A3.6.3, a proof that the modulus of continuity of the trajectory $s_n(t)$ converges to zero in probability (for more details, see e.g. [1]). This could be proved using the Kolmogorov inequality from Corollary 11.2.1.

3.6.2 Convergence in Total Variation

So, to consider weak convergence of distributions in spaces $\langle \mathcal{X}, \mathfrak{B} \rangle$ of a general nature, one has to introduce a topology in the space, which is not always convenient and feasible. There exists another type of convergence of distributions on $\langle \mathcal{X}, \mathfrak{B} \rangle$ which does not require the introduction of topologies. This is convergence in total variation.

Definition A3.6.5 Let γ be a finite signed measure on $\langle \mathcal{X}, \mathfrak{B} \rangle$. The total variation of γ (or the total variation norm $\|\gamma\|$) is the quantity

$$\|\gamma\| = \sup_{f:|f|\le 1} \left| \int f(x)\, d\gamma(x) \right|, \qquad \text{(A3.6.5)}$$

where the supremum is taken over the class of all $\mathfrak{B}$-measurable functions $f(x)$ such that $|f(x)| \le 1$ for all $x \in \mathcal{X}$.

The supremum in (A3.6.5) is clearly attained on such functions f for which, roughly speaking, $f(x) = 1$ at points x such that $d\gamma(x) > 0$, and $f(x) = -1$ at points x for which $d\gamma(x) < 0$. Therefore (A3.6.5) can be written in the form

$$\|\gamma\| = \int \left| d\gamma(x) \right|. \qquad \text{(A3.6.6)}$$

An exact meaning to this expression can be given using the Hahn decomposition theorem (see Corollary A3.5.1), which implies

$$\|\gamma\| = \gamma^+(\mathcal{X}) + \gamma^-(\mathcal{X}). \qquad \text{(A3.6.7)}$$

The right-hand side of this equality may be taken as a definition of $\int |d\gamma(x)|$.

Lemma A3.6.2 *If* $\gamma(\mathcal{X}) = 0$, *then* $\|\gamma\| = 2\sup_{B\in\mathfrak{B}} \gamma(B)$.

Proof From (A3.6.5) it follows that, for any B ($\overline{B}$ is the complement of B, $\gamma(B) \cup \gamma(\overline{B})=0$),

$$\|\gamma\| \ge \left|\gamma(B)\right| + \left|\gamma(\overline{B})\right| = 2\left|\gamma(B)\right|.$$

Therefore $\|\gamma\| \ge 2\sup_{B\in\mathfrak{B}} |\gamma(B)|$.

To obtain the converse inequality, we will make use of Corollary A3.5.1 of the Hahn decomposition theorem. As we have already noted, according to that theorem (for the definition of the set $D^{\pm}$ see the Hahn theorem),

$$\|\gamma\| = \gamma^+(\mathcal{X}) + \gamma^-(\mathcal{X}) = \gamma^+(D^+) + \gamma^-(\overline{D^+})$$
$$= \gamma(D^+) - \gamma(\overline{D^+}) = 2\gamma(D^+) \le 2 \sup_{B \in \mathfrak{B}} \gamma(B).$$

The lemma is proved. □

Definition A3.6.6 Let $\mathbf{P}$ be a distribution and $\mathbf{P}_n$, $n = 1, 2, \ldots$, a sequence of distributions given on $\langle \mathcal{X}, \mathfrak{B} \rangle$. We will say that $\mathbf{P}_n$ *converges to* $\mathbf{P}$ *in total variation*: $\mathbf{P}_n \xrightarrow{TV} \mathbf{P}$, if $\|\mathbf{P}_n - \mathbf{P}\| \to 0$ as $n \to \infty$.

Convergence in total variation is a very strong form of convergence. If $\langle \mathcal{X}, \mathfrak{B} \rangle$ is a metric space and $\mathbf{P}_n \xrightarrow{TV} \mathbf{P}$, then $\mathbf{P}_n \Rightarrow \mathbf{P}$. Indeed, since any functional $f \in C_b(\mathcal{X})$ is bounded: $|f(x)| < b$, we have

$$\left| \int f(d\mathbf{P}_n - d\mathbf{P}) \right| \le b \int \left| d(\mathbf{P}_n - \mathbf{P}) \right| = b\|\mathbf{P}_n - \mathbf{P}\| \to 0.$$

Thus in that case

$$\int f\, d\mathbf{P}_n \to \int f\, d\mathbf{P}$$

even without assuming the continuity of f.

The converse assertion about convergence $\mathbf{P}_n \xrightarrow{TV} \mathbf{P}$ if $\mathbf{P}_n \Rightarrow \mathbf{P}$ is not true. Let, for example, $\mathcal{X} = [0, 1]$, $\mathbf{P}_n$ be the uniform distribution on the set of $n + 1$ points $\{0, 1/n, \ldots, n/n\}$, and $\mathbf{P} = \mathbf{U}_{0,1}$. It is clear that all $\mathbf{P}_n$ are concentrated on the countable set $\mathcal{N}$ of all rational numbers. Therefore $\mathbf{P}_n(\mathcal{N}) = 1$, $\mathbf{P}(\mathcal{N}) = 0$, and $\|\mathbf{P}_n - \mathbf{P}\| = \mathbf{P}_n(\mathcal{N}) + \mathbf{P}(\mathcal{X} \setminus \mathcal{N}) = 2$. At the same time, clearly $\mathbf{P}_n \Rightarrow \mathbf{P}$.

Now let the distribution $\mathbf{P}$ have a density p with respect to a measure μ (one could take, in particular, $\mu = \mathbf{P}$, in which case $p(x) \equiv 1$). Denote by p_n the density (with respect to μ) of the absolutely continuous (with respect to μ) component $\mathbf{P}_n^a$ of the distribution $\mathbf{P}_n$.

Theorem A3.6.4 *A necessary and sufficient condition for convergence* $\mathbf{P}_n \xrightarrow{TV} \mathbf{P}$ *is that* p_n *converges to* p *in measure* μ, *i.e., for any* $\varepsilon > 0$,

$$\mu\left\{x : \left|p_n(x) - p(x)\right| > \varepsilon\right\} \to 0 \quad \text{as } n \to \infty.$$

Proof We have

$$\int \left| d(\mathbf{P}_n - \mathbf{P}) \right| = \int \left| p_n(x) - p(x) \right| \mu(dx) + \left\| \mathbf{P}_n^s \right\|,$$

where $\mathbf{P}_n^s$ is the singular component of $\mathbf{P}_n$ with respect to the measure μ.

Let $\|\mathbf{P}_n - \mathbf{P}\| \to 0$. Then

$$\int |p_n - p|\, d\mu \to 0, \tag{A3.6.8}$$

and hence

$$\mu\{x : |p_n(x) - p(x)| > \varepsilon\} \le \varepsilon^{-1} \int |p_n - p|\, d\mu \to 0.$$

Now let $p_n \xrightarrow{\mu} p$. Put

$$B_\varepsilon = \{x : p(x) \ge \varepsilon\}, \qquad A_{n,\varepsilon} = \{x : |p_n(x) - p(x)| \le \varepsilon^2\}.$$

Then

$$1 \ge \int_{B_\varepsilon} p\, d\mu \ge \varepsilon \mu(B_\varepsilon), \quad \mu(B_\varepsilon) \le \frac{1}{\varepsilon}.$$

Consider

$$\int |p_n - p|\, d\mu = \int_{B_\varepsilon A_{n,\varepsilon}} + \int_{\overline{B_\varepsilon A_{n,\varepsilon}}}. \tag{A3.6.9}$$

Here the first integral on the right-hand side does not exceed ε. Since

$$\lim_{\varepsilon \to 0} \int_{B_\varepsilon} p\, d\mu \to 1,$$

we will have, for a given $\delta > 0$ and sufficiently small ε, the inequality

$$\int_{B_\varepsilon} p\, d\mu > 1 - \delta$$

and, for n large enough,

$$\int_{B_\varepsilon A_{n,\varepsilon}} p\, d\mu > 1 - 2\delta, \quad \int_{B_n A_{n,\varepsilon}} p_n\, d\mu > 1 - 3\delta. \tag{A3.6.10}$$

It follows from these two inequalities that the second integral in (A3.6.9) does not exceed 5δ, which proves (A3.6.8). Furthermore, (A3.6.10) implies that $\|\mathbf{P}_n^a\| > 1 - 3\delta$ and $\|\mathbf{P}_n^s\| < 3\delta$. The theorem is proved. □

The theorem implies that if $\mathbf{P}_n \xrightarrow{TV} \mathbf{P}$ then the absolutely continuous with respect to $\mu = \mathbf{P}$ component $\mathbf{P}_n^a$ of the distribution $\mathbf{P}_n$ has a density $p_n(x) \xrightarrow{p} 1$, $\mathbf{P}_n^a(\mathcal{X}) \to 1$.

Appendix 4
The Helly and Arzelà–Ascoli Theorems

In this appendix we will prove Helly's theorem and the Arzelà–Ascoli theorem. The former theorem was used in Sect. 6.3, and both theorems will be used in the proof of the main theorem of Appendix 9.

Let $\mathfrak{F}$ be the class of all distribution functions, and $\mathfrak{G}$ the class of functions G possessing properties F1 and F2 from Sect. 3.2 (monotonicity and left continuity), and the properties $G(-\infty) \ge 0$ and $G(\infty) \le 1$. We will write $G_n \Rightarrow G$ as $n \to \infty$, $G \in \mathfrak{G}$, if $G_n(x) \to G(x)$ at all points of continuity of the function G.

Theorem A4.1 (Helly) *Any sequence $F_n \in \mathfrak{F}$ contains a convergent subsequence $F_{nn} \Rightarrow F \in \mathfrak{G}$.*

We will need the following.

Lemma A4.1 *A sufficient condition for convergence $F_n \Rightarrow F \in \mathfrak{G}$ is that*

$$F_n(x) \to F(x), \quad x \in D,$$

as $n \to \infty$ on some everywhere dense set D of the reals.

Proof Let x be an arbitrary point of continuity of $F(x)$. For arbitrary $x', x'' \in D$ such that $x' \le x \le x''$, one has

$$F_n(x') \le F_n(x) \le F_n(x'').$$

Consequently,

$$\lim_{n\to\infty} F_n(x') \le \liminf_{n\to\infty} F_n(x) \le \limsup_{n\to\infty} F_n(x) \le \lim_{n\to\infty} F_n(x'').$$

From here and the conditions of the lemma we obtain

$$F(x') \le \liminf_{n\to\infty} F_n(x) \le \limsup_{n\to\infty} F_n(x) \le F(x'').$$

A.A. Borovkov, *Probability Theory*, Universitext,
DOI 10.1007/978-1-4471-5201-9, © Springer-Verlag London 2013

Letting $x' \uparrow x$ and $x'' \downarrow x$ along the set D and taking into account that x is a point of continuity of F, we get

$$\lim_{n\to\infty} F_n(x) = F(x).$$

The lemma is proved. □

Proof of Theorem A4.1 Let $D = \{x_n\}$ be an arbitrary countable everywhere dense set of real numbers. The numerical sequence $\{F_n(x_1)\}$ is bounded and hence contains a convergent sequence $\{F_{1n}(x_1)\}$. Denote the limit of this sequence by $F(x_1)$. Consider now the numerical sequence $\{F_{1n}(x_2)\}$. It also contains a convergent subsequence $\{F_{2n}(x_2)\}$ with a limit $F(x_2)$. Moreover,

$$\lim_{n\to\infty} F_{2n}(x_1) = F(x_1).$$

Continuing this process, we will obtain, for any number k, k sequences

$$\big\{F_{kn}(x_i)\big\}, \quad i = 1, \dots, k,$$

such that $\lim_{n\to\infty} F_{kn}(x_i) = F(x_i)$.

Consider the diagonal sequence of the distribution functions $\{F_{nn}(x)\}$. For any $x_k \in D$, only $k-1$ first elements of the numerical sequence $\{F_{nn}(x_k)\}$ may not belong to the sequence $F_{kn}(x_k)$. Therefore

$$\lim_{n\to\infty} F_{nn}(x_k) = F(x_k).$$

It is clear that $F(x)$ is a non-decreasing bounded function given on D. It can easily be extended by continuity from the left to a non-decreasing function on the whole real line. Now we see that the sequence $\{F_{nn}\}$ and the function F satisfy the conditions of Lemma A4.1. The theorem is proved. □

The conditions of Helly's theorem can be weakened. Namely, instead of $\mathcal{F}$ we could consider a wider class $\mathcal{H}$ of non-decreasing left continuous (i.e., satisfying properties F1 and F3) functions H majorised by a fixed function: for any x, $|H(x)| \le N(x) < \infty$, where N is a given function characterising the class $\mathcal{H}$. We do not exclude the case when $|H(x)|$ (or $N(x)$) grow unboundedly as $|x| \to \infty$. The following generalised version of Helly's theorem is true.

Theorem A4.2 (Generalised Helly theorem) *Any sequence $H_n \in \mathcal{F}$ contains a subsequence H_{nn} which converges to a function $H \in \mathcal{H}$ at each point of continuity of H.*

The Proof repeats the above proof of Helly's theorem. □

To each function $H_n \in \mathcal{H}$ we can associate a measure $\boldsymbol{\mu}_n$ by putting

$$\boldsymbol{\mu}_n\big([a, b)\big) = H_n(b) - H_n(a).$$

The generalised Helly theorem will then mean that, for any sequence of measures μ_n generated by functions from $\mathcal{H}$, there exists a subsequence μ_{nn} converging weakly on each finite interval of which the endpoints are not atoms of the limiting measure μ_n.

We give one more analogue of Helly's theorem which refers to a collection of equicontinuous functions g_n. Recall that a sequence of functions $\{g_n\}$ is said to be *equicontinuous* if, for any $\varepsilon > 0$, there exists a $\delta > 0$ such that $|x_1 - x_2| < \delta$ implies $|g_n(x_1) - g_n(x_2)| < \varepsilon$ for all n.

Theorem A4.3 (Arzelà–Ascoli) *Let $\{g_n\}$ be a sequence of uniformly bounded and equicontinuous functions of a real variable. Then there exists a subsequence g_{n_k} converging to a continuous limit g uniformly on each finite interval.*

Proof Choose again a countable everywhere dense subset $\{x_n\}$ of the real line, and a subsequence $\{g_{n_k}\}$ converging at the points $x_1, x_2, \ldots$ Denote its limit at the point x_j by $g(x_j)$. We have

$$\begin{aligned} \left|g_{n_k}(x) - g_{n_r}(x)\right| \le \left|g_{n_k}(x) - g_{n_k}(x_j)\right| + \left|g_{n_r}(x) - g_{n_r}(x_j)\right| \\ + \left|g_{n_k}(x_j) - g_{n_r}(x_j)\right|. \end{aligned} \tag{A4.1}$$

By assumption, the last term on the right-hand side tends to 0 as $n_k \to \infty$, $n_r \to \infty$. By virtue of equicontinuity, for any point x there exists a point x_j such that, for all n,

$$\left|g_n(x) - g_n(x_j)\right| < \varepsilon. \tag{A4.2}$$

In any given finite interval I there exists a finite collection of points x_j such that (A4.2) will hold for all points $x_j \in I$. This implies that the right-hand side of (A4.1) will be less than 3ε for all sufficiently large n_k, n_r uniformly over $x_j \in I$. Thus there exists the limit $g(x) = \lim g_{n_k}(x)$, for which by (A4.2) we have $|g(x) - g(x_j)| \le \varepsilon$, which implies that g is continuous. The theorem is proved. □

Appendix 5
The Proof of the Berry–Esseen Theorem

In this appendix we prove the following assertion stated in Sect. 8.5.

Theorem A5.1 (Berry–Esseen) *Let ξ_k be independent identically distributed random variables,*

$$\mathbf{E}\xi_k = 0, \qquad \mathrm{Var}(\xi_k) = 1, \qquad \mu = \mathbf{E}|\xi_k|^3 < \infty, \qquad S_n = \sum_{k=1}^{n} \xi_k, \qquad \zeta_n = \frac{S_n}{\sqrt{n}}.$$

Then, for all n,

$$\Delta_n := \sup_x \big|\mathbf{P}(\zeta_n < x) - \Phi(x)\big| < \frac{c\mu}{\sqrt{n}},$$

where Φ is the standard normal distribution function and c is an absolute constant.

Proof We will make use of the composition method. As in Sect. 8.5, we will bound Δ_n based on estimates of smallness of $\mathbf{E}g(\zeta_n) - \mathbf{E}g(\eta)$, $\eta \in \mathbf{\Phi}_{0,1}$, for smooth g. To get a bound for Δ_n in Sect. 8.5, we chose g to be a function constant outside a small interval. The next lemma shows that such a choice is not obligatory. Let G be a distribution function and $\gamma \in G$ be independent of ζ_n and η. Put

$$g(z) := G\left(\frac{x-z}{\varepsilon}\right),$$

so that

$$\mathbf{E}g(\zeta_n) = \mathbf{E}G\left(\frac{x-\zeta_n}{\varepsilon}\right) = \mathbf{P}\left(\gamma < \frac{x-\zeta_n}{\varepsilon}\right) = \mathbf{P}(\zeta_n + \varepsilon\gamma < x),$$
$$\mathbf{E}g(\eta) = \mathbf{P}(\eta + \varepsilon\gamma < x).$$

Set

$$\Delta_{n,\varepsilon} := \sup_x \left|\mathbf{E}G\left(\frac{x-\zeta_n}{\varepsilon}\right) - \mathbf{E}G\left(\frac{x-\eta}{\varepsilon}\right)\right|$$

DOI 10.1007/978-1-4471-5201-9,

$$= \sup_x \big| \mathbf{P}(\zeta_n + \varepsilon\gamma < x) - \mathbf{P}(\eta + \varepsilon\gamma < x) \big|$$
$$= \sup_x \bigg| \int dG(y) \big[\mathbf{P}(\zeta_n < x - \varepsilon y) - \mathbf{P}(\eta < x - \varepsilon y) \big] \bigg|.$$

Clearly, $\Delta_{n,\varepsilon} \le \Delta_n$. Our aim will be to obtain a converse inequality for Δ_n.

Lemma A5.1 *Let $v > 0$ be such that $G(v) - G(-v) \ge 3/4$. Then, for any $\varepsilon > 0$,*

$$\Delta_n \le 2\Delta_{n,\varepsilon} + \frac{3v\varepsilon}{\sqrt{2\pi}}.$$

Proof Assume that the $\sup_x$ in the definition of Δ_n is attained on a positive value $\Delta_n(x) := F_n(x) - \Phi(x)$ (the case of a negative value $\Delta_n(x)$ is similar) and that, for a given $\delta > 0$, the value x_δ is such that

$$\Delta_n(x_\delta) = F_n(x_\delta) - \Phi(x_\delta) \ge \Delta_n - \delta,$$

where F_n is the distribution function of ζ_n. When the argument increases, the value of $\Delta_n(x_\delta)$ varies little in the following sense. Let $|y| < v$. Then $v - y > 0$ and

$$\begin{aligned} \Delta_n\big(x_\delta + \varepsilon(v-y)\big) &= F_n\big(x_\delta + \varepsilon(v-y)\big) - \Phi\big(x_\delta + \varepsilon(v-y)\big) \\ &\ge F_n(x_\delta) - \Phi(x_\delta) - \big[\Phi\big(x_\delta + \varepsilon(v-y)\big) - \Phi(x_\delta)\big]. \end{aligned}$$

Here the difference in the brackets does not exceed $\varepsilon(v-y)\Phi'(0) \le 2v\varepsilon/\sqrt{2\pi}$, and hence

$$\Delta_n\big(x_\delta + \varepsilon(v-y)\big) \ge \Delta_n - \delta - \frac{2v\varepsilon}{\sqrt{2\pi}}.$$

Therefore

$$\begin{aligned} \Delta_{n,\varepsilon} &\ge \int dG(y)\Delta_n(x_\delta + \varepsilon v - \varepsilon y) = \int_{|y|<v} + \int_{|y|\ge v} \\ &\ge \frac{3}{4}\bigg(\Delta_n - \delta - \frac{2v\varepsilon}{\sqrt{2\pi}}\bigg) - \frac{1}{4}\Delta_n = \frac{\Delta_n}{2} - \frac{3}{4}\bigg(\delta + \frac{2v\varepsilon}{\sqrt{2\pi}}\bigg). \end{aligned}$$

Since δ is arbitrary, the assertion of the lemma follows. □

Corollary A5.1 *For $G = \Phi(\gamma \in \boldsymbol{\Phi}_{0,1})$ the value $v = 6/5$ satisfies the condition of Lemma* A5.1, *and*

$$\Delta_n \le 2(\Delta_{n,\varepsilon} + \varepsilon). \tag{A5.1}$$

At the next stage of the proof we bound $\Delta_{n,\varepsilon}$, and it is at that stage where the composition method will be used. Put

$$u(n) := \max_{k \le n} \Delta_k \frac{\sqrt{k}}{\mu}, \qquad \alpha^2 := \varepsilon^2 n.$$

By letters c (with or without indices) we will denote absolute constants, not necessarily the same ones.

Lemma A5.2 *For* $\alpha \ge 1$,

$$\Delta_{n,\varepsilon} \le c\mu\left(\frac{1}{\sqrt{n}} + \frac{\mu u(n-1)}{\alpha\sqrt{n}}\right). \tag{A5.2}$$

Proof Set $H_n := \sum_{k=1}^n \eta_k$, where $\eta_k \in \mathbf{\Phi}_{0,1}$ are independent of each other and of H_n and γ. The composition method is based on the following identity (cf. Theorem 8.5.1 and identity (8.5.3), $\eta \in \mathbf{\Phi}_{0,1}$):

$$\begin{aligned}
&\mathbf{P}(\zeta_n + \varepsilon\gamma < x) - \mathbf{P}(\eta + \varepsilon\gamma < x) = \mathbf{P}(S_n + \alpha\gamma < x\sqrt{n}) - \mathbf{P}(H_n + \alpha\gamma < x\sqrt{n}) \\
&\quad = \sum_{m=1}^{n}\big[\mathbf{P}\big(S_{m-1} + (H_n - H_m) + \xi_m + \alpha\gamma < x\sqrt{n}\big) \\
&\quad - \mathbf{P}\big(S_{m-1} + (H_n - H_m) + \eta_m + \alpha\gamma < x\sqrt{n}\big)\big].
\end{aligned}$$

Since for $\gamma \in \mathbf{\Phi}_{0,1}$ one has $H_n - H_m + \alpha\gamma \in \mathbf{\Phi}_{0,n-m+\alpha^2}$, the last sum is equal to $\sum_{m=1}^n D_m$, where

$$\begin{aligned}
D_m &:= \mathbf{E}\left[\Phi\left(\frac{x\sqrt{n} - S_{m-1} - \xi_m}{d_m}\right) - \Phi\left(\frac{x\sqrt{n} - S_{m-1} - \eta_m}{d_m}\right)\right] \\
&= \mathbf{E}\left[\Phi\left(T_m - \frac{\xi_m}{d_m}\right) - \Phi\left(T_m - \frac{\eta_m}{d_m}\right)\right], \\
d_m^2 &:= n - m + \alpha^2, \qquad T_m := \frac{x\sqrt{n} - S_{m-1}}{d_m}.
\end{aligned}$$

To bound D_m we will adopt the same approach as in Lemma 8.5.1. Because the first two moments of ξ_m and η_m coincide, expanding Φ into a series yields

$$|D_m| \le \frac{2\mu}{d_m^3} \sup_t \mathbf{E}\phi''(T_m + t),$$

where $\phi(x) = \Phi'(x)$ and $\phi'' = \Phi'''$. Since the function ϕ'' is bounded,

$$|D_m| \le \frac{c\mu}{d_m^3}. \tag{A5.3}$$

We will also need another bound for D_m. To obtain it, consider the quantity

$$\begin{aligned}
R_m &:= \sup_t \big|\mathbf{E}\phi''(T_m + t)\big| \\
&\le \sup_t \big|\mathbf{E}\big[\phi''(T_m + t) - \phi''(V_m + t)\big]\big| + \sup_t \big|\mathbf{E}\phi''(V_m + t)\big|,
\end{aligned} \tag{A5.4}$$

where V_m is defined in the same way as T_m but with S_{m-1} replaced by H_{m-1}. Integrating by parts yields

$$\begin{aligned}\big|\mathbf{E}[\phi''(T_m+t)-\phi''(V_m+t)]\big| &= \left|\int \phi''(u+t)\,d\big[\mathbf{P}(T_m<u)-\mathbf{P}(V_m<u)\big]\right| \\ &= \left|\int \phi'''(u+t)\big[\mathbf{P}(T_m<u)-\mathbf{P}(V_m<u)\big]\,du\right| \\ &\le \Delta_{m-1}\int \big|\phi'''(u)\big|\,du = c\Delta_{m-1},\end{aligned}$$

since $|\mathbf{P}(T_m<u)-\mathbf{P}(V_m<u)| \le \Delta_{m-1}$ (the variables T_m and V_m are obtained from $S_{m-1}/\sqrt{m-1}$ and $H_{m-1}/\sqrt{m-1}$, respectively, by one and the same linear transformation).

To bound the second summand on the right-hand side of (A5.4), note that

$$\mathbf{E}\phi''(V_m+t) = \int \phi''(u+t)\frac{1}{r_m}\phi\left(\frac{u-a_m}{r_m}\right)du, \tag{A5.5}$$

where

$$a_m = x\frac{\sqrt{n}}{d_m}, \qquad r_m = \sqrt{\frac{m-1}{n-m+\alpha^2}},$$

so that $\frac{1}{r_m}\phi(\frac{u-a_m}{r_m})$ is the density of $V_m = \frac{(x\sqrt{n}-H_{m-1})}{d_m}$. Integrating the right-hand side of (A5.5) twice by parts, we obtain

$$\big|\mathbf{E}\phi''(V_m+t)\big| = \frac{1}{r_m^3}\left|\int \phi(u+t)\phi''\left(\frac{u-a_m}{r_m}\right)du\right| \le \frac{c}{r_m^3}.$$

Thus,

$$R_m \le c\left(\Delta_{m-1}+\frac{1}{r_m^3}\right), \qquad D_m \le c\mu\left(\frac{\Delta_{m-1}}{d_m^3}+\frac{1}{(m-1)^{3/2}}\right).$$

The bounds derived for D_m do not depend on x. Therefore, using the bound just obtained for $m>n/2$, and bound (A5.3) for $m \le n/2$ (the latter bound implies then that $|D_m| \le c\mu/n^{3/2}$), we get

$$\Delta_{n,\varepsilon} \le c\mu\left[\sum_{m\le n/2} n^{-3/2} + \sum_{m>n/2}^{n}\frac{\Delta_{m-1}}{d_m^3} + \sum_{m>n/2}^{n}\frac{1}{(m-1)^{3/2}}\right]. \tag{A5.6}$$

Here the first sum does not exceed $(n/2)n^{-3/2} = 1/(2\sqrt{n})$ and the last sum does not exceed

$$\int_{n/2-1}^{n}\frac{ds}{s^{3/2}} \le \frac{c}{\sqrt{n}}.$$

It remains to bound the middle sum. Setting $(u(n) := \max_{k \le n} (\Delta_k \sqrt{k})/\mu)$, we have

$$\sum_{m>n/2}^{n} \frac{\Delta_{m-1}}{d_m^3} \le \mu u(n-1)\sqrt{\frac{2}{n}} \sum_{m>n/2}^{n} \frac{1}{(n-m+\alpha^2)^{3/2}}.$$

The last sum does not exceed

$$\sum_{k=0}^{\infty} \frac{1}{(k+\alpha^2)^{3/2}} \le \frac{1}{\alpha^3} + \int_0^{\infty} \frac{dt}{(t+\alpha^2)^{3/2}} = \frac{1}{\alpha^3} + \frac{1}{2\alpha} \le \frac{3}{2\alpha},$$

provided that $\alpha \ge 1$. Collecting (A5.6) and the above estimates together, we obtain the assertion of the lemma. □

We now turn directly to the proof of the theorem. By virtue of (A5.1) and (A5.2),

$$v(n) := \frac{\Delta_n \sqrt{n}}{\mu} \le \frac{2}{\mu} \sqrt{n} \Delta_{n,\varepsilon} + \frac{2\alpha}{\mu} \le 2c + \frac{2c\mu u(n-1)}{\alpha} + \frac{2\alpha}{\mu}.$$

Put here $\alpha := \max(4c\mu, 1)$. Then $(\mu \ge 1)$

$$v(n) \le c_1 + \frac{u(n+1)}{2}.$$

This implies that $u(n) \le 2c_1$ for all n. To verify this, we make use of induction. Clearly, $u(1) = v(1) \le 1 \le 2c_1$. Let $u(n-1) \le 2c_1$. Then $v(n) \le 2c_1$ and $u(n) = \max(v(n), u(n-1)) \le 2c_1$. The theorem is proved. □

Appendix 6
The Basic Properties of Regularly Varying Functions and Subexponential Distributions

The properties of regularly varying functions and subexponential distributions were used in Sects. 8.8, 9.4–9.6 and 12.7 and will be used in Appendices 7 and 8.

6.1 General Properties of Regularly Varying Functions

Definition A6.1.1 A positive measurable function $L(t)$ is called a *slowly varying function* (s.v.f.) as $t \to \infty$ if, for any fixed $v > 0$,

$$\frac{L(vt)}{L(t)} \to 1 \quad \text{as } t \to \infty. \tag{A6.1.1}$$

A function $V(t)$ is called a *regularly varying function* (r.v.f.) (with exponent $-\beta \in \mathbb{R}$) as $t \to \infty$ if it can be represented as

$$V(t) = t^{-\beta} L(t), \tag{A6.1.2}$$

where $L(t)$ is an s.v.f. as $t \to \infty$. We will denote the class of all r.v.f.s by $\mathfrak{R}$.

The definitions of an s.v.f. and r.v.f. as $t \downarrow 0$ are quite similar. In what follows, the term s.v.f. (r.v.f.) will (unless specified otherwise) always refer to a slowly (regularly) varying function at infinity.

It is easy to see that, similarly to (A6.1.1), a characteristic property of regularly varying functions is the convergence, for any fixed $v > 0$,

$$\frac{V(vt)}{V(t)} \to v^{-\beta} \quad \text{as } t \to \infty. \tag{A6.1.3}$$

Thus, an s.v.f. is an r.v.f. with exponent zero.

Typical representatives of the class of s.v.f.s are the logarithmic function and its powers $\ln^{\gamma} t$, $\gamma \in \mathbb{R}$, their linear combinations, multiple logarithms, functions with

A.A. Borovkov, *Probability Theory*, Universitext,
DOI 10.1007/978-1-4471-5201-9,

the property $L(t) \to L = \text{const} \neq 0$ as $t \to \infty$, etc. As an example of a *bounded oscillating* s.v.f. one can give

$$L_0(t) = 2 + \sin(\ln \ln t), \quad t > 1.$$

We will need the following two basic properties of s.v.f.s.

Theorem A6.1.1 (Uniform convergence theorem) *If $L(t)$ is an s.v.f. as $t \to \infty$ then convergence* (A6.1.1) *holds uniformly in v on any segment* $[v_1, v_2]$, $0 < v_1 < v_2 < \infty$.

The theorem implies that the uniform convergence (A6.1.1) on the segment $[1/M, M]$ also takes place in the case when, as $t \to \infty$, the quantity $M = M(t)$ grows unboundedly slowly enough.

Theorem A6.1.2 (Integral representation) *A function $L(t)$ is an s.v.f. as $t \to \infty$ if and only if, for some $t_0 > 0$, one has*

$$L(t) = c(t) \exp\left\{ \int_{t_0}^{t} \frac{\varepsilon(u)}{u} du \right\}, \quad t \geq t_0, \tag{A6.1.4}$$

where the functions $c(t)$ and $\varepsilon(t)$ are measurable and such that $c(t) \to c \in (0, \infty)$ and $\varepsilon(t) \to 0$ as $t \to \infty$.

For instance, for $L(t) = \ln t$ representation (A6.1.4) is valid with $c(t) = 1$, $t_0 = e$ and $\varepsilon(t) = (\ln t)^{-1}$.

Proof of Theorem A6.1.1 Put

$$h(x) := \ln L(e^x). \tag{A6.1.5}$$

Then property (A6.1.1) of s.v.f.s is equivalent, for each $u \in \mathbb{R}$, to the condition that the convergence

$$h(x + u) - h(x) \to 0 \tag{A6.1.6}$$

takes place as $x \to \infty$. To prove the theorem, we need to show that this convergence is uniform in $u \in [u_1, u_2]$ for any fixed $u_i \in \mathbb{R}$. In order to do that, it suffices to verify that convergence (A6.1.6) is uniform on the segment [0, 1]. Indeed, from the obvious inequality

$$\big|h(x + u_1 + u_2) - h(x)\big| \leq \big|h(x + u_1 + u_2) - h(x + u_1)\big| + \big|h(x + u_1) - h(x)\big| \tag{A6.1.7}$$

we have

$$\big|h(x + u) - h(x)\big| \leq (u_2 - u_1 + 1) \sup_{y \in [0,1]} \big|h(x + y) - h(x)\big|, \quad u \in [u_1, u_2].$$

For a given $\varepsilon \in (0,1)$ and an $x > 0$, set $I_x := [x, x+2]$,

$$I_x^* := \{u \in I_x : |h(u) - h(x)| \ge \varepsilon/2\}, \quad I_{0,x}^* := \{u \in I_0 : |h(x+u) - h(x)| \ge \varepsilon/2\}.$$

Clearly, the sets I_x^* and $I_{0,x}^*$ are measurable and differ from each other by a translation by x, so that $\mu(I_x^*) = \mu(I_{0,x}^*)$, where μ is the Lebesgue measure. By (A6.1.6) the indicator function of the set $I_{0,x}^*$ converges, at each point $u \in I_0$, to 0 as $x \to \infty$. Therefore, by the dominated convergence theorem, the integral of this function, being equal to $\mu(I_{0,x}^*)$, converges to 0, so that $\mu(I_x^*) < \varepsilon/2$ for $x \ge x_0$, where x_0 is large enough.

Further, for $s \in [0,1]$, the segment $I_x \cap I_{x+s} = [x+s, x+2]$ has length $2-s \ge 1$, so that, for $x \ge x_0$, the set

$$(I_x \cap I_{x+s}) \setminus (I_x^* \cup I_{x+s}^*)$$

has measure $\ge 1 - \varepsilon > 0$ and hence is non-empty. Let y be a point from this set. Then

$$|h(x+s) - h(x)| \le |h(x+s) - h(y)| + |h(y) - h(x)| < \varepsilon/2 + \varepsilon/2 = \varepsilon$$

for $x \ge x_0$, which proves the required uniformity on $[0,1]$ and hence on any fixed segment. The theorem is proved. □

Proof of Theorem A6.1.2 The fact that the right-hand side of (A6.1.4) is an s.v.f. is almost obvious: for any fixed $v \ne 1$,

$$\frac{L(vt)}{L(t)} = \frac{c(vt)}{c(t)} \exp\left\{\int_t^{vt} \frac{\varepsilon(u)}{u} du\right\}, \tag{A6.1.8}$$

where $c(vt)/c(t) \to c/c = 1$ and, as $t \to \infty$,

$$\int_t^{vt} \frac{\varepsilon(u)}{u} du = o\left(\int_t^{vt} \frac{du}{u}\right) = o(\ln v) = o(1). \tag{A6.1.9}$$

We now prove that any s.v.f. admits the representation (A6.1.4). The required representation in terms of the function (A6.1.5) is equivalent (after substituting $t = e^x$) to the relation

$$h(x) = d(x) + \int_{x_0}^{x} \delta(y)\, dy, \tag{A6.1.10}$$

where $d(x) = \ln c(e^x) \to d \in \mathbb{R}$ and $\delta(x) = \varepsilon(e^x) \to 0$ as $x \to \infty$, $x_0 = \ln t_0$. Therefore it suffices to establish representation (A6.1.10) for the function $h(x)$.

First of all note that $h(x)$ (as well as $L(t)$) is a "locally bounded" function. Indeed, Theorem A6.1.1 implies that, for x_0 large enough and all $x \ge x_0$,

$$\sup_{0 \le y \le 1} |h(x+y) - h(x)| < 1.$$

Hence, for any $x > x_0$, we have by virtue of (A6.1.7) the bound

$$\left|h(x) - h(x_0)\right| \le x - x_0 + 1.$$

Further, the local boundedness and measurability of the function h mean that it is locally integrable on $[x_0, \infty)$ and hence can be represented for $x \ge x_0$ as

$$h(x) = \int_{x_0}^{x_0+1} h(y)\,dy + \int_0^1 \big(h(x) - h(x+y)\big)\,dy + \int_{x_0}^{x} \big(h(y+1) - h(y)\big)\,dy. \tag{A6.1.11}$$

The first integral in (A6.1.11) is a constant, which will be denoted by d. The second integral, by virtue of Theorem A6.1.1, converges to zero as $x \to \infty$, so that

$$d(x) := d + \int_0^1 \big(h(x) - h(x+y)\big)\,dy \to d, \quad x \to \infty.$$

As for the third integral in (A6.1.11), by the definition of an s.v.f., the integrand satisfies

$$\delta(y) := h(y+1) - h(y) \to 0$$

as $y \to \infty$, which completes the proof of representation (A6.1.10). □

6.2 The Basic Asymptotic Properties

In this section we will obtain a number of consequences of Theorems A6.1.1 and A6.1.2 that are related to the asymptotic behaviour of s.v.f.s and r.v.f.s.

Theorem A6.2.1 (i) *If L_1 and L_2 are s.v.f.s then $L_1 + L_2$, $L_1 L_2$, L_1^b and $L(t) := L_1(at + b)$, where $a \ge 0$ and $b \in \mathbb{R}$, are also s.v.f.s*

(ii) *If L is an s.v.f. then, for any $\delta > 0$, there exists a $t_\delta > 0$ such that*

$$t^{-\delta} \le L(t) \le t^{\delta} \quad \textit{for all } t \ge t_\delta. \tag{A6.2.1}$$

In other words, $L(t) = t^{o(1)}$ as $t \to \infty$.

(iii) *If L is an s.v.f. then, for any $\delta > 0$ and $v_0 > 1$, there exists a $t_\delta > 0$ such that, for all $v \ge v_0$ and $t \ge t_\delta$,*

$$v^{-\delta} \le \frac{L(vt)}{L(t)} \le v^{\delta}. \tag{A6.2.2}$$

(iv) (*Karamata's theorem*) *If an r.v.f. V in* (A6.1.2) *has exponent $-\beta$, $\beta > 1$, then*

$$V^I(t) := \int_t^\infty V(u)\,du \sim \frac{tV(t)}{\beta - 1} \quad \textit{as } t \to \infty. \tag{A6.2.3}$$

If $\beta < 1$ then

$$V_I(t) := \int_0^t V(u)\,du \sim \frac{tV(t)}{1-\beta} \quad \text{as } t \to \infty. \tag{A6.2.4}$$

If $\beta = 1$ then

$$V_I(t) = tV(t)L_1(t) \tag{A6.2.5}$$

and

$$V^I(t) = tV(t)L_2(t) \quad \text{if } \int_0^\infty V(u)\,du < \infty, \tag{A6.2.6}$$

where $L_i(t) \to \infty$ as $t \to \infty$, $i = 1, 2$, are s.v.f.s.

(v) *For an r.v.f. V with exponent $-\beta < 0$, put*

$$b(t) := V^{(-1)}(1/t) = \inf\{u : V(u) < 1/t\}.$$

Then $b(t)$ is an r.v.f. with exponent $1/\beta$:

$$b(t) = t^{1/\beta} L_b(t), \tag{A6.2.7}$$

where L_b is an s.v.f. If the function L possesses the property

$$L\big(tL^{1/\beta}(t)\big) \sim L(t) \tag{A6.2.8}$$

as $t \to \infty$ then

$$L_b(t) \sim L^{1/\beta}\big(t^{1/\beta}\big). \tag{A6.2.9}$$

Similar assertions hold for functions slowly/regularly varying as $t \downarrow 0$.

Note that Theorem A6.1.1 and inequality (A6.2.2) imply the following property of s.v.f.s: *for any $\delta > 0$ there exists a $t_\delta > 0$ such that, for all t and v satisfying the inequalities $t \ge t_\delta$ and $vt \ge t_\delta$, we have*

$$(1-\delta)\min\{v^\delta, v^{-\delta}\} \le \frac{L(vt)}{L(t)} \le (1+\delta)\max\{v^\delta, v^{-\delta}\}. \tag{A6.2.10}$$

Proof of Theorem A6.2.1 Assertion (i) is evident (just note that, in order to prove the last part of (i), one needs Theorem A6.1.1).

(ii) This property follows immediately from representation (A6.1.4) and the bound

$$\left|\int_{t_0}^t \frac{\varepsilon(u)}{u}\,du\right| = \left|\int_{t_0}^{\ln t} + \int_{\ln t}^t\right| = O\left(\int_{t_0}^{\ln t} \frac{du}{u}\right) + o\left(\int_{\ln t}^t \frac{du}{u}\right) = o(\ln t)$$

as $t \to \infty$.

(iii) In order to prove this property, notice that on the right-hand side of (A6.1.8), for any fixed $\delta > 0$ and $v_0 > 1$ and all t large enough, we have

$$v^{-\delta/2} \le v_0^{-\delta/2} \le \frac{c(vt)}{c(t)} \le v_0^{\delta/2} \le v^{\delta/2}, \quad v \ge v_0,$$

and

$$\left|\int_t^{vt} \frac{\varepsilon(u)}{u} du\right| \le \frac{\delta}{2} \ln v$$

(by virtue of (A6.1.9)). This implies (A6.2.2).

(iv) By the dominated convergence theorem, we can choose an $M = M(t) \to \infty$ as $t \to \infty$ such that the convergence in (A6.1.1) will be uniform in $v \in [1, M]$. Changing the variable $u = vt$, we obtain

$$V^I(t) = t^{-\beta+1} L(t) \int_1^\infty v^{-\beta} \frac{L(vt)}{L(t)} dv = t^{-\beta+1} L(t) \left[\int_1^M + \int_M^\infty\right]. \tag{A6.2.11}$$

If $\beta > 1$ then, as $t \to \infty$,

$$\int_1^M \sim \int_1^M v^{-\beta} dv \to \frac{1}{\beta - 1},$$

whereas by property (iii), for $\delta = (\beta - 1)/2$, we have

$$\int_M^\infty < \int_M^\infty v^{-\beta+\delta} dv = \int_M^\infty v^{-(\beta+1)/2} dv \to 0.$$

These relations together imply

$$V^I(t) \sim \frac{t^{-\beta+1}}{\beta - 1} L(t) = \frac{tV(t)}{\beta - 1}.$$

The case $\beta < 1$ can be treated quite similarly, but taking into account the uniform in $v \in [1/M, 1]$ convergence in (A6.1.1) and the equality

$$\int_0^1 v^{-\beta} dv = \frac{1}{1 - \beta}.$$

If $\beta = 1$ then the first integral on the right-hand side of (A6.2.11) is

$$\int_1^M \sim \int_1^M v^{-1} dv = \ln M,$$

so that if

$$\int_0^\infty V(u)\, du < \infty \tag{A6.2.12}$$

then

$$V^I(t) \ge \big(1 + o(1)\big) L(t) \ln M \gg L(t) \tag{A6.2.13}$$

and hence

$$L_2(t) := \frac{V^I(t)}{tV(t)} = \frac{V^I(t)}{L(t)} \to \infty \quad \text{as } t \to \infty.$$

Note now that, by property (i), the function L_2 will be an s.v.f. whenever $V^I(t)$ is an s.v.f. But, for $v > 1$,

$$V^I(t) = V^I(vt) + \int_t^{vt} V(u)\,du,$$

where the last integral clearly does not exceed $(v-1)L(t)(1+o(1))$. By (A6.2.13) this implies that $V^I(vt)/V^I(t) \to 1$ as $t \to \infty$, which completes the proof of (A6.2.6).

That relation (A6.2.5) is true in the subcase when (A6.2.12) holds is almost obvious, since

$$V_I(t) = tV(t)L_1(t) = L(t)L_1(t) = \int_0^t V(u)\,du \to \int_0^\infty V(u)\,du,$$

so that, firstly, L_1 is an s.v.f. by property (i) and, secondly, $L_1(t) \to \infty$ because $L(t) \to 0$ by (A6.2.13).

Now let $\beta = 1$ and $\int_0^\infty V(u)\,du = \infty$. Then, as $M = M(t) \to \infty$ slowly enough, similarly to (A6.2.11) and (A6.2.13), by the uniform convergence theorem we have

$$V_I(t) = \int_0^1 v^{-1}L(vt)\,dv \ge \int_{1/M}^1 v^{-1}L(vt)\,dv \sim L(t)\ln M \gg L(t).$$

Therefore $L_1(t) := V_I(t)/L(t) \to \infty$ as $t \to \infty$. Further, also similarly to the above, we have, as $v \in (0,1)$,

$$V_I(t) = V_I(vt) + \int_{vt}^t V(u)\,du,$$

where the last integral does not exceed $(1-v)L(t)(1+o(1)) \ll V_I(t)$, so that $V_I(t)$ (as well as $L_1(t)$ by virtue of property (i)) is an s.v.f. This completes the proof of property (iv).

(v) Clearly, by the uniform convergence theorem the quantity $b = b(t)$ is a solution to the "asymptotic equation"

$$V(b) \sim \frac{1}{t} \quad \text{as } t \to \infty \tag{A6.2.14}$$

(where the symbol $\sim$ can be replaced by the equality sign if the function V is continuous and monotonically decreasing). Substituting $t^{1/\beta}L_b(t)$ for b, we obtain an equivalent relation

$$L_b^{-\beta}L\big(t^{1/\beta}L_b\big) \sim 1, \tag{A6.2.15}$$

where clearly

$$t^{1/\beta}L_b \to \infty \quad \text{as } t \to \infty. \tag{A6.2.16}$$

Fix an arbitrary $v > 0$. Substituting vt for t in (A6.2.15) and setting, for brevity's sake, $L_2 = L_2(t) := L_b(vt)$, we get the relation

$$L_2^{-\beta} L\big(t^{1/\beta} L_2\big) \sim 1, \qquad \text{(A6.2.17)}$$

since $L(v^{1/\beta} t^{1/\beta} L_2) \sim L(t^{1/\beta} L_2)$ by virtue of (A6.2.16) (with L_b replaced with L_2). Now we will show by contradiction that (A6.2.15)–(A6.2.17) imply that $L_b \sim L_2$ as $t \to \infty$, which obviously means that L_b is an s.v.f.

Indeed, the contrary assumption means that there exist $v_0 > 1$ and a sequence $t_n \to \infty$ such that

$$u_n := L_2(t_n)/L_b(t_n) > v_0, \quad n = 1, 2, \ldots \qquad \text{(A6.2.18)}$$

(the possible alternative case can be dealt with in the same way). Clearly, $t_n^* := t_n^{1/\beta} L_b(t_n) \to \infty$ by (A6.2.16), so we obtain from (A6.2.15)–(A6.2.16) and property (iii) with $\delta = \beta/2$ that

$$1 \sim \frac{L_2^{-\beta}(t_n) L(t_n^{1/\beta} L_2(t_n))}{L_b^{-\beta}(t_n) L(t_n^{1/\beta} L_b(t_n))} = u_n^{-\beta} \frac{L(u_n t_n^*)}{L(t_n^*)} \le u_n^{-\beta/2} < v_0^{-\beta/2} < 1.$$

We get a contradiction.

Note that the above argument proves the uniqueness (up to asymptotic equivalence) of the solution to Eq. (A6.2.14).

Finally, relation (A6.2.9) can be proved by a direct verification of (A6.2.14) for $b := t^{1/\beta} L^{1/\beta}(t^{1/\beta})$: using (A6.2.8), we have

$$V(b) = b^{-\beta} L(b) = \frac{L(t^{1/\beta} L^{1/\beta}(t^{1/\beta}))}{t L(t^{1/\beta})} \sim \frac{L(t^{1/\beta})}{t L(t^{1/\beta})} = \frac{1}{t}.$$

The required assertion follows now by the aforementioned uniqueness of the solution to the asymptotic equation (A6.2.14). Theorem A6.2.1 is proved. □

6.3 The Asymptotic Properties of the Transforms of R.V.F.s (Abel-Type Theorems)

For an r.v.f. $V(t)$, its Laplace transform

$$\psi(\lambda) := \int_0^\infty e^{-\lambda t} V(t)\, dt < \infty$$

is defined for all $\lambda > 0$. The following asymptotic relations hold true for the transform.

Theorem A6.3.1 *Assume that* $V(t) \in \mathfrak{R}$ *(i.e.* $V(t)$ *has the form* (A6.1.2)).

(i) *If* $\beta \in [0, 1)$ *then*

$$\psi(\lambda) \sim \frac{\Gamma(1-\beta)}{\lambda} V(1/\lambda) \quad \text{as } \lambda \downarrow 0. \tag{A6.3.1}$$

(ii) *If* $\beta = 1$ *and* $\int_0^\infty V(t)\,dt = \infty$ *then*

$$\psi(\lambda) \sim V_I(1/\lambda) \quad \text{as } \lambda \downarrow 0, \tag{A6.3.2}$$

where $V_I(t) = \int_0^t V(u)\,du \to \infty$ *is an s.v.f. such that* $V_I(t) \gg L(t)$ *as* $t \to \infty$.

(iii) *In any case,* $\psi(\lambda) \uparrow V_I(\infty) = \int_0^\infty V(t)\,dt \le \infty$ *as* $\lambda \downarrow 0$.

Assertions (i) and (ii) are called *Abelian* theorems.

If we resolve relation (A6.3.1) for V then we obtain

$$V(t) \sim \frac{\psi(1/t)}{t\Gamma(1-\beta)} \quad \text{as } t \to \infty.$$

Relations of this type will also be valid in the case when, instead of the regularity of the function V, one requires the monotonicity of V and assumes that $\psi(\lambda)$ is an r.v.f. as $\lambda \downarrow 0$. Statements of such type are called *Tauberian* theorems. We will not need these theorems and so will not dwell on them.

Proof of Theorem A6.3.1 (i) For any fixed $\varepsilon > 0$ we have

$$\psi(\lambda) = \int_0^{\varepsilon/\lambda} + \int_{\varepsilon/\lambda}^\infty,$$

where, for the first integral on the right-hand side, for $\beta < 1$, by virtue of (A6.2.4) we have the following relation

$$\int_0^{\varepsilon/\lambda} e^{-\lambda t} V(t)\,dt \le \int_0^{\varepsilon/\lambda} V(t)\,dt \sim \frac{\varepsilon V(\varepsilon/\lambda)}{\lambda(1-\beta)} \quad \text{as } \lambda \downarrow 0. \tag{A6.3.3}$$

Changing the variable $\lambda t = u$, we can rewrite the second integral in the above representation for $\psi(\lambda)$ as

$$\int_{\varepsilon/\lambda}^\infty = \frac{V(1/\lambda)}{\lambda} \int_\varepsilon^\infty e^{-u} u^{-\beta} \frac{L(u/\lambda)}{L(1/\lambda)}\,du = \frac{V(1/\lambda)}{\lambda}\left[\int_\varepsilon^2 + \int_2^\infty\right]. \tag{A6.3.4}$$

Each of the integrals on the right-hand side converges, as $\lambda \downarrow 0$, to the corresponding integral of $e^{-u}u^{-\beta}$: the former integral converges by the uniform convergence theorem (the convergence $L(u/\lambda)/L(1/\lambda) \to 1$ is uniform in $u \in [\varepsilon, 2]$), and the latter converges by virtue of (A6.1.1) and the dominated convergence theorem, since by Theorem A6.2.1(iii), for all λ small enough, we have $L(u/\lambda)/L(1/\lambda) < u$ for $u \ge 2$. Therefore,

$$\int_{\varepsilon/\lambda}^\infty \sim \frac{V(1/\lambda)}{\lambda} \int_\varepsilon^\infty u^{-\beta} e^{-u}\,du. \tag{A6.3.5}$$

Now note that, as $\lambda \downarrow 0$,

$$\frac{\varepsilon V(\varepsilon/\lambda)}{\lambda} \Big/ \frac{V(1/\lambda)}{\lambda} = \varepsilon^{1-\beta}\frac{L(\varepsilon/\lambda)}{L(1/\lambda)} \to \varepsilon^{1-\beta}.$$

Since $\varepsilon > 0$ can be chosen arbitrarily small, this relation together with (A6.3.3) and (A6.3.5) completes the proof of (A6.3.1).

(ii) Integrating by parts and changing the variable $\lambda t = u$, we obtain, for $\beta = 1$ and $M > 0$, that

$$\begin{aligned}\psi(\lambda) &= \int_0^\infty e^{-\lambda t}\,dV_I(t) = -\int_0^\infty V_I(t)\,de^{-\lambda t}\\ &= \int_0^\infty V_I(u/\lambda)e^{-u}\,du = \int_0^{1/M} + \int_{1/M}^M + \int_M^\infty. \end{aligned} \tag{A6.3.6}$$

By Theorem A6.2.1(iv), $V_I(t) \gg L(t)$ is an s.v.f. as $t \to \infty$. Therefore, by the uniform convergence theorem, for $M = M(\lambda) \to \infty$ slowly enough as $\lambda \to 0$, the middle integral on the right-hand side of (A6.3.6) is

$$V_I(1/\lambda)\int_{1/M}^M \frac{V_I(u/\lambda)}{V_I(1/\lambda)}e^{-u}\,du \sim V_I(1/\lambda)\int_{1/M}^M e^{-u}\,du \sim V_I(1/\lambda).$$

The remaining two integrals are negligibly small: since $V_I(t)$ is an increasing function, the first integral does not exceed $V_I(1/\lambda M)/M = o(V_I(1/\lambda))$, while for the last integral we have by Theorem A6.2.1(iii) that

$$V_I(1/\lambda)\int_M^\infty \frac{V_I(u/\lambda)}{V_I(1/\lambda)}e^{-u}\,du \le V_I(1/\lambda)\int_M^\infty ue^{-u}\,du = o\big(V_I(1/\lambda)\big).$$

Hence (ii) is proved. Assertion (iii) is evident. □

6.4 Subexponential Distributions and Their Properties

Let $\xi, \xi_1, \xi_2, \ldots$ be independent identically distributed random variables with distribution $\mathbf{F}$, and let the *right tail of this distribution*

$$F_+(t) := \mathbf{F}\big([t,\infty)\big) = \mathbf{P}(\xi \ge t), \quad t \in \mathbb{R},$$

be an r.v.f. as $t \to \infty$ of the form (A6.1.2), which we will denote by $V(t)$. Recall that *we denoted the class of all such distributions by* $\mathcal{R}$.

In this section we will introduce one more class of distributions, which is substantially wider than $\mathcal{R}$.

Let $\zeta \in \mathbb{R}$ be a random variable with distribution $\mathbf{G}$: $\mathbf{G}(B) = \mathbf{P}(\zeta \in B)$ for any Borel set B (recall that in this case we write $\zeta \Subset \mathbf{G}$). Denote by $G(t)$ the right tail of the distribution of the random variable ζ:

$$G(t) := \mathbf{P}(\zeta \ge t), \quad t \in \mathbb{R}.$$

The convolution of tails $G_1(t)$ and $G_2(t)$ is the function

$$G_1 * G_2(t) := -\int G(t-y)dG_2(y) = \int G_1(t-y)\mathbf{G}_2(dy) = \mathbf{P}(Z_2 \ge t),$$

where $Z_2 = \zeta_1 + \zeta_2$ is the sum of independent random variables $\zeta_i \in\!\!\!= \mathbf{G}_i$, $i = 1, 2$. Clearly, $G_1 * G_2(t) = G_2 * G_1(t)$. Denote by $G^{2*}(t) := G * G(t)$ the convolution of the tail $G(t)$ with itself and put $G^{(n+1)*}(t) := G * G^{n*}(t)$, $n \ge 2$.

Definition A6.4.1 A distribution $\mathbf{G}$ on $[0, \infty)$ belongs to the class $\mathcal{S}_+$ of *subexponential distributions on the positive half-line* if

$$G^{2*}(t) \sim 2G(t) \quad \text{as } t \to \infty. \tag{A6.4.1}$$

A distribution $\mathbf{G}$ on the whole line $(-\infty, \infty)$ belongs to the class $\mathcal{S}$ of *subexponential distributions* if the distribution $\mathbf{G}^+$ of the positive part $\zeta^+ = \max\{0, \zeta\}$ of the random variable $\zeta \in\!\!\!= \mathbf{G}$ belongs to $\mathcal{S}_+$. A random variable is called subexponential if its distribution is subexponential.

As we will see below (Theorem A6.4.3), the subexponentiality property of a distribution $\mathbf{G}$ is essentially the property of the asymptotics of the tail $G(t)$ as $t \to \infty$. Therefore we can also speak about *subexponential functions*.

A nondecreasing function $G_1(t)$ on $(0, \infty)$ is called *subexponential* if a distribution $\mathbf{G}$ with the tail $G(t) \sim cG_1(t)$ as $t \to \infty$ with some $c > 0$ is subexponential. (For example, distributions with the tails $G(t) = G_1(t)/G_1(0)$ or $G(t) = \min(1, G_1(t))$).

Remark A6.4.1 Since we obviously always have

$$\begin{aligned}(G^+)^{2*}(t) &= \mathbf{P}(\zeta_1^+ + \zeta_2^+ \ge t) \ge \mathbf{P}(\{\zeta_1^+ \ge t\} \cup \{\zeta_2^+ \ge t\}) \\ &= \mathbf{P}(\zeta_1 \ge t) + \mathbf{P}(\zeta_2 \ge t) - \mathbf{P}(\zeta_1 \ge t, \zeta_2 \ge t) \\ &= 2G(t) - G^2(t) = 2G^+(t)\big(1 + o(1)\big)\end{aligned}$$

as $t \to \infty$, subexponentiality is equivalent to the following property:

$$\limsup_{t\to\infty} \frac{(G^+)^{2*}(t)}{G^+(t)} \le 2. \tag{A6.4.2}$$

Note also that, since relation (A6.4.1) makes sense only when $G(t) > 0$ for all $t \in \mathbb{R}$, the support of any subexponential distribution is unbounded from the right.

We show that regularly varying distributions are subexponential, i.e., that $\mathcal{R} \subset \mathcal{S}$. Let $\mathbf{F} \in \mathcal{R}$ and $\mathbf{P}(\xi \ge t) = V(t)$ be r.v.f.s. We need to show that

$$\begin{aligned}\mathbf{P}(\xi_1 + \xi_2 \ge x) &= V^{2*}(x) := V * V(x) \\ &= -\int_{-\infty}^{\infty} V(x-t)\,dV(t) \sim 2V(x).\end{aligned} \tag{A6.4.3}$$

In order to do that, we introduce events $A := \{\xi_1 + \xi_2 \geq x\}$ and $B_i := \{\xi_i < x/2\}$, $i = 1, 2$. Clearly,

$$\mathbf{P}(A) = \mathbf{P}(AB_1) + \mathbf{P}(AB_2) - \mathbf{P}(AB_1 B_2) + \mathbf{P}(A\overline{B}_1\overline{B}_2),$$

where $\mathbf{P}(AB_1B_2) = 0$, $\mathbf{P}(A\overline{B}_1\overline{B}_2) = \mathbf{P}(\overline{B}_1\overline{B}_2) = V^2(x/2)$ (here and in what follows, $\overline{B}$ denotes the event complementary to B) and

$$\mathbf{P}(AB_1) = \mathbf{P}(AB_2) = \int_{-\infty}^{x/2} V(x-t)\,\mathbf{F}(dt).$$

Therefore

$$V^{2*}(x) = 2\int_{-\infty}^{x/2} V(x-t)\,\mathbf{F}(dt) + V^2(x/2). \tag{A6.4.4}$$

(The same result can be obtained by integrating the convolution in (A6.4.3) by parts.) It remains to note that $V^2(x/2) = o(V(x))$ and

$$\int_{-\infty}^{x/2} V(x-t)\,\mathbf{F}(dt) = \int_{-\infty}^{-M} + \int_{-M}^{M} + \int_{M}^{x/2}, \tag{A6.4.5}$$

where, as one can easily see, for any $M = M(x) \to \infty$ as $x \to \infty$ such that $M = o(x)$, we have

$$\int_{-M}^{M} \sim V(x) \quad \text{and} \quad \int_{-\infty}^{-M} + \int_{M}^{x/2} = o\big(V(x)\big),$$

which proves (A6.4.3).

The same argument is valid for distributions with a right tail of the form

$$F_+(t) = e^{-t^\beta L(t)}, \quad \beta \in (0,1), \tag{A6.4.6}$$

where $L(t)$ is an s.v.f. as $t \to \infty$ satisfying a certain smoothness condition (for instance, that L is differentiable with $L'(t) = o(L(t)/t)$ as $t \to \infty$).

One of the basic properties of subexponential distributions $\mathbf{G}$ is that their tails $G(t)$ are asymptotically locally constant in the following sense.

Definition A6.4.2 We will call a function $G(t) > 0$ (asymptotically) *locally constant* (l.c.) if, for any fixed v,

$$\frac{G(t+v)}{G(t)} \to 1 \quad \text{as } t \to \infty. \tag{A6.4.7}$$

In the literature, distributions with l.c. tails are often referred to as long-tailed distributions; however, we feel that the term "locally constant function" better reflects the meaning of the concept. Denote the class of all distributions $\mathbf{G}$ with l.c. tails $G(t)$ by $\mathcal{L}$.

For future reference, we will state the basic properties of l.c. functions as a separate theorem.

Theorem A6.4.1 (i) *For an l.c. function $G(t)$ the convergence in* (A6.4.7) *is uniform in v on any fixed finite interval.*

(ii) *A function $G(t)$ is l.c. if and only if, for some $t_0 > 0$, it admits a representation of the form*

$$G(t) = c(t)\exp\left\{\int_{t_0}^{t} \varepsilon(u)\,du\right\}, \quad t \ge t_0, \tag{A6.4.8}$$

where the functions $c(t)$ and $\varepsilon(t)$ are measurable and such that $c(t) \to c \in (0, \infty)$ and $\varepsilon(t) \to 0$ as $t \to \infty$.

(iii) *If $G_1(t)$ and $G_2(t)$ are l.c. functions then $G_1(t)+G_2(t)$, $G_1(t)G_2(t)$, $G_1^b(t)$, and $G(t) := G_1(at+b)$, where $a \ge 0$ and $b \in \mathbb{R}$, are also l.c.*

(iv) *If $G(t)$ is an l.c. function then, for any $\varepsilon > 0$,*

$$e^{\varepsilon t} G(t) \to \infty \quad as\ t \to \infty.$$

In other words, any l.c. function $G(t)$ can be represented as

$$G(t) = e^{-l(t)}, \quad l(t) = o(t) \quad as\ t \to \infty. \tag{A6.4.9}$$

(v) *Let*

$$G^I(t) := \int_t^\infty G(u)\,du < \infty$$

and at least one of the following conditions be satisfied:

(a) *$G(t)$ is an l.c. function; or*
(b) *$G^I(t)$ is an l.c. function and $G(t)$ is monotone.*

Then

$$G(t) = o\big(G^I(t)\big) \quad as\ t \to \infty. \tag{A6.4.10}$$

(vi) *If $\mathbf{G} \in \mathcal{L}$ then $G^{2*}(t) \sim (G^+)^{2*}(t)$ as $t \to \infty$.*

Remark A6.4.2 Assertion (i) of the theorem implies that the uniform convergence in (A6.4.7) on the interval $[-M, M]$ persists in the case when, as $t \to \infty$, $M = M(t)$ grows unboundedly slowly enough.

Proof of Theorem A6.4.1, (i)–(iii) It is clear from Definitions A6.4.1 and A6.4.2 that $G(t)$ is l.c. if and only if $L(t) := G(\ln t)$ is an s.v.f. Having made this observation, assertion (i) follows directly from Theorem A6.1.1 (on uniform convergence of s.v.f.s), while assertions (ii) and (iii) follow from Theorems A6.1.2 and A6.2.1(i), respectively.

Assertion (iv) follows from the integral representation (A6.4.8).

(v) If (a) holds then, for any $M > 0$ and all t large enough,

$$G^I(t) > \int_t^{t+M} G(u)\,du > \frac{1}{2} M G(t).$$

Since M is arbitrary, $G^I(t) \gg G(t)$. Further, if (b) holds then

$$\frac{G(t)}{G^I(t)} \le \frac{1}{G^I(t)} \int_{t-1}^{t} G(u)\,du = \frac{G^I(t-1)}{G^I(t)} - 1 \to 0$$

as $t \to \infty$.

(vi) Let ζ_1 and ζ_2 be independent copies of a random variable ζ, $Z_2 := \zeta_1 + \zeta_2$, $Z_2^{(+)} := \zeta_1^+ + \zeta_2^+$. Clearly, $\zeta_i \le \zeta_i^+$, so that

$$G^{2*}(t) = \mathbf{P}(Z_2 \ge t) \le \mathbf{P}\big(Z_2^{(+)} \ge t\big) = \big(G^+\big)^{2*}(t). \tag{A6.4.11}$$

On the other hand, for any $M > 0$,

$$G^{2*}(t) \ge \mathbf{P}(Z_2 \ge t, \zeta_1 > 0, \zeta_2 > 0) + \sum_{i=1}^{2} \mathbf{P}\big(Z_2 \ge t, \zeta_i \in [-M, 0]\big),$$

where the first term on the right-hand side is equal to $\mathbf{P}(Z_2^{(+)} \ge t,\ \zeta_1^+ > 0,\ \zeta_2^+ > 0)$, and the last two terms can be bounded as follows: since $\mathbf{G} \in \mathcal{L}$, then, for any $\varepsilon > 0$ and M and t large enough,

$$\begin{aligned}
\mathbf{P}\big(Z_2 \ge t, \zeta_1 \in [-M, 0]\big) &\ge \mathbf{P}\big(\zeta_2 \ge t + M,\ \zeta_1 \in [-M, 0]\big) \\
&= G(t)\frac{G(t+M)}{G(t)}\big[\mathbf{P}(\zeta_1 \le 0) - \mathbf{P}(\zeta_1 < -M)\big] \\
&\ge (1-\varepsilon)G(t)\mathbf{P}\big(\zeta_1^+ = 0\big) = (1-\varepsilon)\mathbf{P}\big(Z_2^{(+)} \ge t,\ \zeta_1^+ = 0\big).
\end{aligned}$$

Thus we obtain for $G^{2*}(t)$ the lower bound

$$\begin{aligned}
G^{2*}(t) &\ge \mathbf{P}\big(Z_2^{(+)} \ge t, \zeta_1^+ > 0, \zeta_2^+ > 0\big) + (1-\varepsilon)\sum_{i=1}^{2} \mathbf{P}\big(Z_2^{(+)} \ge t, \zeta_i^+ = 0\big) \\
&\ge (1-\varepsilon)\mathbf{P}\big(Z_2^{(+)} \ge t\big) = (1-\varepsilon)\big(G^*\big)^{2*}(t).
\end{aligned}$$

Therefore (vi) is proved, as ε can be arbitrarily small. The theorem is proved. □

We return now to our discussion of subexponential distributions. First of all, we turn to the relationship between the classes $\mathcal{S}$ and $\mathcal{L}$.

Theorem A6.4.2 *We have $\mathcal{S} \subset \mathcal{L}$, and hence all the assertions of Theorem* A6.4.1 *are valid for subexponential distributions as well.*

Remark A6.4.3 The coinage of the term "subexponential distribution" was apparently due mostly to the fact that the tail of such a distribution decreases as $t \to \infty$ slower than any exponential function $e^{-\varepsilon t}$, as shown in Theorems A6.4.1(iv) and A6.4.2.

Remark A6.4.4 In the case when the distribution $\mathbf{G}$ is *not concentrated* on $[0, \infty)$, the tails' additivity condition (A6.4.1) alone is not sufficient for the function $G(t)$ to be l.c. (and hence for ensuring the "subexponential decay" of the distribution tail, cf. Remark A6.4.3). This explains the necessity of defining subexponentiality in the general case in terms of condition (A6.4.1) on the distribution $\mathbf{G}^+$ of the random variable ζ^+. Actually, as we will see below (Corollary A6.4.1), the subexponentiality of a distribution $\mathbf{G}$ on $\mathbb{R}$ *is equivalent* to the combination of conditions (A6.4.1) (on $\mathbf{G}$ itself) and $\mathbf{G} \in \mathcal{L}$.

The next example shows that, for random variables *assuming values of both signs*, condition (A6.4.1), generally speaking, does not imply the subexponential behaviour of $G(t)$.

Example A6.4.1 Let $\mu > 0$ be fixed and the right tail of the distribution $\mathbf{G}$ have the form

$$G(t) = e^{-\mu t} V(t), \tag{A6.4.12}$$

where $V(t)$ is an r.v.f. vanishing as $t \to \infty$ and such that

$$g(\mu) := \int_{-\infty}^{\infty} e^{\mu y} \mathbf{G}(dy) < \infty.$$

Similarly to (A6.4.4) and (A6.4.5), we have

$$G^{2*}(t) = 2 \int_{-\infty}^{t/2} G(t-y)\,\mathbf{G}(dy) + G^2(t/2),$$

where

$$\begin{aligned}\int_{-\infty}^{t/2} G(t-y)\,\mathbf{G}(dy) &= e^{-\mu t} \int_{-\infty}^{t/2} e^{\mu y} V(t-y)\,\mathbf{G}(dy) \\ &= e^{-\mu t}\left[\int_{-\infty}^{-M} + \int_{-M}^{M} + \int_{M}^{t/2}\right].\end{aligned}$$

One can easily see that, for $M = M(t) \to \infty$ slowly enough as $t \to \infty$, we have

$$\int_{-M}^{M} e^{\mu y} V(t-y)\,\mathbf{G}(dy) \sim g(\mu) V(t), \qquad \int_{-\infty}^{-M} + \int_{M}^{t/2} = o\big(G(t)\big),$$

while

$$G^2(t/2) = e^{-\mu t} V^2(t/2) \le c e^{-\mu t} V^2(t) = o\big(G(t)\big).$$

Thus, we obtain

$$G^{2*}(t) \sim 2g(\mu) e^{-\mu t} V(t) = 2g(\mu) G(t), \tag{A6.4.13}$$

and it is clear that we can always find a distribution $\mathbf{G}$ (with a negative mean) such that $g(\mu) = 1$. In that case relation (A6.4.1) from the definition of subexponentiality will be satisfied, although $G(t)$ decreases exponentially fast and hence is not an l.c. function.

On the other hand, note that the class of distributions satisfying relation (A6.4.1) only is an extension of the class $\mathcal{S}$. Distributions in the former class possess many of the properties of distributions from $\mathcal{S}$.

Proof of Theorem A6.4.2 We have to prove that $\mathcal{S} \subset \mathcal{L}$. Since the definitions of both classes are given in terms of the right distribution tails, we can assume without loss of generality, that $\mathbf{G} \in \mathcal{S}_+$ (or just consider the distribution $\mathbf{G}^+$). For independent (nonnegative) $\zeta_i \in\!\!\!= \mathbf{G}$ we have, for $t > 0$,

$$G^{2*}(t) = \mathbf{P}(\zeta_1 + \zeta_2 \ge t) = \mathbf{P}(\zeta_1 \ge t) + \mathbf{P}(\zeta_1 + \zeta_2 \ge t, \zeta_1 < t)$$

$$= G(t) + \int_0^t G(t-y)\,\mathbf{G}(dy). \tag{A6.4.14}$$

Since $G(t)$ is non-increasing and $G(0) = 1$, it follows that, for $t > v > 0$,

$$\frac{G^{2*}(t)}{G(t)} = 1 + \int_0^v \frac{G(t-y)}{G(t)}\,\mathbf{G}(dy) + \int_v^t \frac{G(t-y)}{G(t)}\,\mathbf{G}(dy)$$

$$\ge 1 + \bigl[1 - G(v)\bigr] + \frac{G(t-v)}{G(t)}\bigl[G(v) - G(t)\bigr].$$

Therefore, for t large enough (such that $G(v) - G(t) > 0$),

$$1 \le \frac{G(t-v)}{G(t)} \le \frac{1}{G(v) - G(t)}\left[\frac{G^{2*}(t)}{G(t)} - 2 + G(v)\right].$$

Since $\mathbf{G} \in \mathcal{S}_+$, the right-hand side of the last formula converges as $t \to \infty$ to the quantity $G(v)/G(v) = 1$ and hence $\mathbf{G} \in \mathcal{L}$. The theorem is proved. □

The next theorem contains several important properties of subexponential distributions.

Theorem A6.4.3 *Let* $\mathbf{G} \in \mathcal{S}$.

(i) *If* $G_i(t)/G(t) \to c_i$ *as* $t \to \infty$, $c_i \ge 0$, $i = 1, 2$, $c_1 + c_2 > 0$, *then*

$$G_1 * G_2(t) \sim G_1(t) + G_2(t) \sim (c_1 + c_2)G(t).$$

(ii) *If* $G_0(t) \sim cG(t)$ *as* $t \to \infty$, $c > 0$, *then* $\mathbf{G}_0 \in \mathcal{S}$.

(iii) *For any fixed* $n \ge 2$,

$$G^{n*}(t) \sim nG(t) \quad \text{as } t \to \infty. \tag{A6.4.15}$$

(iv) *For any* $\varepsilon > 0$ *there exists a* $b = b(\varepsilon) < \infty$ *such that*

$$\frac{G^{n*}(t)}{G(t)} \le b(1+\varepsilon)^n$$

for all $n \ge 2$ *and* t.

In addition to assertions (i) and (ii) of the theorem, we can also show that if $\mathbf{G} \in \mathcal{S}$ and the function $m(t) \in \mathcal{L}$ possesses the property

$$0 < m_1 \le m(t) \le m_2 < \infty$$

then $G_1(t) = m(t)G(t) \in \mathcal{S}$.

Theorems A6.4.1(vi), A6.4.2 and A6.4.3(iii) imply the following simple statement elucidating the subexponentiality condition for random variables taking values of both signs.

Corollary A6.4.1 *A distribution* $\mathbf{G}$ *belongs to* $\mathcal{S}$ *if and only if* $\mathbf{G} \in \mathcal{L}$ *and* $G^{2*}(t) \sim 2G(t)$ *as* $t \to \infty$.

Remark A6.4.5 Evidently the asymptotic relation $G_1(t) \sim G_2(t)$ as $t \to \infty$ is an *equivalence relation* on the set of distributions on $\mathbb{R}$. Theorem A6.4.3(ii) means that the class $\mathcal{S}$ *is closed with respect to that equivalence.* One can easily see that in each of the equivalence subclasses of the class $\mathcal{S}$ with respect to this relation *there is always a distribution with an arbitrarily smooth tail* $G(t)$.

Indeed, let $p(t)$ be an infinitely differentiable probability density on $\mathbb{R}$ vanishing outside $[0, 1]$ (we can take, e.g., $p(x) = c \cdot e^{-1/(x(1-x))}$ if $x \in (0, 1)$ and $p(x) = 0$ if $x \notin (0, 1)$). Now we "smooth" the function $l(t) := -\ln G(t)$, $\mathbf{G} \in \mathcal{S}$, putting

$$l_0(t) := \int p(t-u)l(u)\,du, \quad \text{and let } G_0(t) := e^{-l_0(t)}. \tag{A6.4.16}$$

Clearly, $G_0(t)$ is an infinitely differentiable function and, since $l(t)$ is nondecreasing and we actually integrate over $[t-1, t]$ only, one has $l(t-1) \le l_0(t) \le l(t)$ and hence by Theorem A6.4.2

$$1 \le \frac{G_0(t)}{G(t)} \le \frac{G(t-1)}{G(t)} \to 1 \quad \text{as } t \to \infty.$$

Thus, the distribution $\mathbf{G}_0$ is equivalent to the original $\mathbf{G}$. A simpler smoothing procedure leading to a less smooth asymptotically equivalent tail consists of replacing the function $l(t)$ with its linear interpolation with nodes at points $(k, l(k))$, k being an integer.

Therefore, *up to a summand* $o(1)$, we can always assume the function $l(t) = -\ln G(t)$, $\mathbf{G} \in \mathcal{S}$, to be arbitrarily smooth.

The aforesaid is clearly applicable to the class $\mathcal{L}$ as well: it is also closed with respect to the introduced equivalence, and each of its equivalence subclass contains arbitrarily smooth representatives.

Remark A6.4.6 Theorem A6.4.3(ii) and (iii) immediately implies that if $\mathbf{G} \in \mathcal{S}$ then also $\mathbf{G}^{n*} \in \mathcal{S}$, $n = 2, 3, \ldots$. Moreover, if we denote by $\mathbf{G}^{n\vee}$ the distribution of the maximum of independent identically distributed random variables $\zeta_1, \dots, \zeta_n \Subset \mathbf{G}$, then the evident relation

$$G^{n\vee}(t) = 1 - \big(1 - G(t)\big)^n \sim nG(t) \quad \text{as } t \to \infty \tag{A6.4.17}$$

and Theorem A6.4.3(ii) imply that $\mathbf{G}^{n\vee}$ also belongs to $\mathcal{S}$.

Relations (A6.4.17) and (A6.4.15) show that, in the case of a subexponential $\mathbf{G}$, the tail $G^{n*}(t)$ of the distribution of the sum of a fixed number n of independent identically distributed random variables $\zeta_i \Subset \mathbf{G}$ is asymptotically equivalent (as $t \to \infty$) to the tail $G^{n\vee}(t)$ of the maximum of these random variables, i.e., the "large" values of this sum are mainly due to by the presence of one "large" term ζ_i in the sum. It is easy to see that this property is characteristic of subexponentiality.

Remark A6.4.7 Note also that an assertion converse to what was stated at the beginning of Remark A6.4.6 is also valid: if $\mathbf{G}^{n*} \in \mathcal{S}$ for some $n \ge 2$ then $\mathbf{G} \in \mathcal{S}$ as well. That $\mathbf{G}^{n\vee} \in \mathcal{S}$ implies $\mathbf{G} \in \mathcal{S}$ evidently follows from (A6.4.17) and Theorem A6.4.3(ii).

Proof of Theorem A6.4.3 (i) First assume that $c_1 c_2 > 0$ and that both distributions $\mathbf{G}_i$ are concentrated on $[0, \infty)$. Fix an arbitrary $\varepsilon > 0$ and choose M large enough to have $G_i(M) < \varepsilon$, $i = 1, 2$, and $G(M) < \varepsilon$, and such that, for $t > M$,

$$(1-\varepsilon)c_i < \frac{G_i(t)}{G(t)} < (1+\varepsilon)c_i, \quad i = 1, 2, \qquad 1 - \varepsilon < \frac{G(t-M)}{G(t)} < 1 + \varepsilon \tag{A6.4.18}$$

(the last inequality holds by virtue of Theorem A6.4.2).

Let $\zeta \Subset \mathbf{G}$ and $\zeta_i \Subset \mathbf{G}_i$, $i = 1, 2$, be independent random variables. Then, for $t > 2M$, we have the representation

$$G_1 * G_2(t) = P_1 + P_2 + P_3 + P_4, \tag{A6.4.19}$$

where

$$\begin{aligned}
P_1 &:= \mathbf{P}\big(\zeta_1 \ge t - \zeta_2, \zeta_2 \in [0, M)\big),\\
P_2 &:= \mathbf{P}\big(\zeta_2 \ge t - \zeta_1,\ \zeta_1 \in [0, M)\big),\\
P_3 &:= \mathbf{P}\big(\zeta_2 \ge t - \zeta_1,\ \zeta_1 \in [M, t - M)\big),\\
P_4 &:= \mathbf{P}(\zeta_2 \ge M, \zeta_1 \ge t - M)
\end{aligned}$$

(see Fig. A.1).

We show that the first two terms on the right-hand side of (A6.4.19) are asymptotically equivalent to $c_1 G(t)$ and $c_2 G(t)$, respectively, while the last two terms are negligibly small compared with $G(t)$. Indeed, for P_1 we have the obvious two-sided bounds

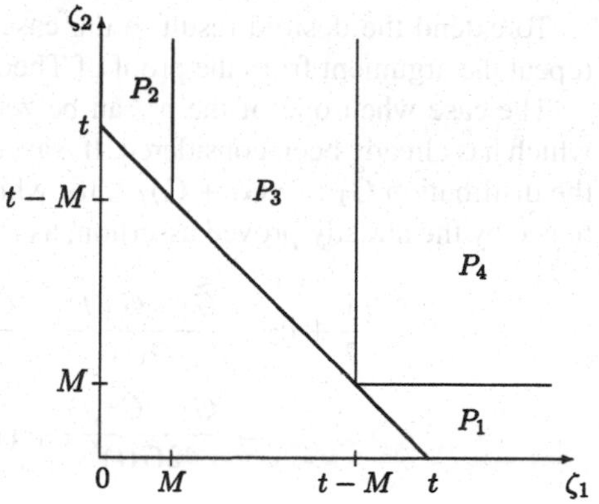

Fig. A.1 Illustration to the proof of Theorem A6.4.3, showing the regions P_i, $i = 1, 2, 3, 4$

$$\begin{aligned}(1-\varepsilon)^2 c_1 G(t) < G_1(t)\big(1 - G_2(M)\big) &= \mathbf{P}\big(\zeta_1 \ge t,\, \zeta_2 \in [0, M)\big)\\ &\le P_1 \le \mathbf{P}(\zeta_1 \ge t - M) = G_1(t-M) \le (1+\varepsilon)^2 c_1 G(t)\end{aligned}$$

by (A6.4.18); the term P_2 can be bounded in a similar way. Further,

$$P_4 = \mathbf{P}(\zeta_2 \ge M,\, \zeta_1 \ge t - M) = G_2(M) G_1(t-M) < \varepsilon(1+\varepsilon)^2 c_2 G(t).$$

It remains to estimate P_3 (note that it is here that we will need the condition $\mathbf{G} \in \mathcal{S}$; so far we have only used the fact that $\mathbf{G} \in \mathcal{L}$). We have

$$P_3 = \int_{[M,t-M)} G_2(t-y)\, \mathbf{G}_1(dy) \le (1+\varepsilon) c_2 \int_{[M,t-M)} G(t-y)\, \mathbf{G}_1(dy), \tag{A6.4.20}$$

where it is clear that, by (A6.4.18), the last integral is equal to

$$\begin{aligned}&\mathbf{P}\big(\zeta + \zeta_1 \ge t,\, \zeta_1 \in [M, t-M)\big)\\ &\quad= \mathbf{P}\big(\zeta \ge t - M,\, \zeta_1 \in [M, t-M)\big) + \mathbf{P}\big(\zeta + \zeta_1 \ge t, \zeta \in [M, t-M)\big)\\ &\quad= G(t-M)\, G_1\big([M, t-M)\big) + \int_{[M,t-M)} G_1(t-y)\, \mathbf{G}(dy)\\ &\quad\le \varepsilon(1+\varepsilon) G(t) + (1+\varepsilon) c_1 \int_{[M,t-M)} G(t-y) \mathbf{G}(dy). \end{aligned}\tag{A6.4.21}$$

Now note that similarly to the above argument we can easily obtain (setting $G_1 = G_2 = G$) that

$$G^{2*}(t) = (1+\theta_1\varepsilon) 2G(t) + \int_{[M,t-M)} G(t-y) \mathbf{G}(dy) + \varepsilon(1+\theta_2\varepsilon) G(t),$$

where $|\theta_i| \le 1$, $i = 1, 2$. Since $G^{2*}(t) \sim 2G(t)$ by virtue of $\mathbf{G} \in \mathcal{S}_+$, this equality means that the integral on the right-hand side is $o(G(t))$. Now (A6.4.21) immediately implies that also $P_3 = o(G(t))$, and hence the required assertion is established for the case $\mathbf{G} \in \mathcal{S}_+$.

To extend the desired result to the case of distributions G_i on $\mathbb{R}$, it suffices to repeat the argument from the proof of Theorem A6.4.1(vi).

The case when one of the c_i can be zero can be reduced to the case $c_1c_2 > 0$, which has already been considered. If, say, $c_1 = 0$ and $c_2 > 0$, then we can introduce the distribution $\widetilde{\mathbf{G}}_1 := (\mathbf{G}_1 + \mathbf{G})/2$, for which clearly $\widetilde{G}_1(t)/G(t) \to \widetilde{c}_1 = 1/2$, and hence by the already proved assertion, as $t \to \infty$,

$$\frac{1}{2} + c_2 \sim \frac{\widetilde{G}_1 * G_2(t)}{G(t)} = \frac{G_1 * G_2(t) + G * G_2(t)}{2G(t)}$$
$$= \frac{G_1 * G_2(t)}{2G(t)} + \big(1 + o(1)\big)\frac{1 + c_2}{2},$$

so that $G_1 * G_2(t)/G(t) \to c_2 = c_1 + c_2$.

(ii) Denote by $\mathbf{G}_0^+$ the distribution of the random variable ζ_0^+, where $\zeta_0 \Subset \mathbf{G}_0$. Since $G_0^+(t) = G_0(t)$ for $t > 0$, it follows immediately from (i) with $\mathbf{G}_1 = \mathbf{G}_2 = \mathbf{G}_0^+$ that $(G_0^+)^{2*}(t) \sim 2G_0^+(t)$, i.e. $\mathbf{G}_0 \in \mathcal{S}$.

(iii) If $\mathbf{G} \in \mathcal{S}$ then by Theorems A6.4.1(vi) and A6.4.2 we have, as $t \to \infty$,

$$G^{2*}(t) \sim \big(G^+\big)^{2*}(t) \sim 2G(t).$$

Now relation (A6.4.15) follows immediately from (i) by induction.

(iv) Similarly to (A6.4.11), we have $G^{n*}(t) \le G_+^{n*}(t)$, $n \ge 1$. Therefore it is clear that it suffices to consider the case $\mathbf{G} \in \mathcal{S}_+$. Put

$$\alpha_n := \sup_{t \ge 0} \frac{G^{n*}(t)}{G(t)}.$$

Similarly to (A6.4.14), for $n \ge 2$, we have

$$G^{n*}(t) = G(t) + \int_0^t G^{(n-1)*}(t - y)\mathbf{G}(dy),$$

and hence, for each $M > 0$,

$$\alpha_n \le 1 + \sup_{0 \le t \le M} \int_0^t \frac{G^{(n-1)*}(t - y)}{G(t)}\mathbf{G}(dy)$$
$$+ \sup_{t > M} \int_0^t \frac{G^{(n-1)*}(t - y)}{G(t - y)} \frac{G(t - y)}{G(t)}\mathbf{G}(dy)$$
$$\le 1 + \frac{1}{G(M)} + \alpha_{n-1} \sup_{t > M} \frac{G^{2*}(t) - G(t)}{G(t)}.$$

Since $\mathbf{G} \in \mathcal{S}$, for any $\varepsilon > 0$ there exists an $M = M(\varepsilon)$ such that

$$\sup_{t > M} \frac{G^{2*}(t) - G(t)}{G(t)} < 1 + \varepsilon$$

and hence

$$\alpha_n \le b_0 + \alpha_{n-1}(1+\varepsilon), \quad b_0 := 1 + 1/G(M), \quad \alpha_1 = 1.$$

This recurrently implies

$$\alpha_n \le b_0 + b_0(1+\varepsilon) + \alpha_{n-2}(1+\varepsilon)^2 \le \cdots \le b_0 \sum_{j=0}^{n-1} (1+\varepsilon)^j \le \frac{b_0}{\varepsilon}(1+\varepsilon)^n.$$

The theorem is proved. □

Appendix 7
The Proofs of Theorems on Convergence to Stable Laws

In this appendix we will prove Theorems 8.8.1–8.8.4.

7.1 The Integral Limit Theorem

In this section we will prove Theorem 8.8.1 on convergence of the distributions of normalised sums $S_n = \sum_{k=1}^{n} \xi_k$ to stable laws. Recall the basic notation:

$$F_+(t) := \mathbf{P}(\xi \ge t), \qquad F_-(t) := \mathbf{P}(\xi < -t),$$
$$F_0(t) := F_+(t) + F_-(t) = \mathbf{P}\bigl(\xi \notin [-t, t)\bigr).$$

The main condition used in the theorem has this form:

$[\mathbf{R}_{\beta,\rho}]$ *The total tail* $F_0(x) = F_-(x) + F_+(x)$ *is a r.v.f. as* $x \to \infty$, *i.e., can be represented as*

$$F_0(x) = t^{-\beta} L_{F_0}(x), \quad \beta \in (0, 2], \tag{A7.1.1}$$

where $L_{F_0}(x)$ *is an s.v.f., and there exists the limit*

$$\rho_+ := \lim_{x\to\infty} \frac{F_+(x)}{F_0(x)} \in [0, 1], \qquad \rho := 2\rho_+ - 1. \tag{A7.1.2}$$

In the case $\beta < 2$ we put

$$b(n) := F_0^{(-1)}(1/n), \tag{A7.1.3}$$

while for $\beta = 2$ we set

$$b(n) := Y^{(-1)}(1/n), \tag{A7.1.4}$$

where

$$Y(t) := 2t^{-2} \int_0^t y F_0(y)\, dy = t^{-2}\mathbf{E}\bigl(\xi^2; -t \le \xi < t\bigr) = t^{-2} L_Y(t), \tag{A7.1.5}$$

A.A. Borovkov, *Probability Theory*, Universitext,
DOI 10.1007/978-1-4471-5201-9, © Springer-Verlag London 2013

$L_Y(t)$ is an s.v.f., so that (see Theorem A6.2.1(v) of Appendix 6)

$$b(n) = n^{1/\alpha} L_b(n), \quad L_b \text{ is an s.v.f.}$$

In the case when $F_+(t)$ and $F_-(t)$ are regularly varying functions (for instance, when condition $[\mathbf{R}_{\beta,\rho}]$ is satisfied and $\rho = 0$), we will denote these functions by $V(t)$ and $W(t)$, respectively, and put

$$V_I(t) := \int_0^t V(y)\,dy, \qquad V^I(t) := \int_t^\infty V(y)\,dy;$$

the same notational convention will be used for W.

If $F_+(t) = o(F_0(t))$ as $t \to \infty$ ($\rho = -1$), then $F_+(t)$ is not necessarily a regularly varying function, but everything we say below regarding the sums $V(t) + W(t)$ and $V^I(t) + W^I(t)$ remains valid if we understand by their first summands quantities negligibly small compared to the second summands (the first summands can also be replaced by zeros). This is also true for the sums $V_I(t) + W_I(t)$, except for the case when $\mathbf{E}\max(0, \xi)$ exists and $V_I(t)$ has to be replaced by $\mathbf{E}(\xi;\ \xi \ge 0) + o(1)$.

Theorem A7.1.1 *Let condition* $[\mathbf{R}_{\beta,\rho}]$ *be satisfied and* $\zeta_n := \frac{S_n}{b(n)}$.

(i) *For* $\beta \in (0, 2)$, $\beta \ne 1$, *and scaling factor* (A7.1.3), *as* $n \to \infty$,

$$\zeta_n \Rightarrow \zeta^{(\beta,\rho)}, \tag{A7.1.6}$$

where the distribution $\mathbf{F}_{\beta,\rho}$ *of the random variable* $\zeta^{(\beta,\rho)}$ *depends on the parameters* β *and* ρ *only and has ch.f.*

$$\varphi^{(\beta,\rho)}(t) := \mathbf{E}e^{it\zeta^{(\beta,\rho)}} = \exp\bigl\{|t|^\beta B(\beta, \rho, \vartheta)\bigr\}, \tag{A7.1.7}$$

where $\vartheta := \operatorname{sign} t$,

$$B(\beta, \rho, \vartheta) := \Gamma(1 - \beta)\left[i\rho\vartheta \sin\frac{\beta\pi}{2} - \cos\frac{\beta\pi}{2}\right] \tag{A7.1.8}$$

and, for $\beta \in (1, 2)$, *we assume that* $\Gamma(1 - \beta) = \Gamma(2 - \beta)/(1 - \beta)$.

(ii) *When* $\beta = 1$, *for the sequence* ζ_n *with scaling factor* (A7.1.3) *to converge to a limiting law the former, generally speaking, needs to be centred. More precisely, we have, as* $n \to \infty$,

$$\zeta_n - A_n \Rightarrow \zeta^{(1,\rho)}, \tag{A7.1.9}$$

where

$$A_n := \frac{n}{b(n)}\bigl[V_I\bigl(b(n)\bigr) - W_I\bigl(b(n)\bigr)\bigr] - \rho C, \tag{A7.1.10}$$

$C \approx 0.5772$ *is the Euler constant, and*

$$\varphi^{(1,\rho)}(t) := \mathbf{E}e^{it\zeta^{(1,\rho)}} = \exp\left\{-\frac{\pi|t|}{2} - i\rho t \ln|t|\right\}. \tag{A7.1.11}$$

If $n[V_I(b(n)) - W_I(b(n))] = o(b(n))$, *then* $\rho = 0$ *and we can put* $A_n = 0$.
If there exists $\mathbf{E}\xi = 0$ *then*

$$A_n = \frac{n}{b(n)}\big[W^I(b(n)) - V^I(b(n))\big] - \rho C.$$

If $\mathbf{E}\xi = 0$ *and* $\rho \neq 0$ *then* $\rho A_n \to -\infty$ *as* $n \to \infty$.
(iii) *For* $\beta = 2$ *and scaling factor* (A7.1.4), *as* $n \to \infty$,

$$\zeta_n \Rightarrow \zeta^{(2,\rho)}, \qquad \varphi^{(2,\rho)}(t) := \mathbf{E}e^{it\zeta} = e^{-t^2/2},$$

so that $\zeta^{(2,\rho)}$ *has the standard normal distribution which does not depend on* ρ.

Proof We will use the same approach as in the proof of the central limit theorem using relation (8.8.1). We will study the asymptotic properties of the ch.f. $\varphi(t) = \mathbf{E}e^{it\xi}$ in the vicinity of zero (more precisely, the asymptotics of

$$\varphi\left(\frac{\mu}{b(n)}\right) - 1 \to 0$$

as $b(n) \to \infty$) and show that, under condition $[\mathbf{R}_{\beta,\rho}]$, for each $\mu \in \mathbb{R}$, we have

$$n\left(\varphi\left(\frac{\mu}{b(n)}\right) - 1\right) \to \ln \varphi^{(\beta,\rho)}(\mu) \quad \text{as } n \to \infty \tag{A7.1.12}$$

(or some modification of this relation, see (A7.1.48)). This will imply that, for $\zeta_n = S(n)/b(n)$, as $n \to \infty$, there holds the relation (cf. Lemma 8.3.2)

$$\varphi_{\zeta_n}(\mu) \to \varphi^{(\beta,\rho)}(\mu). \tag{A7.1.13}$$

Indeed,

$$\varphi_{\zeta_n}(\mu) = \varphi^n\left(\frac{\mu}{b(n)}\right).$$

Since $\varphi(t) \to 1$ as $t \to 0$, one has

$$\begin{aligned}\ln \varphi_{\zeta_n}(\mu) &= n \ln \varphi\left(\frac{\mu}{b(n)}\right) \\ &= n \ln\left[1 + \left(\varphi\left(\frac{\mu}{b(n)}\right) - 1\right)\right] = n\left[\varphi\left(\frac{\mu}{b(n)}\right) - 1\right] + R_n,\end{aligned}$$

where $|R_n| \le n|\varphi(\mu/b(n)) - 1|^2$ for all n large enough, and hence $R_n \to 0$ by virtue of (A7.1.12). It follows that (A7.1.12) implies (A7.1.13).

So first we will study the asymptotics of $\varphi(t)$ as $t \to 0$ and then establish (A7.1.12).

(i) *First let* $\beta \in (0,1)$. We have

$$\varphi(t) = -\int_0^\infty e^{itx}\, dV(x) - \int_0^\infty e^{-itx}\, dW(x). \tag{A7.1.14}$$

Consider the former integral

$$-\int_0^\infty e^{itx}\, dV(x) = V(0) + it\int_0^\infty e^{itx} V(x)\, dx, \tag{A7.1.15}$$

where the substitution $|t|x = y$, $|t| = 1/m$ yields

$$I_+(t) := it\int_0^\infty e^{itx} V(x)\, dx = i\vartheta \int_0^\infty e^{i\vartheta y} V(my)\, dy, \tag{A7.1.16}$$

$\vartheta = \operatorname{sign} t$ (we will henceforth exclude the trivial case $t = 0$).

Assume for the present that $\rho_+ > 0$. Then $V(x)$ is an r.v.f. as $x \to \infty$ and, for each y, by virtue of the properties of s.v.f.s we have, as $|m| \to 0$,

$$V(my) \sim y^{-\beta} V(m).$$

Therefore it is natural to expect that, as $|t| \to 0$,

$$I_+(t) \sim i\vartheta V(m)\int_0^\infty e^{i\vartheta y} y^{-\beta}\, dy = i\vartheta V(m) A(\beta, \vartheta), \tag{A7.1.17}$$

where

$$A(\beta, \vartheta) := \int_0^\infty e^{i\vartheta y} y^{-\beta}\, dy. \tag{A7.1.18}$$

Assume that relation (A7.1.17) holds and similarly (in the case when $\rho_- > 0$)

$$-\int_0^\infty e^{-itx}\, dW(x) = W(0) + I_-(t), \tag{A7.1.19}$$

where

$$\begin{aligned} I_-(t) &:= -it\int_0^\infty e^{-itx} W(x)\, dx \sim -i\vartheta W(m)\int_0^\infty e^{-i\vartheta y} y^{-\beta}\, dy \\ &= -i\vartheta W(m) A(\beta, -\vartheta). \end{aligned} \tag{A7.1.20}$$

Since $V(0) + W(0) = 1$, relations (A7.1.14)–(A7.1.20) mean that, as $t \to 0$,

$$\varphi(t) = 1 + F_0(m) i\vartheta \big[\rho_+ A(\beta, \vartheta) - \rho_- A(\beta, -\vartheta)\big]\big(1 + o(1)\big). \tag{A7.1.21}$$

We can find an explicit form of the integral $A(\beta, \vartheta)$. Observe that the integral along the boundary of the positive quadrant (closed as a contour) in the complex plane of the function $e^{iz}z^{-\beta}$, which, as $|t| \to 0$, is equal to zero. From this it is not hard to obtain that

$$A(\beta, \vartheta) = \Gamma(1-\beta)e^{i\vartheta(1-\beta)\pi/2}, \quad \beta > 0. \tag{A7.1.22}$$

(Note also that (A7.1.18) is a table integral and its value can be found in handbooks, see, e.g., integrals 3.761.4 and 3.761.9 in [18].)

Thus, in (A7.1.21) one has

$$\begin{aligned}
i\vartheta\bigl[\rho_+ A(\beta,\vartheta) - \rho_- A(\beta,-\vartheta)\bigr] &= i\vartheta\,\Gamma(1-\beta)\biggl[\rho_+ \cos\frac{(1-\beta)\pi}{2} \\
&\quad + i\vartheta\rho_+ \sin\frac{(1-\beta)\pi}{2} - \rho_- \cos\frac{(1-\beta)\pi}{2} + i\vartheta\rho_- \sin\frac{(1-\beta)\pi}{2}\biggr] \\
&= \Gamma(1-\beta)\biggl[i\vartheta(\rho_+ - \rho_-)\cos\frac{(1-\beta)\pi}{2} - \sin\frac{(1-\beta)\pi}{2}\biggr] \\
&= \Gamma(1-\beta)\biggl[i\vartheta\rho\sin\frac{\beta\pi}{2} - \cos\frac{\beta\pi}{2}\biggr] = B(\beta,\rho,\vartheta),
\end{aligned}$$

where $B(\beta, \rho, \vartheta)$ is defined in (A7.1.8). Hence, as $t \to 0$,

$$\varphi(t) - 1 = F_0(m)B(\beta,\rho,\vartheta)\bigl(1+o(1)\bigr). \tag{A7.1.23}$$

Putting $t = \mu/b(n)$ (so that $m = b(n)/|\mu|$), where $b(n)$ is defined in (A7.1.3), and taking into account that $F_0(b(n)) \sim 1/n$, we obtain

$$n\biggl[\varphi\biggl(\frac{\mu}{b(n)}\biggr) - 1\biggr] = nF_0\biggl(\frac{b(n)}{|\mu|}\biggr)B(\beta,\rho,\vartheta)\bigl(1+o(1)\bigr) \sim |\mu|^\beta B(\beta,\rho,\vartheta). \tag{A7.1.24}$$

We have established the validity of (A7.1.12) and therefore that of assertion (i) of the theorem in the case $\beta < 1$, $\rho_+ > 0$.

If $\rho_+ = 0$ ($\rho_- = 0$) then, as was already mentioned, the above argument remains valid if we replace $V(m)$ ($W(m)$) by zero. This follows from the fact that in this case $F_+(t)$ ($F_-(t)$) admits a regularly varying majorant $V^*(t) = o(W(t))$ ($W^*(t) = o(V(t))$).

It remains only to justify the asymptotic equivalence in (A7.1.17). To do that, it is sufficient to verify that the integrals

$$\int_0^\varepsilon e^{i\vartheta y} V(my)\,dy, \qquad \int_M^\infty e^{i\vartheta y} V(my)\,dy \tag{A7.1.25}$$

can be made arbitrarily small compared to $V(m)$ by choosing appropriate ε and M. Note first that by Theorem A6.2.1(iii) of Appendix 6 (see (A6.1.2) in Appendix 6), for any $\delta > 0$, there exists an $x_\delta > 0$ such that, for all $v \le 1$ and $vx \ge x_\delta$, we have

$$\frac{V(vx)}{V(x)} \le (1+\delta)v^{-\beta-\delta}.$$

Therefore, for $\delta < 1-\beta$ and $x > x_\delta$,

$$\begin{aligned}\int_0^x V(u)\,du &\le x_\delta + \int_{x_\delta}^x V(u)\,du = x_\delta + xV(x)\int_{x_\delta/x}^1 \frac{V(vx)}{V(x)}\,dv \\ &\le x_\delta + xV(x)(1+\delta)\int_0^1 v^{-\beta-\delta}\,dv \\ &= x_\delta + \frac{xV(x)(1+\delta)}{1-\beta-\delta} \le cxV(x) \qquad \text{(A7.1.26)}\end{aligned}$$

since $xV(x)\to\infty$ as $x\to\infty$. It follows that

$$\left|\int_0^\varepsilon e^{i\vartheta y}V(my)\,dy\right| \le \frac{1}{m}\int_0^{\varepsilon m} V(u)\,du \le c\varepsilon V(\varepsilon m) \sim c\varepsilon^{1-\beta}V(m).$$

Since $\varepsilon^{1-\beta}\to 0$ as $\varepsilon\to 0$, the first assertion in (A7.1.25) is proved. The second integral in (A7.1.25) is equal to

$$\begin{aligned}\int_M^\infty e^{i\vartheta y}V(my)\,dy &= \frac{1}{i\vartheta}e^{i\vartheta y}V(my)\Big|_M^\infty - \frac{1}{i\vartheta}\int_M^\infty e^{i\vartheta y}\,dV(my) \\ &= -\frac{1}{i\vartheta}e^{i\vartheta M}V(mM) - \frac{1}{i\vartheta}\int_{mM}^\infty e^{i\vartheta u/m}\,dV(u),\end{aligned}$$

so its absolute value does not exceed

$$2V(mM)\sim 2M^{-\beta}V(m) \qquad \text{(A7.1.27)}$$

as $m\to\infty$. Hence the value of the second integral in (A7.1.25) can also be made arbitrarily small compared to $V(m)$ by choosing an appropriate M. Relation (A7.1.17) together with the assertion of the theorem in the case $\beta<1$ are proved.

Let now $\beta\in(1,2)$ and hence there exist a finite expectation $\mathbf{E}\xi$ which, according to our condition, will be assumed to be equal to zero. In this case,

$$\varphi(t)-1 = \vartheta\int_0^{|t|}\varphi'(\vartheta u)\,du, \quad \vartheta = \operatorname{sign} t, \qquad \text{(A7.1.28)}$$

and we have to find the asymptotic behaviour of

$$\varphi'(t) = -i\int_0^\infty xe^{itx}\,dV(x) + i\int_0^\infty xe^{-itx}\,dW(x) =: I_+^{(1)}(t) + I_-^{(1)}(t) \qquad \text{(A7.1.29)}$$

as $t\to 0$. Since $x\,dV(x) = d(xV(x)) - V(x)\,dx$, integration by parts yields

$$\begin{aligned}I_+^{(1)}(t) &:= -i\int_0^\infty xe^{itx}\,dV(x) = -i\int_0^\infty e^{itx}\,d\big(xV(x)\big) + i\int_0^\infty e^{itx}V(x)\,dx \\ &= -t\int_0^\infty xV(x)e^{itx}\,dx + iV^I(0) - t\int_0^\infty V^I(x)e^{itx}\,dx\end{aligned}$$

$$= iV^I(0) - t\int_0^\infty \widetilde{V}(x)e^{itx}\,dx, \tag{A7.1.30}$$

where, by Theorem A6.2.1(iv) of Appendix 6, both functions

$$V^I(x) := \int_x^\infty V(u)\,du \sim \frac{xV(x)}{\beta-1} \quad \text{as } x\to\infty, \quad V^I(0)<\infty,$$

and

$$\widetilde{V}(x) := xV(x) + V^I(x) \sim \frac{\beta x V(x)}{\beta-1}$$

are regularly varying.

Letting, as before, $m = 1/|t|$, $m\to\infty$ (cf. (A7.1.16), (A7.1.17)), we get

$$-t\int_0^\infty \widetilde{V}(x)e^{itx}\,dx = -\vartheta\widetilde{V}(m)\int_0^\infty \widetilde{V}(my)e^{i\vartheta y}\,dy$$

$$\sim -\vartheta\int_0^\infty y^{-\beta+1}e^{i\vartheta y}\,dy = -\frac{\beta V(m)}{t(\beta-1)}A(\beta-1,\vartheta),$$

$$I_+^{(1)}(t) = iV^I(0) - \frac{\beta\rho_+F_0(m)}{t(\beta-1)}A(\beta-1,\vartheta)\big(1+o(1)\big), \tag{A7.1.31}$$

where the function $A(\beta,\vartheta)$ defined in (A7.1.18) is equal to (A7.1.22).

Similarly,

$$\begin{aligned} I_-^{(1)}(t) &:= i\int_0^\infty te^{-itx}\,dW(x) \\ &= -t\int_0^\infty xW(x)e^{-itx}\,dx - iW^I(0) - t\int_0^\infty W^I(x)e^{-itx}\,dx \\ &= -iW^I(0) - t\int_0^\infty \widetilde{W}(x)e^{-itx}\,dx, \end{aligned}$$

where

$$W^I(x) := \int_x^\infty W(u)\,du, \qquad \widetilde{W}(x) := xW(x) + W^I(x) \sim \frac{\beta xW(x)}{\beta-1},$$

and

$$-t\int_0^\infty \widetilde{W}(x)\,e^{-itx}\,dx \sim -\frac{\beta W(m)}{t(\beta-1)}A(\beta-1,-\vartheta).$$

Therefore

$$I_-^{(1)}(t) = iW^I(0) - \frac{\beta\rho_-F_0(m)}{t(\beta-1)}A(\beta-1,-\vartheta)\big(1+o(1)\big)$$

and hence, by virtue of (A7.1.29), (A7.1.31), and the equality $V^I(0) - W^I(0) = \mathbf{E}\xi = 0$, we have

$$\varphi'(t) = -\frac{\beta F_0(m)}{t(\beta-1)}\big[\rho_+ A(\beta-1,\vartheta) + \rho_- A(\beta-1,-\vartheta)\big]\big(1+o(1)\big).$$

We return now to relation (A7.1.28). Since

$$\int_0^{|t|} u^{-1} F_0\big(u^{-1}\big)\,du \sim \beta^{-1} F_0\big(|t|^{-1}\big) = \beta^{-1} F_0(m)$$

(see Theorem A6.2.1(iii) of Appendix 6), we obtain, again using (A7.1.22) and an argument similar to the one in the proof for the case $\beta < 1$, that

$$\begin{aligned}
\varphi(t) - 1 &= -\frac{1}{\beta-1} F_0(m)\big[\rho_+ A(\beta-1,\vartheta) + \rho_- A(\beta-1,-\vartheta)\big]\big(1+o(1)\big) \\
&= -\frac{\Gamma(2-\beta)}{\beta-1} F_0(m)\bigg[\rho_+\bigg(\cos\frac{(2-\beta)\pi}{2} + i\vartheta\sin\frac{(2-\beta)\pi}{2}\bigg) \\
&\quad + \rho_-\bigg(\cos\frac{(2-\beta)\pi}{2} - i\vartheta\sin\frac{(2-\beta)\pi}{2}\bigg)\bigg]\big(1+o(1)\big) \\
&= \frac{\Gamma(2-\beta)}{\beta-1} F_0(m)\bigg[\cos\frac{\beta\pi}{2} - i\vartheta\rho\sin\frac{\beta\pi}{2}\bigg]\big(1+o(1)\big) \\
&= F_0(m) B(\beta,\rho,\vartheta)\big(1+o(1)\big).
\end{aligned} \tag{A7.1.32}$$

We arrive once again at relation (A7.1.23) which, by virtue of (A7.1.24), implies the assertion of the theorem for $\beta \in (1,2)$.

(ii) *Case* $\beta = 1$. In this case, the computation is somewhat more complicated. We again follow relations (A7.1.14)–(A7.1.16), according to which

$$\varphi(t) = 1 + I_+(t) + I_-(t). \tag{A7.1.33}$$

Rewrite expression (A7.1.16) for $I_+(t)$ as

$$I_+(x) = i\vartheta \int_0^\infty e^{i\vartheta y} V(my)\,dy = i\vartheta \int_0^\infty V(my)\cos y\,dy - \int_0^\infty V(my)\sin y\,dy, \tag{A7.1.34}$$

where the first integral on the right-hand side can be represented as the sum of two integrals:

$$\int_0^1 V(my)\,dy + \int_0^\infty g(y)V(my)\,dy, \tag{A7.1.35}$$

$$g(y) = \begin{cases} \cos y - 1 & \text{if } y \le 1, \\ \cos y & \text{if } y > 1. \end{cases} \tag{A7.1.36}$$

Note that (see, e.g., integral 3.782 in [18]) the value of the integral

$$-\int_0^\infty g(y)y^{-1}\,dy = C \approx 0.5772 \tag{A7.1.37}$$

is the Euler constant. Since $V(ym)/V(m) \to y^{-1}$ as $m \to \infty$, similarly to the above argument we obtain for the second integral in (A7.1.35) the relation

$$\int_0^\infty g(y)V(my)\,dy \sim -CV(m). \tag{A7.1.38}$$

Consider now the first integral in (A7.1.35):

$$\int_0^1 V(my)\,dy = m^{-1}\int_0^m V(u)\,du = m^{-1}V_I(m), \tag{A7.1.39}$$

where

$$V_I(x) := \int_0^t V(u)\,du \tag{A7.1.40}$$

can easily be seen to be an s.v.f. in the case $\beta = 1$ (see Theorem A6.2.1(iv) of Appendix 6). Here if $\mathbf{E}|\xi| = \infty$ then $V_I(x) \to \infty$ as $x \to \infty$, and if $\mathbf{E}|\xi| < \infty$ then $V_I(x) \to V_I(\infty) < \infty$.

Thus, for the first term on the right-hand side of (A7.1.34) we have

$$\operatorname{Im} I_+(t) = \vartheta\big(-CV(m) + m^{-1}V_I(m)\big) + o\big(V(m)\big). \tag{A7.1.41}$$

Now we will determine how $V_I(vx)$ depends on v as $x \to \infty$. For any fixed $v > 0$,

$$V_I(vx) = V_I(x) + \int_x^{vx} V(u)\,du = V_I(x) + xV(x)\int_1^v \frac{V(yx)}{V(x)}\,dy.$$

By Theorem A6.2.1 of Appendix 6,

$$\int_1^v \frac{V(yx)}{V(x)}\,dy \sim \int_1^v \frac{dy}{y} = \ln v,$$

so that

$$V_I(vx) = V_I(x) + \big(1 + o(1)\big)xV(x)\ln v =: A_V(v,x) + xV(x)\ln v, \tag{A7.1.42}$$

where evidently

$$A_V(v,x) = V_I(x) + o\big(xV(x)\big) \quad \text{as } x \to \infty \tag{A7.1.43}$$

and $V_I(x) \gg xV(x)$ by Theorem A6.2.1(iv) of Appendix 6.

Therefore, for $t = \mu/b(n)$ (so that $m = b(n)/|\mu|$ and hence $V(m) \sim \rho_+|\mu|/n$), we obtain from (A7.1.41) and (A7.1.42) (where one has to put $x = b(n)$, $v = 1/|\mu|$) that the following representation is valid as $n \to \infty$:

$$\begin{aligned}\operatorname{Im} I_+(t) &= -C\frac{\rho_+\mu}{n} + \frac{\mu}{b(n)}\left[A_V\left(|\mu|^{-1}, b(n)\right) - \frac{\rho_+\mu}{n}\ln|\mu|\right] + o\left(n^{-1}\right)\\ &= \frac{\mu}{b(n)}A_V\left(|\mu|^{-1}, b(n)\right) - \frac{\rho_+\mu}{n}\left(C + \ln|\mu|\right) + o\left(n^{-1}\right). \qquad \text{(A7.1.44)}\end{aligned}$$

For the second term on the right-hand side of (A7.1.34) we have

$$\operatorname{Re} I_+(t) = -\int_0^\infty V(my)\sin y\,dy \sim -V(m)\int_0^\infty y^{-1}\sin y\,dy.$$

Because $\sin y \sim y$ as $y \to 0$, the last integral converges. Since $\Gamma(\gamma) \sim 1/\gamma$ as $\gamma \to 0$, the value of this integral can be found to be (see (A7.1.22) and (A7.1.22))

$$\lim_{\gamma\to 0}\Gamma(\gamma)\sin\frac{\gamma\pi}{2} = \frac{\pi}{2}. \qquad \text{(A7.1.45)}$$

Thus, for $t = \mu/b(n)$,

$$\operatorname{Re} I_+(t) = -\frac{\pi|\mu|}{2n} + o\left(n^{-1}\right). \qquad \text{(A7.1.46)}$$

In a similar way we can find an asymptotic representation for the integral $I_-(t)$ (see (A7.1.14)–(A7.1.20)):

$$\begin{aligned}I_-(t) &:= -i\vartheta\int_0^\infty W(my)e^{-i\vartheta y}\,dy\\ &= -i\vartheta\int_0^\infty W(my)\cos y\,dy - \int_0^\infty W(my)\sin y\,dy.\end{aligned}$$

Comparing this with (A7.1.34) and the subsequent computation of $I_+(t)$, we can immediately conclude that, for $t = \mu/b(n)$ (cf. (A7.1.44), (A7.1.46)),

$$\begin{aligned}\operatorname{Im} I_-(t) &= -\frac{-\mu A_W\left(|\mu|^{-1}, b(n)\right)}{b(n)} + \frac{\rho_-\mu}{n}\left(C + \ln|\mu|\right) + o\left(n^{-1}\right),\\ \operatorname{Re} I_-(t) &= -\frac{\pi|\mu|\rho_-}{2n} + o\left(n^{-1}\right).\end{aligned} \qquad \text{(A7.1.47)}$$

Thus we obtain from (A7.1.33), (A7.1.44) and (A7.1.46) that (A7.1.47) imply

$$\begin{aligned}\varphi\left(\frac{\mu}{b(n)}\right) - 1 &= -\frac{\pi|\mu|}{n} - \frac{i\rho\mu}{n}\left(C + \ln|\mu|\right)\\ &\quad + \frac{i\mu}{b(n)}\left[A_V\left(|\mu|^{-1}, b(n)\right) - A_W\left(|\mu|^{-1}, b(n)\right)\right] + o\left(n^{-1}\right).\end{aligned}$$

It follows from (A7.1.43) that the penultimate term here is equal to

$$\frac{i\mu}{b(n)}\big[V_I\big(b(n)\big) - W_I\big(b(n)\big)\big] + o\big(n^{-1}\big),$$

so that finally,

$$\varphi\left(\frac{\mu}{b(n)}\right) - 1 = -\frac{\pi|\mu|}{2n} - \frac{i\rho\mu}{n}\ln|\mu| + i\mu\frac{A_n}{n} + o\big(n^{-1}\big), \tag{A7.1.48}$$

where

$$A_n = \frac{n}{b(n)}\big[V_I\big(b(n)\big) - W_I\big(b(n)\big)\big] - \rho C.$$

Therefore, similarly to (A7.1.12) and (A7.1.13), we obtain

$$\begin{aligned}\varphi_{\zeta_n - A_n}(\mu) &= e^{-i\mu A_n}\varphi^n\left(\frac{\mu}{b(n)}\right) = \exp\left\{-i\mu A_n + n\ln\left[1 + \left(\varphi\left(\frac{\mu}{b(n)}\right) - 1\right)\right]\right\} \\ &= \exp\left\{-i\mu A_n + n\left(\varphi\left(\frac{\mu}{b(n)}\right) - 1\right) + nO\left(\left|\varphi\left(\frac{\mu}{b(n)}\right) - 1\right|^2\right)\right\}.\end{aligned}$$

As, for $\beta = 1$, by Theorem A6.2.1(iv) of Appendix 6, the functions V_I and W_I are slowly varying, by (A7.1.48) one has

$$n\left|\varphi\left(\frac{\mu}{b(n)}\right) - 1\right|^2 \le c\left(\frac{1}{n} + \frac{A_n^2}{n}\right) \le c_1\left(\frac{1}{n} + \frac{1}{b(n)}\big[V_I\big(b(n)\big)^2 + W_I\big(b(n)\big)^2\big]\right) \to 0.$$

Since clearly

$$-i\mu A_n + n\left(\varphi\left(\frac{\mu}{b(n)}\right) - 1\right) \to -\frac{\pi|\mu|}{2} - i\rho\mu\ln|\mu|,$$

we have

$$\varphi_{\zeta_n - A_n}(\mu) \to \exp\left\{-\frac{\pi|\mu|}{2} - i\rho\mu\ln|\mu|\right\},$$

so relation (A7.1.9) is proved. The subsequent assertions regarding the centring sequence $\{A_n\}$ are evident. □

(iii) It remains to consider the case $\beta = 2$. We will follow representations (A7.1.28)–(A7.1.30), according to which we have to find, as $m = 1/|t| \to \infty$, the asymptotics of

$$\varphi'(t) = I_+^{(1)}(t) + I_-^{(1)}(t), \tag{A7.1.49}$$

where

$$I_+^{(1)}(t) := iV^I(0) - t\int_0^\infty \widetilde{V}(x)e^{itx}\,dx = iV^I(0) - \vartheta\int_0^\infty \widetilde{V}(my)e^{i\vartheta y}\,dy \tag{A7.1.50}$$

and, by Theorem A6.2.1(iv) of Appendix 6,

$$V^I(x) = \int_x^\infty V(y)\,dy \sim xV(x), \quad \widetilde{V}(x) = xV(x) + V^I(x) \sim 2xV(x) \quad \text{(A7.1.51)}$$

as $x \to \infty$. Further,

$$\int_0^\infty \widetilde{V}(my)\,e^{i\vartheta y}dy = \int_0^\infty \widetilde{V}(my)\cos y\,dy + \vartheta \int_0^\infty \widetilde{V}(my)\sin y\,dy. \quad \text{(A7.1.52)}$$

Here the second integral on the right-hand side is asymptotically equivalent, as $m \to \infty$, to (see (A7.1.45))

$$\widetilde{V}(m)\int_0^\infty y^{-1}\sin y\,dy = \frac{\pi}{2}\,\widetilde{V}(m).$$

The first integral on the right-hand side of (A7.1.52) is equal to

$$\int_0^1 \widetilde{V}(my)\,dy + \int_0^\infty g(y)\widetilde{V}(my)\,dy,$$

where the function $g(y)$ was defined in (A7.1.35), and

$$\int_0^1 \widetilde{V}(my)\,dy = \frac{1}{m}\int_0^m \widetilde{V}(u)\,du = \frac{1}{m}\,\widetilde{V}_I(m),$$

$\widetilde{V}_I(x) := \int_0^x \widetilde{V}(u)\,du$ being an s.v.f. by (A7.1.51). Since

$$\int_0^x uV(u)\,du = \frac{x^2V(x)}{2} - \frac{1}{2}\int_0^x u^2dV(u),$$

$$\int_0^x V^I(u)\,du = xV^I(x) + \int_0^x uV(u)\,du$$

and $V^I(x) \sim xV(x)$, we have

$$\begin{aligned}
\widetilde{V}_I(x) &= \int_0^x \big(uV(u) + V^I(u)\big)\,du \\
&= xV^I(x) + x^2V(x) - \int_0^x u^2\,dV(u) \\
&= -\int_0^x u^2\,dV(y) + O\big(x^2V(x)\big),
\end{aligned} \quad \text{(A7.1.53)}$$

where the last term is negligibly small, because

$$\int_0^x uV(u)\,du \gg x^2V(x)$$

(see Theorem A6.2.1(iv) of Appendix 6).

It is also clear that, as $x \to \infty$,

$$\widetilde{V}_I(x) \to \widetilde{V}_I(\infty) = \mathbf{E}\big(\xi^2; \xi > 0\big) \in (0, \infty].$$

As a result, we obtain (see also (A7.1.38))

$$\begin{aligned} I_+^{(1)}(t) &= i V^I(0) - \frac{i\pi}{2} \widetilde{V}(m) - t \widetilde{V}_I(m) + \vartheta C \widetilde{V}(m) + o\big(\widetilde{V}(m)\big) \\ &= i V^I(0) - t \widetilde{V}_I(m)\big(1 + o(1)\big) \end{aligned}$$

since $\widetilde{V}_I(x) \gg t \widetilde{V}(x)$.

Quite similarly we get

$$I_-^{(1)}(t) = -i W^I(0) - t \widetilde{W}_I(m)\big(1 + o(1)\big),$$

where $\widetilde{W}_I$ is an s.v.f. which is obtained from the function W in the same way as $\widetilde{V}_I$ from V. Since $V^I(0) = W^I(0)$, relation (A7.1.49) now yields that

$$\varphi'(t) = -t\big[\widetilde{V}_I(m) + \widetilde{W}_I(m)\big]\big(1 + o(1)\big).$$

Hence from (A7.1.28) we obtain the representation

$$\begin{aligned} \varphi(t) - 1 &= \vartheta \int_0^{1/m} \varphi'(\vartheta u)\, du = -\int_0^{1/m} u\big[\widetilde{V}_I(1/u) + \widetilde{W}_I(1/u)\big]\, du \\ &\sim -\frac{1}{2m^2}\big[\widetilde{V}_I(m) + \widetilde{W}_I(m)\big] \sim -\frac{1}{2m^2}\mathbf{E}\big(\xi^2;\ -m \le \xi < m\big) \end{aligned}$$

by virtue of (A7.1.53) and a similar relation for $\widetilde{W}_I$. Turning now to the definition of the function $Y(x) = x^{-2} L_Y(x)$ in (A7.1.5) and putting

$$b(n) := Y^{(-1)}(1/n), \quad t = \mu/b(n),$$

we get

$$n\big(\varphi(t) - 1\big) \sim -\frac{n}{2} Y\big(b(n)/|\mu|\big) \sim -\frac{n\mu^2}{2} Y\big(b(n)\big) \to -\frac{\mu^2}{2}.$$

The theorem is proved. □

7.2 The Integro-Local and Local Limit Theorems

In this section we will prove Theorems 8.8.2–8.8.4. We will begin with the integro-local theorem.

Theorem A7.2.1 (Integro-local Stone's theorem) *Let ξ be a non-lattice random variable and the conditions of Theorem* A7.1.1 *be satisfied. Then, for each fixed* $\Delta > 0$,

$$\mathbf{P}\big(S_n \in \Delta[x)\big) = \frac{\Delta}{b(n)} f^{(\beta,\rho)}\left(\frac{x}{b(n)}\right) + o\left(\frac{1}{b(n)}\right) \quad \text{as } n \to \infty,$$

where the remainder term $o(\frac{1}{b(n)})$ is uniform in x.

Proof of Theorem A7.2.1 The Proof is analogous to the proof of Theorem 8.7.1. We will again use the smoothing approach and consider, along with the sums S_n, the sums

$$Z_n = S_n + \theta\eta,$$

where $\theta = \text{const}$ and η is chosen so that its ch.f. is equal to 0 outside a finite interval. For instance, we can choose η as in Sect. 8.7.3, i.e., with the ch.f. $\varphi_\eta(t) = \max(0, 1 - |t|)$. Then equality (8.7.19) will still be valid with the same decomposition of the integral on its right-hand side into the subintegral I_1 over the domain $|t| < \gamma$ and I_2 over the domain $\gamma \le |t| \le 1/\theta$. Here estimating I_2 can be done in the same way as in Theorem 8.7.1.

For the sake of brevity, put $\widehat{\varphi}(t) := \varphi_{\eta_\Delta}(t)\varphi_{\theta\eta}(t)$. Then, for the integral I_1 with $x = vb(n)$, we have

$$I_1 = \int_{|t|<\gamma} e^{-itx}\varphi^n(t)\widehat{\varphi}(t)\,dt = \frac{1}{b(n)} \int_{|u|<\gamma b(n)} e^{-iuv}\varphi^n\left(\frac{u}{b(n)}\right)\widehat{\varphi}\left(\frac{u}{b(n)}\right)du. \tag{A7.2.1}$$

As was shown in the proof of Theorem 8.1.1, for each u we have

$$\varphi^n\left(\frac{u}{b(n)}\right) \to \varphi^{(\beta,\rho)}(u) \quad \text{as } n \to \infty,$$

and, moreover, for some $c > 0$ and $\gamma > 0$ small enough, by, virtue of, say, (A7.1.23) and (A7.1.32), we have

$$\operatorname{Re}\big(\varphi(t) - 1\big) \le -c\,F_0\left(\frac{1}{|t|}\right),$$

and, for any $\varepsilon > 0$ and all n large enough,

$$n\operatorname{Re}\left(\varphi\left(\frac{u}{b(n)}\right) - 1\right) \le -cnF_0\left(\frac{b(n)}{|u|}\right) \le -c\,|u|^{\beta-\varepsilon}.$$

Here we used the properties of the r.v.f. F_0. Moreover,

$$\widehat{\varphi}\left(\frac{u}{b(n)}\right) \to 1 \quad \text{as } n \to \infty, \qquad \big|\widehat{\varphi}(u)/b(n)\big| \le 1.$$

The above also implies that, for all u such that $|u| < \gamma b(n)$,

$$\left|\varphi^n\left(\frac{u}{b(n)}\right)\right| \le e^{-c|u|^{\beta-\varepsilon}}. \tag{A7.2.2}$$

The obtained relations mean that we can use the dominated convergence theorem in (A7.2.1) which implies

$$\lim_{n\to\infty} b(n)I_1 = \int e^{-iuv}\varphi^{(\beta,\rho)}(u)\,du \tag{A7.2.3}$$

uniformly in v, since the right-hand side of (A7.2.1) is uniformly continuous in v. On the right-hand side of (A7.2.3) is the result of the application of the inversion formula (up to the factor $1/2\pi$) to the ch.f. $\varphi^{(\alpha,\rho)}$. This means that

$$\lim_{n\to\infty} b(n)I_1 = 2\pi f^{(\beta,\rho)}(v).$$

We have established that, for $x = vb(n)$, as $n \to \infty$,

$$\mathbf{P}\big(Z_n \in \Delta[x)\big) = \frac{\Delta}{b(n)} f^{(\beta,\rho)}\left(\frac{x}{b(n)}\right) + o\left(\frac{1}{b(n)}\right)$$

uniformly in v (and hence in x).

To prove the theorem it remains to use Lemma 8.7.1.

The theorem is proved. □

The proofs of the local Theorems 8.8.3 and 8.8.4 can be obtained by an obvious similar modification of the proofs of Theorems 8.7.2 and 8.7.3 under the conditions of Theorem 8.8.1.

Appendix 8
Upper and Lower Bounds for the Distributions of the Sums and the Maxima of the Sums of Independent Random Variables

Let $\xi, \xi_1, \xi_2, \dots$ be independent identically distributed random variables,

$$S_n = \sum_{i=1}^{n} \xi_i, \qquad \overline{S}_n = \max_{1\le k\le n} S_k.$$

The main goal of this appendix is to obtain upper and lower bounds for the probabilities $\mathbf{P}(S_n \ge x)$ and $\mathbf{P}(\overline{S}_n \ge x)$. These bounds were used in Sect. 9.5 to find the asymptotics of the probabilities of large deviations for S_n and $\overline{S}_n$.

8.1 Upper Bounds Under the Cramér Condition

In this section we will assume that the following one-sided Cramér condition is met:

[C] *There exists a* $\lambda > 0$ *such that*

$$\psi(\lambda) = \mathbf{E}e^{\lambda\xi} < \infty. \tag{A8.1.1}$$

The following analogue of the exponential Chebyshev inequality holds true for $\mathbf{P}(S_n \ge x)$.

Theorem A8.1.1 *For all* $n \ge 1$, $x \ge 0$ *and* $\lambda \ge 0$, *we have*

$$\mathbf{P}(\overline{S}_n \ge x) \le e^{-\lambda x} \max\bigl(1, \psi^n(\lambda)\bigr). \tag{A8.1.2}$$

Proof As $\eta(x) := \inf\{k \ge 1 : S_k \ge x\} \le \infty$ is a Markov time, the event $\{\eta(x) = k\}$ is independent of the random variables $S_n - S_k$. Therefore

$$\psi^n(\lambda) = \mathbf{E}\,e^{\lambda S_n} \ge \sum_{k=1}^{n} \mathbf{E}\left(e^{\lambda S_n}; \eta(x) = k\right) \ge \sum_{k=1}^{n} \mathbf{E}\left(e^{\lambda(x+S_n-S_k)}; \eta(x) = k\right)$$

A.A. Borovkov, *Probability Theory*, Universitext,
DOI 10.1007/978-1-4471-5201-9, © Springer-Verlag London 2013

$$= e^{\lambda x} \sum_{k=1}^{n} \psi^{n-k}(\lambda) \mathbf{P}\big(\eta(x) = k\big) \ge e^{\lambda x} \min\big(1, \psi^n(\lambda)\big) \mathbf{P}(\overline{S}_n \ge x).$$

This immediately implies (A8.1.2). The theorem is proved. □

If $\psi(\lambda) \ge 1$ for $\lambda \ge 0$ (this is always the case if there exists $\mathbf{E}\xi \ge 0$) then the right-hand side of (A8.1.2) is equal to $e^{-\lambda x}\psi^n(\lambda)$, and the equality (A8.1.2) itself can also be obtained as a consequence of the well-known Kolmogorov–Doob inequality for submartingales (see Theorem 15.3.4, where one has to put $X_n := S_n$).

Thus, if $\mathbf{E}\xi \ge 0$ then

$$\mathbf{P}(\overline{S}_n \ge x) \le e^{-\lambda x + n \ln \psi(\lambda)}.$$

Choosing the best possible value of λ we obtain the following inequality.

Corollary A8.1.1 *If* $\mathbf{E}\xi \ge 0$ *then, for all* $n \ge 1$ *and* $x \ge 0$, *we have*

$$\mathbf{P}(\overline{S}_n \ge x) \le e^{-n\Lambda(\alpha)},$$

where

$$\alpha := \frac{x}{n}, \qquad \Lambda(\alpha) := \sup_{\lambda}\big(\lambda\alpha - \ln \psi(\lambda)\big).$$

The function $\Lambda(\alpha)$ is the rate function introduced in Sect. 9.1. Its basic properties were stated in that section. In particular, for $\mathbf{E}\xi = 0$ and $\mathbf{E}\xi^2 = \sigma^2 < \infty$, the asymptotic equivalence $\Lambda(\alpha) \sim \frac{\alpha^2}{2\sigma^2}$ as $\alpha \to 0$ takes place, which yields that, for $x = o(n)$,

$$\mathbf{P}(\overline{S}_n \ge x) \le \exp\left\{-\frac{x^2}{2n\sigma^2}\big(1 + o(1)\big)\right\}. \tag{A8.1.3}$$

8.2 Upper Bounds when the Cramér Condition Is Not Met

In this section we will assume that

$$\mathbf{E}\xi = 0, \qquad \mathbf{E}\xi^2 = \sigma^2 < \infty. \tag{A8.2.1}$$

For simplicity's sake, without losing generality, in what follows we will put $\sigma = 1$. The bounds will be obtained for the deviation zone $x > \sqrt{n}$ which is adjacent to the zone of "normal deviations" where

$$\mathbf{P}(S_n \ge x) \sim 1 - \Phi\left(\frac{x}{\sqrt{n}}\right) \tag{A8.2.2}$$

(uniformly in $x \in (0, N_n\sqrt{n})$, where $N_n \to \infty$ slowly enough as $n \to \infty$; see Sect. 8.2). Moreover, it was established in Sect. 19.1 that, in the normal deviations zone,

$$\mathbf{P}(\overline{S}_n \ge x) \sim 2\left(1 - \Phi\left(\frac{x}{\sqrt{n}}\right)\right). \qquad \text{(A8.2.3)}$$

To derive upper bounds in the zone $x > \sqrt{n}$ when the Cramér condition **[C]** is not met, we will need additional conditions on the behaviour of the right tail $F_+(t) = \mathbf{P}(\xi \ge t)$ of the distribution **F**.

Namely, we will assume that the following condition is satisfied.

[<] *For the right tail $F_+(t) = \mathbf{P}(\xi \ge t)$ there exists a regularly varying (see Appendix* 6) *majorant $V(t)$:*

$$F_+(t) \le V(t) := t^{-\beta} L(t) \quad \text{for all } t > 0,$$

where $\beta > 2$ and L is a slowly varying function (s.v.f., see Appendix 6).

By virtue of (A8.2.2) and (A8.2.3), for deviations $x < N_n\sqrt{n}$, $n \to \infty$, it would be natural to expect upper bounds with an exponential right-hand side $e^{-x^2/(2n)}$ (cf. (A8.1.3)). On the other hand, Theorem A6.4.3(iii) of Appendix 6 implies that, for $F_+(t) = V(t) \in \mathcal{R}$ and any fixed n we have, as $x \to \infty$,

$$\mathbf{P}(S_n \ge x) \sim nV(x). \qquad \text{(A8.2.4)}$$

This relation clearly holds true if $n \to \infty$ slowly enough (as $x \to \infty$).

The asymptotics (A8.2.2) and (A8.2.4) merge with each other remarkably as follows:

$$\mathbf{P}(S_n \ge x) \sim \left(1 - \Phi\left(\frac{x}{\sqrt{n}}\right)\right) + nV(x) \qquad \text{(A8.2.5)}$$

as $n \to \infty$ for all $x > \sqrt{n}$ (for more details see, e.g., [8] and the bibliography therein). Relation (A8.2.5) allows us to "guess" the threshold values of $x = b(n)$ for which asymptotics (A8.2.2) changes to asymptotics (A8.2.4). To find such x it suffices to equate the logarithms of the right-hand sides of (A8.2.2) and (A8.2.4):

$$-\frac{x^2}{2n} = \ln nV(x) = \ln n - \beta \ln x + o(\ln x).$$

The main part $b(n)$ of the solution to this equation, as it is not hard to see, has the form

$$b(n) = \sqrt{(\beta - 2)n \ln n}$$

(we exclude the trivial case $n = 1$).

In what follows, we will represent deviations x as $x = sb(n)$. Based on the above, it is natural to expect (and it can be easily verified) that the first term will dominate on the right-hand side of (A8.2.5) if $s < 1$, while the second will dominate if $s > 1$. Accordingly, for small s (but such that $x > \sqrt{n}$), we will have the above-mentioned exponential bounds for $\mathbf{P}(S_n \ge x)$, while for large s there will hold bounds of the form $nV(x)$ (note that $nV(x) \to 0$ for $x > b(n)$ and $\beta > 2$).

The above claim is confirmed by the assertions below. Along with x introduce deviations

$$y = \frac{x}{r},$$

where $r \ge 1$ is fixed, and put

$$B_j := \{\xi_j < y\}, \qquad B := \bigcap_{j=1}^{n} B_j.$$

Theorem A8.2.1 *Let conditions* (A8.2.1) *and* [<] *be satisfied.*

(1) *For any fixed* $h > 1$, $s_0 > 0$, $x = sb(n)$, $s \ge s_0$ *and all* $\Pi := nV(x)$ *small enough, we have*

$$P := \mathbf{P}(\overline{S}_n \ge x;\, B) \le e^r \left(\frac{\Pi(y)}{r}\right)^{r-\theta}, \tag{A8.2.6}$$

where

$$\Pi(y) := nV(y), \qquad \theta := \frac{hr^2}{4s^2}\left(1 + b\frac{\ln s}{\ln n}\right), \qquad b := \frac{2\beta}{\beta - 2}.$$

(2) *For any fixed* $h > 1$, $\tau > 0$, *for* $x = sb(n) > \sqrt{n}$, $s^2 < (h - \tau)/2$, *and all* n *large enough, we have*

$$P \le e^{-x^2/(2nh)}. \tag{A8.2.7}$$

Corollary A8.2.1 (a) *If* $s \to \infty$ *then*

$$\mathbf{P}(\overline{S}_n \ge x) \le nV(x)\big(1 + o(1)\big). \tag{A8.2.8}$$

(b) *If* $s^2 \ge s_0^2$ *for some fixed* $s_0 > 1$ *then, for all* $nV(x)$ *small enough,*

$$\mathbf{P}(\overline{S}_n \ge x) \le cnV(x), \quad c = \text{const}. \tag{A8.2.9}$$

(c) *For any fixed* $h > 1$, $\tau > 0$, *for* $s^2 < (h - \tau)/2$, $x > \sqrt{n}$, *and all* n *large enough,*

$$\mathbf{P}(\overline{S}_n \ge x) \le e^{-x^2/(2nh)}. \tag{A8.2.10}$$

Remark A8.2.1 It is not hard to verify (see the proofs of Theorem A8.2.1 and Corollary A8.2.1) that there exists a function $\varepsilon(t) \downarrow 0$ as $t \uparrow \infty$ such that one has, along with (A8.2.8), the relation

$$\sup_{x:s\ge t} \frac{\mathbf{P}(\overline{S}_n > x)}{nV(x)} \le 1+\varepsilon(t).$$

Proof of Corollary A8.2.1 The proof is based on the inequality

$$\mathbf{P}(\overline{S}_n > x) \le \mathbf{P}(\overline{B}) + \mathbf{P}(\overline{S}_n \ge x;\ B) \le nV(y) + P. \tag{A8.2.11}$$

Since $\theta \to 0$ as $s \to \infty$, we see that, for any fixed $\varepsilon > 0$ and all $\Pi = nV(x)$ small enough, we have $P \le c(nV(y))^{r-\varepsilon}$. Putting $r := 1+2\varepsilon$, we obtain from (A8.2.11) and (A8.2.6) that

$$\mathbf{P}(\overline{S}_n \ge x) \le nV(y) + c\big(nV(y)\big)^{1+\varepsilon} \sim n(1+2\varepsilon)^{-\beta}V(x).$$

Since the left-hand side of this inequality does not depend on ε, relation (A8.2.8) follows.

We now prove (b). If $s \to \infty$ then (b) follows from (a). If s is bounded then necessarily $n \to \infty$ (since $nV(x) \to 0$) and hence

$$r - \theta = r - \frac{hr^2}{4s^2}\left(1 + b\frac{\ln s}{\ln n}\right) = \psi(r,s) + o(1),$$

where the function

$$\psi(r,s) := r - \frac{hr^2}{4s^2}$$

attains its maximum $\psi(r_0, s) = s^2/h$ in r at the point $r_0 = 2s^2/h$. Moreover, $\psi(r,s)$ strictly decreases in s. Therefore, for $r_0 = 2s^2/h$, we obtain

$$\psi(r_0, s) = \frac{s^2}{h}, \tag{A8.2.12}$$

$$r_0 - \theta \ge \frac{s^2}{h} + o(1) \quad \text{as } n \to \infty. \tag{A8.2.13}$$

Choose h so close to 1 and $\tau > 0$ so small that $h + \tau \le s_0^2$. Putting $r := r_0$, for $s^2 \ge s_0^2 \ge h + \tau$ and as $n \to \infty$, we get from (A8.2.6), (A8.2.12) and (A8.2.13) that

$$\mathbf{P}(\overline{S}_n \ge x) \le nV(y) + c\big(nV(y)\big)^{1+\tau/2} \sim nV\left(\frac{x}{r_0}\right) \sim r_0^\beta nV(x).$$

This proves (b).

Relation (c) for $y = x$ follows from the inequality (see (A8.2.7) and (A8.2.11))

$$\mathbf{P}(\overline{S}_n \ge x) \le nV(x) + e^{-x^2/(2nh)}, \tag{A8.2.14}$$

where, for $s^2 < (h-\tau)/2$,

$$e^{-x^2(2nh)} > \exp\left\{-\frac{(h-\tau)}{2}\frac{(\beta-2)n\ln n}{2nh}\right\} > n^{-(\beta-2)/4}.$$

On the other hand, we have $x > \sqrt{n}$,

$$nV(x) \le nV(\sqrt{n}) = n^{-(\beta-2)/2}L^*(n),$$

where L^* is a s.v.f. Therefore the second term dominates on the right-hand side of (A8.2.14). Slightly changing h if necessary, we obtain (c). Corollary A8.2.1 is proved. □

Remark A8.2.2 One can see from the proof of the corollary that the main contribution to the bound for the probability $\mathbf{P}(\overline{S}_n \ge x)$ under the conditions of assertions (a) and (b) comes from the event $\overline{B} = \{\max_{j\le n}\xi_j \ge y\}$ with y close to x, so that the most probable trajectory of $\{S_k\}_{k=1}^n$ that reaches the level x contains at least one jump ξ_j of size comparable to x.

Proof of Theorem A8.2.1 In our case, the Cramér condition **[C]** is not met. In order to use Theorem A8.1.1 in such a situation, we introduce "truncated" random variables with distributions that coincide with the conditional distribution of ξ given $\{\xi < y\}$ for some level y the choice of which will be at our disposal. Namely, we introduce independent identically distributed random variables $\xi_j^{(y)}$, $j = 1, 2, \ldots$, with the distribution function

$$\mathbf{P}\big(\xi_j^{(y)} < t\big) = \mathbf{P}(\xi < t \,|\, \xi < y) = \frac{\mathbf{P}(\xi < t)}{\mathbf{P}(\xi < y)}, \qquad t \le y,$$

and put

$$S_n^{(y)} := \sum_{j=1}^n \xi_j^{(y)}, \qquad \overline{S}_n^{(y)} := \max_{k\le n} S_k^{(y)}.$$

Then

$$P = \mathbf{P}(\overline{S}_n \ge x, B) = \big(\mathbf{P}(\xi < y)\big)^n \mathbf{P}\big(\overline{S}_n^{(y)} \ge x\big). \tag{A8.2.15}$$

Applying Theorem A8.1.1 to the variables $\xi_j^{(y)}$, we obtain that, for any $\lambda \ge 0$,

$$\mathbf{P}\big(\overline{S}_n^{(y)} \ge x\big) \le e^{-\lambda x}\big[\max\big\{1, \mathbf{E}\, e^{\lambda\xi^{(y)}}\big\}\big]^n.$$

Since

$$\mathbf{E}e^{\lambda\xi^{(y)}} = \frac{R(\lambda, y)}{F(y)}, \quad \text{where } R(\lambda, y) := \int_{\infty}^{y} e^{\lambda t}\mathbf{F}(dt),$$

we arrive at the following basic inequality. For $x, y, \lambda \ge 0$,

$$P = \mathbf{P}(\overline{S}_n \ge x, B) \le e^{-\lambda x}\big[\max\big\{\mathbf{P}(\xi < y), R(\lambda, y)\big\}\big]^n$$

$$\le e^{-\lambda x}\max\{1, R^n(\lambda, y)\}. \tag{A8.2.16}$$

Thus, the main problem is to bound the integral $R(\lambda, y)$. Put

$$M(v) := \frac{v}{\lambda}$$

and represent $R(\lambda, y)$ as

$$R(\lambda, y) = I_1 + I_2,$$

where, for a fixed $\varepsilon > 0$,

$$I_1 := \int_{-\infty}^{M(\varepsilon)} e^{\lambda t}\mathbf{F}(dt) = \int_{-\infty}^{M(\varepsilon)} \left(1 + \lambda t + \frac{\lambda^2 t^2}{2} e^{\lambda\theta(t)}\right)\mathbf{F}(dt), \quad 0 \le \frac{\theta(t)}{t} \le 1. \tag{A8.2.17}$$

Here

$$\int_{-\infty}^{M(\varepsilon)} \mathbf{F}(dt) = 1 - V\big(M(\varepsilon)\big) \le 1,$$

$$\int_{-\infty}^{M(\varepsilon)} t\mathbf{F}(dt) = -\int_{M(\varepsilon)}^{\infty} t\mathbf{F}(dt) \le 0, \tag{A8.2.18}$$

$$\int_{-\infty}^{M(\varepsilon)} t^2 e^{\lambda\theta(t)}\mathbf{F}(dt) \le e^{\varepsilon}\int_{-\infty}^{M(\varepsilon)} t^2\mathbf{F}(dt) \le e^{\varepsilon} =: h. \tag{A8.2.19}$$

Therefore,

$$I_1 \le 1 + \frac{\lambda^2 h}{2}. \tag{A8.2.20}$$

Estimate now

$$I_2 := -\int_{M(\varepsilon)}^{y} e^{\lambda t}\, dF_+(t) \le V\big(M(\varepsilon)\big)e^{\varepsilon} + \lambda\int_{M(\varepsilon)}^{y} V(t)e^{\lambda t}\, dt. \tag{A8.2.21}$$

First consider, for $M(\varepsilon) < M(2\beta) < y$, the subintegral

$$I_{2,1} := \lambda\int_{M(\varepsilon)}^{M(2\beta)} V(t)e^{\lambda t}\, dt.$$

For $t = v/\lambda$, as $\lambda \to 0$, we have

$$V(t)e^{\lambda t} = V\left(\frac{v}{\lambda}\right)e^{v} \sim V\left(\frac{1}{\lambda}\right)f(v), \tag{A8.2.22}$$

where the function

$$f(v) := v^{-\beta}e^{v}$$

is convex on $(0, \infty)$. Therefore

$$I_{2,1} \le \frac{\lambda}{2}\big(M(2\beta) - M(\varepsilon)\big) V\left(\frac{1}{\lambda}\right)\big(f(\varepsilon) + f(2\beta)\big)\big(1 + o(1)\big) \le cV\left(\frac{1}{\lambda}\right). \quad \text{(A8.2.23)}$$

We now proceed to estimating the remaining subintegral

$$I_{2,2} := \lambda \int_{M(2\beta)}^{y} V(t)e^{\lambda t}\,dt.$$

For brevity's sake, put $M(2\beta) =: M$. We will choose λ so that

$$\mu = \lambda y \to \infty \quad (y \gg 1/\lambda) \quad \text{(A8.2.24)}$$

as $x \to \infty$. Substituting the variable $(y - t)\lambda =: u$ we obtain

$$\lambda I_{2,2} = e^{\lambda y} V(y) \int_0^{(y-M)\lambda} V\left(y - \frac{u}{\lambda}\right) V^{-1}(y) e^{-u}\,du. \quad \text{(A8.2.25)}$$

Consider the integral on the right-hand side of (A8.2.25). Since $1/\lambda \ll y$, the integrand

$$r_{y,\lambda}(u) := \frac{V(y - u/\lambda)}{V(y)}$$

converges to 1 for each fixed u. In order to use the dominated convergence theorem which implies that the integral on the right-hand side of (A8.2.25) converges, as $y \to \infty$, to

$$\int_0^\infty e^{-u}\,du = 1, \quad \text{(A8.2.26)}$$

it remains to estimate the growth rate of the function $r_{y,\lambda}(u)$ as u increases. By the properties of r.v.f.s (see Theorem A6.2.1(iii) in Appendix 6), for all λ small enough (or M large enough; recall that $y - u/\lambda \ge M$ in the integrand in (A8.2.25)), we have

$$r_{y,\lambda}(u) \le \left(1 - \frac{u}{\lambda y}\right)^{-2\beta/2} =: g(u).$$

Since $g(0) = 1$ and $\lambda y - u \ge M\lambda = 2\beta$, in this domain

$$\big(\ln g(u)\big)' = \frac{3\beta}{2(\lambda y - u)} \le \frac{3\beta}{4\beta} = \frac{3}{4},$$

$$\ln g(u) \le \frac{3u}{4}, \qquad r_{y,\lambda}(u) \le e^{3u/4}.$$

This means that the integrand in (A8.2.25) is dominated by the exponential $e^{-u/4}$, and the use of the dominated convergence theorem is justified. Therefore, due to the convergence of the integral in (A8.2.25) to the limit (A8.2.26), we obtain

$$\lambda I_{2,2} \sim e^{\lambda y} V(y) \int_0^\infty e^u \, du = e^\mu V(y),$$

and it is not hard to find a function $\varepsilon(\mu) \downarrow 0$ as $\mu \uparrow \infty$ such that

$$\lambda I_{2,2} \le e^{\lambda y} V(y)\big(1+\varepsilon(\mu)\big). \tag{A8.2.27}$$

Summarising (A8.2.20)–(A8.2.23) and (A8.2.27), we obtain

$$R(\lambda, y) \le 1 + \frac{\lambda^2 h}{2} + cV\left(\frac{1}{\lambda}\right) + V(y)e^{\lambda y}\big(1+\varepsilon(\mu)\big), \tag{A8.2.28}$$

$$R^n(\lambda, y) \le \exp\left\{\frac{n\lambda^2 h}{2} + cnV\left(\frac{1}{\lambda}\right) + nV(y)e^{\lambda y}\big(1+\varepsilon(\mu)\big)\right\}. \tag{A8.2.29}$$

First take λ to be the value

$$\lambda = \frac{1}{y} \ln T$$

that "almost minimises" the function $-\lambda x + nV(y)e^{\lambda y}$, where $T := \frac{r}{nV(y)}$, so that $\mu = \ln T$. Note that, for such a choice of μ (or of $\lambda = y^{-1}\ln(r/\Pi(y))$), for $\Pi(y) \to 0$ we have that $\mu = \lambda y \sim -\ln \Pi(y) \to \infty$ and hence that the assumption $y \gg 1/\lambda$ we made in (A8.2.24) holds true. For such λ,

$$R^n(\lambda, y) \le \exp\left\{\frac{n\lambda^2 h}{2} + cnV\left(\frac{1}{\lambda}\right) + r\big(1+\varepsilon(\mu)\big)\right\}, \tag{A8.2.30}$$

where, by the properties of r.v.f.s,

$$nV\left(\frac{1}{\mu}\right) \sim nV\left(\frac{y}{\ln T}\right) \sim cnV\left(\frac{y}{|\ln nV(y)|}\right) \le cnV(y)\big|\ln nV(y)\big|^{\beta+\delta} \to 0,$$
$$\delta > 0, \tag{A8.2.31}$$

as $nV(y) \to 0$. Therefore

$$\begin{aligned}\ln P &\le -r\ln T + r + \frac{nh}{2y^2}\ln^2 T + \varepsilon_1(T)\\ &= \left[-r + \frac{nh}{2y^2}\ln T\right]\ln T + r + \varepsilon_1(T),\end{aligned} \tag{A8.2.32}$$

where $\varepsilon_1(T) \downarrow 0$ as $T \uparrow \infty$. If $x = sb(n)$, $b(n) = \sqrt{(\beta-2)n\ln n}$, and $nV(x) \to 0$ then

$$\ln T = -\ln nV(x) + O(1) = -\ln n + \beta \ln s + \frac{\beta}{2}\ln n + O\big(\ln L\big(s\sigma(n)\big)\big) + O(1)$$

$$= \frac{\beta - 2}{2} \ln n \left[1 + b \frac{\ln s}{\ln n} \right] (1 + o(1)), \tag{A8.2.33}$$

where $b = \frac{2\beta}{\beta - 2}$ (the term $o(1)$ in the last equality appears because in our case either $n \to \infty$ or $s \to \infty$.) Hence, by (A8.2.32),

$$\frac{nh}{2y^2} \ln T = \frac{hr^2}{4s^2} \left[1 + b \frac{\ln s}{\ln n} \right] (1 + o(1)),$$

$$\ln P \le r - \left[r - \frac{h' r^2}{4s^2} \left(1 + b \frac{\ln s}{\ln n} \right) \right] \ln T$$

for any $h' > h > 1$ and $nV(x)$ small enough. This proves the first assertion of the theorem.

We now prove the second assertion of the theorem for "small" values of s such that, for some $\tau > 0$,

$$s^2 < \frac{h - \tau}{2}.$$

Since we always assume that $x > \sqrt{n}$, we also have

$$s = \frac{x}{b(n)} > \frac{1}{\sqrt{(\beta - 2) \ln n}}$$

and we can assume that $s^2 \ge n^{-\gamma}$ for some $\gamma > 0$ to be chosen below. This corresponds to the following domain of the values of x^2:

$$cn^{1-\gamma} \ln n < x^2 < \frac{(h - \tau)(\beta - 2)}{2} n \ln n. \tag{A8.2.34}$$

For such x, as will be shown below, the main contribution to the exponent on the right-hand side of (A8.2.29) comes from the quadratic term $n\lambda^2 h/2$, and we will set

$$\lambda := \frac{x}{nh}.$$

Then, for $y = x$ ($r = 1$, $\mu = x^2/(nh)$),

$$\ln P \le -\lambda x + \frac{n\lambda^2 h}{2} + cnV\left(\frac{1}{\lambda}\right) + nV(y) e^{\lambda y} (1 + \varepsilon(\mu))$$

$$= -\frac{x^2}{2nh} + cnV\left(\frac{nh}{x}\right) + nV(x) e^{\frac{x^2}{nh}} (1 + \varepsilon(\mu)). \tag{A8.2.35}$$

We show that the last two terms on the right-hand side are negligibly small as $n \to \infty$. Indeed, by the second inequality in (A8.2.34),

$$nV\left(\frac{nh}{x}\right) \le cnV\left(\sqrt{\frac{n}{\ln n}}\right) \to 0 \quad \text{as } n \to \infty.$$

Further, by the first inequality in (A8.2.34),

$$nV(x) \le n^{(2-\beta)/2+\gamma'},$$

where we can choose γ'. Moreover, by (A8.2.34),

$$\frac{x^2}{nh} \le \frac{(h-\tau)(\beta-2)\ln n}{2h} = \left[\frac{\beta-2}{2} - \frac{\tau(\beta-2)}{2h}\right]\ln n.$$

Therefore

$$nV(x)e^{x^2/(nh)} \le n^{-\tau(\beta-2)/(2h)+\gamma'} \to 0$$

for $\gamma' < \frac{\tau(\beta-2)}{2h}$ as $n \to \infty$.

Thus,

$$\ln P \le -\frac{x^2}{2nh} + o(1).$$

Since $x^2/n > 1$, the term $o(1)$ in the last relation can be omitted by slightly changing $h > 1$. (Formally, we proved that, for any $h > 1$ and all n large enough, inequality (A8.2.7) is valid with the h on its right-hand side replaced with $h' > h$, where we can take, for instance, $h' = h + (h-1)/2$. Since $h' > 1$ can also be made arbitrarily close to 1 by the choice of h, the obtained relation is equivalent to the one from Theorem A8.2.1.) This proves (A8.2.7).

The theorem is proved. □

Comparing the assertions of Theorem A8.2.1 and Corollary A8.1.1, we see that, roughly speaking, for $s < 1/2$ and for $s > 1$ one can obtain quite satisfactory and, in a certain sense, unimprovable upper bounds for the probabilities P and $\mathbf{P}(\overline{S}_n > x)$.

8.3 Lower Bounds

In this section we will again assume that conditions (A8.2.1) are satisfied. The lower bounds for $\mathbf{P}(S_n \ge x)$ (they will clearly hold for $\mathbf{P}(\overline{S}_n \ge x)$ as well) can be obtained in a much simpler way than the upper bounds and need essentially no assumptions.

Theorem A8.3.1 *Let* $\mathbf{E}\xi_j = 0$ *and* $\mathbf{E}\xi_j^2 = 1$. *Then, for* $y = x + t\sqrt{n-1}$,

$$\mathbf{P}(S_n \ge x) \ge nF_+(y)\left[1 - t^{-2} - \frac{n-1}{2}F_+(y)\right]. \tag{A8.3.1}$$

Proof Put $G_n := \{S_n \ge x\}$ and $B_j := \{\xi_j < y\}$. Then

$$\mathbf{P}(S_n \ge x) \ge \mathbf{P}\left(G_n;\ \bigcup_{j=1}^n \overline{B}_j\right) \ge \sum_{j=1}^n \mathbf{P}(G_n\overline{B}_j) - \sum_{i<j\le n} \mathbf{P}(G_n\overline{B}_i\overline{B}_j)$$

$$\geq \sum_{j=1}^{n} \mathbf{P}(G_n \overline{B}_j) - \frac{n(n-1)}{2} F_+^2(y).$$

Here, for $y = x + t\sqrt{n-1}$,

$$\mathbf{P}(G_n \overline{B}_j) = \int_y^\infty \mathbf{P}(S_{n-1} \geq x - u)\mathbf{F}(du) \geq \mathbf{P}(S_{n-1} \geq x - y)F_+(y)$$
$$= \mathbf{P}(S_{n-1} \geq -t\sqrt{n-1})F_+(y) = \big(1 - \mathbf{P}(S_{n-1} < -t\sqrt{n-1})\big)F_+(t),$$

where, by the Chebyshev inequality,

$$\mathbf{P}(S_{n-1} < -t\sqrt{n-1}) \leq t^{-2}.$$

As a result we get

$$\mathbf{P}(S_n \geq x) \geq nF_+(t)\big(1 - t^{-2}\big) - \frac{n(n-1)}{2} F_+^2(t),$$

which is equivalent to (A8.3.1).

The theorem is proved. □

Corollary A8.3.1 *If $x \to \infty$ and $x \gg \sqrt{n}$ then, as $t \to \infty$,*

$$\mathbf{P}(S_n \geq x) \geq nF_+(y)\big(1 + o(1)\big). \quad \text{(A8.3.2)}$$

If, moreover, $F_+(u) \geq V(u) \in \mathcal{R}$ then

$$\mathbf{P}(S_n \geq x) \geq nV(x)\big(1 + o(1)\big).$$

Proof Since $y \geq x$, we have

$$nF_+(y) \ll ny^{-2} < nx^{-2} = o(1).$$

This together with (A8.3.1) implies the first assertion of the corollary as $t \to \infty$. To obtain the second one, in (A8.3.2) one should take $t \to \infty$ such that $t = o(x/\sqrt{n})$. Then $y \sim x$ and $V(y) \sim V(x)$.

The corollary is proved. □

Appendix 9
Renewal Theorems

The main goal of the present section is to prove Theorem 10.4.1, the key renewal theorem in the non-arithmetic case (in the terminology of Chap. 10). We will also consider some refinements and extensions of the theorem.

First consider *positive* independent identically distributed random variables $\tau_j \stackrel{d}{=} \tau$ with distribution function F and finite mean $a := \mathbf{E}\tau < \infty$. Here it will be more convenient to understand by the renewal function its left-continuous version

$$H(t) := \sum_{k=0}^{\infty} F^{*k}(t), \quad t \geq 0,$$

where F^{*k} is the k-fold convolution of the distribution F with itself, which is the distribution function of the sum $T_k = \tau_1 + \cdots + \tau_k$. We first prove the following key assertion.

Theorem A9.1 *If g is a directly integrable function and τ_j are non-arithmetic (see Chap. 10) then, as $t \to \infty$,*

$$\int_0^t g(t-u)\,dH(u) \to \frac{1}{a}\int_0^{\infty} g(u)\,du.$$

The proof of the theorem mostly follows the argument suggested in [13] and will need several auxiliary assertions.

Lemma A9.1 *Let g be a bounded measurable function. The integral*

$$G(t) = \int_0^t g(t-u)\,dH(u) =: g * H(t) \tag{A9.1}$$

is the unique solution of the equation

A.A. Borovkov, *Probability Theory*, Universitext,
DOI 10.1007/978-1-4471-5201-9, © Springer-Verlag London 2013

$$G(t) = g(t) + \int_0^t G(t-u)\,dF(u) = g(t) + G * F(t) \qquad \text{(A9.2)}$$

in the class of functions bounded on finite intervals.

The function $G = H$ is the solution of (A9.2) *when $g \equiv 1$. The function $G \equiv 1$ is the solution of* (A9.2) *when $g = 1 - F$.*

Equation (A9.2) is called the *renewal equation*.

As we already noted in Theorem 10.4.1, one can associate, in an obvious way, measures **H** and **F** with the functions H and F, and write the integrals in (A9.1) and (A9.2) as integrals with respect to the measures:

$$\int_0^t g(t-u)\mathbf{H}(du) \quad \text{and} \quad \int_0^t G(t-u)\mathbf{F}(du),$$

respectively.

Proof of Lemma A9.1 Put

$$H_n(t) := \sum_{k=0}^{n} F^{*k}(t).$$

The functions $G_n = g * H_n$ satisfy the equation $G_{n+1} = g + G_n * F$ and form an increasing sequence $G_n \uparrow$ which is bounded by Lemma 10.2.3. Therefore $G_n \uparrow G$, and passing to the limit in the equation for G_n we obtain that G satisfies (A9.1). To prove uniqueness note that the difference $V = G^{(1)} - G^{(2)}$ of two solutions $G^{(1)}$ and $G^{(2)}$ must satisfy the homogeneous equation $V = V * F$ and therefore also the relations $V = V * (F^{k*})$ or, which is the same,

$$V(t) = \int_0^t V(t-u)\,dF^{*k}(u).$$

But $F^{*k}(u) \to 0$ as $k \to \infty$ for $u \in [0, t]$. Since by the assumption $|V(u)| < c$ on $[0, t]$, we have $V(t) \to 0$ as $k \to \infty$. But V does not depend on k, so that $V(t) \equiv 0$. The last assertion of the lemma can be verified directly. The lemma is proved. □

Note that if we considered functions g of bounded variation, the assertion of Lemma A9.1 would immediately follow from the equation for the Laplace–Stieltjes transform $\widetilde{G}(\lambda) = \int_0^\infty e^{-\lambda t}\,dG(t)$ of G which follows from (A9.2):

$$\widetilde{G}(\lambda) = \widetilde{g}(\lambda) + \widetilde{G}(\lambda)\psi(\lambda), \qquad \text{(A9.3)}$$

where

$$\widetilde{g}(\lambda) := \int_0^\infty e^{-\lambda t}\,dg(t), \qquad \psi(\lambda) := \int_0^\infty e^{-\lambda t}\,dF(t).$$

Indeed, it follows from (A9.3) that

$$\widetilde{G}(\lambda) = \frac{\widetilde{g}(\lambda)}{1-\psi(\lambda)},$$

which is equivalent to (A9.1).

A point t is said to be a *point of growth* of the distribution function F provided that $F(t+\varepsilon) - F(t) > 0$ for any $\varepsilon > 0$.

Lemma A9.2 *Let the distribution F be non-arithmetic and Z be the set of all points of growth of H, i.e. points of growth of the functions $F, F^{*2}, F^{*3}, \ldots$. Then Z is "asymptotically dense at infinity", i.e., for any given $\varepsilon > 0$ and all x large enough, the intersection $(x, x+\varepsilon) \cap Z$ is non-empty.*

Proof Observe first that if t_1 is a point of growth of the distribution F_1 of a random variable τ, and t_2 is a point of growth of the distribution F_2 of a random variable ζ which is independent of τ, then $t = t_1 + t_2$ will be a point of growth of the distribution $F_1 * F_2$ of the variable $\tau + \zeta$. Indeed,

$$\mathbf{P}(t \le \tau + \zeta < t + \varepsilon) \ge \mathbf{P}\left(t_1 \le \tau < t_1 + \frac{\varepsilon}{2}\right)\mathbf{P}\left(t_2 \le \zeta < t_2 + \frac{\varepsilon}{2}\right).$$

Let, further, $x < y$ be two points of the set Z, and $\Delta := y - x$. The following alternative takes place: either

(1) for any $\varepsilon > 0$ there exist x and y such that $\Delta < \varepsilon$, or

(2) there exists a $\delta > 0$ such that $\Delta \ge \delta$ for all x and y from Z.

Put $I_n := [xn, yn]$. If $n\Delta > x$ then that interval contains $[nx, (n+1)x]$ as a subset, and therefore any point $v > v_0 = x^2/\Delta$ belongs to at least one of the intervals $I_1, I_2, \ldots$.

By virtue of the above observation, the $n+1$ points $nx + k\Delta = (n-k)x + ky$, $k = 0, \ldots, n$, belong to Z and divide I_n into n subintervals of length Δ. This means that, for any point $v > v_0$, the distance between v and the points from Z is at most $\Delta/2$. This implies the assertion of the lemma when (1) holds.

If (2) is true, we can assume that x and y are chosen so that $\Delta < 2\delta$. Then the points of the form $nx + k\Delta$ exhaust all the points from Z lying inside I_n. Since the point $(n+1)x$ is among these points, the value x is a multiple of Δ, and all the points of Z lying inside I_n are multiples of Δ. Now let z be an arbitrary point of growth of F. For sufficiently large n, the interval I_n contains a point of the form $z + k\Delta$, and since the latter belongs to Z, the value z is also a multiple of Δ. Thus F is an arithmetic distribution, so that case (2) cannot take place. The lemma is proved. □

Lemma A9.3 *Let $q(x)$ be a bounded uniformly continuous function given on $(-\infty, \infty)$ such that, for all x, $q(x) \le q(0)$ for all x, and*

$$q(x) = \int_0^\infty q(x-y)\, dF(y). \tag{A9.4}$$

Then $q(x) \equiv q(0)$.

Proof Equation (A9.4) means that $q = q * F = \cdots = q * F^{*k}$ for all $k \ge 1$. The right-hand side of (A9.4) does not exceed $q(0)$, and hence, for $x = 0$, the equality (A9.4) is only possible if $q(-y) = q(0)$ for all $y \in Z_k$, where Z_k is the set of points of growth of F^{*k}, and therefore $q(-y) = q(0)$ for all $y \in Z$. By Lemma A9.2 and the uniform continuity of q this means that $q(-y) \to q(0)$ as $y \to \infty$. Further, for an arbitrarily large N we can choose k such that $q(x)$ will be arbitrarily close to $\int_N^\infty q(x-y)\,dF^{*k}(y)$, since $F^{*k}(N) \to 0$ as $k \to \infty$. This means, in turn, that $q(x)$ will be close to $q(0)$. Since $q(x)$ depends neither on N nor on k, we have $q(x) = q(0)$. The lemma is proved. □

Lemma A9.4 *Let g be a continuous function vanishing outside segment* $[0, b]$. *Then the solution G of the renewal equation* (A9.2) *is uniformly continuous and, for any u,*

$$G(x+u) - G(x) \to 0 \tag{A9.5}$$

as $x \to \infty$.

Proof By virtue of Lemma 10.2.3,

$$\begin{aligned}\bigl|G(x+\delta) - G(x)\bigr| &= \left|\int_{x-b}^{x+\delta} \bigl(g(x+\delta-y) - g(x-y)\bigr)\,dH(y)\right| \\ &\le \max_{0\le x\le b+\delta} \bigl|g(x+\delta) - g(x)\bigr|\bigl(c_1 + c_2(b+\delta)\bigr).\end{aligned} \tag{A9.6}$$

This means that the uniform continuity of g implies that of G.

Now assume that g has a continuous derivative g'. Then G' exists and satisfies the renewal equation

$$G'(x) = g'(x) + \int_0^x G'(x-y)\,dF(y).$$

Therefore the derivative G' is bounded and uniformly continuous. Let

$$\limsup_{x\to\infty} G'(x) = s.$$

Choose a sequence $t_n \to \infty$ such that $G'(t_n) \to s$. The family of functions q_n defined by the equalities

$$q_n(x) = G'(t_n + x)$$

is equicontinuous, and

$$q_n(x) = g'(t_n + x) + \int_0^{x+t_n} q_n(x-y)\,dF(y) = g'(t_n + x) + \int_0^\infty q_n(x-y)\,dF(y). \tag{A9.7}$$

By the Arzelà–Ascoli theorem (see Appendix 4) there exists a subsequence t_{n_r} such that q_{n_r} converges to a limit q. From (A9.7) it follows that this limit satisfies the conditions of Lemma A9.3, and therefore $q(x) = q(0) = s$ for all x. Thus $G'(t_{n_r}+x) \to s$ for all x, and hence

$$G(t_{n_r} + x) - G(t_{n_r}) \to sx.$$

Since the last relation holds for any x and the function g is bounded, we get $s = 0$.

We have proved the lemma for continuously differentiable g. But an arbitrary continuous function g vanishing outside $[0, b]$ can be approximated by a continuously differentiable function g_1 which also vanishes outside that interval. Let G_1 be the solution of the renewal equation corresponding to the function g_1. Then $|g - g_1| < \varepsilon$ implies $|G - G_1| < c\varepsilon$, $c = c_1 + c_2 b$ (see Lemma 10.2.3), and therefore

$$\big|G(x + u) - G(x)\big| < (2c + 1)\varepsilon$$

for all sufficiently large x. This proves (A9.5) for arbitrary continuous functions g. The lemma is proved. □

Proof of Theorem A9.1 Consider an arbitrary sequence $t_n \to \infty$ and the measures μ_n generated by the functions

$$H_{(n)}(u) = H(t_n + u) - H(t_n) \qquad \big(\mu_n\big([u, v)\big) = H_{(n)}(v) - H_{(n)}(u)\big).$$

These functions satisfy the conditions of the generalised Helly theorem (see Appendix 4). Therefore there exists a subsequence t_{nn}, the respective subsequence of measures μ_{nn}, and the limiting measure μ such that μ_{nn} converges weakly to μ on any finite interval as $n \to \infty$.

Now let g be a continuous function vanishing outside $[0, b]$. Then

$$\begin{aligned} G(t_{nn} + x) &= \int_{-b}^{0} g(-u)\, dH(t_{nn} + x + u) \\ &= \int_{-b}^{0} g(-u)\, d\big(H(t_{nn} + x + u) - H(t_{nn})\big) \to \int_{0}^{b} g(u)\mu(x + du). \end{aligned}$$

By Lemma A9.4, the sequence $G(t_{nn} + y)$ will have the same limit. This means that the measure $\mu(x + du)$ does not depend on x, and therefore $\mu([u, v))$ is proportional to the length of the interval (u, v):

$$\mu\big((u, v)\big) = c(v - u), \quad \mu(du) = c\, du.$$

Thus, we have proved that

$$G(t_{nn} + x) \to c \int_{0}^{\infty} g(u)\, du \tag{A9.8}$$

for any continuous function g vanishing outside $[0, b]$. But for any Riemann integrable function g on $[0, b]$ and given $\varepsilon > 0$ there exist continuous functions g_1 and g_2, $g_1 < g < g_2$, which are equal to 0 outside $[0, b+1]$ and such that

$$\int_0^b (g_2 - g_1)\,du < \varepsilon.$$

This means that convergence (A9.8) also holds for any Riemann integrable function vanishing outside $[0, b]$.

Now consider an arbitrary directly integrable function g. By property (2) of such functions (see Definition 10.4.1) one can choose a $b > 0$ such that for the function

$$g_{(b)}(u) = \begin{cases} g(u) & \text{if } u \le b, \\ 0 & \text{if } u > b, \end{cases}$$

the left- and right-hand sides of (A9.8) will be arbitrarily close to the respective expressions corresponding to the original function g (for the right-hand side it is obvious, while for the left-hand side it follows from the convergence

$$\left| \int_0^t g(t-s)\,dH(s) - \int g_{(b)}(t-s)\,dH(s) \right|$$
$$= \left| \int_0^{t-b} g(t-s)\,dH(s) \right| \le \sum_{k>b-1} (c_1 + c_2) g_k \to 0$$

as $b \to \infty$ (see Lemma 10.2.3)). Therefore (A9.8) is proved for any directly integrable function g. Putting $g := 1 - F$ we obtain from Lemma A9.1

$$1 = c \int_0^\infty \bigl(1 - F(u)\bigr)\,du = ac, \quad c = \frac{1}{a}.$$

Thus the limit in (A9.8) is one and the same for any initial sequence t_n. From this it follows that, as $t \to \infty$,

$$G(t) \to \frac{1}{a} \int_0^\infty g(u)\,du.$$

The theorem is proved. □

Theorem 10.4.1 is a simple consequence of Theorem A9.1 and the argument used in the proof of Theorem 10.2.3 that extends the key renewal theorem in the arithmetic case was extended to the setting where τ_j, $j \ge 2$, can assume values of different signs, while τ_1 is arbitrary. We will leave it to the reader to apply the argument in the non-arithmetic case.

Now we will give several further consequences of Theorem A9.1. In Sect. 10.4 we obtained a refinement of the renewal theorem in the case when $m_2 := \mathbf{E}\tau_j^2 < \infty$. Approaches developed while proving Theorem A9.1 enable one to obtain an alternative proof of the following assertion coinciding with Theorem 10.4.4.

Theorem A9.2 *Let the conditions of Theorem* A9.1 *be met and* $m_2 < \infty$. *Then*

$$0 \le H(t) - \frac{t}{a} \to \frac{m_2}{2a^2} \quad \text{as } t \to \infty.$$

Proof The function $G(t) := H(t) - t/a$ is the solution of the renewal equation (A9.2) corresponding to the function

$$g(t) := \frac{1}{a} \int_t^\infty \bigl(1 - F(u)\bigr)\, du.$$

Since g is directly integrable, we have

$$G(t) \to \frac{1}{a} \int_0^\infty \int_v^\infty \bigl(1 - F(u)\bigr)\, du\, dv = \frac{m_2}{2a^2}.$$

The theorem is proved. □

Theorem A9.3 (The local renewal theorem for densities) *Assume that* F *has a density* $f = F'$ *and this density is directly integrable. Then* H *has a density* $h = H'$, *and*

$$h(t) \to \frac{1}{a} \quad \text{as } t \to \infty.$$

Proof Denote by $f_n(x)$ the density of the sum $T_n = \tau_1 + \cdots + \tau_n$. We have

$$h(t) = H'(t) = \sum_{n=1}^\infty f_n(t) = f(t) + \int h(t-u) f(u)\, du = f(t) + h * F(t).$$

This means that $h(t)$ satisfies the renewal equation with the function $g = f$. Therefore by Theorem A9.1,

$$h(t) \to \frac{1}{a} \int_0^\infty f(u)\, du = \frac{1}{a}.$$

The theorem is proved. □

Consider now some extensions of Theorem A9.1. A function g given on the whole line $(-\infty, \infty)$ is said to be directly integrable if both functions $g(t)$ and $g(-t)$, $t \ge 0$, are directly integrable.

Theorem A9.4 *If the conditions of Theorem* A9.1 *are met and* g *is directly integrable, then*

$$G(t) = \int_0^\infty g(t-u)\mathbf{H}(du) \to \frac{1}{a} \int_{-\infty}^\infty g(u)\, du \quad \text{as } t \to \infty.$$

The Proof can be obtained by making several small and quite obvious modifications to the argument in the demonstration of Theorem A9.1. The main change is that instead of functions g vanishing outside $[0, b]$ one should now consider functions vanishing outside $[-b, b]$.

Another extension refers to the second version of the renewal function

$$U(t) := \sum_{k=0}^{\infty} F^{*k}(t), \quad -\infty < t < \infty,$$

in the case when τ_j can assume values of different signs.

Theorem A9.5 *If g is directly integrable and $\mathbf{E}\tau_j = a > 0$, then*

$$G(t) = \int_{-\infty}^{\infty} g(t-u)\mathbf{U}(du) \to \frac{1}{a}\int_{-\infty}^{\infty} g(u)\,du \quad \textit{as } t \to \infty,$$

and, for any fixed u, $U(t+u) - U(t) \to 0$ as $t \to \infty$.

The proof is also obtained by modifying the argument proving Theorem A9.1 (see [13]).

References

1. Billingsley, P.: Convergence of Probability Measures. Wiley, New York (1968)
2. Billingsley, P.: Probability and Measure. Anniversary edn. Wiley, Hoboken (2012)
3. Borovkov, A.A.: Stochastic Processes in Queueing Theory. Springer, New York (1976)
4. Borovkov, A.A.: Convergence of measures and random processes. Russ. Math. Surv. **31**, 1–69 (1976)
5. Borovkov, A.A.: Probability Theory. Gordon & Breach, Amsterdam (1998)
6. Borovkov, A.A.: Ergodicity and Stability of Stochastic Processes. Wiley, Chichester (1998)
7. Borovkov, A.A.: Mathematical Statistics. Gordon & Breach, Amsterdam (1998)
8. Borovkov, A.A., Borovkov, K.A.: Asymptotic Analysis of Random Walks. Heavy-Tailed Distributions. Cambridge University Press, Cambridge (2008)
9. Cramér, H., Leadbetter, M.R.: Stationary and Related Stochastic Processes. Willey, New York (1967)
10. Dudley, R.M.: Real Analysis and Probability. Cambridge University Press, Cambridge (2002)
11. Feinstein, A.: Foundations of Information Theory. McGraw-Hill, New York (1995)
12. Feller, W.: An Introduction to Probability Theory and Its Applications, vol. 1. Wiley, New York (1968)
13. Feller, W.: An Introduction to Probability Theory and Its Applications, vol. 2. Wiley, New York (1971)
14. Gikhman, I.I., Skorokhod, A.V.: Introduction to the Theory of Random Processes. Saunders, Philadelphia (1969)
15. Gnedenko, B.V.: The Theory of Probability. Chelsea, New York (1962)
16. Gnedenko, B.V., Kolmogorov, A.N.: Limit Distributions for Sums of Independent Random Variables. Addison-Wesley, Reading (1968)
17. Gradsteyn, I.S., Ryzhik, I.M.: Table of Integrals, Series, and Products. Academic Press, New York (1965). Fourth edition prepared by Ju.V. Geronimus and M.Ju. Ceitlin. Translated from the Russian by Scripta Technica, Inc. Translation edited by Alan Jeffrey
18. Grenander, U.: Probabilities on Algebraic Structures. Almqvist & Wiskel, Stockholm (1963)
19. Halmos, P.R.: Measure Theory. Van Nostrand, New York (1950)
20. Ibragimov, I.A., Linnik, Yu.V.: Independent and Stationary Sequences of Random Variables. Wolters-Noordhoff, Croningen (1971)
21. Khinchin, A.Ya.: Ponyatie entropii v teorii veroyatnostei (The concept of entropy in the theory probability). Usp. Mat. Nauk **8**, 3–20 (1953) (in Russian)
22. Kifer, Yu.: Ergodic Theory of Random Transformations. Birkhäuser, Boston (1986)
23. Kolmogorov, A.N.: Markov chains with a countable number of possible states. In: Shiryaev, A.N. (ed.) Selected Works of A.N. Kolmogorov, vol. 2, pp. 193–208. Kluwer Academic, Dordrecht (1986)

A.A. Borovkov, *Probability Theory*, Universitext,
DOI 10.1007/978-1-4471-5201-9,

24. Kolmogorov, A.N.: The theory of probability. In: Aleksandrov, A.D., et al. (eds.) Mathematics, Its Content, Methods, and Meaning, vol. 2, pp. 229–264. MIT Press, Cambridge (1963)
25. Lamperti, J.: Probability: A Survey of the Mathematical Theory. Wiley, New York (1996)
26. Loeve, M.: Probability Theory. Springer, New York (1977)
27. Meyn, S.P., Tweedie, R.L.: Markov Chains and Stochastic Stability. Springer, New York (1993)
28. Natanson, I.P.: Theory of Functions of a Real Variable. Ungar, New York (1961)
29. Nummelin, E.: General Irreducible Markov Chains and Nonnegative Operators. Cambridge University Press, New York (1984)
30. Petrov, V.V.: Sums of Independent Random Variables. Springer, New York (1975)
31. Shiryaev, A.N.: Probability. Springer, New York (1984)
32. Skorokhod, A.V.: Random Processes with Independent Increments. Kluwer Academic, Dordrecht (1991)
33. Tyurin, I.S.: An improvement of the residual in the Lyapunov theorem. Theory Probab. Appl. **56**(4) (2011)

Index of Basic Notation

Spaces and σ-algebras

$\mathfrak{F}$—a σ-algebra, 14
$\langle \Omega, \mathfrak{F} \rangle$—a measurable space, 14
$\mathbb{R}$—the real line, 17
$\mathbb{R}^n$—n-dimensional Euclidean space, 18
$\mathfrak{B}$—the σ-algebra of Borel-measurable subsets of $\mathbb{R}$, 17
$\mathfrak{B}^n$—the σ-algebra of Borel-measurable subsets of $\mathbb{R}^n$, 18
$\langle \Omega, \mathfrak{F}, \mathbf{P} \rangle$—the probability space, 17
(Note that Ω and $\mathfrak{F}$ can take specific values, i.e. $\mathbb{R}$ and $\mathfrak{B}$, respectively.)

Distributions[1]

$\mathbf{F}_\xi$, $\mathbf{F}$—the distribution of the random variable ξ, 32, 32
$\mathbf{I}_a$—the degenerate distribution (concentrated at the point a), 37
$\mathbf{U}_{a,b}$—the uniform distribution on $[a, b]$, 37
$\mathbf{B}_p$, $\mathbf{B}_p^n$—the binomial distributions, 37
multinomial distributions, 47
$\boldsymbol{\Phi}_{\alpha,\sigma^2}$—the normal (Gaussian) distribution with parameters (α, σ^2), 37, 48
$\phi_{\alpha,\sigma^2}(x)$—the density of the normal law with parameters (α, σ^2), 41
$\mathbf{F}_{\beta,\rho}$—the stable distribution with parameters β, ρ, 231, 233
$f^{(\beta,\rho)}(x)$—the density of the stable distribution with parameters $\mathbf{F}_{\beta,\rho}$, 235
$\varphi^{(\beta,\rho)}(t)$—the characteristic function of distribution $\mathbf{F}_{\beta,\rho}$, 231
$\mathbf{K}_{\alpha,\sigma}$—the Cauchy distribution with parameters (α, σ), 38
$\boldsymbol{\Gamma}_\alpha$—the exponential distribution with parameter α, 38, 177
$\boldsymbol{\Gamma}_{\alpha,\lambda}$—the gamma-distribution with parameters (α, λ), 176
$\boldsymbol{\Pi}_\lambda$—the Poisson distribution with parameter λ, 39
$\boldsymbol{\chi}^2$—the $\boldsymbol{\chi}^2$-distribution, 177
$\Lambda(\alpha)$—the large deviation rate function, 244

[1](All distributions and measures are denoted by bold letters).

A.A. Borovkov, *Probability Theory*, Universitext,
DOI 10.1007/978-1-4471-5201-9, © Springer-Verlag London 2013

Relations

Conditions

Subject Index

A.A. Borovkov, *Probability Theory*, Universitext,
DOI 10.1007/978-1-4471-5201-9, © Springer-Verlag London 2013

S

T